THE PHYSICAL WORLD

The Physical World

An Inspirational Tour of Fundamental Physics

Nicholas Manton

*Professor at the Department of Applied Mathematics and Theoretical Physics,
University of Cambridge*

Nicholas Mee

Trinity College, Cambridge

OXFORD
UNIVERSITY PRESS

OXFORD
UNIVERSITY PRESS

Great Clarendon Street, Oxford, OX2 6DP,
United Kingdom

Oxford University Press is a department of the University of Oxford.
It furthers the University's objective of excellence in research, scholarship,
and education by publishing worldwide. Oxford is a registered trade mark of
Oxford University Press in the UK and in certain other countries

First Edition published in 2017

Impression: 5

Reprinted 2017 (three times, twice with corrections), 2018

Published in the United States of America by Oxford University Press
198 Madison Avenue, New York, NY 10016, United States of America

British Library Cataloguing in Publication Data

Data available

Library of Congress Control Number: 2017934959

ISBN 978–0–19–879593–3 (hbk.)
ISBN 978–0–19–879611–4 (pbk.)

DOI 10.1093/oso/9780198795933.001

Printed and bound by
CPI Group (UK) Ltd, Croydon, CR0 4YY

Preface

Writing this book has been a great pleasure. It has given us the opportunity to think deeply about all the major branches of physics and how they unite to form a coherent picture of the physical world. The result is a panoramic survey of the subject that is both concise and comprehensive. Mathematics is the natural language for describing so much of the world around us, as many commentators such as Richard Feynman and Jim Al Khalili have stressed, and is essential for a true understanding of physics. Our account is therefore necessarily mathematical. We provide clear explanations of the mathematical reasoning that underpins modern fundamental physics and have not shied away from presenting the key equations and their solutions.

There has never been a greater appetite for scientific explanations of the universe. Our aim is to offer an inspirational tour of fundamental physics that is accessible to anyone who has taken at least high school physics and mathematics. For those just leaving school, it can be read as a taster of much of the physics covered in a university course. This book should also suit those who have taken a maths course for scientists or engineers at university and desire a survey of fundamental physics up to the frontiers of current research. It might also appeal to mathematicians and computer scientists who wish to know more about physics.

Explaining the physical world is the work of many generations. The best theories of today are built on the great theories of the past, so it is scarcely possible to appreciate modern physics without a good understanding of the established physics of Newton, Maxwell, Einstein and the many others who have contributed to our understanding of the universe. For this reason, much of the book may look historical. This material is, however, presented in a modern style that may be very different from its original formulation. A key feature of our approach is a unifying idea that runs as a theme throughout *The Physical World*. This is the variational principle, of which the most important example is the *principle of least action*. Almost all successful theories of physics can be formulated using this idea and it lies at the heart of modern physics. Our tour also takes us around the highlights of recent years, including the surveys of the cosmic microwave background by WMAP and Planck, the discovery of the Higgs boson at the Large Hadron Collider and the discovery of gravitational waves by LIGO.

We would like to record here our thanks to the many friends, colleagues and relations who have encouraged us to write this book. We especially thank John Barrow and Jonathan Evans for their encouragement and advice.

Nick Manton thanks Anthony Charlesworth, Helena Aitta, Roger Heumann and Alan Smith for discussions and their interest in this book. He is also grateful to the physics societies of Dulwich College, his old school, and Charterhouse for opportunities to present the principle of least action to a sixth-form audience. He particularly thanks Alasdair Kennedy, who was until recently at Dulwich. He also thanks Anneli and Ben for their encouragement, and their patience with all the rounds of writing and editing.

Nick Mee would like to thank his parents for their enduring support. He is very grateful to John Eastwood for his assistance at numerous book signing events, including the October 2013 Federation of Astronomical Societies Convention in Cambridge which led to a chance encounter with Nick Manton that resulted in the writing of this book. He would also like to thank Jonathan Evans for his hospitality in Gonville and Caius College during this event. He also thanks Mark Sheeky and Debra Nightingale for their help and inspiration. He extends a warm thank you to all the members of his newsletter group and the readers of his blog for their encouragement and enthusiasm. He especially thanks Angie for her patience and fortitude throughout another long project.

We thank Sonke Adlung and Ania Wronski of Oxford University Press for their personal involvement with bringing this book to publication, and Suganiya Karumbayeeram and colleagues for their role in the production of the book. We also thank Mhairi Gray for compiling the index. We would like to express our gratitude to the many authors whose work we consulted while writing the book. At the end of each chapter, we have included a Further Reading section that lists some key books and papers.

Contents

0
Introduction

We live in a fascinating world, bursting with remarkable phenomena on every scale. Our expanding universe is filled with trillions of galaxies, and a supermassive black hole inhabits the heart of each. Exploding stars seed the galaxies with the dust of life, and eight minutes away a blazing nuclear furnace releases the energy that keeps the Earth green, vibrant and full of life. Our marbled, watery globe may be unique, or one of many where sentient beings have evolved. At a smaller scale, all visible matter is made of just a few fundamental particle types, but these combine into more than a hundred different atoms that bond together in countless ways.

Perhaps what is most surprising is the amount that it is possible for us to know of this, and the precision with which it is now understood. Using both natural and artificial materials, we can construct amazing devices that transform our lives and help us delve ever deeper into the mysteries of the cosmos. Some important physical phenomena are at the limit of what we can observe. Occasional collisions of black holes create tiny ripples in the fabric of space that we detect with our most sensitive instruments. Among the fundamental particles are the elusive neutrinos that pass through us every day in immense quantities, but are detected only very occasionally in vast underground chambers. For most of human history, people pursued their daily lives with little knowledge of the physical world. We are fortunate, indeed, to live in an age when so many of its secrets are being revealed.

As physicists, we are compelled by unceasing curiosity to seek out and explain the inner workings of nature. This was traditionally the domain of the philosopher, but subtle reasoned argument can only take us so far. Genuine insight is founded on the twin pillars of experimental investigation and elegant mathematical models. As Richard Feynman put it: 'If you want to learn about nature, to appreciate nature, it is necessary to understand the language that she speaks in.' For this reason, we have adopted a style in *The Physical World* that is unashamedly mathematical.

We will explore the theoretical foundations of physics and present a grand vision of the essential unity of the subject. Our aim is to give a broad overview of physics that will provide the necessary background and motivation to delve deeper into the topics covered in each chapter. We comprehensively cover the laws of physics, but are selective in illustrating these laws with relatively simple applications. In choosing our material, we have blended several strands of information: the fundamental laws and the philosophical principles on which they are based, the mathematical description of the laws, the experimental basis for the laws, the historical development of the laws, the shortcomings of our current understanding, and outstanding questions that remain to be answered. We have followed Albert Einstein's dictum that 'explanations should be made as simple as possible, but no simpler,' to give an

The Physical World. Nicholas Manton and Nicholas Mee, Oxford University Press (2017).
© Nicholas Manton and Nicholas Mee. DOI 10.1093/oso/9780198795933.001.0001

honest account of how modern physicists understand their subject. Our aim is to present an engaging portrait of physics with concise derivations of the important results in a style where every step is clearly explained. The level of the mathematics is broadly similar to that of the Feynman Lectures on Physics. We assume familiarity with algebra, including matrices and their determinants, geometry using Cartesian and polar coordinates, basic calculus, and complex numbers.

In Chapter 1 we cover some introductory ideas: vectors, the use of variational principles in physics, and partial differentiation. We have included an accessible introduction to partial derivatives, as they are an essential ingredient of most of the fundamental equations of physics, including Maxwell's electromagnetic field equations, the Schrödinger and Dirac equations of quantum mechanics, and the Einstein equation of general relativity. Chapter 2 covers Newtonian dynamics, and the application of Newton's law of gravitation to the motion of bodies in the solar system. Chapter 3 is mainly about electromagnetic fields as described by Maxwell's equations.

Particles and fields are the key concepts in classical physics, but Newton's laws and Maxwell's equations are not fully consistent with each other. Einstein resolved this issue when he devised his special theory of relativity, as described in Chapter 4. In special relativity, space and time are unified by introducing the novel idea of spacetime; but then a further problem arises, as special relativity is not compatible with Newton's theory of gravity. Einstein completed his revolution by showing that a consistent theory of particles, fields and gravitational forces requires spacetime to be curved, as described by his general theory of relativity. Chapter 6 is devoted to this theory and its remarkable consequences, such as the existence of black holes. To motivate some of the ideas in this chapter, our presentation is preceded by a discussion of curved space in Chapter 5, including some physical applications. Curved space is easier to visualize than curved spacetime.

Chapters 7 and 8 are about quantum mechanics, the other revolutionary idea of the 20th century, essential for understanding all kinds of phenomena at the atomic scale. Chapter 9 applies quantum mechanics to the structure and properties of materials and explains the fundamental principles of chemistry and solid state physics. Chapter 10 is about thermodynamics, which is built on the concepts of temperature and entropy. We discuss a number of examples including the analysis of black body radiation that led to the quantum revolution. Chapter 11 surveys the atomic nucleus, its properties and behaviour. Investigation of the nucleus with high energy particle beams leads even deeper into the structure of matter and the quest for its ultimate building blocks. Chapter 12 explores particle physics and includes a short description of quantum field theory, the Standard Model and the Higgs mechanism.

Throughout the book, we have carefully selected applications that give insight into fundamental physics. Chapter 13 on stars is an extended example of this, bringing together ideas from many branches of physics: gravity, quantum mechanics, thermodynamics, nuclear and particle physics. Chapter 14 on cosmology surveys the structure and evolution of the universe as a whole, starting from the Big Bang. We conclude with Chapter 15, which discusses some of the fundamental problems that remain, such as the interpretation of quantum mechanics, and the ultimate nature of particles. Some exciting but speculative ideas for better understanding particles and for unifying gravity with the forces of particle physics are also briefly explored. These include supersymmetry, solitons and string theory.

In *The Physical World*, emphasis is placed on the use of variational principles in physics, and in particular the *principle of least action*, an approach that lies at the heart of modern theoretical physics, but has been neglected in most introductory accounts of the subject. We think this concept should be better known. It was one of the great achievements of Feynman to bring the action to the forefront of theoretical physics. We offer a simple and somewhat novel explanation of how the principle of least action underlies Newton's laws of motion, and briefly explain the role of the action in electromagnetic theory and general relativity. Our treatment of quantum mechanics is the traditional one based on the Schrödinger equation, but we also discuss the role of the action, and how this leads to Feynman's path integral approach to quantum mechanics.

Although this book is largely about well established physics, we describe the latest advances in many key areas, such as astrophysics, relativity, nuclear and particle physics. These include gravitational lenses, supermassive black holes, graphene, Bose–Einstein condensates, superheavy nuclei, dark matter, neutrino oscillations, and the discovery of the Higgs boson and gravitational waves.

0.1 Further Reading

R.P. Feynman, *Feynman Lectures on Physics (New Millenium ed.)*, New York: Basic, 2010.

M. Longair, *Theoretical Concepts in Physics: An Alternative View of Theoretical Reasoning in Physics (2nd ed.)*, Cambridge: CUP, 2003.

R. Penrose, *The Road to Reality: A Complete Guide to the Laws of the Universe*, London: Vintage, 2005.

1
Fundamental Ideas

1.1 Variational Principles

Many of our activities in everyday life are directed towards optimizing some quantity. We often try to perform tasks with minimal effort, or as quickly as possible. Here is a simple example: we may plan a road journey to minimize the travel time, taking a longer route in order to go faster along a section of highway. Figure 1.1 is a schematic road map between towns A and B. Speed on the ordinary roads is 50 mph, and on the highway passing through F, G and H it is 70 mph. The minimum journey time is 1 hr 24 mins along route AFGB, even though this is not the shortest route.

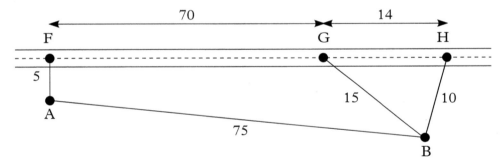

Fig. 1.1 Road map with distances in miles. The speed on ordinary roads is 50 mph, and on the highway 70 mph.

Remarkably, many natural processes can similarly be seen as optimizing some quantity. We say that they satisfy a *variational principle*. An elastic band stretched between two points lies along a straight line; this is the shortest path and also minimizes the elastic band's energy. We can understand why a straight line is the shortest path as follows. First we need to assume that a shortest path *does exist*. In the current situation this is obvious, but there are more complicated optimization problems where there is no optimal solution. Now assume that the shortest path has a curved segment somewhere along it. Any curved segment can be approximated by part of a circle, as shown in Figure 1.2, and using a little trigonometry, we can check that the straight segment CD is shorter than the circular arc CD. In fact, the circular arc has length $2R\alpha$, and the straight segment has length $2R\sin\alpha$, which is shorter. So the assumption that the shortest path is somewhere curved is contradictory. Therefore the shortest path is straight.

The Physical World. Nicholas Manton and Nicholas Mee, Oxford University Press (2017).
© Nicholas Manton and Nicholas Mee. DOI 10.1093/oso/9780198795933.001.0001

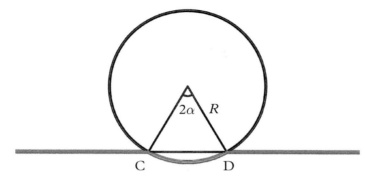

Fig. 1.2 Any curved section of a path can be approximated by part of a circle. A straight chord across this circle is shorter than the curved path.

A soap film is another familiar, physical example of energy optimization. Although it might initially be vibrating, the soap film will eventually settle into a state in which it is at rest. Its energy is then the product of its constant surface tension and its area, so the energy is minimized when the area is minimized. For any smooth surface in 3-dimensional space, there are two principal radii of curvature, r_1 and r_2; for a surface of minimal area the two radii of curvature are equal, but point in opposite directions. Every region of the surface is saddle-like, as shown in Figure 1.3. We can understand physically why the surface tension has this effect. On each small element of the surface, the two curvatures produce forces. If they are equal in magnitude and opposite in direction then they cancel, and the surface element is in equilibrium. We therefore have an intimate connection between the physical ideas of energy and force and the geometrical concept of minimal area. We will discuss the geometry of curved surfaces further in Chapter 5.

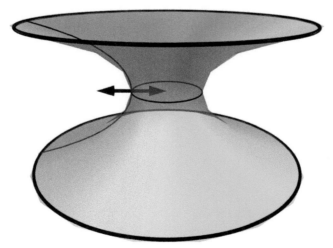

Fig. 1.3 A soap film is a surface of minimal area. The two radii of curvature are equal, but the curvatures are in opposite directions. The force due to the curvature in one direction balances the force due to the curvature in the other direction.

1.1.1 Geometrical optics—reflection and refraction

Fermat's principle in the field of optics was the first optimization principle to be discovered in physics. It was described by Pierre de Fermat in 1662. Geometrical optics is the study of idealized, infinitesimally thin beams of light, known as light rays. In the real world, narrow beams of light that are close to ideal rays can be obtained using parabolic mirrors or by projecting light through a screen containing narrow slits. Even if the light is not physically restricted like this, it can still be considered as a collection of rays travelling in different directions.

Fermat's principle says that the path taken by a light ray between two given points, A and B, is the path that minimizes the total travel time. The path may be straight, or it may be bent or even curved as it passes through various media. A fundamental assumption is that in a given medium, a light ray has a definite, finite speed. In a uniform medium, for example air or water or a vacuum, the travel time equals the length of the path divided by the light speed. Since the speed is constant, the path of minimal time is also the shortest path, and this is the straight line path from A to B. So light rays are straight in a uniform medium, as is readily verified. A light ray heading off in the correct direction from a source at A will arrive at B, and even though the source may emit light in all directions, a small obstacle anywhere along the line between A and B will prevent light reaching B, and will cast a shadow there.

Fermat's principle can be used to understand two basic laws of optics, the laws of *reflection* and *refraction*. First, let's consider reflection. Suppose we have a long flat mirror in a uniform medium, and a light source at A. Let B be the light receiving point, on the same side of the mirror, as shown in Figure 1.4. Consider all the possible light rays from A to B that bounce off the mirror once. If the time for the light to travel from A to B is to be minimized, the segments before and after reflection must be straight. What we'd like to know is the position of the reflection point X.

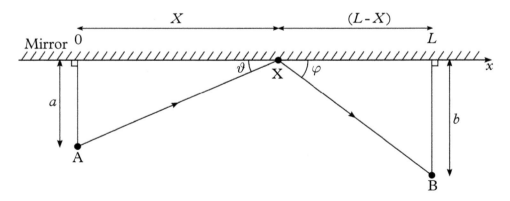

Fig. 1.4 Reflection of a light ray from a mirror.

The coordinates in the figure show the x-axis along the mirror, and the reflection point X is at $x = X$. Consider the various lengths in the figure, and ignore the angles ϑ and φ for the moment. Using Pythagoras' theorem to determine the path lengths, we find that the time for the light to travel from A to B via X is

$$T = \frac{1}{c}\left(\sqrt{a^2 + X^2} + \sqrt{b^2 + (L - X)^2}\right), \tag{1.1}$$

where c is the speed of the light along both straight segments. The derivative of T with respect to X is

$$\frac{dT}{dX} = \frac{1}{c}\left(\frac{X}{\sqrt{a^2 + X^2}} - \frac{L - X}{\sqrt{b^2 + (L - X)^2}}\right), \tag{1.2}$$

and the travel time is minimized when this derivative vanishes, giving the equation for X,

$$\frac{X}{\sqrt{a^2 + X^2}} = \frac{L - X}{\sqrt{b^2 + (L - X)^2}}. \tag{1.3}$$

Now the angles come in handy, as equation (1.3) is equivalent to

$$\cos\vartheta = \cos\varphi, \tag{1.4}$$

as can be seen from Figure 1.4. Therefore ϑ and φ are equal. We haven't explicitly found X, but that doesn't matter. The important result is that the incoming and outgoing light rays meet the mirror surface at equal angles. This is the fundamental law of reflection. In fact, by simplifying equation (1.3) or by considering the equation $\cot\vartheta = \cot\varphi$, we obtain $\frac{X}{a} = \frac{(L-X)}{b}$, and then X is easily found.

Refraction isn't very different. Here, light rays pass from a medium where the speed is c_1 into another medium where the speed is c_2. The geometry of refraction is different from that of reflection, but not very much, and we use similar coordinates (see Figure 1.5). By Fermat's principle, the path of the actual light ray from A to B, or from B to A, is the one that minimizes the time taken. Note that, unless $c_1 = c_2$, this is definitely not the same as the *shortest* path from A to B, which is the straight line between them. The path of minimum time has a kink, just like the route via the highway that we considered earlier.

The rays from A to X and from X to B must be straight, because each of these segments is wholly within a single medium and traced out at a single speed. The total time for the light to travel from A to B is therefore

$$T = \frac{1}{c_1}\sqrt{a^2 + X^2} + \frac{1}{c_2}\sqrt{b^2 + (L - X)^2}. \tag{1.5}$$

The time T is again minimized when the derivative of T with respect to X vanishes, that is,

$$\frac{dT}{dX} = \frac{1}{c_1}\frac{X}{\sqrt{a^2 + X^2}} - \frac{1}{c_2}\frac{L - X}{\sqrt{b^2 + (L - X)^2}} = 0. \tag{1.6}$$

This gives the equation for X,

$$\frac{1}{c_1}\frac{X}{\sqrt{a^2 + X^2}} = \frac{1}{c_2}\frac{L - X}{\sqrt{b^2 + (L - X)^2}}. \tag{1.7}$$

We do not really want to solve this, but rather to express it more geometrically. In terms of the angles ϑ and φ in Figure 1.5, the equation becomes

$$\frac{1}{c_1}\cos\vartheta = \frac{1}{c_2}\cos\varphi, \tag{1.8}$$

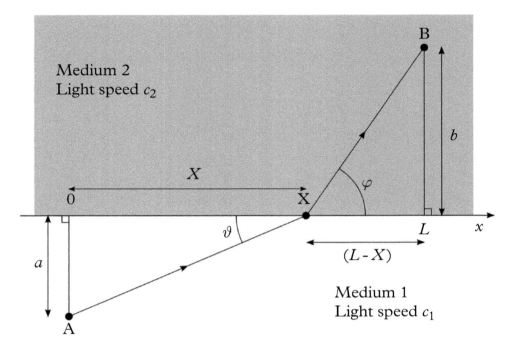

Fig. 1.5 The refraction of a light ray. c_2, the light speed in medium 2, is less than c_1, the speed in medium 1.

or more usefully

$$\cos \varphi = \frac{c_2}{c_1} \cos \vartheta . \tag{1.9}$$

This is Willebrord Snell's law of refraction.[1] It relates the angles of the light rays to the ratio of the light speeds c_2 and c_1. Snell's law can be tested experimentally even if the light speeds are unknown. To do this, the angle at which the light beam hits the surface must be varied, so that A and B are no longer fixed. When $\cos \varphi$ is plotted against $\cos \vartheta$, the resulting graph is a straight line through the origin.

Suppose the light passes from air into water. The speed of light in water is less than its speed in air, so c_2 is less than c_1, and $\cos \varphi$ is less than $\cos \vartheta$. Therefore φ is greater than ϑ. The result, as is easily verified, is that light rays are bent into the water towards the normal to the surface, as shown in Figure 1.5.

Snell's law has many interesting consequences. It is key to applications such as light focussing and lens systems. It also accounts for the phenomenon of total internal reflection. This occurs when a light ray originating at B, in the medium where the light speed is less, hits the surface at a small angle φ for which $\cos \varphi$ is close to 1 and therefore $\cos \vartheta > 1$. There is then no solution for the angle ϑ, so the ray cannot cross the surface into medium 1, and the entire ray is reflected internally. The critical angle of incidence φ_c for total internal reflection depends on the ratio of the light speeds in the two media. Equation (1.9) shows

---

[1] Snell's law may be more familiar in terms of the angles $\varphi' = \frac{\pi}{2} - \varphi$ and $\vartheta' = \frac{\pi}{2} - \vartheta$ between the light ray and the normal (the perpendicular line) to the surface, in which case it takes the form $\sin \varphi' = \frac{c_2}{c_1} \sin \vartheta'$.

that $\cos \varphi_c = \frac{c_2}{c_1}$. This result is important for applications such as the transmission of light signals down fibre optic cables.

Originally, the law of refraction was expressed in terms of a ratio of refractive indices on the right-hand side of equation (1.9). It was by considering Fermat's principle that physicists realised that the ratio could be understood as a ratio of light speeds. Later, when the speed of light in various media could be directly measured, it was found that light travels at its maximal speed in a vacuum, and only slightly slower in air. In denser materials such as water or glass, however, its speed is considerably slower, by about 20–40%. The speed of light in a vacuum is an absolute constant, $299,792,458 \text{ m s}^{-1}$, which is often approximated as $3 \times 10^8 \text{ m s}^{-1}$. In dense media the speed may depend on the wavelength of the light, so in passing from air into glass or water, rays of different colours bend through different angles, which is why a refracted beam of white light splits up when entering a glass prism or water droplet.

1.1.2 The scope of variational principles

We have given a brief flavour of how some mathematical laws of nature can be formulated in terms of variational principles. These principles are actually much more general, and occur throughout physics. Be it the motion of particles, the waveforms of fields, quantum states, or even the shape of spacetime itself, we find that natural processes always optimize some physical quantity. Usually this means that the quantity is minimized or maximized, but it may be at a saddle point.[2] The most important such quantity is known as an *action*, and many laws of physics can be formulated as a *principle of least action*. The appropriate mathematics for analysing these principles is called the *calculus of variations*. It is an extension of ordinary calculus, with its own additional tools that we will explain later.

As long ago as the 18th century, Jean le Rond D'Alembert, Leonhard Euler and Joseph-Louis Lagrange realized that Newton's laws of motion could be derived from a principle of least action. This approach was perfected by William Rowan Hamilton in the 1830s. We now know that Maxwell's equations for electric and magnetic fields also arise from an electromagnetic action principle, and in 1915 David Hilbert showed that Einstein's newly discovered equations describing gravity as curved spacetime arise from an action principle. Even the relationship between classical physics and quantum mechanics is best understood in terms of an action principle. This idea was pioneered by Paul Dirac, and perfected by Feynman. Today, the action principle is seen as the best method of encapsulating the behaviour of particles and fields.

One advantage of formulating physical theories in this way is that the principle of least action is concise and easy to remember. For example, in Maxwell's original formulation of electromagnetism, there were 20 equations for electromagnetic fields. In modern vector notation, due to Josiah Willard Gibbs, there are four Maxwell equations, supplemented by the Lorentz force law for charged particles. The action, on the other hand, is a single quantity constructed from the electromagnetic fields and the trajectories of charged particles, as we will describe in Chapter 3. This economy is essential when developing the more complicated gauge theories of elementary particles, discussed in Chapter 12, and even more esoteric theories, such as string theory.

---

[2] A saddle point in a landscape is a stationary point of the height, like a mountain pass, but is neither a maximum nor a minimum.

In Chapter 2 we will return to these ideas and show how Newtonian mechanics can be understood in terms of the principle of least action. By considering all possible infinitesimal variations in the motion of a physical body through space, we will derive Newton's laws of motion. First, however, we must describe mathematically the arena in which this motion takes place.

1.2   Euclidean Space and Time

Familiar 3-dimensional Euclidean space, known as 3-space for short and often denoted by $\mathbb{R}^3$, is the stage on which the drama of the physical world is played out. This drama takes place in time, but time and space are not unified in non-relativistic physics, so we will not require a geometrical description of time as yet. 3-space has the Euclidean symmetries of rotations and translations, where a translation is a rigid motion without rotation. The most fundamental geometrical concept is the distance between two points, and this is unchanged by translations and rotations. It is natural to express the laws of physics in a way that is independent of position and orientation. Then their form does not change when the entire physical system is translated or rotated. This gives the laws a geometrical significance.

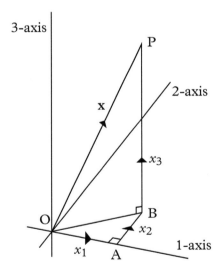

Fig. 1.6 Representation of a point P by a vector **x**.

A point in space is most easily described using Cartesian coordinates. For this one needs to pick an origin O, and a set of axes that are mutually orthogonal, meaning at right angles. Every point P is uniquely represented by three real numbers, collectively written as a *vector* $\mathbf{x} = (x_1, x_2, x_3)$. Often, we will not distinguish a point from the vector representing it. To get from O to P one moves a distance x_1 along the 1-axis, to A, then a distance x_2 parallel to the 2-axis, to B, and finally a distance x_3 parallel to the 3-axis, to P, as shown in Figure 1.6. O itself is represented by the vector $\mathbf{0} = (0, 0, 0)$.

The length or magnitude of **x** is the distance from O to P, and is denoted by $|\mathbf{x}|$. This distance can be calculated using Pythagoras' theorem. OAB is a right angle triangle, so the distance from O to B is $\sqrt{x_1^2 + x_2^2}$, and since OBP is also a right angle triangle, the distance from O to P is $\sqrt{(x_1^2 + x_2^2) + x_3^2}$. The square of the distance is therefore

$$|\mathbf{x}|^2 = x_1^2 + x_2^2 + x_3^2, \tag{1.10}$$

which is the 3-dimensional version of Pythagoras' theorem. If one performs a rotation about O, the distance $|\mathbf{x}|$ remains the same.

The rotation sending $\mathbf{x}$ to $\mathbf{x}'$ may be an active one, making $\mathbf{x}'$ and $\mathbf{x}$ genuinely different points. Alternatively, the rotation may be a passive one, by which we mean that the axes are rotated, but the point $\mathbf{x}$ does not actually change. All that happens is that $\mathbf{x}$ acquires a new set of coordinates $\mathbf{x}' = (x_1', x_2', x_3')$ relative to the rotated axes. In both cases $|\mathbf{x}'| = |\mathbf{x}|$.

The square of the distance between points $\mathbf{x}$ and $\mathbf{y}$ is

$$|\mathbf{x} - \mathbf{y}|^2 = (x_1 - y_1)^2 + (x_2 - y_2)^2 + (x_3 - y_3)^2. \tag{1.11}$$

This distance is unaffected by both rotations and translations. A translation shifts all points by a fixed vector $\mathbf{c}$, so $\mathbf{x}$ and $\mathbf{y}$ are shifted to $\mathbf{x} + \mathbf{c}$ and $\mathbf{y} + \mathbf{c}$. The difference $\mathbf{x} - \mathbf{y}$ is unchanged, and so is $|\mathbf{x} - \mathbf{y}|$.

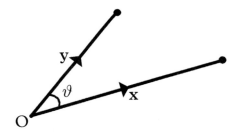

Fig. 1.7 The dot product of two vectors is $\mathbf{x} \cdot \mathbf{y} = |\mathbf{x}||\mathbf{y}| \cos \vartheta$.

When considering a pair of vectors $\mathbf{x}$ and $\mathbf{y}$, it is useful to introduce their *dot product*

$$\mathbf{x} \cdot \mathbf{y} = x_1 y_1 + x_2 y_2 + x_3 y_3. \tag{1.12}$$

A special case of this is $\mathbf{x} \cdot \mathbf{x} = x_1^2 + x_2^2 + x_3^2 = |\mathbf{x}|^2$, expressing the squared length of $\mathbf{x}$ as the dot product of $\mathbf{x}$ with itself. It is not immediately obvious whether $\mathbf{x} \cdot \mathbf{y}$ is affected by a rotation. However, if we expand out the terms on the right-hand side of equation (1.11), we find that

$$|\mathbf{x} - \mathbf{y}|^2 = |\mathbf{x}|^2 + |\mathbf{y}|^2 - 2\mathbf{x} \cdot \mathbf{y}, \tag{1.13}$$

and as $|\mathbf{x}|$, $|\mathbf{y}|$ and $|\mathbf{x} - \mathbf{y}|$ are all unaffected by a rotation, $\mathbf{x} \cdot \mathbf{y}$ must also be unaffected. We can use this result to find a more convenient expression for the dot product of $\mathbf{x}$ and $\mathbf{y}$. When applied to a triangle with edges of length $|\mathbf{x}|$, $|\mathbf{y}|$ and $|\mathbf{x} - \mathbf{y}|$, as shown in Figure 1.7, we can rearrange the expression (1.13), and then use the cosine rule to obtain

$$\mathbf{x} \cdot \mathbf{y} = \frac{1}{2}(|\mathbf{x}|^2 + |\mathbf{y}|^2 - |\mathbf{x} - \mathbf{y}|^2) = |\mathbf{x}||\mathbf{y}| \cos \vartheta, \tag{1.14}$$

where ϑ is the angle between the vectors $\mathbf{x}$ and $\mathbf{y}$.

It follows that if $\mathbf{x} \cdot \mathbf{y} = 0$, and the lengths of the vectors $\mathbf{x}$ and $\mathbf{y}$ are non-zero, then $\cos \vartheta = 0$ so the angle between $\mathbf{x}$ and $\mathbf{y}$ is $\vartheta = \pm\frac{\pi}{2}$, and the two vectors are orthogonal. For example, the basis vectors along the Cartesian axes, $(1,0,0)$, $(0,1,0)$ and $(0,0,1)$ are all of unit length, and the dot product of any pair of them vanishes, so they are orthogonal.

Critically, in Euclidean 3-space, the lengths of vectors and the angles between them are invariant under any rotation of all the vectors together, and this is why the dot product is a useful construction. Quantities such as $\mathbf{x} \cdot \mathbf{y}$ that are unaffected by rotations are called *scalars*.

There is a further, equally useful construction. From two vectors $\mathbf{x}$ and $\mathbf{y}$ one may construct a third vector, their *cross product* $\mathbf{x} \times \mathbf{y}$, as shown in Figure 1.8. This has components

$$\mathbf{x} \times \mathbf{y} = (x_2 y_3 - x_3 y_2, \, x_3 y_1 - x_1 y_3, \, x_1 y_2 - x_2 y_1) \,. \tag{1.15}$$

The cross product is useful, because if both $\mathbf{x}$ and $\mathbf{y}$ are rotated around an arbitrary axis, then $\mathbf{x} \times \mathbf{y}$ rotates with them. (If one invented another vector product of $\mathbf{x}$ and $\mathbf{y}$ with components $(x_2 y_3, x_3 y_1, x_1 y_2)$, say, then it would not have this rotational property and would have little geometrical significance.) Unlike the dot product $\mathbf{x} \cdot \mathbf{y}$, the cross product $\mathbf{x} \times \mathbf{y}$ is not *invariant* under rotations. We say that it transforms *covariantly* with $\mathbf{x}$ and $\mathbf{y}$ under rotations. 'Co-variant' means 'varying with' or 'transforming in the same way as', and this is an idea that occurs frequently in physics.

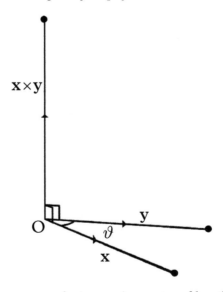

Fig. 1.8 The cross product $\mathbf{x} \times \mathbf{y}$ is a vector of length $|\mathbf{x}||\mathbf{y}| \sin \vartheta$.

We can check this rotational covariance of $\mathbf{x} \times \mathbf{y}$ by considering the dot product of $\mathbf{x} \times \mathbf{y}$ with a third vector $\mathbf{z}$. Using equations (1.15) and (1.12), we find

$$(\mathbf{x} \times \mathbf{y}) \cdot \mathbf{z} = x_2 y_3 z_1 - x_3 y_2 z_1 + x_3 y_1 z_2 - x_1 y_3 z_2 + x_1 y_2 z_3 - x_2 y_1 z_3 \,. \tag{1.16}$$

This is generally non-zero, but if either $\mathbf{z} = \mathbf{x}$ or $\mathbf{z} = \mathbf{y}$, it is easy to see that the six terms above cancel out in pairs, and the result is zero. This means that $\mathbf{x} \times \mathbf{y}$ is orthogonal to $\mathbf{x}$ and orthogonal to $\mathbf{y}$, as shown in Figure 1.8. When subject to a rotation, the directions of $\mathbf{x} \times \mathbf{y}$, $\mathbf{x}$ and $\mathbf{y}$ must therefore all rotate together. Now we just need to check that the length of $\mathbf{x} \times \mathbf{y}$ is invariant under rotations. In terms of its components, the squared length of $\mathbf{x} \times \mathbf{y}$ is

$$|\mathbf{x} \times \mathbf{y}|^2 = (x_2 y_3 - x_3 y_2)^2 + (x_3 y_1 - x_1 y_3)^2 + (x_1 y_2 - x_2 y_1)^2, \qquad (1.17)$$

and this can be re-expressed, after a little algebra, as

$$|\mathbf{x} \times \mathbf{y}|^2 = (\mathbf{x} \cdot \mathbf{x})(\mathbf{y} \cdot \mathbf{y}) - (\mathbf{x} \cdot \mathbf{y})^2. \qquad (1.18)$$

The right-hand side only includes quantities that are rotationally invariant, so $|\mathbf{x} \times \mathbf{y}|$ is similarly invariant. The right-hand side can be expressed in terms of lengths and angles as $|\mathbf{x}|^2 |\mathbf{y}|^2 - |\mathbf{x}|^2 |\mathbf{y}|^2 \cos^2 \vartheta$, which simplifies to $|\mathbf{x}|^2 |\mathbf{y}|^2 \sin^2 \vartheta$. The length of the vector $\mathbf{x} \times \mathbf{y}$ is therefore $|\mathbf{x}||\mathbf{y}| \sin \vartheta$.

Under the exchange of $\mathbf{x}$ and $\mathbf{y}$, the two quantities $\mathbf{x} \cdot \mathbf{y}$ and $\mathbf{x} \times \mathbf{y}$ have opposite symmetry properties. $\mathbf{x} \cdot \mathbf{y} = \mathbf{y} \cdot \mathbf{x}$, but $\mathbf{x} \times \mathbf{y} = -(\mathbf{y} \times \mathbf{x})$, as is clear from equations (1.12) and (1.15). The latter relation implies that $\mathbf{x} \times \mathbf{x} = \mathbf{0}$ for any $\mathbf{x}$.

From three vectors $\mathbf{x}$, $\mathbf{y}$ and $\mathbf{z}$, there are two useful geometrical quantities that can be constructed. One is the scalar $(\mathbf{x} \times \mathbf{y}) \cdot \mathbf{z}$. This has several nice symmetry properties that can be verified using equation (1.16), in particular

$$(\mathbf{x} \times \mathbf{y}) \cdot \mathbf{z} = \mathbf{x} \cdot (\mathbf{y} \times \mathbf{z}). \qquad (1.19)$$

The other geometrical quantity is the double cross product $(\mathbf{x} \times \mathbf{y}) \times \mathbf{z}$, which is a vector. This can be expressed in terms of dot products through the important identity

$$(\mathbf{x} \times \mathbf{y}) \times \mathbf{z} = (\mathbf{x} \cdot \mathbf{z})\mathbf{y} - (\mathbf{y} \cdot \mathbf{z})\mathbf{x}. \qquad (1.20)$$

This identity, which is covariant under rotations, is easily checked using the cross product definition (1.15). To gain some intuition into its form, note that $\mathbf{x} \times \mathbf{y}$ is orthogonal to the plane spanned by $\mathbf{x}$ and $\mathbf{y}$, and taking the cross product with $\mathbf{z}$ gives a vector orthogonal to $\mathbf{x} \times \mathbf{y}$ and therefore back in this plane. $(\mathbf{x} \times \mathbf{y}) \times \mathbf{z}$ must therefore be a linear combination of $\mathbf{x}$ and $\mathbf{y}$. This vector must also be orthogonal to $\mathbf{z}$ and this is clearly true of the right-hand side of the identity, as

$$((\mathbf{x} \cdot \mathbf{z})\mathbf{y} - (\mathbf{y} \cdot \mathbf{z})\mathbf{x}) \cdot \mathbf{z} = (\mathbf{x} \cdot \mathbf{z})(\mathbf{y} \cdot \mathbf{z}) - (\mathbf{y} \cdot \mathbf{z})(\mathbf{x} \cdot \mathbf{z}) = 0. \qquad (1.21)$$

We have gone into these properties of $\mathbf{x} \cdot \mathbf{y}$ and $\mathbf{x} \times \mathbf{y}$ in some detail, because the laws of physics need to be expressed in a way that doesn't change when the entire physical system is rotated or translated. Even more importantly, the laws of physics should not change if one passively rotates the axes or shifts the origin. Dot products and cross products therefore occur frequently in physical contexts, for example, in formulae for energy and angular momentum. In the next section we will meet a vector of partial derivatives, denoted by ∇, and should not be surprised that it appears in electromagnetic theory in expressions such as $\nabla \cdot \mathbf{E}$ and $\nabla \times \mathbf{E}$, where $\mathbf{E}$ is the electric field vector. We will define and use these quantities in Chapter 3.

Geometrically speaking, there is not much to add concerning time until we discuss relativity in Chapter 4. In non-relativistic physics we use a further Cartesian coordinate t to represent time. Given times t_1 and t_2, it is the interval between them, $t_2 - t_1$, that is physically meaningful. Physical phenomena are unaffected by a time shift. If a process can start at t_1 and end at t_2 then it can equally well start at $t_1 + c$ and end at $t_2 + c$. Suppose some system starts at $t = 0$ and ends in the same state at $t = T$. Then it will repeat, and come back to the same state at $t = 2T$, $t = 3T$, and so on. This has an application that is very familiar to us in the guise of a clock.

1.3   Partial Derivatives

Physics in 3-dimensional space often involves functions of several variables. When a function depends on more than one variable, we need to consider the derivatives with respect to all of these. Suppose $\phi(x_1, x_2, x_3)$ is a smooth function defined in Euclidean 3-space. The *partial derivative* $\frac{\partial \phi}{\partial x_1}$ is just the ordinary derivative with respect to x_1, with x_2 and x_3 treated as fixed, or constant. It can be evaluated at any point $\mathbf{x} = (x_1, x_2, x_3)$. By taking x_2 and x_3 fixed, one is really just thinking of ϕ as a function of x_1 along the line through $\mathbf{x}$ parallel to the 1-axis, and the partial derivative $\frac{\partial \phi}{\partial x_1}$ is the ordinary derivative along this line. The partial derivatives $\frac{\partial \phi}{\partial x_2}$ and $\frac{\partial \phi}{\partial x_3}$ are similarly defined at $\mathbf{x}$ by differentiating along the lines through $\mathbf{x}$ parallel to the 2-axis and 3-axis.

It is easy to calculate the partial derivatives of functions that are known explicitly. For example, if $\phi(x_1, x_2, x_3) = x_1^3 x_2^4 x_3$, then $\frac{\partial \phi}{\partial x_1}$ is found by differentiating x_1^3 and treating $x_2^4 x_3$ as a constant, and similarly for $\frac{\partial \phi}{\partial x_2}$ and $\frac{\partial \phi}{\partial x_3}$. Therefore

$$\frac{\partial \phi}{\partial x_1} = 3x_1^2 x_2^4 x_3 \,, \qquad \frac{\partial \phi}{\partial x_2} = 4x_1^3 x_2^3 x_3 \,, \qquad \frac{\partial \phi}{\partial x_3} = x_1^3 x_2^4 \,. \tag{1.22}$$

Recall that by using the ordinary derivative of a function $f(x)$, denoted by $f'(x)$, we can obtain an approximate value for $f(x + \delta x)$ when δx is small:

$$f(x + \delta x) \simeq f(x) + f'(x)\delta x \,. \tag{1.23}$$

Similarly, by using the partial derivative $\frac{\partial \phi}{\partial x_1}$, we obtain

$$\phi(x_1 + \delta x_1, x_2, x_3) \simeq \phi(x_1, x_2, x_3) + \frac{\partial \phi}{\partial x_1} \delta x_1 \,. \tag{1.24}$$

By combining the three partial derivatives of ϕ at $\mathbf{x}$, we obtain the more powerful result

$$\phi(x_1 + \delta x_1, x_2 + \delta x_2, x_3 + \delta x_3)$$
$$\simeq \phi(x_1, x_2, x_3) + \frac{\partial \phi}{\partial x_1} \delta x_1 + \frac{\partial \phi}{\partial x_2} \delta x_2 + \frac{\partial \phi}{\partial x_3} \delta x_3 \,. \tag{1.25}$$

This provides an approximation for ϕ at any point $\mathbf{x} + \delta\mathbf{x}$ close to $\mathbf{x}$.

There is an implicit assumption here, which is that $\frac{\partial \phi}{\partial x_2}$ is essentially the same at the point $(x_1 + \delta x_1, x_2, x_3)$ as it is at (x_1, x_2, x_3), and similarly for $\frac{\partial \phi}{\partial x_3}$. This is why we supposed earlier that ϕ was smooth.

The collection of partial derivatives of ϕ forms a vector, denoted by $\nabla \phi$:

$$\nabla \phi = \left(\frac{\partial \phi}{\partial x_1}, \frac{\partial \phi}{\partial x_2}, \frac{\partial \phi}{\partial x_3} \right) . \tag{1.26}$$

Similarly $\delta \mathbf{x} = (\delta x_1, \delta x_2, \delta x_3)$ is a vector. Equation (1.25) can be written more concisely as

$$\phi(\mathbf{x} + \delta \mathbf{x}) \simeq \phi(\mathbf{x}) + \nabla \phi \cdot \delta \mathbf{x}, \tag{1.27}$$

a result we will use repeatedly. On the right-hand side is a genuine dot product that is unchanged if one rotates the axes. $\nabla \phi$ is called the *gradient* of ϕ.

A good way to think about a function is in terms of its contours. For a function ϕ in 3-space, the contours are the surfaces of constant ϕ. If $\delta \mathbf{x}$ is any vector tangent to the contour surface through $\mathbf{x}$, then $\phi(\mathbf{x} + \delta \mathbf{x}) - \phi(\mathbf{x}) \simeq 0$ to linear order in $\delta \mathbf{x}$, so $\nabla \phi \cdot \delta \mathbf{x} = 0$. Therefore $\nabla \phi$ is orthogonal to $\delta \mathbf{x}$, implying that $\nabla \phi$ is a vector orthogonal to the contour surface, as shown in Figure 1.9. In fact, $\nabla \phi$ is in the direction of steepest ascent of ϕ, and its magnitude is the rate of increase of ϕ with distance in this direction. This justifies the name 'gradient'.

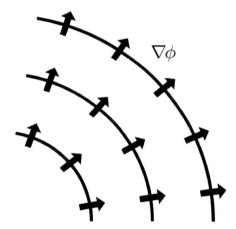

Fig. 1.9 The curves represent the contours of constant ϕ. The arrows represent the gradient $\nabla \phi$.

There can be points $\mathbf{x}$ where all three partial derivatives vanish, and $\nabla \phi = 0$. $\mathbf{x}$ is then a stationary point of ϕ. Whether the stationary point is a minimum, a maximum, or a saddle point depends on the second partial derivatives of ϕ at $\mathbf{x}$.

There are nine possible second partial derivatives of ϕ; these include $\frac{\partial^2 \phi}{\partial x_1^2}$, $\frac{\partial^2 \phi}{\partial x_1 \partial x_2}$, $\frac{\partial^2 \phi}{\partial x_2 \partial x_1}$ and $\frac{\partial^2 \phi}{\partial x_2^2}$. The mixed partial derivative $\frac{\partial^2 \phi}{\partial x_1 \partial x_2}$ is obtained by first differentiating ϕ with respect to x_2, and then differentiating the result by x_1, whereas for $\frac{\partial^2 \phi}{\partial x_2 \partial x_1}$ the order of differentiation is reversed. For example, for the function $\phi(x_1, x_2, x_3) = x_1^3 x_2^4 x_3$, one has

$$\frac{\partial^2 \phi}{\partial x_1^2} = 6 x_1 x_2^4 x_3, \qquad \frac{\partial^2 \phi}{\partial x_1 \partial x_2} = 12 x_1^2 x_2^3 x_3,$$

$$\frac{\partial^2 \phi}{\partial x_2 \partial x_1} = 12 x_1^2 x_2^3 x_3, \qquad \frac{\partial^2 \phi}{\partial x_2^2} = 12 x_1^3 x_2^2 x_3. \tag{1.28}$$

Notice that the mixed partial derivatives are actually the same. This is an important and completely general result.

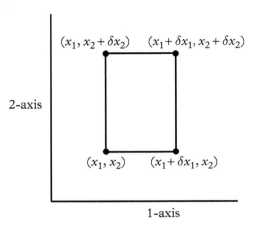

Fig. 1.10 Infinitesimal rectangle showing four positions at which the function ϕ can be evaluated.

To prove this result, one needs to think about the rectangle of values of ϕ shown in Figure 1.10 and estimate in two ways the expression

$$\phi(x_1 + \delta x_1, x_2 + \delta x_2, x_3) - \phi(x_1 + \delta x_1, x_2, x_3) - \phi(x_1, x_2 + \delta x_2, x_3) + \phi(x_1, x_2, x_3). \quad (1.29)$$

One estimate is the difference of the differences along the vertical edges,

$$\{\phi(x_1 + \delta x_1, x_2 + \delta x_2, x_3) - \phi(x_1 + \delta x_1, x_2, x_3)\} - \{\phi(x_1, x_2 + \delta x_2, x_3) - \phi(x_1, x_2, x_3)\}$$
$$\simeq \frac{\partial \phi}{\partial x_2}(x_1 + \delta x_1, x_2, x_3)\,\delta x_2 - \frac{\partial \phi}{\partial x_2}(x_1, x_2, x_3)\,\delta x_2$$
$$\simeq \frac{\partial^2 \phi}{\partial x_1 \partial x_2}(x_1, x_2, x_3)\,\delta x_1 \delta x_2. \quad (1.30)$$

The other estimate, with the bracketing reorganized, is the difference of the differences along the horizontal edges,

$$\{\phi(x_1 + \delta x_1, x_2 + \delta x_2, x_3) - \phi(x_1, x_2 + \delta x_2, x_3)\} - \{\phi(x_1 + \delta x_1, x_2, x_3) - \phi(x_1, x_2, x_3)\}$$
$$\simeq \frac{\partial \phi}{\partial x_1}(x_1, x_2 + \delta x_2, x_3)\,\delta x_1 - \frac{\partial \phi}{\partial x_1}(x_1, x_2, x_3)\,\delta x_1$$
$$\simeq \frac{\partial^2 \phi}{\partial x_2 \partial x_1}(x_1, x_2, x_3)\,\delta x_1 \delta x_2. \quad (1.31)$$

As the left-hand sides of these two expressions (1.30) and (1.31) are the same, the mixed partial derivatives must be equal. This result is called the *symmetry of mixed (second) partial derivatives*, because there is a symmetry under exchange of the order of differentiation. We shall make use of this later, for example, when investigating Maxwell's equations, and when deriving various thermodynamic relationships.

There is a particularly important combination of the second partial derivatives of ϕ, called the *Laplacian* of ϕ, and denoted by $\nabla^2 \phi$. This is

$$\nabla^2 \phi = \frac{\partial^2 \phi}{\partial x_1^2} + \frac{\partial^2 \phi}{\partial x_2^2} + \frac{\partial^2 \phi}{\partial x_3^2}, \quad (1.32)$$

and it is a scalar, unchanged if the axes are rotated. The scalar property is evident if one regards $\left(\frac{\partial}{\partial x_1}, \frac{\partial}{\partial x_2}, \frac{\partial}{\partial x_3}\right)$ as a vector of derivatives and writes

$$\nabla^2 \phi = \left(\frac{\partial}{\partial x_1}, \frac{\partial}{\partial x_2}, \frac{\partial}{\partial x_3}\right) \cdot \left(\frac{\partial \phi}{\partial x_1}, \frac{\partial \phi}{\partial x_2}, \frac{\partial \phi}{\partial x_3}\right), \tag{1.33}$$

or more compactly $\nabla^2 \phi = \nabla \cdot \nabla \phi$. More formally still, $\nabla^2 = \nabla \cdot \nabla$. For our familiar example $\phi = x_1^3 x_2^4 x_3$,

$$\begin{aligned} \nabla^2 (x_1^3 x_2^4 x_3) &= \frac{\partial^2}{\partial x_1^2}(x_1^3 x_2^4 x_3) + \frac{\partial^2}{\partial x_2^2}(x_1^3 x_2^4 x_3) + \frac{\partial^2}{\partial x_3^2}(x_1^3 x_2^4 x_3) \\ &= 6 x_1 x_2^4 x_3 + 12 x_1^3 x_2^2 x_3, \end{aligned} \tag{1.34}$$

a typical, non-zero result. However, there are plenty of functions whose Laplacian vanishes, for example, $x_1^2 - x_2^2$ and $x_1 x_2 x_3$.

In 3-space, one often needs to find the gradient or the Laplacian of a function $f(r)$ that depends only on the radial distance r from O. Here, $r^2 = x_1^2 + x_2^2 + x_3^2 = \mathbf{x} \cdot \mathbf{x}$. These calculations can be a bit fiddly, because r involves a square root, but are simpler if one works with r^2.

Let's find the gradient first. By the chain rule,

$$\nabla(r^2) = 2r\left(\frac{\partial r}{\partial x_1}, \frac{\partial r}{\partial x_2}, \frac{\partial r}{\partial x_3}\right) = 2r\nabla r. \tag{1.35}$$

On the other hand, by direct partial differentiation of $x_1^2 + x_2^2 + x_3^2$,

$$\nabla(r^2) = (2x_1, 2x_2, 2x_3) = 2\mathbf{x}. \tag{1.36}$$

Comparing these expressions we see that

$$\nabla r = \frac{\mathbf{x}}{r} = \hat{\mathbf{x}}. \tag{1.37}$$

$\mathbf{x}$ is a vector of magnitude r, and $\hat{\mathbf{x}}$ is the unit vector that points radially outwards at every point (except O). One can also understand equation (1.37) by noting that the contours of r are spheres centred at O, and the rate of increase of r with distance from O is unity everywhere. Equation (1.35) is easily generalized. For a general function $f(r)$, the chain rule gives

$$\nabla(f(r)) = f'(r)\nabla r = f'(r)\frac{\mathbf{x}}{r} = f'(r)\hat{\mathbf{x}}. \tag{1.38}$$

The most important example of this result is

$$\nabla\left(\frac{1}{r}\right) = -\frac{1}{r^2}\hat{\mathbf{x}}, \tag{1.39}$$

which is useful when considering the inverse square law forces of electrostatics and gravity.

Next, let's find the Laplacian of $f(r)$. We have $\nabla(f(r)) = \frac{1}{r}f'(r)\mathbf{x}$, so

$$\nabla^2(f(r)) = \nabla \cdot \nabla(f(r)) = \nabla \cdot \left(\frac{1}{r}f'(r)\mathbf{x} \right). \tag{1.40}$$

By the usual Leibniz rule, there are two contributions to the last expression. In one, ∇ acts on the function $\frac{1}{r}f'(r)$ to give the contribution

$$\left(\frac{1}{r}f''(r) - \frac{1}{r^2}f'(r) \right) \frac{\mathbf{x}}{r} \cdot \mathbf{x} = f''(r) - \frac{1}{r}f'(r), \tag{1.41}$$

where we have applied the result (1.38) again. The other is a dot product, in which the components $\left(\frac{\partial}{\partial x_1}, \frac{\partial}{\partial x_2}, \frac{\partial}{\partial x_3} \right)$ of ∇ act respectively on the three components (x_1, x_2, x_3) of $\mathbf{x}$ to give the number 3, so the second contribution is $\frac{3}{r}f'(r)$. Adding these, the result is

$$\nabla^2(f(r)) = f''(r) + \frac{2}{r}f'(r). \tag{1.42}$$

The most important example is

$$\nabla^2 \left(\frac{1}{r} \right) = \frac{2}{r^3} + \frac{2}{r}\left(\frac{-1}{r^2} \right) = 0. \tag{1.43}$$

This equation is valid everywhere except at O. $\frac{1}{r}$ is infinite at O, so its gradient is not defined there, and neither is its Laplacian. One says that $\frac{1}{r}$ is singular at O. The most general function just of the variable r whose Laplacian vanishes (except possibly at O) is $\frac{C}{r} + D$, where C and D are constants.

1.4 e, π and Gaussian Integrals

The transcendental numbers e and π appear throughout mathematics and physics, and will be used frequently in what follows. The exponential function e^x, often written as $\exp x$, and its complex counterpart e^{ix} will also appear frequently. There are two remarkable relations between e and π. One is the famous Euler relation

$$e^{i\pi} = -1, \tag{1.44}$$

and the other is the Gaussian integral formula

$$\int_{-\infty}^{\infty} e^{-x^2}\, dx = \sqrt{\pi}. \tag{1.45}$$

We shall explain these in this section and also describe two basic physical applications of the real and complex exponential functions.

   The exponential function is defined by the infinite series

$$e^x = 1 + x + \frac{1}{2}x^2 + \frac{1}{6}x^3 + \cdots + \frac{1}{n!}x^n + \cdots, \tag{1.46}$$

and is positive for all x. Obviously $e^0 = 1$. Euler's constant e is defined as e^1, the sum of the series for $x = 1$. Numerically, $e = 2.718....$ By expanding out, one can verify that

$$e^{x+y} = e^x\, e^y, \tag{1.47}$$

which is the key property of the exponential function. This property makes it consistent to identify e^x (as a series) with the x-th power of e. As an illustration, e^2 (as a series) equals

$e^1 e^1$ (the product of two series) so $e^2 = e \times e$. Differentiating the series (1.46) term by term, one easily sees that

$$\frac{d}{dx}(e^x) = e^x. \tag{1.48}$$

The importance of this simple formula is illustrated in section 1.4.1.

The extension of the exponential function to imaginary arguments is defined using the same series expansion,

$$e^{ix} = 1 + ix - \frac{1}{2}x^2 - \frac{i}{6}x^3 + \cdots + \frac{i^n}{n!}x^n + \cdots, \tag{1.49}$$

where $i^2 = -1$. The real and imaginary parts of this expansion are the well known series expansions of $\cos x$ and $\sin x$,

$$\cos x = 1 - \frac{1}{2}x^2 + \frac{1}{24}x^4 + \cdots, \tag{1.50}$$

$$\sin x = x - \frac{1}{6}x^3 + \cdots, \tag{1.51}$$

so

$$e^{ix} = \cos x + i \sin x. \tag{1.52}$$

Now, $\cos \pi = -1$ and $\sin \pi = 0$, so if we substitute the value $x = \pi$ into this expression we obtain the Euler relation, $e^{i\pi} = -1$. Raising the relation to the power of $2n$, we see that one consequence is that $e^{2ni\pi} = 1$ for any integer n.

1.4.1 Radioactive decay

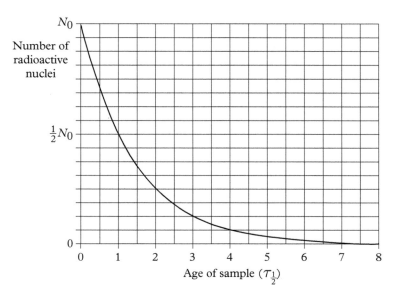

Fig. 1.11 Radioactive decay.

Radioactivity was discovered in 1896 by Henri Becquerel. When a radioactive nucleus decays it changes into a different nucleus. The rate of change of the number of radioactive nuclei N is described by the following law:

$$\frac{dN}{dt} = -\lambda N, \tag{1.53}$$

where λ is known as the decay constant. The radioactivity decays exponentially, as the solution of the differential equation (1.53) is

$$N = N_0 e^{-\lambda t}, \tag{1.54}$$

where N_0 is the initial number of radioactive nuclei when $t = 0$. The solution is shown in Figure 1.11. Taking logarithms, we obtain

$$\ln\left(\frac{N}{N_0}\right) = -\lambda t. \tag{1.55}$$

The time $\tau_{\frac{1}{2}}$ taken for half the nuclei to decay is known as the *half-life* of the radioactive substance. It is given by $\ln\left(\frac{1}{2}\right) = -\lambda \tau_{\frac{1}{2}}$, so

$$\tau_{\frac{1}{2}} = \frac{\ln 2}{\lambda}. \tag{1.56}$$

We can also work out the mean lifetime $\bar{t}$ of the radioactive nucleus. All N_0 nuclei eventually decay so we can average over the times of decay, finding

$$
\begin{aligned}
\bar{t} &= \frac{1}{N_0} \int_0^{N_0} t \, dN \\
&= -\frac{1}{\lambda N_0} \int_0^{N_0} \ln\left(\frac{N}{N_0}\right) dN \\
&= -\frac{1}{\lambda N_0} \Big[N \ln N - N - N \ln N_0 \Big]_0^{N_0} \\
&= \frac{1}{\lambda},
\end{aligned}
\tag{1.57}
$$

where in the second line we have substituted for t from equation (1.55).

Radioactivity provides an extremely useful tool for dating artefacts. If we know that a sample material originally contained N_0 radioactive nuclei, and now contains N of these, then we can determine the time t that has elapsed since the material formed,

$$t = \frac{1}{\lambda} \ln\left(\frac{N_0}{N}\right) = \frac{\tau_{\frac{1}{2}}}{\ln 2} \ln\left(\frac{N_0}{N}\right). \tag{1.58}$$

Different radioactive nuclei may be used depending on the relevant timescales. For instance, uranium-238, with a half-life of about 4.5 billion years, has been used to date meteorites and thereby determine the age of the solar system, and carbon-14, with a half-life of 5730 years, is used to date archaeological remains.

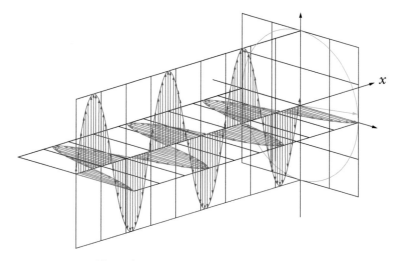

Fig. 1.12 The plane wave $e^{i(kx-\omega t)}$ travels in the x-direction at velocity $\frac{\omega}{k}$. As time passes, the amplitude of the wave at a fixed position remains constant and the phase of the wave rotates around a circle. The wave is shown decomposed into its real and imaginary components, which are two perpendicular sine waves with a relative phase shift of $\frac{\pi}{2}$.

1.4.2 Waves and periodic functions

We can represent a wave varying with position x and time t, travelling towards the positive x-direction, as $e^{i(kx-\omega t)}$ with k and ω positive, as shown in Figure 1.12. Because of the Euler relation, the wave will be the same at positions for which kx differs by an integer multiple of 2π, so the wavelength is $\frac{2\pi}{k}$. Similarly, the wave will be the same at times for which ωt differs by an integer multiple of 2π, so the period is $\frac{2\pi}{\omega}$. k and ω are called, respectively, the *wavenumber* and *angular frequency* of the wave.

The phase of the wave remains constant where $kx - \omega t = $ constant. The phase is therefore constant at a point x that moves along at velocity $\frac{\omega}{k}$, and this is the velocity of the wave. If k is negative, and ω still positive, the wave moves in the opposite direction.

The real and imaginary parts of the wave are $\cos(kx - \omega t)$ and $\sin(kx - \omega t)$. These are called sine waves, but one is phase shifted by $\frac{\pi}{2}$ relative to the other. Many types of wave, for example, electromagnetic waves and the waves on the surface of a fluid, are real, but in quantum mechanics the wavefunction of a freely moving particle is a complex wave.

1.4.3 The Gaussian integral

The integral of the Gaussian function e^{-x^2}, shown in Figure 1.13, cannot be expressed in terms of standard functions, so the indefinite integral from $-\infty$ to X is not elementary. On the other hand, the definite integral from $-\infty$ to ∞ has the value

$$I = \int_{-\infty}^{\infty} e^{-x^2}\, dx = \sqrt{\pi}\,. \tag{1.59}$$

This is the simplest *Gaussian integral*. It often arises in physics, and we will make use of it later.

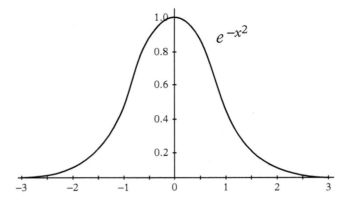

Fig. 1.13 The Gaussian function.

I can be evaulated using a rather surprising trick. We begin by considering its square,

$$I^2 = \int_{-\infty}^{\infty} e^{-x_1^2} \, dx_1 \int_{-\infty}^{\infty} e^{-x_2^2} \, dx_2 \,. \tag{1.60}$$

This can be expressed as the 2-dimensional integral

$$I^2 = \int_{\mathbb{R}^2} e^{-x_1^2 - x_2^2} \, d^2x \,, \tag{1.61}$$

where the integral is over the whole plane, $\mathbb{R}^2$. Now convert to polar coordinates. Let r be the radial coordinate and ϑ the angular coordinate. Then, by Pythagoras' theorem $r^2 = x_1^2 + x_2^2$, and the integration measure is $d^2x = r \, dr \, d\vartheta$. So

$$I^2 = \int_0^{2\pi} \int_0^{\infty} e^{-r^2} r \, dr \, d\vartheta = 2\pi \int_0^{\infty} e^{-r^2} r \, dr \,. \tag{1.62}$$

The range of ϑ is 2π because, geometrically, 2π is the length of the unit circle. The extra factor of r makes the integral over r elementary, and the result is

$$I^2 = 2\pi \left[-\frac{1}{2} e^{-r^2} \right]_0^{\infty} = \pi \,, \tag{1.63}$$

so $I = \sqrt{\pi}$, as claimed.

A more general Gaussian integral is

$$I(\alpha) = \int_{-\infty}^{\infty} e^{-\alpha x^2} \, dx = \frac{1}{\sqrt{\alpha}} \int_{-\infty}^{\infty} e^{-y^2} \, dy = \sqrt{\frac{\pi}{\alpha}} \,, \tag{1.64}$$

where we have used the substitution $y = \sqrt{\alpha} \, x$. Another useful trick allows us to evaluate a sequence of integrals where the Gaussian function is multiplied by an even power of x. Differentiating the integral $I(\alpha)$ with respect to α brings down a factor of $-x^2$, so

$$\int_{-\infty}^{\infty} x^2 e^{-\alpha x^2} \, dx = -\frac{dI(\alpha)}{d\alpha} = -\frac{d}{d\alpha} \sqrt{\frac{\pi}{\alpha}} = \frac{1}{2\alpha} \sqrt{\frac{\pi}{\alpha}} \,. \tag{1.65}$$

Differentiating a second time with respect to α gives

$$\int_{-\infty}^{\infty} x^4 e^{-\alpha x^2}\, dx = -\frac{d}{d\alpha}\left(\frac{1}{2\alpha}\sqrt{\frac{\pi}{\alpha}}\right) = \frac{3}{4\alpha^2}\sqrt{\frac{\pi}{\alpha}}. \tag{1.66}$$

We can continue differentiating with respect to α to evaluate all integrals of the form $\int_{-\infty}^{\infty} x^{2n} e^{-\alpha x^2}\, dx$.

If the Gaussian is multiplied by an odd power of x, the integrand is an odd function, antisymmetric under $x \to -x$, so $\int_{-\infty}^{\infty} x^{2n+1} e^{-\alpha x^2}\, dx = 0$. When the lower limit is 0, these integrals can be evaluated by substituting $y = x^2$, then integrating by parts, to give

$$\begin{aligned}
\int_0^{\infty} x^{2n+1} e^{-x^2}\, dx &= \frac{1}{2}\int_0^{\infty} y^n e^{-y}\, dy \\
&= \left[-\frac{1}{2}y^n e^{-y}\right]_0^{\infty} + \frac{1}{2}n\int_0^{\infty} y^{n-1} e^{-y}\, dy \\
&= \frac{1}{2}n\int_0^{\infty} y^{n-1} e^{-y}\, dy.
\end{aligned} \tag{1.67}$$

Repeating this procedure n times, we find $\int_0^{\infty} y^n e^{-y}\, dy = n!\int_0^{\infty} e^{-y}\, dy = n!$, and therefore

$$\int_0^{\infty} x^{2n+1} e^{-x^2}\, dx = \frac{1}{2}n!. \tag{1.68}$$

The basic Gaussian integral and these variants are useful in many areas of physics, especially quantum mechanics and quantum field theory.

We can also find interesting geometrical results by considering the nth power of I,

$$I^n = \int_{-\infty}^{\infty} e^{-x_1^2}\, dx_1 \int_{-\infty}^{\infty} e^{-x_2^2}\, dx_2 \cdots \int_{-\infty}^{\infty} e^{-x_n^2}\, dx_n, \tag{1.69}$$

which can be re-expressed as an n-dimensional integral

$$I^n = \int_{\mathbb{R}^n} e^{-x_1^2 - x_2^2 - \cdots - x_n^2}\, d^n x. \tag{1.70}$$

Now convert to spherical polar coordinates r, Ω in n dimensions, where Ω denotes collectively the $n-1$ angular coordinates. By Pythagoras' theorem in n dimensions, $r^2 = x_1^2 + x_2^2 + \cdots + x_n^2$, and the integration measure $d^n x$ becomes $r^{n-1}\, dr\, d\Omega$, where $d\Omega$ denotes the volume element of the unit sphere in n dimensions—the unit $(n-1)$-sphere. So

$$I^n = \int\int_0^{\infty} e^{-r^2} r^{n-1}\, dr\, d\Omega. \tag{1.71}$$

The integral of $d\Omega$ is the total volume of the unit $(n-1)$-sphere, and the remaining radial integral is one of the Gaussian integrals considered above.

For example, in the case of I^3 the radial integral has the same form as the integral (1.65), but with the lower limit 0 (and $\alpha = 1$). It equals $\frac{1}{4}\sqrt{\pi}$, half the full Gaussian integral, so

$$I^3 = \frac{1}{4}\sqrt{\pi} A \tag{1.72}$$

where A is the area of the unit 2-sphere, the sphere we are familiar with. We know that $I = \sqrt{\pi}$, so $I^3 = \pi\sqrt{\pi}$, and therefore $A = 4\pi$, the well known result for the area of the sphere. Note that in this calculation, using a Gaussian integral, we have not needed to make an explicit choice of angular coordinates to find A.

By a similar calculation we can find a less well known result, the volume of the unit sphere in four dimensions, the 3-sphere. Just as the 2-sphere is a 2-dimensional surface enclosing a 3-dimensional ball, so the 3-sphere is a 3-dimensional volume enclosing a ball of 4-dimensional space. Equation (1.71) becomes

$$I^4 = V \int_0^\infty e^{-r^2} r^3 \, dr \,, \tag{1.73}$$

where V is the volume of the unit 3-sphere. Using $I^4 = \pi^2$ and the integral (1.68) in the case $n = 1$, namely $\int_0^\infty e^{-r^2} r^3 \, dr = \frac{1}{2}$, we find $V = 2\pi^2$.

1.4.4 The method of steepest descents

In many physics applications we arrive at integrals that cannot be evaluated exactly, where the integrand is a product of a variant of a Gaussian function and another function. We will see an example of this in Chapter 11, when considering nuclear fusion. In such cases, the basic Gaussian integral can be used to give estimates of these more complicated integrals. Suppose $g(x)$ has a single maximum between a and b at x_0; then, since $g'(x_0) = 0$ and $g''(x_0) < 0$, we can use the Taylor expansion $g(x) \simeq g(x_0) - \frac{1}{2}|g''(x_0)|(x - x_0)^2$ near x_0. This means that the integral

$$I = \int_a^b F(x) \exp(g(x)) \, dx \tag{1.74}$$

can be approximated by

$$I \simeq \exp(g(x_0)) \int_a^b F(x) \exp\left(-\frac{1}{2}|g''(x_0)|(x - x_0)^2\right) dx \,. \tag{1.75}$$

If, further, $F(x)$ varies slowly near x_0, then it can be treated as a constant $F(x_0)$ and pulled out of the integral to give

$$I \simeq F(x_0) \exp(g(x_0)) \int_a^b \exp\left(-\frac{1}{2}|g''(x_0)|(x - x_0)^2\right) dx \,. \tag{1.76}$$

As the integrand is concentrated around the point x_0, we can extend the limits of integration to $\pm\infty$ without significantly affecting the value of the integral, so

$$
\begin{aligned}
I &\simeq F(x_0) \exp(g(x_0)) \int_{-\infty}^\infty \exp\left(-\frac{1}{2}|g''(x_0)|(x - x_0)^2\right) dx \\
&= F(x_0) \exp(g(x_0)) \sqrt{\frac{2\pi}{|g''(x_0)|}} \,,
\end{aligned}
\tag{1.77}
$$

where in the last step we used the Gaussian integral (1.64).

This is known as the steepest descents approximation. It is accurate provided the second derivative $g''(x_0)$ has a large magnitude and higher order terms in the Taylor expansions of g and F around x_0 can be neglected.

1.5 Further Reading

For a survey of variational principles and their history, see

D.S. Lemons, *Perfect Form: Variational Principles, Methods, and Applications in Elementary Physics*, Princeton: PUP, 1997.

H.H. Goldstine, *A History of the Calculus of Variations: from the 17th through the 19th Century*, New York: Springer, 1980.

For comprehensive coverage of the mathematics used in this book, consult

K.F. Riley, M.P. Hobson and S.J. Bence, *Mathematical Methods for Physics and Engineering (3rd ed.)*, Cambridge: CUP, 2006.

2

Motions of Bodies—Newton's Laws

2.1 Introduction

In this city-dwelling age, we rarely witness the full beauty of the heavens and it may seem that stargazing is just an expensive and amusing, but ultimately worthless pastime. We should not forget, however, that science began with astronomy. During the final third of the 16th century, Tycho Brahe raised astronomy to a new level of precision. He designed and built large instruments that enabled him to undertake a systematic and accurate survey of the night sky plotting the positions of the planets over a period of several decades, and introduced many procedures now routinely used by scientists when gathering data, such as seeking sources of error and estimating their size. Following Tycho's death in 1601, Johannes Kepler immersed himself in a painstaking analysis of these observations in search of a model that would explain the planetary motions. After several years of intense struggle, in 1609 Kepler published a new and concise description of how the planets move around the Sun. His conclusions are summarized in three laws. The first describes the shape of a planet's orbit. It is an ellipse, with the Sun at one focus of the ellipse. The second describes the relative rate at which the planet travels around this ellipse as it gets closer to and further from the Sun. The third relates the orbital period of the planet to its distance from the Sun.

Kepler's laws are purely descriptive and he was unable to find a true causal explanation. His best guess was that somehow the rotation of the Sun swept the planets around. This issue remained unresolved for most of the 17th century. It was the desire to find a mechanical explanation of Kepler's laws that motivated Isaac Newton to develop his system of mechanics published in the *Principia* in 1687. Newton built on the work of others, most notably Kepler, Galileo Galilei and Jeremiah Horrocks, but his personal achievement was monumental. Newton produced the first rational mechanics and stimulated the development of science as a whole. This brought about a revolution that has ultimately led to the creation of the modern world.

Although Newton was the first to understand calculus, his *Principia* was written in the language of classical geometry. We will not dwell on Newton's original presentation, however, but will use a style of mathematics developed long after Newton. For instance, Newton was the first to recognize that for velocities, accelerations and forces, their directions are just as important as their magnitudes, so they must be treated as vectors. However, the vector notation that we will use was not developed until the late 19th century.

We start with a survey of Newton's laws of motion, and show how these laws can be derived from the principle of least action. We will then consider some important examples of

The Physical World. Nicholas Manton and Nicholas Mee, Oxford University Press (2017). © Nicholas Manton and Nicholas Mee. DOI 10.1093/oso/9780198795933.001.0001

the motion of bodies in three dimensions and show that Kepler's laws follow from Newton's laws of motion if we assume, following Newton, that the attractive force between the Sun and a planet falls off with the inverse square of the distance between them.

2.2 Newton's Laws of Motion

Newton's laws describe the motion of one or more massive bodies. A single body has a definite mass, m. The body's internal structure and shape can often be neglected, and then the body can be treated as a point particle with a definite position $\mathbf{x}$. As it moves, its position traces out a curve in space, $\mathbf{x}(t)$. Later we will show that composite bodies, despite their finite size, can be treated as having a single central position called the *centre of mass*.

Newton's first law states that the motion of a body at constant velocity is self-sustaining and no force is needed. The velocity $\mathbf{v}$ is the time derivative of the position $\mathbf{x}(t)$ of the body,

$$\mathbf{v} = \frac{d\mathbf{x}}{dt}. \tag{2.1}$$

In the absence of a force, the velocity is a constant $\mathbf{v}_0$, so $\frac{d\mathbf{x}}{dt} = \mathbf{v}_0$, and the body's position as a function of time is

$$\mathbf{x}(t) = \mathbf{x}(0) + \mathbf{v}_0 t, \tag{2.2}$$

where $\mathbf{x}(0)$ is the position at the initial time $t = 0$. The body moves along a straight line at constant speed $|\mathbf{v}_0|$ and if the velocity is zero, the body is at rest.

Newton's second law defines what we mean by a force. It states that if a force acts on a body of mass m, then the body accelerates. The acceleration $\mathbf{a}$ and force $\mathbf{F}$ are parallel vectors, and the relation between them is

$$m\mathbf{a} = \mathbf{F}. \tag{2.3}$$

This is the starting point for most calculations involving forces in Newtonian mechanics.

Newton's second law is intimately tied to calculus. The acceleration is the time derivative of velocity and hence the second time derivative of position,

$$\mathbf{a} = \frac{d\mathbf{v}}{dt} = \frac{d^2\mathbf{x}}{dt^2}. \tag{2.4}$$

With a given force, equation (2.3) becomes a second order differential equation for the position of the body as a function of time,

$$m\frac{d^2\mathbf{x}}{dt^2} = \mathbf{F}. \tag{2.5}$$

If there is no force, then the acceleration is zero and the velocity is constant, which is a restatement of the first law, so the first law may be regarded as a special case of the second.

Equation (2.5) is the key to the success of Newtonian mechanics. It has enormous predictive power, but we require some independent information about the form taken by the force $\mathbf{F}$ in order to make use of it. We have this for the electric and magnetic forces on charged particles, using the notions of electric and magnetic fields, as we will discuss in Chapter 3. Forces due to springs, and various types of contact forces, describing collisions and friction, may also be represented by simple algebraic expressions. In the case of gravity,

Newton showed that Kepler's laws can only be explained if the force between the Sun and a planet obeys an inverse square law and we will come to this later.

Newton's law of gravity simplifies for bodies *near* the Earth's surface, and their motion can then be readily found. The force exerted by the Earth on a body of mass m is downwards, with magnitude mg, where $g = 9.81\,\mathrm{m\,s^{-2}}$ is a positive constant combining Newton's *universal gravitational constant* G with the mass and radius of the Earth. In this case Newton's second law reduces to

$$ma = -mg \tag{2.6}$$

where a is the upward acceleration. m cancels, which is always true for acceleration due to gravity, so $a = -g$. The acceleration a is negative, and is of course downwards. g is called *the* acceleration due to gravity. It is the same for all bodies and in this simplified case is independent of their position.

Let us assume that the motion is purely vertical and look more closely at equation (2.6) as a differential equation. After cancelling m, equation (2.6) takes the form

$$\frac{d^2z}{dt^2} = -g\,, \tag{2.7}$$

with z the height of the body above some reference level. The solution is

$$z(t) = -\frac{1}{2}gt^2 + u_0 t + z_0\,, \tag{2.8}$$

where z_0 and u_0 are the height and upward velocity at time $t = 0$. The graph of z against t, for any z_0 and u_0, is a parabola, or part of a parabola if the time interval is finite, as shown in Figure 2.1.

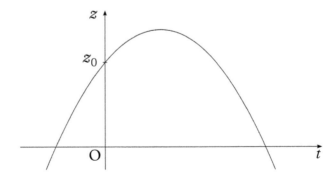

Fig. 2.1 Motion under gravity.

We can consider non-vertical motion too, for example, the motion of a projectile like a cannon ball. The body moves in a vertical plane, with z the vertical coordinate and x the horizontal coordinate. Since there is no horizontal component to the gravitational force, the body has no horizontal acceleration, so x is linearly related to t. With the origin of x chosen appropriately, x is just a constant multiple of t, the multiple being the constant x-component of the velocity. We will assume this multiple is non-zero. On the other hand, the vertical part of the motion is given by equation (2.8), as before. Rather than plotting z against t, we can now plot z against x. This just requires a rescaling of the t-axis, because x is a multiple of t. Figure 2.1 then shows the parabolic trajectory of the body in the (x, z)-plane, rather than the height as a function of time.

Newton's third law states that for every force there is a reaction force acting in the opposite direction. If one body exerts a force $\mathbf{F}$ on a second body, then at the same time the second body exerts a force $-\mathbf{F}$ on the first body. This can be observed in collisions of billiard balls and in the motion of celestial bodies of comparable mass, such as binary stars. Indeed, this is one of the methods by which exoplanets have been discovered around nearby stars, as the apparent position of a star will oscillate as it is orbited by an unseen planet. Similarly, when a body of mass m is near the Earth's surface, and the Earth exerts a gravitational force mg downward, then the body exerts a gravitational force on the Earth of the same magnitude upward, although this may be too small to measure. If the mass hangs from a spring, however, then the spring exerts an upward force mg on the mass that prevents it from falling, and the mass exerts a downward force mg on the spring, which stretches it and allows m to be measured.

We may have solid observational evidence for Newton's third law, but it is not immediately obvious why it should be true. We shall see below, in the context of the principle of least action, that the third law follows from a simple geometrical idea.

2.3   The Principle of Least Action

The motions of massive bodies all have one thing in common. Whether it is a heavy ball thrown through the air or a planet's motion around the Sun, there is a quantity related to the body's energy, known as the *action*, which takes its smallest possible value when evaluated along the path taken by the body. The fact that the action is minimized along the body's trajectory is known as the *principle of least action*. In practice, the principle is used to derive the equations of motion, which are the same as those derived by more standard methods, as we will soon see. The principle of least action says that, in some sense, the actual motion that we observe is the *optimum* out of all conceivable motions that could have occurred. It might appear that nature is working in an efficient way, with minimal effort, to some kind of plan. Of course, nature isn't consciously 'trying' to optimize its performance and there isn't a plan. In fact, no foresight is necessary, as only local information is relevant, which is why the conditions for a trajectory to be optimal can be re-expressed as differential equations. The principle of least action is actually much more basic than Newtonian mechanics and its applicability extends far beyond Newtonian physics. Essentially all the laws of physics, describing everything from the smallest elementary particle to the motion of galaxies in the expanding universe, can be understood using some version of the principle of least action. Indeed, we might regard the ultimate goal of theoretical physicists and applied mathematicians to be the discovery of the correct form taken by the action in each branch of physics.

It is not *essential* to consider the principle of least action. One could instead just use the equations of motion. This is the traditional approach that has been used throughout physics, but surprisingly, the principle of least action seems to be more fundamental than the equations of motion. The argument for this is made, with characteristic enthusiasm, in one of the most famous Feynman lectures. A key part of the argument is that the principle of least action is not just a technique for obtaining the classical equations of motion of particles and fields. It also plays a central role in the relationship between classical and quantum theory.

There are several advantages to using the principle of least action. The first is the conceptual one, in that it appears to be a fundamental and unifying principle in all areas of physical science. A second is that its mathematical formulation is based on the geometry of space and time, and on the key concepts of velocity and energy, whereas the variables that occur in Newton's second law, acceleration and force, become secondary, derived concepts. This is useful as velocity is simpler than acceleration, and energy is easier to understand intuitively than forces. When using Newton's laws, there is always the question of how the forces arise and what determines their form. A third advantage is that there are fewer action principles than equations of motion. All the equations of motion for a system of bodies follow from a single principle. Similarly, all four of Maxwell's equations for electromagnetic fields follow from one action principle. A final advantage is that the action can be written down using any coordinate system, making it easier to understand certain kinds of motion. Converting equations of motion from Cartesian to polar coordinates, for example, is rather fiddly, but the equations in polar coordinates can be obtained relatively easily if one starts with the principle of least action.

So what are the disadvantages? Well, more sophisticated mathematical technology is required. The action is a combination of energy contributions, integrated over time, and the standard method by which one derives the equations of motion is the calculus of variations, which is calculus in function space, not elementary calculus. Also, the equations of motion derived from the principle of least action are differential equations, which still need to be solved.

There is also an apparent physics issue, in that equations of motion derived from a principle of least action have no friction terms, which implies that energy is conserved and the motion continues forever. Friction must be added separately, but this actually represents more of a gain than a loss. At a fundamental level, it expresses the fact that energy really is conserved. Friction terms are a phenomenological way of dealing with energy dissipation, the transfer of energy to microscopic degrees of freedom outside the system being considered.

The calculus of variations may sound formidable, but fortunately, the principle of least action and its consequences can be made more readily accessible. In Chapter 1 we started down this road by showing that in some applications of Fermat's principle involving light rays, the calculus of variations is not required. We obtained physically important results using geometry coupled with elementary calculus. Shortly, we will present the principle of least action for a body moving in one dimension and rederive Newton's second law of motion. For the simple example of motion in a linear potential, which corresponds to a force that is constant, we can again use elementary calculus. Extending the argument, we can go on to derive the equation of motion in a general potential. For completeness, we also give the calculus of variations derivation.

In section 2.4 we discuss the principle of least action for two interacting bodies, which leads to Newton's third law and the law of momentum conservation. We also show that for a composite body with two or more parts, there is a natural notion of its centre of mass. This emerges by considering the body's total momentum.

2.3.1 Motion in one dimension

Let us see how the principle of least action can be used to derive Newton's second law of motion. It is simplest to consider the motion of a single body in one dimension, say along the x-axis. Let $x(t)$ be a *possible* path of the body, but not necessarily the actual one that is taken. The body's velocity is

$$v = \frac{dx}{dt},$$ (2.9)

which is also a function of t.

To set up the principle of least action, we postulate that a moving body has two types of energy. The first is *kinetic energy* due to its velocity. This does not depend on the direction of motion, so it is the same for velocity v and velocity $-v$, which suggests that kinetic energy is a multiple of v^2. What else does it depend on? Intuitively, the kinetic energy of several bodies is the sum of the kinetic energies of the individual bodies. A group of N equal bodies, moving together with the same velocity, has N times the kinetic energy of one body. It also has N times the mass. So kinetic energy is proportional to mass, and the squared velocity. The postulate is that the kinetic energy K of a body of mass m and velocity v is

$$K = \frac{1}{2}mv^2 = \frac{1}{2}m\left(\frac{dx}{dt}\right)^2,$$ (2.10)

where the factor of $\frac{1}{2}$ is convenient for connecting with Newton's laws.

The second type of energy of a body is *potential energy*. This is due to the environment and is independent of velocity. It depends on the presence of other bodies and the way they interact with each other, electrically, gravitationally or otherwise. The postulate is that the potential energy of a body is a function of its position, $V(x)$. We only really need to know V at the location of the body at each moment in time t, which is $x(t)$, so to be precise we write $V(x(t))$. However, it is important that V is defined everywhere that the body *might* be, which is all x in some range. We often say that the body is moving in the potential V.

The form of the potential energy $V(x)$ depends on the physical situation and it must be known in order to perform calculations, just as with Newton's second law the force must be known in order to work out the body's motion. $V(x)$ sometimes has a simple form. For example, if the body is free and has no significant interaction with the environment, then V is independent of position and is just a constant V_0. We shall see later that the value of this constant has no physical effect. For a body near the Earth's surface we know intuitively that it takes energy to lift the body, so the body's potential energy increases with height. Lifting a body through a height h requires a certain energy, and lifting it through a further height h requires the same energy again. Also, lifting two bodies of mass m through height h requires twice the energy needed to lift one body of mass m. This leads to the assertion that the increase in potential energy when a body is raised by height h is mgh, proportional to mass and height, and multiplied by a constant g, which as we will see later, is the acceleration due to gravity. The complete potential energy of the body at height x above some reference level is therefore

$$V(x) = V_0 + mgx,$$ (2.11)

where the constant V_0 again has no effect. (In this section, for consistency, we use x as the coordinate denoting height, rather than z as before.) For a body attached to a stretched spring, the potential is $V(x) = \frac{1}{2}kx^2$, a quadratic function of x, and there are other cases where the form of V is known or can be postulated.

We now consider the body moving between an initial position x_0 at time t_0 and a final position x_1 at a later time t_1. We will adopt Hamilton's definition of the action, which is now standard, although historically there were other definitions. For each possible motion, the action S is defined to be

$$S = \int_{t_0}^{t_1} \left(\frac{1}{2}m \left(\frac{dx}{dt} \right)^2 - V(x(t)) \right) dt \,. \tag{2.12}$$

The integrand is the kinetic energy *minus* the potential energy of the body at time t. The minus sign is important and explains why we spoke earlier about two types of energy. They are distinguished because one depends on velocity and the other does not. The action is sometimes written in the condensed form

$$S = \int_{t_0}^{t_1} (K - V) \, dt \,, \tag{2.13}$$

or more concisely still as

$$S = \int_{t_0}^{t_1} L \, dt \,, \tag{2.14}$$

where $L = K - V$ is called the *Lagrangian*. The action is the time integral of the Lagrangian, and this is the case not just for one body moving in one dimension, but much more generally.

The principle of least action now states that among all the possible paths $x(t)$ that connect the fixed endpoints, the actual path $X(t)$ taken by the body is the one for which the action S takes its minimum value.[1]

Note that we are not minimizing with respect to just a single quantity, such as the position of the body at the mid-time $\frac{1}{2}(t_0 + t_1)$. Rather, we are minimizing with respect to the infinite number of variables that characterize all possible paths, with all their possible wiggles, which is a much more subtle problem. In order to proceed, we must make the physically reasonable assumption that the paths $x(t)$ have some smoothness. In other words, acceptable paths are those for which the acceleration remains finite, so the velocity is continuous. Some typical acceptable paths are shown in Figure 2.2.

We can now see why V_0, either as a constant potential, or as an additive contribution to a non-constant potential such as in equation (2.11), has no effect. When substituted into the integral (2.14), its contribution to the action S is simply $-(t_1 - t_0)V_0$, which is itself a constant, independent of the path $x(t)$. Finding the path $X(t)$ that minimizes S is unaffected by this constant contribution to S. So we will usually simply drop V_0.

2.3.2 A simple example and a simple method

A simple example where the principle of least action can be applied is where the potential energy $V(x)$ is linear in x, so $V(x) = kx$ with k constant. We shall determine the motion of the body over the time interval $-T \leq t \leq T$, assuming that the initial position is $x(-T) = -X$ and the final position is $x(T) = X$. This choice of initial and final times and

---

[1] This is usually the case, but sometimes the action is stationary rather than minimal. The equation of motion is unaffected by this distinction.

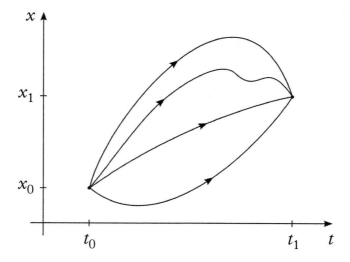

Fig. 2.2 Possible paths $x(t)$.

positions may look rather contrived, but it is always possible to simplify the calculations by choosing the origin of t and of x to be midway between the initial and final times and positions, as they are here. Making such a choice takes advantage of the Euclidean symmetry of space and time.

Next, consider a very limited class among the possible paths $x(t)$ from the initial to final positions. Assume that the graph of $x(t)$ is a parabola passing through the given endpoints, as shown in Figure 2.3, so $x(t)$ is a quadratic function of the form $At^2 + Bt + C$. There are three parameters in this expression, but with two endpoint constraints, only one parameter is *free*. To satisfy the constraints, the form taken by $x(t)$ must be

$$x(t) = \frac{X}{T}t + \frac{1}{2}a(t^2 - T^2).\tag{2.15}$$

$\frac{X}{T}$ is the average velocity, which is detemined by the values of x and t at the endpoints. a is the free parameter and it is equal to the (constant) acceleration, as $\frac{d^2x}{dt^2} = a$. The term proportional to a vanishes at the endpoints, so $x(-T) = -X$ and $x(T) = X$, as required.

For the path given by equation (2.15), the velocity is

$$\frac{dx}{dt} = \frac{X}{T} + at\tag{2.16}$$

and therefore the kinetic energy is $K = \frac{1}{2}m\left(\frac{X}{T} + at\right)^2$. The potential energy is $kx(t)$ at time t, and is k times the expression (2.15). Combining the kinetic and potential energies we obtain the action

$$S = \int_{-T}^{T}\left\{\frac{1}{2}m\left(\frac{X}{T} + at\right)^2 - k\left(\frac{X}{T}t + \frac{1}{2}a(t^2 - T^2)\right)\right\}dt\,,\tag{2.17}$$

the integral of a quadratic function of t. The linear terms integrate to zero, because the integral is from $-T$ to T.

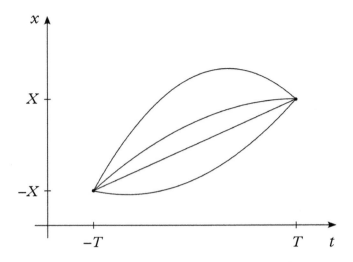

Fig. 2.3 Parabolic paths with various accelerations.

With these terms removed, we have

$$S = \int_{-T}^{T} \left\{ \frac{1}{2}m \left(\frac{X^2}{T^2} + a^2 t^2 \right) - \frac{1}{2}ka(t^2 - T^2) \right\} dt$$

$$= m\frac{X^2}{T} + \frac{1}{3}ma^2 T^3 + \frac{2}{3}kaT^3 \,. \tag{2.18}$$

To satisfy the principle of least action, we must find the value of a that minimizes S. This is standard calculus. Differentiating S with respect to a, we obtain

$$\frac{dS}{da} = \frac{2}{3}maT^3 + \frac{2}{3}kT^3 \,, \tag{2.19}$$

and setting this to zero gives the relation

$$ma = -k \,. \tag{2.20}$$

The acceleration a that minimizes S is therefore $-\frac{1}{m}k$, and substituting into equation (2.15) gives the motion of the body as

$$X(t) = \frac{X}{T}t - \frac{k}{2m}(t^2 - T^2) \,. \tag{2.21}$$

(For this value of a, the action is $S = \frac{mX^2}{T} - \frac{k^2 T^3}{3m}$, but this is of less interest.)

We can interpret equation (2.20) as follows. The linear potential $V(x) = kx$ produces a force $-k$, and equation (2.20) is Newton's second law, with the acceleration a being constant and equal to $-\frac{1}{m}k$. The result would be the same for the potential $V(x) = V_0 + kx$.

Our method has determined the true motion in this simple example. However, the method looks quite incomplete, because we have not minimized S over all paths through the endpoints, but only over the subclass of parabolic paths, which have constant acceleration. The next step is to show that the method is better than it appears, and will lead us to the correct equation of motion for completely general potentials $V(x)$.

2.3.3 Motion in a general potential and Newton's second law

Let us now consider the principle of least action for 1-dimensional motion in a general potential $V(x)$. The action S is given in equation (2.12), still with the endpoint conditions $x(t_0) = x_0$ and $x(t_1) = x_1$. We assume there exists a path $X(t)$ that satisfies these conditions and minimizes S.

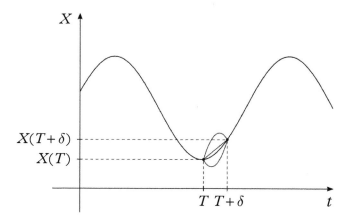

Fig. 2.4 Possible parabolic paths over a very short time interval.

The motion $X(t)$ necessarily minimizes the action over any smaller subinterval of time between t_0 and t_1. If it did not, we could modify the path inside that subinterval, and reduce the total action. Let us therefore focus on a subinterval between times T and $T + \delta$, where δ is very small, and minimize the action on this subinterval. Suppose the true motion is from $X(T)$ to $X(T + \delta)$ in this interval, with $X(T + \delta)$ very close to $X(T)$. As these intervals in time and space are very small, we can make some approximations. The simplest would be to say that the potential V is constant, and X varies linearly with t. But this is too simple and we don't learn anything. A more refined approximation is to say that $V(x)$ varies linearly with x between $X(T)$ and $X(T + \delta)$, and that the path $X(t)$ varies quadratically with t, so its graph is a parabola, as illustrated in Figure 2.4. With $V(x)$ linear it has a definite slope $\frac{dV}{dx}$, which can be regarded as constant between $X(T)$ and $X(T + \delta)$. With $X(t)$ quadratic in this interval, and its graph a parabola, the motion generally has some acceleration.

Now we can make use of the simple calculation in the last section. There we showed that if the potential is linear with slope k, as described by $V(x) = V_0 + kx$, then among the *parabolic* paths, the path that minimizes the action is the one for which ma, the mass times acceleration, equals $-k$. When applied to the small time interval from T to $T + \delta$, this implies that ma equals $-\frac{dV}{dx}$, the negative of the slope of the potential evaluated at $X(T)$. This is the key result. By approximating the graph of $X(t)$ by a parabola over a short time interval, we have found the acceleration. Although we have minimized the action in just one short interval, the same analysis applies to any other short interval. In general $-\frac{dV}{dx}$ will vary from one interval to another, so the acceleration will vary too.

If we write the acceleration as the second time derivative of x, we find the general equation of motion

$$m\frac{d^2x}{dt^2} = -\frac{dV}{dx}\,,$$ (2.22)

obtained from the principle of least action. The true motion, $X(t)$, is a solution of this.

Equation (2.22) has the form of Newton's second law of motion. We identify the force F that acts on the body as $-\frac{dV}{dx}$. This is, in fact, the main lesson from the principle of least action. The potential $V(x)$ is the fundamental input, and the force $F(x)$ is derived from it. The force is minus the derivative of the potential. It is a function of x and needs to be evaluated at the position where the body is, namely $x(t)$.

The expression $F = -\frac{dV}{dx}$ may be familiar in the related form $F\Delta x = -\Delta V$. $F\Delta x$ is the work done, ΔW, when the body moves a short distance Δx. In the absence of friction, the body's kinetic energy increases by ΔW. For us $\Delta W = -\Delta V$, so as the body accelerates, the increase in kinetic energy equals the decrease in potential energy.

Newton's first law follows as a special case of equation (2.22). If the potential V is a constant V_0 then its derivative vanishes, so there is no force and the equation of motion is

$$m\frac{d^2x}{dt^2} = 0\,,$$ (2.23)

which implies that $\frac{dx}{dt}$ is constant and the motion is at constant velocity. Even in a non-constant potential $V(x)$, at any points $\tilde{x}$ where $\frac{dV}{dx}$ vanishes, there is no force. These are possible equilibrium points in the potential where the body can remain at rest. Such equilibrium points may or may not be stable.

We argued earlier that the gravitational potential of a body close to the Earth's surface is $V(x) = V_0 + mgx$, but had not confirmed the interpretation of g. For this potential, $-\frac{dV}{dx} = -mg$, which is precisely the gravitational force on a body of mass m that appeared in equation (2.6), so g is the acceleration due to gravity.

Because equation (2.22) is general, we can now go back and check whether the simplifications we made in section 2.3.2, when dealing with the linear potential $V(x) = kx$, gave the right or wrong answer. In fact, the equation of motion we obtained, (2.20), is correct. This is because the force is the constant $-k$, so the acceleration is constant. The true motion $X(t)$ is therefore quadratic in t and has a parabolic graph.

2.3.4   The calculus of variations

We have derived the equation of motion (2.22), starting from the principle of least action, but our method, based on calculations involving parabolas, is not the most rigorous one and does not easily extend to more complicated problems. For completeness, we show here how to minimize the action S using the calculus of variations. As before, this method gives the differential equation obeyed by the true path $X(t)$, which is Newton's second law of motion. One must still solve this differential equation to find $X(t)$.

Recall that for a general path $x(t)$, running between the fixed endpoints $x(t_0) = x_0$ and $x(t_1) = x_1$,

$$S = \int_{t_0}^{t_1} \left(\frac{1}{2}m\left(\frac{dx}{dt}\right)^2 - V(x(t)) \right) dt\,.$$ (2.24)

As before, assume that there exists a smooth path $x(t) = X(t)$ that minimizes the action. Let S_X denote the action of this optimal path, so

$$S_X = \int_{t_0}^{t_1} \left(\frac{1}{2}m \left(\frac{dX}{dt} \right)^2 - V(X(t)) \right) dt. \tag{2.25}$$

Now suppose that $x(t) = X(t) + h(t)$ is a path infinitesimally close to $X(t)$. As $h(t)$ is infinitesimal, we can ignore quantities quadratic in $h(t)$. $h(t)$ is called the *path variation*, and $X(t) + h(t)$ the varied path. For the varied path the velocity is

$$\frac{dx}{dt} = \frac{dX}{dt} + \frac{dh}{dt} \tag{2.26}$$

and the kinetic energy is

$$K = \frac{1}{2}m \left(\frac{dX}{dt} + \frac{dh}{dt} \right)^2. \tag{2.27}$$

Discarding terms quadratic in h, we obtain

$$K = \frac{1}{2}m \left(\frac{dX}{dt} \right)^2 + m\frac{dX}{dt}\frac{dh}{dt}. \tag{2.28}$$

Next we do a similar analysis of the potential energy. For the varied path, the potential energy at time t is $V(X(t) + h(t))$. We use the usual approximation of calculus (as in equation (1.23))

$$V(X(t) + h(t)) = V(X(t)) + V'(X(t))\,h(t). \tag{2.29}$$

Here V is a function of just one variable (originally x), and we differentiate with respect to x to produce V'.

Combining these results for K and V, we find that the action for the varied path S_{X+h} is

$$S_{X+h} = \int_{t_0}^{t_1} \left(\frac{1}{2}m \left(\frac{dX}{dt} \right)^2 + m\frac{dX}{dt}\frac{dh}{dt} - V(X(t)) - V'(X(t))\,h(t) \right) dt. \tag{2.30}$$

The first and third terms on the right that contain neither h nor $\frac{dh}{dt}$ are exactly those found in expression (2.25) for S_X, so

$$S_{X+h} = S_X + \int_{t_0}^{t_1} \left(m\frac{dX}{dt}\frac{dh}{dt} - V'(X(t))\,h(t) \right) dt. \tag{2.31}$$

We now integrate the first term in the integral by parts, to make both terms have a common factor of $h(t)$. Integrating $\frac{dh}{dt}$ and differentiating $m\frac{dX}{dt}$, we find

$$S_{X+h} = S_X + \left[m\frac{dX}{dt}h(t) \right]_{t_0}^{t_1} - \int_{t_0}^{t_1} \left(m\frac{d^2X}{dt^2} + V'(X(t)) \right) h(t)\,dt. \tag{2.32}$$

$h(t)$ is a very general (infinitesimal) function, but it must vanish at both t_0 and t_1, because the principle of least action applies to paths with fixed endpoints at t_0 and t_1. Consequently, $m\frac{dX}{dt}h(t)$ vanishes at both endpoints, so

$$S_{X+h} - S_X = -\int_{t_0}^{t_1} \left(m\frac{d^2X}{dt^2} + V'(X(t)) \right) h(t)\,dt. \tag{2.33}$$

Between the endpoints, $h(t)$ is unconstrained. (We could even flip its sign if we chose to.)

It follows that if the bracketed expression multiplying $h(t)$ in the integral were non-zero, we could find some $h(t)$ for which $S_{X+h} - S_X$ is negative,[2] and S_{X+h} would therefore be less than S_X, contrary to our assumption that the path $X(t)$ minimizes the action. S_X is therefore the minimum of the action only if the bracketed expression vanishes at all times t between t_0 and t_1. In other words, the principle of least action requires that

$$m\frac{d^2x}{dt^2} + V'(x(t)) = 0 \,. \tag{2.34}$$

This is the differential equation that the actual path $x(t) = X(t)$ must satisfy, and it is the same as equation (2.22). In the context of the calculus of variations it is called the *Euler–Lagrange equation* associated with the action S.

As we said previously, equation (2.34) is a version of Newton's second law in which the force is given by

$$F(x) = -V'(x) \,. \tag{2.35}$$

We have derived Newton's second law from the principle of least action. However, it is no longer the force that is the fundamental quantity. Instead, it is the potential V.

2.3.5 The unimportance of the endpoints

An apparent problem with the principle of least action is that it seems to require the initial and final times, t_0 and t_1, to be specified in advance and the endpoint conditions to be imposed on the path at these times. However, this is not really the case. Usually, there is nothing special about t_0 and t_1. The motion may have started before t_0 and it may continue after t_1. Let us assume, in fact, that the motion takes place for all time, and that it satisfies the equation of motion (2.34). The choice of fixed endpoints can be avoided in the following way. Let us define the action formally, as

$$S = \int \left(\frac{1}{2}m \left(\frac{dx}{dt} \right)^2 - V(x(t)) \right) dt \,. \tag{2.36}$$

No endpoints have been specified. We cannot choose the endpoints as $-\infty$ and ∞ because generally S would then be infinite. Now consider a path variation which replaces the actual motion $x(t) = X(t)$ by $x(t) = X(t) + h(t)$, with h infinitesimal and non-zero only on some finite, but arbitrary, time interval I. h should also be continuous, so it does not jump at the moment it first becomes non-zero, nor at the last moment that it is non-zero. Consider the action S, integrated over any larger time interval I' which contains I. The principle of least action requires that S, defined over this larger interval I', is unchanged to first order by any variation in h, and this implies that the actual motion obeys the equation of motion throughout the smaller interval I, because this is the only interval where h is non-zero. The calculation is exactly the same as in the previous section, and goes through because h vanishes at the endpoints of I. In turn, this implies that the equation of motion is obeyed for all time, because the interval I can be selected freely, and there is always a larger interval I' that contains I. The freedom in the choice of I implies that we have not broken time-translational invariance.

---

[2] This claim is not completely obvious, but can be proved rigorously if it is assumed that the bracketed expression is continuous and somewhere non-zero.

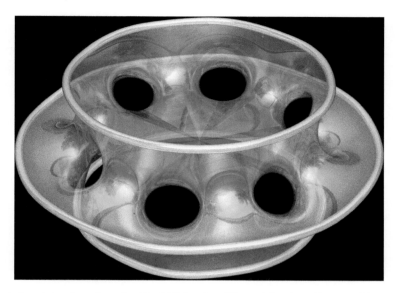

Fig. 2.5 Costa's minimal surface, by Paul Nylander. The minimal surface would extend outwards to infinity, but for clarity it is shown here bounded by three rings.

Escaping from pre-set boundary conditions is also useful for other variational problems. We can model soap films as infinite minimal surfaces stretching through 3-space. The most obvious such surface is a flat planar film, but there are many more unusual ones. One cannot say that these films have minimal total area, because their area is infinite. Rather, the surface is minimal in the sense that if we consider an infinitesimal deformation of the film that is continuous, and non-zero only in a finite region Σ, then the area of the part of the film in a larger finite region Σ' does not change to first order as a result of the deformation. This implies that everywhere in Σ the film obeys the curvature condition that we stated at the end of section 1.1, namely that the two principal curvatures of the surface are of equal magnitude, but in opposite directions. As Σ is arbitrarily chosen, the whole film obeys the curvature condition. An example of such a surface is shown in Figure 2.5.

2.4 The Motion of Several Bodies and Newton's Third Law

We will now use the principle of least action to derive Newton's third law for a system of two bodies moving in one dimension and interacting through a potential. The action is still a single quantity, but now it involves both bodies. Let the possible paths of the bodies be $x^{(1)}(t)$ and $x^{(2)}(t)$ and their masses $m^{(1)}$ and $m^{(2)}$. The kinetic energy is

$$K = \frac{1}{2}m^{(1)}\left(\frac{dx^{(1)}}{dt}\right)^2 + \frac{1}{2}m^{(2)}\left(\frac{dx^{(2)}}{dt}\right)^2 . \tag{2.37}$$

For the potential energy V we suppose the background environment is uniform and has no dynamical effect. V is then some function depending only on the *separation* of the bodies $l = x^{(2)} - x^{(1)}$. This is because of Euclidean symmetry, which in one dimension reduces to translational symmetry along the x-axis. So the potential energy is $V(l) = V(x^{(2)} - x^{(1)})$. (Usually, V only depends on the magnitude $|x^{(2)} - x^{(1)}|$, but this is not essential.)

The action for the pair of bodies is then

$$S = \int_{t_0}^{t_1} \left(\frac{1}{2} m^{(1)} \left(\frac{dx^{(1)}}{dt} \right)^2 + \frac{1}{2} m^{(2)} \left(\frac{dx^{(2)}}{dt} \right)^2 - V \left(x^{(2)}(t) - x^{(1)}(t) \right) \right) dt. \qquad (2.38)$$

The possible paths of the two bodies are independent, but the endpoints of the paths $x^{(1)}(t_0), x^{(2)}(t_0)$ and $x^{(1)}(t_1), x^{(2)}(t_1)$ must be specified in advance. The principle of least action says that the true paths of both bodies, which we denote by $X^{(1)}(t)$ and $X^{(2)}(t)$, minimize S. As before, this principle leads to the equations of motion. The equations are found by requiring that the minimized action has no first order variation under independent path variations $X^{(1)}(t) \rightarrow X^{(1)}(t) + h^{(1)}(t)$ and $X^{(2)}(t) \rightarrow X^{(2)}(t) + h^{(2)}(t)$. Following the same analysis as for a single body that led to equation (2.34), we find that the equations have the form of Newton's second law,

$$m^{(1)} \frac{d^2 x^{(1)}}{dt^2} + \frac{\partial V}{\partial x^{(1)}} = 0,$$

$$m^{(2)} \frac{d^2 x^{(2)}}{dt^2} + \frac{\partial V}{\partial x^{(2)}} = 0. \qquad (2.39)$$

Partial derivatives occur, because V depends on both $x^{(1)}$ and $x^{(2)}$, but V is really a function of just the single quantity $l = x^{(2)} - x^{(1)}$. Let V' denote the derivative $\frac{dV}{dl}$. Then, by the chain rule, $\frac{\partial V}{\partial x^{(2)}} = V'$ and $\frac{\partial V}{\partial x^{(1)}} = -V'$. The equations of motion for the two bodies (2.39) therefore simplify to

$$m^{(1)} \frac{d^2 x^{(1)}}{dt^2} - V'(x^{(2)} - x^{(1)}) = 0,$$

$$m^{(2)} \frac{d^2 x^{(2)}}{dt^2} + V'(x^{(2)} - x^{(1)}) = 0. \qquad (2.40)$$

For body 1 the force is $V'(x^{(2)} - x^{(1)})$, whereas for body 2 it is $-V'(x^{(2)} - x^{(1)})$. The forces are equal and opposite. So we have derived Newton's third law and have seen that it is a consequence of the translational invariance of the potential, which in turn follows from the Euclidean symmetry of space.

This brings us to *momentum*, which is the product of mass and velocity. The form of Newton's second law for one body suggests that it is useful to define the momentum p as

$$p = mv = m \frac{dx}{dt}. \qquad (2.41)$$

The equation of motion for one body, (2.34), now takes the form

$$\frac{dp}{dt} + V'(x(t)) = 0. \qquad (2.42)$$

As $-V'$ is the force on the body, equation (2.42) says that the force is equal to the rate of change of the body's momentum. If V' is zero and there is no force, then p is constant and we say that momentum is *conserved*.

Momentum is more useful when there are two or more bodies. Suppose we add the two equations (2.40). The force terms cancel, leaving

$$m^{(1)}\frac{d^2 x^{(1)}}{dt^2} + m^{(2)}\frac{d^2 x^{(2)}}{dt^2} = 0 \,. \tag{2.43}$$

Integrating this once, we find

$$m^{(1)}\frac{dx^{(1)}}{dt} + m^{(2)}\frac{dx^{(2)}}{dt} = \text{constant} \,. \tag{2.44}$$

In terms of the momenta $p^{(1)}$ and $p^{(2)}$ of the two bodies,

$$p^{(1)} + p^{(2)} = \text{constant} \,. \tag{2.45}$$

This is an important result. Even though the relative motion of the two bodies may be complicated, the total momentum $P_{\text{tot}} = p^{(1)} + p^{(2)}$ does not change with time; it is conserved. This follows from our assumption that led to Newton's third law, that space is uniform, so the bodies have no interaction with their environment, but only with each other.

One interpretation is that the two bodies act as a composite single body whose total momentum is the sum of the momenta of its constituents. The total momentum of the composite body is conserved, which is what is expected for a single body not subject to an external force. We can go further and identify, for a composite body, the equivalent central position that it would have as a single body. Note that

$$P_{\text{tot}} = m^{(1)}\frac{dx^{(1)}}{dt} + m^{(2)}\frac{dx^{(2)}}{dt} = \frac{d}{dt}\left(m^{(1)}x^{(1)} + m^{(2)}x^{(2)}\right), \tag{2.46}$$

and that the total mass of the composite body is $M_{\text{tot}} = m^{(1)} + m^{(2)}$. So let us write

$$P_{\text{tot}} = M_{\text{tot}}\frac{d}{dt}\left(\frac{m^{(1)}}{M_{\text{tot}}}x^{(1)} + \frac{m^{(2)}}{M_{\text{tot}}}x^{(2)}\right). \tag{2.47}$$

This expresses the total momentum in one-body form as the product of the total mass and a velocity $\frac{dX_{\text{CM}}}{dt}$, which is the time derivative of the central position

$$X_{\text{CM}} = \frac{m^{(1)}}{M_{\text{tot}}}x^{(1)} + \frac{m^{(2)}}{M_{\text{tot}}}x^{(2)} \,. \tag{2.48}$$

X_{CM} is called the *centre of mass*. It is the average of the positions of the constituents weighted by their masses, and reduces to the ordinary average if the masses are equal. Because of the conservation of total momentum, X_{CM} moves at a constant velocity. Essentially, the composite body obeys Newton's first law, irrespective of the internal motion of its constituents.

This analysis extends to N bodies. If the N bodies interact through a potential V that depends on their various positions, then one can derive from a single principle of least action the equations of motion for all the bodies. Each equation has the form of Newton's second law of motion for that body. If the whole system is isolated from the environment, then the system is translation invariant and V only depends on the relative positions of the bodies. In this case, the sum of the forces acting on the N bodies vanishes, i.e. $F^{(1)} + F^{(2)} + \cdots + F^{(N)} = 0$. This is a more general version of Newton's third law, but it implies the usual third law. For example, the force on the first body $F^{(1)}$, produced by all the other bodies, is equal and opposite to the total force on the other bodies combined, $F^{(2)} + \cdots + F^{(N)}$.

One can define momenta for each of the bodies, $p^{(1)} = m^{(1)} \frac{dx^{(1)}}{dt}$, $p^{(2)} = m^{(2)} \frac{dx^{(2)}}{dt}$, etc., and a total momentum $P_{\text{tot}} = p^{(1)} + p^{(2)} + \cdots + p^{(N)}$. For an isolated system, where $F^{(1)} + F^{(2)} + \cdots + F^{(N)} = 0$, P_{tot} is conserved. It follows that if we define, for N bodies, the total mass as $M_{\text{tot}} = m^{(1)} + m^{(2)} + \cdots + m^{(N)}$, and the centre of mass as

$$X_{\text{CM}} = \frac{m^{(1)}}{M_{\text{tot}}} x^{(1)} + \frac{m^{(2)}}{M_{\text{tot}}} x^{(2)} + \cdots + \frac{m^{(N)}}{M_{\text{tot}}} x^{(N)} , \tag{2.49}$$

then the centre of mass has a constant velocity. We can think of the N bodies as forming a single composite body, characterized by its total mass and a simple centre of mass motion. The total momentum is

$$P_{\text{tot}} = M_{\text{tot}} \frac{dX_{\text{CM}}}{dt} . \tag{2.50}$$

If this composite body is *not* isolated from the environment, then the sum of the forces F_{tot} will not vanish. The equation of motion for the centre of mass is then

$$M_{\text{tot}} \frac{d^2 X_{\text{CM}}}{dt^2} = F_{\text{tot}} . \tag{2.51}$$

This simple result helps us to understand the motion of composite systems, such as the Earth and Moon, orbiting the Sun together.

2.5   Motion of One Body in Three Dimensions

In most real problems we need to consider motion in three dimensions. The potential energy V depends on the positions of all the bodies in question, so for N bodies, V is a function of $3N$ variables. We need to be able to differentiate with respect to any of these, requiring further use of partial derivatives.

Let us just consider one body for the moment. Its trajectory is $\mathbf{x}(t)$ and its velocity is $\mathbf{v}(t)$. As in one dimension, the body's kinetic energy K is proportional to its mass and to the squared magnitude of its velocity,

$$K = \frac{1}{2} m \mathbf{v} \cdot \mathbf{v} = \frac{1}{2} m \frac{d\mathbf{x}}{dt} \cdot \frac{d\mathbf{x}}{dt} . \tag{2.52}$$

Because of the dot product, K is unchanged if the direction of $\mathbf{v}$ changes. The body also has potential energy $V(\mathbf{x})$.

The action for the body moving on a trajectory $\mathbf{x}(t)$ between initial and final points $\mathbf{x}_0$ and $\mathbf{x}_1$, at times t_0 and t_1, is

$$S = \int_{t_0}^{t_1} \left(\frac{1}{2} m \frac{d\mathbf{x}}{dt} \cdot \frac{d\mathbf{x}}{dt} - V(\mathbf{x}(t)) \right) dt$$

$$= \int_{t_0}^{t_1} \left(\frac{1}{2} m \left(\frac{dx_1}{dt} \right)^2 + \frac{1}{2} m \left(\frac{dx_2}{dt} \right)^2 + \frac{1}{2} m \left(\frac{dx_3}{dt} \right)^2 \right.$$

$$\left. - V\Big(x_1(t), x_2(t), x_3(t) \Big) \right) dt \,. \tag{2.53}$$

This has a similar form to equation (2.38), but the interpretation is quite different. Here, $(x_1(t), x_2(t), x_3(t))$ are the three components of the position of one body, whereas previously, $x^{(1)}(t)$ and $x^{(2)}(t)$ were the positions (in one dimension) of two bodies. Mathematically, this makes no difference, and the principle of least action, applied to the action (2.53), gives the equations of motion

$$m \frac{d^2 x_1}{dt^2} + \frac{\partial V}{\partial x_1} = 0 \,, \quad m \frac{d^2 x_2}{dt^2} + \frac{\partial V}{\partial x_2} = 0 \,, \quad m \frac{d^2 x_3}{dt^2} + \frac{\partial V}{\partial x_3} = 0 \,. \tag{2.54}$$

We may combine these into the vector equation

$$m \frac{d^2 \mathbf{x}}{dt^2} + \nabla V = 0 \,, \tag{2.55}$$

using the definition (1.26) of the gradient ∇. This is Newton's second law (2.3), with the force $\mathbf{F} = -\nabla V$. $\mathbf{F}$ is not an arbitrary function of position, because not every vector function $\mathbf{F}(\mathbf{x})$ can be expressed as minus the gradient of a scalar function $V(\mathbf{x})$. Equation (2.55) has a suitably geometrical interpretation. Recall that ∇V is in the direction of steepest ascent of the potential V. The force is therefore in the direction of steepest descent, and its magnitude is proportional to the slope.

The momentum of a body in three dimensions is $\mathbf{p} = m\mathbf{v} = m\frac{d\mathbf{x}}{dt}$. An alternative version of equation (2.55) is therefore

$$\frac{d\mathbf{p}}{dt} + \nabla V = 0 \,, \tag{2.56}$$

implying again that the force is the rate of change of momentum.

2.5.1 The harmonic oscillator

The vector differential equation (2.55) is generally only solvable by numerical integration, but there are important exceptions. One analytically solvable case is the harmonic oscillator, whose potential V is a quadratic function of x_1, x_2 and x_3. The general quadratic function $V(x_1, x_2, x_3) = \frac{1}{2} A x_1^2 + \frac{1}{2} B x_2^2 + \frac{1}{2} C x_3^2 + D x_1 x_2 + E x_1 x_3 + F x_2 x_3$ is a bit awkward to deal with. We can simplify it by taking advantage of the rotational invariance of the kinetic energy K. It is always possible to choose a new primed set of axes[3] and retain the same expression for K, but lose the mixed terms in V, so that $V(x_1, x_2, x_3) = \frac{1}{2} A x_1^2 + \frac{1}{2} B x_2^2 + \frac{1}{2} C x_3^2$.

---

[3] Strictly speaking, we should use primed coordinates and coefficients here, and then drop the primes.

Let us assume A, B and C are positive, so the potential has a minimum at the origin O. And to simplify this example, let us assume that $m = 1$. The action is

$$S = \frac{1}{2} \int_{t_0}^{t_1} \left(\left(\frac{dx_1}{dt} \right)^2 + \left(\frac{dx_2}{dt} \right)^2 + \left(\frac{dx_3}{dt} \right)^2 - Ax_1^2 - Bx_2^2 - Cx_3^2 \right) dt, \qquad (2.57)$$

with no terms mixing the coordinates. Minimizing S gives us the equations of motion

$$\frac{d^2x_1}{dt^2} + Ax_1 = 0, \quad \frac{d^2x_2}{dt^2} + Bx_2 = 0, \quad \frac{d^2x_3}{dt^2} + Cx_3 = 0. \qquad (2.58)$$

These are three decoupled 1-dimensional harmonic oscillators, with the general solution

$$
\begin{aligned}
x_1(t) &= \alpha_1 \cos \sqrt{A}\,t + \beta_1 \sin \sqrt{A}\,t \\
x_2(t) &= \alpha_2 \cos \sqrt{B}\,t + \beta_2 \sin \sqrt{B}\,t \\
x_3(t) &= \alpha_3 \cos \sqrt{C}\,t + \beta_3 \sin \sqrt{C}\,t,
\end{aligned}
\qquad (2.59)
$$

describing oscillations around the stable equilibrium at O with frequencies $\sqrt{A}$, $\sqrt{B}$ and $\sqrt{C}$.

The 3-dimensional harmonic oscillator is an important and useful example. Even when the potential V is not quadratic, if it has a stable equilibrium point at $\widetilde{\mathbf{x}}$, one can often use a quadratic approximation to V and treat oscillations of *small amplitude* around $\widetilde{\mathbf{x}}$ as harmonic.

Solutions of equation (2.58) are also straightforward to find if one or more of the coefficients A, B, C is negative or zero. For example, if $A < 0$ the first equation has the general solution $x_1(t) = \alpha_1 \exp \sqrt{-A}\,t + \beta_1 \exp -\sqrt{-A}\,t$, and if $A = 0$, $x_1(t) = \alpha_1 t + \beta_1$. These describe motion near an unstable and a neutral equilibrium point, respectively.

A special case of the harmonic oscillator potential is

$$V(x_1, x_2, x_3) = \frac{1}{2} A(x_1^2 + x_2^2 + x_3^2) = \frac{1}{2} A r^2. \qquad (2.60)$$

This is rotationally invariant, or *isotropic*. There is now just one frequency $\sqrt{A}$, and the general oscillatory solution can be written in vector form as

$$\mathbf{x}(t) = \boldsymbol{\alpha} \cos \sqrt{A}\,t + \boldsymbol{\beta} \sin \sqrt{A}\,t. \qquad (2.61)$$

The motion is along an elliptical orbit in the plane spanned by $\boldsymbol{\alpha}$ and $\boldsymbol{\beta}$, with the centre of the ellipse at O. By choosing the initial time appropriately, one can arrange for $\boldsymbol{\alpha}$ to be the semi-major axis and $\boldsymbol{\beta}$ the semi-minor axis. Horrocks and others compared the Keplerian orbits to the elliptical orbits of a pendulum bob, which for small oscillations are described by a harmonic oscillator potential. The crucial difference is that Kepler found that the Sun is positioned at a focus of the elliptical orbit, not its centre.

2.6 Central Forces

We come now to a key question that was addressed by Newton in the *Principia*. Let us consider the motion of a body in a general potential $V(r)$ that only depends on r, the radial distance from O. Recall the general form of $\nabla V(r)$ from equation (1.38). Using this, the equations of motion (2.54) become

$$m\frac{d^2x_1}{dt^2} + V'(r)\frac{x_1}{r} = 0\,, \quad m\frac{d^2x_2}{dt^2} + V'(r)\frac{x_2}{r} = 0\,, \quad m\frac{d^2x_3}{dt^2} + V'(r)\frac{x_3}{r} = 0\,, \quad (2.62)$$

or in vector form,

$$m\frac{d^2\mathbf{x}}{dt^2} + \frac{1}{r}V'(r)\mathbf{x} = 0\,. \tag{2.63}$$

This implies that the acceleration is proportional to the radial vector $\mathbf{x}$, directly towards or away from O, so O acts as a centre of force. For this reason, $V(r)$ is called a *central potential* and the resulting force a *central force*. The isotropic harmonic oscillator is an example, with $V(r) = \frac{1}{2}Ar^2$ and $V'(r) = Ar$.

For a body moving in a central potential, both the kinetic energy and potential energy are unchanged by a rotation around O, and an important consequence is the existence of a conserved quantity—the *angular momentum* of the body. This is a vector $\mathbf{l}$ defined using the cross product of the position $\mathbf{x}$ and velocity $\mathbf{v}$:

$$\mathbf{l} = m\mathbf{x} \times \mathbf{v} = m\mathbf{x} \times \frac{d\mathbf{x}}{dt}\,. \tag{2.64}$$

An alternative expression for angular momentum is $\mathbf{l} = \mathbf{x} \times \mathbf{p}$, where $\mathbf{p} = m\mathbf{v}$ is the ordinary linear momentum of the body.

To show that $\mathbf{l}$ is constant we differentiate equation (2.64), applying the Leibniz rule, then use the equation of motion (2.63), and finally recall that the cross product of any vector with itself vanishes:

$$\begin{aligned}
\frac{d\mathbf{l}}{dt} &= m\frac{d\mathbf{x}}{dt} \times \frac{d\mathbf{x}}{dt} + m\mathbf{x} \times \frac{d^2\mathbf{x}}{dt^2} \\
&= m\frac{d\mathbf{x}}{dt} \times \frac{d\mathbf{x}}{dt} - \frac{1}{r}V'(r)\mathbf{x} \times \mathbf{x} \\
&= 0\,.
\end{aligned} \tag{2.65}$$

The angular momentum $\mathbf{l}$ is therefore conserved.

One direct result of angular momentum conservation is that orbits are planar. Recall that $\mathbf{l} = m\mathbf{x} \times \mathbf{v}$ implies that $\mathbf{l}$ is orthogonal to $\mathbf{x}$ and to $\mathbf{v}$. Think of $\mathbf{l}$, $\mathbf{x}$ and $\mathbf{v}$ as vectors pointing out from the origin O (by a parallel shift of $\mathbf{v}$ if necessary). As $\mathbf{l}$ is constant, $\mathbf{x}$ and $\mathbf{v}$ must be in the fixed plane through O that is orthogonal to $\mathbf{l}$, so provided no other force acts, $\mathbf{x}$ will remain in this plane, and so will $\mathbf{v}$. If we choose axes so that the 3-axis is along $\mathbf{l}$, then the motion is in the plane through O orthogonal to the 3-axis, with Cartesian coordinates that we now denote as x and y, so $\mathbf{x}(t) = (x(t), y(t), 0)$. One of Kepler's earliest findings when analysing the motion of the planets was that their orbits remain fixed in a plane that intersects the Sun. This is sometimes known as Kepler's zeroth law. Here we have seen that it holds for a planet moving in any central potential with the Sun as the centre of force.

It is also useful to consider angular momentum conservation using polar coordinates, where $x = r\cos\varphi$ and $y = r\sin\varphi$. Using these coordinates, the body's changing position is

$$\mathbf{x}(t) = r(t)\Big(\cos\varphi(t), \sin\varphi(t), 0\Big)\,. \tag{2.66}$$

Therefore, using the Leibniz rule,

$$\frac{d\mathbf{x}}{dt} = \frac{dr}{dt}\Big(\cos\varphi(t), \sin\varphi(t), 0\Big) + r(t)\frac{d\varphi}{dt}\Big(-\sin\varphi(t), \cos\varphi(t), 0\Big). \tag{2.67}$$

Only the second term on the right contributes to $\mathbf{l}$, as the first term is in the direction of $\mathbf{x}$, so its cross product with $\mathbf{x}$ is zero. We then find

$$\mathbf{l} = m\mathbf{x} \times \frac{d\mathbf{x}}{dt} = m\left(0, 0, r^2(t)\frac{d\varphi}{dt}\right). \tag{2.68}$$

As expected, $\mathbf{l}$ is in the direction of the 3-axis, and as $\mathbf{l}$ is conserved, it has a constant magnitude l. Equation (2.68) implies that

$$mr^2(t)\frac{d\varphi}{dt} = l. \tag{2.69}$$

For motion in a central potential, therefore, the angular velocity $\frac{d\varphi}{dt}$ is inversely proportional to the squared distance from O, so when the body is far from O it has a smaller angular velocity than when it is near to O.

There is a neat geometrical interpretation of this result. The area of the portion of the orbit swept out by the radius vector up to time t is

$$\frac{1}{2}\int r^2\,d\varphi = \frac{1}{2}\int_0^t r^2(t')\frac{d\varphi}{dt'}\,dt' = \frac{l}{2m}t. \tag{2.70}$$

The rate at which area is swept out, the time derivative of this, therefore has a constant value, $\frac{l}{2m}$. This is Kepler's second law. It holds for the orbits in any central potential, as a consequence of angular momentum conservation.

Typically, the general orbit in an attractive central potential, with $\mathbf{l}$ non-zero, takes the form shown in Figure 2.6. The motion is along a planar trajectory that does not close to form a repeating figure. Even so, there is a kind of periodicity, referred to as *precession*. The motion from B to C repeats the motion from A to B, but rotated through some angle φ_0. This repetition continues indefinitely, with each return to the outermost point of the orbit rotated through a further angle of φ_0.

As long as $\mathbf{l} \neq 0$, the orbit cannot pass through O, because at O, equation (2.64) implies that $\mathbf{l}$ would be zero (as $\mathbf{x} = 0$ and $\mathbf{v}$ is finite). Note, however, that if $\mathbf{l}$ is zero, then $\mathbf{x} \times \mathbf{v} = 0$, so $\mathbf{x}$ and $\mathbf{v}$ are parallel. In this case the motion is along some fixed radial line, and may pass through O.

2.6.1   Circular orbits

When an orbit is circular, the radial distance r and angular velocity $\frac{d\varphi}{dt}$ remain constant, so this is the simplest case to consider for motion with a central force. Let us find the relation between the angular velocity and the strength of the force.

Assume the orbit is

$$\mathbf{x}(t) = r\Big(\cos\varphi(t), \sin\varphi(t), 0\Big), \tag{2.71}$$

with r and $\frac{d\varphi}{dt}$ constant.

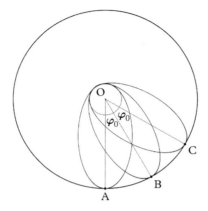

Fig. 2.6 Precession of an orbit.

The velocity is

$$\mathbf{v} = \frac{d\mathbf{x}}{dt} = r\frac{d\varphi}{dt}\left(-\sin\varphi(t), \cos\varphi(t), 0\right),\tag{2.72}$$

a vector orthogonal to $\mathbf{x}$ and therefore tangent to the orbit. The acceleration is

$$\mathbf{a} = \frac{d^2\mathbf{x}}{dt^2} = -r\left(\frac{d\varphi}{dt}\right)^2\left(\cos\varphi(t), \sin\varphi(t), 0\right),\tag{2.73}$$

which is a negative multiple of the position vector $\mathbf{x}$, and therefore a vector towards O. This establishes the important result that along a circular orbit of radius r and at angular velocity $\frac{d\varphi}{dt}$, the acceleration is towards the centre of the circle with magnitude

$$|\mathbf{a}| = r\left(\frac{d\varphi}{dt}\right)^2.\tag{2.74}$$

For a central potential $V(r)$, the magnitude of the force towards O is $V'(r)$, so a circular orbit satisfies the equation of motion $m\mathbf{a} = \mathbf{F}$ provided

$$mr\left(\frac{d\varphi}{dt}\right)^2 = V'(r).\tag{2.75}$$

If the force is attractive, then $V'(r)$ is positive and there is a solution at any fixed radius r with angular velocity

$$\frac{d\varphi}{dt} = \pm\left(\frac{1}{mr}V'(r)\right)^{\frac{1}{2}},\tag{2.76}$$

where the sign determines the direction of motion around the circle. The magnitude of the angular momentum is

$$l = mr^2\frac{d\varphi}{dt} = \left(mr^3V'(r)\right)^{\frac{1}{2}}.\tag{2.77}$$

The range of φ is 2π, so equation (2.76) implies that the period of the orbit at this radius is

$$T = 2\pi \left(\frac{1}{mr}V'(r)\right)^{-\frac{1}{2}}.$$
(2.78)

For the isotropic harmonic oscillator $V'(r) = Ar$, so T is independent of the radius of the orbit, which agrees with the well known fact that for small oscillations, the period of a pendulum is independent of the amplitude of oscillation. However, this rules out the harmonic oscillator as a model for the planetary orbits, as Kepler's third law states that for a circular orbit of radius r, the period T is proportional to $r^{\frac{3}{2}}$. Kepler's law implies that $\left(\frac{1}{mr}V'(r)\right)^{-\frac{1}{2}} \propto r^{\frac{3}{2}}$ or $V'(r) \propto \frac{1}{r^2}$. This was the key that enabled Newton to deduce that the force between the Sun and planets diminishes with an inverse square law.

2.7   The Attractive Inverse Square Law Force

Physically, the most important example of a central force is the attractive inverse square law force, which arises from a potential $V(r) = -\frac{C}{r}$ with C positive. The equation of motion for a body in this potential is

$$m\frac{d^2\mathbf{x}}{dt^2} + \frac{C}{r^3}\mathbf{x} = 0.$$
(2.79)

The body is subject to an inverse square law force of magnitude $\frac{C}{r^2}$ towards O. Newton famously demonstrated that it is only possible to account for all of Kepler's laws of planetary motion if the gravitational force between two massive spherically symmetric bodies takes this form. If the masses are $m^{(1)}$ and $m^{(2)}$ and the separation is r, the magnitude of the force is

$$\frac{Gm^{(1)}m^{(2)}}{r^2}$$
(2.80)

where G is Newton's universal gravitational constant. We will discuss the motion of two bodies under the influence of their mutual gravitational attraction in section 2.10, but for now we consider the simpler situation where the first body is much more massive than the second. Then we can regard the first to be at rest at O and the second as in orbit around it. This is a reasonable first approximation for the motion of a planet around the Sun.[4]

Let the mass of the body at O be $m^{(1)} = M$, and the mass of the orbiting body be $m^{(2)} = m$. The gravitational force is $\frac{GMm}{r^3}\mathbf{x}$, and the equation of motion (2.79) for the orbiting body becomes

$$\frac{d^2\mathbf{x}}{dt^2} + \frac{GM}{r^3}\mathbf{x} = 0.$$
(2.81)

m has cancelled out, as we expect when the force is gravitational. The orbit is independent of m, although quantities like the angular momentum and energy do depend on m. For convenience, in the rest of this section we set $m = 1$.

Let us determine the general orbit for the equation of motion (2.81). As with motion in any central potential, the angular momentum vector $\mathbf{l} = \mathbf{x} \times \mathbf{v}$ is conserved and the motion

---

[4] This is only an approximation to motion in the solar system, as the gravitational attraction of the other planets must also be taken into account. If the body is non-spherical, further corrections are necessary.

is in the plane through O orthogonal to l. Uniquely in the case of the inverse square law force, there is an additional conserved vector known as the *Runge–Lenz vector*,

$$\mathbf{k} = \mathbf{l} \times \mathbf{v} + \frac{GM}{r}\mathbf{x} , \tag{2.82}$$

which lies in the plane of the motion.

Verifying that $\mathbf{k}$ is conserved is a little more involved than verifying the conservation of l. Taking its time derivative, we find

$$\frac{d\mathbf{k}}{dt} = -\frac{GM}{r^3}\mathbf{l} \times \mathbf{x} + GM \left(\nabla \left(\frac{1}{r} \right) \cdot \mathbf{v} \right) \mathbf{x} + \frac{GM}{r}\mathbf{v} , \tag{2.83}$$

where the first term comes from differentiating $\mathbf{v}$ and using the equation of motion (2.81), the second term comes from the fact that the time derivative of any function f of the space variables is $\nabla f \cdot \frac{d\mathbf{x}}{dt} = \nabla f \cdot \mathbf{v}$ (by the chain rule) and the final term comes simply from the time derivative of $\mathbf{x}$. Now we replace $\mathbf{l}$ by $\mathbf{x} \times \mathbf{v}$ and substitute for the gradient term using equation (1.39) to obtain

$$\frac{d\mathbf{k}}{dt} = -\frac{GM}{r^3}(\mathbf{x} \times \mathbf{v}) \times \mathbf{x} - \frac{GM}{r^3}(\mathbf{x} \cdot \mathbf{v})\mathbf{x} + \frac{GM}{r}\mathbf{v} . \tag{2.84}$$

Finally we make use of the double cross product identity (1.20) in the form $(\mathbf{x} \times \mathbf{v}) \times \mathbf{x} = (\mathbf{x} \cdot \mathbf{x})\mathbf{v} - (\mathbf{x} \cdot \mathbf{v})\mathbf{x} = r^2\mathbf{v} - (\mathbf{x} \cdot \mathbf{v})\mathbf{x}$, and see that all the terms on the right-hand side cancel. So $\frac{d\mathbf{k}}{dt} = 0$.

The conservation of the Runge–Lenz vector $\mathbf{k}$ is a key fact about motion subject to an inverse square law force. The consequences are that the direction of $\mathbf{k}$ is fixed, so there is no precession, and more importantly, bounded orbits close up to form ellipses. To show this we must first review the geometry of the ellipse. The standard equation for an ellipse in the (X, Y)-plane, with its centre at the origin, is

$$\frac{X^2}{a^2} + \frac{Y^2}{b^2} = 1 , \tag{2.85}$$

where $a > b$. The ellipse is oriented so that a and b are the lengths of the semi-major and semi-minor axes, as shown in Figure 2.7. The eccentricity e, defined through $b^2 = (1 - e^2)a^2$, specifies how far the ellipse deviates from a circle. The two foci of the ellipse are on the X-axis, at $X = \pm ea$. For any point on the ellipse, the sum of the distances to the two foci is $2a$.

We need to shift the focus at $X = ea$ to the origin and find the equation of the ellipse in this position. So let $x = X - ea$ and $y = Y$. Substituting in equation (2.85), multiplying through by b^2 and then replacing b^2 by $(1 - e^2)a^2$, we obtain

$$(1 - e^2)(x + ea)^2 + y^2 = (1 - e^2)a^2 , \tag{2.86}$$

and therefore, expanding out,

$$(1 - e^2)x^2 + 2e(1 - e^2)ax + (1 - e^2)e^2a^2 + y^2 = (1 - e^2)a^2 . \tag{2.87}$$

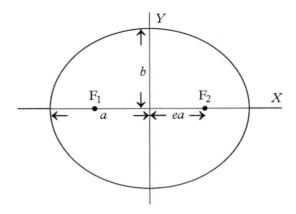

Fig. 2.7 Ellipse: F_1 and F_2 are the foci of the ellipse, a is the length of the semi-major axis and b is the length of the semi-minor axis. e is the eccentricity.

Subtracting the term $(1 - e^2)e^2a^2$ from both sides gives

$$(1 - e^2)x^2 + 2e(1 - e^2)ax + y^2 = (1 - e^2)^2 a^2 \,, \tag{2.88}$$

which can be rearranged as

$$x^2 + y^2 = \left(ex - (1 - e^2)a\right)^2 \,. \tag{2.89}$$

Now we introduce polar coordinates, $x = r\cos\varphi$ and $y = r\sin\varphi$, and take the square root of equation (2.89), to give the equation for the ellipse

$$r = -er\cos\varphi + (1 - e^2)a \,, \tag{2.90}$$

which rearranges into the final form

$$r(1 + e\cos\varphi) = (1 - e^2)a \,. \tag{2.91}$$

This is the polar equation for an ellipse with one focus at the origin.

We can now show that our orbit subject to an inverse square law force is an ellipse, with one focus at the centre of force. Recall that the orbit is in the (x, y)-plane, as is the conserved Runge–Lenz vector $\mathbf{k}$. The equation (2.82) defining the Runge–Lenz vector involves the velocity and position, but remarkably, from this we can directly obtain an equation for the orbit that only depends on the position vector $\mathbf{x}$. To do this we take the dot product of the two sides of equation (2.82) with $\mathbf{x}$, finding

$$\mathbf{k} \cdot \mathbf{x} = (\mathbf{l} \times \mathbf{v}) \cdot \mathbf{x} + GMr \,. \tag{2.92}$$

Now, using the identity $(\mathbf{l} \times \mathbf{v}) \cdot \mathbf{x} = \mathbf{l} \cdot (\mathbf{v} \times \mathbf{x})$, and replacing $(\mathbf{v} \times \mathbf{x})$ by $-\mathbf{l}$, we obtain

$$\mathbf{k} \cdot \mathbf{x} = -|\mathbf{l}|^2 + GMr \,, \tag{2.93}$$

an equation of the desired form, where the velocity has dropped out.

If we orient axes so that $\mathbf{k}$ is along the negative x-axis, then $\mathbf{k} \cdot \mathbf{x} = -kx = -kr \cos \varphi$, where k is the magnitude of $\mathbf{k}$. Substituting this into equation (2.93) produces the result $-kr \cos \varphi = -l^2 + GMr$, which rearranges to

$$r \left(1 + \frac{k}{GM} \cos \varphi \right) = \frac{l^2}{GM}. \tag{2.94}$$

This orbit equation is precisely the polar equation for an ellipse with one focus at the origin, given in equation (2.91). The eccentricity is $e = \frac{k}{GM}$, and the length parameter a is given by $(1 - e^2)a = \frac{l^2}{GM}$. These quantities are determined by the magnitudes of the Runge–Lenz vector and angular momentum vector, respectively, which in turn are determined by the initial conditions. The centre of force is at the origin, which is one focus of the ellipse. Thus, the orbits for an attractive inverse square law force have exactly the form found by Kepler when studying the planets. Kepler's first law states that a planetary orbit is an ellipse with the Sun at one focus.

We have derived the geometry of the orbit, but have not explicitly found the rate at which the body traces it out. Rearranging equation (2.94) gives

$$\frac{1}{r} = \frac{GM}{l^2} \left(1 + \frac{k}{GM} \cos \varphi \right), \tag{2.95}$$

and from the formula for angular momentum in polar coordinates (2.69) we have $\frac{d\varphi}{dt} = \frac{l}{r^2}$. So squaring equation (2.95) and multiplying by l, gives us the differential equation

$$\frac{d\varphi}{dt} = \frac{G^2 M^2}{l^3} \left(1 + \frac{k}{GM} \cos \varphi \right)^2 \tag{2.96}$$

for the angular motion. This equation is not easily solved.

The total orbital period, however, has a simple form. From the polar equation for the ellipse (2.91) we see that, geometrically, the point on the orbit furthest from the origin occurs where $\cos \varphi = -1$, at a distance $r_{\max} = (1 + e)a$, and the point closest to the origin occurs where $\cos \varphi = 1$, at a distance of $r_{\min} = (1 - e)a$. For a Keplerian orbit, we read off from equation (2.94) that $r_{\max} = \frac{l^2}{GM - k}$ and $r_{\min} = \frac{l^2}{GM + k}$. Therefore

$$\frac{1}{2}(r_{\max} + r_{\min}) = a = \frac{GMl^2}{G^2 M^2 - k^2}, \tag{2.97}$$

and

$$r_{\max} r_{\min} = (1 - e^2)a^2 = b^2 = \frac{l^4}{G^2 M^2 - k^2}. \tag{2.98}$$

The area of the orbit is given by the formula for the area of an ellipse, $A = \pi ab$, and is therefore

$$A = \pi \frac{GMl^2}{G^2 M^2 - k^2} \left(\frac{l^4}{G^2 M^2 - k^2} \right)^{\frac{1}{2}} = \pi \frac{GMl^4}{(G^2 M^2 - k^2)^{\frac{3}{2}}} = \pi \frac{l}{(GM)^{\frac{1}{2}}} a^{\frac{3}{2}}. \tag{2.99}$$

From equation (2.70), the orbital period T is equal to the area of the orbit A, divided by the rate at which the area is swept out, $\frac{1}{2}l$. So

$$T = \frac{2\pi}{(GM)^{\frac{1}{2}}} a^{\frac{3}{2}} \,. \tag{2.100}$$

This is Kepler's third law for a general elliptical orbit: The square of the orbital period is proportional to the cube of the length of the semi-major axis of the orbit.

In the solar system, with the Sun as the dominant gravitating body, M is the mass of the Sun $M_\odot$. The constant multiplying $a^{\frac{3}{2}}$ in the third law is therefore $\frac{2\pi}{(GM_\odot)^{\frac{1}{2}}}$, and is the same for all the planets, asteroids and other bodies orbiting the Sun. For a circular orbit the semi-major axis a is simply the radius.

2.8 G and the Mass of the Earth

To determine the masses of the Earth, other planets, and the Sun, one needs an independent determination of Newton's constant G. This is possible if one can measure the gravitational force between bodies of known mass and known separation in a terrestrial experiment. Newton himself thought that such a measurement was too difficult, but by the end of the 18th century an accurate result had been obtained.

In 1774, the Royal Society commissioned the Astronomer Royal Nevil Maskelyne to organize an expedition to the Scottish mountain Schiehallion to measure the gravitational attraction of the material forming the mountain. A pendulum was constructed in the vicinity of the mountain and the angle between vertical, as determined by the stars, and the pendulum wire was measured on either side of the mountain. Schiehallion was chosen for its simple shape, which makes its mass easy to estimate, and because it has a relatively isolated location, so the gravitational effect of any neighbouring mountains can be neglected. Even so, the expedition was unable to determine a very accurate value for G.

In 1798, Henry Cavendish used a method devised by John Michell to obtain a much better result. He used a torsion balance to measure the gravitational force between lead balls, in the set-up shown in Figure 2.8. There are two small, fixed spherical masses $m^{(2)}$, and between them is a beam suspended from a fine wire, with a large spherical mass $m^{(1)}$ at each end. Each mass $m^{(1)}$ is pulled towards the nearer fixed mass $m^{(2)}$. This twists the wire until it reaches an equilibrium position where an equal restoring force is produced. There is a linear relationship between this restoring force and the angle of twist, $F = c\vartheta$. If the value of c is known, then the gravitational force is easily obtained. The clever part of the experiment is that the direction of the beam oscillates around its equilibrium position and the period of oscillation allows the constant c to be determined. As the forces are so small, the period of oscillation is long—around 20 minutes.

With the value of c in hand, the gravitational force can now be obtained. A mirror on the suspension wire allows the beam direction to be accurately determined, so the deflection angle ϑ at the equilibrium position can be measured. This gives the restoring force, which balances the gravitational force $\frac{Gm^{(1)}m^{(2)}}{d^2}$ between the masses $m^{(1)}$ and $m^{(2)}$, where d is the distance between the masses at equilibrium. Since d and the masses are known, G can be calculated.

Cavendish performed the experiment in the sitting room of his house in central London. Rather remarkably, he obtained a value within 1% of today's best value, which is

$$G = 6.67 \times 10^{-11} \, \text{m}^3 \, \text{kg}^{-1} \, \text{s}^{-2} \,. \tag{2.101}$$

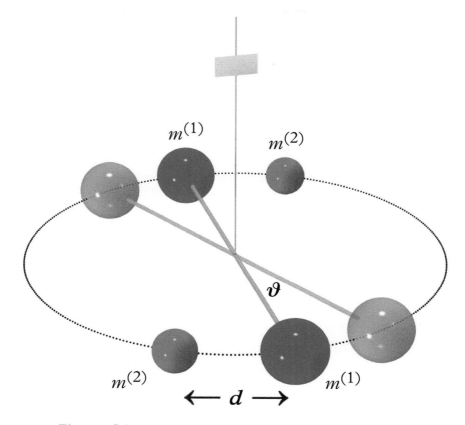

Fig. 2.8 Schematic set-up of the Michell–Cavendish experiment.

Using this result, one can obtain the mass of the Earth. The downward acceleration of a body close to the Earth's surface is $g = \frac{GM}{R^2}$, where M is the mass of the Earth and R is its radius. The radius of the Earth has been known since antiquity, so with g and G known, M can be calculated. The Earth's mass is $M \simeq 5.97 \times 10^{24}$ kg, and its average density is 5.51×10^3 kg m^{-3}. This is much denser than most rocks found at the surface, but consistent with the Earth having a dense metallic core, as indicated by seismological data and the Earth's magnetic field.

2.9 Composite Bodies and Centre of Mass Motion

When a single body moves in a central potential, angular momentum is conserved, but ordinary linear momentum is not, as the centre of force spoils the translational invariance. On the other hand, for a system of N bodies interacting with each other but free from external forces, both total linear momentum and total angular momentum are conserved. We will here explore the contribution of the centre of mass motion to both these conserved quantities.

Let the N bodies have time-dependent positions $\mathbf{x}^{(1)}, \ldots, \mathbf{x}^{(N)}$ and velocities $\mathbf{v}^{(1)}, \ldots, \mathbf{v}^{(N)}$, where the superscript is the label of the body. The potential of the system is some function $V(\mathbf{x}^{(1)}, \ldots, \mathbf{x}^{(N)})$. By Euclidean symmetry, the potential is unchanged by a translation of all the bodies by an arbitrary vector $\mathbf{c}$, so

$$V(\mathbf{x}^{(1)} + \mathbf{c}, \ldots, \mathbf{x}^{(N)} + \mathbf{c}) = V(\mathbf{x}^{(1)}, \ldots, \mathbf{x}^{(N)}). \tag{2.102}$$

For infinitesimal $\mathbf{c}$, using equation (1.27), this implies that

$$\mathbf{c} \cdot \nabla^{(1)} V + \cdots + \mathbf{c} \cdot \nabla^{(N)} V = 0, \tag{2.103}$$

where $\nabla^{(k)}$ is the gradient associated with the position variable $\mathbf{x}^{(k)}$. It follows that

$$\nabla^{(1)} V + \cdots + \nabla^{(N)} V = 0, \tag{2.104}$$

as $\mathbf{c}$ is arbitrary. Similarly, V is unchanged by an infinitesimal rotation of all the bodies about O. An infinitesimal rotation moves $\mathbf{x}$ to $\mathbf{x} + \boldsymbol{\alpha} \times \mathbf{x}$, where the rotation is around the axis in the direction $\boldsymbol{\alpha}$ through the infinitesimal angle $|\boldsymbol{\alpha}|$. Invariance of V implies that

$$\boldsymbol{\alpha} \times \mathbf{x}^{(1)} \cdot \nabla^{(1)} V + \cdots + \boldsymbol{\alpha} \times \mathbf{x}^{(N)} \cdot \nabla^{(N)} V = 0, \tag{2.105}$$

which can be re-expressed (using equation (1.19)) in the form

$$\boldsymbol{\alpha} \cdot \left[\mathbf{x}^{(1)} \times \nabla^{(1)} V + \cdots + \mathbf{x}^{(N)} \times \nabla^{(N)} V \right] = 0. \tag{2.106}$$

As $\boldsymbol{\alpha}$ is arbitrary,

$$\mathbf{x}^{(1)} \times \nabla^{(1)} V + \cdots + \mathbf{x}^{(N)} \times \nabla^{(N)} V = 0. \tag{2.107}$$

Let us now consider the consequences of the invariance properties (2.104) and (2.107). The equations of motion are

$$m^{(k)} \frac{d^2 \mathbf{x}^{(k)}}{dt^2} + \nabla^{(k)} V = 0, \quad k = 1, \ldots, N. \tag{2.108}$$

Adding these, and using equation (2.104), we find

$$m^{(1)} \frac{d^2 \mathbf{x}^{(1)}}{dt^2} + \cdots + m^{(N)} \frac{d^2 \mathbf{x}^{(N)}}{dt^2} = 0. \tag{2.109}$$

Integrating once gives

$$m^{(1)} \frac{d \mathbf{x}^{(1)}}{dt} + \cdots + m^{(N)} \frac{d \mathbf{x}^{(N)}}{dt} = \text{constant}. \tag{2.110}$$

This constant vector is the total momentum $\mathbf{P}_{\text{tot}}$, the sum of the momenta $\mathbf{p}^{(k)} = m^{(k)} \frac{d\mathbf{x}^{(k)}}{dt}$ of all the bodies. $\mathbf{P}_{\text{tot}}$ is directly related to the centre of mass motion, because

$$\begin{aligned} \mathbf{P}_{\text{tot}} &= \frac{d}{dt} \left(m^{(1)} \mathbf{x}^{(1)} + \cdots + m^{(N)} \mathbf{x}^{(N)} \right) \\ &= M_{\text{tot}} \frac{d}{dt} \left(\frac{m^{(1)}}{M_{\text{tot}}} \mathbf{x}^{(1)} + \cdots + \frac{m^{(N)}}{M_{\text{tot}}} \mathbf{x}^{(N)} \right) \\ &= M_{\text{tot}} \frac{d\mathbf{X}_{\text{CM}}}{dt}, \end{aligned} \tag{2.111}$$

where

$$\mathbf{X}_{\text{CM}} = \frac{m^{(1)}}{M_{\text{tot}}} \mathbf{x}^{(1)} + \cdots + \frac{m^{(N)}}{M_{\text{tot}}} \mathbf{x}^{(N)} \tag{2.112}$$

is the centre of mass. Equation (2.111) is the 3-dimensional analogue of equation (2.50). The conserved total momentum is the total mass times the velocity of the centre of mass, so this velocity is constant. There is no contribution to the total momentum from the relative motion of the bodies.

To find the total angular momentum we take the cross product of the equation of motion (2.108) for the kth body with the position $\mathbf{x}^{(k)}$, and again add, obtaining

$$m^{(1)}\mathbf{x}^{(1)} \times \frac{d^2\mathbf{x}^{(1)}}{dt^2} + \cdots + m^{(N)}\mathbf{x}^{(N)} \times \frac{d^2\mathbf{x}^{(N)}}{dt^2} = 0 \tag{2.113}$$

after using equation (2.107). This equation can be expressed as

$$\frac{d}{dt}\left(m^{(1)}\mathbf{x}^{(1)} \times \frac{d\mathbf{x}^{(1)}}{dt} + \cdots + m^{(N)}\mathbf{x}^{(N)} \times \frac{d\mathbf{x}^{(N)}}{dt}\right) = 0\,, \tag{2.114}$$

as all the terms $\frac{d\mathbf{x}^{(k)}}{dt} \times \frac{d\mathbf{x}^{(k)}}{dt}$ are zero. Integrating gives

$$m^{(1)}\mathbf{x}^{(1)} \times \frac{d\mathbf{x}^{(1)}}{dt} + \cdots + m^{(N)}\mathbf{x}^{(N)} \times \frac{d\mathbf{x}^{(N)}}{dt} = \text{constant}\,. \tag{2.115}$$

This constant vector is the conserved total angular momentum $\mathbf{L}_{\text{tot}}$, which has the alternative expressions

$$\begin{aligned} \mathbf{L}_{\text{tot}} &= m^{(1)}\mathbf{x}^{(1)} \times \mathbf{v}^{(1)} + \cdots + m^{(N)}\mathbf{x}^{(N)} \times \mathbf{v}^{(N)} \\ &= \mathbf{x}^{(1)} \times \mathbf{p}^{(1)} + \cdots + \mathbf{x}^{(N)} \times \mathbf{p}^{(N)}\,. \end{aligned} \tag{2.116}$$

$\mathbf{L}_{\text{tot}}$ is the sum of the angular momentum contributions of all N bodies.

We can now see how the motion of the centre of mass contributes to the total angular momentum $\mathbf{L}_{\text{tot}}$ (recall that it is responsible for all of $\mathbf{P}_{\text{tot}}$). Suppose first that the centre of mass is at rest at O and $\mathbf{P}_{\text{tot}}$ is zero. Because of the relative motion of the bodies, $\mathbf{L}_{\text{tot}}$, as given by equation (2.116), is generally not zero. Now, if we combine this relative motion with a centre of mass motion by shifting $\mathbf{x}^{(k)}$ to $\mathbf{x}^{(k)} + \mathbf{X}_{\text{CM}}$ and $\mathbf{v}^{(k)}$ to $\mathbf{v}^{(k)} + \mathbf{V}_{\text{CM}}$, where $\mathbf{V}_{\text{CM}}$ is constant and $\frac{d\mathbf{X}_{\text{CM}}}{dt} = \mathbf{V}_{\text{CM}}$, then the new conserved angular momentum is

$$\begin{aligned} \mathbf{L}'_{\text{tot}} &= \sum_1^N m^{(k)}(\mathbf{x}^{(k)} + \mathbf{X}_{\text{CM}}) \times (\mathbf{v}^{(k)} + \mathbf{V}_{\text{CM}}) \\ &= \mathbf{L}_{\text{tot}} + \mathbf{X}_{\text{CM}} \times \left(\sum_1^N m^{(k)}\mathbf{v}^{(k)}\right) + \left(\sum_1^N m^{(k)}\mathbf{x}^{(k)}\right) \times \mathbf{V}_{\text{CM}} \\ &\quad + M_{\text{tot}}\mathbf{X}_{\text{CM}} \times \mathbf{V}_{\text{CM}}\,. \end{aligned} \tag{2.117}$$

The vector $\sum_1^N m^{(k)}\mathbf{v}^{(k)}$ is the original total momentum, which was zero, and from the definition of the centre of mass given in equation (2.112), we see that $\sum_1^N m^{(k)}\mathbf{x}^{(k)}$ is M_{tot} times the original centre of mass position, which was also the zero vector. Therefore

$$\mathbf{L}'_{\text{tot}} = \mathbf{L}_{\text{tot}} + M_{\text{tot}}\mathbf{X}_{\text{CM}} \times \mathbf{V}_{\text{CM}} = \mathbf{L}_{\text{tot}} + \mathbf{X}_{\text{CM}} \times \mathbf{P}_{\text{tot}}\,. \tag{2.118}$$

The motion of the centre of mass contributes $M_{\text{tot}}\mathbf{X}_{\text{CM}} \times \mathbf{V}_{\text{CM}}$ to the total angular momentum and this is constant, as its time derivative only involves $\mathbf{V}_{\text{CM}} \times \mathbf{V}_{\text{CM}}$, which is zero.

We conclude that for a generic centre of mass motion, the total angular momentum $\mathbf{L}'_{\text{tot}}$ of a system of N bodies has two parts, each of which is time independent. The contribution of the centre of mass motion is not very significant, as it depends on the point O that we choose as origin. The interesting part is the original quantity $\mathbf{L}_{\text{tot}}$, which is the angular momentum *relative* to the centre of mass. We call this the intrinsic angular momentum of the system, or the *spin* of the system. When we come to discuss quantum mechanics, we will discover that the spin of particles and atoms is quantized, meaning that it takes discrete values proportional to Planck's constant $\hbar$. This spin is unaffected by the overall motion of the centre of mass.

A system of bodies in relative motion, for example a galaxy of stars or a solid body composed of many atoms, can be regarded as a rotating composite body. This is a particularly good interpretation if the system rotates rigidly, as a solid body does. The spin angular momentum of a solid body is related to the angular velocity of the body as a whole and its moment of inertia.

2.10 The Kepler 2-Body Problem

Here, briefly, we will explain how the motion of two bodies, under their mutual gravitational attraction, reduces to the central force problem for one body that we considered in section 2.7.

As earlier, let the bodies have masses $m^{(1)}$ and $m^{(2)}$. Their equations of motion are

$$m^{(1)}\frac{d^2\mathbf{x}^{(1)}}{dt^2} + \frac{Gm^{(1)}m^{(2)}}{|\mathbf{x}^{(2)}-\mathbf{x}^{(1)}|^3}(\mathbf{x}^{(1)}-\mathbf{x}^{(2)}) = 0\,,$$

$$m^{(2)}\frac{d^2\mathbf{x}^{(2)}}{dt^2} + \frac{Gm^{(1)}m^{(2)}}{|\mathbf{x}^{(2)}-\mathbf{x}^{(1)}|^3}(\mathbf{x}^{(2)}-\mathbf{x}^{(1)}) = 0\,, \qquad (2.119)$$

where the inverse square law forces are of equal magnitude but in opposite directions.

Adding these equations, we confirm that the centre of mass has a constant velocity, whilst cancelling the repeated mass factors and subtracting, we find

$$\frac{d^2(\mathbf{x}^{(2)}-\mathbf{x}^{(1)})}{dt^2} + \frac{G(m^{(1)}+m^{(2)})}{|\mathbf{x}^{(2)}-\mathbf{x}^{(1)}|^3}(\mathbf{x}^{(2)}-\mathbf{x}^{(1)}) = 0\,, \qquad (2.120)$$

the equation for the relative motion. The separation vector $\mathbf{x}^{(2)}-\mathbf{x}^{(1)}$ obeys a central force equation with an attractive inverse square law force, as in equation (2.81), but the constant GM is replaced here by $GM_{\text{tot}} = G(m^{(1)}+m^{(2)})$. The separation vector therefore moves in an elliptical orbit obeying Kepler's three laws.

Relative to the centre of mass, the path taken by the second body is $\mathbf{x}^{(2)} - \mathbf{X}_{\text{CM}}$, where $\mathbf{X}_{\text{CM}}$ is defined in equation (2.112). This simplifies to

$$\mathbf{x}^{(2)} - \frac{m^{(1)}\mathbf{x}^{(1)} + m^{(2)}\mathbf{x}^{(2)}}{m^{(1)}+m^{(2)}} = \frac{m^{(1)}}{m^{(1)}+m^{(2)}}(\mathbf{x}^{(2)} - \mathbf{x}^{(1)})\,, \qquad (2.121)$$

so the second body's motion relative to the centre of mass is a scaled-down version of the motion of the separation vector. Suppose the separation vector moves on an ellipse of semi-major axis a. Then the second body moves along an ellipse of semi-major axis $a^{(2)} = \frac{m^{(1)}}{M_{\text{tot}}}a$ with the centre of mass at a focus of this ellipse. By swapping labels (1) and (2), we see that the first body also has an elliptical orbit, with the centre of mass at a focus, but of semi-major axis $a^{(1)} = \frac{m^{(2)}}{M_{\text{tot}}}a$. Combining these expressions gives

$$a = a^{(1)} + a^{(2)}, \qquad \frac{a^{(1)}}{a^{(2)}} = \frac{m^{(2)}}{m^{(1)}}. \tag{2.122}$$

Kepler's third law now takes the form

$$T = \frac{2\pi a^{\frac{3}{2}}}{(GM_{\text{tot}})^{\frac{1}{2}}} = \frac{2\pi(a^{(1)} + a^{(2)})^{\frac{3}{2}}}{G^{\frac{1}{2}}(m^{(1)} + m^{(2)})^{\frac{1}{2}}}. \tag{2.123}$$

These relationships have proved very useful for astrophysicists. The orbits of two such bodies are illustrated in Figure 2.9. This figure shows that when both bodies are considered, both foci of the ellipse traced by the separation vector have a dynamical role.

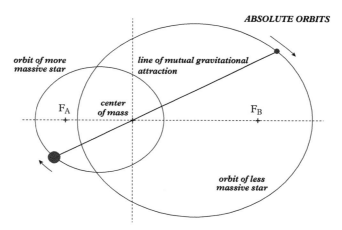

Fig. 2.9 Binary star system.

2.10.1 Binary stars

Stars are often found in binary systems. Many such systems have been observed for decades or even centuries and the changes in their relative positions in the sky over time have been mapped. If the stars orbit in a plane that is perpendicular to our line of sight and we know the distance to the system,[5] then it is possible to determine the mass of each star. With the distance to the system known, the actual size of an orbit is readily determined by measuring the apparent size of the orbit in the sky. If we know the length of the semi-major axis of each orbit, then the ratio of the stellar masses is $\frac{m^{(2)}}{m^{(1)}} = \frac{a^{(1)}}{a^{(2)}}$. We can then use Kepler's third law (2.123) to obtain the sum of the masses M_{tot} from the observed orbital period and the summed lengths of the semi-major axes, $a = a^{(1)} + a^{(2)}$. We can therefore determine each mass separately.

---

[5] The position of a star shifts slightly during the course of a year due to the changing position of the Earth as it orbits the Sun. This is known as *parallax* and it enables the distance to the star to be measured.

The only drawback is that most binary star orbits are not conveniently aligned face-on to us, which introduces some uncertainty into the method. Figure 2.10 shows the orbits of Sirius A, which is the brightest star in the night sky, and its faint companion Sirius B, as measured telescopically over the course of many years. From the Earth these orbits are viewed at an oblique angle, so, although the orbits appear elliptical, we do not see the centre of mass of the system (which is at the origin in the figure) at the foci of the ellipses.

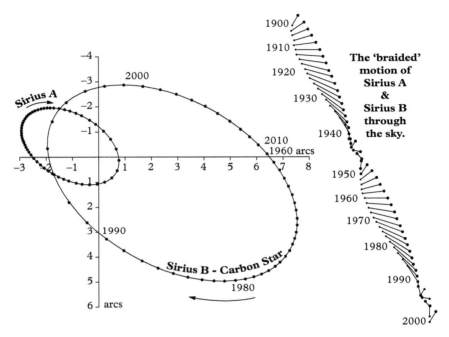

Fig. 2.10 The orbits of Sirius A and B. (The units are arcseconds.)

The determination of the masses of stars in nearby binary systems has been extremely important in enabling astrophysicists to construct accurate theories of the stars. We will explore this fascinating subject in Chapter 13.

2.11 Lagrangian Points

The 3-body problem for gravitating bodies is, in general, insoluble analytically. However, when two of the masses are much greater than the third, there are five points at which the third negligible mass may be placed such that its position remains fixed relative to the other two. These points are named the *Lagrangian points* after the 18th century mathematician Joseph-Louis Lagrange and are denoted L_1 to L_5. They are shown in Figure 2.11. L_1 to L_3 are unstable fixed points, whereas L_4 and L_5 are stable fixed points.

We will assume that the orbits of masses $m^{(1)}$ and $m^{(2)}$ are circular and that $m^{(1)} \gg m^{(2)}$. The separation of these bodies is then constant, and the angular velocities are the same and constant. A test mass $m^{(3)}$ positioned at a Lagrangian point of this system will orbit the centre of mass at the same angular velocity as the other two masses, so $\frac{d\varphi^{(3)}}{dt} = \frac{d\varphi^{(2)}}{dt} = \frac{d\varphi^{(1)}}{dt}$.

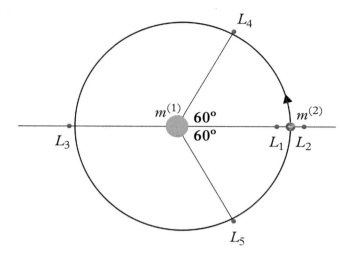

Fig. 2.11 The Lagrangian points in a system with $m^{(1)} \gg m^{(2)}$. In the case of the Sun and Earth, the centre of mass is well inside the body of the Sun.

Lagrangian points L_1 and L_2

The location of L_2 is on the far side of $m^{(2)}$ in a direct line from $m^{(1)}$, as shown in Figure 2.12. Its position can be understood as follows. By Kepler's third law, a test mass further from $m^{(1)}$ than $m^{(2)}$ would normally have a longer orbital period than $m^{(2)}$. However, at L_2 the gravitational attraction of $m^{(2)}$ adds to the attraction of $m^{(1)}$ and thereby decreases the orbital period of the test mass there. At just the right distance r outside the orbit of $m^{(2)}$, the orbital period of the test mass exactly matches that of $m^{(2)}$.

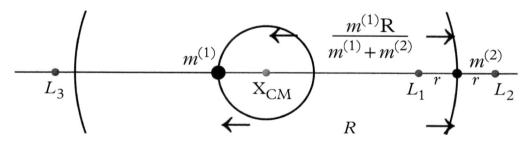

Fig. 2.12 The Lagrangian points L_1 and L_2, indicating the distances between $m^{(1)}, m^{(2)}, \mathbf{X}_{\mathrm{CM}}$ and L_1 and L_2.

L_1 is on the same line, but between $m^{(1)}$ and $m^{(2)}$. In this case, $m^{(2)}$ offsets some of the gravitational attraction of $m^{(1)}$ and thereby increases the orbital period of the test mass. Again, at just the right distance r inside the orbit of $m^{(2)}$ (with r not necessarily the same, although in fact it is), the orbital period of the test mass again exactly matches that of $m^{(2)}$.

We will now determine the distances r for L_2 and L_1. The calculation is very similar in both cases, so we treat them together. If $R = |\mathbf{x}^{(2)} - \mathbf{x}^{(1)}|$ is the distance between $m^{(1)}$ and

$m^{(2)}$, then the distances of $m^{(1)}$ and $m^{(2)}$ from the centre of mass $\mathbf{X}_{\mathrm{CM}}$ are $a^{(1)} = \frac{m^{(2)}R}{m^{(1)}+m^{(2)}}$ and $a^{(2)} = \frac{m^{(1)}R}{m^{(1)}+m^{(2)}}$, and $R = a^{(1)} + a^{(2)}$. The distance of the test mass from $\mathbf{X}_{\mathrm{CM}}$ is therefore $a^{(3)} = \frac{m^{(1)}R}{m^{(1)}+m^{(2)}} + \kappa r$, where $\kappa = 1$ at L_2 and $\kappa = -1$ at L_1. The test mass follows a circular orbit around $\mathbf{X}_{\mathrm{CM}}$, so it satisfies equation (2.75), with $F = V'(r)$ equal to the sum of the gravitational forces on the test mass. This gives

$$\frac{Gm^{(1)}m^{(3)}}{(R+\kappa r)^2} + \kappa \frac{Gm^{(2)}m^{(3)}}{r^2} = m^{(3)} \left(\frac{m^{(1)}R}{m^{(1)}+m^{(2)}} + \kappa r \right) \left(\frac{d\varphi^{(3)}}{dt} \right)^2 . \tag{2.124}$$

The first term is the gravitational force towards $m^{(1)}$, the second is the gravitational force towards $m^{(2)}$, and the term on the right is the mass times the acceleration required for the circular motion.

At the Lagrangian points, the angular velocity of the test mass equals the angular velocities of $m^{(1)}$ and $m^{(2)}$. Kepler's third law for the 2-body problem (2.123), applied to a circular orbit, gives the angular velocity as

$$\left(\frac{d\varphi^{(3)}}{dt} \right)^2 = \left(\frac{d\varphi^{(1)}}{dt} \right)^2 = \left(\frac{d\varphi^{(2)}}{dt} \right)^2 = \left(\frac{2\pi}{T} \right)^2 = \frac{G(m^{(1)}+m^{(2)})}{R^3} . \tag{2.125}$$

If we substitute this into equation (2.124) and cancel factors of $Gm^{(3)}$, we obtain

$$\frac{m^{(1)}}{(R+\kappa r)^2} + \frac{\kappa m^{(2)}}{r^2} = \left(\frac{m^{(1)}R}{m^{(1)}+m^{(2)}} + \kappa r \right) \frac{m^{(1)}+m^{(2)}}{R^3} . \tag{2.126}$$

With $m^{(1)} \gg m^{(2)}$, this reduces to

$$\frac{m^{(1)}}{(R+\kappa r)^2} + \frac{\kappa m^{(2)}}{r^2} \simeq (R+\kappa r)\frac{m^{(1)}}{R^3} , \tag{2.127}$$

and after gathering and rearranging terms, to

$$m^{(1)} \left[\frac{1}{R^2} \left(1 + \frac{\kappa r}{R} \right)^{-2} - \frac{1}{R^2} - \frac{\kappa r}{R^3} \right] \simeq -\frac{\kappa m^{(2)}}{r^2} . \tag{2.128}$$

For $m^{(1)} \gg m^{(2)}$, clearly $R \gg r$, so $(1 + \frac{\kappa r}{R})^{-2} \simeq 1 - \frac{2\kappa r}{R} + \ldots$ and therefore

$$-m^{(1)} \left(\frac{3\kappa r}{R^3} \right) \simeq -\frac{\kappa m^{(2)}}{r^2} . \tag{2.129}$$

κ cancels and the distance r is

$$r \simeq \left(\frac{m^{(2)}}{3m^{(1)}} \right)^{\frac{1}{3}} R , \tag{2.130}$$

the same for both L_2 and L_1.

The mass of the Sun is 1.99×10^{30} kg and the mass of the Earth is 5.97×10^{24} kg, which gives a value for r of $0.01R$. The average distance R between Earth and Sun is around 1.5×10^8 km, so r is 1.5×10^6 km, about four times the average distance of the Moon from

Fig. 2.13 The Moon crossing the face of the Earth, as viewed from the Earth–Sun L_1 point. The photograph was taken by NASA's Deep Space Climate Observatory (DSCOVR).

the Earth. L_1 is at the distance r inside the Earth's orbit in a direct line towards the Sun and L_2 is the same distance away from the Sun, on the opposite side of the Earth. These are suitable places to site various types of space probes. For instance, the Solar and Heliospheric Observatory (SOHO) is positioned at L_1 and the Wilkinson Microwave Anisotropy Probe (WMAP) was positioned at L_2 to minimize the microwave radiation received from the Earth, Moon and Sun. Figure 2.13 is a photograph of the Moon and Earth taken from L_1.

Lagrangian point L_3

L_3 is situated diametrically opposite $m^{(2)}$, on the far side of its orbit, as shown in Figure 2.11. L_3 lies just outside the orbit of $m^{(2)}$, although the distance from L_3 to $m^{(1)}$ is less than R. This is possible because the orbital radius of $m^{(2)}$ is $\frac{m^{(1)}R}{m^{(1)}+m^{(2)}} < R$. If the distance between L_3 and $m^{(1)}$ is $R-r$, then the distance between L_3 and $m^{(2)}$ is $2R-r$. The distance from L_3 to the 2-body centre of mass $\mathbf{X}_{\mathrm{CM}}$ equals the distance to $m^{(1)}$ plus the distance from $m^{(1)}$ to $\mathbf{X}_{\mathrm{CM}}$, which is $R-r+\frac{m^{(2)}R}{m^{(1)}+m^{(2)}}$. To find r, we match the gravitational forces on a test mass at L_3 to the force required to keep the mass in a circular orbit at the same angular velocity as before, thus

$$\frac{m^{(1)}}{(R-r)^2} + \frac{m^{(2)}}{(2R-r)^2} = \left(R-r+\frac{m^{(2)}R}{m^{(1)}+m^{(2)}}\right)\frac{m^{(1)}+m^{(2)}}{R^3}. \qquad (2.131)$$

Making the same approximations as previously, $m^{(1)} \gg m^{(2)}$ and $R \gg r$, we keep terms proportional to $m^{(2)}R$ and $m^{(1)}r$, but neglect terms proportional to $m^{(2)}r$, obtaining

$$\frac{m^{(1)}}{R^2}\left(1 + \frac{2r}{R}\right) + \frac{m^{(2)}}{4R^2} \simeq \left((R-r)(m^{(1)} + m^{(2)}) + m^{(2)}R\right)\frac{1}{R^3}. \tag{2.132}$$

Gathering terms, we find the result

$$\frac{3m^{(1)}r}{R^3} \simeq \frac{7m^{(2)}}{4R^2}, \quad \text{and therefore} \quad r \simeq \frac{7}{12}\frac{m^{(2)}}{m^{(1)}}R. \tag{2.133}$$

The orbital radius at L_3, which is the distance from L_3 to $\mathbf{X}_{\text{CM}}$, is therefore

$$a^{(3)} \simeq R - \frac{7}{12}\frac{m^{(2)}}{m^{(1)}}R + \frac{m^{(2)}R}{m^{(1)} + m^{(2)}} \simeq R + \frac{5}{12}\frac{m^{(2)}}{m^{(1)}}R. \tag{2.134}$$

Lagrangian points L_4 and L_5

The points L_4 and L_5 are each situated at a vertex of an equilateral triangle with $m^{(1)}$ and $m^{(2)}$ forming the other two vertices, as shown in Figure 2.11. By symmetry, the same considerations apply to the positions of both points. We will consider L_4. Its location can be understood by considering Figure 2.14. $\mathbf{a}^{(2)}$ is the acceleration of $m^{(2)}$ due to the gravitational attraction of $m^{(1)}$. Similarly, $\mathbf{a}^{(1)}$ is the acceleration of $m^{(1)}$ due to the gravitational attraction of $m^{(2)}$, so $\frac{|\mathbf{a}^{(1)}|}{|\mathbf{a}^{(2)}|} = \frac{m^{(2)}}{m^{(1)}}$. The masses $m^{(1)}$ and $m^{(2)}$ orbit their centre of mass $\mathbf{X}_{\text{CM}}$. The orbital radius of $m^{(1)}$ is $a^{(1)} = \frac{m^{(2)}R}{m^{(1)} + m^{(2)}}$ and the orbital radius of $m^{(2)}$ is $a^{(2)} = \frac{m^{(1)}R}{m^{(1)} + m^{(2)}}$. The ratio of these radii is

$$\frac{a^{(1)}}{a^{(2)}} = \frac{m^{(2)}}{m^{(1)}} = \frac{|\mathbf{a}^{(1)}|}{|\mathbf{a}^{(2)}|}. \tag{2.135}$$

This is the critical relationship for understanding the position of L_4, as it means that the length of the acceleration vectors $|\mathbf{a}^{(i)}|$ is proportional to the length of the displacement vectors $a^{(i)}$, as shown in Figure 2.14.

The distance from L_4 to $m^{(2)}$ is the same as the distance from $m^{(1)}$ to $m^{(2)}$, so the acceleration $\mathbf{a}^{(3)}(2)$ of a test particle at L_4 due to $m^{(2)}$ is equal in magnitude to the acceleration of $m^{(1)}$ due to $m^{(2)}$, i.e. $|\mathbf{a}^{(3)}(2)| = |\mathbf{a}^{(1)}|$. Similarly, the distance from L_4 to $m^{(1)}$ is the same as the distance from $m^{(2)}$ to $m^{(1)}$, so $|\mathbf{a}^{(3)}(1)| = |\mathbf{a}^{(2)}|$. Because the acceleration vectors are proportional to the displacement vectors, the resultant acceleration $\mathbf{a}^{(3)}$ of the test particle is towards the centre of mass of the 2-body system, as Figure 2.14 shows.

Furthermore, the angular velocity of the test particle at L_4 is the same as the angular velocity of $m^{(1)}$ and $m^{(2)}$, as we will now demonstrate. $m^{(2)}$ is in a circular orbit, so its acceleration has magnitude

$$|\mathbf{a}^{(2)}| = a^{(2)}\left(\frac{d\varphi^{(2)}}{dt}\right)^2, \tag{2.136}$$

Similarly, $|\mathbf{a}^{(1)}| = a^{(1)}\left(\frac{d\varphi^{(1)}}{dt}\right)^2$ and $|\mathbf{a}^{(3)}| = a^{(3)}\left(\frac{d\varphi^{(3)}}{dt}\right)^2$, where $a^{(3)}$ is the distance of L_4 from the centre of mass.

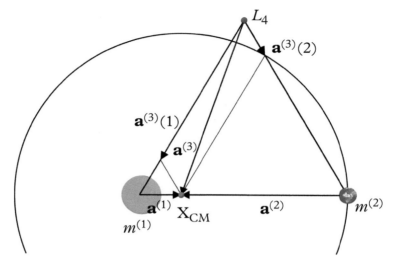

Fig. 2.14 The Lagrangian point L_4.

From the figure, we can see that

$$\frac{|\mathbf{a}^{(1)}|}{a^{(1)}} = \frac{|\mathbf{a}^{(2)}|}{a^{(2)}} = \frac{|\mathbf{a}^{(3)}|}{a^{(3)}}, \quad \text{therefore} \quad \frac{d\varphi^{(1)}}{dt} = \frac{d\varphi^{(2)}}{dt} = \frac{d\varphi^{(3)}}{dt}. \tag{2.137}$$

A number of asteroids are found at the L_4 and L_5 points of the Sun–Jupiter system. These bodies are known as *trojans*. There are also thought to be numerous trojan asteroids at the Sun–Neptune L_4 and L_5 points.

2.12 Conservation of Energy

An important issue that we have so far neglected is the total energy and its conservation. It is a little tricky to get insight into energy conservation directly from the principle of least action. Instead, it is easier to use the equations of motion. Let us start with the example of a single body moving in one dimension, with equation of motion

$$m\frac{d^2x}{dt^2} + \frac{dV}{dx} = 0. \tag{2.138}$$

Multiplying by $\frac{dx}{dt}$, we obtain

$$m\frac{d^2x}{dt^2}\frac{dx}{dt} + \frac{dV}{dx}\frac{dx}{dt} = 0, \tag{2.139}$$

an equation that can be expressed as a total derivative,

$$\frac{d}{dt}\left(\frac{1}{2}m\left(\frac{dx}{dt}\right)^2 + V(x(t))\right) = 0. \tag{2.140}$$

Therefore

$$\frac{1}{2}m\left(\frac{dx}{dt}\right)^2 + V(x(t)) = \text{constant}. \tag{2.141}$$

This constant is the conserved total energy of the body, and is denoted by E. Note that the total energy is the sum of the kinetic and potential energies, $E = K + V$. The sign here is plus, and not the minus sign that occurs in the Lagrangian $L = K - V$.

For a body subject to a general force $\mathbf{F}(\mathbf{x})$ in three dimensions, the energy is not necessarily conserved. However, if the force is due to a potential $V(\mathbf{x})$—which is always the case if the equation of motion is derived from the principle of least action—then the force has the form $\mathbf{F}(\mathbf{x}) = -\nabla V(\mathbf{x})$. In this case, there is again a conserved total energy $E = K + V$. For this reason, any force that can be expressed as $\mathbf{F} = -\nabla V$ is called conservative.

The proof of energy conservation is hardly different from the 1-dimensional case. We take the dot product of the equation of motion (2.55) with $\frac{d\mathbf{x}}{dt}$, and find

$$m\frac{d^2\mathbf{x}}{dt^2}\cdot\frac{d\mathbf{x}}{dt} + \nabla V\cdot\frac{d\mathbf{x}}{dt} = 0. \tag{2.142}$$

The first term is the time derivative of the kinetic energy $K = \frac{1}{2}m\frac{d\mathbf{x}}{dt}\cdot\frac{d\mathbf{x}}{dt}$, and the second is the total time derivative of the potential energy $V(\mathbf{x})$. This follows from

$$V(\mathbf{x}(t+\delta t)) \simeq V\left(\mathbf{x}(t)+\frac{d\mathbf{x}}{dt}\delta t\right) \simeq V(\mathbf{x}(t)) + \nabla V\cdot\frac{d\mathbf{x}}{dt}\delta t. \tag{2.143}$$

So the time derivative of the total energy

$$E = K + V = \frac{1}{2}m\frac{d\mathbf{x}}{dt}\cdot\frac{d\mathbf{x}}{dt} + V(\mathbf{x}(t)) \tag{2.144}$$

is zero.

Total energy is also conserved for a system of N interacting bodies, provided the forces are obtained from a single potential energy function V, which again, is precisely the condition for the equations of motion to be derivable from the principle of least action. The conserved energy is simply the sum of the kinetic energies of the N bodies and the system's potential energy,

$$E = \sum_{1}^{N}\frac{1}{2}m^{(k)}\frac{d\mathbf{x}^{(k)}}{dt}\cdot\frac{d\mathbf{x}^{(k)}}{dt} + V(\mathbf{x}^{(1)},\ldots,\mathbf{x}^{(N)}). \tag{2.145}$$

As in the cases of momentum and angular momentum conservation, it is instructive to determine the contribution of centre of mass motion to the total energy. So suppose that originally the centre of mass is at rest at O, and the total energy is the expression (2.145). Now supplement the velocities of the bodies by a centre of mass velocity $\mathbf{V}_{\text{CM}}$. The potential energy V is unaffected by centre of mass motion because it only depends on the relative positions of the bodies. The new total energy is

$$\begin{aligned}
E' &= \sum_{1}^{N}\frac{1}{2}m^{(k)}\left(\frac{d\mathbf{x}^{(k)}}{dt}+\mathbf{V}_{\text{CM}}\right)\cdot\left(\frac{d\mathbf{x}^{(k)}}{dt}+\mathbf{V}_{\text{CM}}\right) + V(\mathbf{x}^{(1)},\ldots,\mathbf{x}^{(N)}) \\
&= E + \frac{1}{2}M_{\text{tot}}\mathbf{V}_{\text{CM}}\cdot\mathbf{V}_{\text{CM}},
\end{aligned} \tag{2.146}$$

where the terms with a single factor of $\mathbf{V}_{\text{CM}}$ vanish, because the sum of the original momenta

$\sum_1^N m^{(k)} \frac{d\mathbf{x}^{(k)}}{dt}$ is zero if the centre of mass is originally at rest.

We see that the conserved energy (2.146) is the sum of two parts, each of which is separately time independent. The second part is the kinetic energy of the composite body as a whole. The first is the total energy relative to the centre of mass. This energy is called the *internal energy* of the system of bodies. When we come to discuss energy in the context of thermodynamics, it will be the internal energy that will be of interest, and the centre of mass motion will have no thermodynamic significance. For example, the temperature of a gas of molecules depends on the internal energy of the gas, and is unaffected by centre of mass motion.

2.13 Friction and Dissipation

Bodies moving in a vacuum, such as planets and spacecraft in the solar system or elementary particles in a particle accelerator, are subject to negligible friction, but bodies falling through the atmosphere, bodies sliding or rolling on tables, and motorized vehicles all experience friction.

Friction is a complicated force that usually acts on the surfaces of bodies, and has the effect of dissipating their mechanical energy. The sum of the kinetic and potential energies of the bodies is no longer conserved, as some energy is lost as heat, both within the bodies and in the surrounding media. We will not discuss this energy dissipation in any detail, although heat as a form of energy will be discussed in Chapter 10. Here we only explore the simplest way that friction can affect the motion.

The frictional force on a body depends on the body's velocity relative to the medium in contact with its surface. Suppose the medium is at rest. In the simplest model, the frictional force is then proportional to the body's velocity and acts in the opposite direction. The equation of motion (2.22) for the body moving in one dimension, subject to friction, becomes

$$m\frac{d^2x}{dt^2} = -\frac{dV}{dx} - \mu\frac{dx}{dt} \, . \tag{2.147}$$

μ is a positive constant, called the *coefficient of friction*. This simple model is valid for a limited range of velocities. At high velocities, frictional forces often increase more rapidly with velocity, and at extremely low velocities new kinds of sticky surface forces dominate.

We can easily solve the equation of motion (2.147) in special cases. If V is constant, then in the absence of friction the body moves at constant velocity, but with friction present the solution is

$$x(t) = x_0 + \frac{mu_0}{\mu}\left(1 - e^{-\frac{\mu}{m}t}\right) , \tag{2.148}$$

where x_0 and u_0 are the position and velocity at $t = 0$. The body stops at position $x_0 + \frac{mu_0}{\mu}$, but this takes an infinite time. In reality, the body stops after a finite time because of the sticky forces. Another example is a body falling through the atmosphere under gravity. In this case $-\frac{dV}{dx} = -mg$ and the body quickly approaches a terminal velocity, where it no longer accelerates. The terminal velocity is $-\frac{mg}{\mu}$.

There is a rather general result for the rate of energy dissipation due to friction. Consider N bodies in three dimensions, interacting through a potential V, and suppose each body is subject to a frictional force proportional to its velocity. The equations of motion are the modified version of equations (2.108),

$$m^{(k)}\frac{d^2\mathbf{x}^{(k)}}{dt^2} + \nabla^{(k)}V = -\mu\frac{d\mathbf{x}^{(k)}}{dt}\,, \quad k = 1,\ldots,N\,. \tag{2.149}$$

Taking the dot product of each of these with the velocity $\frac{d\mathbf{x}^{(k)}}{dt}$ and adding, we find that

$$\frac{dE}{dt} = -\mu\sum_1^N \frac{d\mathbf{x}^{(k)}}{dt}\cdot\frac{d\mathbf{x}^{(k)}}{dt} \tag{2.150}$$

where $E = K + V$ is the mechanical energy (2.145). So the mechanical energy E is always decreasing, so long as any of the bodies are still moving. The right-hand side of equation (2.150) is not very differerent from the total kinetic energy K. In fact, if all N bodies have the same mass m, then the rate of energy dissipation can be expressed as

$$\frac{d}{dt}(K + V) = -\frac{2\mu}{m}K\,. \tag{2.151}$$

2.14 Further Reading

J.B. Barbour, *The Discovery of Dynamics*, Oxford: OUP, 2001.

T.W.B. Kibble and F.H. Berkshire, *Classical Mechanics (5th ed.)*, London: Imperial College Press, 2004.

L.D. Landau and E.M. Lifshitz, *Mechanics: Course of Theoretical Physics, Vol. 1*, Oxford: Butterworth-Heinemann, 1981.

For a hands-on tool for minimizing the action of a particle moving in one dimension, take a look at

E.F. Taylor and S. Tuleja, *Principle of Least Action Interactive*, which is available here: www.eftaylor.com/software/ActionApplets/LeastAction.html

3

Fields—Maxwell's Equations

3.1 Fields

The fundamental physical ingredients in the previous chapter were space and time, and a set of moving bodies. These bodies were treated as particles, located at a finite set of points, and nothing physical existed in the space between them. Space was completely empty, but despite this, the particles interacted. This is known as *action at a distance*.

From early times, however, it was regarded as rather implausible that particles could interact without anything physical between them. René Descartes and others suggested that it was only reasonable to assume that forces are transmitted by direct contact or through fluids occupying the space between the particles. The modern idea is that space is occupied by *fields* of various types, and these are responsible for the potential energies and forces experienced by particles. At first, the field description of forces was seen as a mathematical reformulation of the Newtonian action at a distance, and perhaps as a physical fiction— but then it was realized that the fields obey their own dynamical equations, and that it is possible for there to be dynamical fields without any particles, at least in some large regions of space.

The great breakthrough in this approach was James Clerk Maxwell's treatment of *electric and magnetic fields*. Previously, these fields had been associated exclusively with charges and currents, but Maxwell's equations allowed additionally for dynamical electric and magnetic fields in the absence of sources. These could be interpreted as light waves. Light is clearly physical, and therefore so are the fields.

Fields now have a central place in physical thinking. We believe that space is everywhere filled with a large number of different kinds of field. In addition to the electromagnetic field, there are Yang–Mills gauge fields and Higgs fields. Even a quintessential particle, like the electron, has an associated field, called the Dirac field. These fields are dynamical. They carry energy and momentum that change as the field transmits forces to particles. There is also a gravitational field, with which one can reformulate the gravitational forces discussed in the previous chapter. Most remarkably, Einstein showed that the only consistent way to describe the gravitational field dynamically is to interpret it as deforming the geometry of spacetime itself.

So, through fields, there is a physical unification of particles, of the forces between them, and of the underlying geometry of spacetime. A dream, yet to be fully realized, is that all physical phenomena derive from a purely geometrical theory of fields.

The common notion of a field is a plot of land with something growing in it. By analogy, physicists, beginning with Michael Faraday, have adopted this term and it is not such a bad one. When crops grow, they have properties that vary from place to place. With large

The Physical World. Nicholas Manton and Nicholas Mee, Oxford University Press (2017).
© Nicholas Manton and Nicholas Mee. DOI 10.1093/oso/9780198795933.001.0001

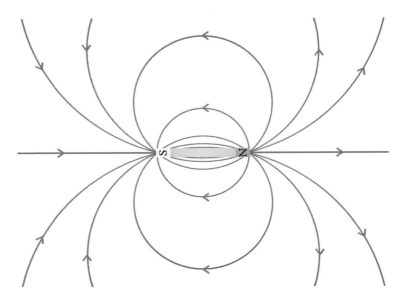

Fig. 3.1 The magnetic field around a magnet.

numbers of individual plants, we can look at average quantities, like the density of plants. This is the number of plants per unit area, which is essentially a continuously varying quantity, both in space and time. The average height of the plants is another quantity that varies continuously in space and time. A farmer may aim for a uniform density of plants, and uniform height, but more commonly these quantities will vary spatially. The height will definitely vary with time.

In physics, the meaning of field is like the quantities described above—plant density and height. Fields are physical quantities that vary in space and time. Usually they are smooth functions, meaning that we can differentiate them with respect to space and time variables as often as we wish. A field is more than just a mathematical function of space and time, because it has some physical reality. The relation between a field and a function is analogous to the relation between a particle trajectory and a geometrical curve.

The earliest examples of fields in physics arose in the description of fluids. We now know that fluids are made of countless particles, atoms or molecules, but they appear to us as continua of matter. Key quantities are the density, the mass of fluid per unit volume, and the fluid velocity. The fluid velocity is assumed to exist at every point in the fluid and to vary smoothly from point to point. The velocity is therefore a function $\mathbf{v}(\mathbf{x}, t)$ defined in the entire region occupied by the fluid, and the density is similarly a function $\rho(\mathbf{x}, t)$. ρ has a single value at each point, and is unaffected by rotations of the spatial axes. Such a quantity is called a *scalar field*. Fluid velocity is a *vector field*.

In this chapter we will first discuss a scalar field, using the principle of least action to find its dynamical equation. The scalar field equation can be used to describe sound waves. We then discuss electric and magnetic fields, and their dynamical equations, the Maxwell equations. Static magnetic fields are rather familiar. We can visualize the field surrounding a magnet, illustrated in Figure 3.1, through the alignment of iron filings sprinkled on a piece of paper covering the magnet. Later, we will consider electrically charged particles and electric currents, which act as sources for electric and magnetic fields, and the equations of motion for the charged particles as they are influenced by the fields. A current is ultimately due to charged particles in motion, but it is often convenient to treat it as an independent concept. Among the most important solutions of Maxwell's equations are the electromagnetic waves describing light.

The dynamical framework of the electromagnetic field interacting with numerous charged particles comes close to being a complete and consistent theory of all electromagnetic phenomena. The dynamical equations for the fields and particles can be derived from a single principle of least action. However, a couple of issues require further exploration. One is the need to modify Newton's equations of motion for particles moving at high speeds comparable to the speed of light. We will deal with this in Chapter 4, on special relativity. The second issue is the idealization of charged particles as point-like. This is problematic if the particles experience very large accelerations and radiate a lot of energy. The equations of motion of point-like particles are then unclear. Guidance from experiment is not yet available, as such accelerations require extremely strong electric or magnetic fields, which have not yet been achieved in a laboratory setting.

It is tempting to discuss the Newtonian gravitational field as a scalar field—but this is an approximation, and the only consistent treatment of gravity as a dynamical theory is through Einstein's equations of general relativity, so we postpone further discussion of gravity until Chapter 6. Quantized field theory and its relation to elementary particle physics is discussed in Chapter 12.

3.2 The Scalar Field Equation

A scalar field is simpler than the electromagnetic field. It is a single-component real function $\psi(\mathbf{x}, t)$, defined throughout space. The field at a given time is called a field configuration, and the dynamical field evolution can be thought of as a smooth trajectory through the (infinite-dimensional) space of field configurations. An action S needs to be defined for any trajectory of the field connecting a given initial field configuration $\psi(\mathbf{x}, t_0)$ at time t_0 to a final configuration $\psi(\mathbf{x}, t_1)$ at time t_1. By applying the principle of least action to S we can derive the dynamical, scalar field equation, which is a type of wave equation.

As for particles, the Lagrangian L is a combination of kinetic energy and potential energy. The kinetic energy is half the square of the partial time derivative of ψ, integrated over space,

$$K = \int \frac{1}{2} \left(\frac{\partial \psi}{\partial t} \right)^2 d^3x \,. \tag{3.1}$$

(Spatial integrals, throughout this chapter, are over 3-space $\mathbb{R}^3$.) K is analogous to the kinetic energy of a particle of unit mass, but because it is an integral over space, the field at all points contributes. K is unaffected by rotations of the axes or translations of the origin. In other words, K is Euclidean invariant.

There is more choice for the potential energy. One possible contribution is the integral of some function of ψ, which we denote $U(\psi)$. As ψ is a function of $\mathbf{x}$, it is more precise to write $U(\psi(\mathbf{x}))$, and integrate this. U can be any familiar function, for example a sine function or an exponential function, but usually in three dimensions it is a polynomial in ψ. This contribution is analogous to the potential energy $V(\mathbf{x})$ of a particle. A further feature of a field is its gradient $\nabla\psi$, incorporating its partial space derivatives. This gives a second possible contribution, the integral over space of $\frac{1}{2}c^2\nabla\psi \cdot \nabla\psi$, where c is a non-zero constant. The total potential energy is then

$$V = \int \left\{ \frac{1}{2}c^2\nabla\psi \cdot \nabla\psi + U(\psi) \right\} d^3x \,, \qquad (3.2)$$

and this is again Euclidean invariant. Notice that V only depends on the field configuration at each instant of time, and this is why it is called potential energy. One could contemplate further terms in V, for example, higher powers of the gradient of ψ. Contributions involving the product of ψ at different locations, integrated over space, are called non-local terms—they lead to dynamical equations where the field evolution at one point produces instantaneous effects at other points, nullifying our reasons for introducing fields. So we do not allow them. One of the main motivations for fields was that physical signals should propagate at finite speed to avoid action at a distance.

We need to say something about the overall shape of the function U and about boundary conditions. To produce a satisfactory stable field theory, $U(\psi)$ should have a minimum at some finite value of ψ. We shall assume, in this chapter, that U has a unique minimum at $\psi = 0$. Examples of such functions are $U(\psi) = \frac{1}{2}\mu^2\psi^2$ with μ non-zero, and $U(\psi) = \frac{1}{2}\mu^2\psi^2 + \frac{1}{4}\nu\psi^4$ with ν positive. A theory with a unique minimum at another value of ψ is essentially equivalent, because by a field redefinition $\psi \rightarrow \psi+\text{constant}$, the minimum can be shifted to $\psi = 0$. We shall assume that the minimum value of U is zero, as it is in these two examples. This ensures that the total potential energy is zero, rather than infinite, for the field configuration $\psi = 0$. Finally, we impose as a boundary condition that $\psi \rightarrow 0$ as $|\mathbf{x}| \rightarrow \infty$. In other words, at spatial infinity, the field minimizes U, and the potential energy density vanishes. At any finite point $\mathbf{x}$, where $\psi \neq 0$ or the gradient is non-zero, the potential energy density is positive.

The configuration $\psi = 0$ everywhere in space and for all time is called the classical vacuum. The field is still present everywhere in spacetime, but it has zero potential energy and zero kinetic energy, the minimum possible values. Such a field carries no energy and no momentum.

The total action S for the scalar field ψ is the integral over time of the Lagrangian $L = K - V$, that is,

$$S = \int_{t_0}^{t_1} \int \left\{ \frac{1}{2}\left(\frac{\partial\psi}{\partial t}\right)^2 - \frac{1}{2}c^2\nabla\psi \cdot \nabla\psi - U(\psi) \right\} d^3x \, dt \,. \qquad (3.3)$$

The integrand, the quantity in curly brackets, is called the *Lagrangian density*, and denoted by $\mathcal{L}(\mathbf{x}, t)$. It depends only on the field value at $(\mathbf{x}, t)$, and the field values in an infinitesimal neighbourhood of $(\mathbf{x}, t)$, which contribute to the time and space derivatives. So $\mathcal{L}$ is said to be local. The Lagrangian L is the integral over space of $\mathcal{L}$, and a further time integration gives the action.

The principle of least action determines the dynamics of the field. The field equation of ψ is the condition for S to be stationary, and to find this we need to use the calculus of variations. Let us fix the field configurations at the initial and final times, t_0 and t_1, and suppose that S is stationary for some trajectory $\psi(\mathbf{x}, t) = \Psi(\mathbf{x}, t)$, where Ψ satisfies the boundary condition $\Psi \rightarrow 0$ as $|\mathbf{x}| \rightarrow \infty$ along with the given endpoint conditions at t_0 and t_1. Now consider a variation of the field $\psi(\mathbf{x}, t) = \Psi(\mathbf{x}, t) + h(\mathbf{x}, t)$, preserving the boundary and endpoint conditions, and require S to be unchanged to first order in h. For the first and third terms in S the calculation is essentially the same as that leading to equation (2.31) for particle motion. For the second term we need to use the expansion

$$\frac{1}{2}c^2\nabla\psi\cdot\nabla\psi = \frac{1}{2}c^2\nabla\Psi\cdot\nabla\Psi + c^2\nabla\Psi\cdot\nabla h \tag{3.4}$$

(keeping only terms up to first order in h). The analogue of equation (2.31) is

$$S_{\Psi+h} = S_\Psi + \int_{t_0}^{t_1}\int\left\{\frac{\partial\Psi}{\partial t}\frac{\partial h}{\partial t} - c^2\nabla\Psi\cdot\nabla h - U'(\Psi)h\right\}d^3x\,dt\,. \tag{3.5}$$

Integration by parts, both in the time and space directions, converts the terms with derivatives of h into terms depending simply on h. The boundary and endpoint terms all vanish, to give the result

$$S_{\Psi+h} = S_\Psi + \int_{t_0}^{t_1}\int\left\{-\frac{\partial^2\Psi}{\partial t^2} + c^2\nabla^2\Psi - U'(\Psi)\right\}h\,d^3x\,dt\,. \tag{3.6}$$

As Ψ makes S stationary, $S_{\Psi+h}$ must equal S_Ψ for any variation $h(\mathbf{x},t)$, so the quantity in curly brackets multiplying h must vanish everywhere. This gives us the field equation (which we write in terms of the field ψ, rather than Ψ)

$$\frac{\partial^2\psi}{\partial t^2} - c^2\nabla^2\psi + U'(\psi) = 0\,. \tag{3.7}$$

For a general function U, this partial differential equation is a nonlinear wave equation, and is difficult to solve. It is important that the gradient term in the action is not absent; if it were, the field would evolve completely independently at each spatial point.

As the action S is invariant under all Euclidean symmetries, there are several conserved quantities. There is a conserved momentum for the field, and also a conserved angular momentum. These are integrals over space of certain densities depending on the time and space derivatives of the field. There is also a conserved energy, which is simply $E = K + V$, or in other words

$$E = \int\left\{\frac{1}{2}\left(\frac{\partial\psi}{\partial t}\right)^2 + \frac{1}{2}c^2\nabla\psi\cdot\nabla\psi + U(\psi)\right\}d^3x\,. \tag{3.8}$$

In the simplest version of scalar field theory, U is absent. Physically, this is acceptable, and the boundary condition that ψ vanishes at spatial infinity can still be imposed. Equation (3.7) then simplifies to the linear wave equation

$$\frac{\partial^2\psi}{\partial t^2} - c^2\nabla^2\psi = 0\,. \tag{3.9}$$

3.3 Waves

The basic solution of the wave equation (3.9) is a 3-dimensional plane wave of the form

$$\psi(\mathbf{x}, t) = e^{i(\mathbf{k}\cdot\mathbf{x}-\omega t)}, \tag{3.10}$$

where $\mathbf{k}$ is the wavevector and ω is the (angular) frequency. The wavelength is $\frac{2\pi}{|\mathbf{k}|}$. For all $\mathbf{x}$ on any spatial plane orthogonal to $\mathbf{k}$ (at a fixed time) the phase of ψ satisfies

$$\mathbf{k}\cdot\mathbf{x} - \omega t = \text{constant}, \tag{3.11}$$

which is why this is called a plane wave. The double time derivative in equation (3.9) brings down from the exponent of ψ two factors of $-i\omega$, i.e. a factor of $-\omega^2$, and the Laplacian $\nabla^2 = \frac{\partial^2}{\partial x_1^2} + \frac{\partial^2}{\partial x_2^2} + \frac{\partial^2}{\partial x_3^2}$ brings down $-k_1^2 - k_2^2 - k_3^2 = -\mathbf{k}\cdot\mathbf{k} = -|\mathbf{k}|^2$, so the plane wave (3.10) satisfies the linear wave equation provided

$$\omega^2 = c^2|\mathbf{k}|^2. \tag{3.12}$$

Therefore $\mathbf{k}$ is an arbitrary constant vector, but ω must take the value $c|\mathbf{k}|$ or $-c|\mathbf{k}|$. A plane wave is shown in Figure 3.2.

The velocity $\mathbf{c}$ of the plane wave is determined by the condition that at a point $\mathbf{x}$ moving at velocity $\mathbf{c}$ in the direction of $\mathbf{k}$, the phase of the wave remains constant in time. Differentiating equation (3.11) with respect to time, and setting $\frac{d\mathbf{x}}{dt} = \mathbf{c}$, we find $\mathbf{k}\cdot\mathbf{c} - \omega = 0$. So $|\mathbf{c}|$, the wave speed, is $\frac{|\omega|}{|\mathbf{k}|}$, and this is simply c, the parameter in the action and field equation. The wave speed is independent of both frequency and the direction of the wavevector.

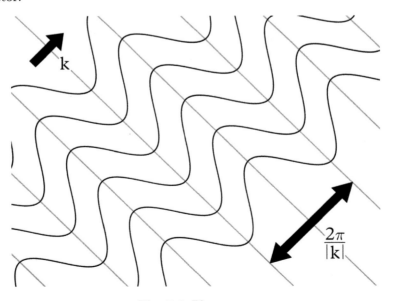

Fig. 3.2 Plane wave.

The basic plane wave solution is not real, nor does it satisfy the boundary condition of vanishing at spatial infinity—but because the wave equation is linear, its general solution is a *linear superposition* of the basic solutions with $\omega = \pm c|\mathbf{k}|$ of the form

$$\psi(\mathbf{x}, t) = \int \left(C(\mathbf{k})e^{i(\mathbf{k}\cdot\mathbf{x}-c|\mathbf{k}|t)} + D(\mathbf{k})e^{i(\mathbf{k}\cdot\mathbf{x}+c|\mathbf{k}|t)} \right) d^3k \,. \tag{3.13}$$

By suitable constraints on the complex functions $C(\mathbf{k})$ and $D(\mathbf{k})$, ψ becomes real and satisfies the boundary condition. Transforming between ψ as a function of $\mathbf{x}$, and C and D as functions of $\mathbf{k}$, is an example of a Fourier transform.

There is a rather beautiful general solution of the linear wave equation in one space dimension, which can also be used for waves in three dimensions provided all the contributing wavevectors $\mathbf{k}$ are in the same direction. In one dimension, where the coordinates are x and t, the wave equation is

$$\frac{\partial^2 \psi}{\partial t^2} - c^2 \frac{\partial^2 \psi}{\partial x^2} = 0 \,. \tag{3.14}$$

This can be written in the factorized form

$$\left(\frac{\partial}{\partial t} + c\frac{\partial}{\partial x} \right) \left(\frac{\partial}{\partial t} - c\frac{\partial}{\partial x} \right) \psi = 0 \,, \tag{3.15}$$

and the order of the factors can be reversed if one wishes. The second operator gives zero acting on any function of $x + ct$, as

$$\frac{\partial}{\partial t} f(x + ct) = cf'(x + ct) = c\frac{\partial}{\partial x} f(x + ct) \,, \tag{3.16}$$

(where f' is the derivative of f) and the first operator gives zero acting on any function of $x - ct$. Therefore the general solution of the wave equation (3.14) is

$$\psi(x, t) = f(x + ct) + g(x - ct) \tag{3.17}$$

where f and g are arbitrary (smooth) functions. These functions are determined by the initial data: ψ and $\frac{\partial \psi}{\partial t}$ at $t = 0$.

A function $f(x + ct)$ is unchanged if t increases by a and simultaneously x decreases by ca. Therefore this function is a wave profile moving in the direction of negative x at speed c. Similarly $g(x - ct)$ is a wave profile moving in the direction of positive x at speed c. These waves are called left-moving and right-moving, respectively. If the initial wave is localized in some finite interval of space, and is zero outside, then it is some combination of left- and right-moving waves, and at later times these waves separate. The field value between the separating waves is uniform and constant, but not necessarily zero.

It is easy to have a wave that is moving purely in one direction, say right-moving. This is the kind of wave that describes a directed flash of light, whose contributing wavevectors are in the direction of the beam, and whose wave fronts are orthogonal to the beam. The 1-dimensional approximation is sensible provided the beam width is much greater than the wavelength.

Another version of the scalar wave equation in three dimensions is worth mentioning. Suppose the function U does not vanish, but has the form $U(\psi) = \frac{1}{2}\mu^2\psi^2$. The field equation, which is still linear, is known as the *Klein–Gordon equation* and is

$$\frac{\partial^2\psi}{\partial t^2} - c^2\nabla^2\psi + \mu^2\psi = 0. \tag{3.18}$$

As before, plane wave solutions have the exponential form (3.10), but the relation of ω to the wavevector $\mathbf{k}$ is

$$\omega^2 = c^2|\mathbf{k}|^2 + \mu^2. \tag{3.19}$$

The waves now have a minimum frequency $\omega = \mu$, and the wave speed depends on frequency. Fourier analysis can again be used to understand more general solutions, with localized, real profiles.

Scalar field theory has some applications. One of these is to sound waves. The density of a gas is a scalar. Small disturbances ψ of the density are described by the action (3.3) with U vanishing. The constant, uniform equilibrium density doesn't appear, because both terms in the action involve derivatives. (There is also an overall constant multiplying S, but this doesn't affect the field equation.) The wave equation (3.9) is then the equation for the sound waves and c is the speed of sound. c depends on the compressibility of the gas and its equilibrium density.

The Klein–Gordon equation will appear again when we consider relativistic scalar fields in the context of particle physics.

3.4 Divergence and Curl

In 3-space, we have seen that it is useful to combine the three partial derivatives into a vector operator

$$\nabla = \left(\frac{\partial}{\partial x_1}, \frac{\partial}{\partial x_2}, \frac{\partial}{\partial x_3}\right). \tag{3.20}$$

Its action on a scalar field ψ gives the gradient, $\nabla\psi$.

∇ can act on a vector field $\mathbf{V}(\mathbf{x}) = (V_1(\mathbf{x}), V_2(\mathbf{x}), V_3(\mathbf{x}))$ in two geometrically natural ways, analogous to the two products $\mathbf{x} \cdot \mathbf{y}$ and $\mathbf{x} \times \mathbf{y}$. The first is $\nabla \cdot \mathbf{V}$, which is called the *divergence* of $\mathbf{V}$, or 'div $\mathbf{V}$', and the second is $\nabla \times \mathbf{V}$, called the *curl* of $\mathbf{V}$, or 'curl $\mathbf{V}$'. Under a rotation of axes, the components of ∇ and $\mathbf{V}$ rotate in the same way, so $\nabla \cdot \mathbf{V}$ is a scalar quantity, invariant under rotations, whereas $\nabla \times \mathbf{V}$ is a vector, rotating with $\mathbf{V}$ and other vectors.

Explicitly, the definition of the divergence of $\mathbf{V}$ is

$$\nabla \cdot \mathbf{V} = \frac{\partial V_1}{\partial x_1} + \frac{\partial V_2}{\partial x_2} + \frac{\partial V_3}{\partial x_3}. \tag{3.21}$$

Note the analogy with the dot product definition (1.12). As $\nabla \cdot \mathbf{V}$ is a function of $\mathbf{x}$, it is a scalar field. If $\nabla \cdot \mathbf{V}$ is positive in some region, then that region is a source for $\mathbf{V}$, and $\mathbf{V}$ tends to point outwards. If $\nabla \cdot \mathbf{V}$ is negative then $\mathbf{V}$ tends to point inwards.

The definition of curl $\mathbf{V}$ is

$$\nabla \times \mathbf{V} = \left(\frac{\partial V_3}{\partial x_2} - \frac{\partial V_2}{\partial x_3}, \frac{\partial V_1}{\partial x_3} - \frac{\partial V_3}{\partial x_1}, \frac{\partial V_2}{\partial x_1} - \frac{\partial V_1}{\partial x_2}\right). \tag{3.22}$$

Again, note the analogy with the cross product definition (1.15). $\nabla \times \mathbf{V}$ is a vector field with three components, and it is a measure of how $\mathbf{V}$ circulates.

A few results about the divergence and curl of general vector fields are important for us. First, if $\mathbf{V}$ can be expressed as $-\nabla\Phi$ for some scalar field Φ, then $\nabla \times \mathbf{V} = -\nabla \times \nabla\Phi = 0$. (The minus sign could be absorbed into Φ, but making it explicit connects to the relation $\mathbf{F} = -\nabla V$ between force and potential.) This is easy to check. For example, the first component of $\nabla \times \mathbf{V}$ is

$$-\frac{\partial}{\partial x_2}\frac{\partial\Phi}{\partial x_3} + \frac{\partial}{\partial x_3}\frac{\partial\Phi}{\partial x_2}, \tag{3.23}$$

and this is zero, by the symmetry of mixed partial derivatives. A deeper result is the converse of this: if $\nabla \times \mathbf{V} = 0$ in some (simply connected) region of space, then there is a scalar field Φ in that region such that $\mathbf{V} = -\nabla\Phi$, and Φ is unique apart from the addition of a constant.

Second, if $\mathbf{V}$ can be expressed as $\nabla \times \mathbf{W}$ for some vector field $\mathbf{W}$, then $\nabla \cdot \mathbf{V} = 0$. This is easy to check too, as

$$\begin{aligned}\nabla \cdot (\nabla \times \mathbf{W}) &= \frac{\partial}{\partial x_1}\left(\frac{\partial W_3}{\partial x_2} - \frac{\partial W_2}{\partial x_3}\right) + \frac{\partial}{\partial x_2}\left(\frac{\partial W_1}{\partial x_3} - \frac{\partial W_3}{\partial x_1}\right) \\ &\quad + \frac{\partial}{\partial x_3}\left(\frac{\partial W_2}{\partial x_1} - \frac{\partial W_1}{\partial x_2}\right) = 0.\end{aligned} \tag{3.24}$$

The result is zero, because by the symmetry of mixed partial derivatives, the terms cancel in pairs. Again, it is the converse that is more profound: if $\nabla \cdot \mathbf{V} = 0$ in some region, then there is a vector field $\mathbf{W}$ such that $\mathbf{V} = \nabla \times \mathbf{W}$. This field $\mathbf{W}$ is unique apart from the addition of the gradient of a scalar, $\nabla\lambda$ (whose curl is identically zero so it doesn't contribute to $\mathbf{V}$). If one wants to pin down $\mathbf{W}$, one can impose a further condition, like $\nabla \cdot \mathbf{W} = 0$, but this isn't always desirable.

In Maxwell theory there are two vector fields, the electric field $\mathbf{E}$ and the magnetic field $\mathbf{B}$. We will see that $\nabla \cdot \mathbf{B}$ is always zero, so $\mathbf{B}$ can be expressed as $\nabla \times \mathbf{A}$. $\mathbf{A}$ is called the *vector potential*. $\nabla \times \mathbf{E}$ is sometimes zero, and if so, $\mathbf{E}$ can be expressed as $-\nabla\Phi$. Φ is called the *scalar potential*. Even when $\nabla \times \mathbf{E}$ is not zero, there is a scalar potential that plays an important role.

3.5 Electromagnetic Fields and Maxwell's Equations

Many electric and magnetic phenomena have been known since antiquity. These include static electricity produced by rubbing amber and other materials, naturally occurring magnetic rocks known as lodestones, the electric shocks given by organisms such as electric eels, and the phenomenon of lightning. However, understanding this diverse array of phenomena and recognizing their connections took a long time. An early breakthrough was when Benjamin Franklin realized in the middle of the 18th century that objects can acquire an electric charge, and that this may be positive or negative. A second fundamental breakthrough was the invention of the battery by Alessandro Volta in 1800, as this gave researchers a ready supply of electricity for their experiments, and studying the electric currents produced by such batteries showed that current is a flow of electric charge.

The next major discovery was the hint of a connection between electricity and magnetism, made by Hans Christian Ørsted in 1820. Ørsted observed the effect on a nearby magnetic compass of an electric current flowing in a wire, as illustrated in Figure 3.3. This small effect would eventually lead to one of the great unifications in the history of science—the theory of electromagnetism. A key conceptual step along the way was Michael Faraday's proposal

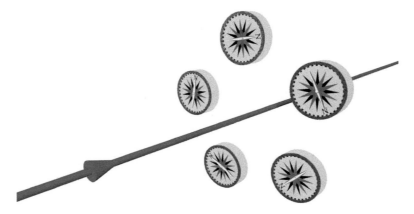

Fig. 3.3 The magnetic field around a wire.

that both electric and magnetic phenomena are best described by postulating the existence throughout space of an electric field $\mathbf{E}$ and a magnetic field $\mathbf{B}$. These are each vectors, with components $\mathbf{E} = (E_1, E_2, E_3)$ and $\mathbf{B} = (B_1, B_2, B_3)$, and they are functions of position $\mathbf{x}$ and time t. $\mathbf{E}$ and $\mathbf{B}$ can be measured by test charges and test magnets (see Figure 3.4). If a charge q is placed at a point $\mathbf{x}$, it experiences an electric force of strength $q\mathbf{E}$. If a small magnet is placed at $\mathbf{x}$, it aligns along the direction of $\mathbf{B}$, and the strength of $\mathbf{B}$ affects how fast this happens. More precisely, the torque or twisting force on the magnet is proportional to the strength of $\mathbf{B}$. It is presumed that $\mathbf{E}$ and $\mathbf{B}$ are present *even if the test devices are removed*. Although once controversial, the power of this idea eventually overcame the doubts of the sceptics.

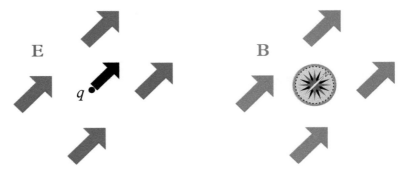

Fig. 3.4 Electric and magnetic forces.

A nagging question is how to incorporate the field due to the test charge into the total field. If the test charge is small, but not infinitesimal, it makes a contribution to the total field. However, in most circumstances, this contribution can be neglected, and it is the field produced by all the other charges and currents that affects the test charge. Only if the test charge is very rapidly accelerating need one worry about the interaction of the test charge with its own self-field.

It is also rather difficult to understand the fields inside materials like conducting metals. Our modern view of materials makes life simpler. Materials have various kinds of charged particles moving about inside them. Fundamentally, therefore, we need a theory of **E** and **B** and of their interactions with moving point particles. The field equations of macroscopic media, such as conductors, dielectric insulators or ferromagnets, can be obtained by averaging the fields produced by their constituent particles.

Building on over a century's experimental work by numerous others, such as Charles de Coulomb, Jean-Baptiste Biot and Félix Savart, Ørsted, André-Marie Ampère, and especially Faraday, Maxwell found the final form of the equations satisfied by **E** and **B**. The sources for these fields are the electric charge density ρ and current density **j**, which are functions of **x** and t. Maxwell wrote his equations in component form, and as a consequence, there were 20 equations in his definitive paper on electromagnetism in 1865. These were later rewritten in a more compact and elegant form by Oliver Heaviside[1] in 1884 using vector notation. It is in this form that the Maxwell equations are usually known. They are

$$\nabla \cdot \mathbf{E} = \rho, \tag{3.25}$$

$$\nabla \times \mathbf{E} = -\frac{\partial \mathbf{B}}{\partial t}, \tag{3.26}$$

$$\nabla \cdot \mathbf{B} = 0, \tag{3.27}$$

$$\nabla \times \mathbf{B} = \mathbf{j} + \frac{\partial \mathbf{E}}{\partial t}. \tag{3.28}$$

The Maxwell equations (3.25)–(3.28) often include the constant parameters ε_0 and μ_0. We have chosen the Heaviside–Lorentz system of units, in which both these constants are unity. Even in these units, the speed of light c usually appears in the equations, but we have made the further choice of working in spacetime units where $c = 1$. This is not the standard SI system of units, but our choice simplifies the mathematics considerably, and is most helpful when we discuss relativity and quantum field theory. (Readers who are uncomfortable with this should consult the many electromagnetism textbooks that discuss units, and work in the SI system.)

The unit of charge is defined in terms of the electric force between two charges at unit distance apart. The unit of current is defined in terms of the magnetic force between two parallel current-carrying wires separated by unit distance. As currents consist of charges in motion, it is sensible to ask about the ratio of the magnetic force to the electric force for moving charges, as depicted in Figure 3.5. At what speed are the forces comparable? The answer is a speed that is present in electromagnetic theory as a fundamental parameter. This turns out to be the speed of light.

There is a good reason for setting the speed of light to unity. Historically, time and length units were defined terrestrially, by a clock measuring a second as 1/86,400 of the length of a day. A metal bar held in Paris was the standard by which a length of one metre was defined. The length of this bar is approximately 10^{-7} times the distance from the North Pole to the equator. The speed of light was then a quantity to be measured, and its value changed over the years, as experimental techniques improved. More recently it was decided to define the time unit in terms of the frequency of the photons emitted in a particular atomic energy level transition, and to define the length unit in terms of the wavelength of

---

[1] This was also done by Heinrich Hertz and Josiah Willard Gibbs at about the same time.

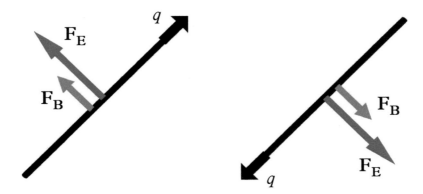

Fig. 3.5 Moving charges produce fields and forces.

the same photons. As a result, the speed of light is now defined by convention to have a value of exactly $c = 299,792,458$ m s^{-1}, the integer giving the best fit to the historical notions of metre and second. As c is a pure integer, of no fundamental significance, it makes more sense to set $c = 1$. The time unit can still be regarded as a second, s, but the length unit is now the light-second, exactly 299,792,458 m.

There is a physical rationale behind this too. We do not normally choose length and time units so that the speed of sound in a gas is unity; this is because sound speed is not universal, but depends on the composition of the gas, and its temperature. However, the speed of light in vacuum is now known to be universal. It is independent of wavelength, so all photons and other elementary particles with negligible mass, such as neutrinos, travel at essentially the same speed. All the field equations of particle physics and the formulae of special and general relativity use the same factor c to relate length to time, so it makes sense to set this universal factor to unity. It was Einstein who had the original insight that the speed of light is the ultimate speed limit and therefore an exceptional quantity.

3.5.1   What Maxwell's equations tell us

Some electromagnetism textbooks spend many chapters discussing the phenomena that motivate Maxwell's equations. Others start with these equations, and spend many chapters solving them and finding their consequences. Here we shall give a brief overview of what each equation tells us, although really it is necessary to consider all the equations together to reach these conclusions.

The first equation, $\nabla \cdot \mathbf{E} = \rho$, says that the charge density ρ is the source for the electric field $\mathbf{E}$. If ρ is positive, $\mathbf{E}$ is directed outwards from the source and weakens in strength as one moves further away. The charge density can be localized at a point, and this models a charged point particle.

The third equation, $\nabla \cdot \mathbf{B} = 0$, says that there is no magnetic analogue of a charge density, so there are no magnetically charged particles, usually referred to as magnetic monopoles. The magnetic field $\mathbf{B}$ behaves like the velocity $\mathbf{v}$ of an incompressible fluid, which obeys $\nabla \cdot \mathbf{v} = 0$ and has no sources or sinks. In fact, over any complete, closed surface, the net outward flux of $\mathbf{B}$ is zero. A magnetic dipole, such as a bar magnet, may appear to have magnetic poles of opposite strength at each end, but actually $\mathbf{B}$ circulates, passing from one end of the magnet to the other in the space outside, and then returning through the material of the magnet. If this were not the case, then one could break the magnet into two pieces, one of which would be a source and the other a sink of magnetic flux. The source of the field **B** produced by a magnet is in fact the electric current **j** present in the material of the magnet, rather than the poles near its ends.

The second Maxwell equation, $\nabla \times \mathbf{E} = -\frac{\partial \mathbf{B}}{\partial t}$, says that **E** circulates around any region where the magnetic field **B** is changing with time. Suppose C is a fixed closed curve bounding a surface through which there is a flux of **B**. When the flux increases or decreases, an electric field **E** is generated that tends to point one way along C. If the geometrical curve C is replaced by a physical, conducting wire then the electric field generates a current along the wire. This is Faraday's law of induction and is depicted in Figure 3.6. It is essential for the generation of electricity. At a power station, mechanical power drives the motion of magnets (actually, electromagnets), and the time-varying magnetic fields generate electric currents which are then distributed over large distances along power cables, for use in all our electrical machinery and devices.

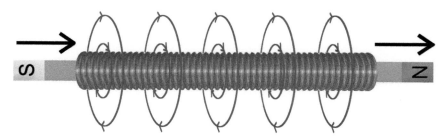

Fig. 3.6 Electromagnetic induction: an electric field is generated by the moving magnet.

An electric field generates a current in a metal wire as follows. The wire is electrically neutral and typically consists of positively charged ions and negatively charged electrons. The electric field produces a force on both of these, but the ions do not move because of the mechanical forces that maintain the integrity of the solid metal. The electrons, however, are free to move, and accelerate in the electric field. They do not accelerate indefinitely, but reach a top speed proportional to **E**, because current flow is limited by the conductor's *resistance*. The outcome is that the current density **j** is proportional to the applied field **E**, which is a version of *Ohm's law*. This picture of current flow is known as Drude theory. It had some success in the early years following the discovery of the electron, but turns out to be rather naïve. In reality, the behaviour of electrons in conductors can be accurately modelled only in the context of quantum theory. We will take a look at the quantum theory of electrons in solids in Chapter 9.

The fourth Maxwell equation, $\nabla \times \mathbf{B} = \mathbf{j} + \frac{\partial \mathbf{E}}{\partial t}$, describes the circulation of **B** around a current, the phenomenon observed by Ørsted. The current is usually flowing in a wire, but it could be a beam of charged particles. In fact, the simplified equation $\nabla \times \mathbf{B} = \mathbf{j}$

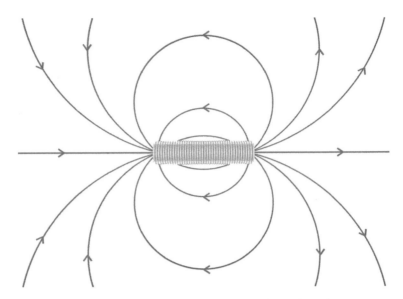

Fig. 3.7 The magnetic field around a solenoid.

describes this quite well and is called *Ampère's law*. Ampère's law is valid for currents in closed circuits, produced by batteries and most electricity generating networks. It is also sufficient for understanding the magnetic field produced by a current running through a coiled wire known as a solenoid, which is very similar to the field of a bar magnet, as shown in Figure 3.7. However, Maxwell realized that Ampère's law by itself is incorrect when a circuit is not closed. For example, current will flow in the set-up shown in Figure 3.8. The current can be driven (briefly) by a battery or (for a longer time) by a time-varying magnetic flux through the interior of the incomplete wire loop. As the current flows, opposite charges build up on the plates at the top that together form a capacitor, and because of the first Maxwell equation, an electric field also builds up between the plates. Maxwell noticed that the magnetic field should behave smoothly as one moves from the region surrounding the wire to the region surrounding the gap between the plates. The charges on the plates do not directly produce a magnetic field, but the time-dependent electric field between the plates does. The fourth Maxwell equation takes into account that **B** is produced by both the current density **j** and the time derivative of **E**.

This second source of **B**, the changing electric field, was not discovered experimentally by earlier scientists, because capacitors joined by wires to a battery tend to charge up rapidly and then stabilize, so there is insufficient time to see the effect. On the other hand, a slowly changing electric field, where there is time to see the effect, only generates a very small magnetic field.

Maxwell's equations are consistent with the conservation of electric charge. Charge can flow, but cannot be created or destroyed. The charge/current conservation equation is

$$\nabla \cdot \mathbf{j} + \frac{\partial \rho}{\partial t} = 0 \,. \tag{3.29}$$

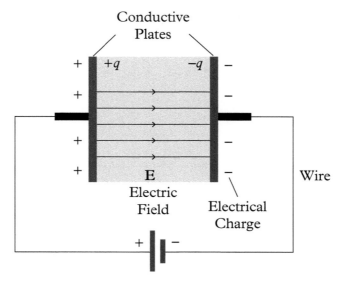

Fig. 3.8 Current flow and charge build-up on a capacitor.

This says that $\rho(\mathbf{x}, t)$, the charge density at $\mathbf{x}$, can change with time, but it can only increase if there is a net flow of current density $\mathbf{j}$ into $\mathbf{x}$, or decrease if there is a net flow out.

Maxwell's equations imply that charge conservation must hold. To see this, take the time derivative of the first Maxwell equation (3.25) and reverse the order of the derivatives $\frac{\partial}{\partial t}$ and ∇ (which is permissible because of the symmetry of mixed partial derivatives) to obtain

$$\nabla \cdot \frac{\partial \mathbf{E}}{\partial t} = \frac{\partial \rho}{\partial t}. \tag{3.30}$$

Using the fourth Maxwell equation (3.28) to substitute for $\frac{\partial \mathbf{E}}{\partial t}$, this becomes

$$\nabla \cdot (\nabla \times \mathbf{B} - \mathbf{j}) = \frac{\partial \rho}{\partial t}, \tag{3.31}$$

which simplifies to the charge/current conservation equation (3.29) because $\nabla \cdot (\nabla \times \mathbf{B}) = 0$ automatically (recall equation (3.24) and the related discussion). Notice that there would have been an inconsistency without the extra term that Maxwell added in equation (3.28). The simpler Ampère law is correct only if $\nabla \cdot \mathbf{j} = 0$, which holds for currents in closed circuits, but not if the charge density is time-varying somewhere in space.

3.6 Electrostatic Fields

The simplest solutions of Maxwell's equations are electrostatic fields. These occur when there are no currents and no magnetic field, and the charge density ρ is static. The electric field $\mathbf{E}$ is then also static. The Maxwell equations (3.27) and (3.28) are trivially satisfied, and the remaining equations are

$$\nabla \cdot \mathbf{E} = \rho(\mathbf{x}), \qquad \nabla \times \mathbf{E} = 0. \tag{3.32}$$

The second of these implies that $\mathbf{E}$ can be expressed as $-\nabla\Phi$, as explained in section 3.4, in which case the first becomes $\nabla \cdot \nabla\Phi = -\rho(\mathbf{x})$. Now recall that the operator $\nabla \cdot \nabla$ is the Laplacian ∇^2, so the basic equation of electrostatics is

$$\nabla^2\Phi = -\rho(\mathbf{x})\,. \tag{3.33}$$

This is *Poisson's equation* for the scalar potential Φ. ρ is the source for Φ but doesn't determine Φ completely, as the Laplace equation $\nabla^2\Phi = 0$ has many solutions. However, if the charge is localized inside a finite region, then we can impose the boundary condition that $\Phi \to 0$ as $|\mathbf{x}| \to \infty$, and then Φ is unique.

To find solutions of Poisson's equation, let us first consider the case where both the charge density ρ and potential Φ are spherically symmetric and smooth, and ρ is zero outside some radius R. The charge density and potential are functions $\rho(r)$ and $\Phi(r)$, where r is the radial coordinate. Using the spherical version of the Laplacian, as in equation (1.42), Poisson's equation simplifies to

$$\frac{d^2\Phi}{dr^2} + \frac{2}{r}\frac{d\Phi}{dr} = -\rho(r)\,. \tag{3.34}$$

This is equivalent to

$$\frac{d}{dr}\left(r^2\frac{d\Phi}{dr}\right) = -r^2\rho(r)\,. \tag{3.35}$$

Integrating, and multiplying both sides by 4π, gives

$$4\pi r^2\frac{d\Phi}{dr} = -\int_0^r 4\pi r'^2\rho(r')\,dr'\,. \tag{3.36}$$

(There is no further integration constant if Φ is to be smooth at the origin.) The right-hand side is minus the charge inside the ball of radius r, which we denote by $Q(r)$. So

$$\frac{d\Phi}{dr} = -\frac{Q(r)}{4\pi r^2}\,, \tag{3.37}$$

and Φ itself can be obtained by integrating once more. The occurrence of 4π here is not obvious from the form of Poisson's equation; it is related to the area of a unit sphere being 4π. The electric field is $\mathbf{E} = -\nabla\Phi = -\frac{d\Phi}{dr}\hat{\mathbf{x}}$, as we saw in equation (1.38), so

$$\mathbf{E}(\mathbf{x}) = \frac{Q(r)}{4\pi r^2}\hat{\mathbf{x}}\,, \tag{3.38}$$

a field of strength $\frac{Q(r)}{4\pi r^2}$ in the radial direction. $\mathbf{E}$ vanishes at the origin, as $Q(r)$ goes to zero more rapidly than r^2 as $r \to 0$.

For r greater than R, the charge density vanishes, so $Q(r) = Q$, where Q is the total charge. Therefore the electric field decays with an inverse square law,

$$\mathbf{E}(\mathbf{x}) = \frac{Q}{4\pi r^2}\hat{\mathbf{x}}\,. \tag{3.39}$$

Here, the potential Φ satisfies Laplace's equation and so must have the form $\Phi(r) = \frac{C}{r} + D$, as we argued at the end of Chapter 1. $\frac{d\Phi}{dr}$ has the right value if $C = \frac{Q}{4\pi}$, and Φ satisfies the boundary condition at infinity if $D = 0$. The potential outside a spherically symmetric charge distribution with total charge Q is therefore

$$\Phi(r) = \frac{Q}{4\pi r}.$$

$$(3.40)$$

A special case is a uniform spherically symmetric charge density ρ_0 in a ball of radius R. The total charge is $Q = \frac{4}{3}\pi R^3 \rho_0$. Outside the ball the potential is $\Phi(r) = \frac{Q}{4\pi r}$ and the electric field is $\mathbf{E}(\mathbf{x}) = \frac{Q}{4\pi r^3}\mathbf{x}$. Inside, $\frac{d\Phi}{dr}$ is given by equation (3.37), where $Q(r) = \frac{4}{3}\pi r^3 \rho_0$. Integrating, one finds that Φ itself is the quadratic expression

$$\Phi(r) = \frac{Q}{8\pi R} \left(3 - \frac{r^2}{R^2} \right),$$

$$(3.41)$$

with the constant of integration fixed so that Φ is continuous at $r = R$, where it equals $\frac{Q}{4\pi R}$. The electric field inside the ball is therefore

$$\mathbf{E}(\mathbf{x}) = \frac{Q}{4\pi R^3}\mathbf{x},$$

$$(3.42)$$

going linearly to zero at the origin.

The most interesting feature of the general results (3.39) and (3.40) is that the exterior electric field and potential depend only on the total charge and not on how it is radially distributed, as sketched in Figure 3.9. This was first established by Newton using a different method, in the context of gravity. He showed that the gravitational attraction of a small test body to a large, spherically symmetric body is the same as if the large body has all its mass located at its centre.

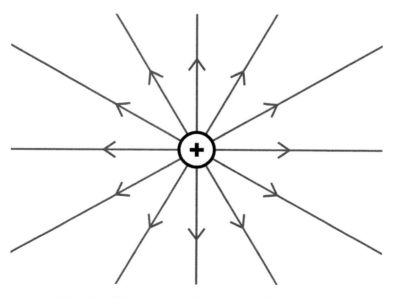

Fig. 3.9 The electric field outside a charged body.

Using formula (3.40) we can go on to find more general solutions of Poisson's equation. To start with, consider the limit where the charge density becomes concentrated at the origin and the total charge[2] is q. Then the potential is

$$\Phi(\mathbf{x}) = \frac{q}{4\pi|\mathbf{x}|}, \tag{3.43}$$

which is finite everywhere except at the origin, where Φ has a singularity. For a point charge at $\mathbf{X}$, we just replace $|\mathbf{x}|$ by $|\mathbf{x} - \mathbf{X}|$. The Poisson equation is linear, so if there are a number of charges $q^{(1)}, \ldots, q^{(N)}$ at locations $\mathbf{x}^{(1)}, \ldots, \mathbf{x}^{(N)}$, the complete potential is obtained by addition, or linear superposition. The potential is

$$\Phi(\mathbf{x}) = \sum_{k=1}^{N} \frac{q^{(k)}}{4\pi|\mathbf{x} - \mathbf{x}^{(k)}|}. \tag{3.44}$$

The solution for a smooth charge density ρ (not necessarily spherically symmetric) is then obtained by replacing the sum in this expression by an integral. The general solution of the original Poisson equation (3.33) is therefore

$$\Phi(\mathbf{x}) = \int \frac{\rho(\mathbf{x}')}{4\pi|\mathbf{x} - \mathbf{x}'|} \, d^3x'. \tag{3.45}$$

This is smooth everywhere.

We have discussed Poisson's equation and its solutions in some detail, because the same equation arises in magnetostatics, as we will see shortly, and also in Newtonian gravity when one considers bodies of finite size.

3.6.1   Charge and dipole moment

Consider a localized but smooth charge density, not necessarily spherically symmetric around the origin, and assume the charge density is zero outside some finite radius R_0. The potential $\Phi(\mathbf{x})$ is the solution (3.45) of Poisson's equation, and outside radius R_0 it automatically satisfies Laplace's equation. Far away, Φ has an expansion in the inverse of the distance r to the origin. From the expansion's first few terms one learns about the main features of the charge distribution, although one cannot resolve all its fine details. The leading term just depends on the total electric charge, and the next term depends on what is known as the *electric dipole moment* of the charge distribution, which is a vector.

To find the charge and dipole moment, and the terms in the potential they produce, we need the expansion of $\frac{1}{|\mathbf{x}-\mathbf{x}'|}$ when $|\mathbf{x}'| \ll |\mathbf{x}|$. This is derived using

$$\begin{aligned}
|\mathbf{x} - \mathbf{x}'|^2 &= (\mathbf{x} - \mathbf{x}') \cdot (\mathbf{x} - \mathbf{x}') \\
&\simeq r^2 - 2\mathbf{x} \cdot \mathbf{x}' \tag{3.46} \\
&= r^2\left(1 - \frac{2\mathbf{x}}{r^2} \cdot \mathbf{x}'\right) \tag{3.47}
\end{aligned}$$

where $r^2 = |\mathbf{x}|^2$ is the squared distance from the origin, and we have dropped the term $|\mathbf{x}'|^2$.

---

[2] We use the notation Q for the total charge of a composite body, and q for the charge of a point-like body or particle.

Inverting and taking the square root gives

$$\frac{1}{|\mathbf{x} - \mathbf{x}'|} \simeq \frac{1}{r}\left(1 + \frac{\mathbf{x}}{r^2} \cdot \mathbf{x}'\right) = \frac{1}{r} + \frac{\mathbf{x}}{r^3} \cdot \mathbf{x}'. \tag{3.48}$$

Substituting this into the solution of Poisson's equation, we find the two leading terms for large r,

$$\Phi(\mathbf{x}) \simeq \frac{1}{4\pi r}\int \rho(\mathbf{x}')\,d^3x' + \frac{\mathbf{x}}{4\pi r^3} \cdot \int \mathbf{x}'\rho(\mathbf{x}')\,d^3x'. \tag{3.49}$$

The integrals here are the total charge

$$Q = \int \rho(\mathbf{x}')\,d^3x', \tag{3.50}$$

and the dipole moment

$$\mathbf{p} = \int \mathbf{x}'\rho(\mathbf{x}')\,d^3x'. \tag{3.51}$$

The leading terms in the potential for large r may then be written more compactly as

$$\Phi(\mathbf{x}) \simeq \frac{Q}{4\pi r} + \frac{\mathbf{p} \cdot \mathbf{x}}{4\pi r^3}, \tag{3.52}$$

and both satisfy Laplace's equation. The charge term decays as $\frac{1}{r}$ and is the same as the potential due to a spherically symmetric charge distribution; the dipole term decays as $\frac{1}{r^2}$ and is a measure of the deviation from spherical symmetry. The electric field is $\mathbf{E} = -\nabla\Phi$, whose dipole part decays as $\frac{1}{r^3}$, with a non-trivial angular dependence.

Under a translation of the charge distribution, say by $\mathbf{a}$, the total charge Q is unchanged, but the dipole moment becomes

$$\tilde{\mathbf{p}} = \int (\mathbf{x}' + \mathbf{a})\rho(\mathbf{x}')\,d^3x' = \mathbf{p} + Q\mathbf{a}. \tag{3.53}$$

So the dipole moment changes if there is a net charge, and can be set to zero by a suitable choice of $\mathbf{a}$, or equivalently, a suitable choice of origin. The dipole moment has no invariant physical significance when Q is non-zero. However, if there is no net charge, then the dipole moment is invariant under translations, and hence more significant. For example, we will see in Chapter 9 that HCl is a polar molecule with no net charge, but with the hydrogen ion positively charged and the chlorine ion negatively charged. The molecule has an electric dipole moment. This rotates as the molecule rotates, but its magnitude is independent of the orientation of the molecule.

The simplest dipole is a point negative charge $-q$ together with a point positive charge q at separation $\mathbf{d}$ away. The dipole moment is $\mathbf{p} = q\mathbf{d}$.

When charges move, the dipole moment of the charge distribution is generally time dependent. An oscillating dipole is the principal source for electromagnetic waves, which we discuss next.

3.7 Electromagnetic Waves

Maxwell's equations involve the fields $\mathbf{E}$ and $\mathbf{B}$. These fields are physical, even though they are not easily visualized, especially if they are time dependent. Surprisingly, the equations simplify if one introduces a new set of fields. These are not directly observable at all, but they seem to be physical too, and fundamental. The new fields are the scalar potential Φ and the vector potential $\mathbf{A}$, briefly mentioned in section 3.4. Φ is a time-dependent version of the potential occurring in electrostatics, and $\mathbf{A}$ is also generally time dependent.

The motivation for introducing Φ and $\mathbf{A}$ is that for any $\mathbf{E}$ and $\mathbf{B}$ satisfying the two sourceless Maxwell equations (3.26) and (3.27), one can always find these new fields, and whereas $\mathbf{E}$ and $\mathbf{B}$ together have six components, Φ and $\mathbf{A}$ have only four.

Whether or not $\mathbf{B}$ is time dependent, it can be expressed as

$$\mathbf{B} = \nabla \times \mathbf{A}, \tag{3.54}$$

because $\nabla \cdot \mathbf{B} = 0$. Differentiating with respect to time gives $\frac{\partial \mathbf{B}}{\partial t} = \nabla \times \frac{\partial \mathbf{A}}{\partial t}$, and substituting this expression into the second Maxwell equation (3.26), we find

$$\nabla \times \left(\mathbf{E} + \frac{\partial \mathbf{A}}{\partial t} \right) = 0. \tag{3.55}$$

Now recall that a vector field with zero curl is the gradient of some scalar field. So we can always write

$$\mathbf{E} + \frac{\partial \mathbf{A}}{\partial t} = -\nabla \Phi. \tag{3.56}$$

This generalizes the electrostatic relation $\mathbf{E} = -\nabla \Phi$. Together, the expressions for $\mathbf{E}$ and $\mathbf{B}$ in terms of Φ and $\mathbf{A}$ are

$$\mathbf{E} = -\frac{\partial \mathbf{A}}{\partial t} - \nabla \Phi, \qquad \mathbf{B} = \nabla \times \mathbf{A}. \tag{3.57}$$

With $\mathbf{E}$ and $\mathbf{B}$ expressed in this way, the Maxwell equations (3.26) and (3.27) are automatically satisfied.

We explained a bit earlier that these two Maxwell equations, and especially Faraday's law of induction (3.26), were experimental discoveries of great physical significance, with important practical consequences. By introducing Φ and $\mathbf{A}$, and expressing $\mathbf{E}$ and $\mathbf{B}$ in terms of them, we seem to have reduced the law of induction to a mathematical triviality, a consequence of the symmetry of mixed partial derivatives. This viewpoint is misleading. A better viewpoint is that Faraday (without realizing it) discovered that the fields Φ and $\mathbf{A}$ have a physical existence even for time-dependent electric and magnetic fields, just as it had earlier been realized that they exist for static fields, where $\mathbf{E} = -\nabla \Phi$ and $\mathbf{B} = \nabla \times \mathbf{A}$.

For given $\mathbf{E}$ and $\mathbf{B}$, the fields Φ and $\mathbf{A}$ are not unique. We can make the replacements

$$\Phi \to \Phi - \frac{\partial \lambda}{\partial t}, \qquad \mathbf{A} \to \mathbf{A} + \nabla \lambda, \tag{3.58}$$

where $\lambda(\mathbf{x}, t)$ is an arbitrary function. $\mathbf{B}$, as defined by equation (3.57), is unaffected because $\nabla \times \nabla \lambda = 0$, and $\mathbf{E}$ is unaffected because in equation (3.57) the additional terms involving λ cancel. The transformation (3.58) is called a *gauge transformation*. (This nomenclature is now standard, but it arose in a different context where there really was a change of measuring gauge, that is, of the length scale.) Fields Φ and $\mathbf{A}$ that differ by a gauge transformation should be regarded as physically equivalent.

We can now substitute the expressions (3.57) for $\mathbf{E}$ and $\mathbf{B}$ into the remaining Maxwell equations (3.25) and (3.28). Equation (3.25) becomes

$$-\nabla \cdot \left(\frac{\partial \mathbf{A}}{\partial t} + \nabla \Phi \right) = \rho \,, \tag{3.59}$$

or, on re-ordering derivatives,

$$-\frac{\partial}{\partial t}(\nabla \cdot \mathbf{A}) - \nabla^2 \Phi = \rho \,. \tag{3.60}$$

Equation (3.28) becomes

$$\nabla \times (\nabla \times \mathbf{A}) = \mathbf{j} - \left(\frac{\partial^2 \mathbf{A}}{\partial t^2} + \nabla \frac{\partial \Phi}{\partial t} \right) \,. \tag{3.61}$$

Using the identity $\nabla \times (\nabla \times \mathbf{A}) = \nabla(\nabla \cdot \mathbf{A}) - \nabla^2 \mathbf{A}$, which is analogous to equation (1.20), equation (3.61) can be re-expressed as

$$\frac{\partial^2 \mathbf{A}}{\partial t^2} - \nabla^2 \mathbf{A} + \nabla \left(\nabla \cdot \mathbf{A} + \frac{\partial \Phi}{\partial t} \right) = \mathbf{j} \,. \tag{3.62}$$

One feature of the transition from $(\mathbf{E}, \mathbf{B})$ to $(\Phi, \mathbf{A})$ becomes clear here. The Maxwell equations only include single time derivatives of the fields, whereas equation (3.62) includes the double time derivative of $\mathbf{A}$.

Equations (3.60) and (3.62) do not look elegant, but a simplification is possible. As Φ and $\mathbf{A}$ are not unique, we may impose one further condition to suit our convenience. This is called a gauge fixing condition. The best choice depends on the circumstances. Sometimes it is $\nabla \cdot \mathbf{A} = 0$, which is known as Coulomb gauge; sometimes it is $\Phi = 0$, which is known as temporal gauge. Here it is best to impose the Lorenz gauge condition, named after Ludvig Lorenz,

$$\nabla \cdot \mathbf{A} + \frac{\partial \Phi}{\partial t} = 0 \,. \tag{3.63}$$

If the potentials $(\Phi, \mathbf{A})$ do not initially satisfy this, we can find a function λ so that following the gauge transformation (3.58) they do.

In Lorenz gauge, equation (3.62) reduces to

$$\frac{\partial^2 \mathbf{A}}{\partial t^2} - \nabla^2 \mathbf{A} = \mathbf{j} \,, \tag{3.64}$$

and we can replace $\nabla \cdot \mathbf{A}$ by $-\frac{\partial \Phi}{\partial t}$ in equation (3.60) to obtain

$$\frac{\partial^2 \Phi}{\partial t^2} - \nabla^2 \Phi = \rho \,. \tag{3.65}$$

This is a remarkable simplification of the Maxwell equations. Both $\mathbf{A}$ and Φ obey wave equations, with $\mathbf{j}$ and ρ as sources, respectively. $\mathbf{A}$ and Φ are not independent, because of the gauge condition (3.63)—but this is reasonable, as $\mathbf{j}$ and ρ are not independent either, as they satisfy the charge/current conservation equation (3.29).

Maxwell's equations in the form of equations (3.64) and (3.65) predict new phenomena, unimagined by Maxwell's predecessors. They imply that oscillating currents and charges generate wave-like electromagnetic fields that propagate through space at the speed of light ($c = 1$) and that visible light only forms a small range of a much broader spectrum of electromagnetic waves. The first experimental confirmation of these ideas was by Heinrich Hertz in 1887. He produced a current in a circuit of wire that generated sparks across a gap in the circuit. Several metres away on the other side of his laboratory Hertz constructed a receiver that responded to the electromagnetic signals generated by the sparks, as shown in Figure 3.10. Generating various types of electromagnetic wave is, of course, commonplace today. Radio waves are typically generated by passing an oscillating current through an aerial or antenna.

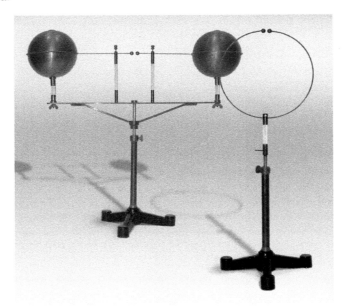

Fig. 3.10 Hertz' experiment for the detection of electromagnetic waves.

Irrespective of how they are generated, electromagnetic waves can be understood by solving the sourceless wave equations. These are equations (3.64) and (3.65) with $\mathbf{j}$ and ρ both set to zero. The simplest solution is a plane wave, a 4-component version of the plane wave solution of scalar field theory. This is of the form

$$\mathbf{A}(\mathbf{x}, t) = \widetilde{\mathbf{A}} e^{i(\mathbf{k} \cdot \mathbf{x} - \omega t)}, \qquad \Phi(\mathbf{x}, t) = \widetilde{\Phi} e^{i(\mathbf{k} \cdot \mathbf{x} - \omega t)}. \tag{3.66}$$

ω is the angular frequency and $\mathbf{k}$ is the wavevector, and the sourceless wave equations imply that these must satisfy

$$\omega^2 = \mathbf{k} \cdot \mathbf{k} = |\mathbf{k}|^2. \tag{3.67}$$

So the wave speed $\frac{|\omega|}{|\mathbf{k}|}$ is unity, which is the speed of light. The vector amplitude $\widetilde{\mathbf{A}}$ and scalar amplitude $\widetilde{\Phi}$ are constants, related by $\mathbf{k} \cdot \widetilde{\mathbf{A}} - \omega\widetilde{\Phi} = 0$ because of the Lorenz gauge condition (3.63).

The wave speed is necessarily the speed of light in our units, but originally the various units and field definitions in the Maxwell equations were based on forces between charges and currents. It was by no means obvious that the speed of the electromagnetic waves derived from the equations would be connected with any other known speed, as no-one knew that light was an electromagnetic phenomenon, although Faraday had found hints that this might be the case in 1845 when he demonstrated that a magnetic field could have a weak but noticeable effect on the polarization of light. Maxwell realized that the waves predicted by his equations have various properties, such as different polarizations, matching the known properties of light, and crucially that they travel at the measured speed of light. The only reasonable conclusion was that light must be an example of an electromagnetic wave. This was one of the most remarkable breakthroughs in the history of science.

The speed of light in vacuum is independent of frequency, or wavelength. This is consistent with the relation (3.67), but inconsistent with a non-zero value for the parameter μ that can occur in the relation $\omega^2 = |\mathbf{k}|^2 + \mu^2$ for a scalar field. As there is no constraint on $\mathbf{k}$, electromagnetic waves may have arbitrarily short or long wavelengths. The wave frequency is determined by the oscillation frequency of the wave generator. As this varies, so does the wavelength. Wavelengths over a vast range, as shown in Figure 3.11, have now been observed, either produced in the laboratory, or through cosmic processes. These include longwave radio signals that have wavelengths around 100 m, modern VHF radio and wireless broadband signals at around 1 m, infrared waves detected in night vision technology at around 10^{-4} m, visible light in the range 7×10^{-7} m (red) to 4×10^{-7} m (violet), X-rays around 10^{-10} m, and laboratory synchrotron 'light' around 10^{-11} m. γ-rays produced in nuclear decays and particle colliders have wavelengths around 10^{-13} m. Electromagnetic waves of such short wavelengths are not well described as classical electromagnetic phenomena and must be considered as composed of individual photons in a quantum theory.

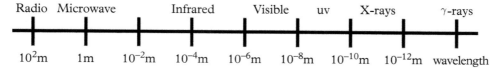

Fig. 3.11 The electromagnetic spectrum.

Let us now consider the geometry of the electric and magnetic fields of an electromagnetic plane wave, as depicted in Figure 3.12. Given $\mathbf{A}$ and Φ, as in equation (3.66), the formulae (3.57) give

$$\mathbf{E} = (i\omega\widetilde{\mathbf{A}} - i\mathbf{k}\widetilde{\Phi})e^{i(\mathbf{k}\cdot\mathbf{x}-\omega t)} \tag{3.68}$$

and

$$\mathbf{B} = i\mathbf{k} \times \widetilde{\mathbf{A}}e^{i(\mathbf{k}\cdot\mathbf{x}-\omega t)} . \tag{3.69}$$

In addition, there is the Lorenz gauge condition, so $\mathbf{k} \cdot \widetilde{\mathbf{A}} - \omega\widetilde{\Phi} = 0$.

The relation between the vectors $\mathbf{k}$ and $\widetilde{\mathbf{A}}$ is not obvious, but we can clarify matters by being more precise about the gauge fixing. For electromagnetic waves, it is possible to gauge fix $\widetilde{\mathbf{A}}$ to be orthogonal to the wavevector $\mathbf{k}$. Then $\mathbf{k} \cdot \widetilde{\mathbf{A}} = 0$ so $\widetilde{\Phi} = 0$. The wave now has an electric field $\mathbf{E}$ of amplitude $\omega\widetilde{\mathbf{A}}$, a vector orthogonal to $\mathbf{k}$, and a magnetic field $\mathbf{B}$ of amplitude $\mathbf{k} \times \widetilde{\mathbf{A}}$, orthogonal to both $\mathbf{E}$ and $\mathbf{k}$. The amplitudes of $\mathbf{E}$ and $\mathbf{B}$ have equal magnitudes in our units, because $|\omega| = |\mathbf{k}|$. The wave is said to be polarized in the direction of $\mathbf{E}$, or equivalently $\widetilde{\mathbf{A}}$. The polarization is transverse because it is orthogonal to the direction of wave propagation, $\mathbf{k}$.

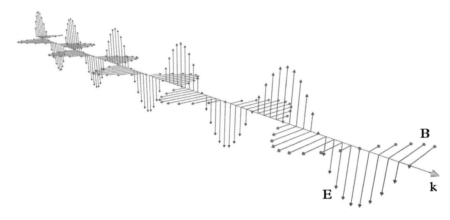

Fig. 3.12 An electromagnetic wave.

With less precise gauge fixing, it remains possible for $\widetilde{\mathbf{A}}$ to have a component parallel to $\mathbf{k}$; this is called the longitudinal polarization component, and if present it is accompanied by a non-zero $\widetilde{\Phi}$. However, this longitudinal component has no physical effect, as it disappears in $\mathbf{E}$ and $\mathbf{B}$, and can be removed by the gauge transformation $\lambda(\mathbf{x}, t) = \frac{i\widetilde{\Phi}}{\omega}e^{i(\mathbf{k}\cdot\mathbf{x}-\omega t)}$, which also makes $\widetilde{\Phi}$ vanish. This gauge transformation preserves the Lorenz gauge condition, because λ satisfies the wave equation.

3.8 Magnetostatics

A steady current produces a static magnetic field. If, additionally, there is no electric charge density and no electric field, then the Maxwell equations reduce to

$$\nabla \cdot \mathbf{B} = 0, \qquad \nabla \times \mathbf{B} = \mathbf{j}. \tag{3.70}$$

The current density $\mathbf{j}$ must satisfy the current conservation equation $\nabla \cdot \mathbf{j} = 0$, the static version of (3.29). The first of equations (3.70) implies that $\mathbf{B}$ can be expressed as $\nabla \times \mathbf{A}$, as before, and the second then becomes

$$\nabla \times (\nabla \times \mathbf{A}) = \mathbf{j}. \tag{3.71}$$

We can again use the identity $\nabla \times (\nabla \times \mathbf{A}) = \nabla(\nabla \cdot \mathbf{A}) - \nabla^2\mathbf{A}$, and now combine this with the Coulomb gauge condition $\nabla \cdot \mathbf{A} = 0$ to fix the gauge. (Coulomb gauge is equivalent to Lorenz gauge if the fields are static.) Then (3.71) simplifies to

$$\nabla^2\mathbf{A} = -\mathbf{j}, \tag{3.72}$$

and this is the basic equation of magnetostatics. It is a vectorial version of the Poisson equation (3.33). Each component of $\mathbf{A}$ satisfies the usual Poisson equation with the corresponding component of $\mathbf{j}$ as source. By taking the divergence, we see that $\nabla^2(\nabla\cdot\mathbf{A}) = 0$, consistent with the Coulomb gauge condition.

We can solve (3.72) by adapting the solution (3.45) of the scalar Poisson equation. The result is

$$\mathbf{A}(\mathbf{x}) = \int \frac{\mathbf{j}(\mathbf{x'})}{4\pi|\mathbf{x} - \mathbf{x'}|}\, d^3x' \,. \tag{3.73}$$

The magnetic field $\mathbf{B}$ is the curl of this expression for $\mathbf{A}$. The derivatives act inside the integral on the variable $\mathbf{x}$, and one finds that

$$\mathbf{B}(\mathbf{x}) = \int \mathbf{j}(\mathbf{x'}) \times \frac{\mathbf{x} - \mathbf{x'}}{4\pi|\mathbf{x} - \mathbf{x'}|^3}\, d^3x' \,, \tag{3.74}$$

which is known as the *Biot–Savart law*.

An alternative way to derive this result is to consider $\nabla \times (\nabla \times \mathbf{B})$. The equations (3.70) then reduce to

$$\nabla^2\mathbf{B} = -\nabla \times \mathbf{j} \,. \tag{3.75}$$

This is Poisson's equation once more, with solution

$$\mathbf{B}(\mathbf{x}) = \int \frac{(\nabla \times \mathbf{j})(\mathbf{x'})}{4\pi|\mathbf{x} - \mathbf{x'}|}\, d^3x' \,. \tag{3.76}$$

The Biot–Savart law is obtained from this by integrating by parts.

The solutions for either $\mathbf{A}$ or $\mathbf{B}$ are not as simple as those in electrostatics. In magnetostatics, the analogue of a spherical charged shell is a circular current loop. This is only circularly, not spherically, symmetric. The magnetic field produced by such a current loop is sketched in Figure 3.13. The field cannot be represented as an elementary function of $\mathbf{x}$ but it can be expressed in terms of *elliptic integrals* (so-named because they are used in calculations of the circumference of an ellipse). A simple expression is available for the field along the symmetry axis through the centre of the loop. The field also simplifies at distances r from the loop that are much greater than the loop radius. Here, the dominant field is a magnetic dipole field whose source is called the *magnetic moment* of the current distribution.

Perhaps the most useful magnetic field is that produced by a solenoid. In practice, a solenoid is a tightly wound, cylindrical coil of wire with a steady current passing along it. Mathematically, it is a finite-length cylinder, with a uniform current density circulating around it (and the current having no component parallel to the cylinder axis). The magnetic field produced by a solenoid is sketched in Figure 3.7. There is no simple, exact formula for the field if the solenoid has finite length, but it is approximately uniform along the interior, and the field emerging from the ends is approximately the same as one would get from opposite magnetic poles placed at the ends. The total field is almost the same as that of a bar magnet, a material object that effectively generates a steady current as a result of quantum effects at the atomic scale, with the same current geometry as in a solenoid.

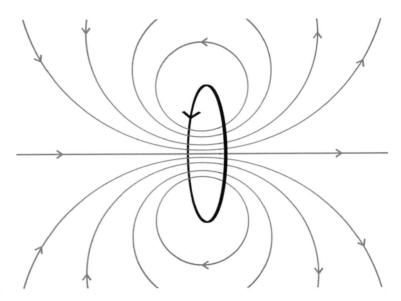

Fig. 3.13 The magnetic field around a current-carrying wire.

3.9   Principle of Least Action for Electromagnetic Fields

Recall that in Newtonian dynamics, the equation of motion for a body moving in one dimension is

$$m\frac{d^2x}{dt^2} = -\frac{dV}{dx}, \qquad (3.77)$$

a single, second order differential equation for x. This equation can be derived from the principle of least action. By introducing the momentum variable p, the equation of motion can be expressed as a first order system

$$\begin{aligned} \frac{dx}{dt} &= \frac{1}{m}p, \\ \frac{dp}{dt} &= -\frac{dV}{dx}. \end{aligned} \qquad (3.78)$$

Equation (3.77) is recovered by eliminating p. The first order system puts x and p on a more symmetrical footing, and is the basis for the Hamiltonian formulation of dynamics.

The Maxwell equations are also a first order dynamical system. $\mathbf{E}$ and $\mathbf{B}$ occur rather symmetrically, and only their first time derivatives appear explicitly. It is not clear which field, if either, is analogous to x, and which is analogous to p. This could make it difficult to find a principle of least action for the electromagnetic fields. The problem is resolved by using the fields $\mathbf{A}$ and Φ. The basic field is $\mathbf{A}$, analogous to a position-like variable in particle dynamics, and the Maxwell equations reduce to a second order dynamical equation for $\mathbf{A}$ that can be derived from a principle of least action.

The field $\mathbf{A}$ changes under a gauge transformation, so only part of $\mathbf{A}$ is gauge invariant and physical. The magnetic field $\mathbf{B} = \nabla \times \mathbf{A}$ captures the local, gauge invariant information held by $\mathbf{A}$ at a given time. Also recall that $\mathbf{E} = -\frac{\partial \mathbf{A}}{\partial t} - \nabla \Phi$. The first term $-\frac{\partial \mathbf{A}}{\partial t}$ is a velocity-type variable, and subtracting off $\nabla \Phi$ picks out the gauge invariant part, the physical electric field **E**. If only **A** and $\frac{\partial \mathbf{A}}{\partial t}$ occurred in the Lagrangian, the Euler–Lagrange equations would completely determine the time evolution of **A**. Because of gauge invariance this cannot be the case. There is some ambiguity in the evolution of **A**, and this is taken into account by including Φ.

After this preamble, we can present the Lagrangian and the action. The charge density ρ and current density **j** are assumed to be externally specified functions of space and time, satisfying the charge/current conservation constraint (3.29). The Lagrangian for the electromagnetic field is then

$$L = \int \left\{ \frac{1}{2} \mathbf{E} \cdot \mathbf{E} - \frac{1}{2} \mathbf{B} \cdot \mathbf{B} + \mathbf{A} \cdot \mathbf{j} - \Phi \rho \right\} d^3 x \,, \tag{3.79}$$

and the action is

$$S = \int_{t_0}^{t_1} L \, dt \,. \tag{3.80}$$

A is the fundamental dynamical field and $\mathbf{E} = -\frac{\partial \mathbf{A}}{\partial t} - \nabla \Phi$ and $\mathbf{B} = \nabla \times \mathbf{A}$ as before. Φ is a further, independent field. For sourceless electromagnetic fields, the integral of $\frac{1}{2} \mathbf{E} \cdot \mathbf{E}$ is the kinetic energy and the integral of $\frac{1}{2} \mathbf{B} \cdot \mathbf{B}$ is the potential energy. The terms $\mathbf{A} \cdot \mathbf{j} - \Phi \rho$ represent the interactions of the fields with the external sources, and they also contribute to the energy.

The Euler–Lagrange equations are obtained by minimizing the action with respect to **A** and Φ. This involves considering small variations of **A** and Φ, and integrating by parts, as usual. The Euler–Lagrange equations (which we will not rederive) reproduce the Maxwell equations in the form of equations (3.60) and (3.62).

There is no term $\frac{\partial \Phi}{\partial t}$ in the action, so Φ is not a dynamical field. Φ is an example of what is called a Lagrange multiplier, or auxiliary field, and the presence of $\nabla \Phi$ terms does not spoil this interpretation. Notice that equation (3.62) includes the double time derivative of **A**, but equation (3.60) only involves a single time derivative. Gauge fixing to simplify the equations is still possible, but Φ should not be removed from the Lagrangian. For example, in Coulomb gauge $\nabla \cdot \mathbf{A} = 0$, equation (3.60) relates Φ to ρ instantaneously. This is a form of constraint equation for Φ, rather than a dynamical equation, but it is still valid physically.

3.10 The Lorentz Force

We have presented charges and currents as sources for electromagnetic fields, but without any detailed consideration of their dynamics. Charges exert forces on each other, and so do currents. For example, to maintain a large current in a coil of wire, the coil has to be held rigidly together in some frame, otherwise it will unwind as a result of the magnetic forces acting between different parts of the coil.

The force between charges at rest is the Coulomb force. The force between current loops of general shape was calculated by Ampère, but is rather complicated. More fundamental than either of these is the force exerted by an electromagnetic field on a charged point particle that is not necessarily at rest. This is known as the *Lorentz force*, after Hendrik Lorentz, and it is

$$\mathbf{F} = q(\mathbf{E} + \mathbf{v} \times \mathbf{B}) \,. \tag{3.81}$$

Being point-like, a particle at **x** experiences a force that only depends on the local field values, $\mathbf{E}(\mathbf{x})$ and $\mathbf{B}(\mathbf{x})$, and on the charge q and velocity **v** of the particle. The electric force

$q\mathbf{E}$ is in the direction of $\mathbf{E}$ and is independent of the particle velocity. The magnetic force $q\mathbf{v} \times \mathbf{B}$ is orthogonal to both the magnetic field and the velocity.

The equation of motion for a particle of charge q and mass m is therefore

$$m\frac{d^2\mathbf{x}}{dt^2} = q\left(\mathbf{E} + \frac{d\mathbf{x}}{dt} \times \mathbf{B}\right), \tag{3.82}$$

where we have written $\frac{d\mathbf{x}}{dt}$ for the velocity. This is Newton's second law for a charged particle, with the Lorentz force on the right-hand side. A combination of acceleration and velocity terms is something we have seen before, for a particle subject to a frictional force, but here there is no friction, because the magnetic force is orthogonal to the velocity and does no work. This can be seen by taking the dot product of equation (3.82) with $\frac{d\mathbf{x}}{dt}$ to give

$$\frac{d}{dt}\left(\frac{1}{2}m\frac{d\mathbf{x}}{dt} \cdot \frac{d\mathbf{x}}{dt}\right) = q\mathbf{E} \cdot \frac{d\mathbf{x}}{dt}. \tag{3.83}$$

We see that the rate of change of the particle's kinetic energy equals the rate at which the electric field does work.

The Coulomb force exerted by one static charge on another may be obtained from the Lorentz force. A charge $q^{(1)}$ at the origin produces an electric field

$$\mathbf{E}(\mathbf{x}) = \frac{q^{(1)}}{4\pi r^2}\hat{\mathbf{x}} = \frac{q^{(1)}}{4\pi r^3}\mathbf{x}. \tag{3.84}$$

Acting on charge $q^{(2)}$ at position $\mathbf{x}$ the force is

$$\mathbf{F} = \frac{q^{(1)}q^{(2)}}{4\pi r^3}\mathbf{x}, \tag{3.85}$$

an inverse square law force. More generally, the Coulomb force exerted by a charge $q^{(1)}$ at $\mathbf{x}^{(1)}$ on a charge $q^{(2)}$ at $\mathbf{x}^{(2)}$ is

$$\mathbf{F} = \frac{q^{(1)}q^{(2)}(\mathbf{x}^{(2)} - \mathbf{x}^{(1)})}{4\pi|\mathbf{x}^{(2)} - \mathbf{x}^{(1)}|^3}. \tag{3.86}$$

The force exerted by charge $q^{(2)}$ on charge $q^{(1)}$ is equal and opposite, consistent with Newton's third law.

The Coulomb potential energy of the pair of charges is

$$V = \frac{q^{(1)}q^{(2)}}{4\pi|\mathbf{x}^{(2)} - \mathbf{x}^{(1)}|}. \tag{3.87}$$

This is analogous to the gravitational potential energy for a pair of masses $m^{(1)}$ and $m^{(2)}$ at these locations, $V = -\frac{Gm^{(1)}m^{(2)}}{|\mathbf{x}^{(2)} - \mathbf{x}^{(1)}|}$, but note that this last quantity is negative, whereas the Coulomb energy is positive for charges of the same sign. Gravity is always attractive, but the Coulomb force is repulsive for charges of the same sign and attractive for charges of opposite sign.

When several charged particles interact, the total force on each particle is to a good approximation the sum of the Coulomb forces exerted by all the others. There are magnetic corrections when the particles are in relative motion, but these are small if the relative speeds are much less than the speed of light.

Knowing the force that acts allows one to calculate the motion of a charged particle in a fixed background electromagnetic field. In general, this is rather complicated, but it simplifies if the background field is uniform and static. We shall describe here the simplest cases. The first is motion in a uniform electric field. If the field is $\mathbf{E} = (0, 0, E)$, then the equation of motion (3.82) implies that the particle has a constant acceleration parallel to the 3-axis, with strength $\frac{qE}{m}$. If the particle starts at rest at the origin at $t = 0$, its position later is $\mathbf{x}(t) = \left(0, 0, \frac{qE}{2m}t^2\right)$. Any motion at constant velocity can be added to this.

More interesting is charged particle motion in a uniform, static magnetic field. The equation of motion is

$$m\frac{d^2\mathbf{x}}{dt^2} = q\frac{d\mathbf{x}}{dt} \times \mathbf{B}\,, \tag{3.88}$$

which can be integrated once to give

$$m\frac{d\mathbf{x}}{dt} = q\mathbf{x} \times \mathbf{B} + \mathbf{u}\,, \tag{3.89}$$

where $\mathbf{u}$ is a constant velocity. The component of the velocity $\frac{d\mathbf{x}}{dt}$ parallel to $\mathbf{B}$ remains constant. For the motion orthogonal to $\mathbf{B}$, the effect of $\mathbf{u}$ can be compensated by a shift of origin. So let's suppose this is done, and set $\mathbf{u}$ to zero. If $\mathbf{B} = (0, 0, B)$ then (using the definition (1.15) of the cross product) the components of the equation of motion projected on the (x_1, x_2)-plane are

$$m\frac{dx_1}{dt} = qBx_2\,, \quad m\frac{dx_2}{dt} = -qBx_1\,. \tag{3.90}$$

These imply that the time derivative of $x_1^2 + x_2^2$ is zero, because

$$\frac{d}{dt}(x_1^2 + x_2^2) = 2x_1\frac{dx_1}{dt} + 2x_2\frac{dx_2}{dt} = \frac{2qB}{m}(x_1x_2 - x_2x_1) = 0\,. \tag{3.91}$$

So $x_1^2 + x_2^2 = R^2$, where R is a constant, and the projected motion is around a circle. If we write $x_1 = R\cos\varphi(t)$, $x_2 = R\sin\varphi(t)$, then equations (3.90) are satisfied if $\frac{d\varphi}{dt} = \frac{-qB}{m}$. The particle therefore moves steadily around the circle at the constant angular frequency $\omega = \frac{qB}{m}$, known as the *cyclotron frequency*. The centre of the circle can be anywhere (when the constant $\mathbf{u}$ is allowed for), and the radius R is arbitrary; the cyclotron frequency is independent of these parameters. In general, when the velocity of the particle includes a component parallel to $\mathbf{B}$, the particle will travel along a helix.

R increases as the particle's speed increases. Consequently if a charged particle passes through a region of uniform magnetic field, its speed can be measured from the radius of curvature of its track. This is utilized in particle detectors, such as the CMS and ATLAS detectors of the Large Hadron Collider at CERN in Geneva.

Circular motion in a magnetic field is also the basis for particle accelerators. In 1932, Ernest Lawrence invented an early accelerator known as the cyclotron, a schematic representation of which is shown in Figure 3.14. The cyclotron is formed of two hollow, D-shaped pieces of metal with a narrow gap between their straight edges. The apparatus is

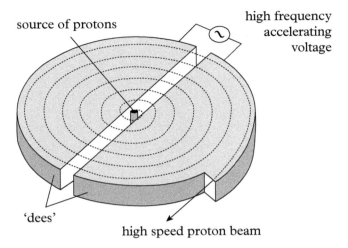

Fig. 3.14 Cyclotron.

placed in a uniform static magnetic field. Charged particles moving at constant speed follow a circular orbit in such a field, as we have seen. The critical design feature of the cyclotron is to apply an oscillating electric field across the gap, with the oscillations timed to coincide with the passage of particles. The particles are thereby accelerated further each time they cross the gap. The result is that particles injected at the centre of the device spiral outwards at ever greater speed until they leave through an opening at the outer edge and are aimed towards a target.

Recent accelerators, like the Large Hadron Collider, have a circular ring of magnets of fixed radius. Again, particles are injected at relatively low energies and are accelerated by electric field pulses. As the particle speed increases the magnetic field must be gradually increased in step to maintain the same radius of curvature and keep the particles in the ring. Because the magnetic field strength is synchronized with the increasing particle speed, these machines are known as synchrotrons. Linear particle accelerators, such as at the Stanford Linear Accelerator Center (SLAC), accelerate particles with a purely electric field. The field must be much stronger than in a circular accelerator, because the particles pass from one end to the other just once.

The particles in all these accelerators travel at close to the speed of light, so the equation of motion (3.82) needs a relativistic modification. We shall look at this in the next chapter.

We have discussed in some detail the force exerted by an electromagnetic field on charged particles. Let us briefly mention the force on a charge or current distribution. Just as a particle with charge q contributes to the charge density ρ, so a moving particle with charge q and velocity $\mathbf{v}$ contributes to the current density $\mathbf{j}$. It follows from the Lorentz force on particles that the electric field exerts a total force

$$\int \rho(\mathbf{x})\mathbf{E}(\mathbf{x}) \, d^3x \tag{3.92}$$

on a charge distribution, and the magnetic field exerts a total force

$$\int \mathbf{j}(\mathbf{x}) \times \mathbf{B}(\mathbf{x}) \, d^3x \tag{3.93}$$

on a current distribution. For a small current loop, the net effect of the magnetic force is to produce a torque on the loop.

3.10.1 The Lorentz force from the principle of least action

The equation of motion (3.82) for a particle of charge q can be derived from a principle of least action. The action is

$$S = \int_{t_0}^{t_1} \left(\frac{1}{2} m \frac{d\mathbf{x}}{dt} \cdot \frac{d\mathbf{x}}{dt} + q\mathbf{A}(\mathbf{x}(t)) \cdot \frac{d\mathbf{x}}{dt} - q\Phi(\mathbf{x}(t)) \right) dt, \qquad (3.94)$$

an extension of the action (2.53) for a particle moving in a potential in three dimensions. Here the particle interacts with both the scalar and vector potentials Φ and $\mathbf{A}$, and these can be time dependent. Unlike in equations (3.79) and (3.80), Φ and $\mathbf{A}$ are background fields, and only the charged particle trajectory $\mathbf{x}(t)$ is varied.

As usual, S is defined for the class of particle paths connecting fixed endpoints $\mathbf{x}(t_0)$ and $\mathbf{x}(t_1)$, and the Euler–Lagrange equation is the condition for S to be a minimum. This equation reproduces the equation of motion (3.82). The integrand of S has a standard kinetic energy term, and a potential energy term $q\Phi(\mathbf{x}(t))$ which is like $V(\mathbf{x}(t))$ in equation (2.53), but the middle term, linear in velocity, is new. When $\mathbf{x}(t)$ is varied, both $\mathbf{A}(\mathbf{x}(t))$ and $\frac{d\mathbf{x}}{dt}$ change. This results in the $\frac{d\mathbf{x}}{dt} \times \mathbf{B}$ term and the $\frac{\partial \mathbf{A}}{\partial t}$ part of the $\mathbf{E}$ term in the equation of motion.

The equation of motion depends only on the gauge invariant quantities $\mathbf{E}$ and $\mathbf{B}$, whereas S appears not to be gauge invariant. A gauge transformation replaces $\mathbf{A}$ and Φ by new potentials $\mathbf{A}' = \mathbf{A} + \nabla\lambda$ and $\Phi' = \Phi - \frac{\partial\lambda}{\partial t}$, and the action changes to

$$S' = S + q\int_{t_0}^{t_1} \left(\nabla\lambda \cdot \frac{d\mathbf{x}}{dt} + \frac{\partial\lambda}{\partial t} \right) dt = S + q\int_{t_0}^{t_1} \frac{d}{dt} (\lambda(x(t))) \, dt. \qquad (3.95)$$

The integrand is the total time derivative of $\lambda(\mathbf{x}(t))$, that is, the time derivative of λ evaluated along the particle trajectory. Integrating gives $S' = S + q\lambda(\mathbf{x}(t_1)) - q\lambda(\mathbf{x}(t_0))$. The additional terms depend only on the values of λ at the endpoints and are independent of the trajectory $\mathbf{x}(t)$ between them, so they do not affect the equation of motion. In this sense the action is gauge invariant.

This example illustrates a more general principle, which is that the fields and even the action are not always strictly gauge invariant, but the physics is gauge invariant. One should regard a change of gauge as something unobservable, affecting the mathematical description of the physics, but not the physics itself.

If both $\mathbf{A}$ and Φ are time independent, then $\mathbf{E} = -\nabla\Phi$ and $\mathbf{B} = \nabla \times \mathbf{A}$, and we can expect the particle to have a conserved energy. It is a general result that a term in a Lagrangian that is linear in velocity does not contribute to the energy. The energy is the sum of the kinetic energy, which is quadratic in velocity, and the potential energy, which is independent of velocity. For the action (3.94) the energy is therefore

$$E = \frac{1}{2} m \frac{d\mathbf{x}}{dt} \cdot \frac{d\mathbf{x}}{dt} + q\Phi(\mathbf{x}(t)). \qquad (3.96)$$

This explains why Φ is called a potential; $\Phi(\mathbf{x})$ is the potential energy of a particle of unit charge at $\mathbf{x}$.

The time derivative of E is zero because

$$
\begin{aligned}
\frac{dE}{dt} &= m\frac{d^2\mathbf{x}}{dt^2}\cdot\frac{d\mathbf{x}}{dt} + q\nabla\Phi(\mathbf{x}(t))\cdot\frac{d\mathbf{x}}{dt} \\
&= q\mathbf{E}\cdot\frac{d\mathbf{x}}{dt} + q\nabla\Phi(\mathbf{x}(t))\cdot\frac{d\mathbf{x}}{dt} \\
&= 0\,.
\end{aligned}
\tag{3.97}
$$

Here, we have used the equation of motion (3.82) to substitute for $m\frac{d^2\mathbf{x}}{dt^2}$, and noted that $\frac{d\mathbf{x}}{dt}\times\mathbf{B}$ is orthogonal to $\frac{d\mathbf{x}}{dt}$. In the second term we have used the chain rule, and finally we have noted that $\mathbf{E} = -\nabla\Phi$ for a static field.

3.11   Field Energy and Momentum

Evaluating energy in electromagnetic theory is not always as straightforward as this. The Maxwell equations specify the dynamics of the fields but do not specify the dynamics of the charge and current sources, beyond requiring charge/current conservation. The charges and currents are subject to mechanical forces and constraints due to the materials in which they reside, in addition to the electromagnetic Lorentz forces. These materials are not necessarily simple, and generally dissipate energy.

The situation is simpler if all the sources are charged point particles, free to move in space. The coupled system of electromagnetic fields and charged particles is a closed one, with a single action, and should have a conserved total energy. Unfortunately, in this case there is a new difficulty, which is that point particles have singularities and their fields appear to have infinite energy. Despite this, energy makes sense in various cases, and can be calculated.

Let's start with electrostatics. For a static charge density and no currents we can assume that the magnetic field $\mathbf{B}$ and the vector potential $\mathbf{A}$ are zero. The field Lagrangian (3.79) simplifies to

$$
L = \int \left\{ \frac{1}{2}\nabla\Phi\cdot\nabla\Phi - \Phi\rho \right\} d^3x\,.
\tag{3.98}
$$

Although $\frac{1}{2}\nabla\Phi\cdot\nabla\Phi$ comes from the electric field contribution, which is normally regarded as kinetic, we can interpret it here as contributing (with opposite sign) to the potential energy. Therefore, in electrostatics there is a potential energy

$$
V = \int \left\{ -\frac{1}{2}\nabla\Phi\cdot\nabla\Phi + \Phi\rho \right\} d^3x\,,
\tag{3.99}
$$

The field action is stationary if V is stationary, and this requires Poisson's equation, $\nabla^2\Phi = -\rho$, to be satisfied.

Provided Φ satisfies Poisson's equation, the two contributions to V in equation (3.99) are closely connected. This is expressed through the *virial relation*

$$
\int (\nabla\Phi\cdot\nabla\Phi - \Phi\rho)\, d^3x = 0\,,
\tag{3.100}
$$

which is easily derived. Suppose we replace Φ by $\mu\Phi$, with μ a real number. V becomes a function of μ, of the form

$$V(\mu) = \int \left\{ -\frac{1}{2}\mu^2 \nabla\Phi \cdot \nabla\Phi + \mu\Phi\rho \right\} d^3x \,, \tag{3.101}$$

and its derivative is

$$\frac{dV}{d\mu} = \int \{ -\mu\nabla\Phi \cdot \nabla\Phi + \Phi\rho \} d^3x \,. \tag{3.102}$$

Now, Poisson's equation is the condition that V is stationary under *all* variations of Φ, including replacing Φ by $\mu\Phi$, so $\frac{dV}{d\mu}$ must be zero when $\mu = 1$, in which case equation (3.102) reduces to the virial relation (3.100).

Using this, one can eliminate either $\Phi\rho$ or $\nabla\Phi \cdot \nabla\Phi$ from V, so V has the alternative expressions

$$V = \frac{1}{2} \int \nabla\Phi \cdot \nabla\Phi \, d^3x \,, \tag{3.103}$$

or

$$V = \frac{1}{2} \int \Phi\rho \, d^3x \,. \tag{3.104}$$

The first integral expresses the energy entirely in terms of the electric field, as $\nabla\Phi{\cdot}\nabla\Phi = \mathbf{E}{\cdot}\mathbf{E}$. If we use the solution (3.45) of Poisson's equation, the second integral becomes

$$V = \int\int \frac{\rho(\mathbf{x})\rho(\mathbf{x}')}{8\pi|\mathbf{x} - \mathbf{x}'|} d^3x \, d^3x' \,, \tag{3.105}$$

which expresses V entirely in terms of the charge density.

The potential energy of a smooth charge distribution is finite, but for a point charge q at the origin, the electric field it produces is given by equation (3.39), and

$$V = \int_0^\infty \frac{q^2}{32\pi^2 r^4} 4\pi r^2 \, dr \,, \tag{3.106}$$

which is a divergent integral representing the charge's self-energy. For a collection of static or slowly moving point charges, one can subtract off an infinite constant to get an effective potential energy representing the finite interaction energy of the charges, but for rapidly moving and accelerating charges, this is not possible. The divergences are no longer simply those of electrostatic fields.

Let us return to dynamical electromagnetic fields. In the absence of sources the energy E is the usual sum of the kinetic energy and potential energy in the field. From the Lagrangian (3.79) we read off that

$$E = \frac{1}{2} \int (\mathbf{E} \cdot \mathbf{E} + \mathbf{B} \cdot \mathbf{B}) \, d^3x \,, \tag{3.107}$$

and can check this is conserved using the sourceless Maxwell equations. Taking the dot products of equation (3.28) with $\mathbf{E}$ and equation (3.26) with $\mathbf{B}$, and subtracting, we find

$$\frac{1}{2}\frac{\partial}{\partial t}(\mathbf{E} \cdot \mathbf{E} + \mathbf{B} \cdot \mathbf{B}) + \nabla \cdot (\mathbf{E} \times \mathbf{B}) = 0 \,, \tag{3.108}$$

so the time derivative of the energy E is the integral of the total derivative $-\nabla \cdot (\mathbf{E} \times \mathbf{B})$, and this integral is automatically zero for fields that decay rapidly enough at infinity.

The field energy density is therefore $\frac{1}{2}(\mathbf{E} \cdot \mathbf{E} + \mathbf{B} \cdot \mathbf{B})$, and the interpretation of equation (3.108) is that the vector $\mathbf{E} \times \mathbf{B}$ is the energy current density. The field simultaneously carries momentum, and the vector $\mathbf{E} \times \mathbf{B}$ is also the field's momentum density. An electromagnetic wave is composed of orthogonal fields $\mathbf{E}$ and $\mathbf{B}$, so $\mathbf{E} \times \mathbf{B}$ is non-zero. It carries both energy and momentum in the direction of the wavevector $\mathbf{k}$.

3.12   Dynamics of Particles and Fields

This almost completes our survey of electromagnetic theory. We have presented Maxwell's equations, which relate the electric and magnetic fields to charge and current sources. These sources may be macroscopic, like currents in wires, or they may be moving point particles. The fields are not completely determined by these sources, because there are independent electromagnetic wave solutions that require no sources. We have also presented the equations of motion for charged particles in an electromagnetic field.

The electric fields of static, spherically symmetric charge distributions, including point charges, are particularly simple, but we haven't explained how to find the field produced by moving charged particles. This is quite technical and leads into conceptually deep waters. In principle, one can determine the charge density ρ and current density $\mathbf{j}$ associated with a point particle of charge q moving on a trajectory $\mathbf{x}(t)$. The charge density is not a smooth function, but is highly localized. There is similarly a localized current density, proportional to the particle velocity, and the conservation equation (3.29) is satisfied provided q is unchanging.

The Maxwell equations determine the fields surrounding the particle. The electric field is a modification of the field of a charged point particle at rest, and the particle velocity results in a magnetic field. In addition, the particle acceleration generates a field far from the particle that is an outgoing electromagnetic wave. This part of the field decreases inversely with distance from the particle, and so dominates the inverse square law decay associated with the other parts of the field. It also carries away some energy and momentum. Given these fields, one can study the complete dynamics of several interacting charged particles. Each particle is mainly affected by the fields produced by the other particles, and not by its own self-field.

There is a total action for N charged particles and the electromagnetic field. This is essentially the sum of the actions for the particles and for the fields $\mathbf{A}$ and Φ, with the interaction terms appearing just once. The Lagrangian is

$$L = \sum_{k=1}^{N} \left(\frac{1}{2} m^{(k)} \frac{d\mathbf{x}^{(k)}}{dt} \cdot \frac{d\mathbf{x}^{(k)}}{dt} + q^{(k)} \mathbf{A}(\mathbf{x}^{(k)}(t)) \cdot \frac{d\mathbf{x}^{(k)}}{dt} - q^{(k)} \Phi(\mathbf{x}^{(k)}(t)) \right)$$
$$+ \frac{1}{2} \int (\mathbf{E} \cdot \mathbf{E} - \mathbf{B} \cdot \mathbf{B}) \, d^3x . \qquad (3.109)$$

where $\mathbf{E} = -\frac{\partial \mathbf{A}}{\partial t} - \nabla\Phi$ and $\mathbf{B} = \nabla \times \mathbf{A}$ as usual. The kth particle here has mass $m^{(k)}$, charge $q^{(k)}$, and trajectory $\mathbf{x}^{(k)}(t)$. The interaction terms coupling the particles to the vector and scalar potentials are the same as in equation (3.94), but they are also the same as in (3.79), because the current and charge densities $\mathbf{j}$ and ρ have the highly localized forms associated with N point particles, and the integrals of $\mathbf{A} \cdot \mathbf{j}$ and $\Phi\rho$ in equation (3.79) reduce to the sums in equation (3.109).

The principle of least action applied to this total system gives the Maxwell equations for **E** and **B** with charged particle sources, together with the equation of motion for each particle. Awkwardly, the electric and magnetic fields acting on each particle include the contribution of the particle's self-field, which is singular at the particle position. The dominant part of the electric self-field does not produce a net force, because it is spherically symmetric and averages to zero over a spherically symmetric charged particle, but some subdominant parts of the self-field are not spherically symmetric and do produce a force.

Just dropping the self-force leads to a contradiction if the particle accelerates and emits electromagnetic waves. Because the electromagnetic radiation carries away energy, it causes the particle itself to lose kinetic energy and slow down. The radiation also carries away momentum, and there needs to be a compensating force on the particle if momentum is to be conserved overall.

The rate of energy radiation from an accelerating charged particle was estimated by Joseph Larmor to be

$$\frac{1}{6\pi}q^2\left|\frac{d^2\mathbf{x}}{dt^2}\right|^2. \tag{3.110}$$

It is proportional to the square of the particle's acceleration. An effective self-force on the particle can be introduced that produces an equivalent loss of kinetic energy, at least when averaged over time. It is proportional to the time derivative of the particle acceleration. However, this is only an approximation, valid while the acceleration is neither very large nor rapidly varying. It is an approximation because the radiated energy must be defined globally and can only be calculated by considering the fields on a large sphere surrounding the particle. There is a delay between the particle accelerating and the radiation reaching this large sphere, which introduces some uncertainty in the timing of the back-reaction and its instantaneous strength.

Attempts were made to resolve these uncertainties around 1900, by Max Abraham and Lorentz and others, and they have continued. A key idea is to give a charged particle, such as an electron, a finite-sized structure. The electron would need a radius of around 10^{-15} m, which is comparable to the size of an atomic nucleus. Unfortunately, an electron with such a structure would explode, due to the Coulomb repulsion between its parts, unless held together by stronger, unknown forces with a non-electromagnetic origin.

Investigations into the possible internal structure of an electron have until now been mainly theoretical. There has been no guidance from experiments. This is because the proposed radius is very small, and the corresponding time for electromagnetic waves to travel this distance is very short. Experiments would need to generate very strong fields of high frequency to produce electron accelerations large enough to require a significant modification of the Lorentz force. The most powerful focussed laser fields currently available do not quite reach this regime, but with the next generation of lasers it may be possible to investigate corrections to the Lorentz force.

In summary, a completely consistent treatment of fields interacting with charged point particles does not seem possible, although the problems are not so severe as to undermine particle accelerator design and operation. Modern thinking is that all matter should be described by fields. Particles are emergent phenomena, and not truly point-like. There is a mathematically appealing theory of particle structure, which we discuss briefly at the end of the book. This is where a particle is modelled as a *soliton*, which is a smooth localized structure in a nonlinear classical field. Although solitons do occur in nature, there is not yet much evidence that they can describe elementary particles like the electron.

This is where classical electromagnetic theory ends and particle physics begins. High energy particle collisions probe the internal structure of particles like electrons, protons and neutrons. These show that a proton has a substructure of smaller, electrically charged quarks, whereas there is no evidence yet for substructure of the electron. We shall return to these aspects of particle physics in Chapter 12, but they are not part of the classical theory of electromagnetism. Quantum field theory is needed to understand particle physics, and it was thought at one point that a quantum theory would completely eliminate the difficulties associated with point-like particles. Despite the successes of quantum field theory in particle physics, this is not really the case.

3.13 Further Reading

P. Lorrain and D.R. Corson, *Electromagnetism: Principles and Applications (2nd ed.)*, New York: Freeman, 1990.

J.D. Jackson, *Classical Electrodynamics (3rd ed.)*, Chichester: Wiley, 1999.

L.D. Landau and E.M. Lifschitz, *The Classical Theory of Fields: Course of Theoretical Physics, Vol. 2 (4th ed.)*, Oxford: Butterworth-Heinemann, 1975.

4

Special Relativity

4.1 Introduction

Before discussing Newton's laws of motion, we considered some of the geometry of Euclidean 3-space. Newtonian physics rests on the idea that space is absolute, and the distance between two points is absolute, but the coordinate system is a matter of choice. Observers can set up Cartesian coordinates with respect to different origins, and the axes can be differently oriented. The coordinates of a point form a vector, but such a vector is not absolute, because it is relative to the origin and choice of axes. However, Newton's laws of motion have a vectorial form that is the same for different observers.

Let's put this more simply. Two observers, located right next to each other but looking in *different directions*, agree on what's going on in the world. If they see a bird pulling a worm from the ground and eating it, they agree that it happened at the same location and agree on how long it took, but if they set up individual spatial coordinate systems, with the origin where they are, their 1-axis forward, their 2-axis to the left, and their 3-axis up, then the location of the bird has different coordinates according to the two observers and when the bird flies off it has different velocity vectors and acceleration vectors. However, the *relationship* between the forces that act on the bird and the bird's acceleration are the same for both observers. In other words, the observers agree on the laws of motion, even though they describe the motion differently.

To discuss dynamics we need to consider 4-dimensional *spacetime*. A point in spacetime is called an *event*, and occurs at a time t and spatial location $\mathbf{x} = (x_1, x_2, x_3)$. These can be brought together into the combination $X = (t, \mathbf{x})$, the *position 4-vector* of the event.

In special relativity, it is spacetime and not space that is absolute. Different observers are generally moving relative to one another, and they will set up different coordinate systems. An observer is not an event, but persists for all time. The most important observers are those experiencing no forces. They are called *inertial observers*, and are equivalent to the bodies in Newtonian dynamics that travel through space at constant velocity. The laws of physics are assumed to be the same for all inertial observers, but the time and space coordinates of events will be relative to each observer. Time is no longer an absolute quantity and neither is 3-dimensional distance. But there is an absolute notion of the separation or *interval* between two events in spacetime, replacing the absolute notion of distance in Euclidean 3-space.

The other key feature of special relativity is that the speed of light is an absolute constant, the same for all inertial observers even if they are in relative motion. We shall see that the interval between two events, one where a light flash is emitted and the other where it is received, is zero, and all observers agree on this.

The Physical World. Nicholas Manton and Nicholas Mee, Oxford University Press (2017).
© Nicholas Manton and Nicholas Mee. DOI 10.1093/oso/9780198795933.001.0001

Let us define the notion of interval for one (inertial) observer. The origin O of spacetime is the event $(0, \mathbf{0})$ at time $t = 0$ and location $\mathbf{x} = \mathbf{0}$. Suppose another event is at time and location $X = (t, \mathbf{x})$. The squared interval τ^2 between these events is defined as $\tau^2 = t^2 - x_1^2 - x_2^2 - x_3^2$, or equivalently

$$\tau^2 = t^2 - \mathbf{x} \cdot \mathbf{x}. \tag{4.1}$$

Notice that τ^2 can be positive, negative or zero, so τ itself can be real or imaginary. If $t^2 > \mathbf{x} \cdot \mathbf{x}$ then τ is real and is taken to be positive if t is positive, and negative if t is negative. The squared interval between two general events $X = (t, \mathbf{x})$ and $Y = (u, \mathbf{y})$ is

$$\begin{aligned} \tau^2 &= (t - u)^2 - (\mathbf{x} - \mathbf{y}) \cdot (\mathbf{x} - \mathbf{y}) \\ &= (t - u)^2 - |\mathbf{x} - \mathbf{y}|^2. \end{aligned} \tag{4.2}$$

Spacetime with this geometry, with the minus sign between the time and space contributions to the squared interval τ^2, is called *Minkowski space*. Alternatively, the geometry is said to be Lorentzian. With a plus sign, τ^2 would be the squared distance between points in 4-dimensional Euclidean space.

Suppose that a second observer sets up a coordinate system with time coordinate t' and space coordinates $\mathbf{x}' = (x_1', x_2', x_3')$. It is assumed that the units along the space axes are calibrated by the second observer with the same type of ruler as used by the first observer, and the time unit is calibrated with the same type of clock. (This is equivalent to the implicit assumption that in Euclidean 3-space, different observers use the same type of ruler to measure distances along their Cartesian axes.)

In special relativity the interval between events is the same for the two observers. If the events are at $X = (t, \mathbf{x})$ and $Y = (u, \mathbf{y})$ for the first observer, and at $X' = (t', \mathbf{x}')$ and $Y' = (u', \mathbf{y}')$ for the second, then

$$(t - u)^2 - (\mathbf{x} - \mathbf{y}) \cdot (\mathbf{x} - \mathbf{y}) = (t' - u')^2 - (\mathbf{x}' - \mathbf{y}') \cdot (\mathbf{x}' - \mathbf{y}'). \tag{4.3}$$

The transformation relating the coordinates of the second observer to those of the first therefore preserves intervals, so it is analogous to a transformation that preserves distances in Euclidean 3-space. Such a transformation generally includes a translation of the spacetime origin, but if it does not, then it is called a *Lorentz transformation* and is analogous to a pure rotation in 3-space.[1]

A Lorentz transformation can be a purely spatial rotation, but it usually mixes the time and space coordinates. This mixing occurs when the first and second observers are moving relative to one another at a constant velocity. Because the fundamental postulate of special relativity is that the laws of physics are the same for all such observers, the laws must be unaffected by a Lorentz transformation, and have a Lorentz covariant form.

Although we did not explicitly discuss it earlier, there is a similar result in Newtonian physics, called Galilean invariance. This says that the laws of motion for a system of bodies are the same for two observers moving relative to one another at constant velocity. In particular, the relative motion of a system of bodies appears the same to the two observers even though the velocity of the centre of mass is different. This explains why we do not

---

[1] Like a rotation, a Lorentz transformation is assumed to be a linear transformation of the coordinates.

notice the large velocity of the Earth around the Sun (which is almost constant on the timescale of a day), and why drinks can be served and consumed on a plane in steady flight as if the plane were not moving. Galilean invariance has its limitations, however. It doesn't apply accurately in electromagnetic theory, and although Galilean invariance puts no upper limit on the relative speed of observers, in practice it is only accurate if the relative speed is much less than the speed of light.

We now look at Lorentz transformations in more detail.

4.2 Lorentz Transformations

Let us focus on two inertial observers whose spacetime origins O agree, as a translation of the origin is not very significant. Consider an event X with coordinates $(t, \mathbf{x})$ for the first observer and $(t', \mathbf{x}')$ for the second. The squared interval between X and O is the same for both observers, so

$$t^2 - \mathbf{x} \cdot \mathbf{x} = t'^2 - \mathbf{x}' \cdot \mathbf{x}' . \tag{4.4}$$

The first observer is at rest in the unprimed coordinates, at $\mathbf{x} = \mathbf{0}$ for all t, and therefore travels along the t-axis in spacetime. This straight line is called the observer's *worldline*. The second observer is similarly at rest in the primed coordinates, at $\mathbf{x}' = \mathbf{0}$, and travels along the t'-axis of spacetime.

There are two basic types of Lorentz transformation. The simpler one is a spatial rotation, with the time coordinate unchanged. The observers are not moving relative to one another, but their spatial axes are differerently oriented. Here, separately, $t'^2 = t^2$ and $\mathbf{x}' \cdot \mathbf{x}' = \mathbf{x} \cdot \mathbf{x}$. More explicitly, suppose that the second observer's axes are rotated by θ relative to the first's in the (x_1, x_2)-plane. The coordinates are then related by

$$\begin{aligned} t' &= t \\ x_1' &= x_1 \cos\theta - x_2 \sin\theta \\ x_2' &= x_1 \sin\theta + x_2 \cos\theta \\ x_3' &= x_3 . \end{aligned} \tag{4.5}$$

Equation (4.4) is satisfied because $\cos^2\theta + \sin^2\theta = 1$, so

$$\begin{aligned} x_1'^2 + x_2'^2 &= (x_1^2 \cos^2\theta - 2x_1 x_2 \cos\theta \sin\theta + x_2^2 \sin^2\theta) \\ &\quad + (x_1^2 \sin^2\theta + 2x_1 x_2 \cos\theta \sin\theta + x_2^2 \cos^2\theta) \\ &= x_1^2 + x_2^2 , \end{aligned} \tag{4.6}$$

and obviously $t'^2 - x_3'^2 = t^2 - x_3^2$. The interval between X and O is therefore preserved. Two observers whose axes only differ by a rotation travel along the same worldline, as their time-axes coincide.

Figure 4.1 shows a spacetime point X relative to the two sets of axes (with t and x_3 suppressed). The angle between the two sets of axes is the rotation angle θ. Each coordinate of X is indicated by where a construction line (in red) parallel to one axis intersects the other axis.[2] The scales along the x_1- and x_2-axes have equal spacing, and the scales along the

---

[2] In three dimensions, the x_3 coordinate of X is given by where the plane parallel to both the x_1- and x_2-axes intersects the x_3-axis.

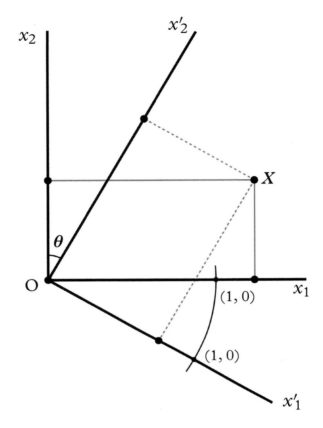

Fig. 4.1 Rotation in the (x_1, x_2)-plane.

x_1'- and x_2'-axes have the same spacing. This is indicated by the circle segment connecting the point $(x_1, x_2) = (1, 0)$ to the point $(x_1', x_2') = (1, 0)$. These points are the same distance from O.

The more interesting type of Lorentz transformation is a *boost*. This takes the coordinate system of one inertial observer to that of a second inertial observer travelling at a constant velocity with respect to the first. If the relative motion is along the x_1-axis, the boost mixes the time coordinate t and the spatial coordinate x_1. The spacetime transformation is a hyperbolic analogue[3] of a rotation in a plane, and again has a parameter θ. It is

$$
\begin{aligned}
t' &= t \cosh\theta - x_1 \sinh\theta \\
x_1' &= -t \sinh\theta + x_1 \cosh\theta \\
x_2' &= x_2 \\
x_3' &= x_3 \, .
\end{aligned}
\tag{4.7}
$$

---

[3] Recall the hyperbolic functions $\cosh\theta = \frac{1}{2}(e^\theta + e^{-\theta})$, $\sinh\theta = \frac{1}{2}(e^\theta - e^{-\theta})$ and $\tanh\theta = \frac{\sinh\theta}{\cosh\theta}$.

This satisfies equation (4.4) because of the identity $\cosh^2 \theta - \sinh^2 \theta = 1$, so

$$
\begin{aligned}
t'^2 - x_1'^2 &= (t^2 \cosh^2 \theta - 2tx_1 \cosh \theta \sinh \theta + x_1^2 \sinh^2 \theta) \\
&\quad - (t^2 \sinh^2 \theta - 2tx_1 \cosh \theta \sinh \theta + x_1^2 \cosh^2 \theta) \\
&= t^2 - x_1^2,
\end{aligned} \tag{4.8}
$$

and obviously $x_2'^2 + x_3'^2 = x_2^2 + x_3^2$. In the context of a rotation, θ is the rotation angle, but here, in the context of a boost, θ is called the *rapidity*. The general Lorentz transformation is a linear transformation of the coordinates obtained by combining rotations and boosts, and is characterized by six parameters. It can be expressed as a 4×4 matrix acting on (t, x_1, x_2, x_3), satisfying equation (4.4).

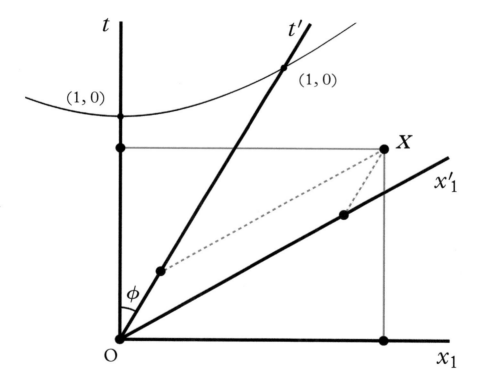

Fig. 4.2 A Lorentz boost transforming the (t, x_1) coordinates to the (t', x_1') coordinates. Here $\tan \phi = \tanh \theta = v$.

The effect of a boost is shown in Figure 4.2. The figure is not intuitively so clear as the figure for a rotation. This is because the (Euclidean) geometry of the figure on the page is not the same as the Lorentzian geometry that it is aiming to represent. The t'- and x_1'-axes are squeezed together relative to the t- and x_1-axes, by the same angle. The coordinates of X are again indicated by where the construction lines (in red) parallel to one axis intersect the other axis. Units along the t- and x_1-axes (in seconds and light-seconds, respectively) have equal spacing, so light rays with speed 1 from O are along the lines at 45° to the axes.

Units along the t'- and x_1'-axes also have equal spacing, but not the same spacing as along the t- and x_1-axes. Nevertheless, light rays for the first observer are still light rays for the second observer.

We can work out the angle ϕ between the t-axis and t'-axis as follows. Along the t-axis, where $x_1 = 0$, the first observer is at rest. Along the t'-axis, where $x_1' = 0$, the second observer is at rest, but moving at some velocity v relative to the first observer. This axis is therefore the line $x_1 = vt$ for the first observer, so the angle ϕ is given by $\tan \phi = v$. From the formulae (4.7) we see that $x_1' = 0$ implies that $x_1 = (\tanh \theta)t$. Therefore

$$\tan \phi = \tanh \theta = v \,. \tag{4.9}$$

Similarly, by setting $x_1 = 0$ in equation (4.7) one can check that according to the second observer, the first is moving at velocity $v = -\tanh \theta$. As θ runs from $-\infty$ to ∞, v runs from -1 to 1, and ϕ runs from $-45°$ to $45°$. The boost velocity v cannot exceed the velocity of light ($c = 1$).

Let us next work out the scale along the t'-axis, relative to the scale along the t-axis. The point $(t', x_1') = (1, 0)$ on the t'-axis has coordinates for the first observer $(t, x_1) = (\cosh \theta, \sinh \theta)$. For both observers the interval to the origin is 1. The point where $t' = 1$ on the t'-axis is marked on the figure. It lies on the hyperbola $t^2 - x_1^2 = 1$, all of whose points are at interval 1 from O.

There is a nice alternative way of understanding why the boost (4.7) preserves the squared interval τ^2. The boost can be re-expressed as

$$\begin{aligned} t' - x_1' &= (t - x_1)e^{\theta} \\ t' + x_1' &= (t + x_1)e^{-\theta} \\ x_2' &= x_2 \\ x_3' &= x_3 \,. \end{aligned} \tag{4.10}$$

Adding and subtracting the first two equations reproduces equations (4.7). Multiplying the first equation by the second shows that $t'^2 - x_1'^2 = t^2 - x_1^2$. The effect of the boost is therefore a coordinate stretching by e^{θ} along one diagonal axis in the (t, x_1)-plane and an equal coordinate compression by $e^{-\theta}$ along the perpendicular diagonal axis, as shown in Figure 4.2.

So far, a boost looks like a purely geometrical construction, a change of coordinates, but the invariance of physics under a boost has physical consequences. One of these is time dilation. The classic example of this is muon decay. A muon is a fundamental particle, rather like an electron but more massive. It always decays in the same way—into an electron, a neutrino and an antineutrino. The decay is quantum mechanical and occurs after a random period of time, but the half-life is a fixed time T, which means that for a muon at rest, the probability that it survives for a time T is $\frac{1}{2}$. For the present discussion we may simply assume the muon lifetime is T.

Muons are produced in particle collisions, or from the decay of other particles. As a result, muons are frequently moving at very high speeds, close to the speed of light. Let us consider a muon produced at the spacetime origin and moving at velocity $v = \tanh \theta$ in the x_1-direction. As before, let the first observer be at rest, and assume the second observer moves at velocity v in the x_1-direction. For the second observer the muon is at rest. In fact, the second observer can be regarded as the muon itself.

For the second observer, the muon decays at time $t' = T$ and at the location $x'_1 = 0$. From the Lorentz boost formulae (4.7) we see that for the first observer, the decay occurs at time t and location x_1, such that

$$T = t \cosh \theta - x_1 \sinh \theta, \quad 0 = -t \sinh \theta + x_1 \cosh \theta. \tag{4.11}$$

The second equation implies that $\frac{x_1}{t} = \tanh \theta$, confirming that the muon has velocity $\tanh \theta$. Eliminating x_1 from the first equation, we find

$$T = t \left(\cosh \theta - \frac{\sinh^2 \theta}{\cosh \theta} \right) = \frac{t}{\cosh \theta}. \tag{4.12}$$

So $t = T \cosh \theta$, and this is later than the time $t = T$, as $\cosh \theta > 1$. The moving particle seen by the first observer therefore has a longer lifetime than the particle at rest seen by the second observer. This is time dilation. The time dilation is by exactly the same factor as shown in Figure 4.2, where the time t of a spacetime event at $(t', x'_1) = (1, 0)$ is $t = \cosh \theta$.

For the first observer, the location at which the particle decays is $x_1 = T \sinh \theta$. This is easily measured, as it is a kink in the muon track, where the muon becomes an electron (with the invisible neutrino and antineutrino carrying away some of the momentum). From this measurement alone, and no independent knowledge of the muon velocity, it would be difficult to confirm the time dilation. However it is possible to find the velocity by a time-of-flight measurement, which is a measurement of the time the muon takes to travel between two detectors that do not significantly slow the muon down.

Note that $T \sinh \theta$, the distance the muon travels, can be much greater than T, the distance the muon would have travelled if it moved essentially at the speed of light, but without time dilation. As a consequence, muons that are produced by cosmic ray collisions in the upper atmosphere frequently strike the ground, even though T is of order 10^{-6} s, and 10^{-6} light-seconds is only about 300 m.

There is nothing special about muons, and the decay of other particles is time dilated in the same way, although this may be more difficult to measure if their lifetime is much shorter or longer than that of a muon. A concrete demonstration of time dilation was provided in 1971, when the very small dilation effect due to motion at fairly slow speeds was confirmed by Joseph Hafele and Richard Keating who travelled around the world on commercial airliners both eastwards and westwards carrying four atomic clocks.

Worldlines at 45° to the time and space axes, as depicted in Figure 4.3, represent light rays. All the light rays through a spacetime point X form the *lightcone* through X. The interval between a pair of events X and Y along a light ray is zero according to formula (4.2), because $|\mathbf{x} - \mathbf{y}| = |t - u|$ for motion at speed 1. Since the interval is Lorentz invariant, all observers agree on the interpretation of light rays, and on the speed of light. This is consistent with Maxwell's equations, which predict an absolute unvarying speed of light.

In the early days of relativity, it was regarded as surprising that the speed of light is unaffected by the motion of the source. Suppose for one observer, a light flash is emitted by a source at rest. A second observer, boosted relative to the first, regards the first observer and the light source as moving, yet the light speed is unaffected by this. This would be paradoxical if light consisted of particles whose velocity depended on the velocity of the source, but classically, light is a wave propagating in an absolute spacetime, so it is quite reasonable for the speed of light to be an absolute constant. The light is not identical as perceived by the two observers, as the light frequency and wavelength are both different.

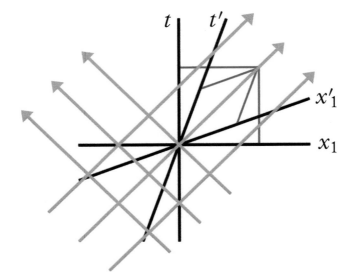

Fig. 4.3 Light worldlines.

Maxwell's equations have a Lorentz covariant form, as we will see later, so all electromagnetic field phenomena, and not just the constancy of the speed of light, are consistent with the principles of special relativity. In fact, Lorentz transformations were originally discovered by studying Maxwell's equations. The evidence in favour of Lorentz covariance was supplemented by Albert Michelson and Edward Morley's measurement of the speed of light. The aim of their experiment was to determine whether the speed of light depends on the direction of the light beam relative to the motion of the Earth. To their great surprise they discovered that light travels at the same speed in all directions and is not affected by the motion of the light source, which varies as the Earth orbits the Sun. The Michelson–Morley apparatus is shown in Figure 4.4.

Einstein's key paper on special relativity was called 'On the Electrodynamics of Moving Bodies'. His critical contribution was to propose that not just electromagnetism, but all of physics should be Lorentz invariant, and to find a modification of Newtonian dynamics that satisfies this requirement. We shall look at relativistic particle dynamics next, and review the Lorentz covariance of Maxwell's equations after that.

4.3 Relativistic Dynamics

The laws of particle dynamics need to be modified to have a Lorentz covariant form. When this is done, the Newtonian laws of motion are recovered for particles moving slowly compared with the speed of light. In special relativity it is easiest to discuss point particles rather than bodies of finite size.

We first need a relativistic notion of velocity, and then acceleration. These are now 4-component quantities rather than the familiar 3-vectors. Recall that a particle at rest has a worldline with $\mathbf{x}$ constant, parallel to the t-axis in spacetime. Two infinitesimally close events on the worldline, $(t, \mathbf{x})$ and $(t + \delta t, \mathbf{x})$, with δt positive, are separated by an interval $\delta\tau = \delta t$. A moving particle has a worldline $X(t) = (t, \mathbf{x}(t))$, so at time t the particle position

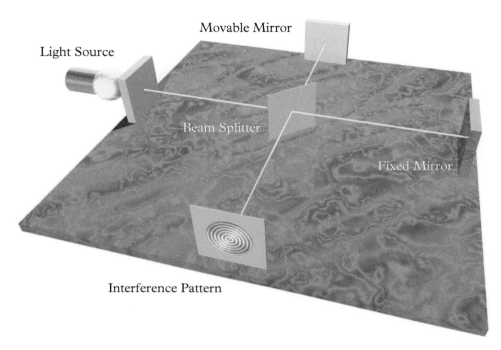

Light Source

Movable Mirror

Beam Splitter

Fixed Mirror

Interference Pattern

Fig. 4.4 A schematic representation of the Michelson–Morley experiment. A beam of light is shone through a half-silvered mirror which splits the beam into two perpendicular beams. These beams are reflected back through the half-silvered mirror and on to a screen where they form an interference pattern. One of the mirrors is fixed, and the other is movable so that the path length can be varied. The whole apparatus can be rotated, so that the orientation of the beams can be changed with respect to the Earth's motion through space.

is $\mathbf{x}(t)$. This worldline is generally curved, as the velocity of the particle may change. The squared interval between two infinitesimally close events along the worldline, $X(t) = (t, \mathbf{x}(t))$ and $X(t + \delta t) = (t + \delta t, \mathbf{x}(t + \delta t))$, is

$$\delta\tau^2 = \delta t^2 - |\mathbf{x}(t + \delta t) - \mathbf{x}(t)|^2 . \tag{4.13}$$

This is positive, as a particle's speed is always less than the speed of light. We can rewrite equation (4.13) as

$$\delta\tau^2 = \left(1 - \left|\frac{(\mathbf{x}(t + \delta t) - \mathbf{x}(t))}{\delta t}\right|^2\right) \delta t^2 . \tag{4.14}$$

The fractional quantity here is just the usual velocity $\mathbf{v}$ of the particle, so

$$\delta\tau = (1 - |\mathbf{v}|^2)^{\frac{1}{2}} \delta t , \tag{4.15}$$

and from this we obtain the useful relation between derivatives

$$\frac{d}{d\tau} = (1 - |\mathbf{v}|^2)^{-\frac{1}{2}} \frac{d}{dt} . \tag{4.16}$$

One might consider the time derivative of the worldline $X(t) = (t, \mathbf{x}(t))$ as a possible Lorentzian analogue of velocity. This is the four-component vector $(1, \mathbf{v})$, However, although the position 4-vector X transforms simply under a Lorentz transformation, t does not, so differentiating with respect to t is not covariant. On the other hand, the parameter τ along the worldline is Lorentz invariant, so the correct procedure is to differentiate X with respect to τ. We therefore define the relativistic 4-velocity V of a particle to be

$$V = \frac{dX}{d\tau} = \frac{d}{d\tau}(t, \mathbf{x}(t)). \tag{4.17}$$

The 4-vector V transforms in the same way as X, and using equation (4.16) we can express V in terms of the usual velocity, finding

$$V = (1 - |\mathbf{v}|^2)^{-\frac{1}{2}}(1, \mathbf{v}). \tag{4.18}$$

The first component of V is called the time component, and the remaining three the space components. Notice that the four components of V depend only on the three independent components of $\mathbf{v}$, so there is one constraint on V that we will make explicit shortly.

The quantity $(1-|\mathbf{v}|^2)^{-\frac{1}{2}}$ occurs frequently in special relativity, so it has its own notation,

$$\gamma(\mathbf{v}) = (1 - |\mathbf{v}|^2)^{-\frac{1}{2}}. \tag{4.19}$$

This is known as the *gamma factor* and is sometimes shortened to γ if the relevant velocity is clear. Then $V = (\gamma, \gamma\mathbf{v})$. If $|\mathbf{v}|$ is small, one can make the approximation that $\gamma(\mathbf{v}) = 1$, or more accurately that $\gamma(\mathbf{v}) = 1 + \frac{1}{2}|\mathbf{v}|^2$. In all cases, cubic terms in $\mathbf{v}$ can be dropped. This is the non-relativistic limit, where relativistic dynamics reduces to Newtonian dynamics. The 4-velocity V in the non-relativistic limit is

$$V = \left(1 + \frac{1}{2}|\mathbf{v}|^2, \mathbf{v}\right). \tag{4.20}$$

This approximation is usually valid provided $|\mathbf{v}|$ is of order 0.01 or less, equivalent to 3×10^6 m s^{-1} in normal units. This is an enormous speed in ordinary life and even in solar system dynamics and space travel.

Note that $\gamma(\mathbf{v}) > 1$ for any positive $|\mathbf{v}|$ less than the speed of light, and that if $|\mathbf{v}| = \tanh\theta$, which is the speed of a particle boosted from rest with rapidity θ, then $\gamma(\mathbf{v}) = \cosh\theta$ and $\gamma(\mathbf{v})|\mathbf{v}| = \sinh\theta$. These quantities all appeared earlier in our discussion of time dilation.

The relativistic acceleration of a particle is defined by taking one further derivative with respect to τ,

$$A = \frac{d^2 X}{d\tau^2} = \frac{d^2}{d\tau^2}(t, \mathbf{x}(t)). \tag{4.21}$$

This 4-acceleration A can be expressed in terms of the usual acceleration $\mathbf{a} = \frac{d^2\mathbf{x}}{dt^2}$, and the velocity $\mathbf{v}$, by differentiating equation (4.18) with respect to t and using equation (4.16), but the formula is a bit complicated and not very enlightening. The important common property of A and V is that they transform covariantly under Lorentz transformations, i.e. in the same way as X, because τ is invariant. This is analogous to the statement that $\mathbf{a}$ and $\mathbf{v}$ are Euclidean 3-vectors transforming in the same way as $\mathbf{x}$ under rotations, because t is rotationally invariant.

In Euclidean 3-space, we defined the rotationally invariant dot product of two vectors. Similarly, in Lorentzian geometry there is a Lorentz invariant inner product of two 4-vectors $X = (t, \mathbf{x})$ and $Y = (u, \mathbf{y})$,

$$X \cdot Y = tu - \mathbf{x} \cdot \mathbf{y}, \tag{4.22}$$

where $\mathbf{x} \cdot \mathbf{y}$ is the usual dot product. This is useful in many ways. The squared interval between X and the spacetime origin O is $X \cdot X = t^2 - \mathbf{x} \cdot \mathbf{x}$, and the squared interval between X and Y is

$$(X - Y) \cdot (X - Y) = X \cdot X - 2X \cdot Y + Y \cdot Y. \tag{4.23}$$

For the 4-velocity V we find

$$V \cdot V = \gamma^2 - \gamma^2 \mathbf{v} \cdot \mathbf{v} = (1 - \mathbf{v} \cdot \mathbf{v})^{-1}(1 - \mathbf{v} \cdot \mathbf{v}) = 1. \tag{4.24}$$

This is the anticipated constraint on the 4-velocity of a particle. Differentiating this constraint with respect to τ we deduce that $A \cdot V = 0$, and this can also be checked using the explicit formulae for A and V.

A particle has a Lorentz invariant property, its mass m, which is positive. This is defined to be the mass measured by conventional means (comparison with a standard mass using a beam balance, say) by an observer for whom the particle is at rest. So there is another 4-vector $P = mV$, called the *4-momentum* of the particle, which Lorentz transforms like V. It has components

$$P = (m\gamma, m\gamma\mathbf{v}). \tag{4.25}$$

The constraint (4.24) on V implies that $P \cdot P = m^2$. Let us write the time and space components of the 4-momentum as

$$P = (E, \mathbf{p}) = (m\gamma, m\gamma\mathbf{v}). \tag{4.26}$$

$E = m\gamma$ is called the relativistic energy, and $\mathbf{p} = m\gamma\mathbf{v}$ the relativistic 3-momentum. The constraint $P \cdot P = m^2$ becomes the important relation

$$E^2 - \mathbf{p} \cdot \mathbf{p} = m^2. \tag{4.27}$$

In particle physics detectors, E and $\mathbf{p}$ can be directly measured, and from the above relation the particle mass can be inferred. A variant is when a particle, for example a Higgs particle, rapidly decays and leaves no track itself. Instead, the decay products leave tracks from which their energies and momenta can be measured. Adding these up gives the energy and momentum of the original decaying particle, from which its mass can be calculated.

In the non-relativistic limit, the 4-momentum P reduces to

$$P = \left(m + \frac{1}{2}m|\mathbf{v}|^2, m\mathbf{v} \right), \tag{4.28}$$

whose spatial part $\mathbf{p} = m\mathbf{v}$ is the ordinary 3-momentum of the particle. The time component is related to the ordinary energy, as it is the sum of the mass of the particle and the usual kinetic energy. We shall return to this.

The 4-vector mA is analogous to the left-hand side in Newton's second law, $m\mathbf{a}$. The relativistic equation of motion for a particle of mass m is therefore

$$mA = F, \qquad (4.29)$$

where F is a 4-vector force. For this to have content one needs to know F in the physical situation of interest. As $A \cdot V = 0$ automatically, any 4-force must satisfy the constraint $F \cdot V = 0$. It is not easy to devise sensible 4-forces. The gravitational force that we considered in Chapter 2 does not have a simple 4-vector equivalent. Also, forces that act instantaneously between spatially separated particles are not compatible with relativity, because different observers do not agree on the time of spatially separated events, nor do they agree on the distance between particles. More importantly, in relativity the maximum speed of signals is the speed of light, so instantaneous action at a distance is ruled out.

One force that can be given a 4-force description is the Lorentz force on a charged particle exerted by electromagnetic fields $\mathbf{E}$ and $\mathbf{B}$. Only the field strengths at the particle's instantaneous position contribute. We shall discuss this below, after reconsidering the electromagnetic fields and Maxwell equations in 4-vector form.

Another case where we can accurately model 4-forces is for two bodies that briefly collide and separate. For point particles, a collision is an event at a single point in spacetime, and all observers agree on its position. The forces produce instantaneous impulses, which change the 4-velocities suddenly. We do not need to know what these changes are in detail, as they depend on the nature of the collision, but the impulse that the first particle exerts on the second is the negative of the impulse that the second exerts on the first. This is analogous to Newton's third law. The important consequence is that the total 4-momentum is conserved in the collision. We reach the same conclusion by supposing that the 4-forces act over the same (infinitesimal) interval, and are opposite, so

$$m^{(1)}A^{(1)} = F, \quad \text{and} \quad m^{(2)}A^{(2)} = -F, \qquad (4.30)$$

where $m^{(1)}$ and $A^{(1)}$ are the mass and 4-acceleration of the first particle, and $m^{(2)}$ and $A^{(2)}$ those of the second. Adding these equations gives $m^{(1)}A^{(1)} + m^{(2)}A^{(2)} = 0$, so the τ derivative of $m^{(1)}V^{(1)} + m^{(2)}V^{(2)}$ is zero. Therefore

$$m^{(1)}V^{(1)} + m^{(2)}V^{(2)} = \text{constant}, \qquad (4.31)$$

confirming the conservation of total 4-momentum. 4-momentum conservation is a fundamental result of relativistic dynamics. It combines the conservation of energy and 3-momentum into a single equation with surprising and significant consequences, as we shall see.

4.3.1   Comparison of Newtonian and relativistic dynamics

So far, our discussion has been principally about spacetime and its Lorentz transformations, and about the relativistic definitions of velocity, acceleration and momentum. This is all rather formal, but there are genuine differences between the predictions of Newtonian and relativistic dynamics. In particular, the conservation of 4-momentum implies that the outcome of a two-particle collision is different according to Newtonian and relativistic dynamics. To illustrate this, we need only consider an elastic collision along a line, say the x_1-axis. We assume that the first and second particles have incoming velocities $u^{(1)}$ and $u^{(2)}$, and that these are known. After the collision, the particle masses are unchanged, and we wish to find the outgoing velocities, $v^{(1)}$ and $v^{(2)}$.

In Newtonian dynamics, momentum and energy conservation require that

$$m^{(1)}v^{(1)} + m^{(2)}v^{(2)} = m^{(1)}u^{(1)} + m^{(2)}u^{(2)} \tag{4.32}$$

and

$$\frac{1}{2}m^{(1)}(v^{(1)})^2 + \frac{1}{2}m^{(2)}(v^{(2)})^2 = \frac{1}{2}m^{(1)}(u^{(1)})^2 + \frac{1}{2}m^{(2)}(u^{(2)})^2. \tag{4.33}$$

Kinetic energy is conserved, because for point particles there are no internal motions that can absorb energy. From these two equations it is possible to determine the unknowns $v^{(1)}$ and $v^{(2)}$ by eliminating one of these velocities and obtaining a quadratic equation for the other. A useful trick is to note that one solution is $v^{(1)} = u^{(1)}$ and $v^{(2)} = u^{(2)}$ (where the particles miss each other), but we are interested in the other solution.

In relativistic dynamics, conservation of 4-momentum requires that

$$m^{(1)}V^{(1)} + m^{(2)}V^{(2)} = m^{(1)}U^{(1)} + m^{(2)}U^{(2)}, \tag{4.34}$$

where $U^{(1)}$ and $U^{(2)}$ are the incoming 4-velocities of the particles, and $V^{(1)}$ and $V^{(2)}$ the outgoing 4-velocities. $U^{(1)}$ has space and time components $\gamma(u^{(1)})u^{(1)}$ and $\gamma(u^{(1)})$, and similarly for the other 4-velocities. So relativistic momentum and energy conservation require

$$m^{(1)}\gamma(v^{(1)})v^{(1)} + m^{(2)}\gamma(v^{(2)})v^{(2)} = m^{(1)}\gamma(u^{(1)})u^{(1)} + m^{(2)}\gamma(u^{(2)})u^{(2)} \tag{4.35}$$

and

$$m^{(1)}\gamma(v^{(1)}) + m^{(2)}\gamma(v^{(2)}) = m^{(1)}\gamma(u^{(1)}) + m^{(2)}\gamma(u^{(2)}). \tag{4.36}$$

It is typical that γ factors, involving square roots, occur in relativistic equations like these. The equations again determine the unknowns $v^{(1)}$ and $v^{(2)}$, but the algebra is now more complicated. As before, one trivial solution is $v^{(1)} = u^{(1)}$ and $v^{(2)} = u^{(2)}$, and knowing this helps to find the other solution.

The relativistic equations agree with the Newtonian ones for particles moving at speeds much less than the speed of light. To see this, it is sufficient in equation (4.35) to make the approximation $\gamma \simeq 1$ in all four terms, and we recover the momentum conservation equation (4.32). In equation (4.36) we need to make the approximation $\gamma(w) \simeq 1 + \frac{1}{2}w^2$ in all four terms, obtaining

$$m^{(1)} + \frac{1}{2}m^{(1)}(v^{(1)})^2 + m^{(2)} + \frac{1}{2}m^{(2)}(v^{(2)})^2 = m^{(1)} + \frac{1}{2}m^{(1)}(u^{(1)})^2 + m^{(2)} + \frac{1}{2}m^{(2)}(u^{(2)})^2. \tag{4.37}$$

This agrees with the energy conservation equation (4.33) after cancelling $m^{(1)} + m^{(2)}$.

For high speed collisions, the equations in the Newtonian and relativistic cases are clearly different, and the predictions for the outgoing velocities $v^{(1)}$ and $v^{(2)}$ are different. One example is sufficient to show this, although we won't go through the algebra. Suppose $m^{(1)} = 2$ and $m^{(2)} = 1$, and $u^{(1)} = \frac{3}{5}$ and $u^{(2)} = 0$. Then in the Newtonian case the outgoing particles have velocities $v^{(1)} = \frac{1}{5}$ and $v^{(2)} = \frac{4}{5}$, whereas in the relativistic case the velocities are different. They are $v^{(1)} = \frac{9}{41}$ and $v^{(2)} = \frac{21}{29}$. (We have chosen unequal masses because if $m^{(1)} = m^{(2)}$, the outgoing velocities are $v^{(1)} = 0$ and $v^{(2)} = \frac{3}{5}$ in both cases.)

The occurrence of only rational numbers (simple fractions) here is a bit surprising. It is easy to show that if $u^{(1)}$ is rational and $u^{(2)} = 0$, and the ratio of masses is rational, then in the Newtonian case $v^{(1)}$ and $v^{(2)}$ are rational. In the relativistic case, $v^{(1)}$ and $v^{(2)}$ can also be shown to be rational, provided both $u^{(1)}$ and $\gamma(u^{(1)})$ are rational, and $u^{(2)} = 0$. That is why we chose $u^{(1)} = \frac{3}{5}$. Because $(3, 4, 5)$ is a Pythagorean triad, $\gamma(u^{(1)}) = \frac{5}{4}$. Similarly, the outgoing relativistic velocities $v^{(1)} = \frac{9}{41}$ and $v^{(2)} = \frac{21}{29}$ are associated to the Pythagorean triads $(9, 40, 41)$ and $(20, 21, 29)$, so $\gamma(v^{(1)}) = \frac{41}{40}$ and $\gamma(v^{(2)}) = \frac{29}{20}$.

In summary, relativistic 4-momentum conservation combines the Newtonian notions of momentum and energy conservation, but in a new way, and it has different consequences in detail for high speed collisions. Experiments involving particle collisions at high energy have shown that the relativistic predictions are correct, and Newtonian dynamics breaks down in this regime.

4.3.2 $E = mc^2$

Now we come to one of the most famous and profound predictions of relativity. We have seen that for a particle, it is the time component of the 4-momentum that is the relativistic version of energy, and this is $m\gamma(\mathbf{v})$, where m is the mass and $\mathbf{v}$ is the usual 3-velocity. As $\gamma(\mathbf{0}) = 1$, a particle at rest has energy $E = m$. This is called the *rest energy* or *rest mass* of the particle. If we had not set the speed of light c to unity, we would have obtained Einstein's famous formula $E = mc^2$. For a particle moving fairly slowly, where $\gamma(\mathbf{v}) \simeq 1 + \frac{1}{2}|\mathbf{v}|^2$, the relativistic energy is

$$E \simeq m + \frac{1}{2}m|\mathbf{v}|^2 \,, \tag{4.38}$$

the sum of the rest energy and the standard, Newtonian kinetic energy.

We saw that in collisions at non-relativistic speeds the rest energies of the particles cancel out, because they appear equally on both sides of the energy conservation equation (4.37). Rest energy can therefore be ignored in Newtonian dynamics. Einstein's belief in relativity and his great insight into physics convinced him that the rest energy of a particle, its mass m, is still physical and that it must be possible to convert it into other forms of energy. This prediction was, of course, correct and has been confirmed in countless ways in the realms of nuclear and particle physics.

For example, a neutron n has a slightly greater mass than a proton p, and it decays, with a half-life of about 10 minutes, through the process

$$n \to p + e^- + \overline{\nu_e} \,, \tag{4.39}$$

where e^- is an electron and $\overline{\nu_e}$ an anti-electron-neutrino, one of several species of neutrino. The electron mass is about one quarter of the difference between the neutron and proton masses, and the antineutrino mass is much smaller still. So although most of the neutron rest energy reappears as the rest energy of the proton and a small amount as the rest energy of the other two particles, there is some energy to spare. This becomes the kinetic energy of the outgoing particles. It has been confirmed that overall, relativistic 4-momentum (i.e. relativistic momentum and energy) is conserved in neutron decay.

Rest energy has great significance for energy generation by nuclear fission, as we will discuss in Chapter 11. A heavy nucleus such as uranium has slightly greater rest energy than its fission fragments. The excess energy appears as kinetic energy of the products,

Fig. 4.5 A collision of two protons inside the ATLAS detector of the Large Hadron Collider. The protons are hidden by the beam pipes and only the outgoing particles are visible.

which can be used to heat water, drive turbines and generate electricity. The kinetic energy released, as in the case of neutron decay, is less than 1% of the mass of the original nucleus, but in everyday terms this is a very large amount of energy. An an illustration, if an outgoing particle has a relativistic energy just one half of 1% above its rest energy, then its speed v is one tenth of the speed of light (as $\frac{1}{2}mv^2 = 0.005m$ if $v = 0.1$), which is enormous in the context of steam driving a turbine. It is also enormous compared to the energy released in the chemical reactions of an equivalent number of atoms. So the mass of fuel needed to run a nuclear power station is far less than that needed to run a coal-, gas- or oil-fired power station.

Conversely, it is also possible to convert the kinetic energy of particles into the rest energy (i.e. mass) of new particles. This happens routinely in high energy collisions of particles in particle accelerators. A collision of two protons at the Large Hadron Collider regularly produces hundreds of new particles, as shown in Figure 4.5. This is possible because the total energy of the incoming protons (which is mainly kinetic) is about 10 TeV, around 10^4 times the rest energy of one proton, so there is enough energy to produce hundreds of new protons and antiprotons, each with substantial kinetic energy. In practice, most of the new particles are pions, electrons and muons, which are less massive than protons.

4.4 More on 4-Vectors

Because of the minus sign occurring in the Lorentzian inner product of 4-vectors (4.22), it is useful to consider for each 4-vector X a second 4-vector $\underline{X}$. This has the same time component as X, but the space components have reversed signs, so if $X = (t, \mathbf{x})$ then $\underline{X} = (t, -\mathbf{x})$.[4] A similar sign-reversal applies to all 4-vectors. The inner product of $X = (t, \mathbf{x})$ and $Y = (u, \mathbf{y})$ can be written as $X \cdot Y$ or $\underline{X} \cdot Y$, both of which are defined to equal $tu - \mathbf{x} \cdot \mathbf{y}$.

---

[4] In many presentations of special relativity, the components of X have an upper (superscript) index and those of $\underline{X}$ have a lower (subscript) index. We will use this notation later.

The convention is that if the inner product involves two 4-vectors with neither underlined, then an explicit minus sign is included in front of the spatial dot product contribution. If one of the 4-vectors is underlined, then there is no explicit minus sign in the inner product and any minus sign comes instead from the spatial components of the underlined 4-vector.

There is a Lorentz transformation rule for $\underline{X}$ that follows from the transformation rule for X. Under a rotation, the transformations of X and $\underline{X}$ are the same because a rotation of $\mathbf{x}$ also rotates $-\mathbf{x}$. However, for a boost with rapidity θ, one needs to reverse the sign of θ in the boost formulae when acting on the components of $\underline{X}$. This is seen most easily from equation (4.10), where an exchange of x_1 and $-x_1$ needs to be accompanied by an exchange of e^θ and $e^{-\theta}$.

In spacetime, it is natural to combine the partial derivatives $\frac{\partial}{\partial t}$ and $\nabla = \left(\frac{\partial}{\partial x_1}, \frac{\partial}{\partial x_2}, \frac{\partial}{\partial x_3}\right)$ into a 4-vector operator, the Lorentzian analogue of ∇. This is an underlined 4-vector

$$\underline{\partial} = \left(\frac{\partial}{\partial t}, \nabla\right) . \tag{4.40}$$

(One must check the effect of a Lorentz transformation to see that $\underline{\partial}$ should be underlined. Roughly, it is because the coordinates appear in the 'denominators' of the partial derivatives.) There is also a regular 4-vector operator $\partial = \left(\frac{\partial}{\partial t}, -\nabla\right)$. The derivatives of a scalar field ψ combine into 4-vectors

$$\underline{\partial}\psi = \left(\frac{\partial\psi}{\partial t}, \nabla\psi\right) \quad \text{and} \quad \partial\psi = \left(\frac{\partial\psi}{\partial t}, -\nabla\psi\right) . \tag{4.41}$$

Another useful operator is the Lorentz invariant wave operator $\partial \cdot \partial = \frac{\partial^2}{\partial t^2} - \nabla^2$. This occurs in the wave equation

$$\frac{\partial^2\psi}{\partial t^2} - \nabla^2\psi = 0 . \tag{4.42}$$

Recall that a plane wave solution is

$$\psi(\mathbf{x}, t) = e^{i(\mathbf{k}\cdot\mathbf{x} - \omega t)} , \tag{4.43}$$

and the wave speed is 1 (the speed of light) because equation (4.42) requires that

$$\omega^2 - \mathbf{k}\cdot\mathbf{k} = 0 . \tag{4.44}$$

The phase in the exponent of ψ is minus the inner product of the 4-vectors $K = (\omega, \mathbf{k})$ and $X = (t, \mathbf{x})$, i.e. $-K \cdot X = \mathbf{k}\cdot\mathbf{x} - \omega t$. Because K transforms like a 4-vector, different observers perceive the wave as having different frequencies ω and spatial wavevectors $\mathbf{k}$. But all observers agree that the speed is 1, because equation (4.44) is the Lorentz invariant condition $K \cdot K = 0$.

4.5 The Relativistic Character of Maxwell's Equations

Some of the ingredients of electromagnetism are clearly 4-vectors. The charge density ρ and current density $\mathbf{j}$ combine into a 4-current density $\mathcal{J} = (\rho, \mathbf{j})$. The conservation equation $\frac{\partial\rho}{\partial t} + \nabla \cdot \mathbf{j} = 0$ can be expressed in 4-vector form simply as

$$\partial \cdot \mathcal{J} = 0 . \tag{4.45}$$

The sign is correct because of the explicit minus sign in $\partial = \left(\frac{\partial}{\partial t}, -\nabla\right)$. $\mathcal{J}$ is a field, defined throughout spacetime, but for a point particle the charge density is singular, and concentrated at the instantaneous particle location. $\mathbf{j}$ is obtained by multiplying ρ by the particle velocity $\mathbf{v}$, so $\mathcal{J} = (\rho, \rho\mathbf{v})$, which is closely related to the particle 4-velocity V. (The absence of explicit gamma factors is because ρ is a density.) The total charge q of a particle, like its mass m, is Lorentz invariant.

The potentials Φ and $\mathbf{A}$ also combine into a 4-vector potential $\mathcal{A} = (\Phi, \mathbf{A})$. Like $\mathcal{J}$, this is a field defined everywhere in spacetime. The Lorenz gauge condition, although not the most fundamental equation of electromagnetism, is simply $\partial \cdot \mathcal{A} = 0$, and hence Lorentz invariant.

Finding a Lorentz covariant formulation of the electric and magnetic fields is more of a challenge. Together, $\mathbf{E}$ and $\mathbf{B}$ have six components, and according to the formulae (3.57) each component is the sum of two terms involving a time or space derivative acting on a potential, which may be Φ or a component of $\mathbf{A}$. The 4-vector version of the fields involves $\partial\mathcal{A}$, without an inner product. As it stands, $\partial\mathcal{A}$ has sixteen components, but if we antisymmetrize then just six distinct components remain. We need a matrix array to show this.

$\partial\mathcal{A}$ is a matrix where each entry is a derivative of a potential:

$$\partial\mathcal{A} = \begin{pmatrix} \frac{\partial\Phi}{\partial t} & \frac{\partial A_1}{\partial t} & \frac{\partial A_2}{\partial t} & \frac{\partial A_3}{\partial t} \\ -\frac{\partial\Phi}{\partial x_1} & -\frac{\partial A_1}{\partial x_1} & -\frac{\partial A_2}{\partial x_1} & -\frac{\partial A_3}{\partial x_1} \\ -\frac{\partial\Phi}{\partial x_2} & -\frac{\partial A_1}{\partial x_2} & -\frac{\partial A_2}{\partial x_2} & -\frac{\partial A_3}{\partial x_2} \\ -\frac{\partial\Phi}{\partial x_3} & -\frac{\partial A_1}{\partial x_3} & -\frac{\partial A_2}{\partial x_3} & -\frac{\partial A_3}{\partial x_3} \end{pmatrix}. \tag{4.46}$$

It has a transposed form $(\partial\mathcal{A})^{\mathrm{T}}$, with the rows and columns exchanged:

$$(\partial\mathcal{A})^{\mathrm{T}} = \begin{pmatrix} \frac{\partial\Phi}{\partial t} & -\frac{\partial\Phi}{\partial x_1} & -\frac{\partial\Phi}{\partial x_2} & -\frac{\partial\Phi}{\partial x_3} \\ \frac{\partial A_1}{\partial t} & -\frac{\partial A_1}{\partial x_1} & -\frac{\partial A_1}{\partial x_2} & -\frac{\partial A_1}{\partial x_3} \\ \frac{\partial A_2}{\partial t} & -\frac{\partial A_2}{\partial x_1} & -\frac{\partial A_2}{\partial x_2} & -\frac{\partial A_2}{\partial x_3} \\ \frac{\partial A_3}{\partial t} & -\frac{\partial A_3}{\partial x_1} & -\frac{\partial A_3}{\partial x_2} & -\frac{\partial A_3}{\partial x_3} \end{pmatrix}. \tag{4.47}$$

The antisymmetrized matrix $\mathcal{F} = \partial\mathcal{A} - (\partial\mathcal{A})^{\mathrm{T}}$ is called the *electromagnetic field tensor*, and is

$$\mathcal{F} = \begin{pmatrix} 0 & \frac{\partial A_1}{\partial t} + \frac{\partial\Phi}{\partial x_1} & \frac{\partial A_2}{\partial t} + \frac{\partial\Phi}{\partial x_2} & \frac{\partial A_3}{\partial t} + \frac{\partial\Phi}{\partial x_3} \\ -\frac{\partial\Phi}{\partial x_1} - \frac{\partial A_1}{\partial t} & 0 & -\frac{\partial A_2}{\partial x_1} + \frac{\partial A_1}{\partial x_2} & -\frac{\partial A_3}{\partial x_1} + \frac{\partial A_1}{\partial x_3} \\ -\frac{\partial\Phi}{\partial x_2} - \frac{\partial A_2}{\partial t} & -\frac{\partial A_1}{\partial x_2} + \frac{\partial A_2}{\partial x_1} & 0 & -\frac{\partial A_3}{\partial x_2} + \frac{\partial A_2}{\partial x_3} \\ -\frac{\partial\Phi}{\partial x_3} - \frac{\partial A_3}{\partial t} & -\frac{\partial A_1}{\partial x_3} + \frac{\partial A_3}{\partial x_1} & -\frac{\partial A_2}{\partial x_3} + \frac{\partial A_3}{\partial x_2} & 0 \end{pmatrix}. \tag{4.48}$$

Each component of $\mathcal{F}$ below the diagonal is the negative of a component above.

The six independent components here are precisely the six components of $\mathbf{E}$ and $\mathbf{B}$, as we can see by comparing $\mathcal{F}$ to the expressions (3.57) and recalling the definition (3.22) of curl. In terms of the electric and magnetic fields, the field tensor is

$$\mathcal{F} = \begin{pmatrix} 0 & -E_1 & -E_2 & -E_3 \\ E_1 & 0 & -B_3 & B_2 \\ E_2 & B_3 & 0 & -B_1 \\ E_3 & -B_2 & B_1 & 0 \end{pmatrix}, \tag{4.49}$$

and it plays the role of the complete electromagnetic field from a spacetime point-of-view.

Under Lorentz transformations, $\partial \mathcal{A}$ transforms doubly as a Lorentz 4-vector, because ∂ and $\mathcal{A}$ each transform as 4-vectors, and $(\partial \mathcal{A})^{\mathrm{T}}$ transforms similarly. $\mathcal{F}$ is said to be a 4-tensor. We will not give all the formulae for Lorentz transformations of $\mathcal{F}$. A rotation just rotates $\mathbf{E}$ and $\mathbf{B}$ individually as 3-vectors, but the effect of the boost (4.7) with rapidity θ is more interesting. It produces new fields

$$E_1' = E_1 \,, \qquad B_1' = B_1$$
$$E_2' = E_2 \cosh\theta - B_3 \sinh\theta \,, \qquad B_2' = B_2 \cosh\theta + E_3 \sinh\theta$$
$$E_3' = E_3 \cosh\theta + B_2 \sinh\theta \,, \qquad B_3' = B_3 \cosh\theta - E_2 \sinh\theta \,, \qquad (4.50)$$

clearly mixing some of the components of the electric and magnetic fields. These formulae can be expressed in terms of the velocity $v = \tanh\theta$ of the boost, by writing $\cosh\theta = \gamma(v)$ and $\sinh\theta = \gamma(v)v$ as before.

In addition to $\mathcal{F}$, there is a second 4-tensor $\widetilde{\mathcal{F}}$ that can be constructed by exchanging $\mathbf{E}$ and $\mathbf{B}$, and changing a sign. This is called the *electromagnetic dual* of $\mathcal{F}$. Its precise form is

$$\widetilde{\mathcal{F}} = \begin{pmatrix} 0 & -B_1 & -B_2 & -B_3 \\ B_1 & 0 & E_3 & -E_2 \\ B_2 & -E_3 & 0 & E_1 \\ B_3 & E_2 & -E_1 & 0 \end{pmatrix} \,, \qquad (4.51)$$

where $(\mathbf{E}, \mathbf{B})$ in $\mathcal{F}$ have been replaced by $(\mathbf{B}, -\mathbf{E})$. Under boosts, $\widetilde{\mathcal{F}}$ Lorentz transforms in the same way as $\mathcal{F}$. One can see this by examining equations (4.50). And $\widetilde{\mathcal{F}}$ transforms in the same way as $\mathcal{F}$ under rotations, since $\mathbf{E}$ and $\mathbf{B}$ transform in the same way under rotations.

A physically interesting consequence of the transformation (4.50) is that what appears as a pure electric field to an observer at rest, appears to a moving observer as a combination of electric and magnetic fields. This is not really surprising. A charged particle at rest produces just an electric field, but to a moving observer the particle appears to be moving in the opposite direction, and therefore carries electric current as well as charge. The moving observer sees a combination of electric and magnetic fields produced by the particle. Similarly, a current loop at rest produces a pure magnetic field, but to a moving observer the magnetic field configuration is swept along through space, and is time dependent, so by the law of induction (3.26) it generates an electric field.

The mixing of electric and magnetic fields affects the interpretation of forces acting on charged particles. For example, a moving charged particle in a purely magnetic field experiences a force and accelerates. But to an observer moving instantaneously at the velocity of the particle, the particle appears to be accelerating from rest, so the force must be due to an electric field (as the magnetic contribution to the Lorentz force vanishes for a particle at rest).

The culmination of all this is that Maxwell's equations have a Lorentz covariant character. The four Maxwell equations combine into just two equations involving the field tensor $\mathcal{F}$ and its dual $\widetilde{\mathcal{F}}$. These are

$$\partial \cdot \mathcal{F} \;=\; \mathcal{J} \,, \qquad (4.52)$$
$$\partial \cdot \widetilde{\mathcal{F}} \;=\; 0 \,. \qquad (4.53)$$

The inner products here are of the row 4-vector operator ∂ acting on each column of the

4-tensors $\mathcal{F}$ and $\widetilde{\mathcal{F}}$. The result is a new row 4-vector, which equals $\mathcal{J}$ in the first equation and zero in the second. The equations are of a manifestly Lorentz covariant form.

Let us check the equivalence of these equations to the earlier forms of Maxwell's equations. Writing equation (4.52) out fully gives

$$\left(\frac{\partial}{\partial t}, -\frac{\partial}{\partial x_1}, -\frac{\partial}{\partial x_2}, -\frac{\partial}{\partial x_3}\right) \cdot \begin{pmatrix} 0 & -E_1 & -E_2 & -E_3 \\ E_1 & 0 & -B_3 & B_2 \\ E_2 & B_3 & 0 & -B_1 \\ E_3 & -B_2 & B_1 & 0 \end{pmatrix} = (\rho, j_1, j_2, j_3) \,. \tag{4.54}$$

We see that the first component is the Maxwell equation $\nabla \cdot \mathbf{E} = \rho$, and the final component is

$$\frac{\partial}{\partial t}(-E_3) + \frac{\partial}{\partial x_1}(B_2) + \frac{\partial}{\partial x_2}(-B_1) = j_3 \,, \tag{4.55}$$

one component of the Maxwell equation (3.28). Similarly, the first component of (4.53) is the Maxwell equation $\nabla \cdot \mathbf{B} = 0$ and the final component is

$$\frac{\partial}{\partial t}(-B_3) + \frac{\partial}{\partial x_1}(-E_2) + \frac{\partial}{\partial x_2}(E_1) = 0 \,, \tag{4.56}$$

one component of the Maxwell equation (3.26). The two middle components in each case give the remaining equations.

The Lorentz force equation can also be modified so that it has a Lorentz covariant form, and this is the correct form for charged particles moving at arbitrary speeds, possibly approaching the speed of light. The original Lorentz force involves the fields $\mathbf{E}$ and $\mathbf{B}$, and the particle velocity $\mathbf{v}$. The relativistic version involves the field tensor $\mathcal{F}$ and the 4-velocity V. We take the inner product of V with each column of $\mathcal{F}$ (just as in $\partial \cdot \mathcal{F}$) and multiply by minus the particle charge q to get the Lorentz 4-force $F = -qV \cdot \mathcal{F}$. The relativistic equation of motion for a particle of mass m and charge q is therefore

$$mA = -qV \cdot \mathcal{F} \,, \tag{4.57}$$

where A is the 4-acceleration. The 4-force $F = -qV \cdot \mathcal{F}$ satisfies the constraint $F \cdot V = 0$, because the double inner product $V \cdot \mathcal{F} \cdot V$ is zero due to the antisymmetry of the matrix $\mathcal{F}$.

The relativistic equation of motion makes different predictions from the Newtonian equation. For example, in a uniform electric field, the speed of a Newtonian charged particle would increase indefinitely. A relativistic particle accelerates too, and its energy continually increases, but its speed is limited to less than the speed of light.

We should check that in the Newtonian limit, when $\mathbf{v}$ is small and $\gamma \simeq 1$, the original Lorentz force law (3.82) emerges. For the final component of equation (4.57), the left-hand side is ma_3, the third component of $m\mathbf{a}$, and as $V \simeq (1, \mathbf{v})$ the right-hand side is $q(E_3 + v_1 B_2 - v_2 B_1)$, the third component of $q(\mathbf{E} + \mathbf{v} \times \mathbf{B})$. The two middle components complete the 3-vector equation of motion. The first component of equation (4.57) is significant too, but not really independent. It states that

$$m\frac{d\gamma}{d\tau} = q\gamma \mathbf{v} \cdot \mathbf{E} \,, \tag{4.58}$$

and equates the rate of change of the particle's relativistic energy, $m\gamma$, to the work done by $\mathbf{E}$ on the particle.

In the Newtonian limit, this reduces to

$$\frac{d}{dt}\left(\frac{1}{2}m|\mathbf{v}|^2\right) = q\mathbf{v}\cdot\mathbf{E}\,,\tag{4.59}$$

which is the energy equation (3.83) associated with the Lorentz force.

The relativistic versions of Maxwell's equations and the Lorentz force law lead to some further insights. From the field tensor and its dual one can construct two independent Lorentz invariant (scalar) quantities, that characterize the type of electromagnetic field at each spacetime point. These are $\mathcal{F}\cdot\mathcal{F}$ and $\mathcal{F}\cdot\widetilde{\mathcal{F}}$, in which one takes the inner product both on the rows and columns. In practice, this means evaluating the sixteen products of the components of the first and second 4-tensors in the same matrix locations, and adding these up, including a minus sign if a product involves components of mixed time/space type (those in the top row or left column). The results are

$$\begin{aligned}\mathcal{F}\cdot\mathcal{F} &= -2(E_1E_1 + E_2E_2 + E_3E_3 - B_1B_1 - B_2B_2 - B_3B_3)\\ &= -2(\mathbf{E}\cdot\mathbf{E} - \mathbf{B}\cdot\mathbf{B})\,,\end{aligned}\tag{4.60}$$

$$\begin{aligned}\mathcal{F}\cdot\widetilde{\mathcal{F}} &= -4(E_1B_1 + E_2B_2 + E_3B_3)\\ &= -4\,\mathbf{E}\cdot\mathbf{B}\,.\end{aligned}\tag{4.61}$$

In the previous chapter we discussed some particular electromagnetic fields. They are special from the point-of-view of these Lorentz invariants. For a purely electrostatic field, $\mathcal{F}\cdot\mathcal{F}$ is negative and $\mathcal{F}\cdot\widetilde{\mathcal{F}} = 0$, whereas for a purely magnetostatic field, $\mathcal{F}\cdot\mathcal{F}$ is positive and $\mathcal{F}\cdot\widetilde{\mathcal{F}} = 0$. Finally, for an electromagnetic wave, where $|\mathbf{E}| = |\mathbf{B}|$ and $\mathbf{E}$ is orthogonal to $\mathbf{B}$, $\mathcal{F}\cdot\mathcal{F} = \mathcal{F}\cdot\widetilde{\mathcal{F}} = 0$.

We also previously considered the motion of a charged particle in constant, uniform fields. By Lorentz invariance, the acceleration we found for a particle in an electric field generalizes to any field with negative $\mathcal{F}\cdot\mathcal{F}$ and $\mathcal{F}\cdot\widetilde{\mathcal{F}} = 0$, an example being an electric field combined with a weaker, perpendicular magnetic field. The circular motion we found for a particle in a magnetic field generalizes to any field with positive $\mathcal{F}\cdot\mathcal{F}$ and $\mathcal{F}\cdot\widetilde{\mathcal{F}} = 0$. We now see that another special case is the motion of a charged particle in a plane electromagnetic wave background, with $\mathcal{F}\cdot\mathcal{F} = \mathcal{F}\cdot\widetilde{\mathcal{F}} = 0$.

4.6   Relativistic Principles of Least Action

In relativistic theories, the action is often Lorentz invariant, and hence observer-independent. This implies that the principle of least action is a particularly elegant way of formulating the dynamics of relativistic fields and particles. We will discuss this briefly, but not rederive the Maxwell equations nor the relativistic equation of motion for a charged particle in an electromagnetic field.

The action (3.80) for the electromagnetic field is the integral over 4-dimensional spacetime of the Lagrangian density

$$\mathcal{L} = \frac{1}{2}\mathbf{E}\cdot\mathbf{E} - \frac{1}{2}\mathbf{B}\cdot\mathbf{B} + \mathbf{A}\cdot\mathbf{j} - \Phi\rho\,,\tag{4.62}$$

with integration element $d^4X = d^3x\,dt$. The integration element is Lorentz invariant, because the matrix of a Lorentz transformation has determinant 1, as is easily checked for the 2×2

matrices of simple rotations and boosts,

$$\begin{pmatrix} \cos\theta & -\sin\theta \\ \sin\theta & \cos\theta \end{pmatrix} \quad \text{and} \quad \begin{pmatrix} \cosh\theta & -\sinh\theta \\ -\sinh\theta & \cosh\theta \end{pmatrix} . \tag{4.63}$$

The Lagrangian density $\mathcal{L}$ can be expressed compactly in terms of the 4-vector potential $\mathcal{A}$, the 4-current $\mathcal{J}$ and the 4-tensor field $\mathcal{F} = \partial\mathcal{A} - (\partial\mathcal{A})^{\mathrm{T}}$ as

$$\mathcal{L} = -\frac{1}{4}\mathcal{F}\cdot\mathcal{F} - \mathcal{A}\cdot\mathcal{J} , \tag{4.64}$$

so the action is

$$S = \int \left(-\frac{1}{4}\mathcal{F}\cdot\mathcal{F} - \mathcal{A}\cdot\mathcal{J} \right) d^4X . \tag{4.65}$$

This is clearly Lorentz invariant as it stands if we drop the initial and final times, t_0 and t_1, and formally integrate over all spacetime.

The principle of least action requires S to be stationary for any smooth variation of the field $\mathcal{A}$ that is non-zero only in some finite region Σ of spacetime. This principle leads to the Maxwell field equations. Different observers will agree on what it means for the action to be stationary, even though they will use different coordinates to specify Σ.

By contrast, the action (3.94) for a charged point particle that we have used so far is not Lorentz invariant, and is only valid when the particle speed is non-relativistic. It needs some modification to allow for particle speeds comparable with the speed of light. For a free particle, the relativistic action is defined to be

$$S = -m \int \frac{1}{\gamma(\mathbf{v})} \, dt , \tag{4.66}$$

with the integral formally along the entire particle worldline. In the Newtonian limit, where $\gamma(\mathbf{v}) \simeq 1 + \frac{1}{2}|\mathbf{v}|^2$, this becomes

$$S \simeq \int \left(-m + \frac{1}{2}m|\mathbf{v}|^2 \right) dt . \tag{4.67}$$

The first part is just a negative constant, and the second is the standard action involving the Newtonian kinetic energy. So the relativistic action has the right Newtonian limit after one drops a constant.

Equation (4.15) then implies that the particle action (4.66) simplifies to

$$S = -m \int d\tau , \tag{4.68}$$

a multiple of the integrated spacetime interval along the particle worldline. In this form it is clearly Lorentz invariant, involving the simplest available quantities. For a time-like worldline—a worldline with speed everywhere less than 1—as shown in Figure 4.6, the action is negative, but it can be arbitrarily close to zero for worldlines made of segments where the particle moves close to the speed of light. The free-particle action is minimized by a straight worldline, where the particle travels at constant velocity.

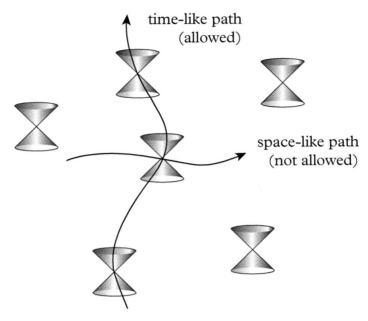

Fig. 4.6 The time-like worldline of a massive particle. The worldline is everywhere inside the lightcone.

For a particle of charge q interacting with a background electromagnetic field, the relativistic action is a combination of the free action (4.66) and the integral over time of the interaction terms $q\mathbf{A}(\mathbf{x}(t)) \cdot \mathbf{v} - q\Phi(\mathbf{x}(t))$ occurring in equation (3.94). The interaction terms need no relativistic modification, because they can be expressed in 4-vector form as $-q\frac{1}{\gamma(\mathbf{v})}V \cdot \mathcal{A}$. Here, the 4-velocity $V = \gamma(\mathbf{v})(1, \mathbf{v})$ and 4-vector potential $\mathcal{A} = (\Phi, \mathbf{A})$ are evaluated at the spacetime points $X = (t, \mathbf{x}(t))$ along the particle worldline. The total relativistic action for the charged particle is therefore the worldline integral

$$S = \int \frac{1}{\gamma(\mathbf{v})}(-m - qV \cdot \mathcal{A})\, dt\,, \tag{4.69}$$

which can be expressed in the manifestly Lorentz invariant form

$$S = \int (-m - qV \cdot \mathcal{A})\, d\tau\,. \tag{4.70}$$

For a dynamical electromagnetic field coupled to relativistic charged particles the action is the sum of the field action (4.65) (with zero $\mathcal{J}$), and a worldline action (4.70) for each particle. The principle of least action gives the formally correct relativistic equations for the fields and the particles, but it does not resolve the difficulties associated with the self-forces and the motion of rapidly accelerating point particles discussed at the end of Chapter 3.

One further example of relativistic dynamics is the action for a point particle of mass m coupled to a background Lorentz scalar field ψ. This action is

$$S = -m \int \exp\left(\frac{1}{m}\psi\right) d\tau\,, \tag{4.71}$$

where the integral is along the particle worldline, and ψ is evaluated at the points X along it. The equation of motion obtained by minimizing S is

$$mA = \partial\psi - (\partial\psi \cdot V)V . \tag{4.72}$$

As required, the 4-force on the right-hand side satisfies the constraint of having zero inner product with V, because $V \cdot V = 1$. Although interesting, this equation has less physical application than the relativistic Lorentz force law (4.57).

4.7 Further Reading

E.F. Taylor and J.A. Wheeler, *Spacetime Physics: Introduction to Special Relativity (2nd ed.)*, New York: Freeman, 2001.

W. Rindler, *Relativity: Special, General and Cosmological (2nd ed.)*, Oxford: OUP, 2006.

For a discussion of relativistic particle collisions, and Pythagorean triples, see

N.S. Manton, *Rational Relativistic Collisions*, arXiv:1406.3014 [physics.pop-ph], 2014.

5
Curved Space

5.1 Spherical Geometry

So far, we have only considered physics in flat Euclidean space and in the flat spacetime of special relativity known as Minkowski space. In this chapter we will take an excursion into the more general geometry of curved space, and consider some of its physical applications. In the next chapter we will use the mathematical technology developed here to describe the curved spacetime geometry of general relativity, Einstein's theory of gravity.

Euclidean space is named after the Greek mathematician Euclid who compiled the most important results of classical geometry into his *Elements* in the 3rd century BC. Euclid began with a short collection of definitions and common notions or axioms. Then step by step he constructed simple results, such as a proof that the interior angles of a triangle are equal to two right angles—so they sum to 180° or π radians—and gradually built up to more complicated results about polygons, circles and the structure of the regular solids. For 2000 years the notion of geometry meant the geometry of Euclid. It was considered so obvious that the results of Euclid must hold in all circumstances that a direct correspondence between Euclid's geometry and the structure of the real world was universally accepted. We now know that this belief was mistaken.

The geometry of Euclid is not even valid on a surface as familiar as the sphere. This is because one of Euclid's axioms—the parallel postulate—does not hold on the sphere. There are a number of equivalent ways to state the parallel postulate. According to one such statement, if we choose a straight line L in a plane and a point P not on L, then we can always draw a *unique* straight line through P that does not intersect L. This is true in the flat plane, but not on a curved surface. On a sphere, the analogue of straight lines are great circles; these include the lines of longitude. Two lines of longitude may look parallel at the equator but they meet at both the North and South Poles. There are *no* parallel lines on a sphere.

Any of Euclid's results that depend on the parallel postulate, such as the proof that the angles of a triangle sum to π, will not hold on the surface of a sphere. This is illustrated in Figure 5.1. On a sphere the angles of a triangle Δ sum to $\Sigma_\Delta = \pi + \frac{A}{a^2}$, where A is the area of the triangle and a is the radius of the sphere. The greater the area of the triangle the larger the angle sum. For instance, the spherical triangle shown in Figure 5.1 covers an octant of the sphere, so its area is $A = \frac{1}{2}\pi a^2$, and therefore the angle sum is $\Sigma_\Delta = \frac{3}{2}\pi$. This may be confirmed by looking at the figure, where it is clear that all three angles are right angles. We know from our experience of living on a globe, that in regions which are small compared to the whole globe, the surface appears approximately flat. The above formula

The Physical World. Nicholas Manton and Nicholas Mee, Oxford University Press (2017).
© Nicholas Manton and Nicholas Mee. DOI 10.1093/oso/9780198795933.001.0001

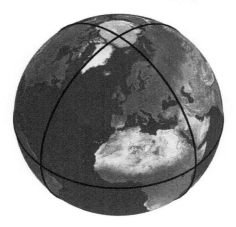

Fig. 5.1 An example of a spherical triangle drawn on the surface of the Earth. All three angles of this triangle are right angles.

ties in with this intuition. For spherical triangles that are small compared to the total area of the sphere, the Euclidean result that the angle sum is π holds to a good approximation.

5.1.1 Geodesics

To understand the geometry of a curved surface we require a more general notion of a straight line. In flat space a straight line is the shortest path between two points. It is natural to extend this property to a surface that may be curved. The Mercator maps of the world that we are familiar with distort our perception of distances. Lines of longitude and latitude both appear straight on these maps, but whereas lines of longitude are the shortest paths between the points they connect, lines of latitude (other than the equator) are certainly not. As is well known to ship and aircraft navigators, the shortest path connecting two points on the globe is part of a great circle, a circle like the equator with the same radius as the sphere itself. These shortest paths are known as *geodesics* and when space is curved they are the equivalent of straight lines. Etymologically, the word geodesic derives from Greek. Geodesy means to survey, literally to divide up the Earth. Lines of longitude are geodesics, because they are segments of great circles, but lines of latitude are not. For instance, as shown in Figure 5.2, the shortest route when flying from London to Tokyo is to follow a great circle route that takes the aircraft well into the Arctic even though Tokyo is much further south than London.

Another important feature of spherical geometry is its uniformity. As in planar geometry, all points on the surface of a sphere are geometrically equivalent, and all directions from a point are equivalent too. The geometry is said to be *homogeneous* and *isotropic*. We can take a geometrical object such as a triangle and move it and rotate it, and its geometrical properties do not change. If two triangles at different locations in a plane have edges of the same lengths then they are congruent, so they also have the same angles at their vertices. Spherical triangles are triangles whose edges are geodesic segments, and they have an equivalent property—triangles in different locations with the same edge lengths automatically have the same angles where the edges meet.

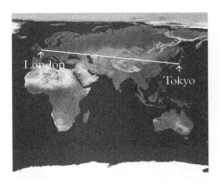

Fig. 5.2 Left: What appears to be a straight path between London and Tokyo on a world map in the Mercator projection. Right: The shortest path between the two cities is actually a great circle route.

5.2   Non-Euclidean, Hyperbolic Geometry

Many mathematicians over the centuries were uneasy about Euclid's parallel postulate, as it seemed to be much more complicated than the other postulates and axioms, appearing more like a theorem. A great deal of effort was expended on attempts to derive it from the other simpler axioms, but without success. By the 1820s the time was ripe for a fundamental reassessment of the situation. The breakthrough was made almost simultaneously by three mathematicians who independently realized that denying the parallel postulate did not lead to any inconsistency. It was still possible to prove geometrical theorems, but the strange geometrical results that they found did not apply to the geometry of Euclid; these theorems described a new *non-Euclidean* geometry. Carl Friedrich Gauss was the world's leading mathematician in the early 19th century. He was the first to discover the possibility of non-Euclidean geometry and privately studied its properties for many years. Only when the work of a young Hungarian mathematician called János Bolyai was brought to his attention did he reveal his own investigations. Bolyai had independently discovered many of the same results as Gauss. It was soon realized that a third mathematician, Nikolai Lobachevsky, had also published very similar results in a Russian journal.

Although spherical geometry is easy to visualize, non-Euclidean geometry is not. It is a completely different 2-dimensional geometry, now known as the geometry of the hyperbolic plane, or simply, hyperbolic geometry. Like a sphere, a hyperbolic plane is a curved surface (and not really a plane) where all points and all directions are geometrically equivalent, and it has a scale parameter a analogous to the radius of a sphere. When we think of curves and surfaces, we usually think of them as embedded in flat 3-dimensional space. However, unlike a sphere, the hyperbolic plane is infinite and cannot be embedded in its entirety as a surface in 3-space, but parts of the surface can be embedded, as shown in Figure 5.3.

The discovery of non-Euclidean geometry generated one of the most profound revolutions in the history of mathematical thought. For two millennia, mathematicians and philosophers had assumed that Euclid's geometry was founded on undeniable truths about the world in which we live. This was expressed most clearly by Immanuel Kant whose philosophy was built on the belief that Euclid's axioms were intuitively obvious atoms of truth about the universe.

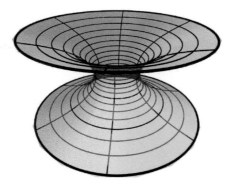

Fig. 5.3 Part of a hyperbolic plane embedded in 3-space. In contrast to the sphere, it is not possible to exhibit the symmetry and uniformity of the hyperbolic plane in such an embedding.

Kant referred to them as *analytic a priori truths*. Kant then argued that as the axioms were necessarily true, the theorems of Euclid—the synthetic truths—must automatically apply to the structure of the universe. The new insight provided by Gauss, Bolyai and Lobachevsky blew a hole in these ideas. It was clear that the axioms of geometry, and by extension the whole of mathematics, were not set in stone. Mathematicians could decide on different sets of axioms and study their implications. Henceforth, the axioms of a system of mathematics would be considered as more like the rules of a game, as agreed before play proceeds. They must be consistent and self-contained, but they need not have any connection to reality. Mathematicians were now set free to explore abstract domains that could be completely independent of any physical basis. The separation between mathematics and physics also raised the question of what the actual geometry of space was. This would be a matter to be determined by experiment and measurement.

Gauss was the first to contemplate the possibility that the space in which we live may not be flat Euclidean 3-space, as had always previously been assumed; it might be some sort of curved 3-dimensional space. To test this, Gauss exploited the theorem that in spherical geometry the sum of the angles of a triangle is greater than π, whereas in hyperbolic geometry it is less than π, with the excess or deficit depending on the size of the triangle (see Figure 5.1). Gauss set up surveying equipment on three peaks in the Harz mountains of central Germany to measure the properties of the triangle whose edges were geodesics connecting the peaks. Since the observations were visual, there was an implicit assumption that light rays travel along geodesics. Gauss measured the angle between the two sides of the triangle meeting at each peak to determine whether the angles sum to π. He found that they did, and concluded that space is Euclidean. (Gauss was, of course, attempting to measure the curvature of the space surrounding the Earth, not the curvature of the Earth's surface.) Gauss's investigations would no doubt have been forgotten long ago, had not the matter been taken up again by Einstein almost a century later.

5.3   Gaussian Curvature

Following Gauss, let's take a look at how mathematicians analyse curvature. At any point P on a curve in Euclidean 3-space, we can find the circle that is most precisely tangent to the curve. Its centre is called the centre of curvature. If there is only mild curvature at P,

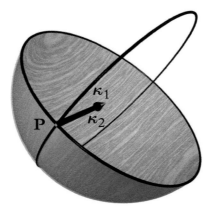

Fig. 5.4 On a sphere, the inverse radii of curvature κ_1 and κ_2 at each point are in the same direction, so the curvature is positive.

then the radius of the circle is large, and if the curvature is strong, the radius is small. So it is natural to use the inverse of the radius of this circle, κ, as a measure of the curvature.

Now consider a point P on a surface in 3-space. There are geodesic curves in all directions through P, and therefore a one-parameter family of circles, depending on direction, which are most precisely tangent to these. The circle curvatures which are maximal and minimal are called the principal curvatures. We will denote them as κ_1 and κ_2. They are known as extrinsic curvatures, as they depend on how the surface is embedded in 3-space. Gauss realized, however, that their product $K = \kappa_1 \kappa_2$ is an intrinsic property of the surface in the neighbourhood of P. K is called the *Gaussian curvature* of the surface at P.

The curves of maximal and minimal curvature cross perpendicularly. In Figure 5.4 two perpendicular curves are drawn through a representative point P on a sphere of radius a. The centre of curvature for both of these is the centre of the sphere, so κ_1 and κ_2 have the same magnitude and sign. Both equal $\frac{1}{a}$, and therefore the Gaussian curvature of the sphere is the positive constant $K = \frac{1}{a^2}$.

Regions of a hyperbolic plane, embedded in 3-space, are shaped like a saddle or curved funnel, as shown in Figure 5.5. The illustration shows a representative point and two perpendicular geodesic curves through it. The centres of curvature are in opposite directions, as indicated by the arrows. In this case κ_1 and κ_2 are regarded as having opposite signs and the Gaussian curvature K is negative. K is a negative constant $-\frac{1}{a^2}$ on the hyperbolic plane, and this is the defining feature of its geometry.

Intrinsically, a hyperbolic plane is circularly symmetric around any of its points, just like a sphere. But this symmetry is necessarily lost when the hyperbolic plane is embedded in 3-space. Although the surface shown in Figure 5.5 has circular symmetry around a vertical axis, no point on this axis belongs to the surface and there is no circular symmetry around any point that is actually on the surface. If the embedded surface was circularly symmetric around one of its points, κ_1 and κ_2 would be the same at that point, and K would be positive or zero. For this reason, the full symmetry of the hyperbolic plane is not apparent in the embedding.

The cylinder offers a third example of a uniform surface. In this case, the distance to one of the centres of curvature is infinite. (One of the curvature circles has degenerated

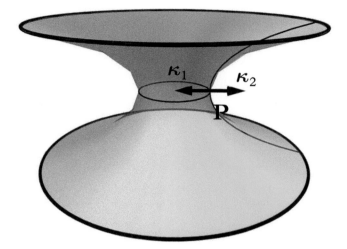

Fig. 5.5 On a hyperbolic plane, the inverse radii of curvature κ_1 and κ_2 are in opposite directions, so the curvature is negative.

into an infinite straight line.) So κ_2 is zero, and the Gaussian curvature is everywhere zero. Geometry on a cylinder is therefore locally the flat planar geometry of Euclid, as we can readily verify by rolling a flat piece of paper into a cylinder.

The Gaussian curvature K is intrinsic to a surface because it can be determined solely in terms of the behaviour of geodesics, and distances within the surface. It is possible to calculate its value at P in the following manner. Consider the spray of geodesics emanating from P. Mark off a small distance ε along each of these. The endpoints, each of distance ε from P, form a closed curve enclosing P. If the surface is flat, then the curve is a circle of length $C(\varepsilon) = 2\pi\varepsilon$. For a curved surface, the length deviates from this result. Positive curvature produces curve shortening and negative curvature produces curve lengthening.

For example, in contrast to the Euclidean plane, on a sphere of radius a in the vicinity of the North Pole, the circle of latitude at distance ε from the pole has circumference

$$C(\varepsilon) = 2\pi a \sin\left(\frac{\varepsilon}{a}\right) = 2\pi a \left(\frac{\varepsilon}{a} - \frac{1}{6}\frac{\varepsilon^3}{a^3} + \dots\right) = 2\pi\left(\varepsilon - \frac{1}{6}K\varepsilon^3 + \dots\right), \qquad (5.1)$$

where the Gaussian curvature K has been substituted for $\frac{1}{a^2}$. This is illustrated in Figure 5.6. Similarly, on a hyperbolic plane the circumference is

$$C(\varepsilon) = 2\pi a \sinh\left(\frac{\varepsilon}{a}\right) = 2\pi a \left(\frac{\varepsilon}{a} + \frac{1}{6}\frac{\varepsilon^3}{a^3} + \dots\right) = 2\pi\left(\varepsilon - \frac{1}{6}K\varepsilon^3 + \dots\right), \qquad (5.2)$$

as $K = -\frac{1}{a^2}$.

In general, the Gaussian curvature is given by the intrinsic expression

$$K = \frac{3}{\pi}\lim_{\varepsilon \to 0}\frac{2\pi\varepsilon - C(\varepsilon)}{\varepsilon^3}, \qquad (5.3)$$

and this defines K at any point P. For most surfaces, the geometry is not uniform, and the curvature varies from point to point. Consider a torus embedded in 3-space, for instance.

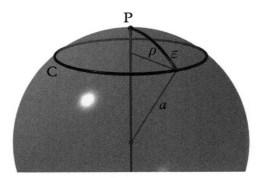

Fig. 5.6 Gaussian curvature: The distance along a geodesic from the North Pole P to the circle C is ε. The circumference of the circle is $2\pi a \sin\left(\frac{\varepsilon}{a}\right)$, where a is the radius of the sphere.

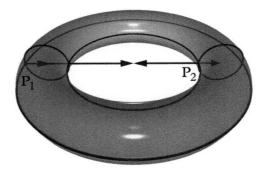

Fig. 5.7 Curvature of a torus embedded in 3-space. At P_1 the Gaussian curvature is positive, and at P_2 it is negative.

There are regions of the surface where the curvature is positive and regions where the curvature is negative, as illustrated in Figure 5.7.

The Gaussian curvature is positive at points on the torus such as P_1 that are on the outer rim of the torus, whereas the curvature is negative at points such as P_2 that are on the inner rim of the torus. If the Gaussian curvature is integrated over the whole surface, the total is zero. This might appear surprising, but can be understood as follows. It is possible for a torus to be completely flat with zero curvature everywhere. Such a surface is intrinsically a rectangle with opposite edges identified, as illustrated in Figure 5.8. The flat torus can be embedded in 3-space only by distorting it, and the distortion changes the Gaussian curvature. However, the distortion does not change the integral of the curvature over the whole surface.[1]

5.4 Riemannian Geometry

The discovery of non-Euclidean, hyperbolic geometry was just the beginning of a new era in geometry. Bernhard Riemann, who was a student of Gauss, developed a general, intrinsic formalism that would enable him to analyse non-uniform smooth geometry in any

---

[1] This result is a special case of the Gauss–Bonnet theorem.

Fig. 5.8 Forming a torus by gluing opposite edges of a rectangle. The long edges are glued to produce a cylindrical tube, then the ends of the tube are bent round and glued to form the torus.

number of dimensions. We shall present here the crux of Riemannian geometry in three space dimensions and later extend it to the 4-dimensional spacetime geometry of Minkowski and Einstein. The key to Riemann's approach is to capture the local distance relationships of a space in a way that generalizes Pythagoras' theorem.

Recall that in Euclidean geometry, points have Cartesian coordinates x_i ($i = 1, 2, 3$) and the square of the infinitesimal distance between points x_i and $x_i + \delta x_i$ is

$$\delta s^2 = \delta x_1^2 + \delta x_2^2 + \delta x_3^2 . \tag{5.4}$$

This expression is called the *Euclidean metric*. δs^2 is positive unless $\delta x_i = 0$ for all i.

3-dimensional Riemannian geometry is similar but generalizes this. It is again assumed that a point can be locally labelled in a unique way by three coordinates, y^i ($i = 1, 2, 3$). There is a change of notation here; from now on, coordinates are denoted with an upper index. This is particularly helpful when the indices are abstract Latin or Greek letters rather than numbers. (It is important not to confuse the coordinates y^2 and y^3 with 'y squared' and 'y cubed'.)

In Riemannian geometry it is assumed that the square of the distance between infinitesimally separated points with coordinates y^i and $y^i + \delta y^i$ is an expression of the form

$$\delta s^2 = g_{ij}(\mathbf{y}) \delta y^i \delta y^j , \tag{5.5}$$

where $g_{ij}(\mathbf{y})$ is a 3×3 matrix that is a smooth function of the coordinates $\mathbf{y} = (y^1, y^2, y^3)$. There is a further notational change here, known as the *summation convention*. In this condensed notation, repeated indices are summed over and an explicit summation symbol is omitted. g_{ij} is taken to be a symmetric matrix, as any antisymmetric part makes no contribution to δs^2, because $\delta y^i \delta y^j$ is necessarily symmetric. In slightly more explicit form (but suppressing the argument $\mathbf{y}$),

$$\delta s^2 = g_{ij}\, \delta y^i \delta y^j = g_{11} \left(\delta y^1\right)^2 + g_{22} \left(\delta y^2\right)^2 + g_{33} \left(\delta y^3\right)^2$$
$$+ 2 g_{12}\, \delta y^1 \delta y^2 + 2 g_{13}\, \delta y^1 \delta y^3 + 2 g_{23}\, \delta y^2 \delta y^3 . \tag{5.6}$$

g_{ij} is called the *metric tensor*, and the expression (5.5) for the infinitesimal squared distance is called the *(Riemannian) metric*. It is assumed that δs^2 is positive for any δy^i

that is not zero for all i. This requires the matrix g_{ij} to be positive definite everywhere. δs is taken to be the positive square root of δs^2 and the distance along a path connecting two points is obtained by integrating δs.

The inverse of the metric tensor g_{ij}, which is denoted by g^{ij}, is the usual 3×3 matrix inverse of g_{ij}. g^{ij} exists everywhere, and is positive definite, because g_{ij} is positive definite. One can write the relation between the metric tensor and its inverse as

$$g^{ij} g_{jk} = \delta^i_k \,, \tag{5.7}$$

where the summation convention applies to the index j. δ^i_k is known as the *Kronecker delta* symbol. It is equal to the unit matrix, so $\delta^i_k = 1$ if the indices i and k are the same and $\delta^i_k = 0$ if i and k are different. The Kronecker delta is unchanged if an index is raised or lowered, so δ_{ik} and δ^{ik} are unit matrices too, with all entries equal to 1 or 0.

One of the fundamental insights of Riemann was that true geometrical quantities do not depend on the choice of coordinate system. Distances and the metric are independent of the coordinates, so under a change of coordinates δs^2 does not change, but y^i and hence g_{ij} do. In fact g_{ij}, evaluated at one point, carries no geometrical information. By a change of coordinates, the metric tensor can be brought at that point into the standard Euclidean form. Therefore, a general Riemannian geometry is indistinguishable from flat geometry in an infinitesimal neighbourhood of a point. To visualize an example of what this means we may consider a spherical triangle, such as that shown in Figure 5.1. For an infinitesimally small triangle the angles add up to π, so locally the surface of the sphere appears planar.

5.4.1   Simple examples of metrics

Let us take a look at the Riemannian metric in a couple of familiar geometries: flat space in spherical polar coordinates and the simplest example of a curved surface, the 2-dimensional sphere or 2-sphere for short. We will also take a look at hyperbolic geometry.

We start with flat 3-space and Cartesian coordinates x^1, x^2, x^3. The Euclidean metric is

$$\delta s^2 = (\delta x^1)^2 + (\delta x^2)^2 + (\delta x^3)^2 \,, \tag{5.8}$$

so the Euclidean metric tensor is $g_{ij} = \delta_{ij}$, or as a matrix,

$$g_{ij} = \begin{pmatrix} 1 & 0 & 0 \\ 0 & 1 & 0 \\ 0 & 0 & 1 \end{pmatrix} \,. \tag{5.9}$$

Now change coordinates to spherical polars, $y^1 = r$, $y^2 = \vartheta$, $y^3 = \varphi$. The formulae for the coordinate changes are

$$x^1 = r \sin \vartheta \cos \varphi, \quad x^2 = r \sin \vartheta \sin \varphi, \quad x^3 = r \cos \vartheta \,. \tag{5.10}$$

The infinitesimal coordinate shifts are found by partial differentiation, giving

$$
\begin{aligned}
\delta x^1 &= \delta r \sin \vartheta \cos \varphi + r \cos \vartheta \, \delta\vartheta \cos \varphi - r \sin \vartheta \sin \varphi \, \delta\varphi \,, \\
\delta x^2 &= \delta r \sin \vartheta \sin \varphi + r \cos \vartheta \, \delta\vartheta \sin \varphi + r \sin \vartheta \cos \varphi \, \delta\varphi \,, \\
\delta x^3 &= \delta r \cos \vartheta - r \sin \vartheta \, \delta\vartheta \,,
\end{aligned}
\tag{5.11}
$$

and substituting these expressions into (5.8) we find the fairly simple metric

$$\delta s^2 = \delta r^2 + r^2 \delta \vartheta^2 + r^2 \sin^2 \vartheta \, \delta \varphi^2 \,. \tag{5.12}$$

Spherical polar coordinates are rather special, in that no cross-terms like $\delta r \delta \vartheta$ occur here. The metric tensor in spherical polars is found by matching expression (5.12) to the general definition of a metric given in equation (5.5). It has components

$$g_{rr} = 1, \quad g_{\vartheta\vartheta} = r^2, \quad g_{\varphi\varphi} = r^2 \sin^2 \vartheta, \quad g_{r\vartheta} = g_{r\varphi} = g_{\vartheta\varphi} = 0, \tag{5.13}$$

which combine to give the matrix

$$g_{ij}(r, \vartheta, \varphi) = \begin{pmatrix} 1 & 0 & 0 \\ 0 & r^2 & 0 \\ 0 & 0 & r^2 \sin^2 \vartheta \end{pmatrix}. \tag{5.14}$$

This has non-trivial functional dependence on the coordinates; nevertheless, the geometry is still flat Euclidean 3-space. (Note that no summation is implied by the repeated indices in equation (5.13). The indices simply represent labels for the components of the metric tensor.)

We can now restrict attention to the surface $r = a$, which is a 2-sphere of radius a, and therefore not flat. r is constant, so the term involving δr in the metric can be dropped. There remain the angular coordinates ϑ and φ, and (5.12) reduces to the sphere metric

$$\delta s^2 = a^2 (\delta \vartheta^2 + \sin^2 \vartheta \, \delta \varphi^2) \,. \tag{5.15}$$

This is a 2-dimensional Riemannian geometry, with metric tensor components

$$g_{\vartheta\vartheta} = a^2, \quad g_{\varphi\varphi} = a^2 \sin^2 \vartheta, \quad g_{\vartheta\varphi} = 0, \tag{5.16}$$

or as a matrix,

$$g_{ij} = \begin{pmatrix} a^2 & 0 \\ 0 & a^2 \sin^2 \vartheta \end{pmatrix}. \tag{5.17}$$

The inverse matrix has components

$$g^{\vartheta\vartheta} = \frac{1}{a^2}, \quad g^{\varphi\varphi} = \frac{1}{a^2 \sin^2 \vartheta}, \quad g^{\vartheta\varphi} = 0. \tag{5.18}$$

The 2-sphere metric looks different if we change coordinates. Instead of the angular coordinate ϑ, let us use the distance ρ from the vertical axis, as illustrated in Figure 5.6, together with the azimuthal angle φ. Then

$$\rho = a \sin \vartheta \quad \text{and} \quad \delta \rho = a \cos \vartheta \, \delta \vartheta \,, \tag{5.19}$$

which can be rearranged to give

$$a^2 \delta \vartheta^2 = \frac{\delta \rho^2}{1 - \frac{\rho^2}{a^2}} = \frac{\delta \rho^2}{1 - K \rho^2} \quad \text{and} \quad a^2 \sin^2 \vartheta \, \delta \varphi^2 = \rho^2 \delta \varphi^2 \,, \tag{5.20}$$

where $K = \frac{1}{a^2}$ is the Gaussian curvature. Therefore, in these coordinates the metric (5.15) becomes

$$\delta s^2 = \frac{\delta \rho^2}{1 - K \rho^2} + \rho^2 \, \delta \varphi^2 \,. \tag{5.21}$$

This formula is singular at $\rho = a$, so strictly speaking it is only valid on the upper hemisphere.

Another interesting change of coordinates on the sphere is

$$x = 2 \tan \frac{\vartheta}{2} \cos \varphi, \quad y = 2 \tan \frac{\vartheta}{2} \sin \varphi. \tag{5.22}$$

The inverse of this transformation is

$$\vartheta = 2 \tan^{-1} \frac{1}{2} (x^2 + y^2)^{\frac{1}{2}}, \quad \varphi = \tan^{-1} \left(\frac{y}{x} \right). \tag{5.23}$$

Differentiating, we find

$$\delta\vartheta = \frac{x\delta x + y\delta y}{\left(1 + \frac{1}{4}(x^2 + y^2) \right) (x^2 + y^2)^{\frac{1}{2}}}, \quad \delta\varphi = \frac{-y\delta x + x\delta y}{x^2 + y^2}. \tag{5.24}$$

We also require the trigonometric identity

$$\sin\vartheta = \frac{2\tan\frac{\vartheta}{2}}{1 + \tan^2\frac{\vartheta}{2}} = \frac{(x^2 + y^2)^{\frac{1}{2}}}{1 + \frac{1}{4}(x^2 + y^2)}. \tag{5.25}$$

When we substitute these quantities into equation (5.15), the cross-terms cancel, and the metric on the sphere in the new coordinates simplifies to

$$\delta s^2 = a^2 \frac{\delta x^2 + \delta y^2}{\left(1 + \frac{1}{4}(x^2 + y^2) \right)^2}. \tag{5.26}$$

This is the planar Euclidean metric $\delta x^2 + \delta y^2$ multiplied by the non-constant function

$$\Omega(x, y) = \frac{a^2}{\left(1 + \frac{1}{4}(x^2 + y^2) \right)^2}. \tag{5.27}$$

A metric that differs from a Euclidean metric by a non-constant, positive factor Ω is called *conformally flat*. The effect of the conformal factor Ω is to rescale distances, but not to change angles. The coordinates x, y range over the entire plane, giving the metric on the entire sphere with the exception of the South Pole, where $\tan\frac{\vartheta}{2}$ is infinite.

The metric on the hyperbolic plane is the hyperbolic counterpart of (5.15),

$$\delta s^2 = a^2(\delta\vartheta^2 + \sinh^2\vartheta\,\delta\varphi^2). \tag{5.28}$$

By a change of coordinates one obtains the metric in the form (5.21), but here $K = -\frac{1}{a^2}$, and ρ has the range $0 \leq \rho < \infty$. The coordinate transformation analogous to (5.22),

$$x = 2\tanh\frac{\vartheta}{2}\cos\varphi, \quad y = 2\tanh\frac{\vartheta}{2}\sin\varphi, \tag{5.29}$$

applied to the metric (5.28), produces the result

$$\delta s^2 = a^2 \frac{\delta x^2 + \delta y^2}{\left(1 - \frac{1}{4}(x^2 + y^2) \right)^2}, \tag{5.30}$$

so, like the sphere, the hyperbolic plane is conformally flat. The coordinates x, y must now be restricted to the interior of the disc $x^2 + y^2 < 4$. This description of the hyperbolic

Fig. 5.9 An infinite tessellation of the Poincaré disc by equilateral triangles whose angles are $\frac{\pi}{4}$.

plane was discovered by Henri Poincaré, and is called the Poincaré disc model. It makes the hyperbolic plane relatively easy to visualize. It can be shown that complete geodesics are segments of circles that cut the boundary orthogonally, and that the true distance to the boundary is infinite. Figure 5.9 shows a tessellation of the hyperbolic plane by equilateral triangles of equal (hyperbolic) size, whose edges are geodesic segments. The diagram does not show the true lengths of the edges, but the angles are correct, as the Poincaré disc model is conformally correct. One can see that the angle sum of each equilateral triangle is less than π, as eight triangles meet at a vertex, so each angle is $\frac{\pi}{4}$. One can also easily see the defining, non-Euclidean nature of the hyperbolic plane. Figure 5.10 shows that there are *infinitely many* geodesics through P that do not intersect the geodesic L. In other words, there are infinitely many 'lines' through P 'parallel' to L. What is not obvious from the diagram is that hyperbolic geometry is uniform and that no point is special.

One property of the metric (5.30) is that in the vicinity of the origin, it approximates to the flat metric $a^2(\delta x^2 + \delta y^2)$. If we use rescaled coordinates $X = ax, Y = ay$, then $\delta s^2 \simeq \delta X^2 + \delta Y^2$ with only quadratic corrections in X and Y, so the metric tensor is $g_{ij} = \delta_{ij}$ $(i, j = 1, 2)$ at the origin, and the first partial derivatives of g_{ij} are zero there too. The metric of the sphere (5.26) has the same property. This is a general feature of Riemannian geometry. In the vicinity of any point P, one can find a coordinate system such that the metric tensor is $g_{ij} = \delta_{ij}$, with only *quadratic* corrections. In other words, all first derivatives of the metric tensor are zero at P. A coordinate system adapted to P in this way is called a *(Riemann) normal coordinate* system. But even in normal coordinates, the second partial derivatives of the metric do not generally vanish. They are closely related to the curvature of the space at P, as we will see.

5.5 Tensors

The natural way to represent objects with a geometrical significance defined locally at a point P is in terms of vectors or more generally *tensors*. Tensors have an innate structure irrespective of the coordinate system. The components of a tensor may take different values in different coordinate systems, just as the components of a vector are different in polar and Cartesian coordinates, but this is simply due to the transformation from one coordinate

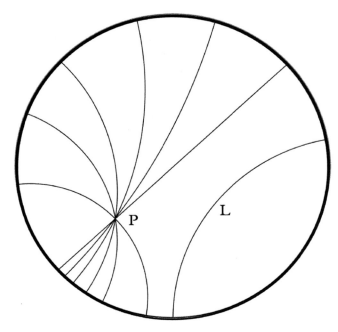

Fig. 5.10 There are infinitely many geodesics through P that do not intersect the geodesic L. Six such geodesics are shown.

system to the other. By extension, the only geometrically meaningful way to express the relationship between objects locally is as a tensor equation. The equations of physics are tensor equations, and are independent of the coordinate system, so if a tensor equation holds in one system of coordinates, it holds in all others. This will prove to be very convenient in what follows.

The basic example of a tensor is a vector V^i. A geometrical object that we can construct from V^i (at a point P, and using coordinates y^i) is the differential operator

$$V^i \frac{\partial}{\partial y^i} , \qquad (5.31)$$

and we regard this as coordinate invariant. So if we change coordinates to z^i, then the vector has new components $\tilde{V}^i$, and

$$\tilde{V}^i \frac{\partial}{\partial z^i} = V^i \frac{\partial}{\partial y^i} . \qquad (5.32)$$

Now, the coordinates z^i are some functions of the coordinates y^j, and this relationship should be invertible, at least locally, so y^j are functions of the coordinates z^i. We can differentiate and find the 3×3 *Jacobian matrix* of partial derivatives $\frac{\partial z^i}{\partial y^j}$. Its entries can be considered as either functions of the old or new coordinates. Then, either formally or by using the chain rule, we see from equation (5.32) that

$$\tilde{V}^i = V^j \frac{\partial z^i}{\partial y^j} . \qquad (5.33)$$

This is the rule for how the components of a vector change under a change of coordinates. The alternative Jacobian matrix $\frac{\partial y^j}{\partial z^i}$ is also useful, and is just the inverse of the original one.

In addition to the vector V^i there is a notion of a covector U_i, with a lower index. By definition, U_i transforms the opposite way to V^i under a coordinate transformation, with the inverse Jacobian matrix. An example of a covector is the gradient of a scalar field,

$$U_i = \frac{\partial \psi}{\partial y^i} \,. \tag{5.34}$$

Using the coordinates z^i, one finds that

$$\tilde{U}_i = \frac{\partial \psi}{\partial z^i} = \frac{\partial \psi}{\partial y^j} \frac{\partial y^j}{\partial z^i} = U_j \frac{\partial y^j}{\partial z^i} \,, \tag{5.35}$$

confirming the opposite transformation rule.

A tensor may have several indices, either upper or lower. For example, $W^{ij}{}_k$ is a 3-index tensor (a tensor of rank 3) with two upper and one lower index, and in three dimensions it has 27 components. Under a coordinate change, there is a Jacobian factor for each index. A tensor equation equates two tensors of the same type (or equivalently, equates a tensor to zero). Under a coordinate change, both sides acquire these Jacobian factors in the same way, so if the equation is satisfied in one coordinate system, it is satisfied in all.

There are some useful constructions involving tensors. A pair of tensors can be multiplied component by component, so, for example, $W^{ij}{}_k U_l$ is a 4-index tensor. Another operation, that of *contracting indices*, produces a tensor with fewer indices. Here, one takes a pair of tensor indices, one upper and one lower, sets them equal and sums the terms. The result may be represented as a tensor with the contracted indices removed. For example, one can contract the indices k and j in $W^{ij}{}_k$ and produce the vector

$$V^i = \sum_{j=1}^{3} W^{ij}{}_j \,. \tag{5.36}$$

(Using the summation convention, this is written as $V^i = W^{ij}{}_j$.) Under coordinate changes, V^i transforms as a vector because two Jacobian factors cancel out when indices are contracted in this way. Another example of index contraction is the product of a covector U_i with a vector V^i (summed over i) giving the scalar $\phi = U_i V^i$, which is unchanged under a coordinate transformation. This is the analogue of the dot product of two vectors, but in a general Riemannian space.

The metric tensor and its inverse, together with index contraction, can also be used to manipulate tensors. From a vector V^i one can create a covector

$$V_i = g_{ij} V^j \,. \tag{5.37}$$

This operation is called index lowering. Similarly, one can raise an index on a covector U_i to produce a vector $U^i = g^{ij} U_j$. Lowering and raising indices are inverse operations. The following quantities are all the same:

$$\phi = U_i V^i = g_{ij} U^j V^i = g^{ij} U_i V_j \,. \tag{5.38}$$

On a general tensor one may raise or lower any of the indices. A tensor with its indices raised or lowered has different component values (unless the metric is Euclidean) but carries essentially the same geometrical information.

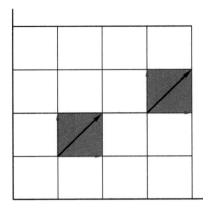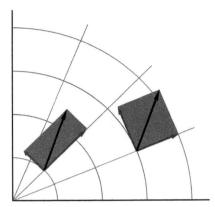

Fig. 5.11 Left: The translation of a vector in Cartesian coordinates does not involve a change of coordinate basis. Right: The translation of a vector in polar coordinates requires a change of basis.

5.5.1 Covariant derivatives and Christoffel symbols

Recall that the fundamental feature of Riemannian geometry is that locally the geometry is Euclidean. In the vicinity of a point there is a normal coordinate system such that the metric tensor is the Kronecker delta, and all first derivatives of the metric tensor vanish. The last property is just as important as the first, as it allows many formulae to be greatly simplified. If we find an equation that holds locally in normal coordinates, we can usually deduce an equivalent tensor equation, and can then be sure that it will hold in any other system of coordinates. This is a powerful method for finding tensor equations.

Whereas a tensor represents a geometrical object at a point in space, a tensor field represents a geometrical object throughout space. To understand how vector and tensor fields vary from point to point, we need to determine how differentiation works in curved space for such fields.

Consider a scalar field in flat, Euclidean space. We can calculate how it varies in each spatial direction by taking its gradient, and in this way generate a (co)vector field. Using Cartesian coordinates, we can perform the same gradient operation on a vector or covector field and generate a tensor field of rank 2. However, in curved space, or even in flat space using general coordinates, differentiation is not tensorial, because it involves taking the difference between tensors at neighbouring points and the coordinate basis at the second point may differ from that at the first point. For example, this is true in the plane in polar coordinates, as illustrated in Figure 5.11. On the left, the figure shows a vector translated between two points in Cartesian coordinates. The same basis is used to decompose the vector at both points, and we see that the vector's components are the same, so infinitesimally, the gradient is zero. On the right, the figure shows the translation of the vector in polar coordinates. The vector decomposes rather differently into radial and angular components at the two points, so naïvely it has a non-zero gradient, but this apparent spatial variation in the vector is simply due to the coordinate system.

We need to find a derivative operator that works covariantly in flat or curved space, such that the covariant derivative of a vector is a tensor of rank 2 in any coordinate system and, in general, the covariant derivative of a tensor of rank n is a tensor of rank $n + 1$. Covariant means to transform the same way, so the basic property of the covariant derivative of a tensor of rank n is that it should transform under coordinate changes in the same way as a tensor of rank $n + 1$.

This requirement is satisfied as follows. At a point where the first derivatives of the metric tensor are zero, the covariant derivative is the usual gradient, whose components are just partial derivatives. If the derivatives of the metric are not zero, then the covariant derivative has a correction term which arises from the change of coordinates needed to make the metric tensor's derivatives vanish locally. One can find this correction explicitly by considering the necessary coordinate transformation; however there is a better way which avoids changing coordinates. Not surprisingly, the correction term involves the derivatives of the metric tensor, as we will see.

The covariant derivative $\frac{D}{Dy^j}$ of a vector field V^i is

$$\frac{DV^i}{Dy^j} = \frac{\partial V^i}{\partial y^j} + \Gamma^i{}_{jk} V^k \,, \tag{5.39}$$

where the second term is the correction term. The *Christoffel symbols* $\Gamma^i{}_{jk}$ have been introduced to compensate for the lack of covariance of the partial derivative $\frac{\partial V^i}{\partial y^j}$. There are three indices, and a sum over k, in order to balance the indices in the equation. As yet, we have not determined $\Gamma^i{}_{jk}$, but will do so shortly.

The covariant derivative of a scalar field is the standard partial derivative, and if we demand that the covariant derivative of the scalar $\phi = U_i V^i$ obeys the Leibniz rule, then the covariant derivative of a covector field U_i must be

$$\frac{DU_i}{Dy^j} = \frac{\partial U_i}{\partial y^j} - \Gamma^k{}_{ji} U_k \,. \tag{5.40}$$

When the covariant derivative acts on the scalar $\phi = U_i V^i$ the Christoffel symbols cancel, as the following calculation shows:

$$
\begin{aligned}
\frac{D(U_i V^i)}{Dy^j} &= \frac{DU_i}{Dy^j} V^i + U_i \frac{DV^i}{Dy^j} \\
&= \frac{\partial U_i}{\partial y^j} V^i - \Gamma^k{}_{ji} U_k V^i + U_i \frac{\partial V^i}{\partial y^j} + U_i \Gamma^i{}_{jk} V^k \\
&= \frac{\partial U_i}{\partial y^j} V^i + U_i \frac{\partial V^i}{\partial y^j} = \frac{\partial(U_i V^i)}{\partial y^j} \,.
\end{aligned}
\tag{5.41}
$$

(Repeated indices are summed over. We can change the label used for such indices without affecting an expression, so in the final term in the second line above, we exchanged index labels i and k. It is then clear that the Γ terms cancel.)

In a similar way, we can extend the operation of the covariant derivative to tensors of higher rank by again demanding that it behaves as a Leibnizian derivative. For example, the covariant derivative of a tensor W_{ij} (which could be the outer product of two covectors, $U_i V_j$) is

$$\frac{DW_{ij}}{Dy^k} = \frac{\partial W_{ij}}{\partial y^k} - \Gamma^l{}_{ki} W_{lj} - \Gamma^l{}_{kj} W_{il} \,, \tag{5.42}$$

with a Γ term for each tensor index. In particular, the covariant derivative of the metric tensor is

$$\frac{Dg_{ij}}{Dy^k} = \frac{\partial g_{ij}}{\partial y^k} - \Gamma^l_{\ ki}g_{lj} - \Gamma^l_{\ kj}g_{il}\,. \tag{5.43}$$

As mentioned previously, in a system of normal coordinates in the vicinity of a point, the metric tensor g_{ij} is equal to the Kronecker delta δ_{ij}, and equally importantly, its derivatives are zero, so

$$\frac{\partial g_{ij}}{\partial y^k} = 0\,. \tag{5.44}$$

This is not a tensor equation, but the covariant derivative must reduce to this in normal coordinates, so we know that the following is an equivalent tensor equation:

$$\frac{Dg_{ij}}{Dy^k} = 0\,. \tag{5.45}$$

This covariant equation is true in one coordinate system (normal coordinates), so it must be true in all coordinate systems, and this is almost enough to determine $\Gamma^i_{\ jk}$.

   For convenience, we will now condense the notation by using commas to indicate partial derivatives, so '$,_i$' signifies differentiation with respect to the coordinate y^i, and '$,_{ij}$' signifies double differentiation with respect to both y^i and y^j. In this new notation

$$\frac{Dg_{ij}}{Dy^k} = g_{ij,k} - \Gamma^l_{\ ki}g_{lj} - \Gamma^l_{\ kj}g_{il} = 0\,. \tag{5.46}$$

Therefore

$$g_{ij,k} = \Gamma^l_{\ ki}g_{lj} + \Gamma^l_{\ kj}g_{il}\,, \tag{5.47}$$

and, permuting indices,

$$g_{ik,j} = \Gamma^l_{\ ji}g_{lk} + \Gamma^l_{\ jk}g_{il}\,, \tag{5.48}$$

$$g_{jk,i} = \Gamma^l_{\ ij}g_{lk} + \Gamma^l_{\ ik}g_{jl}\,. \tag{5.49}$$

We now demand, as a final condition, that $\Gamma^i_{\ jk}$ is symmetric on its two lower indices. Then, adding equations (5.47) and (5.48) and subtracting equation (5.49), we find

$$g_{ij,k} + g_{ik,j} - g_{jk,i} = 2\Gamma^l_{\ jk}g_{il}\,, \tag{5.50}$$

where we have used the assumed symmetry in jk of $\Gamma^i_{\ jk}$. Multiplying by the inverse metric tensor g^{im} and summing over i, we obtain the final expression

$$\Gamma^m_{\ jk} = \frac{1}{2}g^{im}(g_{ij,k} + g_{ik,j} - g_{jk,i})\,, \tag{5.51}$$

determining the Christoffel symbols in terms of derivatives of the metric tensor.

   The Christoffel symbols play a very important role in Riemannian geometry, although they are not the components of a tensor. In a normal coordinate system at P, the derivatives of the metric tensor are zero, so the Christoffel symbols are zero at P too. If they were tensor components, then being zero in one coordinate system would make them zero in all coordinate systems.

5.5.2   Christoffel symbols in plane polar coordinates

To gain some familiarity with Christoffel symbols, let us work them out in the plane, using polar coordinates. The metric is $\delta s^2 = \delta r^2 + r^2 \delta \vartheta^2$, so the metric tensor components are

$$g_{rr} = 1\,, \quad g_{\vartheta\vartheta} = r^2\,, \quad g_{r\vartheta} = 0\,, \tag{5.52}$$

and the only non-zero derivative of the metric tensor is $g_{\vartheta\vartheta,r} = 2r$. In two dimensions there are generally six Christoffel symbols (taking account the symmetry on the lower indices), but here the only non-zero Christoffel symbols are

$$\Gamma^{\vartheta}{}_{r\vartheta} = \Gamma^{\vartheta}{}_{\vartheta r} = \frac{1}{2} g^{\vartheta\vartheta}(g_{\vartheta r,\vartheta} + g_{\vartheta\vartheta,r} - g_{r\vartheta,\vartheta}) = \frac{1}{2}\left(\frac{1}{r^2}\right)2r = \frac{1}{r} \tag{5.53}$$

and

$$\Gamma^{r}{}_{\vartheta\vartheta} = \frac{1}{2} g^{rr}(g_{r\vartheta,\vartheta} + g_{r\vartheta,\vartheta} - g_{\vartheta\vartheta,r}) = \frac{1}{2}(-2r) = -r\,. \tag{5.54}$$

Because the plane is flat, the Christoffel symbols are zero in Cartesian coordinates, but in polar coordinates they are non-zero. This is necessary in order to compensate for the change in the coordinate basis as a vector moves around, as illustrated in Figure 5.11.

5.6   The Riemann Curvature Tensor

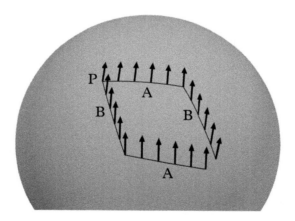

Fig. 5.12 Comparison of a vector transported along two paths in curved space.

Riemann found a geometrical object now known as the *Riemann curvature tensor* (or Riemann tensor, for short) which captures all the information about the curvature at each point, in any number of dimensions. Figure 5.12 illustrates how it is determined. A tangent vector is transported a short distance from P in direction A followed by a short distance in direction B. It is then compared to the same vector transported a short distance in direction B followed by a short distance in direction A. In flat space there would be no difference; the final position and orientation of the vector would be the same either way. However, as illustrated in Figure 5.12, in curved space there is a difference and it depends on the curvature. This is expressed algebraically by the fact that covariant derivatives do not commute in curved space—their ordering matters. The Riemann tensor $R^i{}_{jkl}$ is defined as the *commutator* of covariant derivatives acting on any vector field V^i,

$$R^i{}_{jkl} V^j = \left(\frac{D}{Dy^k} \frac{D}{Dy^l} - \frac{D}{Dy^l} \frac{D}{Dy^k} \right) V^i. \tag{5.55}$$

The commutator is the difference between the double covariant derivative evaluated in one order, and in the other order. (The commutator of standard partial derivatives is zero, because of the symmetry of mixed partial derivatives.)

We can now work out an explicit formula for $R^i{}_{jkl}$. The double covariant derivative of the vector field V^i is the covariant derivative of the single covariant derivative of V^i, the tensor of rank 2 with one upper and one lower index given in equation (5.39). We find (using the condensed notation for partial derivatives)

$$
\begin{aligned}
\frac{D}{Dy^k} \frac{D}{Dy^l} V^i &= \frac{D}{Dy^k} \left(V^i{}_{,l} + \Gamma^i{}_{lj} V^j \right) \\
&= V^i{}_{,lk} + \Gamma^i{}_{lj,k} V^j + \Gamma^i{}_{lj} V^j{}_{,k} + \Gamma^i{}_{kj} V^j{}_{,l} + \Gamma^i{}_{km} \Gamma^m{}_{lj} V^j \\
&\quad - \Gamma^m{}_{kl} V^i{}_{,m} - \Gamma^m{}_{kl} \Gamma^i{}_{mj} V^j.
\end{aligned}
\tag{5.56}
$$

The first, sixth and seventh terms on the right-hand side of this equation are symmetric under the exchange of indices k and l, and the sum of the third and fourth terms is similarly symmetric. (This includes all the terms containing derivatives of V^i.) The only terms that are not symmetric are $\Gamma^i{}_{lj,k} V^j$ and $\Gamma^i{}_{km} \Gamma^m{}_{lj} V^j$. If we now subtract a similar expression for the double covariant derivative in the other order, then all the symmetric terms cancel and we obtain the formula for the Riemann tensor

$$R^i{}_{jkl} = \Gamma^i{}_{lj,k} - \Gamma^i{}_{kj,l} + \Gamma^i{}_{km} \Gamma^m{}_{lj} - \Gamma^i{}_{lm} \Gamma^m{}_{kj}. \tag{5.57}$$

By construction, this tensor is antisymmetric in its indices k and l.

The Riemann curvature tensor is the higher-dimensional generalization of the Gaussian curvature. It involves first derivatives of the Christoffel symbols, and so depends on second derivatives of the metric tensor g_{ij}. We can lower the first index by multiplying by the metric tensor and contracting indices, obtaining the more symmetrical Riemann tensor

$$R_{ijkl} = g_{im} R^m{}_{jkl}. \tag{5.58}$$

Although not obvious, R_{ijkl} has a number of further symmetries. It is antisymmetric in its indices ij, as well as in kl, and it is symmetric under the interchange of the pair of indices ij with the pair kl. It also has a cyclic symmetry on its final three indices

$$R_{ijkl} + R_{iklj} + R_{iljk} = 0, \tag{5.59}$$

known as the *first Bianchi identity*. These symmetries can be shown most easily by expressing R_{ijkl} entirely in terms of the metric tensor and its second derivatives, working in a system of normal coordinates.

For a space of constant (uniform) curvature in all directions, the Riemann tensor simplifies in any coordinate system to

$$R_{ijkl} = C(g_{ik} g_{jl} - g_{il} g_{jk}), \tag{5.60}$$

with C a constant. Notice that this expression is consistent with the index symmetries.

5.6.1 Riemann curvature in plane polar coordinates

We have seen in section 5.5.2 that even in the Euclidean plane, some Christoffel symbols are non-zero in polar coordinates. It is instructive now to take a look at the Riemann curvature tensor in polar coordinates.

The only non-zero derivatives of the Christoffel symbols (5.53) and (5.54) are

$$\Gamma^{\vartheta}{}_{r\vartheta,r} = \Gamma^{\vartheta}{}_{\vartheta r,r} = -\frac{1}{r^2}, \quad \Gamma^{r}{}_{\vartheta\vartheta,r} = -1. \tag{5.61}$$

The Riemann tensor is given by equation (5.57), and all its components are identically zero. For example,

$$\begin{aligned}
R^{r}{}_{\vartheta\vartheta r} &= \Gamma^{r}{}_{r\vartheta,\vartheta} - \Gamma^{r}{}_{\vartheta\vartheta,r} + \Gamma^{r}{}_{\vartheta r}\Gamma^{r}{}_{r\vartheta} + \Gamma^{r}{}_{\vartheta\vartheta}\Gamma^{\vartheta}{}_{r\vartheta} - \Gamma^{r}{}_{rr}\Gamma^{r}{}_{\vartheta\vartheta} - \Gamma^{r}{}_{r\vartheta}\Gamma^{\vartheta}{}_{\vartheta\vartheta} \\
&= 0 + 1 + 0 - r\left(\frac{1}{r}\right) - 0 - 0 = 0.
\end{aligned} \tag{5.62}$$

This confirms what we would expect in a flat space.

5.6.2 Riemann curvature on a sphere

The 2-sphere is an example of a curved space where a non-zero Riemann tensor can readily be calculated. In angular coordinates, the metric tensor components are

$$g_{\vartheta\vartheta} = a^2, \quad g_{\varphi\varphi} = a^2 \sin^2\vartheta, \quad g_{\vartheta\varphi} = 0, \tag{5.63}$$

and their only non-zero derivative is

$$g_{\varphi\varphi,\vartheta} = 2a^2 \sin\vartheta \cos\vartheta. \tag{5.64}$$

The only non-zero Christoffel symbols are therefore

$$\begin{aligned}
\Gamma^{\varphi}{}_{\vartheta\varphi} = \Gamma^{\varphi}{}_{\varphi\vartheta} &= \frac{1}{2}g^{\varphi\varphi}(g_{\varphi\vartheta,\varphi} + g_{\varphi\varphi,\vartheta} - g_{\vartheta\varphi,\varphi}) \\
&= \frac{1}{2}\left(\frac{1}{a^2 \sin^2\vartheta}\right)(2a^2 \sin\vartheta \cos\vartheta) \\
&= \frac{\cos\vartheta}{\sin\vartheta}
\end{aligned} \tag{5.65}$$

and

$$\Gamma^{\vartheta}{}_{\varphi\varphi} = \frac{1}{2}g^{\vartheta\vartheta}(g_{\vartheta\varphi,\varphi} + g_{\vartheta\varphi,\varphi} - g_{\varphi\varphi,\vartheta}) = -\sin\vartheta \cos\vartheta. \tag{5.66}$$

So, omitting terms that are zero,

$$\begin{aligned}
R^{\vartheta}{}_{\varphi\vartheta\varphi} &= \Gamma^{\vartheta}{}_{\varphi\varphi,\vartheta} - \Gamma^{\vartheta}{}_{\varphi\varphi}\Gamma^{\varphi}{}_{\vartheta\varphi} \\
&= -\cos^2\vartheta + \sin^2\vartheta + \sin\vartheta \cos\vartheta\left(\frac{\cos\vartheta}{\sin\vartheta}\right) \\
&= \sin^2\vartheta
\end{aligned} \tag{5.67}$$

and, lowering the first index,

$$R_{\vartheta\varphi\vartheta\varphi} = g_{\vartheta\vartheta}R^{\vartheta}{}_{\varphi\vartheta\varphi} = a^2 \sin^2\vartheta. \tag{5.68}$$

This is essentially the only component of the Riemann tensor on the sphere, as all the other non-zero components are related to this one by index symmetries.

The 2-sphere has constant curvature, because we can write the Riemann tensor in the form of equation (5.60) as

$$R_{\vartheta\varphi\vartheta\varphi} = \frac{1}{a^2}(g_{\vartheta\vartheta}g_{\varphi\varphi} - g_{\vartheta\varphi}g_{\varphi\vartheta}) = a^2\sin^2\vartheta\,. \tag{5.69}$$

The constant C in (5.60) can be identified in two dimensions as the Gaussian curvature, $K = \frac{1}{a^2}$.

5.6.3   The 3-sphere

Another example of a Riemannian space with constant curvature is the 3-dimensional sphere, or 3-sphere for short. This is the sphere of fixed radius a in 4-dimensional Euclidean space. It has important applications in general relativity and cosmology.

Euclidean 4-space has Cartesian coordinates (x^1, x^2, x^3, x^4) and metric $\delta s^2 = (\delta x^1)^2 + (\delta x^2)^2 + (\delta x^3)^2 + (\delta x^4)^2$. Conversion to polar coordinates is through the formulae

$$\begin{aligned} x^1 &= R\sin\chi\sin\vartheta\cos\varphi\,, \quad x^2 = R\sin\chi\sin\vartheta\sin\varphi\,, \\ x^3 &= R\sin\chi\cos\vartheta\,, \quad x^4 = R\cos\chi\,. \end{aligned} \tag{5.70}$$

Then a similar calculation to the coordinate change (5.11) leads to the metric in polar coordinates

$$\delta s^2 = \delta R^2 + R^2\delta\chi^2 + R^2\sin^2\chi(\delta\vartheta^2 + \sin^2\vartheta\,\delta\varphi^2)\,. \tag{5.71}$$

By fixing $R = a$, we obtain the metric on the 3-sphere

$$\delta s^2 = a^2(\delta\chi^2 + \sin^2\chi(\delta\vartheta^2 + \sin^2\vartheta\,\delta\varphi^2))\,. \tag{5.72}$$

An alternative formula for the 3-sphere metric is obtained by the change of coordinate $r = a\sin\chi$. The metric (5.72) becomes

$$\delta s^2 = \frac{\delta r^2}{1 - Kr^2} + r^2(\delta\vartheta^2 + \sin^2\vartheta\,\delta\varphi^2)\,, \tag{5.73}$$

where $K = \frac{1}{a^2}$. This is the counterpart of the 2-sphere metric (5.21). Any equatorial slice of the 3-sphere, for example the slice $\vartheta = \frac{\pi}{2}$, is a 2-sphere with Gaussian curvature K. By taking the limit $K \to 0$, we recover the Euclidean 3-space metric (5.12).

A final version of the 3-sphere metric, which we will not explicitly derive, but which is equivalent to the 2-sphere metric (5.26), is

$$\delta s^2 = a^2\frac{\delta x^2 + \delta y^2 + \delta z^2}{\left(1 + \frac{1}{4}(x^2 + y^2 + z^2)\right)^2} = a^2\frac{\delta\mathbf{x}\cdot\delta\mathbf{x}}{\left(1 + \frac{1}{4}\mathbf{x}\cdot\mathbf{x}\right)^2}\,. \tag{5.74}$$

This formula shows that the 3-sphere is a conformally flat space. The whole sphere except one point is covered as $\mathbf{x}$ ranges over all of (standard) 3-space.

5.7   The Geodesic Equation

Armed with some of the machinery developed by Riemann and his followers, we can now take a look at how particles move through curved space. First we need to understand geometrical paths in a Riemannian space with coordinates y^i and metric tensor g_{ij}. The total length s of a path connecting endpoints G and H is given by the integral[2]

$$s = \int_G^H ds = \int_G^H \sqrt{g_{jk}(\mathbf{y})\, dy^j dy^k}\,. \tag{5.75}$$

If we introduce a parameter λ along the path, then the path is given by a vector function $\mathbf{y}(\lambda) = (y^1(\lambda), y^2(\lambda), y^3(\lambda))$, and we can turn the path length into a regular integral

$$s = \int_{\lambda(G)}^{\lambda(H)} \sqrt{g_{jk}(\mathbf{y}(\lambda)) \frac{dy^j}{d\lambda} \frac{dy^k}{d\lambda}}\, d\lambda\,, \tag{5.76}$$

which is actually independent of the choice of parametrization. Of particular significance is the geodesic or shortest path connecting G to H. This is obtained by minimizing s.

It is natural to expect that just as freely moving particles in flat space move along straight lines, so freely moving (inertial) particles in a curved space follow geodesics. Therefore, we can find the equation for a geodesic by minimizing the action for a particle, rather than minimizing the path length s.

Consider a particle of mass m moving from G to H along the path $\mathbf{y}(t)$, where the parameter t is now the time. The particle's speed is

$$\frac{ds}{dt} = \sqrt{g_{jk}(\mathbf{y}(t)) \frac{dy^j}{dt} \frac{dy^k}{dt}}\,, \tag{5.77}$$

and its kinetic energy is half the mass times the square of the speed,

$$\frac{1}{2} m \left(\frac{ds}{dt} \right)^2 = \frac{1}{2} m\, g_{jk}(\mathbf{y}(t)) \frac{dy^j}{dt} \frac{dy^k}{dt}\,. \tag{5.78}$$

We define the action of the particle to be

$$S = \int_{t_0}^{t_1} \frac{1}{2} m\, g_{jk}(\mathbf{y}(t)) \frac{dy^j}{dt} \frac{dy^k}{dt}\, dt\,, \tag{5.79}$$

by analogy with expression (2.53) for a particle moving in flat space. The integrand, the Lagrangian, is just the kinetic energy. If the particle were influenced by a potential, then there would be a further contribution to S. As usual, the principle of least action requires the true particle motion to be the path $\mathbf{y}(t)$ that minimizes S.

Both the path length and the action are geometrically meaningful in that they do not depend on the coordinate system that is used. This is because they are constructed from a basic geometrical quantity, the infinitesimal distance ds. The action S is simpler to use than the path length s, because it does not involve a square root, but unlike the path length, S does depend on the parametrization. t is not just an arbitrary parameter along the path, but the physical time.

---

[2] In integrals it makes more sense to use ds rather than δs to denote an infinitesimal element of length.

The particle's equation of motion is the Euler–Lagrange equation derived from S (obtained using the calculus of variations), and this is

$$\frac{d}{dt}\left(m\,g_{lj}(\mathbf{y})\frac{dy^j}{dt}\right) - \frac{\partial}{\partial y^l}\left(\frac{1}{2}m\,g_{jk}(\mathbf{y})\frac{dy^j}{dt}\frac{dy^k}{dt}\right) = 0. \tag{5.80}$$

(Note that m cancels out.) Expanding out the derivatives, the equation of motion becomes

$$g_{lj}\frac{d^2y^j}{dt^2} + \left(g_{lj,k} - \frac{1}{2}g_{jk,l}\right)\frac{dy^j}{dt}\frac{dy^k}{dt} = 0. \tag{5.81}$$

The first term in the bracket, which comes from the time dependence of $g_{lj}(\mathbf{y})$ through $\mathbf{y}(t)$, is not symmetric in j and k, but since it multiplies $\frac{dy^j}{dt}\frac{dy^k}{dt}$, which is symmetric, we can explicitly symmetrize this term, obtaining

$$g_{lj}\frac{d^2y^j}{dt^2} + \frac{1}{2}\left(g_{lj,k} + g_{lk,j} - g_{jk,l}\right)\frac{dy^j}{dt}\frac{dy^k}{dt} = 0. \tag{5.82}$$

Multiplying through by the inverse metric g^{il} then gives

$$\frac{d^2y^i}{dt^2} + \frac{1}{2}g^{il}\left(g_{lj,k} + g_{lk,j} - g_{jk,l}\right)\frac{dy^j}{dt}\frac{dy^k}{dt} = 0. \tag{5.83}$$

The second term involves a combination of the inverse metric and first derivatives of the metric that should be familiar. It is a Christoffel symbol, as derived in section 5.5.1, so the equation of motion has the final form

$$\frac{d^2y^i}{dt^2} + \Gamma^i{}_{jk}\frac{dy^j}{dt}\frac{dy^k}{dt} = 0. \tag{5.84}$$

This is the *geodesic equation*. Its solutions describe the motion of a particle along a geodesic.

Equation (5.84) is the analogue for motion in curved space of the equation of motion for a free particle in flat space, which in Cartesian coordinates is $\frac{d^2x^i}{dt^2} = 0$, saying that the particle has no acceleration, and moves along a straight line. If we adopt a normal coordinate system around P, then all Christoffel symbols $\Gamma^i{}_{jk}$ vanish at P, so for any geodesic motion, the acceleration $\frac{d^2y^i}{dt^2}$ at P vanishes too. By considering the whole family of coordinate changes to local normal coordinates along the trajectory of the particle, we see that there is no genuine acceleration anywhere, and therefore no genuine forces are acting. This is characteristic of geodesic motion. However, we usually wish to work with a single coordinate system in an extended region surrounding the particle trajectory, in which case there are non-zero coordinate accelerations, governed by the second term in equation (5.84). We say that these accelerations are produced by *fictitious forces*. Notice that they are quadratic in the velocity components $\frac{dy^i}{dt}$.

Geodesic motion takes place at constant speed, where the speed is given by equation (5.77). This can be understood in several ways. One can directly verify that

$$\frac{d}{dt}\left(g_{jk}(\mathbf{y}(t))\frac{dy^j}{dt}\frac{dy^k}{dt}\right) = 0, \tag{5.85}$$

using the geodesic equation (5.84) and the formula for $\Gamma^i{}_{jk}$ in terms of the metric. More simply, as argued in the previous paragraph, in normal coordinates the speed is obviously constant as the acceleration is zero. Finally, as the energy is purely kinetic, the conservation of energy is equivalent to conservation of speed.

Let λ be the length parameter along the geodesic, not an arbitrary parameter. As geodesic motion proceeds at a constant speed, λ is a constant multiple of t. Therefore the geometric version of the geodesic equation is

$$\frac{d^2 y^i}{d\lambda^2} + \Gamma^i{}_{jk} \frac{dy^j}{d\lambda} \frac{dy^k}{d\lambda} = 0, \tag{5.86}$$

combined with the auxillary expression relating length to the metric,

$$g_{jk}(\mathbf{y}) \frac{dy^j}{d\lambda} \frac{dy^k}{d\lambda} = 1. \tag{5.87}$$

5.7.1 Geodesics in plane polar coordinates

As an illustration, let us consider the geodesic equation in the plane, using polar coordinates r and ϑ.

Circles of constant r are not geodesics. A particle moving along such a circle at constant angular velocity has no coordinate accelerations, but it is subject to a genuine, radially inward force, as we saw when discussing circular orbits in section (2.6). On the other hand, geodesic motion involves coordinate accelerations. In section 5.5.2 we found the non-zero Christoffel symbols to be

$$\Gamma^r{}_{\vartheta\vartheta} = -r, \quad \Gamma^\vartheta{}_{r\vartheta} = \frac{1}{r}, \tag{5.88}$$

and if we insert these into equation (5.84), we obtain the two components of the geodesic equation,

$$\frac{d^2 r}{dt^2} - r \left(\frac{d\vartheta}{dt} \right)^2 = 0, \quad \frac{d^2 \vartheta}{dt^2} + \frac{2}{r} \frac{dr}{dt} \frac{d\vartheta}{dt} = 0. \tag{5.89}$$

The equations include coordinate accelerations due to a *centrifugal force* $mr \left(\frac{d\vartheta}{dt} \right)^2$ and a *Coriolis force* $-\frac{2m}{r} \frac{dr}{dt} \frac{d\vartheta}{dt}$. These are examples of fictitious forces. The centrifugal force is radially outward. It is not obvious, but the geodesics obtained by solving the equations are straight lines in the plane.

5.7.2 The equation of geodesic deviation

If two straight lines in the plane are parallel, they have a fixed separation. If they intersect, then their separation increases or decreases at a constant rate, as illustrated in Figure 5.13. Let λ parametrize distance along both lines, and let $\eta(\lambda)$ be the distance that separates corresponding points on the lines. Then $\frac{d\eta}{d\lambda}$ is a constant. Differentiating a second time gives

$$\frac{d^2 \eta}{d\lambda^2} = 0, \tag{5.90}$$

which is a rather trivial example of the *equation of geodesic deviation*.

Let's now determine the geodesic deviation on a sphere of radius a, where geodesics are great circles. Consider two such geodesics that intersect at the North Pole at an angle φ, as illustrated in Figure 5.14. Their separation increases until they reach the equator, then decreases until they meet again at the South Pole. It is straightforward to calculate η, the

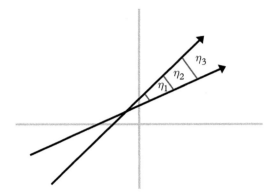

Fig. 5.13 Geodesic deviation in flat space.

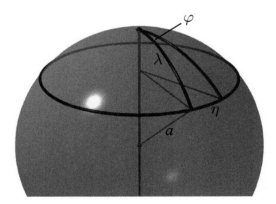

Fig. 5.14 Geodesic deviation on a sphere.

separation of the geodesics. If we again use λ to parametrize the distance along the geodesics, then

$$\eta(\lambda) = \left(a \sin \frac{\lambda}{a} \right) \varphi. \tag{5.91}$$

Differentiating twice with respect to λ, we obtain

$$\frac{d^2\eta}{d\lambda^2} + \frac{1}{a^2}\eta = 0, \tag{5.92}$$

or equivalently

$$\frac{d^2\eta}{d\lambda^2} + K\eta = 0, \tag{5.93}$$

where $K = \frac{1}{a^2}$ is the Gaussian curvature. This is the equation of geodesic deviation on the sphere.

To determine the equation of geodesic deviation on a general Riemannian space we do a similar, but less explicit calculation. We take a geodesic $\mathbf{y}(\lambda)$, and a nearby geodesic $\mathbf{y}(\lambda) + \boldsymbol{\eta}(\lambda)$ separated by a small vector $\boldsymbol{\eta}$.

These satisfy, respectively, the geodesic equation

$$\frac{d^2 y^i}{d\lambda^2} + \Gamma^i{}_{jk}(\mathbf{y})\frac{dy^j}{d\lambda}\frac{dy^k}{d\lambda} = 0 \tag{5.94}$$

and its nearby variant

$$\frac{d^2(y+\eta)^i}{d\lambda^2} + \Gamma^i{}_{jk}(\mathbf{y}+\boldsymbol{\eta})\frac{d(y+\eta)^j}{d\lambda}\frac{d(y+\eta)^k}{d\lambda} = 0\,. \tag{5.95}$$

Expanding equation (5.95) up to linear order in $\boldsymbol{\eta}$, and subtracting equation (5.94), we find the linear equation for $\boldsymbol{\eta}$,

$$\frac{d^2\eta^i}{d\lambda^2} + \Gamma^i{}_{jk,l}\,\eta^l\frac{dy^j}{d\lambda}\frac{dy^k}{d\lambda} + 2\Gamma^i{}_{jk}\frac{dy^j}{d\lambda}\frac{d\eta^k}{d\lambda} = 0\,. \tag{5.96}$$

Here we have dropped terms of quadratic and higher order in $\boldsymbol{\eta}$, and suppressed the $\mathbf{y}$-dependence of the Christoffel symbols. This equation is not obviously geometrical or covariant, but it can be cast in a more elegant form in terms of the Riemann tensor.

To show this we need a notion of the covariant derivative along a curve $\mathbf{y}(\lambda)$. For a field V^i, the covariant derivative along the curve, denoted by $\frac{DV^i}{D\lambda}$, is the projection of the covariant derivative in space in the direction of the curve,

$$\begin{aligned}
\frac{DV^i}{D\lambda} &= \frac{DV^i}{Dy^j}\frac{dy^j}{d\lambda} \\
&= \left(\frac{\partial V^i}{\partial y^j} + \Gamma^i{}_{jk}V^k\right)\frac{dy^j}{d\lambda} \\
&= \frac{dV^i}{d\lambda} + \Gamma^i{}_{jk}V^k\frac{dy^j}{d\lambda}\,.
\end{aligned} \tag{5.97}$$

The last expression involves V^i and its derivative $\frac{dV^i}{d\lambda}$ along the curve. It is therefore particularly useful for vector quantities that are only defined on a curve, and not throughout space. An example of such a vector is the velocity of a particle moving along a geodesic, $\frac{dy^i}{dt}$. The geodesic equation of motion (5.84) can be re-expressed as

$$\frac{D}{Dt}\frac{dy^i}{dt} = 0\,, \tag{5.98}$$

so not only is the speed of the particle constant but also the velocity vector is covariantly constant.

The geodesic deviation vector $\boldsymbol{\eta}$ is only defined on a geodesic. Its covariant derivative along the geodesic is

$$\frac{D\eta^i}{D\lambda} = \frac{d\eta^i}{d\lambda} + \Gamma^i{}_{jk}\eta^k\frac{dy^j}{d\lambda}\,, \tag{5.99}$$

and as this too is a vector along the geodesic, it has a further covariant derivative,

$$
\begin{aligned}
\frac{D^2\eta^i}{D\lambda^2} &= \frac{d}{d\lambda}\left(\frac{d\eta^i}{d\lambda} + \Gamma^i{}_{jk}\eta^k\frac{dy^j}{d\lambda}\right) + \Gamma^i{}_{jk}\left(\frac{d\eta^k}{d\lambda} + \Gamma^k{}_{lm}\eta^m\frac{dy^l}{d\lambda}\right)\frac{dy^j}{d\lambda} \\
&= \frac{d^2\eta^i}{d\lambda^2} + \Gamma^i{}_{jk,l}\frac{dy^l}{d\lambda}\eta^k\frac{dy^j}{d\lambda} + 2\Gamma^i{}_{jk}\frac{d\eta^k}{d\lambda}\frac{dy^j}{d\lambda} + \Gamma^i{}_{jk}\eta^k\frac{d^2y^j}{d\lambda^2} \\
&\quad + \Gamma^i{}_{jk}\Gamma^k{}_{lm}\eta^m\frac{dy^l}{d\lambda}\frac{dy^j}{d\lambda}\,.
\end{aligned}
\tag{5.100}
$$

We can now use equation (5.96) to eliminate the terms $\frac{d^2\eta^i}{d\lambda^2} + 2\Gamma^i{}_{jk}\frac{d\eta^k}{d\lambda}\frac{dy^j}{d\lambda}$, and equation (5.94) to eliminate $\frac{d^2y^j}{d\lambda^2}$. Equation (5.100) then becomes

$$
\frac{D^2\eta^i}{D\lambda^2} = (\Gamma^i{}_{lj,k} - \Gamma^i{}_{kj,l} + \Gamma^i{}_{km}\Gamma^m{}_{lj} - \Gamma^i{}_{lm}\Gamma^m{}_{kj})\frac{dy^j}{d\lambda}\frac{dy^k}{d\lambda}\eta^l\,.
\tag{5.101}
$$

The terms involving the Christoffel symbols are in the precise combination that forms the Riemann curvature tensor, so finally

$$
\frac{D^2\eta^i}{D\lambda^2} = R^i{}_{jkl}\frac{dy^j}{d\lambda}\frac{dy^k}{d\lambda}\eta^l\,.
\tag{5.102}
$$

This tensor equation, which holds in all coordinate systems, is the covariant version of the equation of geodesic deviation. It generalizes the result (5.93) on a sphere, involving the Gaussian curvature, to any higher-dimensional curved space. We will later apply this equation to study tidal forces in general relativity.

5.8 Applications

Configuration space

There are some interesting physical applications of curved Riemannian geometry that have nothing to do with Einstein's theory of curved spacetime and gravity.

At least to an extremely good approximation, space is Euclidean. When we model the motion of N particles in Euclidean space, the geometry of the $3N$ Cartesian coordinates of the particles remains Euclidean. However, the interactions between particles are sometimes so strong that the particles behave collectively as a single object, in a configuration described by a small number of *collective coordinates*. The set of all configurations forms the *configuration space* and often its geometry is curved.

A classic example is a finite-sized rigid body, composed of countless individual particles. The body can have any fixed shape. It may be free to move, like a planet through space, or it may be constrained, like a pendulum that can only swing in a given plane. Treating such a body as rigid is an approximation, valid when the frequency of its motion as a rigid body is small compared with the frequency of elastic, shape-varying vibrations. Needless to say, the forces acting must be much weaker than those that would set it vibrating or break it up.

The maximum number of collective coordinates needed to specify the configuration of a rigid body is six—three for the centre of mass location, and three for the body's orientation. The centre of mass behaves like the position of one particle, and its geometry is the flat geometry of 3-space, so let's suppose the centre of mass is fixed, and ignore it. The orientation is specified by three angles. For example, for the Earth one needs to specify the point on the celestial sphere where the axis of rotation is directed. This point is observed to be close to the Pole Star, and is parametrized by two angular coordinates ϑ, φ. One further angle ψ parametrizes the orientation of the Earth about its rotation axis; this continually increases, with a 24-hour period. The angles ϑ and φ are almost constant, but they do change slowly, over thousands of years.

These three angles ϑ, φ, ψ are called *Euler angles*, and they each have a finite range. They are collective coordinates on the orientational configuration space of the rigid body, which is a curved 3-dimensional Riemannian space, geometrically related to a 2-sphere.

By choosing the body axes carefully, we can be more explicit about the metric on the configuration space. The metric can be rather complicated, but is simpler if the body is symmetrical around an axis. The Earth, with its slightly flattened, oblate spheroid shape, offers an illuminating example. In this case, the metric has the form

$$\delta s^2 = I_1(\delta\vartheta^2 + \sin^2\vartheta\,\delta\varphi^2) + I_3(\delta\psi + \cos\vartheta\,\delta\varphi)^2 \,. \tag{5.103}$$

How have we arrived at this? It is derived by calculating the kinetic energy of the body. When the body is rotating, the three angles are time dependent. All the particles making up the body are moving, and one can calculate the linear instantaneous velocity of each one. Then, by integration, one can find the total kinetic energy. The result depends on how the particles are distributed in the body, and on their masses, but in the end it only involves two constants associated with the matter distribution, I_1 and I_3. (If the body does not have an axis of rotational symmetry, then there is a further, independent constant I_2.) These constants are called the *moments of inertia*. One finds that the total kinetic energy of the body has the form

$$K = \frac{1}{2}I_1\left(\left(\frac{d\vartheta}{dt}\right)^2 + \sin^2\vartheta\left(\frac{d\varphi}{dt}\right)^2\right) + \frac{1}{2}I_3\left(\frac{d\psi}{dt} + \cos\vartheta\frac{d\varphi}{dt}\right)^2 \,. \tag{5.104}$$

Dropping the $\frac{1}{2}$, and the time derivatives, we obtain the metric (5.103).

Calculations of moments of inertia are not hard for bodies with uniform density and some symmetry. They also simplify for thin, 2-dimensional bodies constrained to move in their own plane, where there is just one orientational angle. In all cases one can read off a Riemannian metric from the kinetic energy.

If the body rotates freely, and no forces act, then the motion of the body is geodesic motion on the configuration space with this metric. This is because the kinetic energy K is the complete Lagrangian. Solving the equations of motion for the angles is possible, but there is a simpler intermediate step, which is to solve an equation for the angular velocity about each axis. Particularly simple is motion around the axis of symmetry. Here only one angle is time-varying, and the angular velocity is constant. The general motion is not around an axis of symmetry, and the trajectory does not close up after any finite time. It is still geodesic motion, because any short segment of the trajectory is always the shortest path between the ends of that segment.

The Lagrangian of the rigid body may have a potential term V in addition to the kinetic term K. This produces a force, or more precisely a torque. The standard example is a

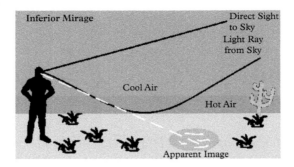

Fig. 5.15 A mirage.

spinning top with its base point fixed on a table. The top's configuration is still specified by three angles, but the height of the centre of mass is variable, so there is a gravitational potential energy depending on one of the angles.

There are extensions of these ideas to bodies that are less rigid, such as a pair of rigid bodies connected by a flexible joint. This compound object might model a molecule. Again there is a Riemannian geometry derived from the kinetic energy. The dimension of the configuration space increases if more coordinates are needed to fully specify a configuration of the system at any particular time. One advantage of this geometrical viewpoint is that the Lagrangian and the resulting dynamics are independent of the choice of coordinates.

Geometrical optics

A rather different application of geodesics and Riemannian geometry is to Fermat's principle of minimal travel time for light rays. Previously, in section 1.1.1, we considered two uniform optical media, with differing light speeds, meeting on a planar surface. Light rays are refracted when passing from one medium to the other. The refractive index $n = \frac{c}{v}$ of a medium is the ratio of the vacuum speed of light, c, to the speed of light in the medium, v. In our units $c = 1$ and $v \leq 1$, so $n \geq 1$ and is only equal to 1 in a vacuum. Suppose now that a medium has a refractive index $n(\mathbf{x})$ which varies spatially in a continuous way. Locally, the speed of light is $\frac{1}{n(\mathbf{x})}$, so the time for light to travel an infinitesimal distance ds in the vicinity of $\mathbf{x}$ is $n(\mathbf{x})ds$.

The total time that light would take to travel from A to B in the medium, along a path $\mathbf{x}(t)$, is

$$T = \int_A^B n(\mathbf{x}(t))\, ds = \int_A^B n(\mathbf{x}(t))\frac{ds}{dt}\, dt. \tag{5.105}$$

Equivalently, introducing the Euclidean metric, it is

$$T = \int_A^B n(\mathbf{x}(t))\sqrt{\delta_{ij}\frac{dx^i}{dt}\frac{dx^j}{dt}}\, dt, \tag{5.106}$$

which can be rewritten as

$$T = \int_A^B \sqrt{n^2(\mathbf{x}(t))\,\delta_{ij}\frac{dx^i}{dt}\frac{dx^j}{dt}}\, dt. \tag{5.107}$$

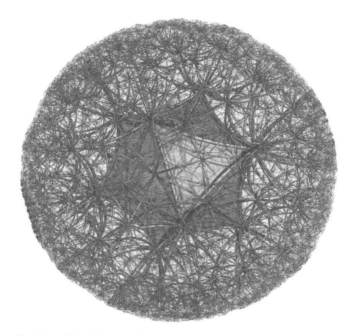

Fig. 5.16 Poincaré ball model of the regular icosahedral honeycomb $\{3, 5, 3\}$ in hyperbolic 3-space.

The time T has become the distance in a geometrically modified space with metric tensor $g_{ij}(\mathbf{x}) = n^2(\mathbf{x})\,\delta_{ij}$. Fermat's principle says that the true light rays are geodesics for this metric. They are generally curved in the ambient Euclidean 3-space.

$g_{ij}(\mathbf{x})$ is conformally flat, as it is the Euclidean metric tensor δ_{ij} scaled by the conformal factor $n^2(\mathbf{x})$, so the Christoffel symbols are relatively simple. They are

$$\Gamma^i{}_{jk} = \frac{1}{n}(n_{,j}\,\delta^i_k + n_{,k}\,\delta^i_j - n_{,m}\,\delta^{im}\delta_{jk})\,, \tag{5.108}$$

and the geodesic equation determining the light rays is

$$\frac{d^2 x^i}{dt^2} + \frac{2}{n}n_{,j}\frac{dx^j}{dt}\frac{dx^i}{dt} - \frac{1}{n}n_{,m}\,\delta^{im}\delta_{jk}\frac{dx^j}{dt}\frac{dx^k}{dt} = 0\,. \tag{5.109}$$

This can also be written in vector notation as

$$\mathbf{a} + \frac{2}{n}(\nabla n \cdot \mathbf{v})\mathbf{v} - \frac{1}{n}(\mathbf{v} \cdot \mathbf{v})\nabla n = \mathbf{0}\,, \tag{5.110}$$

where $\mathbf{v}$ is the velocity and $\mathbf{a}$ is the acceleration of the light. The conserved quantity here is $n^2\,\mathbf{v} \cdot \mathbf{v}$, which equals 1.

With n a function of height above the ground, the resulting geodesics can explain mirages. Near hot ground, the air is less dense than higher up. The refractive index is lower, and the light speed is faster. The gradient ∇n is in the upward direction. Grazing light therefore curves away from the ground, as shown in Figure 5.15. Looking towards the ground far away, one sees the bright sky rather than the dark ground. This looks like the reflection of the sky in a stretch of water.

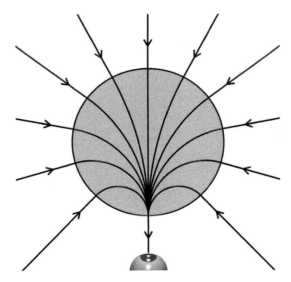

Fig. 5.17 An ideal fisheye lens.

Fig. 5.18 Photograph taken with a fisheye lens.

Examples of conformally flat geometries that we have previously encountered are the 2-sphere and 3-sphere, and the hyperbolic plane. Another example is the hyperbolic 3-space with constant negative curvature, whose metric is the simple generalization of (5.30),

$$\delta s^2 = \frac{\delta x^2 + \delta y^2 + \delta z^2}{\left(1 - \frac{1}{4}(x^2 + y^2 + z^2)\right)^2}, \tag{5.111}$$

where we have set the constant scale factor to unity. The coordinates are restricted to the interior of a ball of radius 2, but the true geometry is infinite in extent. Equatorial slices of this ball are hyperbolic planes, and the geodesics, as before, are segments of circles that meet the boundary sphere orthogonally. Figure 5.16 shows a honeycomb in hyperbolic 3-space, whose edges are all geodesics.

This metric can be interpreted as describing an optical medium that is an ordinary ball in Euclidean 3-space, but whose refractive index is

$$n(x, y, z) = \frac{1}{1 - \frac{1}{4}(x^2 + y^2 + z^2)}. \tag{5.112}$$

The refractive index is 1 at the exact centre, so here the medium is close to the vacuum, but it approaches infinity at the boundary of the ball, so light slows to zero speed at the boundary. Light rays follow the circular geodesics of the hyperbolic 3-space, but they take an infinite time to travel from boundary to boundary.

For a ball slightly smaller than this, we show in Figure 5.17 a set of light rays all ending at an observer's eye near the boundary. The ball allows vision in all directions. In fact it is an idealized fisheye lens. Manufactured fisheye lenses are made of several rounded layers of glass with increasing refractive index towards the outside, and the paths of light rays are not very different from what is shown in Figure 5.17. Figure 5.18 shows a photograph taken with a fisheye lens.

5.9   Further Reading

J.M. Lee, *Riemannian Manifolds: An Introduction to Curvature*, New York: Springer, 1997.

J. Oprea, *Differential Geometry and its Applications (2nd ed.)*, Washington DC: The Mathematical Association of America, 2007.

6

General Relativity

6.1 The Equivalence Principle

Following the development of his special theory of relativity, it was clear to Einstein that a new theory of gravity was necessary to complete his revolution. In Newton's theory, gravity appears to act instantaneously at a distance, whereas the cornerstone of special relativity is that the maximum speed of any interaction is the speed of light. In the years following the publication of special relativity, various attempts were made to incorporate a finite speed of interaction into a theory of gravity, but these early ideas proved to be too simplistic.

Gravity is special among forces, because it acts in the same way on all massive bodies. This observation dates back to Galileo's experiments rolling balls down inclined slopes. Galileo conclusively demonstrated that balls with different masses, released together and allowed to fall under gravity, hit the ground at the same instant if no other forces are acting. This observation is explained in Newtonian physics by the cancellation of mass between Newton's second law (2.3) and Newton's law of gravitation (2.80). In Newton's second law, acceleration equals the applied force divided by the mass of the object being accelerated. In this role, the mass is known as *inertial* mass, as its effect is to resist the change in motion of the body. Remarkably, the gravitational force on an object is also proportional to its mass. In this role, the mass is known as *gravitational* mass. The ratio of inertial mass to gravitational mass could in principle be different for different bodies made of different materials, but experimentally, the ratio is always 1, so we can regard inertial mass as the same as gravitational mass. Motion in a gravitational field is then independent of mass. For example, near the Earth's surface, the equation of motion for a freely falling body is

$$\frac{d^2 z}{dt^2} = -g \,, \tag{6.1}$$

independent of mass. The cancellation of mass is unique to gravity, as the strength of other forces is unrelated to the mass of the body on which they act. The electrostatic force, for instance, is proportional to the electric charge of the body, not its mass.

In Newton's theory, the equality of inertial and gravitational mass seems almost accidental, but in 1907 Einstein realized that this feature of gravitation might be the perfect foundation for a new relativistic theory. He raised his insight to the status of a new principle of physics and named it the *equivalence principle*—the equivalence of gravitational and inertial mass. In special relativity, as in Newtonian mechanics, there is no way to determine one's absolute velocity. Similarly, according to the equivalence principle, when in free fall, there is no way to determine one's absolute acceleration, as all nearby bodies fall with the same acceleration. Einstein postulated that for consistency with the rest of physics, this

The Physical World. Nicholas Manton and Nicholas Mee, Oxford University Press (2017).
© Nicholas Manton and Nicholas Mee. DOI 10.1093/oso/9780198795933.001.0001

principle must extend to all the laws of physics and not just mechanics. He believed that locally it must be impossible to distinguish between free fall in the presence of gravitating bodies, and being in a state of rest in the absence of gravitating bodies.

Einstein illustrated this with a thought experiment. Imagine being in a lift whose cable has snapped. As the lift falls, the occupants feel weightless, just as though gravity did not exist. The reason is that the lift and everything in it, including every part of the occupants' bodies, fall with the same downward acceleration g. We can always find a coordinate system for a falling body, such that there is no instantaneous acceleration. In a uniform gravitational field, the appropriate coordinate change is from (z, t) to (y, t) where

$$y = z + \frac{1}{2}gt^2 \,. \tag{6.2}$$

Then $\frac{d^2y}{dt^2} = \frac{d^2z}{dt^2} + g$, so equation (6.1) is transformed into the equation of motion

$$\frac{d^2y}{dt^2} = 0 \,. \tag{6.3}$$

The coordinate change has eliminated the effect of gravity. This makes the force of gravity reminiscent of the fictitious forces that we considered in section 5.7, which arise from the choice of coordinates. Equation (6.3) has solutions representing motion with any constant velocity. Therefore the relative motion of freely falling bodies has constant velocity, which is exactly the same as for bodies moving freely in the absence of gravity.

We do not feel the force of gravity. We are only aware of it when other forces are acting, such as on the surface of the Earth, where our free fall is stopped by the rigidity of the ground, and our natural frame of reference is non-inertial. In developing general relativity, Einstein would discover a way to model gravity without a gravitational force at all.

6.2 The Newtonian Gravitational Field and Tidal Forces

To understand general relativity, it is useful to first reformulate Newtonian gravity as a field theory. This works very well for massive bodies moving slowly compared to the speed of light, and many of the details are actually very similar to electrostatics. It is important to realize that, in the neighbourhood of the Earth, gravity is weak. For example, a freely falling satellite takes about 90 minutes to orbit the Earth, but a light ray would travel the same distance in about 0.1 seconds, so the satellite motion is comparatively slow.

The Newtonian gravitational force exerted by one body on another is described by an inverse square law force (2.80). It is proportional to the product of the masses of the bodies and inversely proportional to the square of their separation. This is similar to the Coulomb force between two charges (3.86), which is proportional to the product of the charges, and inversely proportional to the square of their separation. Like an electrostatic force, the gravitational force on a body can be interpreted as due to a gravitational field produced by all the other massive bodies.

We saw in section 3.6 that any static distribution of electric charge produces an electric field that is minus the gradient of a potential, and the most significant property of this potential is that away from the charge sources, it satisfies Laplace's equation. Newtonian gravity is very similar. The gravitational field is minus the gradient of a potential $\phi(\mathbf{x})$.

The potential due to a point mass M at the origin is

$$\phi(\mathbf{x}) = -\frac{GM}{r},$$ (6.4)

where G is Newton's universal gravitational constant, and $r = |\mathbf{x}|$ is the distance from the mass.[1] The gradient of this potential is the inverse square law force on a unit mass. Furthermore, the potential satisfies Laplace's equation, $\nabla^2 \phi = 0$, except at the origin.

More generally, the gravitational potential $\phi(\mathbf{x})$ produced by a matter distribution of density $\rho(\mathbf{x})$ satisfies Poisson's equation,

$$\nabla^2 \phi = 4\pi G \rho.$$ (6.5)

For sources that are extended bodies or collections of point masses, $\phi(\mathbf{x})$ is not generally spherically symmetric. The gravitational force on a test body of mass m at $\mathbf{x}$, due to all the other bodies, is

$$\mathbf{F} = -m\nabla\phi,$$ (6.6)

with $\nabla\phi$ evaluated at $\mathbf{x}$. The acceleration of the body is therefore

$$\mathbf{a} = -\nabla\phi.$$ (6.7)

If the sources of the gravitational potential are located in some finite region, then the total potential that they produce becomes uniform at distances that are large compared to their separation. This is usually modelled by imposing the boundary condition $\phi(\mathbf{x}) \to 0$ as $|\mathbf{x}| \to \infty$.

Even in the absence of a test body, the vector field $-\nabla\phi$ may be identified as a physical gravitational field, permeating space. It is often easier to work with the potential ϕ than with the bodies and their component parts that are its sources. For example, to characterize the gravitational field outside the Earth, one need only consider the general solution of Laplace's equation that approaches zero as $|\mathbf{x}| \to \infty$. This solution is an infinite sum of terms that approach zero with increasing inverse powers of the distance from the centre of the Earth. Their coefficients can be determined by observing the motion of orbiting satellites. As the Earth is spherical to a good approximation, the potential ϕ is dominated by the spherically symmetric term $-\frac{GM}{r}$; corrections depend on the deviation of the Earth's shape from spherical, and on the asymmetrical distribution of the Earth's mass. Accurate knowledge of the potential is vital for satellite navigation and GPS systems, and tells us something about the internal structure of the Earth.

Around any point, the potential $\phi(\mathbf{x})$ has a *local* expansion that determines the gravitational field nearby, and the first two or three terms in the expansion are sufficient to model its main effects. Suppose P is a point just above the Earth's surface, chosen to be the origin of Cartesian coordinates (x, y, z). Near P,

$$\phi \simeq \phi_0 + gz + \frac{1}{2}h(x^2 + y^2 - 2z^2)$$ (6.8)

with g and h positive constants. (Note that $x^2 + y^2 - 2z^2$ satisfies Laplace's equation, but x^2, y^2 and z^2 individually do not.) The constant ϕ_0 does not contribute to the gravitational

---

[1] By convention, there is no factor of 4π here.

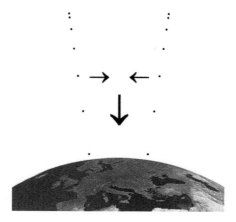

Fig. 6.1 In addition to the downward acceleration on two bodies falling towards the Earth, there is also a relative sideways acceleration.

field. The second term describes the familiar field just above the Earth's surface, where the potential is proportional to height z, and its gradient is the vector $(0, 0, g)$, producing a downward acceleration of magnitude g. However, gravity is not perfectly uniform. Objects that are spatially separated do not feel the same force and will have a relative acceleration. The third term in the potential (6.8) captures this. Its gradient is $(hx, hy, -2hz)$, so the total acceleration is $\mathbf{a} = (-hx, -hy, -g + 2hz)$. The downward gravitational acceleration, $g - 2hz$, is reduced above P and increased below P, and there is a sideways acceleration of magnitude $h\sqrt{x^2 + y^2}$ towards the z-axis. This correctly describes the relative motion of two or more bodies falling towards the Earth's centre, as shown in Figure 6.1.

The linear term in the expansion (6.8) determines the local, approximately uniform gravitational field, whereas the quadratic terms determine its spatial variation. Although the effects of the linear term can always be removed by a change of coordinates, as in equation (6.3), in general the effects of the quadratic terms cannot. These quadratic terms give rise to *tidal effects*. The tides produced by the Moon in the vicinity of the Earth are the defining example and are illustrated in Figure 6.2. The additional acceleration of a body on the side of the Earth facing the Moon, the near side, is compared to the additional acceleration on the far side. The difference is known as a tidal acceleration, and was first invoked by Newton to explain the tides. On the near side, the oceans flow because the pull of the Moon on them is greater than the average pull on the bulk of the Earth, and on the far side, the oceans flow because they are pulled less than the bulk of the Earth. The relative acceleration is away from the Earth's centre. In addition to these effects along the Earth–Moon axis, there are sideways tidal forces in the directions perpendicular to the Earth–Moon axis, as shown in Figure 6.2. The solid Earth is distorted by the pull of the Moon too, but not enough for us to notice it.

The gravitational field due to the Moon is proportional to $\frac{GM}{r^2}$, where r is the distance from the Moon and M is the Moon's mass. The diameter of the Earth is small compared to the Earth–Moon separation, so the difference between accelerations on opposite sides of the Earth is proportional to the derivative of $\frac{GM}{r^2}$. The tidal effects are therefore of magnitude $\frac{GM}{R^3}$, where R is the Earth–Moon separation.

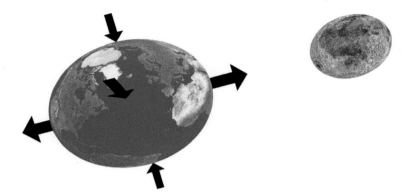

Fig. 6.2 The tidal stretching and squeezing of the Earth in the gravitational field of the Moon (greatly exaggerated).

Einstein realized that, due to tidal forces, the trajectories of two test particles initially travelling freely along parallel Euclidean lines through a gravitational field do not generally remain parallel. This is very similar to the geodesic deviation of particle trajectories in a curved space that was described in section 5.7.2. So Einstein made the astonishing proposal that gravity could be described in terms of curved spacetime. In this picture, massive freely falling bodies follow geodesics through spacetime, and tidal accelerations arise from spacetime curvature.

As a prelude to discussing curved spacetime further, we will describe some of the geometry of flat Minkowski space.

6.3 Minkowski Space

When discussing special relativity in Chapter 4, we saw the advantages of sewing space and time together into the 4-dimensional spacetime known as Minkowski space. The squared infinitesimal interval between events at $(t, \mathbf{x})$ and $(t + dt, \mathbf{x} + d\mathbf{x})$ in Minkowski space is

$$d\tau^2 = dt^2 - d\mathbf{x} \cdot d\mathbf{x}. \tag{6.9}$$

This is analogous to the squared infinitesimal distance $ds^2 = d\mathbf{x} \cdot d\mathbf{x}$ in Euclidean 3-space.[2] The squared interval $d\tau^2$ is Lorentz invariant, which means that it is the same for all inertial observers in uniform relative motion, even though they may have different notions of what the individual time and space coordinates are.

If $d\tau^2$ is positive, then its positive square root $d\tau$ is called the *proper time* separation of the events. Infinitesimal vectors $(dt, d\mathbf{x})$ for which $d\tau^2$ is positive are called time-like and those for which $d\tau^2$ is negative are space-like. Vectors for which $d\tau^2$ is zero are light-like and lie on a double cone, called the lightcone, as shown in Figure 6.3.

---

[2] From here on, we express the infinitesimal interval as $d\tau$, and an infinitesimal distance as ds, rather than using the notation $\delta\tau$ or δs.

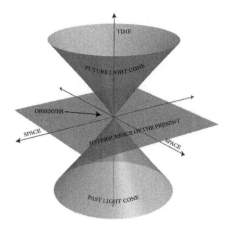

Fig. 6.3 The lightcone. Light rays travel on the lightcone. The trajectories of massive bodies must remain within the local lightcone throughout spacetime.

Consider a curved worldline $X(\lambda) = (t(\lambda), \mathbf{x}(\lambda))$, parametrized by λ, with fixed endpoints $X(\lambda_0)$ and $X(\lambda_1)$. The proper time along the worldline is

$$\tau = \int_{\lambda_0}^{\lambda_1} \sqrt{\left(\frac{dt}{d\lambda}\right)^2 - \frac{d\mathbf{x}}{d\lambda} \cdot \frac{d\mathbf{x}}{d\lambda}} \, d\lambda. \tag{6.10}$$

If $X(\lambda)$ is the path of a massive particle, the quantity under the square root symbol must be positive. The paths that *maximize* τ are time-like geodesics, and they are straight lines in Minkowski space, representing a particle moving at constant velocity. There are also geodesics for which the integrand is zero, in which case τ is also zero. Such geodesics are light-like and correspond to light rays. There are other paths for which the integrand in equation (6.10) is the square root of a negative quantity, in which case τ is imaginary. Such paths are called space-like, and nothing physical can move along them.

The reason that τ is maximized rather than minimized along a particle geodesic is readily understood. In the particle's rest frame, the worldline is a straight line parallel to the time axis, and for such a trajectory the infinitesimal proper time is $d\tau = dt$. Any deviation in the path will introduce negative, spatial contributions to $d\tau^2$, which reduce τ. As τ is invariant under a Lorentz transformation, this result is true for all observers.

We will adopt a uniform notation $x^\mu = (x^0, x^1, x^2, x^3)$ for coordinates in 4-dimensional Minkowski space, where $x^0 = t$ is the time coordinate and (x^1, x^2, x^3) are the space coordinates. These are mixed by Lorentz transformations. Generally, in what follows, Greek indices like μ and ν will range from 0 to 3. A 4-vector has a single Greek index, and a tensor has two or more of these indices. The squared infinitesimal interval (6.9) in Minkowski space can be expressed in terms of a metric tensor $\eta_{\mu\nu}$ as

$$d\tau^2 = \eta_{\mu\nu} dx^\mu dx^\nu = (dx^0)^2 - (dx^1)^2 - (dx^2)^2 - (dx^3)^2. \tag{6.11}$$

This is known as the Minkowski spacetime metric. The metric tensor is diagonal, with components $\eta_{00} = 1$, $\eta_{11} = \eta_{22} = \eta_{33} = -1$, and the off-diagonal components all zero. It is sometimes convenient to write $\eta_{\mu\nu} = \text{diag}(1, -1, -1, -1)$. The inverse metric tensor $\eta^{\mu\nu}$ has identical components. Minkowski space is the appropriate geometry for special relativistic physics, and like Euclidean space, it is flat.

6.4 Curved Spacetime Geometry

Einstein incorporated gravity by transforming spacetime into a dynamical part of the theory, and allowing it to be curved. This more general curved spacetime has four coordinates that we denote in a uniform way by y^μ. A spacetime point with coordinates y^μ is often denoted by y. Locally, the geometrically meaningful quantity is the squared infinitesimal interval between a point with coordinates y^μ and a point with coordinates $y^\mu + dy^\mu$. This has the form

$$d\tau^2 = g_{\mu\nu}(y)dy^\mu dy^\nu \,, \tag{6.12}$$

where $g_{\mu\nu}(y)$ is a symmetric 4×4 matrix varying throughout spacetime, called the spacetime metric tensor. Under any coordinate transformation, the components of $g_{\mu\nu}$ change in such a way that $d\tau^2$ is unchanged.

In 3-dimensional Riemannian geometry, the metric tensor g_{ij} is positive definite everywhere, meaning that by a suitable choice of coordinates it can be brought locally to the form δ_{ij}, whose three entries are $+1$. In spacetime geometry, we require that by a suitable choice of coordinates, the metric tensor $g_{\mu\nu}$ can be brought locally to the Minkowski form $\eta_{\mu\nu} = \text{diag}(1, -1, -1, -1)$. If this property holds, the metric is said to be Lorentzian. A Lorentzian metric tensor $g_{\mu\nu}$ has an inverse $g^{\mu\nu}$ at each point. This is simply the matrix inverse of $g_{\mu\nu}$ so $g^{\lambda\mu}g_{\mu\nu} = \delta^\lambda_\nu$, where the Kronecker delta symbol δ^λ_ν has, as before, the value 1 if the indices are the same and 0 otherwise. At each point in spacetime there are infinitesimal time-like and space-like vectors dy^μ, separated by a local lightcone of light-like vectors.

In Chapter 5 we derived the significant results of Riemannian geometry, such as the form of the Christoffel symbols and the Riemann curvature tensor, for a positive definite metric. However, we made no use of the positive definiteness, only that the metric was invertible, so all these results carry over to the Lorentzian metrics of general relativity. We will assume their validity from here on without further comment. The Christoffel symbols are

$$\Gamma^\beta{}_{\lambda\delta} = \frac{1}{2}g^{\alpha\beta}(g_{\alpha\lambda,\delta} + g_{\alpha\delta,\lambda} - g_{\lambda\delta,\alpha}) \,, \tag{6.13}$$

and the Riemann curvature tensor is

$$R^\alpha{}_{\lambda\gamma\delta} = \Gamma^\alpha{}_{\delta\lambda,\gamma} - \Gamma^\alpha{}_{\gamma\lambda,\delta} + \Gamma^\alpha{}_{\gamma\beta}\Gamma^\beta{}_{\delta\lambda} - \Gamma^\alpha{}_{\delta\beta}\Gamma^\beta{}_{\gamma\lambda} \,, \tag{6.14}$$

where the indices run from 0 to 3. In 2-dimensional space, the curvature is completely determined by a single number at each point, the Gaussian curvature. There are many more curvature components at each point in 4-dimensional spacetime. The Riemann curvature tensor $R_{\alpha\lambda\gamma\delta}$ is antisymmetric in its first two indices, and in its last two indices. This gives $\frac{4 \times 3}{2} = 6$ independent combinations for each pair. It is also symmetric under the interchange of these two pairs of indices, which reduces the number of independent components to $\frac{6 \times 7}{2} = 21$. Finally, the components of the Riemann tensor obey the first Bianchi identity (5.59),

$$R_{\alpha\nu\gamma\lambda} + R_{\alpha\lambda\nu\gamma} + R_{\alpha\gamma\lambda\nu} = 0 \,, \tag{6.15}$$

which reduces the total number of independent curvature components to 20.

It is always possible to transform to local coordinates around a point, such that $g_{\mu\nu}$ has the standard Minkowskian form $\eta_{\mu\nu}$, and in addition the derivatives of $g_{\mu\nu}$ are zero. Then the Christoffel symbols are zero at that point. Such coordinates are called *inertial*, or *freely falling*, and are analogous to normal coordinates in Riemannian geometry. The existence of inertial coordinates is the mathematical counterpart of the equivalence principle. Physically, the Christoffel symbols of general relativity are generalizations of the gravitational field of Newtonian gravity, so the fact that they vanish in an inertial frame ties in with Einstein's observation that we do not feel gravity when in free fall. The first derivatives of the Christoffel symbols involve second derivatives of the metric tensor. In general, they do not vanish. Physically, this is to be expected, as it is these derivatives that determine the spacetime curvature of general relativity and the tidal accelerations in the Newtonian picture.

Let us now consider a particle worldline in curved spacetime, a parametrized time-like path $y(\lambda)$. The integrated interval between fixed endpoints $y(\lambda_0)$ and $y(\lambda_1)$ is

$$\tau = \int_{\lambda_0}^{\lambda_1} \sqrt{g_{\mu\nu}(y(\lambda))\frac{dy^\mu}{d\lambda}\frac{dy^\nu}{d\lambda}}\, d\lambda\,. \tag{6.16}$$

Notice that because of the square root, $d\lambda$ formally cancels, so τ is unchanged if the worldline is reparametrized. Maximizing τ gives the Euler–Lagrange equation[3]

$$\frac{d^2y^\mu}{d\lambda^2} + \Gamma^\mu{}_{\nu\sigma}\frac{dy^\nu}{d\lambda}\frac{dy^\sigma}{d\lambda} = 0\,. \tag{6.17}$$

As in Riemannian geometry, this is the geodesic equation, the analogue of (5.84), and λ is no longer arbitrary but linked to the interval along the worldline.

Suppose a geodesic passes through a point P. Using inertial coordinates where $\Gamma^\mu{}_{\nu\sigma} = 0$, we see that $\frac{d^2y^\mu}{d\lambda^2} = 0$ at P. Each coordinate y^μ of the worldline is therefore locally a linear function of λ, just as for the free motion of a particle in Minkowski space. This is the motion required by the equivalence principle, and implies that a freely falling particle must follow a geodesic through spacetime.

Along a geodesic, the quantity

$$\Xi = g_{\mu\nu}(y(\lambda))\frac{dy^\mu}{d\lambda}\frac{dy^\nu}{d\lambda} \tag{6.18}$$

is independent of λ; in other words, it is conserved. This may be checked by differentiating equation (6.18) with respect to λ, and using the geodesic equation (6.17) together with the formula for the Christoffel symbols. Ξ is positive, zero or negative for time-like, light-like or space-like geodesics, respectively. If the geodesic is time-like we can rescale Ξ to be 1, and then the parameter λ becomes the proper time τ along the geodesic. Only time-like geodesics correspond to the trajectories of physical particles.

Along a light-like geodesic, τ is zero, and λ itself is a better parameter. A light-like geodesic is the path of a light ray in curved spacetime. It describes physical light propagation in the geometrical optics limit, where the wavelength is much less than any length scale associated with the curvature.

---

[3] The same equation occurs if the square root is omitted in the integrand, as optimizing the square root of a function is essentially the same problem as optimizing the function itself.

If the metric of spacetime has symmetries, for example, a rotational symmetry or symmetry under time translation, then geodesic motion has further conservation laws. A continuous symmetry is most simply realized if there is a choice of coordinates y^μ so that the metric tensor is independent of one of the coordinates, say y^α. In that case one can show, again using equation (6.17), that

$$Q = g_{\alpha\mu}(y(\lambda)) \frac{dy^\mu}{d\lambda} \tag{6.19}$$

is conserved along any geodesic. Conservation of Ξ and Q will be useful later when we consider geodesic motion of particles and light in the spacetime surrounding a star or black hole, where there is time translation symmetry and some rotational symmetry.

6.4.1 Weak gravitational fields

According to the equivalence principle, we should be able to model the motion of a freely falling body in a Newtonian gravitational field as a time-like geodesic in a curved spacetime with a suitably defined metric. We know that in a weak gravitational field, Newtonian dynamics works extremely well for bodies moving much slower than the speed of light, so in such circumstances the corresponding metric must be close to Minkowskian. We will therefore use the usual coordinates of Minkowski space, $x^0 = t$ and x^1, x^2, x^3.

If the Newtonian gravitational potential is $\phi(\mathbf{x})$, then the appropriate metric for modelling Newtonian gravity is

$$d\tau^2 = (1 + 2\phi(\mathbf{x})) \, dt^2 - d\mathbf{x} \cdot d\mathbf{x}. \tag{6.20}$$

We can neglect any time dependence of ϕ, as the bodies producing the potential are moving slowly. The only component of the metric tensor that differs from the Minkowski case is $g_{tt} = 1 + 2\phi(\mathbf{x})$, and the difference is small, because in our units, $|\phi| \ll 1$. To verify that this metric has the appropriate geodesics, consider the interval τ along a worldline $X(t) = (t, \mathbf{x}(t))$ parametrized by t, where the velocity $\mathbf{v} = \frac{d\mathbf{x}}{dt}$ is small. The interval is

$$
\begin{aligned}
\tau &= \int_{t_0}^{t_1} \sqrt{(1 + 2\phi(\mathbf{x}(t))) \left(\frac{dt}{dt}\right)^2 - \frac{d\mathbf{x}}{dt} \cdot \frac{d\mathbf{x}}{dt}} \, dt \\
&= \int_{t_0}^{t_1} \sqrt{1 + 2\phi(\mathbf{x}(t)) - \mathbf{v} \cdot \mathbf{v}} \, dt.
\end{aligned} \tag{6.21}
$$

Since ϕ and $\mathbf{v}$ are small, we can approximate the square root, giving

$$\tau \simeq \int_{t_0}^{t_1} \left(1 + \phi(\mathbf{x}(t)) - \frac{1}{2}\mathbf{v} \cdot \mathbf{v}\right) dt. \tag{6.22}$$

The integral of 1 is path independent and can be dropped.

τ is the quantity we should maximize to find particle geodesics, but if we multiply by $-m$, where m is a particle mass, then equivalently we can minimize

$$S = \int_{t_0}^{t_1} \left(\frac{1}{2}m\mathbf{v} \cdot \mathbf{v} - m\phi(\mathbf{x}(t))\right) dt. \tag{6.23}$$

S is the action (2.53) for a non-relativistic particle of mass m, with kinetic energy $\frac{1}{2}m\mathbf{v}\cdot\mathbf{v}$ and potential energy $m\phi$. As we saw in section 2.3, the equation of motion derived by minimizing S is

$$\frac{d^2\mathbf{x}}{dt^2} + \nabla\phi = 0\,, \tag{6.24}$$

which is the defining equation of Newtonian gravity. This shows that in the low velocity limit, a time-like geodesic in the curved spacetime with metric (6.20) reproduces the motion expected in Newtonian gravity.

We can explicitly check the low velocity limit of the geodesic equation (6.17). In this limit, $\tau \simeq t$ so the derivatives with respect to τ can be replaced by derivatives with respect to t. The dominant Christoffel symbol is

$$\Gamma^i{}_{tt} = \frac{1}{2}g^{\alpha i}(g_{\alpha t,t} + g_{\alpha t,t} - g_{tt,\alpha}) = -\frac{1}{2}g^{\alpha i}g_{tt,\alpha} = -\frac{1}{2}g^{ii}g_{tt,i}\,, \tag{6.25}$$

where $g^{ii} = -1$, $g_{tt} = 1 + 2\phi$ and $g_{tt,i} = 2\frac{\partial\phi}{\partial x^i}$, so $\Gamma^i{}_{tt} = \frac{\partial\phi}{\partial x^i}$. (There is no summation over i in the last expression in (6.25).) The space components of (6.17) are therefore

$$\frac{d^2x^i}{dt^2} + \frac{\partial\phi}{\partial x^i} = 0\,, \tag{6.26}$$

again agreeing with the Newtonian equation of motion.

We will see later that the metric (6.20) does not satisfy the Einstein field equation exactly and a further term involving the Newtonian potential ϕ appears in the spatial part of the metric tensor, but this produces a negligible correction to the equation of motion for a slowly moving particle. It is perhaps a little surprising that the most important effect of the Newtonian potential is to distort the term g_{tt} in the spacetime metric tensor. One might have guessed that gravity would curve space. However, a distortion of time is consistent with our earlier finding that free fall in a constant gravitational field appears inertial after the coordinate change (6.2) involving time.

6.5 The Gravitational Field Equation

If we accept the idea that spacetime is curved, then, as we have seen, we can expect massive bodies and light to travel along geodesics—but how does spacetime curvature arise in the first place? What is the gravitational field equation that determines the relationship between matter and spacetime curvature? Einstein assumed that the field equation must satisfy three guiding principles:

   1) it must be generally covariant,
   2) it must be consistent with the equivalence principle,
   3) it must reduce to the equation for the Newtonian gravitational potential, for matter of low density and low velocity.

Principle 1) means that the field equation must be a tensor equation, taking the same form in any coordinate system. Principle 2) had been the idea that initially got the ball rolling. It suggested to Einstein that gravity could be treated as spacetime curvature, because gravity affects all bodies in the same way. Moreover, the equivalence principle implies that even in a gravitational field, physics is indistinguishable in a locally inertial frame from the physics of special relativity. In other words, spacetime is locally Minkowskian.

Principle 2) combined with Principle 1) implies that one side of the field equation must be composed of some form of curvature tensor. Principle 3) provides the constant of proportionality relating the mass density to the curvature and leads to a vital check that the field equation is consistent with well established Newtonian physics.

As mentioned in section 6.2, in the presence of a mass density ρ, the Newtonian potential ϕ obeys Poisson's equation $\nabla^2 \phi = 4\pi G\rho$. The task faced by Einstein was to find the relativistic counterpart of Poisson's equation. This should be a covariant equation relating a tensor describing the curvature of spacetime to a tensor describing the distribution of matter, and it should reduce to Poisson's equation for low mass densities and matter speeds much less than the speed of light.

6.5.1   The energy–momentum tensor

The gravitational source producing the curvature must be a density, like the mass density appearing on the right-hand side of Poisson's equation—but in a relativistic theory, mass in one frame contributes to energy and momentum in another frame, so energy, mass and momentum must all contribute as sources of gravitational curvature.

Energy is the time component of a 4-vector. Under a Lorentz boost it is multiplied by a gamma factor $\gamma = (1 - \mathbf{v} \cdot \mathbf{v})^{-\frac{1}{2}}$. The energy density—the energy per unit volume—acquires a second factor of γ, because a volume element contracts by γ in the direction of the boost. Therefore the energy density transforms as the 00 component of a 2-index tensor. This tensor is known as the *energy–momentum tensor* or *stress–energy tensor* and is denoted by $T^{\mu\nu}$. Although this argument is based on physics in Minkowski space, it also applies in curved spacetime, as the equivalence principle implies that spacetime is always locally Minkowskian.

For pure matter, the density in its rest frame is denoted by ρ and is (by definition) Lorentz invariant. In this frame, $T^{00} = \rho$ is the dominant contribution to $T^{\mu\nu}$. If the matter is moving (and this just depends on the coordinate system that has been chosen) then there is an expression for $T^{\mu\nu}$ that depends on the density ρ and the matter's local 4-velocity v^μ. The components T^{i0} ($i = 1, 2, 3$) give the density of the momentum in the ith direction. T^{ij} is the current, or flux, of the ith component of the momentum density in the jth direction. It has a contribution from the net flow of the matter, and from the random motion of matter particles colliding at the microscopic level, which generates a pressure.

Astrophysicists refer to an idealized fluid of non-interacting free particles, with negligible relative motion between them, and hence negligible pressure, as a *dust*. The energy–momentum tensor of a dust takes the simple form

$$T^{\mu\nu} = \rho v^\mu v^\nu \,, \tag{6.27}$$

where v^μ is the local 4-velocity of the dust. For a more general perfect fluid the energy–momentum tensor includes a pressure term, and has the form

$$T^{\mu\nu} = (\rho + P)\, v^\mu v^\nu - P g^{\mu\nu} \,, \tag{6.28}$$

where ρ is the density and P is the pressure. ρ and P are Lorentz invariants defined in the local rest frame of the fluid, and are related by an *equation of state*. They vary throughout spacetime, so they are fields. More generally, $T^{\mu\nu}$ can also include terms that describe electromagnetic radiation or any other physical phenomena.

Quite generally, $T^{\mu\nu}$ is symmetric under the interchange of its two indices, so of its sixteen components at each point, only ten are independent (four diagonal and six off-diagonal entries). In particular, the current of the energy density T^{0i} equals the momentum density T^{i0}. The components of $T^{\mu\nu}$ are not completely arbitrary functions of space and time, however, as the matter and radiation satisfy their own local field equations. Electromagnetic radiation, for example, obeys Maxwell's equations adapted to a curved spacetime background. For a dilute gas of matter particles, the effectively free particles move along geodesic paths through spacetime. For denser gases of matter, as occur for example in stars, one needs to consider the equations for fluid motion, where pressure plays a role. These dynamical field equations lead to a further constraint on $T^{\mu\nu}$.

Recall that because electric charge is physically conserved in Minkowski space, there is a local electromagnetic 4-current conservation equation (4.45), which says that the spacetime divergence of $\mathcal{J}$ is zero. Using 4-vector notation (and the index notation $_{,\nu}$ for spacetime partial derivatives) this becomes

$$\partial \cdot \mathcal{J} = \mathcal{J}^{\nu}{}_{,\nu} = 0 \,. \tag{6.29}$$

In curved spacetime, the divergence generalizes to the covariant derivative with a contraction of indices, so in coordinates y^{μ}, the electromagnetic current must satisfy the covariant conservation equation

$$\frac{D}{Dy^{\nu}} \mathcal{J}^{\nu} = \mathcal{J}^{\nu}{}_{,\nu} + \Gamma^{\nu}{}_{\nu\alpha} \mathcal{J}^{\alpha} = 0 \,. \tag{6.30}$$

Because of the equivalence principle, we know that locally, even in curved spacetime, energy and the three components of momentum are conserved. There is a corresponding current conservation law—the covariant spacetime divergence of the energy–momentum tensor is zero. In inertial coordinates

$$T^{\mu\nu}{}_{,\nu} = 0 \,, \tag{6.31}$$

and in a general coordinate system this becomes

$$\frac{D}{Dy^{\nu}} T^{\mu\nu} = T^{\mu\nu}{}_{,\nu} + \Gamma^{\mu}{}_{\nu\alpha} T^{\alpha\nu} + \Gamma^{\nu}{}_{\nu\alpha} T^{\mu\alpha} = 0 \,. \tag{6.32}$$

This is the further constraint on $T^{\mu\nu}$. In fact, as μ is a free index running from 0 to 3, there are four local constraints corresponding to the conservation of energy and momentum.

We will now introduce a new shorthand notation. Previously we replaced the partial derivative $\frac{\partial}{\partial y^{\nu}}$ by the comma notation $_{,\nu}$ and will continue to do this. From here on, we will also replace the covariant derivative $\frac{D}{Dy^{\nu}}$, which in curved spacetime includes terms containing Christoffel symbols, with the semi-colon notation $_{;\nu}$. For example,

$$\frac{D}{Dy^{\nu}} V^{\mu} = V^{\mu}{}_{;\nu} = V^{\mu}{}_{,\nu} + \Gamma^{\mu}{}_{\nu\alpha} V^{\alpha} \,. \tag{6.33}$$

In this notation, the vanishing of the covariant divergence of the energy–momentum tensor is rewritten as

$$T^{\mu\nu}{}_{;\nu} = 0 \,. \tag{6.34}$$

6.5.2 The Einstein tensor and the Einstein equation

Einstein realized that the energy–momentum tensor had all the properties required for one side of the gravitational field equation. What he needed was the appropriate tensor, now known as the Einstein tensor, that would sit on the other side of the equation. It would describe the curvature of spacetime and must therefore be related to the Riemann tensor. In empty space, the energy–momentum tensor is zero. This implies that the Einstein tensor on the other side of the equation cannot simply be a multiple of the Riemann tensor, otherwise empty space would be flat and there would be no gravitational effects in the empty space between massive bodies, which is obviously wrong.

The energy–momentum tensor is a symmetric tensor of rank 2, so the Einstein tensor must also have these properties. It is convenient to work with $T_{\mu\nu}$, having lowered its indices using the metric tensor. In Einstein's notebooks he initially wrote the equation as

$$?_{\mu\nu} = \kappa T_{\mu\nu}\,, \tag{6.35}$$

where κ is a constant of proportionality and $?_{\mu\nu}$ is the Einstein tensor, whose form he set out to discover. The energy–momentum tensor has zero divergence, so for consistency, the Einstein tensor must have zero divergence too. The theory then automatically incorporates conservation of energy and momentum.

As we have seen, the Riemann curvature tensor has four indices—it is of rank 4—but there is a closely related rank 2 tensor that derives from it. This is the *Ricci tensor* $R_{\mu\nu}$. It is obtained by contracting indices, that is, summing over selected components of the Riemann tensor, as follows:

$$R_{\mu\nu} = R^\alpha{}_{\mu\alpha\nu} = R^0{}_{\mu 0\nu} + R^1{}_{\mu 1\nu} + R^2{}_{\mu 2\nu} + R^3{}_{\mu 3\nu}\,. \tag{6.36}$$

Due to the symmetry of the Riemann tensor under exchange of the first and second pairs of indices, the Ricci tensor is symmetric in its two indices and therefore has ten independent components. A further contraction of indices is possible, which produces the *Ricci scalar*

$$R = g^{\mu\nu} R_{\mu\nu} = R^\mu{}_\mu\,. \tag{6.37}$$

The Ricci tensor, the metric tensor and the Ricci scalar can be combined into a family of symmetric, rank 2 curvature tensors

$$R_{\mu\nu} - \xi g_{\mu\nu} R\,, \tag{6.38}$$

where ξ is an arbitrary constant. We now show that just one value of ξ gives a tensor whose divergence is identically zero in any spacetime. This divergence may be calculated with the help of an identity involving derivatives of the Riemann tensor. As with many tensor equations, the identity is most easily proved by using local inertial coordinates. The Riemann tensor is expressed in terms of the Christoffel symbols in equation (6.14). In inertial coordinates the Christoffel symbols are zero, so by the Leibniz rule, the contribution of the last two terms of the Riemann tensor, which are products of Christoffel symbols, is still zero after differentiating once. Therefore in inertial coordinates,

$$R^\alpha{}_{\nu\gamma\lambda,\mu} = \Gamma^\alpha{}_{\lambda\nu,\gamma\mu} - \Gamma^\alpha{}_{\gamma\nu,\lambda\mu}\,. \tag{6.39}$$

Similarly, by a permutation of the indices,

$$R^{\alpha}{}_{\nu\mu\gamma,\lambda} = \Gamma^{\alpha}{}_{\gamma\nu,\mu\lambda} - \Gamma^{\alpha}{}_{\mu\nu,\gamma\lambda}, \quad R^{\alpha}{}_{\nu\lambda\mu,\gamma} = \Gamma^{\alpha}{}_{\mu\nu,\lambda\gamma} - \Gamma^{\alpha}{}_{\lambda\nu,\mu\gamma}. \tag{6.40}$$

Adding these three expressions, and using the symmetry of mixed partial derivatives, gives

$$R^{\alpha}{}_{\nu\gamma\lambda,\mu} + R^{\alpha}{}_{\nu\mu\gamma,\lambda} + R^{\alpha}{}_{\nu\lambda\mu,\gamma} = 0. \tag{6.41}$$

This expression is true in inertial coordinates. Replacing the partial derivatives with covariant derivatives produces the tensor identity

$$R^{\alpha}{}_{\nu\gamma\lambda;\mu} + R^{\alpha}{}_{\nu\mu\gamma;\lambda} + R^{\alpha}{}_{\nu\lambda\mu;\gamma} = 0, \tag{6.42}$$

valid in any coordinate system. This is known as the *second Bianchi identity*.

This identity takes us a big step towards finding the Einstein tensor. As we have seen, the Ricci tensor $R_{\mu\nu}$ is formed by taking the trace over the first and third indices of the Riemann tensor. If we contract α with γ in each term in equation (6.42), we obtain

$$R_{\nu\lambda;\mu} - R_{\nu\mu;\lambda} + R^{\alpha}{}_{\nu\lambda\mu;\alpha} = 0, \tag{6.43}$$

where we have used the antisymmetry of the Riemann tensor in its last two indices to obtain the middle term. We can multiply through by $g^{\nu\lambda}$ and contract again, to obtain $R^{\lambda}{}_{\lambda;\mu} - R^{\lambda}{}_{\mu;\lambda} + R^{\alpha\lambda}{}_{\lambda\mu;\alpha} = 0$, and hence, using the antisymmetry of $R^{\alpha\lambda}{}_{\lambda\mu}$ in its first two indices, $R^{\lambda}{}_{\lambda;\mu} - 2R^{\lambda}{}_{\mu;\lambda} = 0$. As $R = R^{\lambda}{}_{\lambda}$ is the Ricci scalar, this becomes

$$R_{;\mu} - 2R^{\lambda}{}_{\mu;\lambda} = 0. \tag{6.44}$$

The metric is covariantly constant, as shown in equation (5.45), so we can multiply by the inverse metric $g^{\mu\nu}$ and pull it through the covariant derivative. After swapping the order of the two terms, and using the same symbol for the repeated indices, this gives

$$\left(R^{\mu\nu} - \frac{1}{2} g^{\mu\nu} R \right)_{;\mu} = 0. \tag{6.45}$$

We have found a symmetric tensor of rank 2 with zero divergence.

We therefore fix the constant ξ in equation (6.38) to be $\frac{1}{2}$, and define the Einstein tensor (with lowered indices) to be

$$G_{\mu\nu} = R_{\mu\nu} - \frac{1}{2} g_{\mu\nu} R. \tag{6.46}$$

The field equation of general relativity follows immediately. Plugging $G_{\mu\nu}$ into equation (6.35), we find

$$G_{\mu\nu} = \kappa T_{\mu\nu}. \tag{6.47}$$

This is the Einstein equation. It has ten components and equates two symmetric, divergence-free rank 2 tensors. Given a particular distribution of matter, energy and momentum it determines the metric $g_{\mu\nu}$, and therefore how spacetime curves. Spacetime is generally curved even in regions that are empty, where $T_{\mu\nu} = 0$, because some components of the Riemann tensor may still be non-zero.

We can find an alternative form of the Einstein equation as follows. If we multiply equation (6.46) through by $g^{\lambda\mu}$ and contract the indices λ and ν, we obtain

$$G^{\nu}{}_{\nu} = R - 2R = \kappa T^{\nu}{}_{\nu}\,, \tag{6.48}$$

as $g^{\mu\nu}g_{\mu\nu} = \delta^{\nu}{}_{\nu} = 4$. If we write $T = T^{\nu}{}_{\nu}$ for the contracted energy–momentum tensor, then

$$-R = \kappa T\,, \tag{6.49}$$

and substituting this back into the Einstein equation (6.47) gives

$$R_{\mu\nu} = \kappa \left(T_{\mu\nu} - \frac{1}{2}g_{\mu\nu}T\right)\,. \tag{6.50}$$

This is the original form in which Einstein presented the field equation.

6.5.3 Determining the constant of proportionality

In the weak field, low velocity limit we should recover Newtonian gravity in the form of Poisson's equation, and we can use the 00 component of the field equation in this limit to determine κ.

In inertial coordinates, the Christoffel symbols are zero, but their derivatives may be non-zero. Contracting the first and third index of the Riemann tensor in equation (6.14) gives the Ricci tensor in inertial coordinates: $R_{\mu\nu} = \Gamma^{\alpha}{}_{\nu\mu,\alpha} - \Gamma^{\alpha}{}_{\alpha\mu,\nu}$. Using the formula (6.13) for the Christoffel symbols, this becomes

$$
\begin{aligned}
R_{\mu\nu} &= \frac{1}{2}g^{\beta\alpha}(g_{\beta\nu,\mu\alpha} + g_{\beta\mu,\nu\alpha} - g_{\nu\mu,\beta\alpha}) - \frac{1}{2}g^{\beta\alpha}(g_{\beta\alpha,\mu\nu} + g_{\beta\mu,\alpha\nu} - g_{\alpha\mu,\beta\nu}) \\
&= \frac{1}{2}g^{\beta\alpha}(g_{\beta\nu,\mu\alpha} - g_{\nu\mu,\beta\alpha} - g_{\beta\alpha,\mu\nu} + g_{\alpha\mu,\beta\nu})\,,
\end{aligned} \tag{6.51}
$$

as the middle terms in the two brackets cancel. For weak fields, the deviations from flatness are small, so we can write

$$g_{\mu\nu} = \eta_{\mu\nu} + h_{\mu\nu}\,, \tag{6.52}$$

where $\eta_{\mu\nu}$ is the metric tensor of the Minkowski space background and $h_{\mu\nu} \ll 1$, and we can choose the coordinates y^{μ} to be the ordinary time and space coordinates ($x^0 = t, x^1, x^2, x^3$). Discarding terms that are quadratic in $h_{\mu\nu}$, the 00 component of the Ricci tensor is then

$$R_{00} = \frac{1}{2}\eta^{\beta\alpha}(h_{\beta0,0\alpha} - h_{00,\beta\alpha} - h_{\beta\alpha,00} + h_{\alpha0,\beta0})\,. \tag{6.53}$$

Slowly moving matter produces a slowly varying metric and therefore we can neglect time derivatives. Every term apart from the second in the expression for R_{00} includes at least one explicit time derivative, so

$$R_{00} = -\frac{1}{2}\eta^{\beta\alpha}h_{00,\beta\alpha}\,. \tag{6.54}$$

Neglecting the remaining time derivatives in this gives

$$R_{00} = -\frac{1}{2}\eta^{ji}h_{00,ji} = \frac{1}{2}h_{00,ii} = \frac{1}{2}\nabla^2 h_{00}\,. \tag{6.55}$$

For slowly moving matter, the rest mass is much greater than the kinetic energy, so the dominant term in the energy–momentum tensor is $T_{00} = \rho$, and for an approximately flat metric, $T^\mu_{\ \mu} = T = \rho$, so the 00 component of the right-hand side of equation (6.50) is $\frac{1}{2}\kappa\rho$. Combining this result with equation (6.55) gives

$$\nabla^2 h_{00} = \kappa\rho \,. \tag{6.56}$$

In the Newtonian limit, $h_{00} = 2\phi$ from equation (6.20) so

$$\nabla^2\phi = \frac{1}{2}\kappa\rho \,, \tag{6.57}$$

which matches Poisson's equation (6.5) if $\kappa = 8\pi G$. This fixes κ and gives the final form of the Einstein equation,

$$G_{\mu\nu} = 8\pi G\, T_{\mu\nu} \,. \tag{6.58}$$

6.6 The Classic Tests of General Relativity

Here we describe three classic tests of general relativity and the historical observations confirming the theory. Detailed calculations of the size of the first two effects are left to subsequent sections.

6.6.1 The perihelion advance of Mercury

Newtonian gravity is described by an inverse square law force. This results in elliptical orbits, which explains Kepler's first law of planetary motion, as we saw in section 2.7. The inverse square law has an enhanced symmetry, resulting in the conserved Runge–Lenz vector that points along the major axis of a planet's orbit, and remains fixed in space. Any small additional force acting on the planet will break this symmetry and the effect will be a gradual precession of the axis of the ellipse, as illustrated in Figure 6.4.

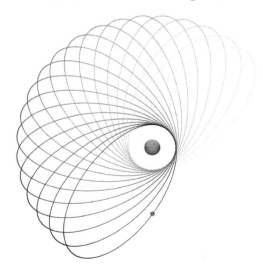

Fig. 6.4 Precession of Mercury's orbit.

Observations in the 19th century showed that Mercury's orbit around the Sun precesses by 574 arcseconds per century. (1 arcsecond is $\frac{1}{60}$ of an arcminute and in turn $\frac{1}{3600}$ of a degree.) In around 225,000 years the axis of Mercury's orbit traces out a complete circuit of the Sun. Most of this can be accounted for by perturbations due to the gravitational attraction of the other planets. The pull of Venus accounts for a shift of 277 arcseconds per century. Jupiter adds another 153 arcseconds. The Earth accounts for 90 arcseconds and the rest of the planets about 11 arcseconds more. These contributions total 531 arcseconds, which leaves 43 arcseconds per century unaccounted for.

During November 1915, Einstein addressed this issue. He performed a calculation of geodesic motion one step beyond the Newtonian approximation, which is perfectly adequate for analysing small effects in the solar system, and discovered that general relativity introduces an additional force that decreases as the inverse fourth power of distance. In the solar system, this extra term is largest in the case of Mercury, because Mercury is closest to the Sun. The additional force due to general relativity causes Mercury's orbit to precess by 43 arcseconds per century, just the right amount to explain the total observed precession. This was the moment when Einstein knew that his theory was a success. Beside himself with joyous excitement, Einstein wrote: 'My wildest dreams have been fulfilled. General covariance. Perihelion motion of Mercury wonderfully exact.'

6.6.2 The deflection of starlight

General relativity predicts that light will be deflected by the curved spacetime around a massive body. Within the solar system, the curvature of spacetime is very small. Even near the Sun, gravity is a weak force. A beam of starlight just grazing the edge of the Sun and following a light-like geodesic is deflected through an angle of just 1.75 arcseconds.

In 1919, a British expedition led by Arthur Eddington and Andrew Crommelin set out to test this prediction by photographing the deflections in the positions of stars located close to the edge of the Sun during a total eclipse. The eclipse would cross Northern Brazil, the Atlantic and Africa on 29 May 1919 and was very favourable for the mission as the duration of totality was six minutes, close to the maximum possible. It was also at an ideal position in the sky, being situated in the open star cluster known as the Hyades, where there are plenty of reasonably bright stars whose positions could be measured. Crommelin's expedition photographed the eclipse from Sobral in Brazil and Eddington's expedition photographed the eclipse from the island of Príncipe off the African coast. The measurements from Sobral gave a shift of 1.98 ± 0.16 arcseconds and the results from Príncipe gave a shift of 1.61 ± 0.4 arcseconds, confirming the prediction of general relativity.

The results of the eclipse expedition were hailed as a great triumph for general relativity. Einstein was thrust into the media spotlight and would be celebrated as an intellectual giant for the rest of his life. Figure 6.5 shows some of the news coverage of the expedition from later that year.

6.6.3 Clocks and gravitational redshift

Another prediction of general relativity is that gravity affects the passage of time. The effect can be understood most easily using the Newtonian approximation (6.20) to the spacetime metric, $d\tau^2 = (1 + 2\phi(\mathbf{x})) \, dt^2 - d\mathbf{x} \cdot d\mathbf{x}$, valid when the potential ϕ is small and vanishes at spatial infinity.

The time measured by a clock is its local proper time τ. The proper time gap between ticks is a constant $\Delta\tau$ that is independent of the position or motion of the clock. As the metric is Minkowskian at infinity, a clock at rest there moves inertially, and measures the

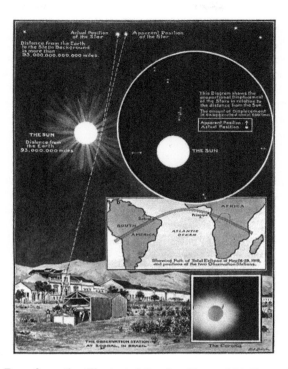

Fig. 6.5 Page from the *Illustrated London News* of 22 November 1919.

coordinate time. The gap between its ticks is $\Delta t = \Delta \tau$. A similar clock at rest at position $\mathbf{x}$, deeper in the gravitational potential, will not be moving inertially; it must be accelerating in order to remain at rest, but we assume that the acceleration has no effect on its time-keeping.[4] As the gap between ticks of the clock at $\mathbf{x}$ is $\Delta \tau$, we deduce using the metric and the approximation $(1 + 2\phi(\mathbf{x}))^{\frac{1}{2}} \simeq 1 + \phi(\mathbf{x})$ that the corresponding gap in coordinate time is $\Delta t = \Delta \tau / (1 + \phi(\mathbf{x}))$.

Suppose the ticks from the clock at $\mathbf{x}$ are signalled out to infinity. There is a time delay, but the gap Δt in coordinate time between ticks is the same at the position of the clock as it is at infinity. This is because the metric has a symmetry under any time shift, so a physical process can be moved forward by Δt throughout spacetime, and remain physical. The ticks arriving at infinity therefore have separation $\Delta \tau / (1 + \phi(\mathbf{x}))$, which is greater than $\Delta \tau$ because $\phi(\mathbf{x})$ is negative. The clock at infinity has ticks separated by $\Delta \tau$, so the clock deeper in the potential appears to an observer at infinity to have slowed down. Conversely, an observer at $\mathbf{x}$ receiving clock signals from infinity will observe the signals speeded up compared to a local clock. In summary, gravity affects clocks—not locally, but when one compares their time-keeping at different points.

A test body of mass m has (negative) potential energy $m\phi(\mathbf{x})$ at a point $\mathbf{x}$, and zero potential energy at infinity. The total energy, including the body's rest energy, is $m(1 + \phi(\mathbf{x}))$ at $\mathbf{x}$ and m at infinity. To approach spatial infinity through free motion, the body would

---

[4] This has been checked for atomic clocks subject to moderate accelerations, but of course fails for clocks that depend on the force of gravity, like pendulum clocks.

need to start with some additional kinetic energy, and this would decrease as the body approached infinity.

Similarly, a photon loses energy as it approaches infinity. Suppose that a photon is emitted from the surface of a massive body such as a star or planet, where the Newtonian potential is ϕ, and is detected by a distant observer where the potential is effectively zero. Suppose the emitted photon has (angular) frequency ω, and the observed frequency is ω_∞. The energy of a photon is initially $E = \hbar\omega$, where $\hbar$ is Planck's constant, as we will discuss in Chapter 7. The photon's energy decreases in the same way as that of a massive body, so

$$\hbar\omega_\infty = \hbar\omega(1 + \phi). \tag{6.59}$$

As ϕ is negative, ω_∞ is less than ω, and we say that the photon has undergone a *redshift* in its climb out of the gravitational well.[5] The pulsing of an electromagnetic wave is an ideal measure of the passage of time. The reduction in energy of the photon between emission and detection can be interpreted as due to a difference in the rate at which time passes at these two points. From the above calculation, we deduce again that a proper time interval at infinity is shorter, by a factor $1 + \phi$, than an equivalent proper time interval signalled out to infinity from a location where the potential is ϕ.

In 1914, Walter Adams described the first member of a new class of stars later named white dwarfs. The following year, the faint companion of the star Sirius was identified as a second such star. These stars are remarkable because they are extremely faint compared to other stars that exhibit similar spectra. As they are in binary systems their masses can be estimated, and turn out to be comparable to the mass of the Sun, $M_\odot$. (The best modern estimate for the mass of Sirius B is $0.98 M_\odot$.) Eddington argued in 1924 that these stars could only be so faint because they are very small compared to a normal star. He estimated that they are similar in size to the Earth and therefore must be incredibly compact objects with an exceptionally high density. He calculated the gravitational redshift of light emitted from Sirius B to be the equivalent of a 20 km s^{-1} Doppler shift. The following year Adams made spectrographic observations of Sirius B and measured the shift in the lines in its spectrum. After accounting for the shift due to the orbital motion of the white dwarf, there remained a redshift equivalent to a Doppler shift of 19 km s^{-1}, just as Eddington had predicted. This was acclaimed by Eddington as another great triumph for general relativity. In reality, however, there was a great deal of uncertainty in both the measurement and Eddington's calculation, so the precise agreement was rather coincidental. The modern figure for the Doppler equivalent of the gravitational redshift of Sirius B is 80.42 ± 4.83 km s^{-1}.

In 1959, a much more accurate measurement of the gravitational frequency shift was undertaken in a classic experiment at Harvard University by Robert Pound and Glen Rebka. Pound and Rebka fired gamma ray photons down the 22.5 metre Jefferson Tower at the university and measured the blueshift in the frequency of the photons at the bottom of the tower due to their fall in the Earth's gravitational field. The initial results agreed with the predictions of general relativity to within 10% accuracy. Subsequent improvements to the experiment made by Pound and Joseph Snider brought the accuracy of the agreement to within 1%.

The effects of gravitational time distortion are now routinely taken into account by the GPS (Global Positioning System) network, which is used daily by millions of people around

---

[5] Red light forms the low frequency end of the visible spectrum. Redshift is the term used to describe the decrease in frequency of electromagnetic radiation, whether or not it is in the visible part of the spectrum.

the world. The GPS system could not function for more than a few minutes if the predictions of general relativity were not incorporated into the system.

6.7   The Schwarzschild Solution of the Einstein Equation

Karl Schwarzschild was a German mathematician and astrophysicist who was stationed on the Eastern Front in 1915. During the War, he began to suffer from a rare and extremely painful autoimmune disease of the skin known as pemphigus. Somehow, in these incredibly difficult circumstances, Schwarzschild found the most important solutions to the Einstein equation, which had only been published a month earlier. His solutions describe spacetime inside and outside a perfectly spherical body, such as a star or a planet. Schwarzschild's skin condition soon worsened and in March 1916 he was removed from the front. Two months later he died.

The Einstein equation is a tensor equation relating second partial derivatives of the metric tensor to the matter and energy density. In empty space, the energy–momentum tensor $T_{\mu\nu}$ vanishes, so, as is clear from the form of equation (6.50), the Einstein equation simplifies to

$$R_{\mu\nu} = 0 \,. \tag{6.60}$$

This is known as the vacuum Einstein equation. The simplest vacuum solution is Minkowski space, the flat spacetime of special relativity, where the entire Riemann tensor vanishes.

Less trivially, the exterior Schwarzschild solution is not flat, and describes the vacuum spacetime around a spherically symmetric body. It is most simply described using polar coordinates $(t, r, \vartheta, \varphi)$. For a body of mass M whose centre is situated at the point $r = 0$, the exterior Schwarzschild metric is

$$d\tau^2 = \left(1 - \frac{2GM}{r}\right) dt^2 - \left(1 - \frac{2GM}{r}\right)^{-1} dr^2 - r^2(d\vartheta^2 + \sin^2\vartheta \, d\varphi^2) \,. \tag{6.61}$$

The non-zero metric tensor components are

$$g_{tt} = 1 - \frac{2GM}{r}, \quad g_{rr} = -\left(1 - \frac{2GM}{r}\right)^{-1}, \quad g_{\vartheta\vartheta} = -r^2, \quad g_{\varphi\varphi} = -r^2\sin^2\vartheta \,. \tag{6.62}$$

The metric tensor is diagonal, and comparing the sign of each component to the Minkowski metric, we see that t should be regarded as a time coordinate and r, ϑ, φ as spatial polar coordinates throughout the region $r > 2GM$.

The Schwarzschild metric is the relativistic counterpart of the Newtonian gravitational field outside a spherically symmetric mass, as described by the gravitational potential $\phi(r) = -\frac{GM}{r}$. Both the Newtonian field and the Schwarzschild metric include a single parameter M. We can see at once that $g_{tt} = 1 + 2\phi$, but the Einstein equation requires that the spatial part of the metric also depends on ϕ. At radii $r \gg GM$, however, the geometry is completely equivalent to the Newtonian gravitational field.

The Schwarzschild metric tensor has two manifest symmetries, as its components are independent of φ and t. There is a rotational symmetry associated with a shift of φ and, as the final terms in the metric are proportional to the 2-sphere metric, this is part of a full spherical symmetry. The metric is also symmetric under time shifts, and is said to be static. It is, in fact, the most general spherically symmetric solution of the vacuum Einstein equation.

This result is known as Birkhoff's theorem, and means that the Schwarzschild exterior metric even applies around matter that is undergoing spherically symmetric collapse or expansion. It also implies that an empty spherical cavity in a spherically symmetric spacetime is described by the Schwarzschild metric with $M = 0$, which is just flat Minkowski space. The equivalent Newtonian result is that the gravitational field vanishes within a spherical shell of matter, because the only spherically symmetric potentials satisfying Laplace's equation are of the form $\phi = \frac{C}{r} + D$, but C must be zero if there is no singularity at the origin, and then the gradient of ϕ is zero.

Fig. 6.6 2-dimensional (r, φ)-slice through the exterior Schwarzschild space.

Figure 6.6 shows a 2-dimensional slice through the space of the exterior Schwarzschild metric, at fixed t and ϑ. The slice ends at the surface of the body, as the exterior metric is no longer applicable inside.

Proving that the exterior Schwarzschild metric satisfies the vacuum Einstein equation (6.60) is not difficult. The algebra is quite laborious, but it is a useful exercise that we will sketch out. The Christoffel symbols can be worked out by plugging the components of the metric into the defining formula (6.13). Most of the terms are zero, so the Christoffel symbols can be computed quite rapidly. For instance,

$$\Gamma^t{}_{tr} = \frac{1}{2} g^{\alpha t}(g_{\alpha t,r} + g_{\alpha r,t} - g_{tr,\alpha}) = \frac{1}{2} g^{\alpha t} g_{\alpha t,r} = \frac{1}{2} g^{tt} g_{tt,r} \,, \tag{6.63}$$

because none of the metric components are time dependent, and all off-diagonal components including g_{tr} are zero. Therefore

$$\Gamma^t{}_{tr} = \frac{1}{2} \left(1 - \frac{2GM}{r}\right)^{-1} \left(\frac{2GM}{r^2}\right) = \frac{GM}{r^2 Z} \,, \tag{6.64}$$

where $Z = 1 - \frac{2GM}{r}$. The only non-zero Christoffel symbols are

$$\Gamma^t{}_{tr} = \Gamma^t{}_{rt} = \frac{GM}{r^2 Z} \,, \quad \Gamma^r{}_{tt} = \frac{GMZ}{r^2} \,, \quad \Gamma^r{}_{rr} = -\frac{GM}{r^2 Z} \,, \quad \Gamma^r{}_{\vartheta\vartheta} = -rZ \,,$$

$$\Gamma^r{}_{\varphi\varphi} = -rZ \sin^2\vartheta \,, \quad \Gamma^\vartheta{}_{r\vartheta} = \Gamma^\vartheta{}_{\vartheta r} = \Gamma^\varphi{}_{r\varphi} = \Gamma^\varphi{}_{\varphi r} = \frac{1}{r} \,,$$

$$\Gamma^\vartheta{}_{\varphi\varphi} = -\sin\vartheta\cos\vartheta \,, \quad \Gamma^\varphi{}_{\varphi\vartheta} = \Gamma^\varphi{}_{\vartheta\varphi} = \cot\vartheta \,, \tag{6.65}$$

and they can be used to compute the components of the Riemann tensor. For instance, from equation (6.14),

$$R^r{}_{trt} = \Gamma^r{}_{tt,r} - \Gamma^r{}_{rt,t} + \Gamma^r{}_{r\beta}\Gamma^\beta{}_{tt} - \Gamma^r{}_{t\beta}\Gamma^\beta{}_{rt}. \tag{6.66}$$

Most terms, such as $\Gamma^r{}_{tr,t}$ and $\Gamma^r{}_{r\vartheta}\Gamma^\vartheta{}_{tt}$, vanish leaving

$$\begin{aligned} R^r{}_{trt} &= \Gamma^r{}_{tt,r} + \Gamma^r{}_{rr}\Gamma^r{}_{tt} - \Gamma^r{}_{tt}\Gamma^t{}_{rt} \\ &= -\frac{2GMZ}{r^3} + \frac{2G^2M^2}{r^4} - \left(\frac{GM}{r^2Z}\right)\left(\frac{GMZ}{r^2}\right) - \left(\frac{GMZ}{r^2}\right)\left(\frac{GM}{r^2Z}\right) \\ &= -\frac{2GMZ}{r^3}, \end{aligned} \tag{6.67}$$

where the first two terms in the second line come from the radial derivative of $\Gamma^r{}_{tt}$. Similar calculations give other components of the Riemann tensor, for instance,

$$R^t{}_{ttt} = 0, \quad R^\vartheta{}_{t\vartheta t} = \frac{GMZ}{r^3}, \quad R^\varphi{}_{t\varphi t} = \frac{GMZ}{r^3}. \tag{6.68}$$

These results combine to give the tt component of the Ricci tensor

$$R_{tt} = R^\alpha{}_{t\alpha t} = -\frac{2GMZ}{r^3} + \frac{GMZ}{r^3} + \frac{GMZ}{r^3} = 0. \tag{6.69}$$

It can similarly be verified that all the other components of the Ricci tensor are zero and therefore the Schwarzschild metric satisfies the vacuum Einstein equation.

The vacuum Einstein equation itself does not include any mass parameter, and therefore the above calculation cannot determine the parameter M that appears in the Schwarzschild metric. The simplest way to show that M is the mass of the gravitating body is to consider the Newtonian limit at large r. Alternatively, it may be established by matching the exterior Schwarzschild metric to the interior Schwarzschild metric at the surface of the body. We will discuss the interior metric in section 6.10.

6.7.1 The Newtonian limit

It is illuminating to look further at the exterior Schwarzschild metric in the Newtonian approximation. We have already noted that this metric corresponds to a Newtonian potential $\phi(r) = -\frac{GM}{r}$ at large r, whose gradient has magnitude $\frac{GM}{r^2}$. This is rather significant. Newton's theory was built around an inverse square law force in order to match the observed motions of the planets. There was no inherent reason why the force had to diminish in this way; the choice was made in order to fit the observations. In Einstein's theory no such choice is possible. The form of the field equation is determined by very general principles, and implies that the Ricci tensor vanishes in empty space. It is a genuine prediction of general relativity that in the Newtonian limit, the gravitational potential around a spherically symmetric body falls off inversely with distance and the force decreases with the inverse square of distance. One of the most significant features of the universe has been deduced from geometrical principles.

We can also gain insight into the Newtonian limit of general relativity by looking at the equation of geodesic deviation (5.102),

$$\frac{D^2\eta^\mu}{D\tau^2} = R^\mu{}_{\nu\rho\lambda}\frac{dy^\nu}{d\tau}\frac{dy^\rho}{d\tau}\eta^\lambda,\tag{6.70}$$

where η^μ is a vector linking points on two nearby, time-like geodesics. In Minkowski space the Riemann tensor vanishes, so

$$\frac{d^2\eta^\mu}{d\tau^2} = 0.\tag{6.71}$$

This is equivalent to Newton's first law of motion applied to the non-relativistic, relative motion of two bodies.

For motion in the radial direction of Schwarzschild spacetime, the r component of equation (6.70) is

$$\frac{D^2\eta^r}{D\tau^2} = R^r{}_{ttr}\frac{dt}{d\tau}\frac{dt}{d\tau}\eta^r.\tag{6.72}$$

From equation (6.67) and the antisymmetry properties of the Riemann tensor we find $R^r{}_{ttr} = \frac{2GMZ}{r^3}$, and for the Schwarzschild metric $\left(\frac{dt}{d\tau}\right)^2 = \frac{1}{Z}$, so

$$\frac{D^2\eta^r}{D\tau^2} = \frac{2GM}{r^3}\eta^r.\tag{6.73}$$

In the Newtonian limit, the factor $\frac{2GM}{r^3}$ is interpreted as a tidal stretch along the line pointing radially out from the mass M. Geodesic deviation in the transverse ϑ and φ directions can be similarly determined. As $R^\vartheta{}_{tt\vartheta} = R^\varphi{}_{tt\varphi} = -\frac{GMZ}{r^3}$,

$$\frac{D^2\eta^\vartheta}{D\tau^2} = -\frac{GM}{r^3}\eta^\vartheta,\quad \frac{D^2\eta^\varphi}{D\tau^2} = -\frac{GM}{r^3}\eta^\varphi.\tag{6.74}$$

The factors $-\frac{GM}{r^3}$ are interpreted as tidal squeezes. Figure 6.2 shows these tidal forces acting on the Earth due to the gravitational field of the Moon.

6.8 Particle Motion in Schwarzschild Spacetime

The spacetime around a massive body such as the Sun is described to a very good approximation by the exterior Schwarzschild metric. A particle freely falling through this spacetime will follow a time-like geodesic as described by equation (6.17). For simplicity, we will suppose this particle has unit mass. As mentioned before, the parameter λ along such a geodesic can be taken to be the proper time τ, and the constant Ξ in equation (6.18) is then 1.

As the metric is spherically symmetric, we can assume the particle's worldline is in the equatorial plane $\vartheta = \frac{\pi}{2}$, without any loss of generality. The symmetry of the metric under the reflection $\vartheta \to \pi - \vartheta$ implies that any worldline starting tangent to this plane remains there. So $\sin\vartheta = 1$ and $d\vartheta = 0$, and the Schwarzschild metric reduces to

$$d\tau^2 = \left(1 - \frac{2GM}{r}\right)dt^2 - \left(1 - \frac{2GM}{r}\right)^{-1}dr^2 - r^2\,d\varphi^2.\tag{6.75}$$

A geodesic is a worldline $(t(\tau), r(\tau), \varphi(\tau))$ satisfying

$$\left(1 - \frac{2GM}{r}\right)\left(\frac{dt}{d\tau}\right)^2 - \left(1 - \frac{2GM}{r}\right)^{-1}\left(\frac{dr}{d\tau}\right)^2 - r^2\left(\frac{d\varphi}{d\tau}\right)^2 = 1, \tag{6.76}$$

the appropriate version of (6.18).

The Schwarzschild metric is static, so the particle has a conserved energy

$$E = \left(1 - \frac{2GM}{r}\right)\frac{dt}{d\tau}, \tag{6.77}$$

as implied by equation (6.19). Similarly, as the metric is symmetric under φ-rotations, the particle has a conserved angular momentum

$$l = r^2\frac{d\varphi}{d\tau}. \tag{6.78}$$

Because of these conserved quantities, equation (6.76) simplifies to

$$\left(1 - \frac{2GM}{r}\right)^{-1}\left(E^2 - \left(\frac{dr}{d\tau}\right)^2\right) - \frac{l^2}{r^2} = 1, \tag{6.79}$$

which rearranges to

$$\frac{1}{2}\left(\frac{dr}{d\tau}\right)^2 + V(r) = \frac{1}{2}E^2, \tag{6.80}$$

where

$$V(r) = \frac{1}{2}\left(1 - \frac{2GM}{r}\right)\left(1 + \frac{l^2}{r^2}\right) = \frac{1}{2} - \frac{GM}{r} + \frac{1}{2}\frac{l^2}{r^2} - \frac{GMl^2}{r^3}. \tag{6.81}$$

The geodesic equation therefore reduces to a 1-dimensional problem of a particle of unit mass and kinetic energy $\frac{1}{2}\left(\frac{dr}{d\tau}\right)^2$ moving in the potential $V(r)$, with total 'energy' $\frac{1}{2}E^2$. The second and third terms in V are the standard Newtonian gravitational potential and centrifugal terms that occur in the analysis of Newtonian orbits, but the final, inverse cubic potential is a new relativistic term that gives rise to an inverse quartic force responsible for the orbit precession.

Let us make the substitution $u = \frac{1}{r}$. Then

$$\frac{dr}{d\tau} = \frac{dr}{du}\frac{du}{d\tau} = -r^2\frac{du}{d\tau} = -r^2\frac{d\varphi}{d\tau}\frac{du}{d\varphi} = -l\frac{du}{d\varphi}. \tag{6.82}$$

Using this change of variable we find from equations (6.80) and (6.81) that

$$\frac{1}{2}l^2\left(\frac{du}{d\varphi}\right)^2 + \frac{1}{2} - GMu + \frac{1}{2}l^2u^2 - GMl^2u^3 = \frac{1}{2}E^2. \tag{6.83}$$

Differentiating with respect to φ and dividing through by $l^2\frac{du}{d\varphi}$ gives

$$\frac{d^2u}{d\varphi^2} + u - \frac{GM}{l^2} = 3GMu^2. \tag{6.84}$$

In the absence of the term on the right-hand side, the solution is

$$u = \frac{GM}{l^2}(1 + e\cos\varphi),$$ (6.85)

or equivalently $r(1+e\cos\varphi) = \frac{l^2}{GM}$, matching the solution (2.95) we found for the Newtonian orbit.[6] e and l are constants determined by the initial conditions.

The additional term $3GMu^2$ produces the relativistic modification to the Newtonian orbit. In the solar system this term is very small, and u can be approximated by substituting the Newtonian solution, giving

$$\frac{d^2u}{d\varphi^2} + u - \frac{GM}{l^2} = \frac{3(GM)^3}{l^4}(1 + e\cos\varphi)^2.$$ (6.86)

The improved solution is then

$$u = \frac{GM}{l^2}(1 + e\cos\varphi) + \frac{3(GM)^3}{l^4}\left\{\left(1 + \frac{e^2}{2}\right) - \frac{e^2}{6}\cos 2\varphi + e\,\varphi\sin\varphi\right\}.$$ (6.87)

The first term in the braces is a small constant and the second is cyclic, producing a small correction that repeats every orbit and does not increase with time. Keeping only the final increasing term, we obtain

$$u = \frac{GM}{l^2} + \frac{GMe}{l^2}\left(\cos\varphi + \frac{3(GM)^2}{l^2}\varphi\sin\varphi\right).$$ (6.88)

The functions of φ on the right-hand side can be combined by using the trigonometric expansion

$$\cos\{(1-\alpha)\varphi\} = \cos\varphi\cos\alpha\varphi + \sin\varphi\sin\alpha\varphi \simeq \cos\varphi + \alpha\varphi\sin\varphi$$ (6.89)

for small α, leading to

$$u = \frac{GM}{l^2} + \frac{GMe}{l^2}\cos\{(1-\alpha)\varphi\},$$ (6.90)

where

$$\alpha = \frac{3(GM)^2}{l^2}.$$ (6.91)

At perihelion, the point of closest approach to the Sun, r reaches its minimum and u its maximum, so $\cos\{(1-\alpha)\varphi\} = 1$, and therefore after N orbits

$$(1-\alpha)\varphi = 2\pi N.$$ (6.92)

The angle φ at perihelion is therefore

$$\varphi \simeq 2\pi N + 2\pi N\alpha,$$ (6.93)

so with each orbit, the perihelion advances by

$$\Delta\varphi = 2\pi\alpha = \frac{6\pi(GM)^2}{l^2} = \frac{6\pi GM}{a(1-e^2)},$$ (6.94)

where we have recalled that for the Newtonian orbit of a unit mass particle, the relation between angular momentum and the semi-major axis is $\frac{l^2}{GM} = a(1-e^2)$.

---

[6] The solution $\sin\varphi$ is not needed if we choose φ to be zero at the maximum of u.

In the solar system the effect is greatest in the case of Mercury, which is closest to the Sun. Mercury also completes its orbits in less time than the other planets so the deviations from Newtonian behaviour accumulate faster.

Newton's gravitational constant is $G = 6.67 \times 10^{-11}$ m^3 kg^{-1} s^{-2} and the speed of light is $c = 3.00 \times 10^8$ m s^{-1}, so in units where the speed of light is 1, Newton's constant is $G = 7.42 \times 10^{-28}$ m kg^{-1}. The mass of the Sun is $M_\odot = 1.99 \times 10^{30}$ kg, so $GM_\odot = 1.48 \times 10^3$ m. The semi-major axis of Mercury's orbit is $a = 5.79 \times 10^{10}$ m and its eccentricity is $e = 0.206$. Plugging these numbers in, we find that the rate of perihelion advance is 5.04×10^{-7} radians per orbit. Mercury has an orbital period of 88.0 days, so there are 415 orbits per century. Therefore, the perihelion advance per century is 2.09×10^{-4} radians, or 43.1 arcseconds.

In 1974 the first binary neutron star system PSR B1913+16 was discovered by Russell Hulse and Joseph Taylor using the giant radio telescope at Arecibo in Puerto Rico. Neutron stars are collapsed stellar remnants that are compressed to nuclear densities. One of the neutron stars in the binary system generates a pulsar, which is a beam of electromagnetic radiation that points in our direction once every revolution of the neutron star. (We will discuss pulsars in section 13.8.1.) The neutron star rotates 17 times a second, so we receive a pulse of radio waves every 59 milliseconds. The radio pulses are received with an incredible regularity, but vary slowly with a period of 7.75 hours due to the Doppler shifts as the neutron star orbits its companion. These Doppler shifts have enabled astronomers to determine the orbital characteristics of this pulsar system with exquisite precision. Many pulsars are known in binary systems, but in most of these systems the companion is a normal star and the transfer of material on to the neutron star complicates the dynamics. By contrast, PSR B1913+16 is a very clean environment in which to study the orbital mechanics. The intense gravitational field in this system, and the precision with which the position of the neutron stars can be calculated, make it the ideal testing ground for general relativity. Astronomers have determined the masses of the two neutron stars to be $1.4411 \pm 0.0007 \, M_\odot$ for the pulsar and $1.3873 \pm 0.0007 \, M_\odot$ for the companion. Their orbit is highly eccentric, with $e = 0.617$, and the length of the semi-major axis is 9.75×10^8 m. At the point of closest approach, the separation of the two neutron stars is just 1.1 solar radii; at their furthest separation it is 4.8 solar radii.

The axis of the orbit advances much more rapidly than for Mercury. Plugging the numbers from the above paragraph into the formula (6.94) produces a precession of 4.2 degrees per year. The observed precession is in perfect agreement with this prediction of general relativity. Each day the orbit shifts by 41.4 arcseconds, almost as much as Mercury's orbit shifts in a century.

In 2003 a double pulsar system of neutron stars known as PSR J0737-3039A and PSR J0737-3039B was discovered at the Parkes observatory in Australia. This remains the only known binary system in which both components are visible pulsars, enabling the system to be accurately monitored. The orbital period is just 2.4 hours, and the axis of the orbit precesses by 16.90 degrees per year, again confirming general relativity's prediction.

6.9   Light Deflection in Schwarzschild Spacetime

To calculate the deflection of light passing through the curved spacetime close to a spherical mass, as illustrated in Figure 6.7, we need to find the light rays in Schwarzschild spacetime. A light ray follows a light-like geodesic and we may again assume it is in the equatorial

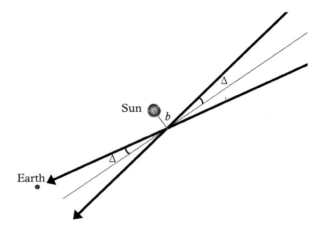

Fig. 6.7 Deflection of light around a massive body.

plane $\vartheta = \frac{\pi}{2}$. In equation (6.18) we set $\Xi = 0$, and the parameter λ is no longer τ. Using the conservation laws for energy and angular momentum and setting $u = \frac{1}{r}$ as before, we find the equation for the light ray

$$\frac{d^2u}{d\varphi^2} + u = 3GMu^2 \,. \tag{6.95}$$

In the solar system the term on the right is again very small. If this term is neglected, the solution is a straight line

$$u = \frac{1}{b}\cos\varphi \,, \tag{6.96}$$

where b is the impact parameter, the distance of closest approach to the central mass. For convenience we have chosen φ to be zero at closest approach, so φ increases from $-\frac{\pi}{2}$ to $\frac{\pi}{2}$ along the line. To find an improved solution, we substitute the straight line solution into the small term on the right-hand side of (6.95), giving

$$\frac{d^2u}{d\varphi^2} + u = \frac{3GM}{b^2}\cos^2\varphi \,, \tag{6.97}$$

whose solution is readily seen to be

$$u = \frac{1}{b}\cos\varphi + \frac{GM}{b^2}(2 - \cos^2\varphi) \,. \tag{6.98}$$

At the ends of the light ray, where $u = 0$,

$$\frac{1}{b}\cos\varphi + \frac{GM}{b^2}(2 - \cos^2\varphi) = 0 \,. \tag{6.99}$$

As φ is close to $\pm\frac{\pi}{2}$, we neglect the $\cos^2\varphi$ term, obtaining

$$\cos\varphi = -\frac{2GM}{b} \,. \tag{6.100}$$

The solution is $\varphi = -\frac{\pi}{2} - \Delta$ in one direction and $\varphi = \frac{\pi}{2} + \Delta$ in the other, where Δ is small. Using the familiar trigonometric formulae $\cos\left(-\frac{\pi}{2} - \Delta\right) = \cos\left(\frac{\pi}{2} + \Delta\right) = -\sin\Delta \simeq -\Delta$, we then find

$$\Delta \simeq \frac{2GM}{b}, \tag{6.101}$$

so the full angular deflection is

$$2\Delta \simeq \frac{4GM}{b}. \tag{6.102}$$

For the Sun, $GM_\odot = 1.48 \times 10^3$ m, and if we take b to be the solar radius, which is 6.96×10^8 m, then a ray of starlight that just grazes the edge of the Sun is deflected by an angle of 8.48×10^{-6} radians or 1.75 arcseconds, as Einstein famously predicted and the eclipse expedition of 1919 confirmed.

The bending of light due to gravity can be seen in gravitational lenses. The light from a galaxy at cosmic distances may be bent around an intervening cluster of galaxies to produce multiple images of the more distant galaxy. Numerous instances of such gravitational lensing systems have been discovered. In the ideal situation where the alignment is exact and the lensing mass is spherically symmetric the image should warp into a circle known as an *Einstein ring*. An example of an almost perfect Einstein ring is shown in Figure 6.8.

Gravitational lenses offer an unambiguous method of determining the mass of a cluster of galaxies. Distances to both the lensing galactic cluster and the more distant galaxy whose distorted image is observed can be determined from their redshifts. (We will look at cosmological redshift in Chapter 14.) The angular size of the ring produced by the gravitational lens can be measured. Combining the distances and the angular size gives the impact parameter b and the total angular deflection 2Δ. Then formula (6.102) can be used to determine the mass of the gravitational lens. Such calculations produce estimates for the amount of material in a cluster of galaxies that greatly exceed what is inferred from the amount of light emitted by the cluster. This suggests that clusters of galaxies are accompanied by a great deal of material that does not emit light and is therefore known as *dark matter*. The identity of this dark matter is, as yet, unknown. The prime candidate is an unknown species of stable particle that would have been produced in vast quantities in the very early universe. We will return to this question in Chapter 14.

6.10 The Interior Schwarzschild Solution

The interior Schwarzschild metric describes spacetime in the interior of a spherically symmetric body of density $\rho(r)$ and pressure $P(r)$ whose centre is situated at $r = 0$. It takes the form

$$d\tau^2 = e^{2\psi(r)}dt^2 - \left(1 - \frac{2GM(r)}{r}\right)^{-1}dr^2 - r^2(d\vartheta^2 + \sin^2\vartheta\,d\varphi^2), \tag{6.103}$$

where

$$M(r) = 4\pi \int_0^r \rho(r')r'^2 dr' \tag{6.104}$$

is the integrated mass from the centre, and $\psi(r)$ is the solution of

$$\frac{d\psi}{dr} = \frac{G(M(r) + 4\pi r^3 P(r))}{r(r - 2GM(r))}, \tag{6.105}$$

satisfying $\psi(\infty) = 0$.

Fig. 6.8 The almost perfect Einstein ring LRG 3-757 photographed by the Hubble Space Telescope Wide Field Camera 3. The ring has a diameter on the celestial sphere of 11 arcseconds (ESA-Hubble and NASA).

In the idealized situation where the density ρ is constant throughout the body, $M(r) = \frac{4}{3}\pi\rho r^3$. In this case, the spatial part of the metric (6.103) is

$$ds^2 = \frac{dr^2}{1 - Kr^2} + r^2(d\vartheta^2 + \sin^2\vartheta\, d\varphi^2) \tag{6.106}$$

with $K = \frac{8}{3}\pi G\rho$. This is the metric (5.73) of a 3-sphere of constant curvature $\frac{8}{3}\pi G\rho$, and hence radius $\left(\frac{3}{8\pi G\rho}\right)^{\frac{1}{2}}$. The interior metric only covers part of the 3-sphere, and only a very small part for a body like the Earth, where $GM(r)$ is everywhere much less than r, so Kr^2 is much less than 1.

Outside a spherically symmetric mass, space is described by the exterior Schwarzschild solution, and has curvature components of both signs. This is similar to the hyperbolic soap film shown in Figure 1.3, where at each point on the surface the curvature components are equal and opposite to produce balancing surface tension forces. In the exterior Schwarzschild geometry there are three spatial dimensions and to satisfy the Einstein equation the inwardly directed curvature in the two angular directions balances the outwardly directed curvature in the radial direction, as given in equation (6.69). In the Newtonian picture, tidal forces stretch bodies radially and squeeze them in the orthogonal directions.

Within the mass, all three spatial curvatures are inwardly directed, so space is positively curved and bodies are squeezed in all three directions. The three curvature components are now balanced in the Einstein equation by the non-gravitational, outward stress exerted by the matter. The positive spatial curvature compresses the material of which the body is composed, and this is resisted by structural forces within the body. In the absence of such forces, which may be electromagnetic or nuclear, the body must collapse.

Figure 6.9 shows a 2-dimensional slice through the exterior and interior spatial metrics corresponding to the space in and around a spherical mass of uniform density. The interior Schwarzschild metric joins the exterior Schwarzschild metric continuously at the surface, where $M(r)$ equals the total mass M. This confirms that the parameter M in the exterior metric is the total mass inside.

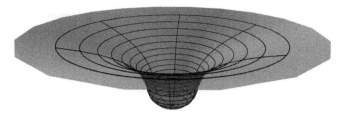

Fig. 6.9 2-dimensional slice through the Schwarzschild exterior and interior geometry. (Note that the portion of the 3-sphere within the spherical body has become part of a 2-sphere in the slice.)

6.11 Black Holes

Something strange appears to happen to the exterior Schwarzschild metric (6.61) at the radius $r_S = 2GM$, which is known as the *Schwarzschild radius*. Here, g_{tt} is zero and g_{rr} is infinite. The Schwarzschild radius of a body is usually irrelevant, as the body is physically much larger than this radius and the exterior Schwarzschild geometry morphs into the interior Schwarzschild geometry at the surface of the body. For example, the Schwarzschild radius of the Sun is about 3 km, but the exterior solution only applies at distances greater than the radius of the Sun, which is about 700,000 km. Inside the Sun, to a very good approximation, the geometry is described by the interior Schwarzschild solution.

The Sun is supported by the pressure that arises from thermal motion of its constituent particles, which is dependent on the continuous release of energy through nuclear fusion, as we will discuss in Chapter 13. When a star has consumed its nuclear fuel, it must collapse under its own gravity. The end result depends on the mass of the star. A mass of up to $1.44M_\odot$ can be supported by electron degeneracy pressure in the form of a white dwarf. More massive stars collapse to form neutron stars which are supported by nuclear forces and neutron degeneracy pressure. They have radii of 10–15 km, perilously close to their Schwarzschild radii. The maximum mass that can be supported as a neutron star is believed to be in the region of 2–3 $M_\odot$. There is no known mechanism to support a star with a greater mass once its nuclear fuel has been consumed.

If a massive body is crushed under its own gravity to the extent that its radius shrinks within its Schwarzschild radius, then nothing can prevent its inexorable collapse. Such a collapsed object is known as a *black hole*, as not even light can escape from inside the Schwarzschild radius. The vacuum spacetime around a black hole of mass M is described by the Schwarzschild metric, from radius $r = 0$ outwards.

The observational evidence for the existence of black holes is now overwhelming. Numerous examples are known of black holes with a mass of order ten solar masses, and supermassive black holes with masses of millions or even billions of solar masses are known to inhabit the central regions of most, if not all, galaxies. The black hole at the centre of the Milky Way has a mass of $4 \times 10^6 M_\odot$, as inferred from the motion of stars seen orbiting it. Recently, detection of gravitational waves, apparently generated in a black hole merger, has provided direct and compelling further evidence for the existence of black holes.

A black hole is very small by cosmic standards. Because of this, it is quite unlikely that much material falls directly into one. Rather, a swirling accretion disc is expected to form around a black hole. Friction causes the material in the accretion disc to gradually lose energy and spiral inwards before finally plummeting into the abyss. A large amount of gravitational energy is released in this process. This heats the accretion disc to extremely high temperatures resulting in the emission of X-rays.

We will now examine the nature of the circular orbits in Schwarzschild spacetime with the aim of shedding some light on the energy release in a black hole accretion disc. The orbits available to unit mass particles in the spacetime described by the exterior Schwarzschild metric are given by solutions to equation (6.80). Stable circular orbits with no radial motion are found at the minimum of the potential (6.81). It is convenient to use again the variable $u = \frac{1}{r}$, so the potential is

$$V(u) = \frac{1}{2} - GMu + \frac{1}{2}l^2u^2 - GMl^2u^3 \,. \tag{6.107}$$

Its stationary points are where $\frac{dV}{du} = 0$, that is, where

$$-GM + l^2u - 3GMl^2u^2 = 0 \,, \tag{6.108}$$

which we can rewrite as

$$\frac{GM}{l^2} = u - 3GMu^2 \,. \tag{6.109}$$

The right-hand side increases from 0 at $u = 0$ to a maximum value $\frac{1}{12GM}$ at $u = \frac{1}{6GM}$, then decreases to 0 at $u = \frac{1}{3GM}$. So for all $l^2 > 12(GM)^2$, there are two solutions for u, one less than and one greater than $\frac{1}{6GM}$. The second derivative of the potential $V(u)$ is $l^2(1 - 6GMu)$, so the solution with $u < \frac{1}{6GM}$ is a minimum of V and is stable, and the other solution is unstable. In terms of the radius, the orbits with $r > 6GM$, three times the Schwarzschild radius, are stable, and those with $6GM \geq r > 3GM$ are unstable.

This means that the inner radius of an accretion disc around a black hole of mass M lies at a distance of $r = 6GM$, and particles there have the critical value of the angular momentum $l = \sqrt{12}\,GM$. We can readily calculate the energy released by any material that reaches this inner edge. Returning to equation (6.80), we see that the energy of a unit mass particle in a circular orbit around the black hole is given by

$$E^2 = (1 - 2GMu)(1 + l^2u^2) \,. \tag{6.110}$$

At the inner edge of the accretion disc, $u = \frac{1}{6GM}$ so $2GMu = \frac{1}{3}$, $l^2u^2 = \frac{1}{3}$ and the particle has energy

$$E = \sqrt{\frac{8}{9}} \,. \tag{6.111}$$

The fraction of the mass of the particle that has therefore been released on its trip to this point is

$$1 - E = 1 - \sqrt{\frac{8}{9}} \approx 0.057 \,. \tag{6.112}$$

From here on, the particle may be expected to rapidly fall into the black hole taking any kinetic energy generated in the final plummet with it. Therefore we can expect a total of about 5.7% of the mass accreted by the black hole to be emitted as energy before the mass disappears into the black hole. This may be compared to the nuclear fusion of hydrogen into helium, which releases around 0.7% of the mass of the hydrogen as energy. We will see shortly that spinning black holes have the potential to release even more energy into their environment.

6.11.1 Eddington–Finkelstein coordinates

The exterior Schwarzschild metric (6.61) is asymptotically Minkowskian. The coordinates $(t, r, \vartheta, \varphi)$ that we have used to describe it are convenient for an observer far from the centre, but from the time component of the metric tensor we see that clocks appear to slow down and stop at the Schwarzschild radius $r_S = 2GM$. This implies a large redshift of any signals detected by a distant observer from an object close to the Schwarzschild radius. The redshift affects both the frequency of any emitted radiation and the period of time between radiation pulses. To a distant observer, an object falling into a black hole disappears just before reaching the Schwarzschild radius.

Something even more alarming appears to happen to the radial component of the metric tensor at the Schwarzschild radius. It blows up there, suggesting that the geometry is singular. However, this is an artifact of the coordinate choice, and in reality the metric as a whole remains smooth and Lorentzian at $r = r_S$. To understand this, we need better coordinates. Useful coordinates were discovered by Eddington in 1924 and independently rediscovered by David Finkelstein in 1958. The Eddington–Finkelstein coordinates retain r, ϑ, φ and replace the time t with a new coordinate v defined by

$$t = v - r - 2GM \log \left| \frac{r}{2GM} - 1 \right| . \tag{6.113}$$

Differentiating this expression gives

$$dt = dv - dr + \frac{dr}{1 - \frac{r}{2GM}} = dv - \frac{dr}{1 - \frac{2GM}{r}} , \tag{6.114}$$

and substituting for dt, the Schwarzschild metric (6.61) is transformed to

$$d\tau^2 = \left(1 - \frac{2GM}{r} \right) dv^2 - 2\, dv\, dr - r^2 \left(d\vartheta^2 + \sin^2 \vartheta\, d\varphi^2 \right) \tag{6.115}$$

in both the regions $r < 2GM$ and $r > 2GM$. The metric is now well behaved even at $r = r_S = 2GM$. The surface at this radius is a sphere with metric

$$(2GM)^2 \left(d\vartheta^2 + \sin^2 \vartheta\, d\varphi^2 \right) . \tag{6.116}$$

This is the boundary between light rays that fall towards the centre of the black hole and those that escape to infinity, and is known as the *event horizon* of the black hole. The event horizon has area $4\pi(2GM)^2$.

For large r, the logarithmic term in equation (6.113) is negligible compared to r, so $t \simeq v - r$ and the metric is approximately

$$d\tau^2 \simeq dv^2 - 2\, dv\, dr - r^2 \left(d\vartheta^2 + \sin^2 \vartheta\, d\varphi^2 \right) , \tag{6.117}$$

which is the flat Minkowski metric, as one sees by changing to coordinates $(v - r, r, \vartheta, \varphi)$.

Light travels along light-like geodesics, on which $d\tau^2 = 0$. We are interested here in the radial light rays, and can make use of the spherical symmetry to set $d\vartheta = d\varphi = 0$. There are two radial rays through each point. In flat space far from the black hole they would travel in opposite directions, one radially inwards and one radially outwards, and would be represented on a time–radius diagram by lines at $45°$ to the vertical.

The radial light rays in Eddington–Finkelstein coordinates are given by

$$\left(1 - \frac{2GM}{r}\right) dv^2 - 2\, dv\, dr = 0\,. \tag{6.118}$$

One solution is $dv = 0$, implying that v is constant, and it represents a light ray going inwards towards the centre of the black hole. We see from equation (6.113) that as t increases, r must decrease if v is to remain constant. This solution behaves as we might expect, but the second solution of (6.118), which satisfies

$$\frac{dr}{dv} = \frac{1}{2}\left(1 - \frac{2GM}{r}\right)\,, \tag{6.119}$$

is more remarkable. When $r > 2GM$, $\frac{dr}{dv}$ is positive, so the ray is outgoing. However, when $r < 2GM$, $\frac{dr}{dv}$ is negative, so the ray is ingoing. This means that once inside the event horizon of a black hole all the light emitted by a radiating body will ultimately fall inwards to the centre of the black hole. Integrating equation (6.119) gives

$$v - 2r - 4GM \log\left|\frac{r}{2GM} - 1\right| = \text{constant}\,. \tag{6.120}$$

In Figure 6.10, $\tilde{t} \equiv v - r$ is plotted against r. The figure shows the radial light-like geodesics around a black hole in Eddington–Finkelstein coordinates. The lightcones appear to tip over as the event horizon is approached. The paths of ingoing light rays, given by our first solution $v = $ constant, are shown as straight lines inclined at $-45°$ to the axes. The curved lines represent our second solution, which, outside the event horizon, are the paths of outgoing light rays, but inside the event horizon are ingoing light rays falling to the centre of the black hole.

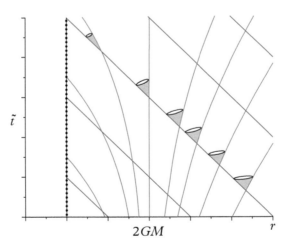

Fig. 6.10 Spacetime diagram in advanced Eddington–Finkelstein coordinates.

The trajectory of any material particle always lies within the light cones, where $d\tau^2 > 0$. Using the lightcone diagram, we can visualize the possible radial trajectories of massive particles.

As we have seen, the apparent metric singularity at $r = r_S$ is innocuous, but the Schwarzschild metric has a singularity of an altogether different character at the point $r = 0$. This singularity cannot be removed by a transformation to different coordinates. It is a point of infinite density and infinite spacetime curvature. As no such object can exist physically, it is believed that the prediction of this singularity indicates that general relativity has been stretched into a regime where it no longer accurately represents the physical world. At some point in the gravitational collapse within a black hole, such incredible densities will be reached that the physics can only be described in terms of a quantum theory of gravity. As yet, we do not have a viable quantum theory of gravity, so the centre of a black hole remains a mystery.

6.11.2   The Kerr metric

Although the exterior Schwarzschild metric is a possible geometry of spacetime around a black hole, there is a sense in which it is not physically realistic. Black holes are the result of the gravitational collapse of rotating objects, and are expected to be rapidly spinning. This has been confirmed by astronomical observations. The Schwarzschild metric, being spherically symmetric, describes the spacetime around a non-spinning spherical mass or black hole. In 1963, Roy Kerr found a more general solution of the vacuum Einstein equation. Outside a spinning body or black hole of mass M and angular momentum J, spacetime is described by the axisymmetric Kerr metric

$$
d\tau^2 = \left(1 - \frac{2GMr}{\rho^2}\right) dt^2 + \frac{4GMar\sin^2\vartheta}{\rho^2}\, dt\, d\varphi - \frac{\rho^2}{r^2 - 2GMr + a^2}\, dr^2
$$
$$
- \rho^2\, d\vartheta^2 - \left(r^2 + a^2 + \frac{2GMa^2 r\sin^2\vartheta}{\rho^2}\right)\sin^2\vartheta\, d\varphi^2, \tag{6.121}
$$

where $a = \frac{J}{M}$ is known as the angular momentum parameter, and $\rho^2 = r^2 + a^2\cos^2\vartheta$.

The body generating the metric rotates steadily, so none of the Kerr metric components are functions of time. However, unlike the Schwarzschild metric, the Kerr metric includes a time-space cross-term $g_{t\varphi}\, dt\, d\varphi$, with a coefficient proportional to J. Time reversal, $t \to -t$, changes the sign of this term and no others. This may be cancelled by the transformation $\varphi \to -\varphi$, so time reversal is equivalent to reversing the direction of rotation of the body, that is, to reversing the sign of J. The Kerr metric is referred to as stationary, but not static. It reduces to the exterior Schwarzschild metric when $J = 0$.

The Kerr metric is almost, but not quite, the most general metric representing a black hole. There is an extension known as the Kerr–Newman metric, which includes an electromagnetic field and describes an electrically charged, spinning black hole. In 1972 Stephen Hawking proved that this is the most general metric of an isolated black hole. So, according to general relativity, all black holes can be described in terms of just three parameters: M, J and Q, where M is the mass, J the angular momentum, and Q the charge. This is known as the *no-hair theorem*. There is no known mechanism for giving a significant charge to a black hole, so it is almost certain that real black holes can be described simply in terms of M and J.

The radius r_+ of the event horizon of a spinning black hole is smaller than in the non-spinning case and is given by

$$r_+ = \frac{1}{2}r_S + \sqrt{\frac{1}{4}r_S^2 - a^2}\,, \tag{6.122}$$

where $r_S = 2GM$ is the Schwarzschild radius. The maximum possible angular momentum parameter of the black hole is $a = \frac{1}{2}r_S = GM$, in which case the angular momentum is $J = GM^2$. In this limit the event horizon radius is $r_+ = GM$, half the Schwarzschild radius. The angular velocity of the event horizon is $\Omega = \frac{a}{2Mr_+}$. This is the rate at which light rays at the event horizon rotate around the black hole.

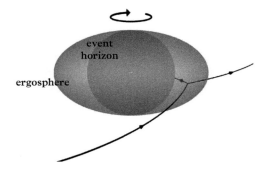

Fig. 6.11 The ergosphere of a rotating black hole.

A freely falling body in the vicinity of a black hole, with no angular momentum, follows a time-like geodesic that takes it inside the event horizon towards the centre of the black hole. To remain static with respect to a distant observer, such that it hovers above the event horizon at a fixed radial coordinate r, the body must be subject to an acceleration, provided by a rocket engine perhaps. For such a body, $dr = d\vartheta = d\varphi = 0$. If we first consider Schwarzschild spacetime, then on such a trajectory all the components of the metric (6.61) vanish except $g_{tt}\, dt^2$. The worldline of a massive particle must be time-like, so $d\tau^2 > 0$, which implies that $g_{tt} > 0$ and therefore $\frac{2GM}{r} < 1$. This is always true outside the event horizon, so given sufficient acceleration it is always possible for the body to remain static. This is not so in the Kerr spacetime around a spinning black hole. In Kerr spacetime there is a region outside the event horizon known as the *ergosphere*, in which it is still possible to escape the black hole, but it is not possible to remain static with respect to a distant observer. Static time-like worldlines are again only possible if g_{tt} is positive and therefore

$$1 - \frac{2GMr}{r^2 + a^2 \cos^2\vartheta} > 0\,. \tag{6.123}$$

This condition only holds outside the ergosphere. The boundary of the ergosphere is determined by the quadratic equation

$$r_{\text{ergo}}^2 + a^2 \cos^2\vartheta = 2GMr_{\text{ergo}} = r_S r_{\text{ergo}}\,, \tag{6.124}$$

so it is the oblate spheroid

$$r_{\text{ergo}}(\vartheta) = \frac{1}{2}r_S + \sqrt{\frac{1}{4}r_S^2 - a^2 \cos^2 \vartheta}\,. \qquad (6.125)$$

Any body inside the ergosphere will necessarily be dragged around by the rotation of the black hole.

The boundary of the ergosphere touches the event horizon at the poles, $\vartheta = 0$ and $\vartheta = \pi$, where the effect of the black hole's spin disappears. The ergosphere was named by Roger Penrose, who showed in 1969 that it is possible to extract rotational energy from a black hole. According to his scheme, material could be sent into the ergosphere where it would divide into two pieces, one of which is fired into the black hole with negative energy, while the other escapes to infinity with greater total energy than the original material that entered the ergosphere. The Penrose process is depicted in Figure 6.11.

According to the Kerr metric, even a massive spinning body such as the Earth drags space around with it. This *frame-dragging* effect, as it is known, was confirmed by Gravity Probe B, which was placed in Earth orbit in 2004. The frame-dragging due to the Earth's rotation was measured to an accuracy of around 15% by a quartet of gyroscopes aboard the probe. The effect was measured to be just 40 milliarcseconds per year, which agrees with the prediction of general relativity. The effect is much greater in the vicinity of a black hole; it implies that the accretion disc of a spinning black hole must lie in the black hole's equatorial plane and rotate in the same direction as the black hole. Orbits in the same direction as the black hole's spin are known as co-rotating, and orbits in the opposite direction are counter-rotating. There are stable co-rotating particle orbits in the Kerr metric that are much closer to the origin than in the Schwarzschild case, so the inner edge of an accretion disc around a rapidly spinning black hole is much deeper in its gravitational well. This greatly increases the binding energy of the closest stable circular orbit. Orbits of unit mass particles in the equatorial plane ($\vartheta = \frac{\pi}{2}$) of the Kerr metric correspond to solutions of

$$\frac{1}{2}\left(\frac{dr}{d\tau}\right)^2 + V(r) = \frac{1}{2}E^2\,. \qquad (6.126)$$

In terms of $u = \frac{1}{r}$, the effective potential here is

$$V(u) = \frac{1}{2} - GMu + \frac{1}{2}(l^2 + a^2(1 - E^2))u^2 - GM(l - aE)^2 u^3\,, \qquad (6.127)$$

where $l > 0$ for co-rotating orbits and $l < 0$ for counter-rotating orbits. V reduces to the Schwarzschild potential (6.107) when $a = 0$. For circular orbits, r and hence u is constant, so $V(u) = \frac{1}{2}E^2$. These orbits are stable if $V(u)$ is a minimum, requiring $\frac{dV}{du} = 0$ and $\frac{d^2V}{du^2} > 0$. The inner edge of the accretion disc lies at the radius r_{min} where $\frac{d^2V}{du^2} = 0$, beyond which there are no stable circular orbits. From the simultaneous equations arising from the conditions $\frac{dV}{du} = \frac{d^2V}{du^2} = 0$, we obtain

$$\frac{1}{2}(l^2 + a^2(1 - E^2))u = GM\,, \quad 3GM(l - aE)^2 u^2 = GM\,. \qquad (6.128)$$

Substituting these expressions into the equation $V(u) = \frac{1}{2}E^2$, we find

$$1 - E^2 = \frac{2}{3}GMu = \frac{2GM}{3r}. \tag{6.129}$$

The inner edge of the accretion disc is closer to the event horizon than for a Schwarzschild black hole. In the limiting case of a maximally spinning black hole, where $a = GM$, the inner edge coincides with the event horizon[7] at $r_{\min} = r_+ = GM$. In this case, equation (6.129) implies $E = \frac{1}{\sqrt{3}}$. To reach this radius, particles must release a large proportion of their rest mass as energy; as $1 - E = 1 - \frac{1}{\sqrt{3}} \simeq 0.42$, a remarkable 42% of the rest mass of material in the accretion disc is converted into other forms of energy prior to the material entering the black hole.

This has important astrophysical consequences. The release of gravitational energy from material falling into a black hole spinning at close to the maximum possible rate will approach 30–40%. For this reason, rapidly spinning supermassive black holes are now generally accepted as the origin of the most energetic phenomena in the universe, such as quasars and active galactic nuclei.

Fig. 6.12 The quasar designated RX J1131-1231 appears as four images here due to a gravitational lens—the three bright spots on the left of the ring and the one on the right. The diameter of the ring is about 3 arcseconds. (Combined image from NASA's Chandra X-ray Observatory and the Hubble Space Telescope.)

A quasar at a distance of six billion light years, discovered in 2008, appears as four images due to the lensing effect of an intervening galaxy, as shown in Figure 6.12. The energy source of the quasar, designated RX J1131-1231, is thought to be a supermassive black hole with a mass of around $10^8 M_\odot$. The quasar images are magnified by a factor of 50 by the gravitational lens. This has enabled astrophysicists to determine the inner radius of the black hole's accretion disc, by measuring the broadening of an emission line in the spectrum of iron atoms in the disc due to their gravitational redshift.[8] It has been

---

[7] The minimum radius of counter-rotating circular orbits is $r = 9GM$ for a maximally spinning black hole.

[8] The line corresponds to the emission of 6.4 keV X-ray photons.

estimated that the inner edge of the accretion disc has a radius less than $3GM$, half that of a non-spinning Schwarzschild black hole, so the black hole must be spinning extremely rapidly. The most likely value for the angular momentum parameter is $a \simeq 0.87\,GM$.

6.12 Gravitational Waves

When an electrically charged object such as an electron is shaken, it emits electromagnetic waves. This is what happens in a radio transmitter. Pulses of the electromagnetic field propagate through spacetime in accordance with Maxwell's equations, and produce oscillating forces when they impinge on test charges. Similarly, according to general relativity, shaking or colliding massive objects generate gravitational waves. These ripples in the gravitational field are propagating distortions of the spacetime metric; they are not simply oscillations of the coordinates, because the curvature oscillates too. The detection of a gravitational wave requires the position of at least two test particles to be monitored. Figure 6.13 shows the effect of a passing gravitational wave on a ring of test particles.

Fig. 6.13 The effect of a gravitational wave of one polarization on a ring of test particles. Five frames are shown from one wave cycle.

Fig. 6.14 The effect of a gravitational wave of the other polarization.

Gravitational waves have no analogue in Newtonian gravity, because the Newtonian potential ϕ is determined by the instantaneous matter density and does not obey a wave equation, so their very existence is a critical test of general relativity. Because gravity is intrinsically so weak, gravitational waves have an incredibly small amplitude. Only the most energetic events in the universe produce waves that could conceivably be detected on Earth. It is thought that the largest gravitational waves incident on the Earth produce fractional changes in distance of the order of 10^{-21}. Despite this, they carry vast amounts of energy distributed over enormous regions.

As gravitational wave amplitudes are so small, we can safely use the linear approximation to the metric tensor,

$$g_{\mu\nu} = \eta_{\mu\nu} + h_{\mu\nu}\,. \tag{6.130}$$

$\eta_{\mu\nu}$ is the metric tensor of flat Minkowski space and $h_{\mu\nu}$ is a small perturbation that corresponds to the gravitational wave. The vacuum Einstein equation for $g_{\mu\nu}$ reduces to a linear wave equation for $h_{\mu\nu}$. In Cartesian coordinates (t, x, y, z), there are two independent gravitational wave solutions for waves propagating in the z-direction. They are both polarized transverse to the z-direction and propagate at the speed of light.

The metric for the polarization shown in Figure 6.13 is

$$d\tau^2 = dt^2 - (1 + f(t-z))\, dx^2 - (1 - f(t-z))\, dy^2 - dz^2\,,\qquad(6.131)$$

and for the polarization shown in Figure 6.14, rotated by 45°, it is

$$d\tau^2 = dt^2 - dx^2 - 2g(t-z)dxdy - dy^2 - dz^2\,.\qquad(6.132)$$

$f(t-z)$ and $g(t-z)$ are arbitrary functions of small amplitude.

6.12.1 The detection of gravitational waves

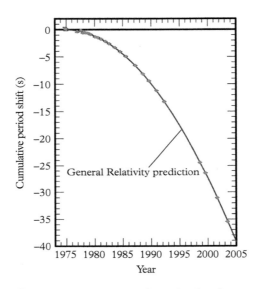

Fig. 6.15 Binary neutron star and gravitational wave emission.

The existence of gravitational waves has been confirmed indirectly by monitoring the Hulse–Taylor binary neutron star system PSR B1913+16, described in section 6.8. The pulsar signal has been observed for several decades and the period of the orbit is gradually diminishing; each year it decreases by 76 microseconds. This can be compared to the expected decrease of the orbital period due to the energy lost through the emission of gravitational radiation, as shown in Figure 6.15. The agreement is a staggeringly good confirmation of general relativity.

The emission of gravitational radiation has now been confirmed in other binary pulsar systems including PSR J0348+0432, a system discovered at the Green Bank observatory in West Virginia in 2007. This remarkable system consists of a neutron star of mass $2M_\odot$ in a tight orbit with a white dwarf of mass $0.17M_\odot$. Their orbital period is just 2 hours 27 minutes, and it decays at the expected rate of 8 microseconds per year.

The detection of gravitational waves on Earth has been an important goal for physicists for several decades. Detectors have been built at various locations around the world. These include LIGO (Laser Interferometer Gravitational-Wave Observatory) which has two facilities separated by a distance of 3000 km at Hanford, Washington and Livingston,

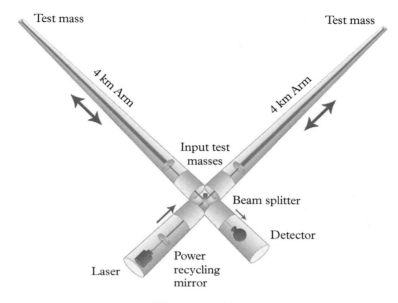

Fig. 6.16 LIGO.

Louisiana in the United States. A schematic set-up of one of these facilities is shown in Figure 6.16. The interferometers are L-shaped with two perpendicular 4 km long arms. The whole apparatus is housed within an ultrahigh vacuum. A laser beam impinges on a beamsplitter that directs half the beam down each arm of the interferometer. The light is then bounced back and forth 400 times between two mirrors in each arm that act as test masses, before passing through the beamsplitter again where the two half-beams are recombined and sent to a photodetector. This makes the arms effectively 1600 km long. If the light travels exactly the same distance down both arms the waves cancel, with the peaks in one beam meeting the troughs in the other, so no signal is detected by the photodetector. However, a passing gravitational wave changes the relative lengths of the arms very slightly, in which case the light waves no longer perfectly cancel and a signal is detected. The sensitivity of the apparatus is extraordinary, as it must be to have any chance of detecting gravitational waves. The latest phase of operation is dubbed Advanced LIGO. The upgraded detectors are now sensitive to gravitational waves with amplitudes as small as 5×10^{-22}. Two widely separated facilities are required to distinguish true gravitational wave events from the inevitable noise from local background disturbances.

Four days before the official start of the Advanced LIGO programme on 14 September 2015, an unmistakable and practically identical signal lasting 0.2 seconds was measured by both detectors within a few milliseconds of each other, as shown in Figure 6.17. This signal was interpreted as a train of gravitational waves produced by the merger of two black holes at a distance of around 1.3 billion light years. It was the first ever detection of a binary black hole system and the most direct observation of black holes ever made. The signal from the event also confirms that gravitational waves travel at the speed of light.

Binary black holes should emit a continuous stream of gravitational waves at twice their orbital frequency. With their emission, the binary system loses energy and the black

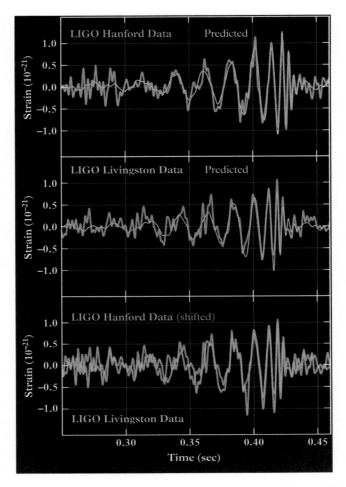

Fig. 6.17 First gravitational wave signal detected by Advanced LIGO.

holes gradually spiral together. In the final moments of inspiral, the amplitude of the waves increases dramatically. Initially, the newly merged black hole is rather asymmetrical, but it rapidly settles down with a final blast of gravitational waves, known as the *ring-down*. The signal detected by Advanced LIGO was produced during the final inspiral and ring-down.

Comparison with computer models of black hole merger processes enables researchers to extract a great deal of information about the observed event. The frequency of the gravitational waves allows the masses of the black holes to be deduced and the amplitude of the waves allows their distance to be estimated. Also, the difference in the arrival time of the waves at the two LIGO facilities determines the direction towards the event, at least roughly. Putting all this information together, we know that in this first event, the signal was produced by the merger of two black holes with masses of around $29M_\odot$ and $36M_\odot$ that coalesced to form a rapidly spinning black hole of $62M_\odot$. In the process, an incredible

$3M_\odot$ was converted into energy in the form of gravitational waves. The resulting black hole has an angular momentum parameter $a \simeq 0.67GM$.

The energy density of an expanding spherical wavefront of electromagnetic radiation emitted by a star decreases with the inverse square of distance from the star. This follows from the conservation of energy. Similarly, the energy density of a gravitational wave decreases with the inverse square of distance from its source. But there is an important difference in how we detect these two types of wave. With electromagnetic waves, we always measure their energy density or intensity, whether the detector is our eye, a CCD camera or a photographic plate. Gravitational wave detectors, on the other hand, directly measure the amplitude of gravitational waves. This is rather advantageous. The energy density of a wave is proportional to the square of its amplitude, so the wave amplitude only decreases inversely with distance from the source. This means that if the sensitivity of Advanced LIGO could be increased by another factor of 10, the volume of space being surveyed would be increased by a factor of 1000. This could increase the rate at which black hole merger and other extreme events are observed by over 1000 times, as they were probably more common in the distant past. We could possibly see black hole mergers like the one described here all the way back to the Big Bang.

There are already plans to increase the sensitivity of Advanced LIGO by a factor of three in the next round of upgrades, and gravitational wave detectors elsewhere are also coming online. The era of gravitational wave astronomy has only just begun.

6.13 The Einstein–Hilbert Action

In section 3.2 we considered the action S of a classical field theory. It is of the form

$$S = \int \mathcal{L}(\mathbf{x}, t) \, d^3x \, dt \,, \tag{6.133}$$

where the Lagrangian density $\mathcal{L}(\mathbf{x}, t)$ is integrated over flat 4-dimensional Minkowski space. In the case of a relativistic scalar field ψ the Lagrangian density is

$$\mathcal{L}(\mathbf{x}, t) = \frac{1}{2} \left(\frac{\partial \psi}{\partial t} \right)^2 - \frac{1}{2} \nabla \psi \cdot \nabla \psi - U(\psi) \,. \tag{6.134}$$

The principle of least action says that for a physical field evolution $\psi(\mathbf{x}, t)$, the action S is stationary under any variation of the field. As we have seen, the field equation may be derived by varying ψ, and equating δS to zero.

We can use the same procedure in general relativity, but now the dynamical field is the spacetime metric itself, so this must be varied, and we cannot simply assume a fixed, flat background spacetime. Within days of Einstein's announcement of the field equation of general relativity in November 1915, Hilbert found an appropriate action for the theory, which is now known as the Einstein–Hilbert action. The Lagrangian density must be a scalar, and the simplest one available is the Ricci scalar R. This is indeed the Lagrangian density, and the Einstein–Hilbert action is

$$S = \int \mathcal{L}(\mathbf{x}, t) \sqrt{-g} \, d^4y = \int R \sqrt{-g} \, d^4y \,. \tag{6.135}$$

Here $\sqrt{-g} \, d^4y$ is the spacetime volume of a coordinate integration element d^4y. It is known as the *measure* of the integral. g is the determinant of the metric $g_{\mu\nu}$, and the minus sign is required for a Lorentzian metric so that $-g$ is positive. $\sqrt{-g}$ is also the determinant of the Jacobian factor when one changes from local normal coordinates to a general coordinate system in the integral.

Let us look at a couple of illustrative examples. In section 5.4.1 we considered the metric tensor

$$g_{ij} = \begin{pmatrix} a^2 & 0 \\ 0 & a^2 \sin^2 \vartheta \end{pmatrix} \tag{6.136}$$

on a 2-sphere of radius a. As there are no off-diagonal terms, the infinitesimal squared distance is

$$ds^2 = a^2 d\vartheta^2 + a^2 \sin^2 \vartheta \, d\varphi^2 \,, \tag{6.137}$$

and an infinitesimal area element is $a \, d\vartheta \times a \sin \vartheta \, d\varphi = a^2 \sin \vartheta \, d\vartheta \, d\varphi = \sqrt{g} \, d\vartheta \, d\varphi$. This is the measure that must be used when integrating over the 2-sphere. The total area of the sphere is

$$\int_0^{2\pi} \int_0^{\pi} a^2 \sin \vartheta \, d\vartheta \, d\varphi = 4\pi a^2 \,. \tag{6.138}$$

With a diagonal 4-dimensional Lorentzian metric, the measure is

$$\sqrt{g_{00}} \, dy^0 \times \sqrt{-g_{11}} \, dy^1 \times \sqrt{-g_{22}} \, dy^2 \times \sqrt{-g_{33}} \, dy^3 = \sqrt{-g} \, dy^0 \, dy^1 \, dy^2 \, dy^3 \,. \tag{6.139}$$

For instance, the measure for the exterior Schwarzschild metric (6.61) is

$$\sqrt{-g_{tt} \, g_{rr} \, g_{\vartheta\vartheta} \, g_{\varphi\varphi}} \, dt \, dr \, d\vartheta \, d\varphi = \sqrt{-g} \, dt \, dr \, d\vartheta \, d\varphi = r^2 \sin \vartheta \, dt \, dr \, d\vartheta \, d\varphi \,. \tag{6.140}$$

The metric is diagonal in these coordinates, so g is the product of the four diagonal entries of the metric tensor. (The factors Z and Z^{-1} in the first two terms of the Schwarzschild metric cancel.) We could change coordinates and in general this would produce off-diagonal terms. However, the appropriate measure is still $\sqrt{-g} \, d^4 y$.

Rather than give a complete derivation of the field equation of general relativity from the Einstein–Hilbert action, which is rather technical and complicated, we will partially sketch it out. The field equation is the condition for the action S to be unchanged to first order when a small change $\delta g^{\mu\nu}$ is made to the (inverse) metric tensor. Both the Ricci scalar and the measure vary as the metric varies. To see what this implies, we require the following results, which we simply quote:

$$\frac{\delta R}{\delta g^{\mu\nu}} = R_{\mu\nu} \,, \quad \frac{\delta \sqrt{-g}}{\delta g^{\mu\nu}} = -\frac{1}{2} \sqrt{-g} \, g_{\mu\nu} \,. \tag{6.141}$$

From these expressions, it follows that

$$\begin{aligned} \delta S &= \int \left(\frac{\delta(R\sqrt{-g})}{\delta g^{\mu\nu}} \right) \delta g^{\mu\nu} \, d^4 y \\ &= \int \left(\sqrt{-g} \frac{\delta R}{\delta g^{\mu\nu}} + R \frac{\delta \sqrt{-g}}{\delta g^{\mu\nu}} \right) \delta g^{\mu\nu} \, d^4 y \\ &= \int \left(R_{\mu\nu} - \frac{1}{2} g_{\mu\nu} R \right) \delta g^{\mu\nu} \sqrt{-g} \, d^4 y \,. \end{aligned} \tag{6.142}$$

The tensor in brackets is the Einstein tensor $G_{\mu\nu}$. According to the principle of least action, $\delta S = 0$ for any infinitesimal variation $\delta g^{\mu\nu}$. This will only be true if the tensor in brackets is zero, which tells us that the vacuum Einstein equation is $G_{\mu\nu} = 0$.

We can also include matter fields in the theory and then the action becomes

$$S = S_G + S_M = \int (R + \alpha \mathcal{L}_M) \sqrt{-g} \, d^4 y \,, \tag{6.143}$$

where α is a constant of proportionality and $\mathcal{L}_M$ is the Lagrangian density for the matter fields. In general the matter Lagrangian will depend on various fields, such as scalar fields or Maxwell fields. If we vary S_M we find

$$
\begin{aligned}
\delta S_M &= \alpha \int \left(\sqrt{-g} \frac{\delta \mathcal{L}_M}{\delta g^{\mu\nu}} + \mathcal{L}_M \frac{\delta \sqrt{-g}}{\delta g^{\mu\nu}} \right) \delta g^{\mu\nu} \, d^4 y \\
&= \alpha \int \left(\frac{\delta \mathcal{L}_M}{\delta g^{\mu\nu}} - \frac{1}{2} g_{\mu\nu} \mathcal{L}_M \right) \delta g^{\mu\nu} \sqrt{-g} \, d^4 y \,.
\end{aligned} \tag{6.144}
$$

The energy–momentum tensor is defined to be[9]

$$T_{\mu\nu} = -2 \frac{\delta \mathcal{L}_M}{\delta g^{\mu\nu}} + g_{\mu\nu} \mathcal{L}_M \,, \tag{6.145}$$

so if we vary the whole action $S_G + S_M$ with respect to both the metric and the matter fields, we find

$$G_{\mu\nu} = \frac{\alpha}{2} T_{\mu\nu} \,, \tag{6.146}$$

together with the field equations of the matter fields in a curved spacetime background. Fixing the constant to be $\alpha = 16\pi G$, we have the Einstein equation in the presence of matter.

The only other term that could be added to the Lagrangian density is a constant, $\mathcal{L}_\Lambda = 2\Lambda$, known as the *cosmological constant* term. (The factor of 2 is conventional.) The variation of the additional action S_Λ is

$$
\begin{aligned}
\delta S_\Lambda &= \int \left(2\Lambda \frac{\delta \sqrt{-g}}{\delta g^{\mu\nu}} \right) \delta g^{\mu\nu} \, d^4 y \\
&= \int (-\Lambda \, g_{\mu\nu}) \, \delta g^{\mu\nu} \sqrt{-g} \, d^4 y \,.
\end{aligned} \tag{6.147}
$$

If this term is included, then the full Einstein equation is

$$G_{\mu\nu} - \Lambda g_{\mu\nu} = 8\pi G \, T_{\mu\nu} \,. \tag{6.148}$$

We will consider the significance of the cosmological constant in Chapter 14.

---

[9] This curved spacetime approach to determining the energy–momentum tensor of matter fields is very convenient, and is consistent with what is found by considering energy and momentum conservation in Minkowski spacetime.

6.14 Further Reading

For an overview of gravitation and an introduction to general relativity, see

M. Begelman and M. Rees, *Gravity's Fatal Attraction: Black Holes in the Universe (2nd ed.)*, Cambridge: CUP, 2010.

N.J. Mee, *Gravity: Cracking the Cosmic Code*, London: Virtual Image, 2014.

For comprehensive coverage of general relativity, see

I.R. Kenyon, *General Relativity*, Oxford: OUP, 1990.

S. Carroll, *Spacetime and Geometry: An Introduction to General Relativity*, San Francisco: Addison Wesley, 2004.

J.B. Hartle, *Gravity: An Introduction to Einstein's General Relativity*, San Francisco: Addison Wesley, 2003.

For an approach to general relativity based on particle dynamics, see

J. Franklin, *Advanced Mechanics and General Relativity*, Cambridge: CUP, 2010.

For a comprehensive treatise on black holes, see

V.P. Frolov and I.D. Novikov, *Black Hole Physics: Basic Concepts and New Developments*, Dordrecht: Kluwer, 1998.

7
Quantum Mechanics

7.1 Introduction

There were physicists towards the end of the 19th century who believed that their subject was essentially complete and further progress would simply be a matter of refining what was already known. In reality, a crisis was looming that would shake physics to its roots, and the implications are still being felt today. As we have seen, revolutionary ideas were required to understand space and time on the largest scale, but an even greater revolution would be required to understand energy and matter on the very short atomic and sub-atomic scales. A new era began in 1900. Max Planck had struggled for some time to explain the observed relationship between the wavelength and intensity of the radiation emitted by a black body. In 1900 he published a formula that accurately described the radiation. (We will derive this formula in Chapter 10.) In doing so, he introduced a new fundamental constant into physics. This constant $\hbar$, Planck's constant, was the first step towards quantum mechanics. It appears wherever quantum mechanical ideas are used, and unifies the subject. Planck originally introduced a constant $h = 2\pi\hbar$, but it is almost always more convenient to use $\hbar$. $\hbar$ has a numerical value of approximately 1.055×10^{-34} J s.

Many metals emit electrons when light shines on them. This is known as the photoelectric effect. According to experiment, the energy of each emitted electron depends on the frequency of the light and not its intensity, an observation that is hard to reconcile with classical physics. In 1905 Einstein wrote a paper that he realized was far more revolutionary than his papers on special relativity published the same year. In this paper, Einstein proposed that electromagnetic radiation is not a continuous wave, but is composed of particles that we now know as photons, and that the energy of these photons is given by a simple formula that includes Planck's constant: $E = \hbar\omega$, where ω is the (angular) frequency of the light. With this far-reaching idea Einstein explained the photoelectric effect. Each photoelectron emitted from a metal results from a single collision with a photon whose energy is given by Einstein's formula.

A couple of years later Einstein applied similar ideas to the thermal vibrations in solids. Based on the assumption that these vibrations are quantized and also obey the equation $E = \hbar\omega$, Einstein derived a formula for the heat capacity of solids.

The next step in the development of quantum theory came when Niels Bohr attempted to explain the structure of the atom following Ernest Rutherford's discovery of the nucleus. Bohr postulated that electrons orbit the nucleus, but that the only possible orbits are those for which each electron's angular momentum is quantized in integer multiples of Planck's constant. This postulate implies that the electron energy levels are discrete, and are separated by finite energy gaps. Many materials emit light when their atoms are placed in a

The Physical World. Nicholas Manton and Nicholas Mee, Oxford University Press (2017). © Nicholas Manton and Nicholas Mee. DOI 10.1093/oso/9780198795933.001.0001

flame, with colours that are very pure, corresponding to precise wavelengths. Bohr realized that these sharp line spectra are due to atomic electrons emitting single photons as they fall from one discrete energy level to another lower energy level, and he used his model to account for them with great accuracy. We will not discuss photons further in this chapter, as a combination of quantum mechanical ideas and relativity is required to understand their behaviour. This chapter will focus on the quantum theory of non-relativistic particles, which is the appropriate theory for atomic and molecular physics.

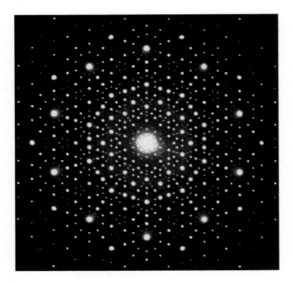

Fig. 7.1 An electron diffraction pattern taken along the tenfold symmetry axis of the $Al_{72}Ni_{20}Co_8$ decagonal quasi-crystal.

For two decades, quantum theory was applied to a range of physical problems, but it developed in an ad hoc way. This approach is now referred to as the old quantum theory. Everything changed with the publication in 1924 of an idea by Louis de Broglie. He had the remarkable insight that if a wave could have a particle-like character, then perhaps a particle could have a wave-like character. He suggested that a particle of momentum $\mathbf{p}$ should have a wavelength $\frac{2\pi\hbar}{|\mathbf{p}|}$. This was confirmed experimentally three years later when electron interference patterns were observed as beams of electrons were fired through the crystalline atomic grid in a metal film. An electron diffraction pattern is shown in Figure 7.1.

From 1925 onwards, the old quantum theory was replaced by a radical new theory—quantum mechanics—which incorporated many of the older ideas but was much more consistent and complete. It gave extraordinarily accurate predictions of atomic structure and dynamical properties, including atomic spectra. This new theory would be applied even more widely than initially anticipated. In 1929, Paul Dirac said that it is 'the mathematical theory of a large part of physics and the whole of chemistry.' Quantum mechanics was rapidly applied to numerous branches of physics, many of which we will discuss in later chapters. It was invoked to explain the properties of atoms and the structure of the

Periodic Table and to understand the chemical bond, which was the key to the whole subject of chemistry. It was applied to the physics of the atomic nucleus, which led to an understanding of nuclear energy and the power source of the stars. Quantum theory was also applied to elementary particles and the forces between them, and to the search for the fundamental constituents of matter. On a larger scale, it was used to explain the structure and properties of solids. Quantum mechanics has also stimulated the discovery of many phenomena that have widespread applications including the components of commonplace devices that are used worldwide. These include lasers, transistors, light-emitting diodes (LEDs), superconductivity, superfluidity and exceptionally strong magnets.

Quantum mechanics is without question more fundamental than classical dynamics, but classical physics has, of course, not been totally discarded. It still has many applications in situations where $\hbar$ is negligible. Classical dynamics remains the best way to understand the motion of macroscopic objects from billiard balls to cars, planets and stars. Most fluids are also well described by classical physics. Even the electromagnetic fields that are so important in power and communication engineering are well described by Maxwell's equations, and one does not need to invoke individual photons to model their behaviour. Roughly speaking, the boundary between the domain where quantum mechanics offers the most useful description of physics and the domain of classical dynamics is the boundary between the atomic length scale and longer length scales, but this boundary is not a sharp one. Indeed, as $\hbar$ is not a unit of length, there are some larger-scale phenomena, whose explanation requires quantum effects to be taken into account.

7.2 Position and Momentum in Quantum Mechanics

In classical Newtonian dynamics, a point particle has a time dependent position $\mathbf{x}(t)$, and by taking its time derivatives we obtain the velocity $\mathbf{v} = \frac{d\mathbf{x}}{dt}$ and acceleration $\mathbf{a} = \frac{d^2\mathbf{x}}{dt^2}$. We are free to specify at any instant $\mathbf{x}$ and $\mathbf{v}$, but $\mathbf{a}$ is determined by the force acting. In fact, there are a number of reasons to prefer $\mathbf{x}$ and $\mathbf{p}$ as the dynamical variables, where $\mathbf{p} = m\mathbf{v}$ is the particle momentum: (i) Newton's second law equates force to the rate of change of $\mathbf{p}$; (ii) when two bodies interact, the sum $\mathbf{p}_1 + \mathbf{p}_2$ is conserved; (iii) angular momentum has a simple expression, $\mathbf{l} = \mathbf{x} \times \mathbf{p}$. Quantum mechanics is formulated in terms of position $\mathbf{x}$ and momentum $\mathbf{p}$, and velocity is less important—but the character of $\mathbf{x}$ and $\mathbf{p}$ is fundamentally different in quantum mechanics than in classical dynamics.

In this chapter, we limit ourselves to 1-dimensional particle motion, so the dynamical variables are x and p. In classical dynamics, x and p are ordinary real numbers taking any values between $-\infty$ and ∞. In quantum mechanics, Werner Heisenberg proposed that x and p were not simply numbers, but *operators* that act on a physical *state* characterizing the particle. The algebra of the operators is postulated to be

$$xp - px = i\hbar\mathbf{1}. \tag{7.1}$$

On the right-hand side, i is $\sqrt{-1}$, $\hbar$ is Planck's constant, and $\mathbf{1}$ is the unit operator, which leaves a state unchanged when it acts. $xp - px$ is called the *commutator* of x and p, and is given the notation $[x, p]$. Equation (7.1) is the fundamental position–momentum *commutation relation* in 1-dimensional quantum mechanics.

We haven't yet said what the operators x and p are, or what values they take, but we will insist that they obey the commutation relation (7.1). The classical limit corresponds to $\hbar = 0$, which implies that $xp = px$ and this is satisfied if x and p are ordinary numbers.

Heisenberg constructed a dynamical theory by giving a rule for how the operators x and p evolve in time, whilst still obeying equation (7.1) at every instant. He also managed to extract physical meaning from x and p and from derived quantities like energy. However, this approach to quantum mechanics is rather abstract and austere.

Matrices do not generally commute when multiplied. For example,

$$\begin{pmatrix} 0 & a \\ 0 & 0 \end{pmatrix} \begin{pmatrix} 0 & 0 \\ b & 0 \end{pmatrix} - \begin{pmatrix} 0 & 0 \\ b & 0 \end{pmatrix} \begin{pmatrix} 0 & a \\ 0 & 0 \end{pmatrix} = \begin{pmatrix} ab & 0 \\ 0 & -ab \end{pmatrix}, \tag{7.2}$$

and this is what sometimes makes matrices a suitable tool for representing operators in quantum mechanics. Heisenberg's quantum mechanics would be simpler if there were a solution of equation (7.1) by square $n \times n$ matrices of real or complex numbers. However, there is no matrix solution $x = X$ and $p = P$, for any finite n. We can show this by taking the trace. (The trace of a matrix M, denoted by $\mathrm{Tr}\, M$, is the sum of the elements on the main diagonal.) Suppose that

$$XP - PX = i\hbar \mathbf{1}_n, \tag{7.3}$$

where $\mathbf{1}_n$ is the unit $n \times n$ matrix. Then

$$\mathrm{Tr}(XP) - \mathrm{Tr}(PX) = i\hbar \mathrm{Tr}\, \mathbf{1}_n = i\hbar n, \tag{7.4}$$

but the trace of a product of matrices does *not* depend on the order in which they are multiplied,[1] so the left-hand side of equation (7.4) is zero, leading to a contradiction. Therefore the original premise of equation (7.3) must be false.

There is a solution of equation (7.1) by *infinite* matrices, discovered by Heisenberg. In this case, the previous argument does not apply, because Tr cannot generally be defined for infinite matrices. Such infinite matrices are messy to write down, so this approach is complicated. Instead we give the Schrödinger point-of-view. Erwin Schrödinger independently developed an approach to quantum mechanics that initially appeared quite different from Heisenberg's, but it was soon realized that they were equivalent, so we now speak of quantum mechanics in the Schrödinger picture and in the Heisenberg picture, having the same physical content. Schrödinger's quantum mechanics focusses more on the state on which the operators x and p act.

In the Schrödinger picture one solves equation (7.1) by representing x and p by differential operators, not matrices, acting on the state ψ. Rather than being a column vector of numbers acted on by a finite matrix, ψ is a function of x (and usually of time t too). A differential operator of first order in the derivative $\frac{d}{dx}$ has the form

$$D = a(x) + b(x)\frac{d}{dx}, \tag{7.5}$$

with $a(x)$ and $b(x)$ ordinary functions, where either $a(x)$ or $b(x)$ may be zero. Acting on ψ,

$$D\psi = a(x)\psi + b(x)\frac{d\psi}{dx}, \tag{7.6}$$

---

[1] The elements of XP are $(XP)_{ab} = \sum_c X_{ac}P_{cb}$ so the diagonal elements are $(XP)_{aa} = \sum_c X_{ac}P_{ca}$ and therefore $\mathrm{Tr}(XP) = \sum_a \sum_c X_{ac}P_{ca}$. Similarly $\mathrm{Tr}(PX) = \sum_a \sum_c P_{ac}X_{ca}$, and this becomes equal to $\mathrm{Tr}(XP)$ if one swaps the labels on the indices a, c and the order of the summations.

so $D\psi$ is a new function[2] of x. There are differential operators with higher order derivatives too, as we will see shortly. The operators representing x and p are each of the first order form (7.5). The operator x is represented by x, i.e. by D with $a(x) = x$ and $b(x) = 0$. There is a notational confusion here that one must live with: x can mean an operator, a function, or a particular real numerical value, but the meaning should be clear from the context. The position operator x acts by multiplication on the function $\psi(x)$ to produce the new function $x\psi(x)$. The momentum operator is represented by

$$p = -i\hbar\frac{d}{dx}, \tag{7.7}$$

i.e. by D with $a(x) = 0$ and $b(x) = -i\hbar$. Finally the unit operator $\mathbf{1}$ is represented by 1, i.e. by D with $a(x) = 1$ and $b(x) = 0$.

It is important to check that equation (7.1) is satisfied for ψ in the Schrödinger picture. This is done as follows:

$$\begin{aligned}
(xp - px)\psi &= x\left(-i\hbar\frac{d}{dx}\right)\psi - \left(-i\hbar\frac{d}{dx}\right)(x\psi) \\
&= -i\hbar x\frac{d\psi}{dx} + i\hbar\left(\psi + x\frac{d\psi}{dx}\right) \\
&= i\hbar\psi \\
&= (i\hbar\mathbf{1})\psi, \tag{7.8}
\end{aligned}$$

and since this result holds for any function ψ, the commutation relation (7.1) is confirmed. Notice that the key step is the Leibniz rule for the derivative of $x\psi$, which is not something that has an obvious analogue in the context of matrices.

In the Schrödinger picture, x and p are represented by differential operators whereas in the Heisenberg picture they are represented by infinite matrices, but this is actually rather a formal difference. A more significant difference is that in the Heisenberg picture the operators change with time, and the state ψ is unchanging. In the Schrödinger picture x and p are unchanging, and ψ changes with time. These pictures are equivalent provided one is careful when extracting physical results from the formalism. In the Schrödinger picture, the operators x and p are universal entities, the same whatever the dynamics of the particle. It is the dynamics of the state ψ that varies between different particles in different situations.

7.3 The Schrödinger Equation

The Newtonian dynamics of a particle in one dimension is controlled by the potential $V(x)$ in which the particle moves. The potential determines the force on the particle. In quantum mechanics too, the dynamics of the particle is controlled by the potential $V(x)$. In the Schrödinger picture, physical information about the particle is carried by the state $\psi(x,t)$, which is also called the *wavefunction* of the particle at time t. We shall shortly discuss what can be deduced from the x-dependence of ψ, but first we discuss the dynamics of ψ—how it evolves in time. This depends on the potential.

---

[2] The notation suppresses the argument x of ψ and $D\psi$, as it is awkward to show this for derivatives.

A new operator H, called the *Hamiltonian*, controls the dynamical evolution of ψ. It is modelled on the classical total energy of the particle, a function of x and p. For a particle moving in a potential $V(x)$, the Hamiltonian is the operator

$$H = \frac{1}{2m}p^2 + V(x)\,, \tag{7.9}$$

the sum of the kinetic energy and the potential energy (these terms are still applicable in quantum mechanics). m is the classical mass of the particle and is a positive constant. We will discuss later why the time evolution is determined by this particular operator.

Since p is represented by $-i\hbar\frac{d}{dx}$, p^2 is simply this operator acting twice. This gives the second order differential operator

$$p^2 = \left(-i\hbar\frac{d}{dx}\right)\left(-i\hbar\frac{d}{dx}\right) = -\hbar^2\frac{d^2}{dx^2}\,. \tag{7.10}$$

On the other hand $V(x)$, which is simply a function of x, just acts by multiplication. Acting on a state $\psi(x,t)$, the derivatives with respect to x become partial derivatives, so

$$H\psi = -\frac{\hbar^2}{2m}\frac{\partial^2\psi}{\partial x^2} + V(x)\psi\,, \tag{7.11}$$

which is a new function of x and t. Fortunately, p^2 is unambiguous as an operator. Some classical quantities, like xp, would have an ordering ambiguity, since xp and px are the same classically, but differ by a constant as operators, because of equation (7.1).

The dynamical principle of quantum mechanics in the Schrödinger picture is that the state $\psi(x,t)$ evolves with time according to the equation

$$i\hbar\frac{\partial\psi}{\partial t} = H\psi\,, \tag{7.12}$$

or, written out fully,

$$i\hbar\frac{\partial\psi}{\partial t} = -\frac{\hbar^2}{2m}\frac{\partial^2\psi}{\partial x^2} + V(x)\psi\,. \tag{7.13}$$

This is the *Schrödinger equation*. As initial data one must specify ψ for all x, and as equation (7.13) is a first order differential equation in time, this is sufficient. $\psi = 0$ is always a solution of the Schrödinger equation, but does not describe a physical state, so from here on, by a solution we always mean one that is non-zero.

Equation (7.13) is a linear partial differential equation, so one approach to constructing solutions is to find a set of special solutions, and then construct the general solution as a linear superposition of the special solutions. More explicitly, if $\psi_0(x,t), \psi_1(x,t), \psi_2(x,t), \ldots$ are independent solutions of (7.13), then

$$\psi(x,t) = a_0\psi_0(x,t) + a_1\psi_1(x,t) + a_2\psi_2(x,t) + \cdots \tag{7.14}$$

is also a solution for any constants $a_0, a_1, a_2, \ldots$ which do not all vanish, and such that the sum converges.

Two things should be spelled out here. The first is that the Schrödinger equation contains an explicit i, so the solutions ψ are generally *complex*. Therefore the constants $a_0, a_1, a_2, \ldots$ are complex numbers. They are called *amplitudes*. The second is that *all* the solutions obtained by linear superposition are physically valid. None is excluded. This is the *superposition principle* of quantum mechanics, and is really just a consequence of having a linear equation. It produces an analogy between the Schrödinger equation and linear wave equations like the Maxwell equations without sources, whose wave solutions can be superposed, and where any solution is physical. There is nothing comparable in classical particle dynamics. Superposition of waves creates interference patterns, and this behaviour, though surprising, has been confirmed experimentally. Formerly, quantum mechanics was called wave mechanics. However, although particle states have wave-like properties, the particles themselves remain local, point-like objects.

A technical issue is now to find a particularly convenient set of independent states $\psi_0, \psi_1, \psi_2, \ldots$, and to decide how many there are. Actually, there are infinitely many. The space of functions ψ, at some given time, is an infinite-dimensional vector space, and the Schrödinger equation produces an evolution of ψ in this space. There is still a preferred set of states, called *stationary states*. These stationary states are not time independent, but they have a particularly simple dependence on time, and most of their physical properties are time independent.

To find stationary states one separates variables. One assumes that ψ is a product of a function of x and a function of t. This implies that the dependence on time is through a simple exponential $e^{-\frac{i}{\hbar}Et}$, with an initially unknown real constant E. The complete wavefunction is then

$$\psi(x,t) = \chi(x)e^{-\frac{i}{\hbar}Et}, \tag{7.15}$$

where $\chi(x)$ and E are yet to be found. On substituting the wavefunction (7.15) into the Schrödinger equation (7.13), the operator $i\hbar\frac{\partial}{\partial t}$ just differentiates the time dependent phase factor, bringing down E. The operator H, involving spatial derivatives, just acts on χ, so we find

$$E\chi e^{-\frac{i}{\hbar}Et} = H\chi e^{-\frac{i}{\hbar}Et}. \tag{7.16}$$

The time dependent factor cancels, leaving

$$H\chi = E\chi. \tag{7.17}$$

More explicitly,

$$-\frac{\hbar^2}{2m}\frac{d^2\chi}{dx^2} + V(x)\chi = E\chi. \tag{7.18}$$

This is the stationary Schrödinger equation. A solution $\chi(x)$ is called a stationary state wavefunction, and E is its energy. Note that the partial derivative $\frac{\partial}{\partial x}$ has become the ordinary derivative $\frac{d}{dx}$ again, because there is no longer any time dependence.

We can now explain why it was reasonable to use the operator H, based on the classical energy of the particle, as the quantum mechanical evolution operator. According to de Broglie's idea, a particle with positive momentum p is described by a wave with wavelength $\frac{2\pi\hbar}{p}$. A wave e^{ikx} has wavelength $\frac{2\pi}{k}$, so this wave (with wavenumber k) describes a particle of momentum $p = \hbar k$. For this wave, the first term of the stationary Schrödinger equation (7.18) is

$$-\frac{\hbar^2}{2m}\frac{d^2}{dx^2}e^{ikx} = \frac{\hbar^2 k^2}{2m}e^{ikx} = \frac{p^2}{2m}e^{ikx}, \tag{7.19}$$

and the coefficient $\frac{p^2}{2m}$ on the right-hand side is the kinetic energy (with p here the classical momentum, not an operator). Suppose now that $V(x)$ is a smooth function which varies with x on a much greater length scale than $\frac{2\pi}{k}$. Then, locally, e^{ikx} approximately solves the stationary Schrödinger equation (7.18) provided

$$\frac{p^2}{2m}e^{ikx} + V(x)e^{ikx} = Ee^{ikx}, \tag{7.20}$$

where $p = \hbar k$ and k varies slowly with x. The coefficients on both sides match if $\frac{p^2}{2m} + V(x) = E$. In this way, the Schrödinger equation is related to the classical energy equation. As E is a constant, energy is conserved, even though p and V are separately varying. This argument, though crude, shows that classical particle motion can be converted into a solution of the stationary Schrödinger equation whenever the quantum particle wavelength is much shorter than the length scale of the externally specified potential V, and shows that it is correct to interpret E, the constant first appearing in equation (7.15), as energy. A better approximation to a solution of equation (7.18) is of the form

$$\chi(x) = A(x)e^{ik(x)x} \tag{7.21}$$

where $\frac{\hbar^2 k^2(x)}{2m} = E - V(x)$, and $A(x)$ has a slowly varying magnitude and phase, and this strengthens the argument.

Let's now return to equation (7.17) and its exact solutions. Equation (7.17), and its explicit version (7.18), is one that is much-studied in the theory of operators. H acts on the function χ but doesn't produce a completely independent function; it just produces a constant multiple of χ. Such functions χ are special, and E is special too. E is called an *eigenvalue* of H, or an energy eigenvalue, and $\chi(x)$ the *eigenfunction* or *eigenstate* of H associated with the eigenvalue E. For physically reasonable potentials $V(x)$, H has infinitely many eigenvalues, E. They may be discrete (like the integers) or a continuum (like all the real numbers) or some combination (some discrete values and a continuum filling one or more intervals), as shown in Figure 7.2. Energy eigenvalues are often referred to in physics as energy levels. They are the only precise energies the particle can have. The eigenfunction associated with the lowest energy level is called the ground state, and those associated with higher levels are called excited states. Two or more states with the same energy are referred to as *degenerate*.

Solving the 1-dimensional Schrödinger equation (7.18) explicitly, and finding the energy eigenvalues, is only possible for certain potentials. We will take a look at a few important examples. Investigation of the spectrum of eigenvalues more generally is part of the theory of Schrödinger operators, a deep and complicated subject.

7.3.1   The free particle

Let us start with the example of a free particle, where V is everywhere zero. The Schrödinger equation for a free particle is

$$i\hbar\frac{\partial\psi}{\partial t} = -\frac{\hbar^2}{2m}\frac{\partial^2\psi}{\partial x^2}. \tag{7.22}$$

By separation of variables this reduces to the stationary Schrödinger equation

$$-\frac{\hbar^2}{2m}\frac{d^2\chi}{dx^2} = E\chi. \tag{7.23}$$

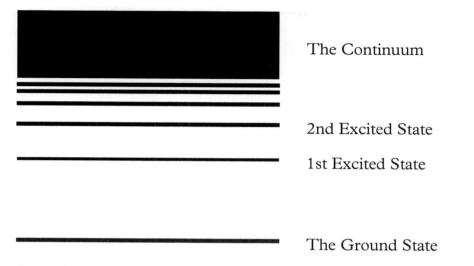

Fig. 7.2 A typical potential may include a set of discrete energy levels as well as a continuum of energy levels.

Equation (7.23) is a second order ordinary differential equation and has two independent real solutions for any positive energy E. These are

$$\chi_1(x) = \cos\left(\frac{1}{\hbar}\sqrt{2mE}\,x\right), \quad \chi_2(x) = \sin\left(\frac{1}{\hbar}\sqrt{2mE}\,x\right). \tag{7.24}$$

The general solution is a linear superposition

$$\chi(x) = A\cos\left(\frac{1}{\hbar}\sqrt{2mE}\,x\right) + B\sin\left(\frac{1}{\hbar}\sqrt{2mE}\,x\right), \tag{7.25}$$

where A and B are real or complex constants.

We now have to consider how $\chi(x)$ can behave as $x \to \pm\infty$. Physical solutions for χ must remain bounded as $x \to \pm\infty$ (that is, the magnitude of χ cannot grow indefinitely large). We will call such solutions acceptable. Unacceptable solutions are ones for which χ grows indefinitely in one or both directions. For this reason, the solution (7.25) is acceptable for any A and B, but if E were negative, then independent solutions would be $\chi_1(x) = e^{\frac{1}{\hbar}\sqrt{2m|E|}\,x}$ and $\chi_2(x) = e^{-\frac{1}{\hbar}\sqrt{2m|E|}\,x}$, which grow exponentially either as $x \to \infty$ or $x \to -\infty$, and are both unacceptable. If $E = 0$ there is just one acceptable solution, $\chi(x) = $ constant. The conclusion is that for a free particle, the allowed energy eigenvalues consist of all the real numbers $E \geq 0$, a continuum of eigenvalues. There are no negative energy levels.

It is often more convenient to work with particular solutions that are complex exponentials. By taking $A = 1$ and $B = \pm i$ in equation (7.25), we obtain independent solutions

$$\chi_+(x) = e^{\frac{i}{\hbar}\sqrt{2mE}\,x}, \quad \chi_-(x) = e^{-\frac{i}{\hbar}\sqrt{2mE}\,x}, \tag{7.26}$$

and the general solution is a superposition of these. The full wavefunctions including their time dependence are

$$\psi_+(x,t) = e^{\frac{i}{\hbar}\left(\sqrt{2mE}\,x - Et\right)}, \quad \psi_-(x,t) = e^{\frac{i}{\hbar}\left(-\sqrt{2mE}\,x - Et\right)}. \qquad (7.27)$$

If we use the de Broglie wavenumber $k = \pm\frac{1}{\hbar}\sqrt{2mE}$, then $E = \frac{\hbar^2 k^2}{2m}$, and a complete set of independent stationary solutions of the Schrödinger equation for a free particle can be written more simply in terms of k as

$$\psi(x,t) = e^{i\left(kx - \frac{\hbar k^2}{2m}t\right)}, \qquad (7.28)$$

with k taking any real value, including negative ones. These are simple wave-like solutions, with spatial wavelength $\frac{2\pi}{|k|}$, representing a particle with momentum $\hbar k$ and energy $\frac{\hbar^2 k^2}{2m}$.

We must not forget that the general solution of the Schrödinger equation is a superposition of stationary states, with their time dependent factors included. For the free particle, the general solution is

$$\psi(x,t) = \int_{-\infty}^{\infty} F(k) e^{i\left(kx - \frac{\hbar k^2}{2m}t\right)} dk, \qquad (7.29)$$

where $F(k)$ is an arbitrary complex function, decaying fast enough as $|k| \to \infty$ so that the integral converges. k is a continuous parameter, which is why the superposition is an integral over k rather than a sum.

7.3.2 The harmonic oscillator

Let's now investigate a second important example, the quantum harmonic oscillator with Schrödinger equation

$$i\hbar \frac{\partial \psi}{\partial t} = -\frac{\hbar^2}{2m} \frac{\partial^2 \psi}{\partial x^2} + \frac{1}{2} m\omega^2 x^2 \psi. \qquad (7.30)$$

The potential is $V(x) = \frac{1}{2} m\omega^2 x^2$, the potential in which a classical particle of mass m oscillates with frequency ω. The stationary Schrödinger equation now takes the form

$$-\frac{\hbar^2}{2m} \frac{d^2\chi}{dx^2} + \frac{1}{2} m\omega^2 x^2 \chi = E\chi. \qquad (7.31)$$

Classically, a particle of finite energy in this potential oscillates in a finite interval. In quantum mechanics we impose the boundary condition that χ must approach zero as $x \to \pm\infty$.

To analyse equation (7.31) it is helpful to simplify the notation by working with scaled length and energy variables. We choose $y = \sqrt{\frac{m\omega}{\hbar}}\,x$ and $\varepsilon = \frac{2}{\hbar\omega}E$. Then $\chi(y)$ satisfies

$$-\frac{d^2\chi}{dy^2} + y^2\chi = \varepsilon\chi. \qquad (7.32)$$

Recall that for a free particle, the energy could take any positive value, or zero. Here, the eigenvalue ε can only take certain discrete values. The reason is as follows. For large $|y|$, the two independent solutions of equation (7.32) are roughly $e^{-\frac{1}{2}y^2}$ and $e^{\frac{1}{2}y^2}$, for any ε. Only the solution that behaves like $e^{-\frac{1}{2}y^2}$ for large positive y is potentially acceptable, as the

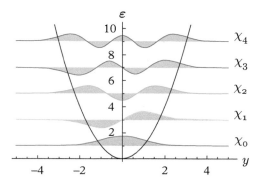

Fig. 7.3 The five lowest energy solutions $\chi_n(y)$ of the harmonic oscillator.

other solution grows. If we take this solution and extend it to large negative y, then it will generally be some combination of the $e^{-\frac{1}{2}y^2}$ and $e^{\frac{1}{2}y^2}$ solutions there, of which the $e^{\frac{1}{2}y^2}$ part will completely dominate. Its growing behaviour makes this solution unacceptable after all. So there is no acceptable solution satisfying both boundary conditions for a general value of ε. Only for certain discrete values of ε is there a solution that decays like $e^{-\frac{1}{2}y^2}$ for both large positive and large negative y, and these special values are the eigenvalues.

What are the eigenvalues? The lowest is $\varepsilon = 1$, and the solution is exactly $\chi(y) = e^{-\frac{1}{2}y^2}$, because (differentiating twice)

$$\left(-\frac{d^2}{dy^2} + y^2\right) e^{-\frac{1}{2}y^2} = \frac{d}{dy}\left(ye^{-\frac{1}{2}y^2}\right) + y^2 e^{-\frac{1}{2}y^2} = e^{-\frac{1}{2}y^2}. \tag{7.33}$$

This is the ground state solution. The complete set of eigenvalues forms the discrete sequence: $\varepsilon = 1, 3, 5, 7, \ldots$. Let us denote this sequence by $\varepsilon_n = 2n + 1$, starting at $n = 0$. The corresponding stationary state wavefunctions are $\chi_n(y)$, and they have the form

$$\chi_n(y) = H_n(y)e^{-\frac{1}{2}y^2} \tag{7.34}$$

where $H_n(y)$ is a polynomial in y of degree n. The first five are shown in Figure 7.3. $\chi_n(y)$ decays almost as fast as $e^{-\frac{1}{2}y^2}$ for large $|y|$, despite the polynomial prefactor. In the original variables, the nth solution is

$$\chi_n(x) = H_n\left(\sqrt{\frac{m\omega}{\hbar}}x\right) e^{-\frac{m\omega}{2\hbar}x^2}, \quad \text{with} \quad E_n = \left(n + \frac{1}{2}\right)\hbar\omega. \tag{7.35}$$

The polynomials in this sequence are known as the *Hermite polynomials*. The first few examples are

$$
\begin{aligned}
H_0(y) &= 1, \\
H_1(y) &= y, \\
H_2(y) &= y^2 - \frac{1}{2}, \\
H_3(y) &= y^3 - \frac{3}{2}y,
\end{aligned} \tag{7.36}
$$

where, by convention, the coefficient of the leading term has been set to 1. A general formula for all these polynomials is

$$H_n(y) = \left(-\frac{1}{2}\right)^n e^{y^2} \frac{d^n}{dy^n} e^{-y^2}. \tag{7.37}$$

The energy eigenvalues E_n ($n = 0, 1, 2, \ldots$) are the energy levels of the harmonic oscillator. The ground state has energy $E_0 = \frac{1}{2}\hbar\omega$ and all states with energies above E_0 are excited states. It is noteworthy that the ground state energy is positive, and that the gaps between neighbouring levels are all equal. $\frac{1}{2}\hbar\omega$ is called the zero-point energy, and is the lowest energy that the quantum harmonic oscillator can have. By contrast, the classical oscillator has minimum energy zero when the particle is at rest at the bottom of the potential, at $x = 0$.

For the harmonic oscillator the general solution is

$$\psi(x, t) = \sum_{n=0}^{\infty} a_n H_n\left(\sqrt{\frac{m\omega}{\hbar}} x\right) e^{-\frac{m\omega}{2\hbar} x^2} e^{-i\left(n+\frac{1}{2}\right)\omega t}. \tag{7.38}$$

The amplitudes a_n must approach zero fast enough as $n \to \infty$ so that the sum converges, but otherwise they are arbitrary.

What about more general potentials? One can consider the Schrödinger equation with, for example, $V(x) = \frac{1}{2}m\omega^2 x^2 + x^4$ or $V(x) = |x|$. These are potentials whose minimum value is zero and which grow to ∞ as $x \to \pm\infty$. It is not easy to determine their energy eigenvalues, but for potentials like these the energies are an infinite discrete set of positive numbers, above the minimum of the potential. Potentials like $V(x) = \frac{1}{2}m\omega^2 x^2 + x^3$, for which V approaches $-\infty$ as $x \to -\infty$, do not lead to physically realistic models in quantum mechanics. There are no energy eigenvalues at all, and the particle dynamics is unstable.

7.4 Interpretation of Wavefunctions—Observables

How should solutions of the Schrödinger equation be interpreted? What is their physical meaning? We know that a stationary state

$$\psi(x, t) = \chi(x) e^{-\frac{i}{\hbar} E t}, \tag{7.39}$$

where $H\chi = E\chi$, is a state with energy E, but there are also states that are superpositions of these, with different values of E. How are they to be understood? Where is the particle and what is its momentum? Does the particle have a definite energy?

These matters were resolved by Max Born, who formulated the standard statistical view of quantum theory. According to Born, the Schrödinger wavefunction gives *probabilistic information* about a particle and its dynamics. This goes against Schrödinger's original intuition that the wavefunction is a measurable object, like an electromagnetic wave or a wave at sea, but it rapidly became a key component of the standard interpretation of quantum mechanics. Although the interpretation of quantum mechanics has still not been resolved to everyone's satisfaction, quantum mechanics provides a recipe for predicting probabilistic results of experiments that has proved incredibly successful. No experiment or observation has ever cast doubt on the fact that quantum mechanics works extremely well.

7.4.1 Position probabilities

We first need to discuss where the particle is. The basic idea is that, at time t,

$$\int_{x_0}^{x_1} |\psi(x,t)|^2 dx \tag{7.40}$$

represents the probability that the particle is located between x_0 and x_1. The integrand is the modulus squared of the wavefunction, $|\psi(x,t)|^2 = \overline{\psi(x,t)}\psi(x,t)$, where $\overline{\psi}$ is the complex conjugate of ψ. $|\psi(x,t)|^2$ is real and non-negative, and represents the probability density for finding the particle at x.

The total probability has to be 1, so for equation (7.40) to make sense, the wavefunction has to be *normalized*, meaning that it satisfies

$$\int_{-\infty}^{\infty} |\psi(x,t)|^2 dx = 1. \tag{7.41}$$

For ψ obeying the Schrödinger equation (7.12) this normalization condition holds for all time, provided it is satisfied at some initial time. If a wavefunction is not normalized, then either the wavefunction needs to be multiplied by a constant so that it is normalized, or equivalently the formula (7.40) is replaced by

$$\frac{\int_{x_0}^{x_1} |\psi(x,t)|^2 dx}{\int_{-\infty}^{\infty} |\psi(x,t)|^2 dx}. \tag{7.42}$$

If ψ is normalized, then $e^{i\alpha}\psi$ (with α a real constant) is still normalized and satisfies the Schrödinger equation. The probability density is not affected by the phase factor $e^{i\alpha}$, and neither are any other physical properties of the particle. So one regards ψ and $e^{i\alpha}\psi$ as physically equivalent wavefunctions.

In a stationary state, $\psi(x,t) = \chi(x)e^{-\frac{i}{\hbar}Et}$, so $|\psi(x,t)|^2 = |\chi(x)|^2$ and the normalization condition on χ is

$$\int_{-\infty}^{\infty} |\chi(x)|^2 dx = 1. \tag{7.43}$$

The probability density $|\chi(x)|^2$ is time independent and this is the reason such a state is called stationary. For a more general wavefunction, the probability of finding the particle between x_0 and x_1 changes with time, so in this sense, the particle is moving.

As an example, let's consider the ground state and first excited state of the harmonic oscillator, using the scaled coordinate y. The normalized stationary states are

$$\chi_0(y) = \left(\frac{1}{\pi}\right)^{\frac{1}{4}} e^{-\frac{1}{2}y^2} \quad \text{and} \quad \chi_1(y) = \left(\frac{4}{\pi}\right)^{\frac{1}{4}} y e^{-\frac{1}{2}y^2}, \tag{7.44}$$

as one can verify using the Gaussian integrals (1.45) and (1.65). Individually, these states have definite energies, and in these states, the probabilities that the particle is between $y = 0$

and $y = 1$ are

$$\left(\frac{1}{\pi}\right)^{\frac{1}{2}} \int_0^1 e^{-y^2}\, dy \simeq 0.421 \quad \text{and} \quad \left(\frac{4}{\pi}\right)^{\frac{1}{2}} \int_0^1 y^2 e^{-y^2}\, dy \simeq 0.214. \tag{7.45}$$

The second of these is smaller, because the state is more spread out in the excited state than in the ground state. This is analogous to the amplitude of the classical oscillation increasing as the energy increases. In a superposition of these states, the time dependent phase factors have different frequencies, and the probability of finding the particle between $y = 0$ and $y = 1$ oscillates with the harmonic oscillator frequency ω.

The probability interpretation can be experimentally tested by measuring whether the particle is between x_0 and x_1. On a single occasion, the answer will be *yes* or *no*, but if the experiment is repeated, with the state being prepared the same way many times, then the fraction of the measurements that give *yes* should approach the predicted probability.

Some physicists find an interpretation that relies on repeated measurements unsatisfactory, and have proposed different interpretations. However, what seems definite is that there is no version of quantum mechanics that carries more information, such that the result of a position measurement is completely certain. Probabilities are an inescapable feature of quantum mechanics.

7.4.2 Other physical quantities—hermitian operators

We have introduced the idea that the basic dynamical variables, position x and momentum p, become operators in quantum mechanics. We have also seen that the Hamiltonian $H = \frac{1}{2m}p^2 + V(x)$ is a key operator; it appears in the Schrödinger equation and is related to the particle energy. Another operator is the kinetic energy alone, $\frac{1}{2m}p^2$. It is a fundamental postulate of quantum mechanics that every *observable*—every physical quantity that one might measure—is represented by an operator. Operators are usually associated with classical dynamical variables, generally functions of x and p, but the spin operators that we will meet later do not have close classical analogues. In quantum mechanics, observables are always represented by *hermitian operators* which, like the Hermite polynomials, are named after the mathematician Charles Hermite. The most significant property of hermitian operators is that they have real eigenvalues, as we shall show, so they are analogous to real dynamical variables.

Mathematically, an operator O is hermitian if it has the following symmetry property, known as *hermiticity*: O is hermitian if, for any complex functions $\phi(x)$ and $\eta(x)$ rapidly approaching zero as $x \to \pm\infty$,

$$\int_{-\infty}^{\infty} \overline{O\eta}\, \phi\, dx = \int_{-\infty}^{\infty} \overline{\eta}\, O\phi\, dx. \tag{7.46}$$

(Remember that ϕ, η, $O\phi$ and $O\eta$ are all functions of x.) An equivalent way of saying this is that $\int_{-\infty}^{\infty} \overline{\phi}\, O\eta\, dx$ is the complex conjugate of $\int_{-\infty}^{\infty} \overline{\eta}\, O\phi\, dx$.

It is not difficult to check hermiticity for particular operators. It usually requires an integration by parts. For example, $\frac{d^2}{dx^2}$ is hermitian because

$$\int_{-\infty}^{\infty} \overline{\frac{d^2\eta}{dx^2}}\, \phi\, dx = \int_{-\infty}^{\infty} \overline{\eta}\, \frac{d^2\phi}{dx^2}\, dx, \tag{7.47}$$

as one can verify by integrating by parts twice. Similarly $i\frac{d}{dx}$ is hermitian—the factor i is essential—because

$$\int_{-\infty}^{\infty} \overline{\left(i\frac{d\eta}{dx}\right)} \phi \, dx = \int_{-\infty}^{\infty} \overline{\eta} \left(i\frac{d\phi}{dx}\right) dx \,. \tag{7.48}$$

Integration by parts gives a minus sign, but $\overline{i} = -i$. It clearly follows that both the Hamiltonian $H = -\frac{\hbar^2}{2m}\frac{d^2}{dx^2} + V(x)$ and the momentum operator $p = -i\hbar\frac{d}{dx}$ are hermitian.

A hermitian operator O generally has an infinite number of independent eigenfunctions, and we assume they can be labelled $k = 0, 1, 2, \dots$. (This is convenient, and true for an entire class of operators, but for some other operators one needs to use a continuous rather than a discrete label.) So O has a discrete spectrum of eigenfunctions and eigenvalues

$$O\phi_k = \lambda_k \phi_k \,, \quad k = 0, 1, 2, \dots \,. \tag{7.49}$$

Two key results follow from the hermiticity of O: (i) each eigenvalue λ_k is *real*; (ii) eigenfunctions $\phi_k(x)$ and $\phi_l(x)$ associated to distinct eigenvalues λ_k and λ_l are *orthogonal*, in the sense that

$$\int_{-\infty}^{\infty} \overline{\phi_l}\,\phi_k \, dx = 0 \,. \tag{7.50}$$

(This is the analogue, for complex functions, of two orthogonal vectors having zero dot product.)

The proofs of results (i) and (ii) are rather similar. We start with a pair of eigenfunctions of O, satisfying the equations

$$O\phi_k = \lambda_k \phi_k \,, \tag{7.51}$$
$$O\phi_l = \lambda_l \phi_l \,, \tag{7.52}$$

and assume that that $\lambda_l \neq \lambda_k$. Then, using the complex conjugate of equation (7.51), the hermiticity of O and equation (7.51) again, we find

$$\overline{\lambda_k} \int_{-\infty}^{\infty} \overline{\phi_k}\,\phi_k \, dx = \int_{-\infty}^{\infty} \overline{O\phi_k}\,\phi_k \, dx = \int_{-\infty}^{\infty} \overline{\phi_k}\,O\phi_k \, dx = \lambda_k \int_{-\infty}^{\infty} \overline{\phi_k}\,\phi_k \, dx \,, \tag{7.53}$$

so $\overline{\lambda_k} = \lambda_k$, and therefore λ_k is real.

Similarly,

$$\lambda_l \int_{-\infty}^{\infty} \overline{\phi_l}\,\phi_k \, dx = \int_{-\infty}^{\infty} \overline{O\phi_l}\,\phi_k \, dx = \int_{-\infty}^{\infty} \overline{\phi_l}\,O\phi_k \, dx = \lambda_k \int_{-\infty}^{\infty} \overline{\phi_l}\,\phi_k \, dx \,, \tag{7.54}$$

where we have used the complex conjugate of equation (7.52), the reality of the eigenvalues of O (which we just proved), the hermiticity of O and equation (7.51). As $\lambda_l \neq \lambda_k$, this chain of equalities implies that $\int_{-\infty}^{\infty} \overline{\phi_l}\,\phi_k \, dx = 0$, which is the orthogonality condition (7.50) we set out to prove.

One can slightly strengthen this orthogonality result to orthonormality, by working with normalized eigenfunctions ϕ_k satisfying

$$\int_{-\infty}^{\infty} \overline{\phi_k}\,\phi_k \, dx = 1 \,. \tag{7.55}$$

The orthonormality condition is then

$$\int_{-\infty}^{\infty} \overline{\phi_l}\, \phi_k \, dx = \delta_{lk} \,, \tag{7.56}$$

where δ_{lk} is the Kronecker delta symbol taking the value 1 if $l = k$, and 0 if $l \neq k$. Equation (7.56) combines the orthogonality condition (7.50) with the normalization condition (7.55). Orthonormality can also be enforced in situations where there is more than one eigenfunction for some of the eigenvalues (i.e. in situations where eigenvalues are degenerate).

A deep theorem in the analysis of hermitian operators is that the eigenfunctions ϕ_k are a *complete* set, meaning that any wavefunction ψ can be expressed as a linear combination of them. (This is probably the most important mathematical theorem required in quantum mechanics and is routinely used in applications throughout atomic and condensed matter physics, as well as theoretical chemistry.) It is a generalization of the idea of Fourier series, that *any* periodic function can be expressed as a linear combination of sine and cosine functions with the same period. Using completeness, we can write

$$\psi(x,t) = \sum_{k=0}^{\infty} c_k(t)\phi_k(x) \,, \tag{7.57}$$

where the amplitudes $c_k(t)$ depend on time t because ψ is a function of t.

If ψ is normalized, then the set of amplitudes is normalized in the sense that

$$\sum_{k=0}^{\infty} |c_k(t)|^2 = 1 \,. \tag{7.58}$$

This is because, for a normalized ψ,

$$
\begin{aligned}
1 = \int_{-\infty}^{\infty} |\psi(x,t)|^2 \, dx &= \sum_{l=0}^{\infty} \sum_{k=0}^{\infty} \overline{c_l(t)}\, c_k(t) \int_{-\infty}^{\infty} \overline{\phi_l}\, \phi_k \, dx \\
&= \sum_{l=0}^{\infty} \sum_{k=0}^{\infty} \overline{c_l(t)}\, c_k(t)\delta_{lk} \\
&= \sum_{k=0}^{\infty} \overline{c_k(t)}\, c_k(t) \\
&= \sum_{k=0}^{\infty} |c_k(t)|^2 \,.
\end{aligned} \tag{7.59}
$$

Here we have used the orthonormality of the eigenfunctions, and the basic property of the Kronecker delta, $\sum_l \alpha_l \,\delta_{lk} = \alpha_k$, which holds because only the term with $l = k$ contributes to the sum.

7.4.3 Measurements of observables

A physical quantity represented by a hermitian operator O is known as an *observable*. We will now consider the measurement of observables. We will need to use the results that O

has real eigenvalues λ_k, and that the normalized eigenfunctions $\phi_k(x)$ of O are a complete, orthonormal set of functions, as just discussed. Let the wavefunction, representing the state of the particle, have the expansion (7.57), $\psi(x,t) = \sum_{k=0}^{\infty} c_k(t)\phi_k(x)$.

It is a basic postulate of quantum mechanics that the *possible* outcomes of a measurement of the quantity represented by the operator O are the eigenvalues λ_k of O, and these values only. If the eigenvalues form a discrete set then there are necessarily gaps between them. The set of eigenvalues depends, of course, only on O and not on the wavefunction. The wavefunction simply determines the *probabilities* of the various outcomes. If the measurement is done at time t, then the outcome is λ_k with probability $|c_k(t)|^2$. The interpretation of the normalization condition (7.58) is that the sum of the probabilities of all possible outcomes is 1, as it must be.

Knowing all the possible outcomes of a measurement, and their probabilities, is as much as one can hope for in quantum mechanics. There are no hidden variables that give more information about the particle than the wavefunction. Sometimes there is less information. Generally there is uncertainty, because various outcomes are possible with various probabilities. The exception is if the wavefunction happens to be (at time t) an eigenfunction of O, say the eigenfunction ϕ_K, with eigenvalue λ_K. In this case we say that the particle has a definite value λ_K for the observable represented by O. The probability that a measurement yields λ_K is then 1.

What about the energy? Suppose the eigenvalues of the Hamiltonian H are E_n ($n = 0, 1, 2, \ldots$) and the normalized eigenfunction associated with E_n is $\chi_n(x)$. Recall that $\chi_n(x)$ is a stationary state wavefunction. The general, non-stationary solution of the Schrödinger equation is expressed in terms of these as

$$\psi(x,t) = \sum_{n=0}^{\infty} a_n \chi_n(x) e^{-\frac{i}{\hbar} E_n t}\,, \tag{7.60}$$

and if $\psi(x,t)$ is normalized, then

$$\sum_{n=0}^{\infty} |a_n|^2 = 1\,. \tag{7.61}$$

If we measure the energy then the outcome will be one of the values E_n, and the probability that the measurement gives E_n is $|a_n|^2$. This is a particular case of what we have said for a general operator O, because equation (7.60) is the expansion of ψ in terms of the eigenfunctions of H. The amplitude $c_n(t)$ in this case is $a_n e^{-\frac{i}{\hbar} E_n t}$, and $|c_n(t)|^2 = |a_n|^2$. Energy is special in that these probabilities do not change with time, even for non-stationary wavefunctions. This is an aspect of energy conservation in quantum mechanics. However, energy is still uncertain in a non-stationary state—only in a stationary state does the energy have a definite value.

We can also consider the measurement of momentum. As we have seen, the momentum operator $p = -i\hbar \frac{d}{dx}$ is hermitian. Its eigenvalue equation is

$$-i\hbar \frac{d}{dx} \phi = \lambda \phi\,, \tag{7.62}$$

with the boundary condition that ϕ should not grow exponentially as $x \to \pm\infty$. The solutions are $\phi_k(x) = e^{ikx}$, with $\lambda = \hbar k$ and k any real constant. The function e^{ikx} neither grows nor decays in magnitude as $x \to \pm\infty$, and this is acceptable. So e^{ikx} (which is a stationary state of a free particle) is an eigenfunction of the momentum operator p with eigenvalue $\hbar k$. Since k takes any real value, a measurement of momentum can have any real-valued outcome. A general wavefunction $\psi(x, t)$ has an expansion in terms of these eigenfunctions of the form

$$\psi(x, t) = \frac{1}{2\pi} \int_{-\infty}^{\infty} \widetilde{\psi}(k, t) e^{ikx} dk, \tag{7.63}$$

and the probability density for finding the momentum value $\hbar k$ is $\frac{1}{2\pi}|\widetilde{\psi}(k, t)|^2$. Equation (7.63) is the formula for the inverse Fourier transform, so $\widetilde{\psi}(k, t)$ is the Fourier transform of $\psi(x, t)$. If the wavefunction ψ is normalized, then

$$1 = \int_{-\infty}^{\infty} |\psi(x, t)|^2 dx = \frac{1}{2\pi} \int_{-\infty}^{\infty} |\widetilde{\psi}(k, t)|^2 dk, \tag{7.64}$$

showing that the momentum probability density is correctly normalized. (In the theory of Fourier transforms, this result is known as Parseval's theorem.)

Momentum has a definite value only for a wavefunction that is an eigenfunction of the momentum operator. If the wavefunction is e^{ikx} at some instant, the momentum is $\hbar k$. This incorporates de Broglie's insight about momentum into the general quantum mechanical framework of observables, hermitian operators and probabilities. Such a wavefunction, with definite momentum, is actually not normalizable. This is typically the case for eigenfunctions of an operator with a continuous set of eigenvalues. To make the analysis physically meaningful, one must restrict the wavefunctions to a large, finite region of space, so that they are normalizable, in which case the momentum is not quite definite. As we will see, this can be interpreted as a manifestation of the uncertainty principle.

7.5   Expectation Values

Here we discuss the mean value of the outcome of a quantum measurement. By definition, the mean value is the average of the outcomes, weighted by the probabilities. For example, if a fair dice with the usual faces 1 to 6 is thrown, the mean value of the outcomes is $3\frac{1}{2}$. In quantum mechanics, the mean value is called the *expectation value*.

Recall the expansion (7.57) of a normalized wavefunction ψ in terms of the eigenfunctions of O, with coefficients $c_k(t)$. The expectation value for a measurement at time t of the observable represented by O is

$$\langle O \rangle = \sum_k |c_k(t)|^2 \lambda_k, \tag{7.65}$$

as the outcome λ_k has probability $|c_k(t)|^2$. Because $c_k(t)$ is related to the wavefunction, there is an elegant alternative formula for $\langle O \rangle$,

$$\langle O \rangle = \int_{-\infty}^{\infty} \overline{\psi(x, t)}\, O\, \psi(x, t)\, dx, \tag{7.66}$$

which we prove shortly. Notice that this formula depends directly on how the operator O acts on the wavefunction, and one does not need explicit knowledge of the individual probabilities to determine $\langle O \rangle$. One might have anticipated that the physical value of a dynamical variable has something to do with how the operator representing this variable acts on the wavefunction, and equation (7.66) confirms this.

If the wavefunction ψ is an eigenfunction of O with eigenvalue λ then the state has a definite value for the operator O, namely λ. The formula (7.66) is consistent with this, because in this case $O\psi = \lambda\psi$ so

$$\langle O \rangle = \lambda \int_{-\infty}^{\infty} \overline{\psi(x,t)}\psi(x,t)\,dx = \lambda\,. \tag{7.67}$$

The proof of the equivalence of equations (7.65) and (7.66) is as follows. Starting with equation (7.66), we use the expansion of the wavefunction in terms of the eigenfunctions of O to find successively

$$
\begin{aligned}
\langle O \rangle &= \sum_k \sum_l \overline{c_k(t)} c_l(t) \int_{-\infty}^{\infty} \overline{\phi_k(x)}\, O\, \phi_l(x)\, dx \\
&= \sum_k \sum_l \overline{c_k(t)} c_l(t) \lambda_l \int_{-\infty}^{\infty} \overline{\phi_k(x)}\phi_l(x)\, dx \\
&= \sum_k \sum_l \overline{c_k(t)} c_l(t) \lambda_l \delta_{kl} \\
&= \sum_k \overline{c_k(t)} c_k(t) \lambda_k \\
&= \sum_k |c_k(t)|^2 \lambda_k\,,
\end{aligned} \tag{7.68}
$$

where the key step (from the second to the third line) is to use the orthonormality condition (7.56) for eigenfunctions of O.

Here are some examples of equation (7.66). The expectation value of the energy is

$$\langle H \rangle = \int_{-\infty}^{\infty} \overline{\psi(x,t)}\, H\, \psi(x,t)\, dx\,, \tag{7.69}$$

which equals $\sum_{n=0}^{\infty} |a_n|^2 E_n$, and is time independent. The expectation value of the position x simplifies, because the operator x just acts by multiplying the wavefunction by x, so

$$
\begin{aligned}
\langle x \rangle &= \int_{-\infty}^{\infty} \overline{\psi(x,t)}\, x\, \psi(x,t)\, dx \\
&= \int_{-\infty}^{\infty} |\psi(x,t)|^2 x\, dx\,.
\end{aligned} \tag{7.70}
$$

This is consistent with $|\psi(x,t)|^2$ being the position probability density. Perhaps the most useful example is the expectation value of momentum. This is

$$\langle p \rangle = -i\hbar \int_{-\infty}^{\infty} \overline{\psi(x,t)} \frac{\partial}{\partial x} \psi(x,t)\, dx\,. \tag{7.71}$$

Even though the momentum probability density involves the Fourier transform of ψ, the formula (7.71) for the expectation value does not.

7.6 After a Measurement

Quantum mechanics has a further postulate, concerning what happens to the wavefunction just after one has performed a measurement. Suppose the dynamical variable represented by O is measured. Recall that the wavefunction has the expansion

$$\psi(x,t) = \sum_k c_k(t)\phi_k(x), \qquad (7.72)$$

where $\phi_k(x)$ is a normalized eigenfunction of O with eigenvalue λ_k. The outcome of the measurement at time t is λ_k with probability $|c_k(t)|^2$. Suppose the result of the measurement is one of these possible values, λ_K. The postulate is that immediately after the measurement the wavefunction is no longer ψ; it is the eigenfunction $\phi_K(x)$ associated with the eigenvalue λ_K. The measurement overrides the Schrödinger equation, and the wavefunction jumps. This jump is called *wavefunction collapse*. If the measurement is repeated immediately, the result will be λ_K again, with probability 1. If no further measurement is done, the wavefunction will evolve from ϕ_K according to the Schrödinger equation.

The collapse of the wavefunction is rather mysterious, especially as it is not described by any dynamical equation. The established Copenhagen interpretation, pioneered by Bohr, is that measurements are made by devices obeying classical physics, and these must have definite values. If the measurement gives the value λ_K, then the observable O has the definite value λ_K and the state *must* now be ϕ_K. But it wasn't beforehand. The Copenhagen interpretation requires a classical world to coexist with a quantum world of atoms, so quantum mechanics by itself, without classical measuring devices, has no meaning.

This is not really satisfactory when one realizes that measuring devices and the records they produce are physical, and not so fundamentally different from the objects being measured. This has become increasingly the case as quantum phenomena are observed in larger and larger systems, and measuring devices and the systems that record measurements get smaller, so the size distinction between macroscopic lab equipment and quantum systems dissolves. For example, the recording of electron positions in particle physics experiments now involves, not pointers or photographs, but other electrons in silicon chips and similar semiconductor devices.

No-one really understands these purported jumps in the wavefunction. One viewpoint is that measurements do not collapse wavefunctions, but set up correlations between the object and the measuring device. This relies on the possibility that a measuring device can be in a superposition of states. As these remarks should indicate, the interpretation of quantum mechanics is still debated by physicists, and no consensus has been reached on exactly what the state of a quantum system means or how to understand wavefunction collapse. We will return to these unresolved issues in the final chapter.

7.7 Uncertainty Relations

If the wavefunction of a particle is an eigenfunction of some hermitian operator O, with eigenvalue λ, then a measurement of O will give the outcome λ with probability 1. The particle has a definite value for O, and there is no uncertainty. For example, in a stationary state—an eigenfunction of the Hamiltonian—the particle has a definite energy. For more general wavefunctions, there is a range of outcomes with various probabilities, so there is uncertainty.

Consider now two observables, represented by O_1 and O_2. Can both of these have definite values, with no uncertainty? A closely related question is whether both can be measured at the same time. The answer depends on whether these operators commute or not. Suppose that O_1 and O_2 do not commute. This means that the commutator is a third operator O_3 that is not identically zero:

$$[O_1, O_2] = O_1 O_2 - O_2 O_1 = O_3 . \qquad (7.73)$$

The classic non-commuting operators are x and p, with the canonical commutation relation $[x, p] = i\hbar\mathbf{1}$, but there are many others, including pairs of spin operators that we will discuss in section 8.5.

A consequence of O_1 and O_2 not commuting is that their eigenfunctions are not all the same. For suppose they have an eigenfunction ϕ in common. Then

$$O_1\phi = \lambda_1\phi, \quad O_2\phi = \lambda_2\phi, \qquad (7.74)$$

where the eigenvalues may be different. Now, operating on the second equation with O_1, and on the first with O_2, gives

$$\begin{aligned} O_1 O_2 \phi &= \lambda_2 O_1 \phi &= \lambda_2 \lambda_1 \phi, \\ O_2 O_1 \phi &= \lambda_1 O_2 \phi &= \lambda_1 \lambda_2 \phi. \end{aligned} \qquad (7.75)$$

Subtracting these equations, we find

$$[O_1, O_2]\phi = 0 , \qquad (7.76)$$

so $O_3\phi = 0$. In general, O_3 does not have zero as an eigenvalue, so this last equation has no solutions, and the conclusion is that O_1 and O_2 have no common eigenfunction.

Exceptionally, O_3 may have one or more eigenfunctions with zero eigenvalue, and such functions can be eigenfunctions of O_1 and O_2 simultaneously. The subspace of eigenfunctions of O_3 with zero eigenvalue is certainly not a complete set of functions; if it were, O_3 would be identically zero. Some eigenfunctions of O_1 are therefore outside this subspace, and these cannot simultaneously be eigenfunctions of O_2. The consequence is that the states for which O_1 and O_2 both have definite values is limited, for such states are simultaneous eigenfunctions of O_1 and O_2. There may be no such states at all, or at most a few.

Excluding these few states, one can say that there is always uncertainty in the combined values of O_1 and O_2. A general state has uncertainty for both of these observables, since it will not be an eigenfunction of either, but even if O_1 has a definite value, then O_2 does not, and if O_2 has a definite value, then O_1 does not. This conclusion is about the uncertainty of the outcome of a measurement, in situations where one can choose to measure O_1 or O_2.

A more physical consequence is that it is actually impossible to simultaneously measure O_1 and O_2 if they do not commute. According to the measurement postulate in quantum mechanics, a simultaneous measurement would produce definite outcomes for both observables, and the wavefunction would collapse to a simultaneous eigenfunction of the two operators. However, we have just seen that the existence of a simultaneous eigenfunction is incompatible with the non-commutativity of the operators (as there is no reason to suppose that the state being measured is an eigenfunction of O_3 with eigenvalue zero).

In fact, one can understand this more physically. The measuring apparatus for observable O_1 would physically obstruct the measuring apparatus for observable O_2. For example, a precise position measurement requires a localized device that can intercept the particle. On the other hand, a precise momentum measurement requires the particle to be able to move freely over a large region. One way to measure momentum is to find the angle of scattering when the particle passes through a region in which there is a uniform magnetic field, but there cannot be a precise position detector inside this region at the same time.

Another example is a particle in a non-trivial potential $V(x)$. The operators p and H do not commute, so momentum and total energy cannot be simultaneously measured. Physically, this is because a particle of definite energy, whose position is constrained by the potential V, is not at the same time free to move through a region sufficiently large for its momentum to be determined.

There is a quantitative uncertainty relation between position and momentum measurements. For a given state there is some distribution of probabilities around the mean value, for either a position or a momentum measurement. The simplest quantities parametrizing these distributions are the standard deviations Δx and Δp. These are related by the *Heisenberg uncertainty relation*

$$\Delta x \Delta p \geq \frac{1}{2}\hbar \,, \tag{7.77}$$

which can be derived using the commutation relation (7.1). A state can have small Δx, but then Δp is large, and vice versa. If the momentum is known with certainty, then the position is completely unknown. Similar results apply to other non-commuting operators. This uncertainty relation allows one to understand the inaccuracy in the tracks left by a particle in a detector. Such tracks do measure the position, but not very precisely. Their curvature measures the particle momentum, again not with absolute precision. The combined uncertainties in position and momentum are compatible with the uncertainty relation.

Let us turn now to the case where O_1 and O_2 do commute, so $[O_1, O_2] = 0$. Suppose that λ_1 is an eigenvalue of O_1 which is non-degenerate. This means there is an eigenfunction ϕ such that

$$O_1 \phi = \lambda_1 \phi \,, \tag{7.78}$$

and the only solutions of this equation are constant multiples of ϕ. Now act with O_2 to obtain $O_2 O_1 \phi = \lambda_1 O_2 \phi$, and therefore, as O_1 and O_2 commute,

$$O_1 O_2 \phi = \lambda_1 O_2 \phi \,. \tag{7.79}$$

This shows that $O_2 \phi$ is an eigenfunction of O_1 with eigenvalue λ_1, and by the non-degeneracy assumption, it is a multiple of ϕ. So $O_2 \phi = \lambda_2 \phi$ for some λ_2, and therefore ϕ is a simultaneous eigenfunction of O_1 and O_2. If all the eigenvalues of O_1 are non-degenerate, then every eigenfunction of O_1 is simultaneously an eigenfunction of O_2, when O_1 and O_2 commute.

An eigenvalue of O_1 may be degenerate, in which case there are two or more independent eigenfunctions with this eigenvalue, and it is then not true that every eigenfunction of O_1 is automatically an eigenfunction of O_2. However, if one carefully chooses the eigenfunctions of O_1 (from inside the subspaces corresponding to each degenerate eigenvalue) then they can be arranged to be simultaneously eigenfunctions of O_1 and O_2. Moreover, these simultaneous eigenfunctions form a complete set of functions. The wavefunction therefore has an expansion in terms of these simultaneous eigenfunctions. Physically, this means that O_1 and O_2 can be simultaneously measured, and the outcome of the measurement is the pair of eigenvalues of O_1 and O_2 for one of the simultaneous eigenfunctions.

Some examples of commuting operators are rather trivial. For example, any power of p commutes with any other power of p. In particular, p commutes with the kinetic energy $\frac{1}{2m}p^2$, so an eigenfunction of momentum is automatically an eigenfunction of kinetic energy. We will find more interesting examples of commuting operators when we discuss quantum mechanics in three dimensions, in Chapter 8.

7.8 Scattering and Tunnelling

Let us leave questions of measurement and interpretation, and return to the nitty-gritty of quantum mechanics—solving the Schrödinger equation. Suppose $V(x)$ is a potential of limited range that approaches zero as $x \to \pm\infty$. The stationary Schrödinger equation for a particle in this potential is

$$-\frac{\hbar^2}{2m}\frac{d^2\chi}{dx^2} + V(x)\chi = E\chi. \tag{7.80}$$

At large $|x|$, the particle does not notice the potential, so it is almost free. There are two independent solutions of the stationary Schrödinger equation for a free particle having a given positive energy. These are

$$\chi_+(x) = e^{ikx}, \quad \chi_-(x) = e^{-ikx}, \tag{7.81}$$

where k is positive. The first solution represents a right-moving particle, with momentum $\hbar k$ and energy $\frac{\hbar^2 k^2}{2m}$, and the second solution represents a left-moving particle, with momentum $-\hbar k$ and energy $\frac{\hbar^2 k^2}{2m}$.

The potential V has the effect of connecting these free-particle solutions on the left ($x \ll 0$) and on the right ($x \gg 0$) in a definite way. One solution of equation (7.80) takes the form

$$\chi(x) = e^{ikx} + Re^{-ikx} \quad (x \ll 0),$$
$$\chi(x) = Te^{ikx} \quad (x \gg 0). \tag{7.82}$$

On the left, this is a superposition of an incoming wave (of unit amplitude) from the left, together with a reflected wave of amplitude R; on the right, it is a purely outgoing wave, with amplitude T. This solution is interpreted quantum mechanically as describing an incoming particle of momentum $\hbar k$ that is scattered by the potential. R is the reflected amplitude and T the transmitted amplitude. Both of these are functions of k. The probability that the particle is reflected is $|R|^2$ and the probability that it is transmitted is $|T|^2$. One can prove that for any real potential, $|R|^2 + |T|^2 = 1$, consistent with the probabilistic interpretation of quantum mechanics. There is no simple formula for the full solution in the central region where V has a significant effect, but it is this full solution (which often has to be found numerically) that determines R and T. A second independent solution represents a particle incoming from the right. This has different, but not totally independent, reflection and transmission amplitudes.

Quantum scattering by a positive potential, known as a *potential barrier*, is quite different from the equivalent classical situation. In one dimension, a classical particle will

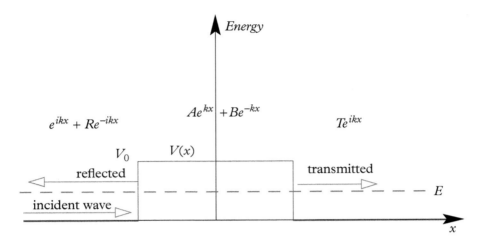

Fig. 7.4 Potential barrier. The form of the wavefunction in each region is indicated at the top.

not be reflected if its initial kinetic energy exceeds the barrier height—the maximum of the potential. The particle will slow down crossing the barrier, but will always be transmitted. Conversely, if the particle's kinetic energy is less than the barrier height, then it cannot cross the barrier and will always be reflected. The quantum mechanical results agree with these classical expectations in limiting cases. For a particle whose energy is much greater than the barrier height, the reflection probability is very small, and for a particle whose energy is much less than the barrier height, the transmission probability is very small. However, at intermediate energies, the classical and quantum behaviour is different. The transmission of a particle whose total energy is somewhat less than the barrier height is known as tunnelling. The tunnelling probability depends on both the amount by which the barrier height exceeds the particle energy, and also on how wide the barrier is. Tunnelling is more likely, for a given particle energy and barrier height, if the barrier is narrow. Tunnelling has particularly important applications in nuclear physics that we will look at in Chapter 11.

R and T can be calculated fairly easily for some simple potentials, such as a step potential that is a non-zero constant on some interval, and otherwise zero. The step potential barrier and step potential well are shown in Figures 7.4 and 7.5.

A curious special case of a potential with an exact particle scattering solution is the smooth potential well

$$V(y) = -\frac{2}{\cosh^2 y}\,. \tag{7.83}$$

The scattering solution for the scaled, stationary Schrödinger equation

$$-\frac{d^2\chi}{dy^2} - \frac{2}{\cosh^2 y}\chi = k^2\chi\,, \tag{7.84}$$

with scaled energy $\varepsilon = k^2$, is[3]

$$\chi(y) = \frac{k + i\tanh y}{k - i}\,e^{iky}\,. \tag{7.85}$$

---

[3] To show this, one needs $\frac{d}{dy}\cosh y = \sinh y$, $\frac{d}{dy}\sinh y = \cosh y$ and $\frac{d}{dy}\tanh y = \frac{1}{\cosh^2 y}$.

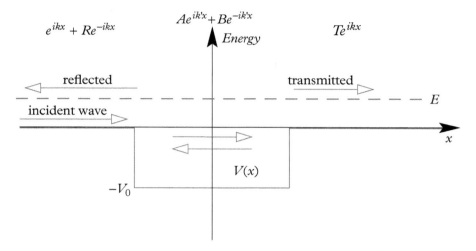

Fig. 7.5 Potential well. The form of the wavefunction in each region is indicated at the top.

Its asymptotic form matches equation (7.82), with $R = 0$ and $T = \frac{k+i}{k-i}$, because $\tanh y \to \pm 1$ as $y \to \pm\infty$. There is no reflection, and the transmission probability $|T|^2$ is 1 at all energies. This is unusual, and $V(y)$ is called a reflectionless potential.

There is also one discrete bound state for the potential (7.83). Its energy is $\varepsilon = -1$ and the eigenfunction is $\chi(y) = \frac{1}{\cosh y}$ as

$$\left(-\frac{d^2}{dy^2} - \frac{2}{\cosh^2 y} \right) \frac{1}{\cosh y} = -\frac{1}{\cosh y} \,. \tag{7.86}$$

The bound state energy $\varepsilon = -1$ corresponds to an unphysical imaginary value, $k = i$, for the scattering problem, and it is not a coincidence that both the scattering solution (7.85) and the transmission amplitude T have a singularity for this value of k.

More generally, $V(y) = -\frac{n(n+1)}{\cosh^2 y}$ is reflectionless for any positive integer n, and it also has bound states with discrete energy eigenvalues $-1, -4, \ldots, -n^2$.

7.9 Variational Principles in Quantum Mechanics

It is possible to derive the Schrödinger equation as an Euler–Lagrange equation from a variational principle. This is because it is a type of wave equation. The quantity one needs to consider is not an action, as for a classical particle or wave field, and it is not Lorentz invariant, so we denote it by I rather than S. I, which involves the wavefunction $\psi(x, t)$ and its complex conjugate $\overline{\psi}(x, t)$, and their first partial derivatives, is

$$I = \int \left\{ \frac{1}{2} i\hbar \overline{\psi} \frac{\partial \psi}{\partial t} - \frac{1}{2} i\hbar \frac{\partial \overline{\psi}}{\partial t} \psi - \frac{\hbar^2}{2m} \frac{\partial \overline{\psi}}{\partial x} \frac{\partial \psi}{\partial x} - V(x) \overline{\psi} \psi \right\} dx dt \,. \tag{7.87}$$

This is a formal expression in the sense discussed in section 2.3.5, with the integral being over all space and time. I is real, and the independent functions here are actually the real and imaginary parts of ψ, but in this and similar cases, one can treat ψ and $\overline{\psi}$ as independent.

Requiring I to be stationary under a localized variation of $\overline{\psi}$ gives an Euler–Lagrange equation of the general form

$$\frac{\partial}{\partial t}\left[\frac{\partial I}{\partial(\partial\overline{\psi}/\partial t)}\right] + \frac{\partial}{\partial x}\left[\frac{\partial I}{\partial(\partial\overline{\psi}/\partial x)}\right] - \frac{\partial I}{\partial\overline{\psi}} = 0. \tag{7.88}$$

For I as defined in (7.87), this is

$$-i\hbar\frac{\partial\psi}{\partial t} - \frac{\hbar^2}{2m}\frac{\partial^2\psi}{\partial x^2} + V(x)\psi = 0, \tag{7.89}$$

a rearrangement of the Schrödinger equation (7.13).

One should also consider a variation of ψ, but this just gives the complex conjugate of the Schrödinger equation, which is automatically satisfied when the Schrödinger equation is satisfied.

It is interesting to see that the Schrödinger equation can be derived in this way, but this approach does not have many applications. We cannot easily turn it into a practical tool for finding solutions. As the Schrödinger equation is first order in the time derivative, we cannot completely specify the wavefunction at both an initial time t_0 and a final time t_1. Instead, it is mathematically consistent to fix the wavefunction's phase (but not magnitude) for all x at times t_0 and t_1, but usually a problem of physical interest is different, and involves a wavefunction whose phase and magnitude are both fixed just at the initial time t_0.

Much more useful is a variational principle that applies to stationary state wavefunctions and especially to the ground state, the state of lowest energy. This is the *Rayleigh–Ritz principle*, that allows one to estimate the ground state energy in situations where it is hard to calculate exactly. We will use this principle to study chemical bonds in Chapter 9.

Suppose a quantum mechanical particle has Hamiltonian H, with the discrete set of energy levels

$$E_0 < E_1 \leq E_2 \leq \ldots \tag{7.90}$$

and corresponding orthonormal stationary states

$$\chi_0(x), \chi_1(x), \chi_2(x), \ldots, \tag{7.91}$$

but neither E_0 nor $\chi_0(x)$ are accurately known. We will assume that the ground state is non-degenerate—there is actually a theorem to this effect that applies to any sensible Hamiltonian. Recall that a general normalized wavefunction at a fixed time can be expanded in terms of the stationary states as

$$\psi(x) = \sum_{n=0}^{\infty} c_n\chi_n(x), \quad \sum_{n=0}^{\infty} |c_n|^2 = 1, \tag{7.92}$$

and that the expectation value of the energy for this wavefunction has two equivalent expressions

$$E = \langle H \rangle = \int_{-\infty}^{\infty} \overline{\psi(x)}H\psi(x)\,dx \tag{7.93}$$

and

$$E = \sum_{n=0}^{\infty} |c_n|^2 E_n \, . \tag{7.94}$$

The second expression is a weighted average of the energies E_n, and is therefore no lower than E_0. The minimum value of E is E_0 and occurs only if $|c_0| = 1$ and $c_1 = c_2 = \ldots = 0$.

So the ground state energy E_0 is also the minimum value of the first expression

$$E = \int_{-\infty}^{\infty} \overline{\psi(x)} H \psi(x) \, dx \tag{7.95}$$

as ψ varies over all normalized functions of x. If ψ is not constrained to be normalized, then E_0 is the minimum value of

$$E = \frac{\int_{-\infty}^{\infty} \overline{\psi(x)} H \psi(x) \, dx}{\int_{-\infty}^{\infty} \overline{\psi(x)} \psi(x) \, dx} \, . \tag{7.96}$$

This minimum value of E is a practical definition of the ground state energy of the particle, avoiding the need to solve the stationary Schrödinger equation. It can easily be generalized to the Hamiltonians of more complicated systems, including a particle in three dimensions, and multi-particle systems.

One can use formulae (7.95) and (7.96) to *estimate* the ground state energy E_0. One just needs to find a trial function $\psi(x)$, not necessarily normalized, that is reasonably close to the ground state wavefunction $\chi_0(x)$. The ratio (7.96), known as the *Rayleigh quotient*, is then the estimate for the ground state energy. A useful tactic is to find a family of functions $\psi(x; \alpha)$ depending on one or more parameters α, calculate the Rayleigh quotient as a function of α, and then find the minimum with respect to α. The last step is often straightforward, using ordinary calculus, or simple numerics.

This estimate for E_0 is usually remarkably accurate, for the following reason. Suppose ψ is normalized, so we can use formula (7.95). Any trial function ψ which is close to the true, normalized ground state χ_0 can be expressed as

$$\psi = \frac{1}{\sqrt{1+\varepsilon^2}} (\chi_0 + \varepsilon \chi_\perp) \tag{7.97}$$

where ε is small. Here $\chi_\perp$ is some normalized linear combination of the excited states $\chi_1, \chi_2, \ldots$. The prefactor normalizes ψ. The ground state energy estimate, using ψ, is

$$\begin{aligned} E &= \frac{1}{1+\varepsilon^2} \int_{-\infty}^{\infty} \overline{(\chi_0 + \varepsilon \chi_\perp)} H(\chi_0 + \varepsilon \chi_\perp) \, dx \\ &= \frac{1}{1+\varepsilon^2} \int_{-\infty}^{\infty} \overline{(\chi_0 + \varepsilon \chi_\perp)} (E_0 \chi_0 + \varepsilon H \chi_\perp) \, dx \, . \end{aligned} \tag{7.98}$$

Now, both $\chi_\perp$ and $H\chi_\perp$ are linear combinations of $\chi_1, \chi_2, \ldots$, so they are automatically orthogonal to χ_0. Therefore both cross-terms, which are proportional to ε in the expression (7.98), are zero. So the expression equals E_0 plus corrections of order ε^2, which come partly from the prefactor $\frac{1}{1+\varepsilon^2}$ and partly from the contribution of $\overline{\chi_\perp} H \chi_\perp$. This means that the error in the energy estimate, of order ε^2, is typically much smaller than the error in the trial wavefunction, which is of order ε.

Let's apply this method to an example where the ground state wavefunction and its energy are not known in closed form, the pure quartic oscillator with Hamiltonian (in simplified units)

$$H = -\frac{d^2}{dy^2} + y^4 \,. \tag{7.99}$$

Like the harmonic oscillator, this has an infinite, discrete set of energy levels. We use as trial function a normalized harmonic oscillator ground state,

$$\psi(y; \alpha) = \left(\frac{\alpha}{\pi}\right)^{\frac{1}{4}} e^{-\frac{1}{2}\alpha y^2} \,. \tag{7.100}$$

α is a width parameter that is readily varied, related to the frequency of the harmonic oscillator. The formula (7.95) estimates the ground state energy E_0 to be

$$E = \int_{-\infty}^{\infty} \left(-\psi \frac{d^2}{dy^2}\psi + y^4 \psi^2\right) dy = \int_{-\infty}^{\infty} \left(\left(\frac{d\psi}{dy}\right)^2 + y^4\psi^2\right) dy \,, \tag{7.101}$$

where we have used the reality of ψ, and integrated by parts to obtain the second expression. For our trial function, $\frac{d\psi}{dy} = -\left(\frac{\alpha}{\pi}\right)^{\frac{1}{4}} \alpha y \, e^{-\frac{1}{2}\alpha y^2}$ and we find, using the Gaussian integrals (1.65) and (1.66), that

$$E = \frac{\alpha}{2} + \frac{3}{4\alpha^2} \,. \tag{7.102}$$

Now we optimize this by varying α. The minimum of E is at $\frac{\partial E}{\partial \alpha} = \frac{1}{2} - \frac{3}{2\alpha^3} = 0$, so $\alpha = 3^{\frac{1}{3}}$. The optimal energy estimate using this family of trial functions is therefore

$$E_0 \simeq \frac{3}{4} 3^{\frac{1}{3}} \simeq 1.08 \,. \tag{7.103}$$

This is about 2% above the true ground state energy of the quartic oscillator, $E_0 \simeq 1.06$, which may be found numerically or by using a more refined class of trial functions.

The variational approach also has something to say about higher energy levels. The Rayleigh quotient has a saddle point at each eigenfunction of the Hamiltonian H. If a trial wavefunction is close to χ_n with an error of order ε, then the Rayleigh quotient will be E_n with an error of order ε^2. However, the error could be positive or negative, and for this reason it is hard to find parameters like α that can be systematically varied. So it is not easy to find saddle points and higher energy levels, even using families of trial functions.

7.10   Further Reading

B.H. Bransden and C.J. Joachain, *Quantum Mechanics (2nd ed.)*, Harlow: Pearson, 2000.

A.I.M. Rae, *Quantum Mechanics (5th ed.)*, Boca Raton FL: Taylor and Francis, 2008.

L.D. Landau and E.M. Lifschitz, *Quantum Mechanics (Non-Relativistic Theory): Course of Theoretical Physics, Vol. 3 (3rd ed.)*, Oxford: Butterworth-Heinemann, 1977.

8

Quantum Mechanics in Three Dimensions

8.1 Introduction

As regards operators and measurements, the principles of quantum mechanics are essentially the same in three dimensions as in one. The Schrödinger equation includes the Laplacian ∇^2 rather than $\frac{d^2}{dx^2}$, and a potential $V(\mathbf{x})$, and is generally more difficult to solve than in one dimension. However, when the potential is spherically symmetric, these difficulties are greatly reduced and finding the particle states and their energies becomes much easier. One physically important example is the Coulomb potential produced by the positively charged nucleus of an atom. The atomic electrons obey the Schrödinger equation in this attractive potential and the solutions occur at a discrete, infinite set of negative energy levels, providing a good explanation for atomic structure. For these electron bound states, the position probability density is concentrated around the nucleus and decays rapidly away from the centre. In a Coulomb potential, there are also positive energy electron scattering states. Scattering is a more complicated topic in three dimensions than in one, because scattered particles can emerge in all directions; they are not just transmitted forwards or reflected backwards. Quantum mechanical scattering theory is important for understanding many types of experiment involving particle beams, but we will not discuss it in detail.

In 3-dimensional quantum mechanics there are operators for angular momentum that have no analogues in one dimension. When the potential is spherically symmetric, angular momentum commutes with the Hamiltonian and stationary states are classified by the eigenvalues of the angular momentum operators, as well as by their energies. The details are quite subtle, because the operators representing the three components of angular momentum do not mutually commute.

In the 1920s it was discovered that particles have an intrinsic quantum *spin* in addition to the angular momentum they carry as they orbit other particles. Spin must be included in the total angular momentum, so even a freely moving particle, such as an electron not bound by a potential, carries some angular momentum. This is a quintessentially quantum mechanical feature of fundamental particles. Most particles, including electrons, protons, neutrons and photons have non-zero spin, although a few particles, including pions and the Higgs particle, have zero spin. Spin can be an integer or half-integer multiple of Planck's constant, $\hbar$.

A phenomenon that has no classical analogue is the remarkable quantum behaviour of systems of identical particles. For example, electrons in more complex atoms than hydrogen obey the quantum mechanical rule known as the *Pauli exclusion principle*, after Wolfgang

The Physical World. Nicholas Manton and Nicholas Mee, Oxford University Press (2017).
© Nicholas Manton and Nicholas Mee. DOI 10.1093/oso/9780198795933.001.0001

Pauli. Rather surprisingly, this is a direct consequence of their half-integer spin, so although electrons obey the Pauli principle, pions and photons do not.

At the end of this chapter we discuss how the classical action of a particle plays a role in quantum theory. It appears in a reformulation of quantum mechanics known as the *path integral* formulation. The path integral approach gives some insight into the classical limit of quantum mechanics, and also into such basic features of quantum mechanics as the de Broglie wavelength of a particle.

8.2   Position and Momentum Operators

A key idea in 1-dimensional quantum mechanics is that the classical dynamical variables are replaced by operators that do not always commute. In three dimensions we need three separate operators to represent the Cartesian position coordinates of a particle, and three operators to represent the particle's momentum components. These operators are denoted by x_i and p_i ($i = 1, 2, 3$), and they are all hermitian, so each represents an observable that can be measured. Collectively, the operators are denoted by vectors $\mathbf{x}$ and $\mathbf{p}$. The position operators mutually commute, which implies that they are simultaneously measurable. Therefore a measurement can pinpoint the location of a particle in 3-space. Similarly, the momentum operators mutually commute, so momentum is also measurable as a vector. However, as in one dimension, the position operators do not all commute with the momentum operators. The precise commutation relations are (with $i, j = 1, 2, 3$)

$$[x_i, x_j] = 0\,, \quad [p_i, p_j] = 0\,, \quad [x_i, p_j] = i\hbar\delta_{ij}\mathbf{1}\,. \tag{8.1}$$

Recall that δ_{ij} is 1 if $i = j$ and zero otherwise, and $\mathbf{1}$ is the unit operator. The position–momentum commutation relations say that, for example, $[x_1, p_1] = i\hbar\mathbf{1}$, which is analogous to the relation (7.1) in one dimension, but that $[x_1, p_2] = 0$. So it is possible to simultaneously measure the position of a particle in the 1-direction (by some extended planar device) and the momentum in the 2-direction. The commutation relations are unchanged by a shift of the origin, or by a rotation of the Cartesian axes, but they would look different if we used non-Cartesian coordinates. Because the position operators commute, there is no problem creating polynomials from them. The most important of these is the squared radius, $r^2 = x_1^2 + x_2^2 + x_3^2$. Also useful is the radius itself, r, which is well defined, despite being the square root of r^2. Both are rotationally invariant, scalar operators.

As in one dimension, we need a convenient representation of the position and momentum operators, to act on wavefunctions. The wavefunction $\psi(\mathbf{x}, t)$ of a particle is a function of its position and also depends on time. The position operators act by multiplication, with x_i acting on $\psi(\mathbf{x}, t)$ to give the new function $x_i\psi(\mathbf{x}, t)$. As the functions $x_i x_j \psi(\mathbf{x}, t)$ and $x_j x_i \psi(\mathbf{x}, t)$ are the same, the commutation relation $[x_i, x_j] = 0$ is satisfied. The momentum operators are multiples of the partial derivatives,

$$p_i = -i\hbar\frac{\partial}{\partial x_i}\,, \tag{8.2}$$

generalizing the 1-dimensional momentum operator (7.7). In vector form, $\mathbf{p} = -i\hbar\nabla$. Partial derivatives mutually commute, a result we have used several times before, so $[p_i, p_j] = 0$.

The position–momentum commutation relations may be verified by acting on a generic wavefunction:

$$[x_i, p_j]\psi = x_i \left(-i\hbar \frac{\partial}{\partial x_j}\right) \psi - \left(-i\hbar \frac{\partial}{\partial x_j}\right) (x_i \psi)$$

$$= -i\hbar x_i \frac{\partial \psi}{\partial x_j} + i\hbar \left(\delta_{ij}\psi + x_i \frac{\partial \psi}{\partial x_j}\right)$$

$$= i\hbar \delta_{ij}\psi, \tag{8.3}$$

where we have used the Leibniz rule, as before, and the result that the partial derivative of x_i with respect to x_j is δ_{ij}.

We can construct other operators from the momentum operators. Analogous to the squared radius, there is the squared momentum,[1] the scalar operator

$$\mathbf{p}^2 = p_1^2 + p_2^2 + p_3^2. \tag{8.4}$$

With the momentum operators represented as partial derivatives, we find

$$\mathbf{p}^2 = -\hbar^2 \left(\frac{\partial^2}{\partial x_1^2} + \frac{\partial^2}{\partial x_2^2} + \frac{\partial^2}{\partial x_3^2}\right) = -\hbar^2 \nabla^2, \tag{8.5}$$

a multiple of the Laplacian.

In classical dynamics, the kinetic energy of a particle of mass m is $\frac{1}{2m}\mathbf{p}^2$, so in quantum mechanics, kinetic energy is represented as $-\frac{\hbar^2}{2m}\nabla^2$. The total Hamiltonian H for a particle is the sum of the kinetic and potential energies, and the potential $V(\mathbf{x})$ is just a function of the spatial position. So the Schrödinger equation for a particle in three dimensions is

$$i\hbar \frac{\partial \psi}{\partial t} = H\psi = -\frac{\hbar^2}{2m}\nabla^2 \psi + V(\mathbf{x})\psi. \tag{8.6}$$

As in one dimension, we assume the potential is not explicitly dependent on time.

The most useful solutions of the Schrödinger equation are the stationary states, with their simple exponential time dependence,

$$\psi(\mathbf{x}, t) = \chi(\mathbf{x})e^{-\frac{i}{\hbar}Et}. \tag{8.7}$$

For these states, the Schrödinger equation simplifies to

$$H\chi = E\chi, \tag{8.8}$$

or explicitly

$$-\frac{\hbar^2}{2m}\nabla^2 \chi + V(\mathbf{x})\chi = E\chi. \tag{8.9}$$

The challenge, as before, is to find the energy eigenvalues E of the Hamiltonian, and the corresponding stationary state wavefunctions $\chi(\mathbf{x})$, which are also the eigenfunctions of H. For physical solutions, $\chi(\mathbf{x})$ should not grow as $|\mathbf{x}| \to \infty$. The simplest case is where

---

[1] From now on, for any vector $\mathbf{v}$, it will be convenient to use the notation $\mathbf{v}^2$ for $\mathbf{v} \cdot \mathbf{v}$.

the potential $V(\mathbf{x})$ vanishes, and we have a free particle. The solution of (8.9) is a pure exponential in the position variables,

$$\chi(\mathbf{x}) = e^{i\mathbf{k}\cdot\mathbf{x}} = e^{ik_1 x_1} e^{ik_2 x_2} e^{ik_3 x_3} . \tag{8.10}$$

This is an eigenfunction of all three momentum operators p_i, with eigenvalues $\hbar k_i$. Equivalently, it is an eigenfunction of $\mathbf{p}$ with eigenvalue $\hbar\mathbf{k}$, and the vector $\mathbf{k}$ is unconstrained. $\chi(\mathbf{x})$ is also an eigenfunction of the Hamiltonian $H = \frac{1}{2m}\mathbf{p}^2 = -\frac{\hbar^2}{2m}\nabla^2$, with energy eigenvalue $E = \frac{\hbar^2 \mathbf{k}^2}{2m}$, where $\mathbf{k}^2 = k_1^2 + k_2^2 + k_3^2$. This stationary state wavefunction is appropriate for a particle in a particle beam, having definite momentum and positive energy.

8.2.1 Particle in a box

A minimal constraint on a particle is to restrict its motion to a box of finite volume. Such a constraint is useful for describing an electron in a sample of metal, and in many other condensed matter contexts. It is also useful for describing a gas molecule in a container.

Mathematically, the most convenient box is a cuboidal one with side lengths L_1, L_2, L_3. We impose periodic boundary conditions on the wavefunction of the particle; this identifies opposite sides of the box, which is not physically realistic, but other boundary conditions lead to similar results. The wavefunction of a free particle still takes the form $\chi(\mathbf{x}) = e^{ik_1 x_1} e^{ik_2 x_2} e^{ik_3 x_3}$, and its energy is still $E = \frac{\hbar^2 \mathbf{k}^2}{2m}$, but now the periodicity condition requires that $e^{ik_1 x_1} = e^{ik_1 (x_1 + L_1)}$, so $e^{ik_1 L_1} = 1$, and similarly for L_2 and L_3. Therefore $\mathbf{k}$ is constrained to satisfy

$$\mathbf{k} = (k_1, k_2, k_3) = \left(\frac{2\pi n_1}{L_1}, \frac{2\pi n_2}{L_2}, \frac{2\pi n_3}{L_3} \right) \tag{8.11}$$

with (n_1, n_2, n_3) integers. There is just one allowed state for each cell in $\mathbf{k}$-space with side lengths $\left(\frac{2\pi}{L_1}, \frac{2\pi}{L_2}, \frac{2\pi}{L_3} \right)$. The cell has volume $\frac{(2\pi)^3}{L_1 L_2 L_3} = \frac{(2\pi)^3}{V}$, where $V = L_1 L_2 L_3$ is the volume of the box. As this result just depends on V, we will ignore the detailed dependence of the energies on the box shape from now on.

Because there is one state per cell of size $\frac{(2\pi)^3}{V}$, the density of states in $\mathbf{k}$-space is $\frac{V}{(2\pi)^3}$. This is relevant for a range of physical wave systems in a box. In quantum mechanics it is more convenient to convert the result to momentum space. As the particle momentum is $\mathbf{p} = \hbar\mathbf{k}$, the constraint on $\mathbf{p}$ involves $2\pi\hbar$, and the density of states in $\mathbf{p}$-space is $\frac{V}{(2\pi\hbar)^3}$.

For particles in a box of macroscopic size this is a very high density, and the states are quasi-continuously distributed in $\mathbf{p}$-space. In the classical limit, particles are characterized by their position as well as momentum, and we can say that the density of particle states makes sense in position and momentum space together (phase space). The density in phase space is $\frac{1}{(2\pi\hbar)^3}$. Integrating this over the spatial box with measure d^3x, we obtain the factor V, and recover the density in momentum space. Although this argument is not rigorous it gives an important indication of the relationship between quantum mechanics and its classical limit.

The density in momentum space can be converted to a density in energy E. The number of states with momentum magnitude between p and $p + dp$ is $4\pi p^2\, dp$ times the density $\frac{V}{(2\pi\hbar)^3}$ in **p**-space, so the density in p is

$$\widetilde{g}(p) = \frac{Vp^2}{2\pi^2\hbar^3}\,. \tag{8.12}$$

By the further change of variable $E = \frac{1}{2m}p^2$, we find the density in E is[2]

$$g(E) = \frac{V}{4\pi^2}\left(\frac{2m}{\hbar^2}\right)^{\frac{3}{2}} E^{\frac{1}{2}}\,. \tag{8.13}$$

The number of states between energies E and $E + dE$ is $g(E)\, dE$. When referring to the *density of states* for a quantum particle in a box, we usually mean this function $g(E)$.

Because the particle states in a box form a quasi-continuum, we can use the density of states to replace any sum over states by an integral. If the discrete states are labelled by n and have energies E_n, then

$$\sum_{\text{states}} f(E_n) \simeq \int_{E_{\min}}^{\infty} g(E)f(E)\, dE\,. \tag{8.14}$$

This is valid for most functions f of the particle energy.

8.3 Angular Momentum Operators

The classical (orbital) angular momentum of a particle is $\mathbf{l} = \mathbf{x} \times \mathbf{p}$, a vector with three components. The component in the 1-direction, for example, is $l_1 = x_2 p_3 - x_3 p_2$. To obtain the quantum, *orbital angular momentum* operators, we just substitute the expressions for the momentum operators p_i as partial derivatives. It is conventional here to leave out the factor of $\hbar$, and define

$$l_1 = -i\left(x_2\frac{\partial}{\partial x_3} - x_3\frac{\partial}{\partial x_2}\right)\,, \quad l_2 = -i\left(x_3\frac{\partial}{\partial x_1} - x_1\frac{\partial}{\partial x_3}\right)\,,$$
$$l_3 = -i\left(x_1\frac{\partial}{\partial x_2} - x_2\frac{\partial}{\partial x_1}\right)\,. \tag{8.15}$$

In vector form, the angular momentum operator is $\mathbf{l} = -i\mathbf{x} \times \nabla$. The physical angular momentum operator is $\hbar$ times this. The advantage of this convention is that l_1, l_2, l_3 are dimensionless, and their eigenvalues are integers, as we will see. Planck's constant $\hbar$ is a unit of action, with dimensions of energy multiplied by time, so, for example, the exponent $-\frac{i}{\hbar}Et$ in the wavefunction (8.7) is dimensionless. Rather coincidentally, $\hbar$ also has the dimensions of angular momentum, so it is natural that in quantum mechanics, angular momentum is a pure number times $\hbar$. We might anticipate that angular momentum is an integer times $\hbar$. Often it is, but the spin of an electron is $\frac{1}{2}\hbar$, and we speak of the electron as having spin $\frac{1}{2}$.

---

[2] The densities are related by $\widetilde{g}(p)\, dp = g(E)\, dE$, where $dE = \frac{1}{m}p\, dp$.

Unlike the momentum operators, the orbital angular momentum operators do not mutually commute. Their commutators are

$$[l_1, l_2] = il_3, \quad [l_2, l_3] = il_1, \quad [l_3, l_1] = il_2. \tag{8.16}$$

The first of these can be verified by calculating

$$\begin{aligned}
[l_1, l_2]\psi &= -\left(x_2\frac{\partial}{\partial x_3} - x_3\frac{\partial}{\partial x_2}\right)\left(x_3\frac{\partial}{\partial x_1} - x_1\frac{\partial}{\partial x_3}\right)\psi \\
&\quad + \left(x_3\frac{\partial}{\partial x_1} - x_1\frac{\partial}{\partial x_3}\right)\left(x_2\frac{\partial}{\partial x_3} - x_3\frac{\partial}{\partial x_2}\right)\psi, \\
&= \left(x_1\frac{\partial}{\partial x_2} - x_2\frac{\partial}{\partial x_1}\right)\psi = il_3\psi, \tag{8.17}
\end{aligned}$$

where all the terms have cancelled except those arising from the operator $\frac{\partial}{\partial x_3}$ acting on x_3. The other two commutators are found by cyclically permuting the labels.

Another useful operator is the scalar, squared angular momentum

$$\mathbf{l}^2 = l_1^2 + l_2^2 + l_3^2. \tag{8.18}$$

One can check that $\mathbf{l}^2$ commutes with each of the individual operators l_1, l_2, l_3. The operators r^2 and ∇^2 also commute with each of the angular momentum operators. The underlying reason is that $\mathbf{l}^2, r^2$ and ∇^2 are scalar, rotationally invariant operators and all such operators must commute with angular momentum.

The final operator we will need is

$$\mathbf{x} \cdot \nabla = x_1\frac{\partial}{\partial x_1} + x_2\frac{\partial}{\partial x_2} + x_3\frac{\partial}{\partial x_3}. \tag{8.19}$$

This is the operator $\mathbf{x} \cdot \mathbf{p}$ with the factor $-i\hbar$ left out, and is again dimensionless. Recall that if $\mathbf{n}$ is any unit vector, then $\mathbf{n} \cdot \nabla$ is the derivative in the direction $\mathbf{n}$. The vector $\mathbf{x}$ has magnitude r and points radially outwards, so the operator (8.19) is r times the derivative in the outward radial direction, which we can write as $r\frac{\partial}{\partial r}$. This operator, first considered by Euler, is the radial scaling operator.

There is a useful relation involving the position and momentum of a classical particle, derived by decomposing $\mathbf{p}$ into its components parallel and orthogonal to $\mathbf{x}$. The unit vector in the $\mathbf{x}$ direction is $\frac{1}{r}\mathbf{x}$. The component of $\mathbf{p}$ in this direction is therefore $\frac{1}{r}\mathbf{x} \cdot \mathbf{p}$ and the orthogonal component is of magnitude $\frac{1}{r}|\mathbf{x} \times \mathbf{p}|$. By Pythagoras' theorem, the sum of the squared lengths of these components is $\mathbf{p}^2$, so classically

$$\mathbf{p}^2 = \frac{1}{r^2}(\mathbf{x} \cdot \mathbf{p})^2 + \frac{1}{r^2}(\mathbf{x} \times \mathbf{p}) \cdot (\mathbf{x} \times \mathbf{p}), \tag{8.20}$$

and therefore

$$r^2\mathbf{p}^2 = (\mathbf{x} \cdot \mathbf{p})^2 + (\mathbf{x} \times \mathbf{p}) \cdot (\mathbf{x} \times \mathbf{p}). \tag{8.21}$$

This relation has a quantum analogue. $\mathbf{p}^2$ is proportional to the kinetic energy operator and hence the Laplacian ∇^2, while $\mathbf{x} \cdot \mathbf{p}$ is proportional to the radial scaling operator $r\frac{\partial}{\partial r}$, and $\mathbf{x} \times \mathbf{p}$ is proportional to angular momentum $\mathbf{l}$. The quantum operator version of equation (8.21) is

$$r^2 \nabla^2 = \left(r \frac{\partial}{\partial r} \right)^2 + r \frac{\partial}{\partial r} - \mathbf{l}^2 \, . \tag{8.22}$$

This is not identical to the classical relation because not all the components of the operators $\mathbf{x}$ and $\mathbf{p}$ commute, and this produces the extra single power of $r \frac{\partial}{\partial r}$. To verify equation (8.22) one has to carefully calculate the operator $\mathbf{l}^2 = l_1^2 + l_2^2 + l_3^2$ using the definitions (8.15), and the square of $r \frac{\partial}{\partial r} = \mathbf{x} \cdot \nabla$ using equation (8.19). The relation (8.22) will be useful, because it connects the squared angular momentum to the Laplacian, and hence to the particle Hamiltonian.

8.3.1 Eigenfunctions of $\mathbf{l}^2$ using Cartesian coordinates

We shall now use the relation (8.22) to find a complete set of eigenfunctions and eigenvalues of the squared angular momentum operator $\mathbf{l}^2$. This is a key step towards solving the Schrödinger equation for a particle in a spherically symmetric potential. The standard approach is to use spherical polar coordinates r, ϑ, φ. Then one finds that each component of angular momentum is an operator that involves derivatives with respect to the angular coordinates ϑ and φ, but no derivative with respect to r. The same is the case for $\mathbf{l}^2$. The eigenfunctions of $\mathbf{l}^2$ are therefore functions of ϑ and φ, and are called *spherical harmonics*.

Here we adopt another approach, working mainly with Cartesian coordinates. Consider all the monomials in x_1, x_2 and x_3. These are products of powers of x_1, x_2, x_3,

$$p^{\{abc\}} = x_1^a x_2^b x_3^c \, , \tag{8.23}$$

where a, b, c are non-negative integers that we use as a label. (In section 1.3, examples of monomials were used to illustrate partial differentiation, and how the Laplacian acts. The discussion here will be similar but more systematic.) Let us denote by l the sum of the powers, $l = a + b + c$. l is called the degree of the monomial, and we should not forget the monomial of degree zero, $p^{\{000\}} = 1$. A polynomial is a finite sum of monomials with arbitrary numerical coefficients, and is said to be of degree l if all the contributing monomials are of degree l.

How many distinct monomials are there, for given l? In other words, how many choices are there for a, b, c? The answer is found by considering $l + 2$ objects in a row and choosing two of them as sticks:

$$\bullet \bullet \ldots \bullet \; \Big| \; \bullet \ldots \bullet \; \Big| \; \bullet \bullet \ldots \bullet \bullet \, . \tag{8.24}$$

This shows a objects on the left, b in the middle, and c on the right. The sticks can be in any two of the $l + 2$ locations: for example, if they are neighbours then $b = 0$, if one is on the far left then $a = 0$, and so on. The number of ways of choosing two positions out of $l + 2$ is $\frac{1}{2}(l+2)(l+1)$, and this is the number of monomials of degree l. The space of polynomials of degree l is the space of linear combinations of the monomials, and is therefore of dimension $\frac{1}{2}(l+2)(l+1)$.

The radial scaling operator is

$$x_1 \frac{\partial}{\partial x_1} + x_2 \frac{\partial}{\partial x_2} + x_3 \frac{\partial}{\partial x_3} \, . \tag{8.25}$$

Acting on $p^{\{abc\}}$, the first term brings down a factor a but otherwise leaves $p^{\{abc\}}$ unchanged. Similarly the second and third terms bring down b and c. Therefore

$$\left(x_1 \frac{\partial}{\partial x_1} + x_2 \frac{\partial}{\partial x_2} + x_3 \frac{\partial}{\partial x_3}\right) p^{\{abc\}} = (a+b+c)p^{\{abc\}} = l\,p^{\{abc\}}. \tag{8.26}$$

The same result is also evident using spherical polars where the operator is expressed as $r\frac{\partial}{\partial r}$. In spherical polars, the monomial $p^{\{abc\}}$ is r^l times an angular function, because

$$(x_1, x_2, x_3) = (r\sin\vartheta\cos\varphi, r\sin\vartheta\sin\varphi, r\cos\vartheta) \tag{8.27}$$

and each of these expressions has one factor of r. Therefore

$$r\frac{\partial}{\partial r} p^{\{abc\}} = l\,p^{\{abc\}}, \tag{8.28}$$

and the operator acts in this same simple way on all polynomials of degree l.

Next, let's consider how the Laplacian $\nabla^2 = \frac{\partial^2}{\partial x_1^2} + \frac{\partial^2}{\partial x_2^2} + \frac{\partial^2}{\partial x_3^2}$ acts on a monomial of degree l. The general action is

$$\nabla^2(x_1^a x_2^b x_3^c) = a(a-1)x_1^{a-2}x_2^b x_3^c + b(b-1)x_1^a x_2^{b-2}x_3^c + c(c-1)x_1^a x_2^b x_3^{c-2}. \tag{8.29}$$

Irrespective of the details, the result is always a polynomial of degree $l-2$. The Laplacian ∇^2 therefore acts on the space of degree l polynomials, a space of dimension $\frac{1}{2}(l+2)(l+1)$, and maps it into the space of degree $l-2$ polynomials, a space of dimension $\frac{1}{2}l(l-1)$. Since the dimension of the second space is smaller than the first, there must be polynomials of degree l on which the action of the Laplacian gives zero. We say that such polynomials are annihilated by the Laplacian. (Less dramatically, they satisfy the Laplace equation.) In fact, the dimension of the space of polynomials of degree l annihilated by the Laplacian is precisely the difference between the dimensions $\frac{1}{2}(l+2)(l+1)$ and $\frac{1}{2}l(l-1)$, which is $2l+1$. It is no greater, because every polynomial of degree $l-2$ can be obtained through the Laplacian acting on some polynomial of degree l.

For example, there are six degree 2 monomials, $x_1^2, x_2^2, x_3^2, x_1x_2, x_2x_3, x_1x_3$, and one degree 0 monomial, the number 1. Therefore, five independent polynomials of degree 2 are annihilated by the Laplacian. They can be chosen as

$$x_1^2 - x_2^2, \; x_1^2 + x_2^2 - 2x_3^2, \; x_1x_2, \; x_2x_3, \; x_1x_3. \tag{8.30}$$

We now reach the payoff of this approach. All the polynomials of degree l that are annihilated by the Laplacian are eigenfunctions of the squared angular momentum operator $\mathbf{l}^2$ with the same, easily found eigenvalue. To see this, let P be such a polynomial. The radial scaling operator $r\frac{\partial}{\partial r}$ acting on P gives lP, so

$$\left(\left(r\frac{\partial}{\partial r}\right)^2 + r\frac{\partial}{\partial r}\right) P = l(l+1)P. \tag{8.31}$$

The Laplacian acting on P gives zero, so from equation (8.22) we obtain

$$\mathbf{l}^2 P = l(l+1)\,P. \tag{8.32}$$

The conclusion is that a polynomial of degree l annihilated by ∇^2 is an eigenfunction of $\mathbf{l}^2$ with eigenvalue $l(l+1)$. Naïvely, one might have expected the eigenvalue to be l^2, but the

+1 in this formula occurs because the relevant operators do not all commute, which is a key feature of quantum mechanics.

As an illustration, all the functions (8.30) are of degree 2, and are therefore eigenfunctions of $\mathbf{l}^2$ with eigenvalue 6. One can check this directly, but recognizing that these functions are annihilated by the Laplacian gives further insight into why this is the case.

Earlier we said that eigenfunctions of $\mathbf{l}^2$ are usually the purely angular functions known as spherical harmonics. The polynomials we have obtained here are equal to these spherical harmonics multiplied by the power r^l. The factor r^l doesn't affect the $\mathbf{l}^2$-eigenvalue. We call these polynomials *harmonics*, a term often used for functions satisfying Laplace's equation. We write a harmonic P whose $\mathbf{l}^2$ eigenvalue is $l(l+1)$ as $P = r^l P_l(\vartheta, \varphi)$, where $P_l(\vartheta, \varphi)$ is a spherical harmonic. Unlike the spherical harmonics, the harmonics are smooth functions at the origin, as is clear from their Cartesian form.

In summary, by considering the polynomials in x_1, x_2, x_3 that are annihilated by the Laplacian, for all degrees l, we obtain a complete set of eigenfunctions of $\mathbf{l}^2$. Alternatively, from the perspective of spherical polar coordinates, we obtain a complete set of spherical harmonics, multiplied by powers of r. Polynomials not annihilated by the Laplacian are related by additional powers of r to spherical harmonics.

While focussing on the operator $\mathbf{l}^2$, and its eigenvalues and eigenfunctions, we have rather lost sight of the original angular momentum operators l_1, l_2, l_3. They each commute with $\mathbf{l}^2$ but not with each other. So we pick one of them, l_3, and rearrange the harmonics we have found so that they are simultaneous eigenfunctions of $\mathbf{l}^2$ and l_3. To do this we need to know how the operator $l_3 = -i \left(x_1 \frac{\partial}{\partial x_2} - x_2 \frac{\partial}{\partial x_1} \right)$ acts on Cartesian coordinates. This is most simply presented as

$$
\begin{aligned}
l_3(x_1 + ix_2) &= x_1 + ix_2\,, \\
l_3 x_3 &= 0\,, \\
l_3(x_1 - ix_2) &= -(x_1 - ix_2)\,.
\end{aligned}
\tag{8.33}
$$

Acting on these combinations of Cartesian coordinates, we see that l_3 has eigenvalues $m = 1, 0, -1$. Because l_3 is a linear differential operator it obeys the Leibniz rule, which means that a product of l_3-eigenfunctions has an l_3-eigenvalue equal to the sum of the eigenvalues of the factors. If a polynomial contains m_1 factors of $x_1 + ix_2$, m_2 factors of $x_1 - ix_2$, and any number of factors of x_3, then it is an l_3-eigenfunction with eigenvalue $m = m_1 - m_2$.

The harmonics of degree l have a basis set with these definite l_3-eigenvalues m. For example, for $l = 2$, we can select the basis

$$
(x_1 + ix_2)^2\,,\ (x_1 + ix_2)x_3\,,\ (x_1 + ix_2)(x_1 - ix_2) - 2x_3^2\,,\ (x_1 - ix_2)x_3\,,\ (x_1 - ix_2)^2\,. \tag{8.34}
$$

We can easily see that these harmonics are eigenfunctions of l_3 with eigenvalues $m = 2, 1, 0, -1, -2$, respectively. Each is a linear combination of the polynomials in the list (8.30), so these harmonics are also eigenfunctions of $\mathbf{l}^2$ with eigenvalue 6. Generally, the $2l + 1$ eigenfunctions of $\mathbf{l}^2$ with eigenvalue $l(l + 1)$ have a basis set labelled by the eigenvalue m of l_3. Each of the $2l + 1$ values of m in the range $l, l - 1, \ldots, -(l - 1), -l$ occurs once. At the top of the range is the eigenfunction $(x_1 + ix_2)^l$ and at the bottom is $(x_1 - ix_2)^l$. The spherical harmonic with definite l and m labels is denoted by $P_l^m(\vartheta, \varphi)$, and the corresponding harmonic is $r^l P_l^m(\vartheta, \varphi)$.

8.4   The Schrödinger Equation with a Spherical Potential

The stationary Schrödinger equation, with a spherically symmetric potential $V(r)$, is

$$-\frac{\hbar^2}{2m}\nabla^2\chi + V(r)\chi = E\chi. \tag{8.35}$$

This can be solved by taking a harmonic $P = r^l P_l^m(\vartheta, \varphi)$ and multiplying by a further radial function $f(r)$,

$$\chi(r, \vartheta, \varphi) = f(r)r^l P_l^m(\vartheta, \varphi). \tag{8.36}$$

Expanding out the radial derivative terms in equation (8.22), we can express the Laplacian as

$$\nabla^2 = \frac{\partial^2}{\partial r^2} + \frac{2}{r}\frac{\partial}{\partial r} - \frac{1}{r^2}\mathbf{l}^2. \tag{8.37}$$

The radial derivatives act solely on $f(r)r^l$, and $\mathbf{l}^2$ acts on P_l^m producing $l(l+1)P_l^m$. Therefore all terms in equation (8.35) are proportional to P_l^m, and after a little simplification it becomes the purely radial equation

$$\left\{ -\frac{\hbar^2}{2m}\left(\frac{d^2}{dr^2} + \frac{2}{r}\frac{d}{dr} - \frac{l(l+1)}{r^2} \right) + V(r) - E \right\} f(r)r^l = 0. \tag{8.38}$$

This depends explicitly on the value of the integer angular momentum label l, and has a set of energy eigenvalues E for each l. Note that the label m does not appear in equation (8.38), so for a given l and given energy, there is always a $(2l + 1)$-fold degeneracy of the states. These states have the same energy and squared angular momentum, but the projection m of the angular momentum along the 3-axis differs. More informally, the states differ in the direction that the angular momentum points. By rotational symmetry, the energy cannot depend on this direction.

If $l \neq 0$, the term $\frac{\hbar^2}{2m}\frac{l(l+1)}{r^2}$ has a strong, positive singularity at the origin, dominating any physical potential $V(r)$ there. The usual effect is for the energy levels E to increase as l increases. For a general potential there is no simple relation between the energy levels for different l-values. Depending on the shape of V, and the value of l, some states may be bound states and others may be scattering states. Sometimes there are no bound states at all, and sometimes no scattering states.[3] Take the spherical harmonic oscillator, for example, whose potential is $V(r) = Ar^2$ with A positive. Here, for each integer l, there is an infinite, discrete set of bound states with positive energies. The particle cannot escape to infinity, so there are no scattering states.

Another type of potential is a well of finite depth. Here $V(r)$ is negative, approaching zero rapidly as $r \to R$, the radial width of the well. Bound states in a well have negative energies, and these may or may not exist, but there are always scattering states for all positive energies. If V is deep enough (actually it is a combination of depth and width that matters) then there will be bound states for small l. However, for large l there will not, as the term $\frac{\hbar^2}{2m}\frac{l(l+1)}{r^2}$ has a repulsive effect that overcomes the attraction of the well. If V is shallow there will be no bound states at all, even for $l = 0$.

---

[3] The relationship between the strength of a potential and the number of bound states is described by Levinson's theorem, which dates back to 1949.

8.4.1 The Coulomb potential

We will now take a more detailed look at the attractive Coulomb potential. This potential is important, because it describes an electron interacting with a proton. The bound states are the states of the simplest atom, the hydrogen atom. As the proton is nearly 2000 times as massive as the electron, we can consider the proton to be at rest at the origin, and solve for the stationary state wavefunction of the electron, $\chi(r, \vartheta, \varphi) = f(r) r^l P_l^m(\vartheta, \varphi)$. The proton has positive charge e and the electron has the opposite, negative charge $-e$. The potential is the electrostatic Coulomb potential $V(r) = -\frac{e^2}{4\pi r}$, whose (negative) gradient gives an attractive inverse square law force. As V is negative and not shallow, we can expect bound states. In fact, bound states exist for all possible values of l, i.e. all non-negative integers, because the potential approaches zero rather slowly as $r \to \infty$, and for large r the $\frac{1}{r}$ attraction dominates the $\frac{1}{r^2}$ repulsion. In addition there are states with positive energy, representing electrons scattering off the proton, but we shall focus on the bound states. The $\frac{1}{r}$ singularity in the potential is rather mild, and wavefunctions remain finite at the origin, both for bound and scattering states. For $l \neq 0$, the wavefunctions vanish at the origin.

With the Coulomb potential, equation (8.38) becomes

$$\left\{ -\frac{\hbar^2}{2m_e} \left(\frac{d^2}{dr^2} + \frac{2}{r}\frac{d}{dr} - \frac{l(l+1)}{r^2} \right) - \frac{e^2}{4\pi r} - E \right\} f(r) r^l = 0, \tag{8.39}$$

where m_e is the electron mass. This is simplified by multiplying by $\frac{2m_e}{\hbar^2}$ and defining $\alpha = \frac{m_e e^2}{2\pi \hbar^2}$ and $\nu^2 = -\frac{2m_e E}{\hbar^2}$. E is negative for bound states, so ν^2 is positive. After explicitly working out the radial derivatives of $f(r) r^l$ using the Leibniz rule, equation (8.39) becomes

$$\left(-\frac{d^2}{dr^2} - \frac{2(l+1)}{r}\frac{d}{dr} - \frac{\alpha}{r} + \nu^2 \right) f(r) = 0. \tag{8.40}$$

The most singular $\frac{l(l+1)}{r^2}$ term in the equation has gone. It has cancelled due to the r^l factor in the wavefunction.

Let us consider how f behaves for large r. The constant ν^2 dominates the two terms with $\frac{1}{r}$ factors, and a suitable asymptotic solution is $f(r) \sim e^{-\nu r}$, which decays rapidly for large r. The full solution can be written as

$$f(r) = g(r) e^{-\nu r}. \tag{8.41}$$

$g(r)$ is required to be finite at the origin, and to grow more slowly than an exponential for large r. This last condition determines the possible values of ν, and for these values, $g(r)$ is a polynomial.

The simplest solution is $g(r) = 1$, so $f(r) = e^{-\nu r}$ exactly. Substituting this into equation (8.40), one finds that $\nu = \frac{\alpha}{2(l+1)}$. Therefore

$$f(r) = e^{-\frac{\alpha}{2(l+1)}r}, \tag{8.42}$$

and

$$E = -\frac{\hbar^2 \nu^2}{2m_e} = -\frac{\hbar^2 \alpha^2}{8m_e} \frac{1}{(l+1)^2} = -\frac{m_e e^4}{32\pi^2 \hbar^2} \frac{1}{(l+1)^2}. \tag{8.43}$$

Because $f(r)$ doesn't pass through zero, this is actually the lowest energy solution for each value of l. Normally, one would next seek higher energy solutions for the same l (i.e. with smaller ν, as this corresponds to less negative E), and there are indeed infinitely many of them, with $g(r)$ being polynomials of ever higher degree. However, the Coulomb potential is very special, and it is easier to describe what happens if the energy is kept fixed, and l is varied.

Let's change notation and set $\nu = \frac{\alpha}{2N}$, so $N = l + 1$ when $g(r) = 1$, and

$$E = -\frac{\hbar^2 \alpha^2}{8m_e} \frac{1}{N^2} = -\frac{m_e e^4}{32\pi^2 \hbar^2} \frac{1}{N^2}, \tag{8.44}$$

with $N \geq 1$, a positive integer. N is called the *principal quantum number*, as it determines the energy. The solution we have been considering so far is

$$f(r) = e^{-\frac{\alpha}{2N}r}, \tag{8.45}$$

and its angular momentum label is $l = N - 1$. Now, with N and the energy fixed, one can show that there is a solution of equation (8.40) for any integer l in the finite range $l = 0, 1, 2, \ldots, N - 1$. The solution is of the form

$$f(r) = g(r)e^{-\frac{\alpha}{2N}r}, \tag{8.46}$$

with g a polynomial of degree $N - l - 1$, called a generalized Laguerre polynomial. For example, for $l = N - 2$, $g(r) = 1 - \frac{\alpha}{2N(N-1)}r$.

8.4.2   Spectroscopy

Figure 8.1 shows the bound state energies of the hydrogen atom. The energies depend solely on the principal quantum number N, and the fixed physical constants. The lowest energy state, with $N = 1$, is unique and has zero angular momentum. It is the ground state of the hydrogen atom. The states with higher N have two further labels, the angular momentum labels l and m, and recall that given l, there are $2l + 1$ allowed values of m. There is more degeneracy here than for a general potential $V(r)$. For each N, there is one state with $l = 0$, three with $l = 1$, and so on, and finally $2N - 1$ states with $l = N - 1$. In total there are $N^2 = \sum_{l=0}^{N-1}(2l + 1)$ states with energy $-\frac{m_e e^4}{32\pi^2 \hbar^2} \frac{1}{N^2}$. This extra degeneracy, special to quantized particle motion in an attractive $\frac{1}{r}$ potential, can be understood using a quantum analogue of the Runge–Lenz vector that we used to study classical Kepler orbits in Chapter 2.

Some of these states correspond to the ones found by Bohr in his early study of the hydrogen atom. In this original quantum mechanical model of an atom, an electron has a circular classical orbit around a proton with angular momentum $N\hbar$, an integer multiple of $\hbar$. The classical equation of motion, with the attractive Coulomb force, implies that the electron energy is $E = -\frac{m_e e^4}{32\pi^2 \hbar^2} \frac{1}{N^2}$. The Bohr model predicts the correct energy levels, but it does not account for the relationship between angular momentum and energy. What the Bohr model misses is that for a given N there are quantum states with angular momentum projections $m\hbar$ and $|m|$ taking any integer value from 0 up to $N - 1$. Arnold Sommerfeld later made an important addition to the Bohr model by considering elliptical orbits and quantizing the l_3 component of angular momentum. The full Bohr–Sommerfeld model of the hydrogen atom, although based on rather ad hoc principles, agrees with the analysis based on the Schrödinger equation given here.

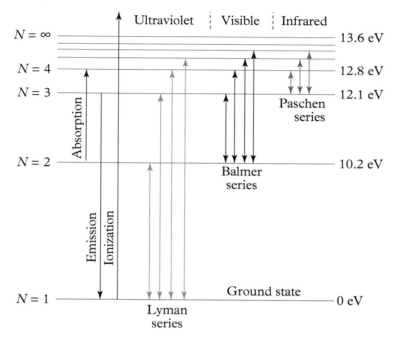

Fig. 8.1 Bound state energies of the hydrogen atom. Electron transitions produces sharp lines in the hydrogen spectrum. Transitions from $N > 1$ to $N = 1$ form the Lyman series. Transitions from $N > 2$ to $N = 2$ form the Balmer series. Transitions from $N > 3$ to $N = 3$ form the Paschen series.

In the ground state of a hydrogen atom the electron has its lowest possible energy. Electrons may be excited into higher energy states in various ways. These include electrical excitations, thermal excitations involving collisions with other atoms, and interactions with particles that may be incident on the atom. In our discussion, the ground and excited states are stationary states, and there are no transitions between them, but this is only an approximation. Because an electron is charged, it also interacts with the electromagnetic field. This interaction is not simple to analyse because one needs to consider quantum aspects of the electromagnetic field, not included in our Schrödinger equation. The most important effect is that electrons in excited states have finite lifetimes and make transitions to lower energy states and eventually the ground state. The energy released is emitted as one or more photons, the quantized states of the electromagnetic field.

The energy of a single emitted photon is the difference in energy between the initial and final (almost) stationary states of an electron. In a transition between states with principal quantum numbers N' and N, with $N' > N$, the photon energy is

$$\frac{m_e e^4}{32\pi^2\hbar^2}\left(\frac{1}{N^2} - \frac{1}{N'^2}\right). \tag{8.47}$$

If a sample of atomic hydrogen emits lots of photons, this is detected as ordinary electromagnetic radiation. The photon energy is proportional to the frequency of the radiation, and hence inversely proportional to the wavelength. In units where the speed of light is unity, these transitions result in emitted radiation of wavelength λ, where

$$\frac{1}{\lambda} = \frac{m_e e^4}{64\pi^3 \hbar^3} \left(\frac{1}{N^2} - \frac{1}{N'^2} \right). \tag{8.48}$$

The allowed wavelengths, and their colour range, are shown in Figure 8.1. The transitions to the $N = 1$ level are all in the ultraviolet, but the transitions to the $N = 2$ level are in the visible light spectrum. The spectral lines are very sharp.

8.5   Spin

The key feature of angular momentum in quantum mechanics is the set of commutation relations (8.16). We derived these starting with the differential operators representing the orbital angular momentum $\mathbf{l} = \mathbf{x} \times \mathbf{p}$. It is reasonable to ask whether the commutation relations are satisfied by any other kind of representation, such as matrices. The answer is yes, and unlike for the position and momentum operators, the matrices are of finite size.

The smallest non-trivial matrices that work are 2×2 matrices. These are called *spin operators*, and have their own notation $\mathbf{s} = (s_1, s_2, s_3)$. The spin operators are

$$s_1 = \frac{1}{2} \begin{pmatrix} 0 & 1 \\ 1 & 0 \end{pmatrix}, \quad s_2 = \frac{1}{2} \begin{pmatrix} 0 & -i \\ i & 0 \end{pmatrix}, \quad s_3 = \frac{1}{2} \begin{pmatrix} 1 & 0 \\ 0 & -1 \end{pmatrix}, \tag{8.49}$$

and they obey the commutation relations $[s_1, s_2] = is_3$ etc., as in equation (8.16). Without the factors of $\frac{1}{2}$, the matrices are called *Pauli matrices*, and are denoted as $\boldsymbol{\sigma} = (\sigma_1, \sigma_2, \sigma_3)$, so $\mathbf{s} = \frac{1}{2}\boldsymbol{\sigma}$. The physical spin operators are $\hbar\mathbf{s} = \frac{\hbar}{2}\boldsymbol{\sigma}$.

The spin operators give an alternative quantum mechanical realization of angular momentum. As the matrices are 2×2, the simplest quantum states with spin have just two complex components. Such states are called *2-component spinors*, and are written as

$$\phi = \begin{pmatrix} \phi_1 \\ \phi_2 \end{pmatrix}. \tag{8.50}$$

To see that the spin operators are not equivalent to the earlier representation of angular momentum as differential operators, consider the squared spin

$$\mathbf{s}^2 = s_1^2 + s_2^2 + s_3^2. \tag{8.51}$$

The square of each Pauli matrix σ_i is the unit 2×2 matrix $\mathbf{1}$, so $\mathbf{s}^2 = \frac{3}{4}\mathbf{1}$. Any 2-component spinor is therefore an eigenstate of $\mathbf{s}^2$, with eigenvalue $\frac{3}{4}$. This is the value of $s(s+1)$ when $s = \frac{1}{2}$, so spin is a manifestation of angular momentum with label $\frac{1}{2}$, rather than the integer label l that we found earlier. One speaks of the states as having spin $\frac{1}{2}$.

The spin operator $s_3 = \begin{pmatrix} \frac{1}{2} & 0 \\ 0 & -\frac{1}{2} \end{pmatrix}$ has two distinct eigenvalues, which are just the diagonal entries $\frac{1}{2}$ and $-\frac{1}{2}$. And as

$$s_3 \begin{pmatrix} 1 \\ 0 \end{pmatrix} = \frac{1}{2} \begin{pmatrix} 1 \\ 0 \end{pmatrix} \quad \text{and} \quad s_3 \begin{pmatrix} 0 \\ 1 \end{pmatrix} = -\frac{1}{2} \begin{pmatrix} 0 \\ 1 \end{pmatrix}, \tag{8.52}$$

the respective eigenstates are the two spinors, $\begin{pmatrix} 1 \\ 0 \end{pmatrix}$ and $\begin{pmatrix} 0 \\ 1 \end{pmatrix}$, shown here. They are called the spin up state and the spin down state (relative to the x_3-axis) with $s = \frac{1}{2}$. The eigenvalues of s_3 run from s to $-s$ in a single integer step, just as the eigenvalues m of l_3 run from l to $-l$ in integer steps.

Interestingly, every 2-component spinor is up with respect to some direction. A spinor ϕ that is up in the direction $\mathbf{n} = (\sin\vartheta\cos\varphi, \sin\vartheta\sin\varphi, \cos\vartheta)$ must satisfy $(\mathbf{n}\cdot\mathbf{s})\phi = \frac{1}{2}\phi$, or explicitly

$$(\sin\vartheta\cos\varphi\, s_1 + \sin\vartheta\sin\varphi\, s_2 + \cos\vartheta\, s_3)\phi = \frac{1}{2}\begin{pmatrix} \cos\vartheta & \sin\vartheta\, e^{-i\varphi} \\ \sin\vartheta\, e^{i\varphi} & -\cos\vartheta \end{pmatrix}\phi = \frac{1}{2}\phi. \quad (8.53)$$

A solution is $\phi = \begin{pmatrix} \cos\frac{1}{2}\vartheta \\ \sin\frac{1}{2}\vartheta\, e^{i\varphi} \end{pmatrix}$. Conversely, $\phi = \begin{pmatrix} \phi_1 \\ \phi_2 \end{pmatrix}$ is spin up in the direction defined by $\tan\frac{1}{2}\vartheta\, e^{i\varphi} = \frac{\phi_2}{\phi_1}$.

8.5.1 The Stern–Gerlach experiment

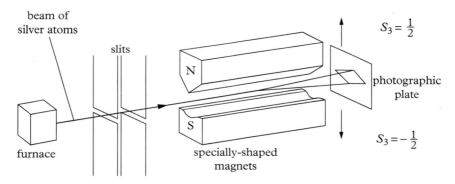

Fig. 8.2 The Stern–Gerlach apparatus.

It is a remarkable fact that physical particles can have spin $\frac{1}{2}$. This was discovered in 1922 when Otto Stern and Walther Gerlach passed a beam of (neutral) silver atoms through a magnet designed to produce a non-uniform magnetic field aligned in the x_3-direction. The magnetic moments of the atoms in the beam interact with the magnetic field as they pass through the Stern–Gerlach device and the atoms are deflected from their original trajectory, as shown in Figure 8.2. The result found by Stern and Gerlach was that the beam of silver atoms splits in two. Each atom has a magnetic moment proportional to its spin, and this is aligned up in the spin up state or down in the spin down state. Passing through the magnetic field, the atom is deflected up or down, depending on whether its magnetic moment is up or down, as shown on the right of Figure 8.3. Classically, we would expect the spin of the atoms to be oriented in any direction, and that the deflection would depend on the angle of orientation. The deflection would be continuous, running between the two extreme values corresponding to perfect alignment and perfect anti-alignment with the magnetic field. The Stern–Gerlach experiment shows that our classical expectations are wrong and that spin can only be understood as a quantum phenomenon. The results can be interpreted as a measurement of s_3, and show that the only possible outcomes are the eigenvalues $\frac{1}{2}$ and $-\frac{1}{2}$.

It is not just silver atoms that have spin $\frac{1}{2}$; the electron, proton and neutron also have spin $\frac{1}{2}$. Electron spin affects the states available to an electron in a hydrogen atom.

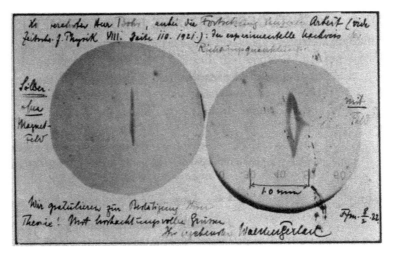

Fig. 8.3 The impression on a photographic plate made by silver atoms after passing through the Stern–Gerlach apparatus. Left: When the magnetic field is switched off, there is no deflection. Right: When the magnetic field is switched on, the silver atoms are deflected through two discrete angles.

A stationary state wavefunction is a position dependent spinor,

$$\phi(\mathbf{x}) = \left(\begin{array}{c} \phi_1(\mathbf{x}) \\ \phi_2(\mathbf{x}) \end{array} \right). \tag{8.54}$$

To a very good approximation, the Schrödinger equation for ϕ reduces to two copies of the Schrödinger equation (8.35) that we looked at previously, one for ϕ_1 and one for ϕ_2. So the states are the same as before, and have the same energies, but there is an additional label that signifies whether the electron has spin up or spin down. The number of independent states with energy (8.44) is now $2N^2$, twice that of the spinless case considered previously.

8.5.2 The Zeeman effect

Placing an atom in a strong magnetic field breaks the spherical symmetry of the Hamiltonian and splits the degeneracy of the electron states. The splitting of the spectral lines is visible when the spectrum is viewed through a diffraction grating. A single spectral line is observed to split as a strong magnet is brought close to a tube containing a gas of excited atoms. This is known as the Zeeman effect. The first effect of the magnetic field is to split the energy of the previously degenerate electron states of given l into $2l+1$ energy levels for the different m values. This *normal* Zeeman splitting is illustrated in Figure 8.4. The $l = 1$ states split into three, and the $l = 2$ states into five. The spectral lines due to the emission and absorption of photons associated with transitions between these energy levels split accordingly. The second effect of the magnetic field is that the spin up electron state is shifted in energy relative to the spin down state, because of the interaction of the electron's spin with the magnetic field. The difference between spin up and spin down is one unit of angular momentum, so we might expect the spin splitting to be the same size as the splitting between consecutive values of m. However, the energy splitting due to the different spin values is almost exactly double that due to one unit of orbital angular momentum. This can be understood using the Dirac equation, the relativistic quantum mechanical equation that we will introduce in Chapter 12.

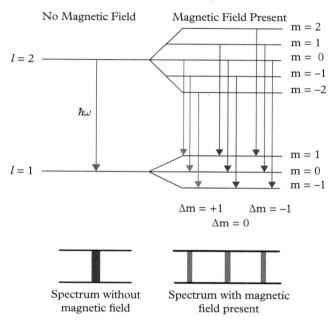

Fig. 8.4 The energies of electron states split when an atom is placed in a magnetic field. This is the basic (normal) Zeeman effect.

We illustrate this in Figure 8.5. For an electron in an $l = 1$ state, there are three m values, and for each of these there are two spin values, up and down, giving six states in total. The energies of the six states are split by a magnetic field as shown on the right. There are four equally spaced states, collectively labelled $P_{\frac{3}{2}}$ and two more states, labelled $P_{\frac{1}{2}}$. This reflects the way that angular momentum adds up, and the splitting is known as the *anomalous* Zeeman effect, although it is not uncommon. The orbital angular momentum is 1 and the spin is $\frac{1}{2}$. If these angular momentum vectors point in the same direction, the total angular momentum is $\frac{3}{2}$. If they point in opposite directions, the total is $\frac{1}{2}$. With combined angular momentum $\frac{3}{2}$ there are four different projections, and with combined angular momentum $\frac{1}{2}$ there are two.

The Zeeman effect is a valuable tool for astronomers as it enables them to study the magnetic fields of stars. For example, through the Zeeman effect we know that the solar sunspot cycle is due to periodic variations in the Sun's magnetic field. It has also enabled astronomers to prove that the typical magnetic field strength of a neutron star is around 10^6 T (where T denotes Tesla). By comparison, the Earth's magnetic field is 10^{-5} T.

8.5.3 Other spin representations

There are larger matrix representations of the angular momentum commutation relations, similar to the spin operators **s**. For example, there is a representation by 3×3 matrices, acting on 3-component spinors. The squared angular momentum $\mathbf{s}^2$ has eigenvalue $s(s + 1) = 2$,

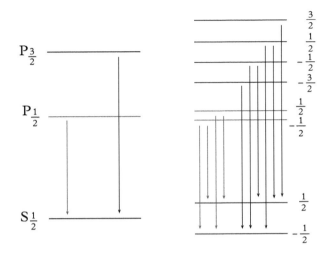

Fig. 8.5 When combined with the two electron spin states, the $l = 0$ angular momentum state (s state) splits into two, and the three $l = 1$ states (p states) split into a total of six states. This is the anomalous Zeeman effect.

so the spin label is $s = 1$. These matrices are equivalent to what one obtains by restricting the action of the usual angular momentum operators (8.15) to the degree 1 polynomials $a_1 x_1 + a_2 x_2 + a_3 x_3$. Here $l = 1$. Despite this equivalence, the matrix representation is useful for describing particles with spin 1. Unlike 2-component spinors, not every 3-component spinor represents a spin 1 particle with spin up in some direction.

The Z boson is a massive spin 1 particle, and its three polarization states are independent states of a 3-component spinor. Conceivably, it is a bound state of two spin $\frac{1}{2}$ particles and its spin arises from the orbital angular momentum and spins of its constituents, but this interpretation would be very difficult to reconcile with a wide range of experimental observations. The Z boson seems to be a fundamental particle without substructure, like the electron, and there are sound theoretical reasons why it can still have spin 1, as we will see in Chapter 12.

There is also a set of 4×4 matrices representing spin. This representation describes a particle of spin $\frac{3}{2}$. Again, such particles exist. The Delta resonances, which are excited states of protons and neutrons, have spin $\frac{3}{2}$. However, spin $\frac{3}{2}$ is less fundamental than spin $\frac{1}{2}$, and the Delta resonances can be modelled in terms of three constituent quarks, each carrying spin $\frac{1}{2}$.

In summary, there are spin representations for spins $0, \frac{1}{2}, 1, \frac{3}{2}, 2, \frac{5}{2}, \ldots$. The values $0, 1, 2, \ldots$ are called integer spins, and $\frac{1}{2}, \frac{3}{2}, \frac{5}{2}, \ldots$ half-integer spins. Of these, spins $0, \frac{1}{2}, 1$ seem the most fundamental.

8.6   Spin $\frac{1}{2}$ as a Quantum Paradigm

Spin $\frac{1}{2}$ particles provide a simple example of the counter-intuitive nature of quantum mechanics and offer a great test of its axioms. Let us ignore the spatial wavefunction, and just treat a spin $\frac{1}{2}$ particle as a two-state system. It is convenient to regard the particle as a component of a beam, whose spin state can be investigated by one or more Stern–Gerlach magnets placed along the beam direction. The spin operators s_1, s_2, s_3 in equation (8.49) are hermitian.[4] They therefore represent observables. If the beam is in the x_1-direction, the spin along any orthogonal direction can be measured by aligning a Stern–Gerlach magnet appropriately.

Suppose the first magnet encountered by the particle has its magnetic field aligned along the x_3-direction. The measurement of s_3 has possible outcomes $\frac{1}{2}$ and $-\frac{1}{2}$, and suppose a particular measurement has outcome $\frac{1}{2}$. Immediately afterwards, the state is $\binom{1}{0}$, the normalized eigenstate of s_3 with eigenvalue $\frac{1}{2}$. If the measurement is repeated with a second magnet having the same alignment, the result is $\frac{1}{2}$ with certainty.

Suppose instead that the second magnet has its field aligned along the x_2-direction, so that s_2 is measured. One might anticipate that a state with spin up along the x_3-direction would have zero component of spin along the x_2-direction. But this is not how quantum mechanics works. The eigenvalues of

$$ s_2 = \frac{1}{2} \begin{pmatrix} 0 & -i \\ i & 0 \end{pmatrix} \tag{8.55} $$

are again $\frac{1}{2}$ and $-\frac{1}{2}$, and the normalized eigenstates are, respectively,

$$ \frac{1}{\sqrt{2}} \begin{pmatrix} 1 \\ i \end{pmatrix} \quad \text{and} \quad \frac{1}{\sqrt{2}} \begin{pmatrix} 1 \\ -i \end{pmatrix}. \tag{8.56} $$

The incoming spin up state $\binom{1}{0}$ can be expressed as a linear superposition of the s_2 eigenstates as

$$ \begin{pmatrix} 1 \\ 0 \end{pmatrix} = \frac{1}{\sqrt{2}} \left\{ \frac{1}{\sqrt{2}} \begin{pmatrix} 1 \\ i \end{pmatrix} \right\} + \frac{1}{\sqrt{2}} \left\{ \frac{1}{\sqrt{2}} \begin{pmatrix} 1 \\ -i \end{pmatrix} \right\}. \tag{8.57} $$

The probabilities for the measurement outcomes are the squares of the coefficients in this expression. So the measurement of s_2 has outcome $\frac{1}{2}$ with probability $\frac{1}{2}$ and outcome $-\frac{1}{2}$ with probability $\frac{1}{2}$. The incoming state is pure, but the outcome of the s_2 measurement is probabilistic and uncertain. This has been confirmed in the laboratory. The fundamental reason for the uncertainty is that s_3 and s_2 do not commute.

A similar analysis can be done with the field of the second magnet aligned at any angle in the (x_2, x_3)-plane. The spin is always measured to be $\frac{1}{2}$ or $-\frac{1}{2}$ but the predicted probabilities of the outcomes depend on the angle and are generally unequal. These probabilities have also been confirmed experimentally.

Strange as these ideas are, they now have technological applications and have been developed into systems that enable messages to be exchanged with perfect security. This is known as quantum cryptography.

8.7 Quantum Mechanics of Several Identical Particles

In classical mechanics, if there are two or more particles, they have positions $\mathbf{x}^{(1)}, \mathbf{x}^{(2)}, \ldots$ and momenta $\mathbf{p}^{(1)}, \mathbf{p}^{(2)}, \ldots$. In quantum mechanics, the state of a multi-particle system is described by a wavefunction that is a function only of the particle positions $\Psi(\mathbf{x}^{(1)}, \mathbf{x}^{(2)}, \ldots)$.

---

[4] A square matrix M_{ab} is hermitian if its transpose (the matrix with rows and columns exchanged) is the same as its complex conjugate, i.e. if $M_{ba} = \overline{M_{ab}}$. The eigenvalues of a hermitian matrix are real.

The wavefunction also depends on time, but we suppress this dependence here. The modulus squared of the wavefunction,

$$|\Psi(\mathbf{x}^{(1)}, \mathbf{x}^{(2)}, \ldots)|^2,\tag{8.58}$$

is the probability density for simultaneously finding the first particle at $\mathbf{x}^{(1)}$, the second at $\mathbf{x}^{(2)}$, and so on. For this to make sense, the wavefunction must satisfy the normalization condition

$$\int |\Psi(\mathbf{x}^{(1)}, \mathbf{x}^{(2)}, \ldots)|^2 \, d^3x^{(1)} \, d^3x^{(2)} \ldots = 1,\tag{8.59}$$

with the integrations over all space.

The position and momentum operators for each particle act on the wavefunction, and just as for a single particle, the position operators act by multiplication and the momenta act by partial differentiation. So, for example, the three components of the total momentum operator are the sums

$$P_1 = -i\hbar \frac{\partial}{\partial x_1^{(1)}} - i\hbar \frac{\partial}{\partial x_1^{(2)}} - \cdots, \quad P_2 = -i\hbar \frac{\partial}{\partial x_2^{(1)}} - i\hbar \frac{\partial}{\partial x_2^{(2)}} - \cdots,$$

$$P_3 = -i\hbar \frac{\partial}{\partial x_3^{(1)}} - i\hbar \frac{\partial}{\partial x_3^{(2)}} - \cdots,\tag{8.60}$$

or in vector form

$$\mathbf{P} = -i\hbar \nabla^{(1)} - i\hbar \nabla^{(2)} - \cdots.\tag{8.61}$$

The total Hamiltonian H is the sum of the kinetic energies, plus a potential that depends on the positions of all the particles. If there are N particles, with masses $m^{(1)}, m^{(2)}, \ldots, m^{(N)}$, then

$$H = -\sum_{k=1}^{N} \frac{\hbar^2}{2m^{(k)}} (\nabla^{(k)})^2 + V(\mathbf{x}^{(1)}, \mathbf{x}^{(2)}, \ldots, \mathbf{x}^{(N)}),\tag{8.62}$$

where $(\nabla^{(k)})^2$ is the Laplacian for the variable $\mathbf{x}^{(k)}$. N-particle stationary states are eigenfunctions of H, and the eigenvalue E is the total energy of the particles.

Recall that if the potential depends only on the relative positions of the particles, then the system is translation invariant and in classical mechanics the total momentum is conserved. Similarly in quantum mechanics, translation invariance implies that the total momentum operator $\mathbf{P}$ commutes with the Hamiltonian H. This means there is a complete set of stationary states that are simultaneously eigenstates of H and $\mathbf{P}$. These states have a definite energy, and a definite total momentum (also denoted by $\mathbf{P}$). For a system of two particles, such states have wavefunctions of the form

$$\Psi(\mathbf{x}^{(1)}, \mathbf{x}^{(2)}) = e^{\frac{i}{\hbar}\mathbf{P}\cdot\mathbf{X}_{\mathrm{CM}}} \psi(\mathbf{x}^{(2)} - \mathbf{x}^{(1)}),\tag{8.63}$$

where $\mathbf{X}_{\mathrm{CM}}$ is the usual centre of mass of the two particles and $\mathbf{x}^{(2)} - \mathbf{x}^{(1)}$ is the particle separation vector.

Suppose now that the N particles are *identical*. (By identical we mean that the particles cannot, even in principle, be distinguished.) The particles could, for example, be the electrons in an atom interacting with each other and with the fixed nucleus, or a gas of identical atoms, where the atoms are treated as point particles and their internal structure is ignored.

Identical particles all have the same mass m, so the Hamiltonian (8.62) has the slightly simpler form

$$H = -\sum_{k=1}^{N} \frac{\hbar^2}{2m}(\nabla^{(k)})^2 + V(\mathbf{x}^{(1)}, \mathbf{x}^{(2)}, \dots, \mathbf{x}^{(N)}). \qquad (8.64)$$

V is unchanged under permutations of the points $\mathbf{x}^{(1)}, \mathbf{x}^{(2)}, \dots, \mathbf{x}^{(N)}$, because a permutation doesn't change the particle configuration, it only changes the particle labels. The kinetic part of the Hamiltonian has the same permutation symmetry. What implications does this have for the wavefunction?

Exchanging the labels in the wavefunction doesn't change a configuration of identical particles. It follows that the probability density $|\Psi(\mathbf{x}^{(1)}, \mathbf{x}^{(2)}, \dots, \mathbf{x}^{(N)})|^2$ must equal the probability density $|\Psi(\mathbf{x}^{(2)}, \mathbf{x}^{(1)}, \dots, \mathbf{x}^{(N)})|^2$, where we have exchanged the first two labels. However, the wavefunction itself could acquire a phase factor under this exchange of labels:

$$\Psi(\mathbf{x}^{(2)}, \mathbf{x}^{(1)}, \dots, \mathbf{x}^{(N)}) = e^{i\alpha}\, \Psi(\mathbf{x}^{(1)}, \mathbf{x}^{(2)}, \dots, \mathbf{x}^{(N)}). \qquad (8.65)$$

If we exchange the first two labels again, we return to the original function, so $e^{2i\alpha} = 1$. There are therefore just two possibilities. Either $e^{i\alpha} = 1$ or $e^{i\alpha} = -1$. In the first case we say the wavefunction is bosonic, and in the second that it is fermionic. If the wavefunction is bosonic, then the particles are called *bosons*; if it is fermionic, then the particles are called *fermions*.

Once a choice is made for the effect of exchanging the first pair of labels, the same choice has to be made for any pair. This is because, as the particles are identical, they must be treated in exactly the same way. Also, a mixture of choices would not be compatible with the symmetry of the Hamiltonian. A bosonic wavefunction is therefore unchanged under exchange of any pair of labels, so it is unchanged under all possible permutations of labels, and is said to be totally symmetric. A fermionic wavefunction changes sign under exchange of any single pair of labels, and is therefore totally antisymmetric. It changes sign under any odd permutation of the labels and is unchanged under an even permutation. By an odd permutation we mean the combined effect of an odd number of pair exchanges. It can be expressed as such a combination in many ways, but the oddness is always the same. This is easy to prove using the properties of the determinant of a square matrix. An odd permutation of the rows changes the sign of the determinant, so it must always be the result of an odd number of row exchanges, each of which changes the sign. Similarly, an even permutation is a combination of an even number of pair exchanges, and an even permutation of the rows of a determinant doesn't change the sign.

These different types of wavefunction are not just algebraically different. They are physically different and so may have different energies, as we can illustrate with a simple example. Consider two identical particles in one dimension interacting through a harmonic oscillator potential. Let $\xi = x^{(2)} - x^{(1)}$ be the separation, so the potential is $V(\xi) = \frac{1}{2}m\omega^2\xi^2$. Notice that V is unchanged by an exchange of particle labels. The centre of mass is $X_{\mathrm{CM}} = \frac{1}{2}(x^{(1)} + x^{(2)})$ and a stationary wavefunction is of the form

$$\chi(x^{(1)}, x^{(2)}) = e^{\frac{i}{\hbar}PX_{\mathrm{CM}}}g(\xi), \qquad (8.66)$$

where P is the total momentum. Under the exchange of labels, ξ changes sign, but X_{CM} is unaffected and the phase factor involving X_{CM} does not change. Therefore, for two fermions,

g must be an odd function of ξ, a function that changes sign when ξ changes sign, whereas for two bosons, g must be an even function. Earlier we studied the stationary states of the harmonic oscillator. The nth state, with energy $E_n = \left(n + \frac{1}{2}\right)\hbar\omega$, is a Hermite polynomial H_n times an exponential factor that is even in ξ. The Hermite polynomials are even functions for even n, and odd for odd n. Therefore for bosons n must be even, whereas for fermions n must be odd. The ground state for two bosons is the $n = 0$ state, but the ground state for two fermions is the $n = 1$ state, which has higher energy.

The fermionic states are functions that vanish at $\xi = 0$. In other words, the two fermions cannot be at the same location, and the probability that they are very close is also small. This result generalizes to an N-fermion wavefunction. A fermionic wavefunction $\Psi(\mathbf{x}^{(1)}, \mathbf{x}^{(2)}, \ldots, \mathbf{x}^{(N)})$ changes sign if any pair of labels is exchanged. Therefore Ψ must be zero when any of the arguments $\mathbf{x}^{(k)}$ and $\mathbf{x}^{(l)}$ are the same, that is, when two particles are at the same location. As the derivatives of the wavefunction are usually finite, Ψ is also small whenever the separation of any pair of particles is small. This result is called the *Pauli exclusion principle*.

Because of the Pauli principle, fermionic particles physically repel each other, but this is not the consequence of a repulsive potential. If N fermions are confined in a box of fixed, finite size, the energy increases with N much more rapidly than N itself, because of this repulsion. For bosons there is no such effect.

A simple model for N identical particles has them interacting with a background potential U, but not interacting directly with each other. The potential in this case is a sum of 1-particle terms,

$$V(\mathbf{x}^{(1)}, \mathbf{x}^{(2)}, \ldots, \mathbf{x}^{(N)}) = U(\mathbf{x}^{(1)}) + U(\mathbf{x}^{(2)}) + \cdots + U(\mathbf{x}^{(N)}), \tag{8.67}$$

a permutation-symmetric function. A solution of the stationary Schrödinger equation is now a product wavefunction

$$\chi(\mathbf{x}^{(1)}, \mathbf{x}^{(2)}, \ldots, \mathbf{x}^{(N)}) = \chi^{(1)}(\mathbf{x}^{(1)})\chi^{(2)}(\mathbf{x}^{(2)}) \cdots \chi^{(N)}(\mathbf{x}^{(N)}), \tag{8.68}$$

where $\chi^{(1)}, \chi^{(2)}, \ldots, \chi^{(N)}$ are solutions of the 1-particle problem. We will from now on use the notation ε to denote a 1-particle energy, and E to denote the total energy of the N-particle system. The 1-particle energies for the wavefunction (8.68) are $\varepsilon^{(1)}, \varepsilon^{(2)}, \ldots, \varepsilon^{(N)}$, and in the absence of inter-particle interactions the total energy is $E = \varepsilon^{(1)} + \varepsilon^{(2)} + \cdots + \varepsilon^{(N)}$.

This wavefunction needs further work, however, to satisfy the permutation symmetry property required for either bosons or fermions. We must symmetrize or antisymmetrize it. The ground state for bosons is particularly simple. We choose $\chi^{(1)}, \chi^{(2)}, \ldots, \chi^{(N)}$ all to be the 1-particle ground state χ_0, with energy ε_0, so

$$\chi(\mathbf{x}^{(1)}, \mathbf{x}^{(2)}, \ldots, \mathbf{x}^{(N)}) = \chi_0(\mathbf{x}^{(1)})\chi_0(\mathbf{x}^{(2)}) \cdots \chi_0(\mathbf{x}^{(N)}). \tag{8.69}$$

This wavefunction is totally symmetric, and has energy $N\varepsilon_0$.

For fermions, an acceptable stationary state wavefunction is more complicated, but may be written as a determinant. We choose N distinct 1-particle wavefunctions $\chi^{(1)}, \ldots, \chi^{(N)}$, and write the complete wavefunction as

$$\chi(\mathbf{x}^{(1)}, \mathbf{x}^{(2)}, \ldots, \mathbf{x}^{(N)}) = \begin{vmatrix} \chi^{(1)}(\mathbf{x}^{(1)}) & \chi^{(2)}(\mathbf{x}^{(1)}) & \cdots & \chi^{(N)}(\mathbf{x}^{(1)}) \\ \chi^{(1)}(\mathbf{x}^{(2)}) & \chi^{(2)}(\mathbf{x}^{(2)}) & \cdots & \chi^{(N)}(\mathbf{x}^{(2)}) \\ \vdots & & \ddots & \vdots \\ \chi^{(1)}(\mathbf{x}^{(N)}) & \chi^{(2)}(\mathbf{x}^{(N)}) & \cdots & \chi^{(N)}(\mathbf{x}^{(N)}) \end{vmatrix}. \tag{8.70}$$

When expanded out this gives a sum (including some minus signs) of $N!$ products of the type on the right-hand side of equation (8.68). χ is totally antisymmetric, because a determinant changes sign under interchange of any pair of rows. It is necessary for each 1-particle wavefunction to be distinct, otherwise two columns of the determinant would be the same, and the whole wavefunction would be zero. The total energy is still the sum of the single particle energies, and the ground state is obtained by choosing the N lowest energy, *distinct* 1-particle states to combine into the determinant. The ground state energy is greater than $N\varepsilon_0$.

This model, in which several particles interact with a background potential but not directly with each other, is often a useful approximation. It is used throughout chemistry, and is known as the independent electron model in solid state physics. Using this approximation, there is the following simple recipe for constructing bosonic and fermionic stationary states for N particles.

One first solves the 1-particle problem in the potential, and labels the states. They can be labelled $0, 1, 2, \ldots$ in order of increasing energy, or labelled by the actual energies $\varepsilon_0, \varepsilon_1, \ldots$. (If an energy level is degenerate one needs additional labels, like angular momentum labels.) Then a bosonic state is specified by giving the occupation numbers $n_0, n_1, \ldots$ of the 1-particle states. From these, the complete wavefunction can be reconstructed as a sum of products of the 1-particle wavefunctions. The total number of particles must be $n_0 + n_1 + \ldots = N$, so only finitely many occupation numbers can be non-zero. There is no other constraint on the occupation numbers. The total energy is the sum of the energies of the occupied states, with the occupation numbers taken into account. The ground state has $n_0 = N$ and all other occupation numbers zero, as shown on the left of Figure 8.6.

A fermionic state is also specified by occupation numbers, but these numbers must be either 0 or 1. No 1-particle state can be multiply occupied. This is another way of stating the Pauli exclusion principle. For N fermions, N 1-particle states are singly occupied, and the rest are empty. The wavefunction is the determinant constructed from the occupied states, and the total energy is the sum of the energies of the occupied states.

Our earlier discussion of the Pauli principle must be refined a little to take account of spin. For N spin $\frac{1}{2}$ fermions, one must include the spin state in the wavefunction. The situation is simplest when the particles are not directly interacting, and each particle can be in either a spin up or a spin down state. 1-particle wavefunctions are written as $\chi(\mathbf{x}) \uparrow$ and $\chi(\mathbf{x}) \downarrow$. The total wavefunction must change sign when both the position and spin labels of a pair of particles are simultaneously exchanged. It doesn't matter whether the energy depends on the spin state or not. The version of the Pauli principle that says that the occupation number of any state is either 0 or 1 remains valid.

The Pauli principle allows two identical spin $\frac{1}{2}$ particles (but no more) to have the same spatial wavefunction χ, as shown on the right of Figure 8.6, provided that one particle has spin up and the other has spin down, and the spin state is antisymmetrized. Such a wavefunction is written

$$\chi(\mathbf{x}^{(1)})\chi(\mathbf{x}^{(2)})\frac{1}{\sqrt{2}}(\uparrow\downarrow - \downarrow\uparrow). \tag{8.71}$$

This is symmetric in space and antisymmetric in spin. Alternatively, the spin state can be symmetric, provided the spatial wavefunction is antisymmetric, for example,

$$(\chi^{(1)}(\mathbf{x}^{(1)})\chi^{(2)}(\mathbf{x}^{(2)}) - \chi^{(2)}(\mathbf{x}^{(1)})\chi^{(1)}(\mathbf{x}^{(2)})) \uparrow\uparrow, \tag{8.72}$$

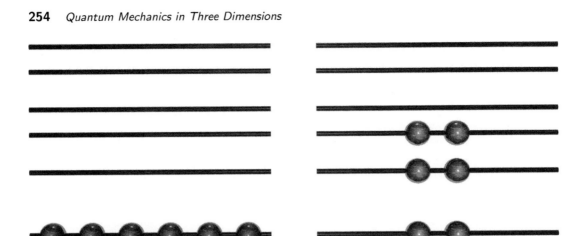

Fig. 8.6 Left: Bosonic ground state. Right: Fermionic ground state.

with $\chi^{(1)}$ and $\chi^{(2)}$ distinct.

For two particles, the antisymmetrized spin state $\frac{1}{\sqrt{2}}(\uparrow\downarrow - \downarrow\uparrow)$ is unique and has total spin zero, but there are three symmetric spin states, $\uparrow\uparrow$, $\frac{1}{\sqrt{2}}(\uparrow\downarrow + \downarrow\uparrow)$ and $\downarrow\downarrow$, and these are the three projections of a state with total spin 1.

When there are genuine interactions between the particles, these wavefunctions are not exact solutions of the stationary Schrödinger equation. However they are useful as approximations, and they have the correct permutation symmetry. Better approximations, especially for the ground states, are obtained by adjusting the background potential and the 1-particle wavefunctions to take some account of the forces that operate between the particles. This is known as the Hartree method (for bosons) or the Hartree–Fock method (for fermions).

8.7.1 The Fermi sphere

Suppose we have N spin $\frac{1}{2}$ fermions in a box, where N is very large, and that these particles are non-interacting. Electrons in a sample of metal are quite a good example, because although the electron–electron Coulomb forces are considerable, they are approximately neutralized by the background positively charged ions in the metal. In the ground state, the fermions occupy the states of lowest available energy. The 1-particle states have occupation number 1 up to some energy ε_F and occupation number 0 for all higher energies. ε_F is called the *Fermi energy*.

We can work out ε_F and the total energy of the ground state using the density of states. For spin $\frac{1}{2}$ particles of mass m in a box of volume V, the density of 1-particle states is $g(\varepsilon) = \frac{V}{2\pi^2}\left(\frac{2m}{\hbar^2}\right)^{\frac{3}{2}} \varepsilon^{\frac{1}{2}}$. This is written in terms of the energy ε, and the additional factor of two compared with the formula (8.13) is because of the two independent spin states. For the occupied states, ε ranges upwards from zero, the 1-particle ground state energy, to the Fermi energy ε_F. The total number of particles N is therefore

$$N = \int_0^{\varepsilon_F} g(\varepsilon)\, d\varepsilon = \frac{V}{2\pi^2}\left(\frac{2m}{\hbar^2}\right)^{\frac{3}{2}} \frac{2}{3} \varepsilon_F^{\frac{3}{2}}. \tag{8.73}$$

Inverting this, we find

$$\varepsilon_{\mathrm{F}} = \frac{\hbar^2}{2m} \left(\frac{3\pi^2 N}{V} \right)^{\frac{2}{3}}, \tag{8.74}$$

showing that the Fermi energy depends only on the spatial density of the particles, $\frac{N}{V}$. The total energy E is a similar integral to that for N but weighted by ε,

$$E = \int_0^{\varepsilon_{\mathrm{F}}} g(\varepsilon)\varepsilon \, d\varepsilon = \frac{V}{2\pi^2} \left(\frac{2m}{\hbar^2} \right)^{\frac{3}{2}} \frac{2}{5} \varepsilon_{\mathrm{F}}^{\frac{5}{2}}. \tag{8.75}$$

This can be expressed in terms of N and V using the formula (8.74), to give

$$E = \frac{3(3\pi^2)^{\frac{2}{3}}}{5} \frac{\hbar^2}{2m} \left(\frac{N}{V} \right)^{\frac{2}{3}} N. \tag{8.76}$$

In **k**-space, or in **p**-space, the occupied states fill the interior of a sphere called the *Fermi sphere*. The radius of the Fermi sphere, k_{F}, is related to the Fermi energy ε_{F} by $\frac{1}{2m}(\hbar k_{\mathrm{F}})^2 = \varepsilon_{\mathrm{F}}$, and is simply $k_{\mathrm{F}} = \left(\frac{3\pi^2 N}{V} \right)^{\frac{1}{3}}$.

8.8 Bosons, Fermions and Spin

Every particle, whether elementary like an electron, or composite like an atom, is either a boson or a fermion. But which is it? Remarkably, it just depends on the particle's spin. Experiments have shown that particles with integer spin $0, 1, \ldots$ are always bosons, and particles with half-integer spin $\frac{1}{2}, \frac{3}{2}, \ldots$ are always fermions. Within quantum mechanics there is really no understanding of why this is the case, and the requirement that a multi-electron wavefunction must be antisymmetric has to be put in by hand when considering physical problems such as atomic structure. However, in the relativistic theory of particles there is a theorem explaining this relation, although it is not elementary.

Electrons have spin $\frac{1}{2}$, as we have seen, and are therefore fermions. The importance of this fact cannot be overstated. It has critical consequences for the behaviour of the electrons in atoms and molecules, and for the countless electrons in materials like metals and semiconductors. We will explore these consequences in Chapter 9, building on our discussion of the Fermi sphere.

An atom is composed of electrons, protons and neutrons, all of which are fermionic particles. The whole atom is then bosonic if the number of fermions it contains is even, because exchanging labels on two such atoms requires an exchange of labels on an even number of fermion pairs. This is consistent with the atom's spin, which will be an integer. (The atom's total spin is a combination of the spins of an even number of interacting half-integer spin fermions, together with some orbital angular momentum, which is always an integer.) By contrast, exchanging labels on the components of two atoms each composed of an odd number of fermions shows that such an atom must be fermionic. Again this is consistent with the atom's spin, which is a combination of an odd number of half-integer spins and is therefore also half-integer.

Neutral atoms may be either fermions or bosons depending on their composition. This can give rise to some very surprising behaviour at low temperatures. For example, atoms of helium-4 contain two electrons, two protons and two neutrons and are therefore bosons. At temperatures below 4.2 K, helium-4 is a liquid. On further cooling, at 2.17 K it transforms into a *superfluid*—a liquid with no viscosity and no resistance to its flow. A container that holds normal liquid helium perfectly well will suddenly spring numerous leaks when it is cooled below this temperature, as superfluid helium seeps out through ultra-microscopic pores in the container. Superfluid helium-4 has many strange and wonderful properties.

On the other hand, helium-3 remains a normal liquid at these temperatures. This is because helium-3 atoms are composed of two electrons, two protons and one neutron, so they are fermions. When helium-3 is cooled much further, something remarkable happens. At a temperature of just 2.49×10^{-3} K the helium-3 atoms pair up. Each helium-3 atom has spin $\frac{1}{2}$, but the spins of a pair of atoms are aligned in the same direction so that the total spin is 1. A helium-3 pair is a boson, and the result is that helium-3 becomes a superfluid. The bonds that hold the helium-3 pairs together are extremely weak and a tiny increase in temperature will shake them apart. That is why helium-3 must be cooled to such an ultra-low temperature before it becomes a superfluid.

Photons have spin 1 and are therefore another important type of boson. They travel at the speed of light and must be described by a relativistic version of quantum theory. The bosonic character of photons makes lasers possible, and also underlies the possibility of treating light as a classical electromagnetic wave, satisfying Maxwell's equations. As we will see in Chapter 12, the electromagnetic force is due to the exchange of photons between charged particles. In general, in quantum theory, the forces of nature are produced by the exchange of bosons between other particles.

8.9 Return to the Action

In everyday life, we are familiar with the idea that particles follow well-defined trajectories. The quantum description of particle motion is quite different. As long ago as 1800, Thomas Young described interference patterns made by light and compared his results to the interference of water waves. We now know that light is composed of photons, and these individual, bosonic particles have wave-like properties. In the 1920s Clinton Davisson and Lester Germer confirmed that electrons behave in a very similar way. This is best illustrated by the double slit experiment, as shown in Figure 8.7. Electrons are fired from a source A and pass through two nearby slits B and C before reaching a detector screen. The probability that an electron is detected at point D on the screen varies periodically across the screen. We conclude that electrons are associated with waves, as suggested by de Broglie, and that these waves add together or cancel depending on their relative phase. As there are two distinct paths available to an electron when travelling from A to D, the amplitude $\Psi = \mu(e^{i\varphi_{\mathrm{B}}} + e^{i\varphi_{\mathrm{C}}})$ for its arrival at D is the sum of the amplitudes for its passage along the two paths, via B or C. The probability that it will be found at D is

$$|\Psi|^2 = \mu^2(e^{i\varphi_{\mathrm{B}}} + e^{i\varphi_{\mathrm{C}}})(e^{-i\varphi_{\mathrm{B}}} + e^{-i\varphi_{\mathrm{C}}}) = 2\mu^2(1 + \cos(\varphi_{\mathrm{C}} - \varphi_{\mathrm{B}})), \qquad (8.77)$$

where μ is a normalization constant.

Experimentally, it is possible to make the electron source so weak that only one electron is passing through the apparatus at any particular time. An interference pattern still emerges, establishing the wave-like nature of individual electrons in quantum mechanics. Also characteristic of quantum mechanics is that the pattern is built up probabilistically, becoming clear only after a large number of individual electrons have been detected.

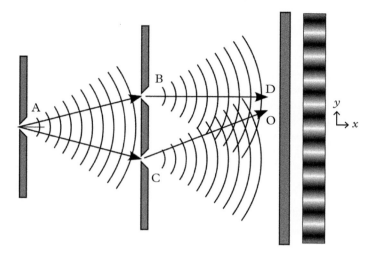

Fig. 8.7 An electron beam fired from source A through two slits in a barrier produces an interference pattern on the screen beyond. The amplitude Ψ for the arrival of an electron at D is the sum of the amplitudes for its passage along the two possible paths.

So far, we have not explained how to calculate the phase difference $\varphi_C - \varphi_B$. We could try to solve the Schrödinger equation for a single electron in this apparatus. However, there is an alternative approach, using the classical particle action. In Chapter 1 we showed how the well known properties of light such as reflection and refraction follow from Fermat's principle, and in Chapter 2 we showed how Newton's laws of motion can be derived from the principle of least action. Indeed, with the correct form for the action, this principle can account for the whole of classical physics. Clearly, the principle of least action has far-reaching consequences, but thus far we have offered no explanation for why it works. The answer is rather surprising and it gives the clearest insight into the relationship between classical and quantum mechanics.

Dirac first discussed the role of the action in quantum mechanics in 1933 and his remarks were followed up by Feynman, who used them to develop an alternative approach to quantum mechanics that is completely equivalent to the earlier versions but offers additional insight into the strangeness and meaning of the quantum world. Feynman's analysis was inspired by the double slit experiment. He decided to take seriously the apparently nonsensical idea that an electron passing through the double slit apparatus follows both available paths, or at least somehow knows of both paths. Following Dirac, he postulated that to determine the amplitude Ψ for an electron to arrive at D on the screen, one must evaluate the action along both paths through the apparatus and include both contributions as phase factors, so

$$\Psi = \mu(e^{i\varphi_B} + e^{i\varphi_C}) = \mu\left(\exp\left(\frac{i}{\hbar}S_B\right) + \exp\left(\frac{i}{\hbar}S_C\right)\right). \tag{8.78}$$

Here S_B is the action for the electron to travel along the path $\mathbf{x}(t)$ from A to D via B, and S_C is the action for the path via C. The action is as in Chapter 2,

$$S[\mathbf{x}(t)] = \int L \, dt = \int (K - V) \, dt \, , \tag{8.79}$$

where K is the kinetic energy and V the potential energy.

It is the relative phase that is important. If the two slits B and C are equidistant from the source A, then the phases are the same for the paths from source to slits, so they cancel in $|\Psi|^2$. We need only consider the paths from slits to detector screen. Let us introduce coordinates as shown in Figure 8.7, with the point O on the screen, equidistant from the slits, as origin, x the distance orthogonal to the screen and y the distance along the screen. Let $2f$ be the separation of the slits, and l the x-separation between the slits and the screen. Suppose the electron passes the slits at time 0 and reaches the screen at time T. (We are summing the contributions to the amplitude at time T, so this time is the same for each path.)

For a free non-relativistic electron, $V = 0$ and the Lagrangian L is the kinetic energy $K = \frac{1}{2}m(v_x^2 + v_y^2)$. The action to arrive at location D on the screen, with coordinate y, is $S = \frac{1}{2}m(v_x^2 + v_y^2)\,T$. The velocities in the x- and y-directions are

$$v_x = \frac{l}{T}\, , \quad v_y = \frac{y \mp f}{T}\, , \tag{8.80}$$

with the upper sign for the path via B and the lower sign for the path via C. Therefore

$$\varphi_{\mathrm{B}} = \frac{1}{\hbar}S_{\mathrm{B}} = \frac{m}{2\hbar}\left(\left(\frac{l}{T}\right)^2 + \left(\frac{(y-f)}{T}\right)^2\right)T = \frac{m}{2\hbar}\left(l^2 + (y-f)^2\right)\frac{v_x}{l}\, , \tag{8.81}$$

and similarly

$$\varphi_{\mathrm{C}} = \frac{m}{2\hbar}\left(l^2 + (y+f)^2\right)\frac{v_x}{l}\, . \tag{8.82}$$

Therefore

$$\varphi_{\mathrm{C}} - \varphi_{\mathrm{B}} = \frac{m}{2\hbar}(4yf)\frac{v_x}{l} = \frac{2fmv_x}{\hbar l}\, y\, , \tag{8.83}$$

and using equation (8.77), we find that the probability that an electron will arrive at D on the detector screen is

$$|\Psi|^2 = 2\mu^2\left(1 + \cos\left(\frac{2fmv_x}{\hbar l}\, y\right)\right)\, . \tag{8.84}$$

This oscillates between $4\mu^2$ and zero, taking its maximum value at $y = 0$, where the two paths are of equal length. The wavelength of the interference pattern on the screen is

$$\lambda = 2\pi\frac{\hbar l}{2fmv_x}\, . \tag{8.85}$$

v_x is determined by l and T, but rather than fix T it is more realistic to fix the energy E of the electron and use the relation $E = \frac{1}{2}mv_x^2$ to find v_x.

Feynman imagined increasing the number of available paths by increasing the number of slits in the barrier. Each such path contributes to the amplitude for the electron to arrive at a particular point on the screen and this total amplitude determines the form taken by the interference pattern. Finally, to model electrons in free space, Feynman imagined increasing the number of slits until the barrier disappears entirely. He then argued that, for consistency, the electrons must still follow paths that pass through all points of the now invisible barrier. Furthermore, the position of the barrier is completely arbitrary; it could have been positioned anywhere between the source and the detector. So we reach the outrageous, but inevitable, result that a free electron in empty space must actually follow all conceivable paths between A and D and that the amplitude for the arrival of the electron at D includes contributions from this vast infinite collection of paths. Feynman concluded that the amplitude for an electron (or other particle) to be emitted at A and detected at D is therefore

$$\Psi = \mu \sum_{\text{paths}} \exp\left(\frac{i}{\hbar} S[\mathbf{x}(t)]\right), \tag{8.86}$$

where the sum is over all possible smooth paths between A and D, over the fixed time interval from 0 to T, and the instantaneous velocity of the particle along each path is not fixed.

This formula is referred to as the *Feynman path integral*. It sums contributions from an uncountably infinite number of paths, most of which include untold wiggles, but it is very democratic in the sense that all paths are treated equally; the only difference between the amplitudes along different paths is their phase. Feynman showed that (allowing the position of D to be variable) the amplitude Ψ satisfies the Schrödinger equation and is completely equivalent to the usual wavefunction. In non-relativistic quantum mechanics it is usually much simpler to solve the Schrödinger equation than to perform the equivalent path integral calculation, but the technique comes into its own when considering quantum field theory, and it becomes essential when working with gauge theories.

Feynman's path integral approach offers a very interesting perspective on the relationship between classical and quantum mechanics. Classical mechanics is applicable in situations where we can treat $\hbar$ as very small. The phase in the expression $\exp(\frac{i}{\hbar} S[\mathbf{x}(t)])$ then varies rapidly, and the phase due to one path may be significantly different to that of a neighbouring path. Let $S_i[\mathbf{x}(t)]$ denote the action evaluated along path i. Then if $S_2[\mathbf{x}(t)] = S_1[\mathbf{x}(t)] + \pi\hbar$, the contribution of path 2 exactly cancels the contribution of path 1, as $\exp(\frac{i}{\hbar} S_2[\mathbf{x}(t)]) = -\exp(\frac{i}{\hbar} S_1[\mathbf{x}(t)])$. In general, the phase takes every value from 0 to 2π when evaluated along paths that differ very slightly from any particular path. When added together, the contributions from these neighbouring paths destructively interfere and do not contribute to the total amplitude for the particle to arrive at D. The dominant paths that do contribute are those for which the action does not change to first order as we move away from the path. These are the paths for which $S[\mathbf{x}(t)])$ is minimal, or stationary, and are known as the paths of stationary phase. The condition of stationary phase in the path integral corresponds exactly to the classical principle of least action—the phase is stationary when the action is stationary. We conclude that in quantum mechanics, all possible paths must be considered, but when the classical approximation is valid, the path integral is dominated by the classical paths for which the action is stationary.

We can ask how far a path must vary in order for the action to change by $\pi\hbar$, and for the contributions of the original path and varied path to cancel in the path integral. For a free non-relativistic particle of mass m and momentum p travelling for a short time Δt, the action is $\Delta S = \frac{p^2}{2m}\Delta t$. If the distance travelled is Δx then the momentum is $p = m\frac{\Delta x}{\Delta t}$, so

$$\Delta S = \frac{p^2}{2m}\Delta t = \frac{1}{2}p\Delta x. \tag{8.87}$$

Therefore $\Delta S = \pi \hbar$ if $\Delta x = \frac{2\pi \hbar}{p}$, which is precisely the de Broglie wavelength. In this way, the path integral picture of quantum mechanics explains why a particle of momentum p behaves like a wave whose wavelength is given by the de Broglie formula.

It also sheds light on when classical mechanics is applicable. The classical trajectory of a particle is a useful concept so long as we are not considering length scales smaller than the particle's de Broglie wavelength. For a billiard ball moving at $10\,\mathrm{m\,s^{-1}}$, this wavelength is an utterly minuscule 10^{-34} m, so it is always safe to describe the motion using classical mechanics. However, the de Broglie wavelength of an electron in an atom is around a nanometre, which is larger than the atom, so when considering the interactions between electrons and atoms, we cannot use a classical mechanical approximation. This will be important in the next chapter. In our discussion of an electron passing through the macroscopic double slit apparatus, however, we are justified in using classical trajectories to determine the action for each path, and take the path integral to be the sum of just two contributions.

8.10   Further Reading

In addition to the suggestions for further reading at the end of Chapter 7, see

E. Merzbacher, *Quantum Mechanics (3rd ed.)*, New York: Wiley, 1998.

For a revised and corrected edition of the original book on Feynman's path integral approach to quantum mechanics, see

R.P. Feynman and A.R. Hibbs, *Quantum Mechanics and Path Integrals: Emended Edition by D.F. Styer*, Mineola NY: Dover, 2010.

9

Atoms, Molecules and Solids

9.1 Atoms

Most matter on Earth is composed of atoms. Each atom contains a tiny nucleus that carries almost all its mass, and this is surrounded by orbiting electrons. The nucleus is composed of protons and neutrons, which have very similar masses, and the electric charge e of a proton is positive, whereas the neutron has no net charge. The number of protons in the nucleus is called the *atomic number*, Z, and as the electron charge is $-e$, a neutral atom must have exactly Z electrons. In most nuclei the number of neutrons is equal to or greater than the number of protons, although there are exceptions. For example, the nucleus of the hydrogen atom is just a single proton. Quantum mechanics is essential for understanding the structure and stability of both the nucleus and the electron orbitals.

Electrons can be stripped from a neutral atom; the atom is then ionized and is referred to as an *ion*. This requires high energies or high temperatures. One or two electrons are often transferred from one atom to another when atoms interact chemically to produce a molecule. This involves much less energy, because overall the molecule remains electrically neutral. Fully ionized atoms only occur at temperatures of about 10^4 K or higher, for example in stars, or in artificial environments like particle accelerators and fusion devices. A gas of fully ionized atoms, consisting of free nuclei and electrons, is called a *plasma*.

Atoms with different values of Z are given distinct names; for example, atoms with $Z = 6$ are carbon atoms. A pure substance consisting of atoms with just one value of Z is called an *element*. The number of neutrons in the atomic nuclei of an element may vary. For instance, naturally occurring carbon is largely composed of carbon-12 atoms with nuclei that each contain six protons and six neutrons, but there is also a small proportion of carbon-13 atoms whose nuclei contain seven neutrons, and these nuclei are also stable. Atoms with the same number of protons, but different numbers of neutrons in their nuclei are referred to as different *isotopes* of the element.

The total number of protons and neutrons in an atom is called the *atomic mass number* and is denoted by A. When specifying a particular isotope, A may be written after the name of the element. To a first approximation, the measured mass of an atom equals the proton mass times A, but this is not exact as neutrons are a bit more massive than protons and the nuclear binding energy reduces the mass slightly; there is also a small contribution from the mass of the electrons.

There are naturally occurring atoms for most values of Z up to $Z = 92$. In most cases these nuclei are stable, but some have unstable nuclei with half-lives as long as billions of years. Almost all the elements up to $Z = 83$ bismuth have stable isotopes. One of the exceptions is element $Z = 43$ technetium, whose name means artificially created. Unstable

The Physical World. Nicholas Manton and Nicholas Mee, Oxford University Press (2017).
© Nicholas Manton and Nicholas Mee. DOI 10.1093/oso/9780198795933.001.0001

nuclei are described as *radioactive*, although this is a little misleading. The name arose because decaying nuclei emit particles that appeared to be similar to intense electromagnetic radiation, such as X-rays, when first detected.

Nuclei are produced through various processes in the cosmos that we will discuss in later chapters. Radioactive elements with isotopes whose half-lives are comparable to the age of the Earth, such as $Z = 92$ uranium, may be relatively abundant on Earth. Their decay products may also be unstable, and if they have short half-lives then they will be relatively scarce. As might be expected, the decay products of uranium, such as $Z = 88$ radium, are found in association with uranium deposits. Radium is much rarer than uranium, because its half-life is of the order of one thousand years. This relatively short half-life makes it powerfully radioactive, as first observed by Marie and Pierre Curie, who isolated small quantities of radium from a uranium ore known as pitchblende.

Nuclei with $Z > 92$ can be created artificially in nuclear reactors or through nuclear collisions, and some have half-lives of tens or hundreds of years. Long before they decay, these nuclei acquire electrons and form neutral atoms. The heaviest known nuclei have Z-values approaching 120, but these are both hard to produce and highly unstable.

The chemical properties of atoms are almost exclusively due to the orbital structure of their electrons, and this is what we will discuss in this chapter. Chapter 11 is devoted to the structure and properties of nuclei. Different isotopes of an element have very similar chemistry, although the different atomic masses do have a small effect on the rate of chemical reactions. However, it is very difficult to separate the isotopes of an element chemically. The elements are organized in the *Periodic Table*, which was originally established by Dmitri Mendeleev. The table's structure largely reflects chemical properties that vary from left to right across each row of the table, but are rather similar within each column. The mass of the atoms increases from left to right and downwards through the table. The elements are ordered in exact accordance with increasing Z, from $Z = 1$ up to $Z = 118$. Understanding the precise structure of the table requires an understanding of the quantum mechanics of many-electron atoms.

In the age of the particle accelerator it has become customary to quote the mass and energy of a particle in electron volts (eV). This is convenient when considering the production of particles in accelerators. One electron volt is the energy that an electron, or any other particle carrying one unit of electric charge $\pm e$, gains when accelerated through a 1-volt electrical circuit. (1 eV is the equivalent of 1.6×10^{-19} joules (J) or 1.8×10^{-36} kg.) Chemical reactions typically involve energies of a few electron volts, and this is why batteries have a voltage in this range. High voltage devices operating at several kilovolts (keV) can strip electrons from atoms and be used to produce electron beams, as in older types of televisions and X-ray machines.

The energy released in a nuclear process is generally quoted in MeV where 1 MeV is a million (mega) electron volts. The mass of an electron is about half an MeV, which is why electron production is possible in the nuclear process of beta decay. A uranium-238 nucleus undergoes alpha decay; the alpha particle that is released is always emitted with the same sharply defined kinetic energy: 4.2 MeV. One billion (giga) electron volts is written as 1 GeV, and is slightly more than the mass of a proton. One trillion (tera) electron volts, or 1000 GeV, is written as 1 TeV. The Large Hadron Collider is a collider of protons that have each been accelerated to an energy of 6.5 TeV. We will use these units throughout this and subsequent chapters.

9.1.1 Atomic orbitals

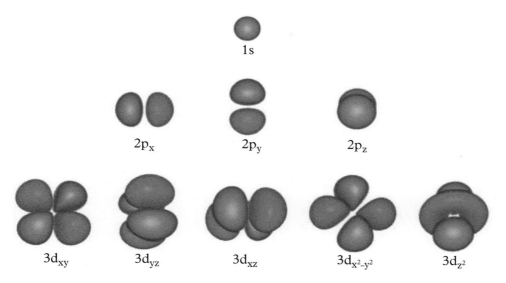

Fig. 9.1 Atomic orbitals.

Each electron in an atom moves in an attractive, hydrogen-like Coulomb potential $-\frac{Ze^2}{4\pi r}$, in which the electron of charge $-e$ interacts with the charge Ze of the nucleus. The electron is also subject to an additional potential due to the other electrons in the atom. The first step in calculating the energy levels is to solve the Schrödinger equation for 1-electron bound states in the attractive Coulomb potential, and ignore the electron–electron interactions.[1] This produces a sequence of energy levels with the degeneracies: 1, 4, 9, 16, ..., N^2, as discussed in section 8.4.2. The energy levels are more strongly negative than in hydrogen, by the factor Z, and the wavefunctions are also more spatially compact, by the same factor. With two electron spin states this gives the following numbers of states at successive energy levels: 2, 8, 18, 32, ..., $2N^2$. As electrons are fermions, they must each occupy a different 1-electron state. For a neutral atom in its ground state, there are Z electrons, filling the 1-electron states in order of increasing energy.

The stationary state wavefunctions, also called *orbitals*, are labelled by the principal quantum number N and the angular momentum l, where $l \leq N - 1$. By convention, the orbitals corresponding to angular momentum 0, 1, 2 and 3 are named s, p, d and f respectively, and the degeneracy of these states (excluding spin) is 1, 3, 5 and 7. For example, when $N = 2$ there are 2s and 2p states and the total degeneracy is 8. The angular momentum corresponds to the number of angular *nodes* in the wavefunction, that is, the number of times the wavefunction changes sign as one moves around an axis through the nucleus. For instance, s orbitals are spherically symmetric and p orbitals have two lobes of opposite sign, as shown in Figure 9.1.

---

[1] The Coulomb potential has an enhanced symmetry and the energy levels have enhanced degeneracies, as we showed in Chapter 8.

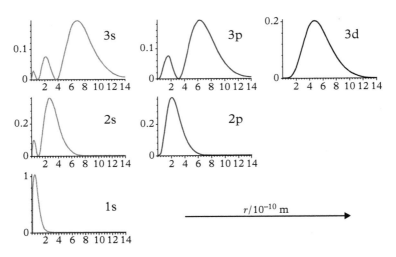

Fig. 9.2 Squared radial functions for atomic orbitals.

The orbitals of different l are orthogonal as they have different numbers of angular nodes, and orbitals of the same l but different N are orthogonal as they have different numbers of radial nodes. With increasing N, the number of radial nodes rises and there is a big increase in orbital radii, as shown in Figure 9.2.

9.1.2 Atomic shell model

For atoms beyond hydrogen, the mutual repulsion of the electrons in the atom must also be taken into account. This breaks some of the symmetry of the Coulomb potential and reduces the degeneracy of the solutions, as is readily understood in qualitative terms. Electrons in the filled inner shells *screen* the electric charge on the nucleus, so the effective nuclear potential seen by the outer electrons approaches zero faster than $-\frac{Ze^2}{4\pi r}$ for large r. The average distance of an electron from the nucleus increases with angular momentum; therefore the average effective nuclear charge seen by an electron decreases with increasing angular momentum. Electrons in s orbitals, for example, have a substantial probability of being very close to the nucleus where they see the full nuclear charge. By constrast, p, d and f orbitals have a node at the nucleus, so electrons in these orbitals do not approach as close to the nucleus. For states corresponding to the same principal quantum number N, those with greater orbital angular momentum l are more loosely bound and therefore higher in energy.

This effect can be represented as an additional spherically symmetric term in the Hamiltonian. It is a mean field perturbation proportional to l^2. For instance, the mean electron field breaks the 32-fold degeneracy of the $N = 4$ energy level of the hydrogen-like atom into two 4s, six 4p, ten 4d and fourteen 4f states, with increasing energy. The way in which the levels split is illustrated in Figure 9.3. Although some of the degeneracy is lost, the states still fall into angular momentum multiplets and *atomic shells* consist of groups of these multiplets with similar energies.

At higher energy levels, the splitting is large enough for some levels to cross over. The screening effect increases the energy of the 3d states above the 4s states. The 3d states may be higher in energy, but the peaks of their orbitals are significantly closer to the nucleus

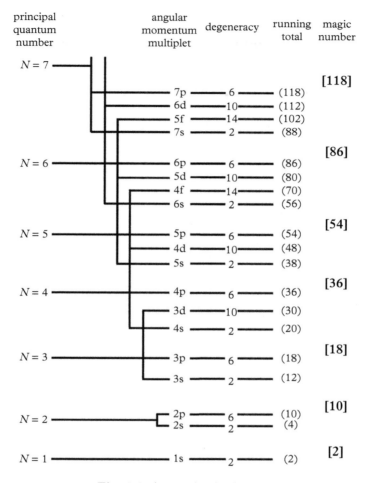

principal quantum number | angular momentum multiplet | degeneracy | running total | magic number

$N = 7$

[118]

7p — 6 — (118)
6d — 10 — (112)
5f — 14 — (102)
7s — 2 — (88)

[86]

$N = 6$
6p — 6 — (86)
5d — 10 — (80)
4f — 14 — (70)
6s — 2 — (56)

[54]

$N = 5$
5p — 6 — (54)
4d — 10 — (48)
5s — 2 — (38)

[36]

$N = 4$
4p — 6 — (36)
3d — 10 — (30)
4s — 2 — (20)

[18]

$N = 3$
3p — 6 — (18)
3s — 2 — (12)

[10]

$N = 2$
2p — 6 — (10)
2s — 2 — (4)

$N = 1$
1s — 2 — (2)

[2]

Fig. 9.3 Atomic-level splitting.

than those of the 4s states, as can be deduced from Figure 9.2. We will see that this has a profound impact on the chemistry of elements with outer electrons in 3d orbitals, and similarly for higher d orbitals.

A shell is complete, or full, when there is a large energy gap to the next available multiplet of levels. An atom with a full-shell configuration is stable as it takes a lot of energy to excite an electron to the next empty energy level. Neutral atoms contain equal numbers of electrons and protons. Therefore nuclei containing Z protons, where Z equals a full-shell number, form atoms that are chemically inert. The numbers of states in successive shells are 2, 8, 8, 18, 18, 32, 32 ..., as shown in Figure 9.3. This gives atomic *magic numbers:*[2] 2, 10, 18, 36, 54 and 86 corresponding to the inert gases: He, Ne, Ar, Kr, Xe and Rn, as set out in Table 9.1. For example, the inert gas krypton has atomic number $Z = 36$. Its electrons fill the first four atomic shells.

---

[2] The term *magic numbers* is borrowed from nuclear physics.

Shell	States	Degeneracy (incl. spin)	Magic numbers
1	1s	2	2
2	2s, 2p	$2 + 6 = 8$	10
3	3s, 3p	$2 + 6 = 8$	18
4	4s, 3d, 4p	$2 + 10 + 6 = 18$	36
5	5s, 4d, 5p	$2 + 10 + 6 = 18$	54
6	6s, 4f, 5d, 6p	$2 + 14 + 10 + 6 = 32$	86
7	7s, 5f, 6d, 7p	$2 + 14 + 10 + 6 = 32$	118

Table 9.1 Table of states for electrons in atoms.

The relationship between the energy levels governs the order in which they are filled and the electrons in the highest incompletely filled shell determine the chemistry of the element. The electrons within an atom comprise an inner *core* of electrons in complete shells plus *valence* electrons that occupy states in an incomplete shell. In section 9.2 we will discuss chemical bonding. In neighbouring atoms, the outer orbitals occupied by the valence electrons may overlap to form covalent or polar bonds. The core electrons lie well within the atomic radius, as shown in Figure 9.2, and so cannot participate in bonds. Also the energy of the core electrons is too low for their participation in ionic bonding, so chemical bonding only involves the valence electrons.

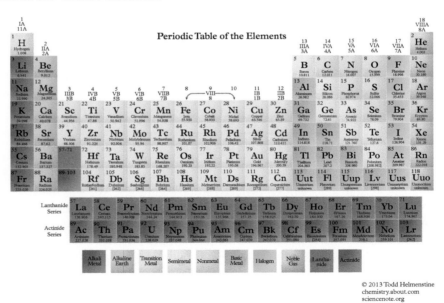

Fig. 9.4 The Periodic Table.

The elements, ordered by their atomic number Z, are presented in the Periodic Table. In its usual arrangement, this has eighteen columns, as shown in Figure 9.4. The chemical properties of an atom are determined by the orbitals occupied by its valence electrons and this is reflected in the layout of the table. There are two s states, ten d states and six p states in the fourth and fifth atomic shells and this explains the number of columns. (There are no p states in the first shell, which corresponds to the first row or *period*, and no d states in the first three shells. This explains the gaps in the top part of the table.) Each column forms a *group* of elements with similar chemical properties. The Group 18 elements in the rightmost column have no valence electrons. They are the inert (or noble) gases. On the left, the Group 1 and Group 2 elements have one or two valence electrons respectively in an s orbital. These are the alkali metals and the alkali earth metals. On the right, in Groups 13 to 17 the valence electrons are in p orbitals. (These groups are often referred to as Groups III to VII in solid state physics.) From the fourth shell onwards there is a sequence of elements in Groups 3 to 12 whose valence electrons are in d orbitals. These are the transition metals. It is significant that the 4s orbitals are lower in energy than the 3d orbitals (and similarly for the 5s and 4d orbitals, and 6s and 5d orbitals, in the subsequent periods of transition metals). Although the 3d orbitals are filled after the 4s orbitals, their orbital radii are much smaller than the radii of the orbitals occupied by the outer electrons. This greatly reduces the overlap between d orbitals on neighbouring atoms compared to the overlap of the outer s and p orbitals.

Set apart beneath the main body of the Periodic Table is a row of fifteen elements known as the lanthanides. Lanthanum has outer electronic structure $6s^2 5d^1$. In the next fourteen elements, from cerium ($6s^2 5d^1 4f^1$) to lutetium ($6s^2 5d^1 4f^{14}$), the 4f orbitals are gradually filled. The radii of the f orbitals are much smaller than the atomic radius, so the electrons in these orbitals do not take part in chemical bonding. All fourteen of these elements therefore have similar chemical properties to lanthanum and they are usually found combined together in their ores. Below the lanthanides is the sequence of actinides that begins with actinium. The pattern of these elements repeats that of the lanthanides. The 5f orbitals are gradually filled, moving through these elements. All of the actinides are radioactive.

9.2 Molecules

There are around 90 different types of naturally occurring atom. It is by combining this limited number of atoms into a myriad *compounds* that the diversity of our world is produced. Millions of compounds are known to chemists and there is no limit to the number of compounds that could, in principle, be formed. The chemical bond that holds atoms together is the key to this diversity. The stability of chemical bonds cannot be explained by classical dynamics. It was not until the discovery of the Schrödinger equation in 1926 that chemists could begin to understand this most fundamental feature of their subject.

9.2.1 Covalent bonding

If two atoms are sufficiently close together, the orbitals occupied by their outer electrons will overlap. These electrons then move in the potential of both atoms. The total energy of the two atoms may be reduced by this orbital overlap, in which case they will reach a stable configuration at an internuclear distance corresponding to the energy minimum and a chemical bond is formed. A bond between two atoms is called *covalent* if the electrons contributing to the bond are shared equally or nearly equally between the atoms. The atoms remain electrically neutral, or nearly so. A collection of two or more atoms bound together by chemical bonds is known as a *molecule*.

There are no exact solutions of the Schrödinger equation for molecules other than the very simplest, the hydrogen molecular ion H_2^+, which consists of two hydrogen nuclei orbited by one electron. To examine the general properties of chemical bonds and the molecules that result, we must make a number of approximations. Nuclei are much more massive than electrons ($m_p \simeq 1836\, m_e$) so we can assume that the nuclei in a molecule are fixed and consider the energy levels available to the electrons. (This is the Born–Oppenheimer approximation. The motion of the nuclei may be treated separately in order to determine the rotational and vibrational spectra of a molecule.) It is convenient to construct the Schrödinger equation for a single electron in an average potential due to the nuclei and all the other electrons in the molecule. This is known as the independent electron model. The single electron problem is also dramatically simplified by assuming that molecular orbitals may be constructed as *linear combinations of atomic orbitals*. This is known as the LCAO approximation. The ultimate justification for this approach is its success when compared to experiment.

We proceed by selecting a suitable collection of real, normalized, atomic stationary state wavefunctions χ_i, and combine them into a molecular orbital $\Psi = c_1\chi_1 + c_2\chi_2 + \dots$ with real coefficents c_i that are to be determined. Consider H_2, the hydrogen molecule. Here the 1s orbitals of the two atoms overlap. If the single-electron Hamiltonian is H, then the energy of an electron with wavefunction $\Psi = c_1\chi_1 + c_2\chi_2$ is

$$E = \frac{\int \Psi H \Psi d^3 x}{\int \Psi^2 \, d^3 x} = \frac{\int (c_1\chi_1 + c_2\chi_2) H (c_1\chi_1 + c_2\chi_2) \, d^3 x}{\int (c_1\chi_1 + c_2\chi_2)^2 \, d^3 x}. \tag{9.1}$$

The numerator simplifies to

$$\int \Psi H \Psi \, d^3 x = c_1^2 \int \chi_1 H \chi_1 \, d^3 x + c_1 c_2 \int (\chi_1 H \chi_2 + \chi_2 H \chi_1) \, d^3 x$$

$$+ c_2^2 \int \chi_2 H \chi_2 \, d^3 x$$

$$= c_1^2 \alpha_1 + 2 c_1 c_2 \beta + c_2^2 \alpha_2, \tag{9.2}$$

where $\alpha_i = \int \chi_i H \chi_i \, d^3 x$ is the energy of the electron in atomic orbital i, which is a negative quantity, and the *matrix element* $\beta = \frac{1}{2} \int (\chi_1 H \chi_2 + \chi_2 H \chi_1) \, d^3 x$, which is also negative, measures the strength of the bonding interaction between orbitals 1 and 2. The denominator is

$$\int \Psi^2 \, d^3 x = c_1^2 \int \chi_1 \chi_1 \, d^3 x + c_1 c_2 \int (\chi_1 \chi_2 + \chi_2 \chi_1) \, d^3 x + c_2^2 \int \chi_2 \chi_2 \, d^3 x$$

$$= c_1^2 + 2 c_1 c_2 S + c_2^2, \tag{9.3}$$

where $S = \int \chi_1 \chi_2 \, d^3 x$ measures the overlap of the orbitals on the neighbouring atoms. S is small and is often neglected in order to simplify calculations.

Now we can use the Rayleigh–Ritz variational principle. To find the optimum combination of atomic orbitals that minimizes the energy, we differentiate

$$E = \frac{c_1^2 \alpha_1 + 2 c_1 c_2 \beta + c_2^2 \alpha_2}{c_1^2 + 2 c_1 c_2 S + c_2^2} \tag{9.4}$$

with respect to each coefficient and set the result equal to zero to produce a set of equations that must be simultaneously satisfied by the coefficients. Rearranging equation (9.4), we obtain

$$E(c_1^2 + 2c_1c_2S + c_2^2) = c_1^2\alpha_1 + 2c_1c_2\beta + c_2^2\alpha_2 \,. \tag{9.5}$$

Differentiating with respect to c_1 gives

$$\frac{\partial E}{\partial c_1}(c_1^2 + 2c_1c_2S + c_2^2) + E(2c_1 + 2c_2S) = 2c_1\alpha_1 + 2c_2\beta \,, \tag{9.6}$$

and setting $\frac{\partial E}{\partial c_1} = 0$ produces $E(2c_1 + 2c_2S) = 2c_1\alpha_1 + 2c_2\beta$, or equivalently

$$c_1(\alpha_1 - E) + c_2(\beta - ES) = 0 \,. \tag{9.7}$$

Similarly, differentiating equation (9.5) with respect to c_2 leads to

$$c_1(\beta - ES) + c_2(\alpha_2 - E) = 0 \,. \tag{9.8}$$

These simultaneous equations are known as the *secular equations*. They have a non-trivial solution if the determinant of the coefficients is zero,

$$\begin{vmatrix} \alpha_1 - E & \beta - ES \\ \beta - ES & \alpha_2 - E \end{vmatrix} = 0 \,. \tag{9.9}$$

In the hydrogen molecule both nuclei are the same, so $\alpha_1 = \alpha_2 = \alpha$. Expanding the determinant gives $(\alpha - E)^2 - (\beta - ES)^2 = 0$, with the two solutions $E - \alpha = \pm(\beta - ES)$, which can be rearranged to give

$$E(1 \pm S) = \alpha \pm \beta \,. \tag{9.10}$$

One solution has energy

$$E_b = \frac{\alpha - |\beta|}{1 + S} \,, \tag{9.11}$$

which is lower than the original atomic energy α. (The energy is expressed in terms of $|\beta|$ as a reminder that β is negative.) It corresponds to a symmetrical combination of atomic orbitals $\Psi_b = \chi_1 + \chi_2$ known as a *bonding orbital*. The second solution has a higher energy,

$$E_a = \frac{\alpha + |\beta|}{1 - S} \,, \tag{9.12}$$

and corresponds to an antisymmetric combination $\Psi_a = \chi_1 - \chi_2$, known as an *antibonding orbital*. The difference in energy between the antibonding and bonding orbitals is $E_a - E_b \simeq 2|\beta|$.

The 1s orbitals of the two hydrogen atoms combine to produce a lower-energy bonding orbital σ and a higher-energy antibonding orbital σ^* as shown in Figure 9.5. Electrons have two spin states and this means that two electrons may occupy each orbital. In the ground state, the bonding orbital will be occupied by one electron from each hydrogen atom. This is shown in Figure 9.6 (left). The total electron energy is therefore lower for the hydrogen molecule than for two well-separated hydrogen atoms. The reduction in energy of

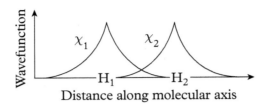

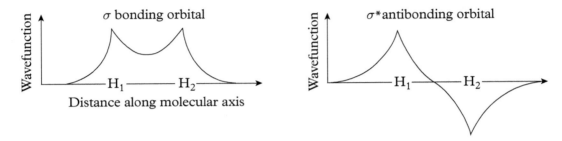

Fig. 9.5 Molecular orbitals in the hydrogen molecule.

the electrons in the bonding orbital is readily explained. This orbital is symmetrical with respect to the two nuclei so the electrons have longer wavelengths than they would have in either the atomic 1s orbitals or the antibonding orbital. This means that, in accordance with the de Broglie relationship, the electrons have less momentum and less kinetic energy.

So far, the separation of the nuclei has not been fixed. The reduction in energy due to the electrons being in the bonding orbital tends to increase as the nuclei get closer, but the repulsion between the positively charged nuclei also increases, and more rapidly at small separations. The total energy has a minimum at a definite separation of the nuclei, the equilibrium bond length. The hydrogen molecule is stable and energy must be put in to dissociate the two hydrogen atoms. The dissociation energy is 4.5 eV.

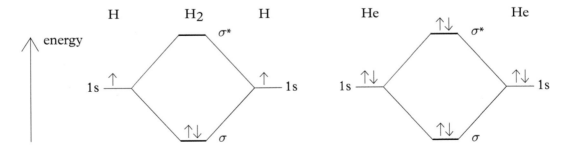

Fig. 9.6 1s orbitals on two adjacent atoms overlap to form a bonding orbital σ and an antibonding orbital σ^*. Left: The electrons from two hydrogen atoms, with spin up and spin down, both enter the σ orbital to form an H_2 molecule. Right: In the case of helium two electrons must also go into the σ^* orbital so a bond will not form.

We can pursue a similar analysis for two helium atoms. In this case there are four electrons, so two electrons must occupy the bonding orbital and two must occupy the antibonding orbital, as also shown in Figure 9.6. Combining equations (9.11) and (9.12) shows that for positive S, $E_a + E_b = \frac{2\alpha + 2|\beta|S}{1 - S^2} > 2\alpha$; therefore the energy cost of putting an electron in an antibonding orbital is greater than the energy gain from putting an electron in a bonding orbital. The total energy rises when the orbitals of two helium atoms overlap, so the helium atoms repel each other and no bond will form.

In the ground state of the first four atoms in the Periodic Table the electrons occupy s orbitals. Beyond $Z = 4$ beryllium the occupation of p orbitals begins, and this opens up new issues when considering the overlap of orbitals. Bonding orbitals are produced by the constructive interference between overlapping wavefunctions; antibonding orbitals are produced by destructive interference. The existence of non-zero overlap between orbitals can be deduced from the symmetry of the orbitals. The single-electron Hamiltonian is symmetrical under rotations around the bond axis so matrix elements between orbitals with different symmetry with respect to the bond axis must vanish. For instance, if the bond axis is the z-axis, then the overlap between an s orbital on atom 1 and a p_x orbital on atom 2 must vanish as any positive overlap in the $+x$-direction will be cancelled by an equal negative overlap in the $-x$-direction. Similarly, there is no overlap between an s orbital and a p_y orbital, but there will be a non-zero overlap between an s orbital and a p_z orbital, which may be positive or negative, as illustrated in Figure 9.7.

Bonding orbitals are classified according to their symmetry with respect to the bond axis, and named by analogy with the names of atomic orbitals according to their angular momentum m around this axis. When m $= 0$, the orbital is known as a σ orbital. The overlap of two s orbitals produces σ orbitals, as shown at the top of Figure 9.7. The overlap between two p_z orbitals also produces σ orbitals. When m $= 1$ (or m $= -1$), the molecular orbitals are known as π orbitals. The overlap between two p_x or two p_y orbitals produces π orbitals, as shown at the bottom of Figure 9.7. If m $= 2$ the molecular orbitals are known as δ orbitals. When atoms from the second and third periods of the Periodic Table bond, there are four orbitals on each atom to consider: one s and three p orbitals. Only m $= 0$ and m $= \pm 1$ occur here.

The overlap between two p_z orbitals is greater than the overlap between two p_x or two p_y orbitals, so the σ-bonding combination of p_z orbitals is lower in energy than the π-bonding combinations of the other two p orbitals, and the σ^*-antibonding combination of p_z orbitals is higher in energy that the π^*-antibonding combinations of the other two p orbitals. Combined with the fact that the atomic p states are higher in energy than the atomic s states, the result is the sequence of molecular energy levels shown in Figure 9.8.

A nitrogen atom has two 1s core electrons, plus five 2s and 2p valence electrons in its second incomplete shell. When two nitrogen atoms bond, a total of ten valence electrons occupy the molecular orbitals. In the ground state these electrons occupy the lowest five orbitals, as shown in Figure 9.8 (left). The four 2s electrons fill both the bonding and antibonding orbitals, but the six 2p electrons only occupy bonding orbitals, so the nitrogen–nitrogen bond is a triple bond. This confers great stability on the nitrogen molecule N_2 relative to isolated nitrogen atoms. Its dissociation energy is 9.79 eV, which makes this the second strongest bond in nature after the bond of the carbon monoxide molecule CO (which also involves ten valence electrons). Unlike carbon monoxide, the nitrogen molecule

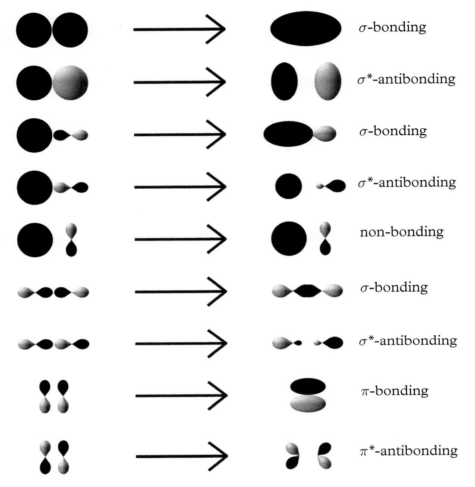

Fig. 9.7 Molecular orbitals. The bonding distorts the atomic orbitals as shown.

is non-polar (it has no electric dipole), so N_2, which forms the bulk of our atmosphere, is chemically inert.

Oxygen atoms have six valence electrons, as shown in Figure 9.8 (right). In the ground state of an oxygen molecule twelve electrons occupy the lowest six molecular orbitals. This means that two electrons must occupy π^*-antibonding orbitals, thus offsetting the energy gain of two electrons in bonding orbitals. Overall, the oxygen–oxygen bond is therefore a double bond. The two electrons in antibonding orbitals have raised energies relative to the oxygen atoms, so they are readily available to form bonds with other atoms. This makes the oxygen molecule O_2 highly reactive.

The π_x^* and π_y^* orbitals are degenerate and this raises the question of whether the two electrons that occupy them in the O_2 molecule are in the same orbital or not. To find the lowest energy state of the oxygen molecule we must move beyond the 1-electron approximation and consider the 2-electron wavefunction. Electrons are fermions so the

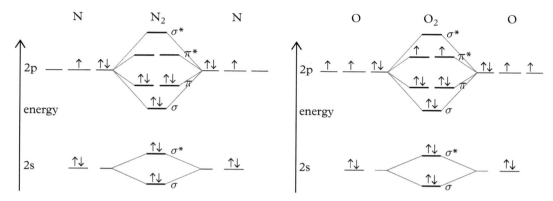

Fig. 9.8 Left: Bonding and antibonding orbitals in the nitrogen molecule. Right: Bonding and antibonding orbitals in the oxygen molecule. Only the valence electron orbitals are shown, not the filled 1s core orbital.

wavefunction is antisymmetric under the exchange of any two electrons, as discussed in section 8.7. This means that if the 2-electron spatial wavefunction is symmetric, then the spin state is antisymmetric and conversely if the spatial wavefunction is antisymmetric, then the spin state is symmetric. For two electrons in the same orbital, the spatial wavefunction is necessarily symmetric, so the spin state must be antisymmetric and the two electrons have opposite spin. Thus each core electron is paired with another electron with opposite spin. If two valence electrons are in different orbitals, then it is possible for the spatial 2-electron wavefunction to be either symmetric or antisymmetric and these two possibilities will not in general be degenerate. The effect of the electrostatic repulsion between the two electrons can be estimated to be

$$\Delta E_{\pm} = \frac{1}{2}(J \pm K),\tag{9.13}$$

where J is the Coulomb repulsion term and K is known as the exchange energy. The plus and minus signs refer to the energies of the symmetric and antisymmetric wavefunctions. For a repulsive potential K is positive, so the antisymmetric spatial wavefunction gives the lower energy state. This is intuitively reasonable, as the separation of two electrons is maximized if their combined wavefunction is antisymmetric. In the O_2 ground state, the 2-electron π^* wavefunction is therefore antisymmetric with one electron in the π_x^* orbital and one electron in the π_y^* orbital. The electron spin state is then symmetric, so the spins of these two electrons are aligned, which gives the oxygen molecule O_2 a magnetic moment. This is an example of Hund's first rule which states that when a number of degenerate states are available to electrons, they preferentially occupy states in which their spins are aligned.

The next element in the second period is fluorine, which also forms a diatomic molecule F_2. The fluorine molecule shares fourteen valence electrons, with four in the antibonding π^* orbitals, which makes it extremely reactive. All the electron spins are paired, so like nitrogen and unlike oxygen, the fluorine molecule does not have a magnetic moment. The fluorine–fluorine bond is a single bond, as just one pair of electrons in a bonding orbital is unmatched by a corresponding pair in an antibonding orbital. In general, the stronger a bond the shorter its length. The fluorine–fluorine single bond is 0.142 nm long, compared to

0.121 nm for the oxygen–oxygen double bond and 0.110 nm for the nitrogen–nitrogen triple bond.

Neon atoms have eight outer electrons. Two neon atoms do not bond at all as the sixteen valence electrons would fill all the bonding and antibonding orbitals making such a bond energetically unfavourable.

9.2.2 Polar bonds

Bonds between different atoms can be analysed in the same way. Consider a molecule HCl formed of a hydrogen atom and a chlorine atom. The bond is formed through the overlap of the 1s orbital of the hydrogen atom χ_1 and the $3p_z$ orbital of chlorine χ_2. The matrix elements $\alpha_1 = \int \chi_1 H \chi_1 \, d^3x$ and $\alpha_2 = \int \chi_2 H \chi_2 \, d^3x$ are now unequal, as illustrated in Figure 9.9, with $\alpha_2 < \alpha_1$.

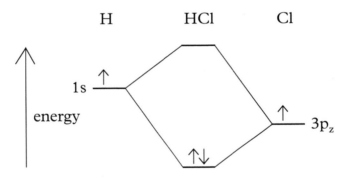

Fig. 9.9 Bonding and antibonding energy levels in a polar bond.

Returning to the secular determinant (9.9), with $\alpha_1 \neq \alpha_2$, the energies of the two molecular orbitals are

$$E_{a,b} = \frac{\{(\alpha_1 + \alpha_2) - 2S\beta\} \pm \sqrt{\{(\alpha_1 + \alpha_2) - 2S\beta\}^2 + 4(1 - S^2)(\beta^2 - \alpha_1\alpha_2)}}{2(1 - S^2)}. \tag{9.14}$$

This simplifies to

$$E_{a,b} = \frac{1}{2}(\alpha_1 + \alpha_2) \pm \frac{1}{2}\sqrt{(\alpha_1 - \alpha_2)^2 + 4\beta^2}, \tag{9.15}$$

if we neglect the overlap integral S. The difference in energy between the antibonding and bonding orbitals is

$$\Delta = E_a - E_b = \sqrt{(\alpha_1 - \alpha_2)^2 + 4\beta^2}, \tag{9.16}$$

where $\alpha_1 - \alpha_2$ is the ionic contribution to the energy gap, equal to the difference in the energy levels of the isolated atoms, and $2|\beta|$ is the covalent contribution, equal to the splitting when the two atoms are identical. We can parametrize the molecular orbitals as

$$\Psi_b = \chi_1 \sin\theta + \chi_2 \cos\theta, \quad \Psi_a = \chi_1 \cos\theta - \chi_2 \sin\theta, \tag{9.17}$$

where

$$\tan 2\theta = \frac{2|\beta|}{\alpha_1 - \alpha_2}. \tag{9.18}$$

If $\alpha_2 < \alpha_1$ then $\tan 2\theta$ is positive, so $0 < \theta < \frac{\pi}{4}$ and $\cos\theta > \sin\theta$. This means that the bonding orbital Ψ_b is concentrated on atom 2 and the antibonding orbital Ψ_a is concentrated on atom 1. In the bonding state, there will be a partial negative electric charge associated with atom 2 and a partial positive charge associated with atom 1, so the bond has an electric dipole. This is known as a *polar* bond.

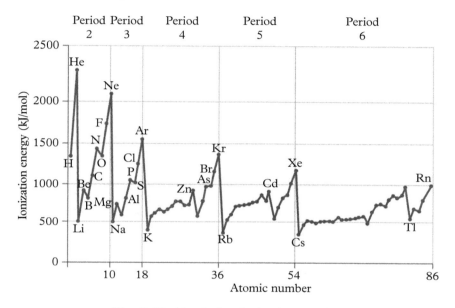

Fig. 9.10 Atomic first ionization energies.

In the limit where $|\beta| \ll \frac{1}{2}|\alpha_1 - \alpha_2|$, then $\Delta \to \alpha_1 - \alpha_2$ and

$$E_b = \frac{1}{2}(\alpha_1 + \alpha_2 - \Delta) \ \to \ \alpha_2, \quad E_a = \frac{1}{2}(\alpha_1 + \alpha_2 + \Delta) \ \to \ \alpha_1. \tag{9.19}$$

Furthermore $\tan 2\theta \to 0$, so $\cos \theta \to 1$ and $\sin \theta \to 0$. Therefore $\Psi_b \to \chi_2$ with energy α_2 and $\Psi_a \to \chi_1$ with energy α_1.[3] In the ground state of the molecule there are two electrons in the bonding orbital, giving a charge of $-e$ to atom 2, and none in the antibonding orbital leaving atom 1 with a charge of $+e$. An electron has been transferred completely from one atom to the other to produce an anion X^- and a cation Y^+. We have a highly polar or *ionic* bond between a negative and a positive ion. The bond that holds the hydrogen atom and the chlorine atom together in the molecule HCl is an example of an ionic bond, with Cl^- the anion and H^+ the cation.

The elements on the left of the Periodic Table readily lose electrons to form positive ions, while the outer electrons of the elements on the right are tightly bound, and they tend to gain extra electrons to form negative ions. The ease with which atoms lose an electron is illustrated in Figure 9.10. As atomic number Z changes, the energy required to remove an electron from an atom reaches a peak at each of the inert gases, which explains why they do not readily form ionic compounds and exist naturally as monatomic gases.

---

[3] This illustrates the important point that significant overlap between orbitals only occurs if the orbitals are relatively close in energy, as well as having a spatial overlap, a point that we will return to when considering solids.

9.2.3 Simple molecules

The familiar idea that the number of bonds formed by atoms arises from the desire of atoms to achieve an inert gas configuration is a handy rule of thumb that holds true for many molecules formed by low mass atoms, but it is not really a fundamental law of chemistry. The key to the formation of molecules is to be found in comparisons between the total energy of different molecular configurations.

Nevertheless, it is useful to attribute a *valency* to the smaller atoms. Carbon atoms have two core 1s electrons and four valence electrons in the second incomplete shell that are available to form covalent bonds. Carbon therefore has a valency of four and will form four covalent bonds to produce molecules such as methane CH_4. All four carbon–hydrogen bonds in methane are identical, giving the molecule tetrahedral symmetry (see Figure 9.11). This minimizes the energy of the molecule by maximizing the distance between the four bonds, thereby reducing the repulsion between the bonding electrons. The angle between the bonds is $2\cos^{-1}(\frac{1}{\sqrt{3}}) \simeq 109.5°$.

The carbon 2s orbital and each of the three 2p orbitals contribute equally to the bonds. Chemists refer to these bonds that mix the s and p orbitals as sp^3 hybrids. Even though the bonds are not truly covalent, but slightly polar, the methane molecule has no net electric dipole because of its tetrahedral symmetry.

Fig. 9.11 Left: methane CH_4. Centre: ammonia NH_3. Right: water H_2O.

A nitrogen atom has five valence electrons in its second incomplete shell. A nitrogen atom will bond with three hydrogen atoms to form a molecule of ammonia NH_3. As in methane, there are eight electrons in four sp^3 hybrid orbitals. The electrons in three of these hybrid orbitals form bonds between the nitrogen and hydrogen atoms. The electrons in the other hybrid orbital are described as a *lone pair*. The distinction between the bonding electrons and the lone pair has a small effect on the tetrahedral arrangement of the hybrid orbitals. The angle between the bonds is reduced to approximately 107.8°. The nitrogen–hydrogen bond is polar giving the nitrogen atom a small negative charge and the hydrogen atoms a small positive charge. The hydrogen atoms are not arranged tetrahedrally, so unlike methane, ammonia is a polar molecule.

With six valence electrons, oxygen atoms will bond to two hydrogen atoms to form water H_2O. Again, there are eight electrons in four sp^3 hybrid orbitals, but now there are two lone pairs on the oxygen atom. The orbitals are still arranged approximately tetrahedrally, but the angle between the bonds is reduced to 104.5°. The polarity of the oxygen–hydrogen bond is even greater than the nitrogen–hydrogen bond and this has a significant effect on the physical properties of water. The interaction between the polar water molecules raises the melting and boiling points of water dramatically. At atmospheric pressure the boiling point of methane is −161°C, that of ammonia is −33°C, while that of water is 100°C. We have the polarity of the oxygen–hydrogen bond to thank for the fact that our planet is covered in liquid water.

9.3 Organic Chemistry

9.3.1 Hückel theory—benzene

Carbon atoms have the remarkable property that they readily form molecular chains and rings. These structures are essential for the existence of life. In 1865 the concept of chemical bonding was in its infancy. Chemists had begun to use stick models to represent molecules, but August Kekulé could not understand the structure of the benzene molecule, which contains six carbon atoms and six hydrogen atoms. Kekulé later recalled that while engrossed in his work he dozed in front of a fire and dreamt of snakes twisting around to bite their own tails. He awoke with the realization that the carbon atoms in benzene must close to form a ring. He proposed a structure where the carbon–carbon atoms alternately form single and double bonds, as shown in Figure 9.12. This was a huge step forward in the elucidation of organic chemistry, but today's chemists know that it is not exactly true as all six carbon–carbon bonds in the benzene ring are identical. The length of carbon–carbon single bonds is 0.154 nm, while the stronger carbon–carbon double bonds are 0.134 nm in length. However, in benzene the carbon–carbon bonds are all the same length, 0.139 nm. So it would appear that these bonds are neither single nor double bonds.

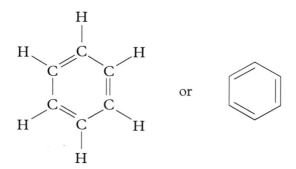

Fig. 9.12 The Kekulé structure of benzene.

In the plane of the benzene ring, each carbon atom forms three σ bonds. One bond is to a hydrogen atom situated radially out from the ring and the other two are to the adjacent carbon atoms in the ring. The angle between these bonds is 120°. The linear superpositions of carbon orbitals that form these bonds are described as sp^2 hybrids. This accounts for three of the four valence electrons of each carbon atom.

The remaining electrons are found in p orbitals that are oriented perpendicular to the plane of the benzene ring. There is one electron per carbon atom in these p orbitals, which overlap on adjacent carbon atoms and provide additional bonding. Chemists refer to such chains of overlapping p orbitals as *conjugated p orbitals*. The bonding in these molecules can be understood by applying molecular orbital theory to their carbon–carbon bonds. We can form molecular orbitals from linear combinations of the conjugated p orbitals, then minimize the energy by varying the coefficients to produce the secular equations, as we did previously for the diatomic molecules.

The problem is greatly simplified by adopting several approximations first introduced by Erich Hückel in 1930. This approach, referred to as Hückel theory, offers a qualitative understanding of some of the chemical properties of planar molecules such as benzene. The electron orbitals Ψ_n in such a system of conjugated p orbitals are approximated by a linear combination of normalized atomic orbitals χ_r, where r labels the atoms, thus

$$\Psi_n = \sum_r c_r^{(n)} \chi_r, \tag{9.20}$$

with constant coefficients $c_r^{(n)}$.

First, we assume that the overlap S between atomic orbitals on different atoms can be neglected. Next, we assume that the atomic environment of each carbon atom is identical, so that all diagonal elements are the same, i.e. $\int \chi_r H \chi_r \, d^3x = \alpha$. This is exactly true for a symmetrical molecule such as benzene. Finally, we assume that all matrix elements are zero except those corresponding to nearest neighbour atoms, i.e. $\int \chi_r H \chi_s \, d^3x = \beta$ if atom r is a nearest neighbour of atom s and zero otherwise, and β is negative. In the case of benzene, this simplifies the secular determinant to

$$\begin{vmatrix} \alpha - E & \beta & 0 & 0 & 0 & \beta \\ \beta & \alpha - E & \beta & 0 & 0 & 0 \\ 0 & \beta & \alpha - E & \beta & 0 & 0 \\ 0 & 0 & \beta & \alpha - E & \beta & 0 \\ 0 & 0 & 0 & \beta & \alpha - E & \beta \\ \beta & 0 & 0 & 0 & \beta & \alpha - E \end{vmatrix} = 0. \tag{9.21}$$

The six secular equations, also known as the *Hückel equations*, may be solved by making the ansatz

$$c_r^{(n)} = e^{\frac{i}{6}(2\pi n r)}. \tag{9.22}$$

Substituting into the Hückel equations gives

$$(\alpha - E_n)e^{\frac{i}{6}(2\pi n r)} + \beta \left\{ e^{\frac{i}{6}(2\pi n(r-1))} + e^{\frac{i}{6}(2\pi n(r+1))} \right\} = 0, \tag{9.23}$$

which simplifies to

$$(\alpha - E_n) + \beta \left\{ e^{-\frac{i}{6}(2\pi n)} + e^{\frac{i}{6}(2\pi n)} \right\} = 0, \tag{9.24}$$

and therefore

$$E_n = \alpha + 2\beta \cos\left\{ \frac{1}{6}(2\pi n) \right\}. \tag{9.25}$$

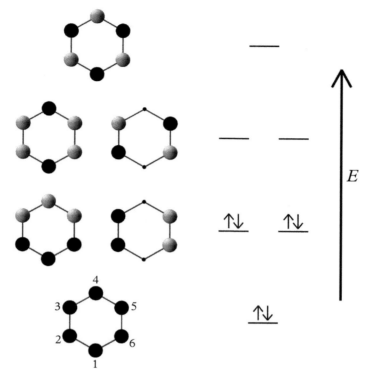

Fig. 9.13 Left: Electron orbitals in benzene formed from linear combinations of conjugated carbon p_z orbitals. The orbitals are oriented perpendicular to the benzene ring. The black and white spheres represent positive and negative lobes. A dot indicates that there is no contribution from the orbital at that atomic site. Right: The arrows represent spin up and spin down electrons occupying the lowest energy orbitals in the benzene ground state.

We can find linear combinations of the degenerate solutions such that all six orthonormal wavefunctions have real coefficients, thus:

$$\Psi_0 = \frac{1}{\sqrt{6}}(\chi_1 + \chi_2 + \chi_3 + \chi_4 + \chi_5 + \chi_6), \qquad E_0 = \alpha - 2|\beta|$$

$$\Psi_1 = \frac{1}{\sqrt{12}}(2\chi_1 + \chi_2 - \chi_3 - 2\chi_4 - \chi_5 + \chi_6), \qquad E_1 = \alpha - |\beta|$$

$$\Psi_2 = \frac{1}{\sqrt{12}}(2\chi_1 - \chi_2 - \chi_3 + 2\chi_4 - \chi_5 - \chi_6), \qquad E_2 = \alpha + |\beta|$$

$$\Psi_3 = \frac{1}{\sqrt{6}}(\chi_1 - \chi_2 + \chi_3 - \chi_4 + \chi_5 - \chi_6), \qquad E_3 = \alpha + 2|\beta|$$

$$\Psi_4 = \frac{1}{2}(\chi_2 - \chi_3 + \chi_5 - \chi_6), \qquad E_4 = \alpha + |\beta|$$

$$\Psi_5 = \frac{1}{2}(\chi_2 + \chi_3 - \chi_5 - \chi_6), \qquad E_5 = \alpha - |\beta|. \qquad (9.26)$$

These orbitals are illustrated in Figure 9.13. The completely in-phase combination of the six atomic orbitals produces the lowest energy state Ψ_0. Electrons in this orbital have the longest wavelength and therefore the lowest kinetic energy. The energy increases with the number of nodes in the wavefunction, as each node reduces the wavelength. In general, the delocalization of electrons allows them to occupy molecular orbitals of greater wavelength and therefore lower kinetic energy. This reduction in energy compared to atomic orbitals increases the stability of molecules.

Setting aside the electrons in the sp^2 hybrid orbitals, as they are unimportant to the argument, we can calculate the extra stability conferred on the benzene ring due to the delocalization of the electrons in the conjugated p$_z$ orbitals. The benzene structure proposed by Kekulé has three single and three double bonds. This structure includes three π bonds with two electrons in each. We have defined $\int \chi_r H \chi_s = \beta$, so the energy of an electron in a π bonding orbital is $\alpha - |\beta|$. In the Kekulé structure, the total energy of these six electrons would be $6(\alpha - |\beta|) = 6\alpha - 6|\beta|$. The Hückel theory shows that in the ground state of benzene there are six electrons in conjugated p$_z$ orbitals, doubly occupying each of the three lowest energy orbitals available, as shown in Figure 9.13. The total energy of these six electrons is $E = 2(\alpha - 2|\beta|) + 4(\alpha - |\beta|) = 6\alpha - 8|\beta|$. Therefore benzene is stabilized by an additional energy of $-2|\beta|$ compared to the Kekulé structure. This extra stability explains the relative chemical unreactivity of benzene.

9.3.2 Polyenes

Compounds of carbon and hydrogen $C_n H_{2n+2}$ built around a chain of single-bonded carbon atoms are known as alkanes, where the suffix 'ane' indicates a chain of carbon–carbon single bonds. A carbon–carbon double bond is signified by the suffix 'ene', so carbon–carbon chains containing conjugated p orbitals are referred to as *polyenes*. Figure 9.14 shows the chemical structure of pentadiene, which is a polyene with five carbon atoms and two double bonds.

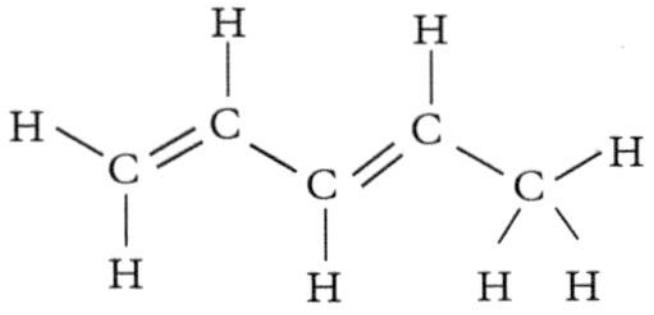

Fig. 9.14 Pentadiene.

Hückel theory can be applied to polyene molecules. The orbitals must vanish beyond the ends of the molecules, so for a linear molecule of length N the boundary conditions are $c_0^{(n)} = c_{N+1}^{(n)} = 0$. We can guess the solution:

$$c_r^{(n)} = \sin\left(\frac{\pi n r}{N+1}\right). \tag{9.27}$$

The orbitals available to the delocalized electrons in the first few polyene molecules are shown in Figure 9.15.

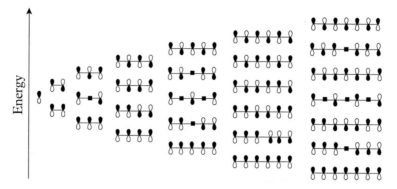

Fig. 9.15 Polyene orbitals in molecules with up to seven conjugated p orbitals (ethene to heptatriene).

Substituting the coefficients $c_r^{(n)}$ into the Hückel equations gives

$$(\alpha - E_n)\sin\left(\frac{\pi n r}{N+1}\right) + \beta\left\{\sin\left(\frac{\pi n(r-1)}{N+1}\right) + \sin\left(\frac{\pi n(r+1)}{N+1}\right)\right\} = 0\,, \qquad (9.28)$$

which after a little algebraic manipulation reduces to

$$(\alpha - E_n) + 2\beta\cos\left(\frac{\pi n}{N+1}\right) = 0\,, \qquad (9.29)$$

and therefore

$$E_n = \alpha + 2\beta\cos\left(\frac{\pi n}{N+1}\right). \qquad (9.30)$$

As $-1 \le \cos\theta \le 1$, the energies of the molecular orbitals lie in the range

$$\alpha - 2|\beta| \le E_n \le \alpha + 2|\beta|\,, \qquad (9.31)$$

as illustrated in Figure 9.16. This range is independent of the length of the polyene, so the average separation of the energy levels decreases as N increases. Each state is available for up to two electrons with opposite spin. As each carbon atom contributes one electron to the conjugated p orbitals, in the ground state only the lower half of the orbitals will be occupied, just as for benzene. Each of these orbitals is lower in energy than the atomic p orbitals, so extra stability is conferred on the polyenes by the delocalized electrons.

   The energy required to promote an electron from the highest occupied energy level to the lowest unoccupied energy level decreases as the polyene grows in length. Ultraviolet photons are required to promote electrons in short alkenes. For larger N, visible light photons can promote electrons into the unoccupied levels. Therefore, long polyenes are coloured. As $N \to \infty$ the separation between the energy levels tends to zero, so there is a half-filled continuous band of energies of width $4|\beta|$, and the system can absorb light over a continuous range of frequencies. A half-filled band is also the characteristic feature of a conductor, as electrons in a partially filled band are free to enter new states and will do so when subject to a small electric field or when thermally excited.

N 2   3   14   26   44

Fig. 9.16 Polyene energy levels. N is the number of carbon atoms in the polyene.

In 2010 Andre Geim and Konstantin Novoselov won the Nobel Prize in Physics for the discovery of graphene which is composed of a 2-dimensional hexagonal lattice of carbon atoms. Weight for weight, graphene is around 100 times as strong as steel. It may be considered as an infinite polycyclic aromatic molecule. Each carbon atom is bonded to three other carbon atoms through planar sp^2 orbitals with a 120° angle between the bonds. For each carbon atom there is also one electron in a p orbital which is perpendicular to this plane. These orbitals form an essentially infinite system of conjugated p orbitals. The orbitals available to the electrons extend throughout graphene and the energy levels form a continuous band. The total number of available states is unchanged when the carbon atoms are bonded into the macromolecule and, as each carbon atom contributes a single electron, only the states in the lower half of the band are filled. So, as expected from our previous discussion of such systems, graphene is an excellent conductor of heat and electricity.

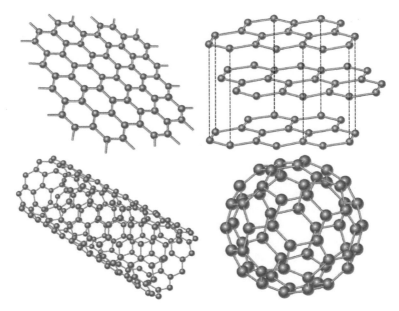

Fig. 9.17 Carbon allotropes. Clockwise from the top left: graphene, graphite, buckyball, carbon nanotube.

In recent years, carbon nanotubes have been created by chemists (Figure 9.17). These are equivalent to graphene molecules that have been wrapped around to form long cylinders. Nanotubes promise to have very important technological applications due to their structural and electrical properties.

Graphite is formed of layers of graphene with half the carbon atoms in each layer positioned above the centre of the hexagons in the previous layer. As in graphene, the p orbitals overlap and form a band that is half-filled with electrons, which results in the shiny (metallic) appearance of graphite and its well known electrical conductivity. Clearly, the arrangement of atoms in a molecule is critical with regard to its physical and chemical properties.

9.4 Solids

Most materials that we come across in everyday life exist as solids formed of vast numbers of atoms. Many of these substances are macromolecules bound together by bonds often characterized as ionic, covalent or metallic. This classification is not mutually exclusive as the bonds may display features midway between these different types, and materials may contain a number of distinct bonds.

The ideas of the previous section can be extended to describe bonding in solids in terms of the overlapping outer atomic orbitals on neighbouring atoms. In isolated atoms, electrons are confined to orbitals with sharply defined energy levels. When packed into a solid, this remains true of the inner core orbitals. They have a small radius and do not significantly overlap with neighbouring atoms. Their energy levels may be different to those of a free atom, but the differences will be very small. By contrast, the outer orbitals generally overlap with those of other atoms and when this happens their energy levels will broaden to form continuous bands just like in the Hückel model of the infinite polyene. The total number of states remains unchanged so that, for instance, a single atomic orbital will give rise to a band containing 10^{23} states if 10^{23} atoms bond together. This must be the case if the states are considered as linear combinations of the atomic orbitals. As electrons obey the exclusion principle, no two can exist in the same state. In the ground state, which is the state of the solid at absolute zero temperature, the electrons fill the energy levels in increasing order of energy, and the highest filled level is at the Fermi energy ε_F. The density of states in the region of the Fermi energy can be determined using techniques such as X-ray spectroscopy. These states play a fundamental role in determining the properties of the solid. Figure 9.18 shows schematic representations of the density of states in various types of substance.

For some materials, there is no energy gap between the highest occupied states and the lowest unoccupied states. Electrons are free to enter new states and may be induced to do so by a small electric field, so these substances conduct electricity. They obey *Ohm's law*, the linear relation between electric current and applied field, and their conductivity decreases with increasing temperature. We know them as *metals*. Metals are opaque and good reflectors of light with a mirror-like surface due to the strong interaction of visible light with their electrons. They are also malleable.

In many substances, including ionic solids such as sodium chloride, there is a large energy gap ΔE between the highest occupied states and the lowest unoccupied states. The band of occupied states is called the *valence band* and the band above is called the *conduction band*. These materials are *insulators* because their electrons cannot respond to a small electric field. They are typically transparent or white as photons of visible light do not

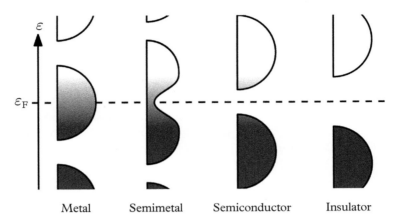

Fig. 9.18 Band gaps in various types of material. The grey strength indicates the filling of the bands, at a small non-zero temperature.

have sufficient energy to promote electrons into higher energy states. Photons pass through, or are scattered,[4] but they are not absorbed.

In some materials the Fermi energy ε_F lies within an energy band, but the density of electron states there is close to zero. These materials are known as *semimetals*. They include arsenic, bismuth and graphite. In bismuth, for instance, there is just one electron in the conduction band for every 10^5 atoms. This is, however, many more than in other materials, such as silicon and germanium, where there is a small energy gap between the valence band, which is full, and the conduction band. If the gap ΔE is less than about 1 eV the material is classed as a *semiconductor*. In such materials, small numbers of electrons are promoted into the conduction band by random thermal vibrations, whose probability is given by a Boltzmann factor[5] $\exp(-\Delta E/k_B T)$. (At room temperature $k_B T \simeq 0.025$ eV, which gives a Boltzmann factor of perhaps 10^{-10} in a semiconductor compared to 10^{-30} to 10^{-40} in an insulator. With 10^{23} electrons in the valence band, there may be 10^{13} electrons promoted into the conduction band of a semiconductor compared to none in an insulator.) Unlike in metals, the conductivity of semiconductors increases with temperature as more electrons are promoted into the conduction band. Electrons may also be excited by exposure to light, as visible light photons have sufficient energy to promote them into vacant states. Incident photons are therefore absorbed, so typically semiconductors look black. The promotion of electrons into the conduction band by the absorption of light increases the conductivity of the semiconductor and this can be used as the basis for a light detector such as a photodiode, one of the many applications of semiconductors. Typically, semiconductors are covalent solids.

9.4.1 Covalent solids

Carbon atoms bond covalently into the macromolecule diamond, whose structure is shown in Figure 9.19. The 2s and 2p orbitals on each carbon atom overlap with those on four others. We can think of the orbitals as sp^3 hybrids. The arrangement of the bonds is tetrahedrally

---

[4] This scattering may be due to impurities or imperfections in the crystal lattice.

[5] The Boltzmann or Gibbs distribution is discussed in section 10.4.

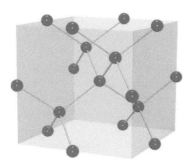 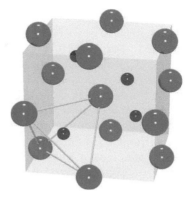

Fig. 9.19 Left: Diamond structure. Right: Zinc sulphide structure.

symmetric around each carbon atom. Because of the symmetry, and the atoms all being the same, none of the atoms acquires a net charge.

The physical properties of diamond follow from its macromolecular structure. Each carbon atom has four nearest neighbours. This number is known as the *coordination number* of the packing. Four is a very low coordination number, and for this reason diamond has a low density. The diamond structure is very stable, held together by strong covalent bonds, and this gives diamond an extremely high melting point (3820 K). The overlapping orbitals form bands of bonding and antibonding orbitals, each of which contain four states for every atom. Every atom contributes four electrons, so the bonding orbitals are filled and the antibonding orbitals are empty. The large gap of 5.5 eV between the filled bonding orbitals (the valence band) and the empty antibonding orbitals (the conduction band) means that diamond is an insulator. It also means that diamond is transparent, as visible light does not have sufficient energy to excite electrons into the conduction band.

The elements silicon and germanium beneath carbon in Group 14 (Group IV) have the same outer electronic structure and not surprisingly they also adopt the diamond structure as solids. The atomic radius increases as one moves down the group, which reduces the strength of the covalent bonds and lowers the melting points of these elements. It also increases the orbital overlap on neighbouring atoms which broadens the bands and reduces the gap between the valence and conduction bands. Whereas diamond is an insulator with a band gap of 5.5 eV, silicon and germanium are semiconductors with band gaps of 1.1 eV and 0.67 eV respectively. Tin, the next element in the group, exists in several allotropes. One of these, grey tin, has the diamond structure but has zero band gap and so is a semimetal. (In lead, the bonding is weak and the diamond structure does not form; lead is also a semimetal.)

Substances with the diamond structure have, as mentioned earlier, low coordination number and low density. When they melt, their atoms become more closely packed and their density increases.[6] The close-packing of the atoms in the liquid state increases the overlap of orbitals on neighbouring atoms. This increases the width of the bands with the result that the band gap disappears. So liquid silicon and liquid germanium are metallic.

---

[6] Water also shows this unusual behaviour—ice floats—and it is for the same reason. Water adopts the diamond structure when it is solid. In ice, the water molecules are tetrahedrally coordinated, with their two electron lone pairs bonded to the hydrogen atoms in neighbouring water molecules.

This is a good example of how the atomic arrangement of substances can have a dramatic effect on their physical properties.

Also shown in Figure 9.19 is the packing known as zinc sulphide (ZnS) structure, that is closely related to diamond structure. Half the carbon atoms are replaced by zinc atoms and half by sulphur atoms, such that each atom is bonded to four atoms of the opposite type. The bonds are slightly polar. This structure is adopted by many useful semiconducting compounds formed of elements from the groups on opposite sides of Group IV. These include the III–V compounds boron nitride, gallium arsenide and indium antimonide, and II–VI compounds such as zinc selenide.

Semiconductors have a wide range of applications that have transformed our world, but we are unable to do justice to this vast subject.

9.5 Band Theory

We discussed molecular orbitals in section 9.2, and similar ideas apply to the electron orbitals within a solid. As in the molecular case, we will assume that electron–electron interactions are not too strong and adopt the independent electron model. Whereas the overlap of two orbitals in a diatomic molecule produces one bonding and one antibonding orbital, within a solid the orbitals split into bands whose energies fall between extreme bonding and antibonding limits. For each atomic orbital there is a corresponding energy band in the solid, whose width depends on the overlap between the orbitals on neighbouring atoms. In materials with close-packed atoms this overlap may be large and the greater the overlap, the broader the band. Broad bands may merge, removing the gaps between the bands. This leads to metallic behaviour. Conversely, when electrons are strongly bound to individual atoms, as in ionic crystals, there is little overlap, so the bands are narrow and there is a large gap between the highest filled band and the next available states. This leads to insulating behaviour.

9.5.1 Atomic lattices

The key to determining the detailed electronic structure of a solid is to take advantage of the fact that its atoms are often arranged in a crystalline lattice. We will consider an idealized infinite and perfectly periodic crystal, neglecting any surface effects and ignoring the defects and impurities that inevitably occur in real crystals.

The centre of each atom in a crystal is situated at a point in the lattice. A 3-dimensional lattice is a regular array of points whose positions are the integer sums of three primitive vectors or generators $\mathbf{a}_i$ ($i = 1, 2, 3$). We can define a unit cell whose edges are these three vectors. The position of a general lattice point is

$$\mathbf{R} = n_1\mathbf{a}_1 + n_2\mathbf{a}_2 + n_3\mathbf{a}_3 , \qquad (9.32)$$

where (n_1, n_2, n_3) are integers. Although one point of the lattice is the origin, all lattice points are geometrically equivalent.

The simplest lattice is the simple cubic lattice, which is generated by the vectors

$$\mathbf{a}_1 = a(1, 0, 0), \quad \mathbf{a}_2 = a(0, 1, 0), \quad \mathbf{a}_3 = a(0, 0, 1), \qquad (9.33)$$

where a is the lattice spacing. The unit cell is a cube of side a. Mathematicians refer to a space-filling collection of polyhedra as a honeycomb. The points in the simple cubic lattice are situated at the vertices of the cubes in a cubic honeycomb, as shown in Figure 9.20 (left). Each atom in a simple cubic lattice has six nearest neighbours, so one says that the coordination number is six.

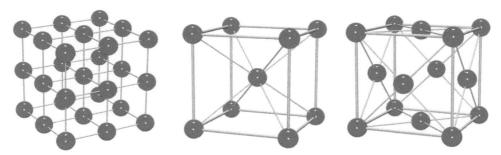

Fig. 9.20 Left: Simple cubic packing. Centre: Body centred cubic (bcc) packing. Right: Face centred cubic (fcc) packing.

When considering waves propagating through a lattice, it is convenient to define the *reciprocal* or *dual lattice*. In three dimensions, each vector in the reciprocal lattice is orthogonal to a plane in the original crystal lattice and vice versa. The reciprocal lattice is generated by three primitive vectors $\mathbf{A}_j$ $(j = 1, 2, 3)$, and the general vector in the reciprocal lattice is

$$\mathbf{K} = k_1\mathbf{A}_1 + k_2\mathbf{A}_2 + k_3\mathbf{A}_3, \tag{9.34}$$

where (k_1, k_2, k_3) are integers. The defining feature of the reciprocal lattice is that its primitive vectors satisfy $\mathbf{a}_i \cdot \mathbf{A}_j = 2\pi\delta_{ij}$. (The factor of 2π is conventional in solid state physics.) $\mathbf{A}_1$ is therefore orthogonal to $\mathbf{a}_2$ and $\mathbf{a}_3$, and hence orthogonal to the lattice plane generated by $\mathbf{a}_2$ and $\mathbf{a}_3$. Similarly for $\mathbf{A}_2$ and $\mathbf{A}_3$. In general,

$$\mathbf{A}_1 = 2\pi\frac{\mathbf{a}_2 \times \mathbf{a}_3}{\mathbf{a}_1 \cdot \mathbf{a}_2 \times \mathbf{a}_3}, \quad \mathbf{A}_2 = 2\pi\frac{\mathbf{a}_3 \times \mathbf{a}_1}{\mathbf{a}_2 \cdot \mathbf{a}_3 \times \mathbf{a}_1}, \quad \mathbf{A}_3 = 2\pi\frac{\mathbf{a}_1 \times \mathbf{a}_2}{\mathbf{a}_3 \cdot \mathbf{a}_1 \times \mathbf{a}_2}, \tag{9.35}$$

which automatically satisfy the defining equations.[7]

The dot product between a vector $\mathbf{K}$ in the reciprocal lattice and a vector $\mathbf{R}$ in the original lattice is 2π times an integer, as

$$\mathbf{K} \cdot \mathbf{R} = (k_1\mathbf{A}_1 + k_2\mathbf{A}_2 + k_3\mathbf{A}_3) \cdot (n_1\mathbf{a}_1 + n_2\mathbf{a}_2 + n_3\mathbf{a}_3) = 2\pi\sum_{i=1}^{3} k_i n_i. \tag{9.36}$$

This has the important consequence that

$$\exp(i\mathbf{K} \cdot \mathbf{R}) = 1. \tag{9.37}$$

---

[7] Note that the denominators are all the same.

It is easy to see that the simple cubic lattice with generators (9.33) has the reciprocal lattice with generators

$$\mathbf{A}_1 = \frac{2\pi}{a}(1,0,0)\,, \quad \mathbf{A}_2 = \frac{2\pi}{a}(0,1,0)\,, \quad \mathbf{A}_3 = \frac{2\pi}{a}(0,0,1)\,, \tag{9.38}$$

a scaled copy of the original lattice. Figure 9.20 shows the cells of two other lattices that occur frequently in solid state physics: the body centred cubic (bcc) lattice, with primitive vectors

$$\mathbf{a}_1 = \frac{a}{2}(-1,1,1)\,, \quad \mathbf{a}_2 = \frac{a}{2}(1,-1,1)\,, \quad \mathbf{a}_3 = \frac{a}{2}(1,1,-1)\,, \tag{9.39}$$

and the face centred cubic (fcc) lattice, with primitive vectors

$$\mathbf{a}_1 = \frac{a}{2}(0,1,1)\,, \quad \mathbf{a}_2 = \frac{a}{2}(1,0,1)\,, \quad \mathbf{a}_3 = \frac{a}{2}(1,1,0)\,. \tag{9.40}$$

Atoms in the bcc lattice have coordination number eight, but they also have six next-nearest neighbours that are not much further away. Atoms in the fcc lattice have coordination number twelve. The nearest neighbours are at the vertices of a cuboctahedron, as shown in Figure 9.21. For the fcc lattice with generators (9.40), the vectors

$$\mathbf{A}_1 = \frac{2\pi}{a}(-1,1,1)\,, \quad \mathbf{A}_2 = \frac{2\pi}{a}(1,-1,1)\,, \quad \mathbf{A}_3 = \frac{2\pi}{a}(1,1,-1) \tag{9.41}$$

generate its reciprocal lattice, as is readily verified by calculating the dot products $\mathbf{a}_i \cdot \mathbf{A}_j$. These reciprocal lattice generators are clearly rescaled bcc lattice generators (9.39), so the fcc and bcc lattices are reciprocal.

We now look at why the reciprocal lattice is a useful concept when describing electron wavefunctions within a periodic, crystalline array of atoms.

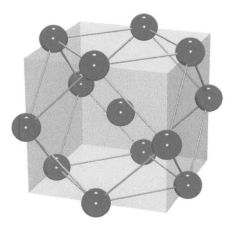

Fig. 9.21 The nearest neighbours of each atom in an fcc lattice are situated at the vertices of a cuboctahedron.

9.5.2 Bloch's theorem

It is reasonable to assume that the Hamiltonian H for an electron in a crystalline solid has the same periodicity as the crystal itself. Therefore, for any lattice vector $\mathbf{R}$, $H(\mathbf{r}) = H(\mathbf{r} + \mathbf{R})$. Let us define, for the vector $\mathbf{R}$, a translation operator $T_{\mathbf{R}}$ that shifts the argument of any function f by $\mathbf{R}$, so

$$T_{\mathbf{R}} f(\mathbf{r}) = f(\mathbf{r} + \mathbf{R}). \tag{9.42}$$

These translation operators satisfy $T_{\mathbf{R}} T_{\mathbf{R}'} = T_{\mathbf{R} + \mathbf{R}'}$, and they all commute, as

$$T_{\mathbf{R}} T_{\mathbf{R}'} f(\mathbf{r}) = f(\mathbf{r} + \mathbf{R} + \mathbf{R}') = T_{\mathbf{R}'} T_{\mathbf{R}} f(\mathbf{r}). \tag{9.43}$$

If $\Psi(\mathbf{r})$ is a wavefunction, then

$$T_{\mathbf{R}} H(\mathbf{r}) \Psi(\mathbf{r}) = H(\mathbf{r} + \mathbf{R}) \Psi(\mathbf{r} + \mathbf{R}) = H(\mathbf{r}) \Psi(\mathbf{r} + \mathbf{R}) = H(\mathbf{r}) T_{\mathbf{R}} \Psi(\mathbf{r}). \tag{9.44}$$

Therefore, the translation operators commute with the Hamiltonian as well as with each other. This means that we can choose the eigenstates of the Hamiltonian to be simultaneous eigenstates of all the lattice translation operators, as discussed in section 7.7.

A simultaneous eigenstate of the translation operators $T_{\mathbf{R}}$ satisfies $T_{\mathbf{R}} \Psi = c(\mathbf{R}) \Psi$, where $c(\mathbf{R})$ is the eigenvalue, and $T_{\mathbf{R}} T_{\mathbf{R}'} = T_{\mathbf{R} + \mathbf{R}'}$ implies that

$$T_{\mathbf{R}} T_{\mathbf{R}'} \Psi(\mathbf{r}) = T_{\mathbf{R}} c(\mathbf{R}') \Psi(\mathbf{r}) = c(\mathbf{R}) c(\mathbf{R}') \Psi(\mathbf{r}) = c(\mathbf{R} + \mathbf{R}') \Psi(\mathbf{r}), \tag{9.45}$$

so $c(\mathbf{R} + \mathbf{R}') = c(\mathbf{R}) c(\mathbf{R}')$. Applying this result repeatedly, we see that for a general lattice vector $\mathbf{R} = n_1 \mathbf{a}_1 + n_2 \mathbf{a}_2 + n_3 \mathbf{a}_3$, with $\mathbf{a}_i$ the primitive lattice vectors,

$$c(\mathbf{R}) = c(\mathbf{a}_1)^{n_1} c(\mathbf{a}_2)^{n_2} c(\mathbf{a}_3)^{n_3}. \tag{9.46}$$

It follows that we can represent $c(\mathbf{R})$ as an exponential function, and as electron wavefunctions do not grow exponentially in any direction, $c(\mathbf{R})$ must have unit modulus and therefore have the form $c(\mathbf{R}) = \exp(i\mathbf{k} \cdot \mathbf{R})$. $\mathbf{k}$ can conveniently be expressed as $\mathbf{k} = k_1 \mathbf{A}_1 + k_2 \mathbf{A}_2 + k_3 \mathbf{A}_3$, where $\mathbf{A}_i$ are the reciprocal lattice generators and (k_1, k_2, k_3) are arbitrary here. In summary, we can find eigenstates Ψ of H such that for every lattice vector $\mathbf{R}$,

$$T_{\mathbf{R}} \Psi(\mathbf{r}) = \Psi(\mathbf{r} + \mathbf{R}) = c(\mathbf{R}) \Psi(\mathbf{r}) = \exp(i\mathbf{k} \cdot \mathbf{R}) \Psi(\mathbf{r}). \tag{9.47}$$

This last equation is solved if Ψ is a *Bloch state*,

$$\Psi(\mathbf{r}) = \exp(i\mathbf{k} \cdot \mathbf{r}) u(\mathbf{r}), \tag{9.48}$$

a product of a function with the periodicity of the lattice, $u(\mathbf{r}) = u(\mathbf{r} + \mathbf{R})$, and a plane wave $\exp(i\mathbf{k} \cdot \mathbf{r})$. Equation (9.47) is satisfied as

$$\begin{aligned} T_{\mathbf{R}} \Psi(\mathbf{r}) &= \exp(i\mathbf{k} \cdot (\mathbf{r} + \mathbf{R})) u(\mathbf{r} + \mathbf{R}) \\ &= \exp(i\mathbf{k} \cdot (\mathbf{r} + \mathbf{R})) u(\mathbf{r}) = \exp(i\mathbf{k} \cdot \mathbf{R}) \Psi(\mathbf{r}). \end{aligned} \tag{9.49}$$

The periodic function $u(\mathbf{r})$ generally varies with $\mathbf{k}$. One says that in the Bloch state, the periodic function $u(\mathbf{r})$ is modulated by the plane wave $\exp(i\mathbf{k} \cdot \mathbf{r})$. A 1-dimensional example is shown in Figure 9.22. In three dimensions, a plane wave fills space and, as the name implies, remains in phase across a plane, oscillating in the direction perpendicular to the plane.

Bloch's theorem says that states of type (9.48), i.e. periodic functions times plane waves, form a complete set of stationary state wavefunctions when the Hamiltonian is periodic. Despite its apparent simplicity, it is a profound result that is the foundation stone for the study of quantum mechanics in crystalline solids.

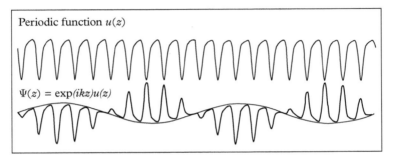

Fig. 9.22 An example of a Bloch state (real part).

$\mathbf{k}$ is known as the *crystal momentum*. Its magnitude is $|\mathbf{k}| = \frac{2\pi}{\lambda}$, where λ is the wavelength of the modulating wave. Although $\mathbf{k}$ may take any value, these different values of $\mathbf{k}$ will not necessarily lead to different Bloch states. The function $\exp(i\mathbf{K} \cdot \mathbf{r})$ has the periodicity of the lattice, for any $\mathbf{K}$ in the reciprocal lattice, because of equation (9.37). Therefore shifting $\mathbf{k}$ by $\mathbf{K}$ doesn't change the Bloch state, as the additional factor $\exp(i\mathbf{K} \cdot \mathbf{r})$ can be absorbed into $u(\mathbf{r})$. To find all the unique solutions of the Schrödinger equation we need only consider values of $\mathbf{k}$ within one unit cell of the reciprocal lattice. It is convenient to take this cell to be the *Wigner–Seitz cell*, which is the region around a chosen reciprocal lattice point that is closer to this point than to any other reciprocal lattice point.[8] The Wigner–Seitz cell is the region bounded by planes that bisect straight lines between the chosen point and its neighbours in the reciprocal lattice. By construction, these cells fill $\mathbf{k}$-space to form a honeycomb. The Wigner–Seitz cell centred at the origin is also known to physicists as the *first Brillouin zone*. In one dimension the first Brillouin zone is obtained by restricting k to values in the range $-\frac{\pi}{a} \le k \le \frac{\pi}{a}$. The first Brillouin zone of an fcc lattice is a rhombic dodecahedron, while for a bcc lattice it is a truncated octahedron, as shown in Figure 9.23.

The opposite faces of a Brillouin zone differ by a reciprocal lattice vector, so the corresponding $\mathbf{k}$-values on these faces give the same Bloch states, with the same energy. In several of the following figures, the equivalence of states on opposite faces of a Brillouin zone can be seen.

9.5.3 Bloch states in a finite crystal

We saw in section (8.2.1) that a free particle in a finite box of volume V has a density of states $\frac{V}{(2\pi)^3}$ in $\mathbf{k}$-space. Similarly, if we have a finite crystal, a cuboidal piece of crystalline solid of volume V, the density of Bloch states in the first Brillouin zone is $\frac{V}{(2\pi)^3}$. This is because periodic boundary conditions applied to a Bloch state $\exp(i\mathbf{k} \cdot \mathbf{r})u(\mathbf{r})$ constrain $\mathbf{k}$

---

[8] This is known as a Voronoi cell to mathematicians.

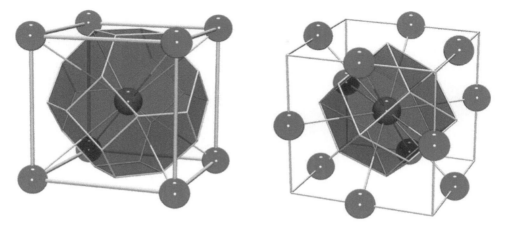

Fig. 9.23 Left: The Wigner–Seitz cell of the bcc lattice is a truncated octahedron. Right: The Wigner–Seitz cell of the fcc lattice is a rhombic dodecahedron.

in the same way as they do for a free particle state $\exp(i\mathbf{k} \cdot \mathbf{r})$, since the function $u(\mathbf{r})$ is automatically periodic. The first Brillouin zone is a region in $\mathbf{k}$-space whose volume is the volume of the reciprocal lattice unit cell, i.e. $\frac{(2\pi)^3}{a^3}$ for a simple cubic lattice. The total number of states in the Brillouin zone is therefore $\frac{V}{a^3}$, the volume of the crystal measured in units of the crystal's unit cell. This is simply the number of unit cells in the finite crystal. As we are assuming that there is one atom for each unit cell, this is just the total number of atoms, N.

Each overlapping atomic orbital gives rise to a band of states, with $\mathbf{k}$-values filling the Brillouin zone. As each atomic orbital may be occupied by two electrons with opposite spin projections, the number of states in the band is $2N$, twice the number of atoms in the crystal. The energy of these states is a function of $\mathbf{k}$.

Not all of these states need be occupied. For instance, sodium has just one valence electron. The sodium s band, which is produced by the overlap of the sodium s orbitals, includes sufficient states for two electrons per atom, but the available electrons only fill the lower half of the band. There is a *Fermi surface*, a surface in $\mathbf{k}$-space within the 3-dimensional, first Brillouin zone. The energy of the highest filled level is the Fermi energy ε_{F}, and the Fermi surface consists of all $\mathbf{k}$-values giving this energy. The value of ε_{F} is such that the interior of the Fermi surface contains half the total volume of the Brillouin zone.

9.5.4 The tight-binding model

When applied to solids, molecular orbital theory in the LCAO approximation is known as the *tight-binding model*. To illustrate the model, consider a 1-dimensional solid formed of an infinite line of equally spaced atoms distance a apart, along the z-axis. The reciprocal lattice is an infinite line of equally spaced points separated by a distance of $\frac{2\pi}{a}$. Electrons within the solid are subject to a periodic potential produced by the array of atoms. According to Bloch's theorem, the stationary state solutions of the Schrödinger equation with a 1-dimensional periodic potential take the form

$$\Psi_k(z) = \exp(ikz)u(z)\,, \quad -\frac{\pi}{a} \leq k \leq \frac{\pi}{a}\,, \tag{9.50}$$

where $u(z)$ is a periodic function with the same periodicity as the potential, so $u(z + a) = u(z)$. $\Psi_k(z)$ is a 1-dimensional Bloch state. It is a periodic function modulated by the phase factor $\exp(ikz)$.

We will use molecular orbital theory to find the electron wavefunctions in the solid. This is very similar to the Hückel analysis of the polyene chains. Consider an infinite sequence of overlapping s orbitals with $u(z)$ equal to an infinite sum of s orbitals, one for each atom. The energy of the states is

$$
\begin{aligned}
E_\mathrm{s} &= \alpha_\mathrm{s} + \beta_\mathrm{s}(\exp(ika) + \exp(-ika)) \\
&= \alpha_\mathrm{s} - 2|\beta_\mathrm{s}| \cos(ka),
\end{aligned}
\tag{9.51}
$$

where, as previously, α_s is the atomic energy level of the s orbitals and $\beta_\mathrm{s} < 0$ is the matrix element due to the overlap of s orbitals on adjacent atoms. The lowest energy state consists of a totally in-phase sum of the s orbitals giving a Bloch state with $k = \frac{2\pi}{\lambda} = 0$. The phase of the s orbital is the same on each atom. The other solutions have the same $u(z)$ in the tight-binding approximation, but k takes different values all the way up to the highest energy state where $|k| = \frac{\pi}{a}$. When $|k| = \frac{\pi}{a}$, the wavelength is equal to twice the interatomic spacing, $2a$. This is the totally out-of-phase solution where the phase is opposite on each adjacent atom. The results are shown in the bottom half of Figure 9.24 (left).

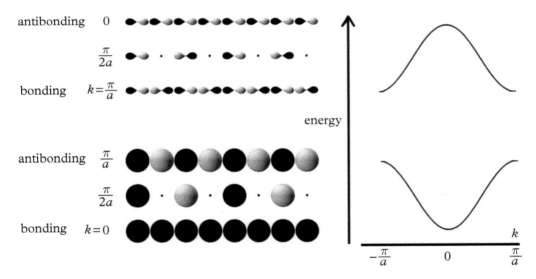

Fig. 9.24 Orbitals in a 1-dimensional solid. Bottom: Overlapping s orbitals. The $k = 0$ state has maximum constructive overlap and therefore lowest energy. The $|k| = \frac{\pi}{a}$ states have maximum destructive overlap and therefore highest energy. Top: Overlapping p_z orbitals. The p_z orbitals have maximum destructive overlap and highest energy when $k = 0$ and maximum constructive overlap and lowest energy when $|k| = \frac{\pi}{a}$.

From equation (9.51) we see that when $k = 0$, $E_\mathrm{s} = \alpha_\mathrm{s} - 2|\beta_\mathrm{s}|$ and when $|k| = \frac{\pi}{a}$, $E_\mathrm{s} = \alpha_\mathrm{s} + 2|\beta_\mathrm{s}|$, so the width of the band is $4|\beta_\mathrm{s}|$. This is largely due to the difference in the kinetic energy between the bottom and the top of the band. The form of the Bloch wavefunctions is shown in the bottom half of Figure 9.24. When $k = 0$, the waves have infinite wavelength. These correspond to bonding states in the molecular language. When $|k| = \frac{\pi}{2a}$, there is no overlap on adjacent atoms and the energy equals the energy of the atomic orbitals. Higher values of $|k|$ produce states with shorter wavelength than the atomic orbitals. These correspond to antibonding states. The highest energy state is reached when $|k| = \frac{\pi}{a}$.

In a solid such as an alkali metal, with one free electron per atom, there are enough free electrons to fill exactly half the states in the s band, so all the bonding orbitals are occupied. In the 1-dimensional tight-binding model, these states fill half the first Brillouin zone, running from $k = 0$ to the Fermi level which is at $|k| = \frac{\pi}{2a}$. The Fermi surface consists of just the two points $k = \pm\frac{\pi}{2a}$. There are free states immediately above the Fermi level, so such a material is metallic. All the occupied states are lower in energy than the atomic s orbitals, so this provides cohesive energy which holds the solid together. The reduction in energy due to the delocalization of electrons throughout a solid is known as metallic bonding.

The p orbitals can be analysed in a similar way. Consider a 1-dimensional chain of overlapping p_z orbitals aligned to form molecular σ orbitals in the solid. The periodic function $u(z)$ is now an infinite sum of p_z orbitals. The lowest energy state occurs when the orientation of the p_z orbitals alternates from atom to atom along the chain, so that lobes of the same sign are adjacent. The maximum constructive overlap of the orbitals is when $|k| = \frac{\pi}{a}$, as shown in the top half of Figure 9.24 (left), and corresponds to a wavelength $\lambda = 2a$, twice the interatomic spacing. There is a band of solutions between this solution and the infinite wavelength solution ($k = 0$) in which all the orbitals are oriented in the same direction, such that the positive lobe on each atom overlaps the negative lobe on the next atom. This is the highest energy maximally antibonding orbital. When $|k| = \frac{\pi}{a}$, $E_p = \alpha_p - 2|\beta_p|$ and when $k = 0$, $E_p = \alpha_p + 2|\beta_p|$, where α_p is the atomic energy level of the p_z orbitals and β_p is the overlap matrix element of orbitals on adjacent atoms.

There is a gap between the energy of the s band state with $k = \frac{\pi}{a}$ and the energy of the p band state with $k = \frac{\pi}{a}$. Both these states have the same wavelength $2a$, but the electron wavefunction in the s band state is concentrated around the site of the atoms, whereas the wavefunction of the p band state is concentrated between the atoms and has nodes at each atom. As the electrons are delocalized in both bands, the remaining atoms on the lattice sites are positively charged ions. The energy at the top of the s band is less than the energy at the bottom of the p band because the electrons are closer to the ions in the s band.

9.5.5 The nearly free electron model

The overlapping molecular orbital model offers a good picture of the s band in real materials. There is a very different complementary picture. The electrons within the solid can be viewed as a free or nearly free electron gas. For electron states whose wavelength is much longer than the interatomic spacing, the potential due to the lattice of ions can be averaged over, to give a constant background potential V_0. This contributes equally to the energy of all the states to give

$$E_k = \frac{\hbar^2 k^2}{2m_s^*} + V_0 \,, \tag{9.52}$$

and the difference between the energy of the states is simply due to the difference in their kinetic energy. m_s^* is the effective mass of the s band electrons in the solid and is not necessarily equal to the electron rest mass. By equating the difference in kinetic energy between the top ($k = \frac{\pi}{a}$) and bottom ($k = 0$) of the band to the width of the band we deduce that

$$\frac{\hbar^2 \pi^2}{2m_s^* a^2} = 4|\beta_s|\,. \tag{9.53}$$

This is intuitively reasonable because the electrons cannot move through the background potential of the solid as freely as they would through empty space; their mobility is constrained by the degree of overlap between atomic orbitals. Equation (9.53) implies that if the overlap $|\beta_s|$ between orbitals is large and the band is wide, then the effective mass is small so the mobility of the electrons is large. Conversely, in a solid such as an ionic solid where the overlap is small, $|\beta_s|$ will be small producing a large effective mass and low mobility, so the electrons will be strongly bound to the ions and unable to move easily through the solid. The effective mass will vary from band to band and may be different in different directions.

9.5.6 Ionic solids

Ionic compounds, such as sodium chloride, condense into regular close-packed arrays of atoms arranged such that the positive ions are surrounded by negative ions and vice versa. They are stabilized by long-range Coulomb interactions. The electrons are tightly bound to individual ions, so there is little overlap between the orbitals on neighbouring ions. This produces narrow bands and a large gap between the valence and conduction bands, so ionic solids tend to be insulators. Ionic crystals are typically hard but brittle, with considerable mechanical strength. For instance, aluminium oxide Al_2O_3 forms the mineral corundum and zirconium oxide ZrO_2 forms the mineral zirconia. Ionic solids are usually transparent unless they include impurities. When transition metal ions are present in ionic solids, gemstones may be formed. The colour of rubies is due to a small number of Cr^{3+} ions replacing aluminium ions in corundum, whereas iron ions produce blue sapphires. These impurity ions produce states within the band gap. Absorption of visible light of the right energy can excite electrons from the valence band into these states, giving the crystal its colour. The precise positions of the energy levels are determined by complicated interactions between the impurity ion and its crystal environment. However, as the visible spectrum runs from red light with photons of energy 1.8 eV to violet light with photons of energy around 3.0 eV, new energy levels anywhere near the middle of the gap will result in light absorption.

Ions differ markedly in size (metal ions are small, non-metal ions are large). This plays an important role in governing the coordination number of the ions and how the ions are packed in the solid. Sodium chloride is the prototype for one common arrangement shown in Figure 9.25. Here, the coordination number of both the sodium and chloride ions is six. If the sodium and chloride ions were replaced by identical atoms then the packing would be simple cubic.

9.5.7 Example of caesium chloride

Caesium chloride offers an example of the structure of electron bands in a real ionic solid. In the caesium chloride packing shown in Figure 9.25 the coordination number is eight. The caesium ions form a simple cubic packing and the chloride ions form a second, diagonally shifted, simple cubic packing. If the caesium and chloride ions were replaced by identical atoms then the packing would be body centred cubic.

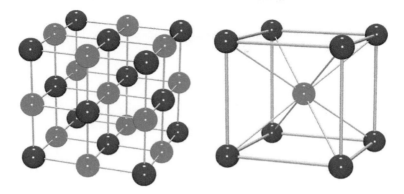

Fig. 9.25 Left: Sodium chloride structure. Right: Caesium chloride structure.

The caesium ion has a closed shell structure, with the electrons in core orbitals. Similarly the electrons in the first and second chlorine shells are in low energy core orbitals that do not overlap those on other ions. The important states responsible for bonding are the 3s and 3p chlorine orbitals. These states are much lower in energy than the lowest available caesium states, so we can neglect any overlap between the chlorine and caesium orbitals and confine our attention to the overlap between the outer orbitals on nearest-neighbour chlorine atoms.[9] These 3s and 3p states will broaden into bands. As discussed in section 9.5.4, the resulting states are Bloch states labelled by the crystal momentum $\mathbf{k}$. As the sublattice formed of chloride ions is simple cubic, its reciprocal lattice is also simple cubic, as shown in Figure 9.26. The crystal momentum is defined throughout the cubic Wigner–Seitz cell of the reciprocal lattice extending from the centre at $\mathbf{k} = (0, 0, 0)$ to the corners at $\mathbf{k} = (\pm\frac{\pi}{a}, \pm\frac{\pi}{a}, \pm\frac{\pi}{a})$. There is one orbital in the 3s band and one orbital in each of the three 3p bands for each value of $\mathbf{k}$ in the Wigner–Seitz cell, and their energies are functions of $\mathbf{k}$.

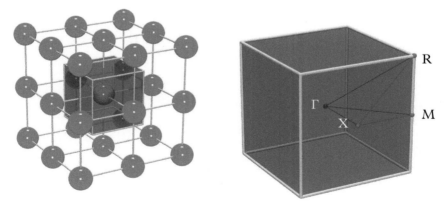

Fig. 9.26 Left: The Wigner–Seitz cell of the simple cubic lattice is a cube. Right: The points of high symmetry of the Wigner–Seitz cell are shown with their conventional labels.

---

[9] Significant overlap between orbitals only occurs for orbitals that are close in energy, as discussed when considering polar bonds in section 9.2.2.

Each chloride ion has six nearest-neighbour chloride ions. Neglecting all but the nearest neighbours and solving the secular equations as previously, the overlap of chlorine s orbitals produces states with energies

$$E(\mathbf{k}) = \alpha_s - |\beta_s| \sum_{j=1}^{6} \exp(i\mathbf{k} \cdot \mathbf{a}_j)$$

$$= \alpha_s - 2|\beta_s|(\cos(k_1 a) + \cos(k_2 a) + \cos(k_3 a)). \tag{9.54}$$

The width of this band is $12|\beta_s|$. (Generally, if we assume that the orbital overlap is only significant for nearest neighbours, then the band width is $2Q|\beta_s|$ where Q is the coordination number. This ties in with the earlier examples, such as the diatomic molecule ($Q = 1$) in equations (9.11) and (9.12), and benzene ($Q = 2$) in equation (9.26).)

Plotting $E(\mathbf{k})$ is tricky as it requires three dimensions for the coordinates of $\mathbf{k}$ and one for $E(\mathbf{k})$. Fortunately, the cubic Wigner–Seitz cell has reflection symmetries which makes much of this information redundant. Solid state physicists define the points of high symmetry, as shown in Figure 9.26 (right), and plot $E(\mathbf{k})$ along straight lines connecting these points. By convention, the point $\mathbf{k} = (0,0,0)$ is labelled Γ. Figure 9.27 plots $E(\mathbf{k})$ for the s band and the three p bands of caesium chloride along two of these lines. ΓX connects $\mathbf{k} = (0,0,0)$ to $\mathbf{k} = (\frac{\pi}{a},0,0)$. From equation (9.54) we see that the energy of the s states increases from $\alpha_s - 6|\beta_s|$ at Γ to $\alpha_s - 2|\beta_s|$ at X.

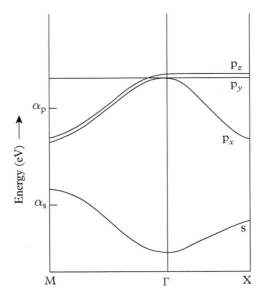

Fig. 9.27 Caesium chloride bands.

Turning to the p orbitals, the p states reach their maximum energy at Γ, as in the 1-dimensional example in section 9.5.4. The p_x states form σ bonds in the k_1-direction. Along ΓX, the energy of the p_x states decreases from $\alpha_p + 6|\beta_p|$ to $\alpha_p + 4|\beta_p|$. The energies of the p_y and p_z states do not change along ΓX, as k_2 and k_3 are both zero along this line.

To the left of Figure 9.27, ΓM connects $\mathbf{k} = (0,0,0)$ to $\mathbf{k} = (\frac{\pi}{a}, \frac{\pi}{a}, 0)$. Along this line $k_1 = k_2$ and $k_3 = 0$, so the p_x and p_y states are degenerate and the energy of the p_z states is constant. The energy of the s band states is $E(\mathbf{k}) = \alpha_s - 2|\beta_s|(2\cos(k_1 a) + 1)$, which increases from $\alpha_s - 6|\beta_s|$ to $\alpha_s + 2|\beta_s|$.

Each chlorine atom contributes seven valence electrons and each caesium atom contributes one. These eight electrons per unit cell fill all the states in the chlorine 3s band and the three chlorine 3p bands. The Fermi level is at the top of the 3p bands. There is a large gap to the next available states which are in the band produced by the overlap of the caesium 6s orbitals, so caesium chloride is an insulator.

9.5.8 Metals

Most elements are metals. Their atoms are close-packed. Many, such as those in Group 1, form bcc structures, while others form fcc structures. Another important packing adopted by many metals is known as hexagonal close packing (hcp). Metals fall into several categories depending on the electronic structure of their atoms. The elements of Groups 1 and 2 plus aluminium are known as simple metals or sp metals. Their outer electrons are in s orbitals. (In the case of aluminium there is also one valence electron in a p orbital.) The elements in Groups 3 to 10, whose valence electrons are in s and d orbitals, are known as transition metals. The outer s and d orbitals are close in energy, so the order in which these orbitals are occupied varies slightly from element to element and depends on the interactions between the outer electrons. Groups 11 and 12 are known as the post-transition metals as their d orbitals are full and their valence electrons are in s orbitals giving them outer electronic configurations $d^{10}s^1$ and $d^{10}s^2$ respectively. The Group 11 elements copper, silver and gold are known as the coinage or noble metals.

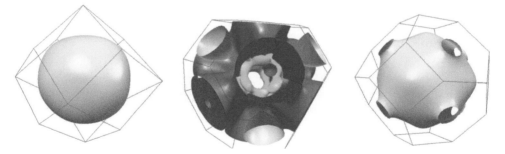

Fig. 9.28 Fermi surfaces. Left to right: Sodium Na (bcc), Manganese Mn (fcc), Copper Cu (fcc).

The physical properties of some metals, notably those in Groups 1 and 2, are well described by treating their electrons as a free electron gas. The free electron model works well because there is a big overlap of the s orbitals on adjacent atoms which gives rise to broad bands. The interaction between the electrons and the ions can be treated as a constant background potential, as described in section 9.5.5. In the ground state, the 1-electron states are filled up to the Fermi level ε_F. There is a filled sphere in momentum space of radius k_F, where

$$\varepsilon_F = \frac{\hbar^2 k_F^2}{2m}. \tag{9.55}$$

The Fermi surface of an alkali metal, such as sodium, is indeed almost spherical, as shown in Figure 9.28 (left).

We can calculate the size of the Fermi sphere. The volume of the Brillouin zone, as discussed in section 9.5.3, is $\left(\frac{2\pi}{a}\right)^3$, and the volume of the Fermi sphere is half this, $\frac{4\pi^3}{a^3}$. Using the formula for the volume of a sphere, we have

$$\frac{4}{3}\pi k_F^3 = \frac{4\pi^3}{a^3} \tag{9.56}$$

so $k_F = (3\pi^2)^{\frac{1}{3}}\frac{1}{a}$. We also have the relation between the volume of metal and the number of its electrons, $V = Na^3$. Therefore

$$k_F = \left(\frac{3\pi^2 N}{V}\right)^{\frac{1}{3}} \tag{9.57}$$

and the Fermi energy is, as in equation (8.74),

$$\varepsilon_F = \frac{\hbar^2}{2m}\left(\frac{3\pi^2 N}{V}\right)^{\frac{2}{3}}, \tag{9.58}$$

depending solely on the electron density in the metal. By identifying ε_F with the kinetic energy of a classical particle $\frac{1}{2}mv_F^2$, we can estimate the electron velocity in a metal, v_F.

For example, in sodium the interionic distance is 0.366 nm. Sodium has a bcc structure, and taking this into account gives $\varepsilon_F \simeq 3.2$ eV, which corresponds to a Fermi velocity in the region of $v_F \simeq 10^6$ m s^{-1}. (The particles of a gas would reach these energies at a temperature of $T_F = \frac{\varepsilon_F}{k_B} \simeq 37,000$ K.)

9.5.9 Example of copper

The free electron model does not work as well for the transition metals. The d orbitals have much smaller radii than the s orbitals so their overlap with the orbitals on neighbouring atoms is much smaller. Consequently the d orbitals give rise to narrow bands that cut through the s band. This complicates the electronic structure of the transition metals, because their Fermi surfaces are in the region of the five partly filled d bands. Computer programs are now available that routinely calculate accurate band structures. The Fermi surface of the transition metal manganese produced by such a program is shown in Figure 9.28 (centre).

In post-transition metals, such as copper, which has outer electronic structure 3d^{10}4s^1, the d bands are fully occupied and there is one valence electron per atom in the 4s band. The atoms in copper, and the other noble metals, adopt the fcc packing. The first Brillouin zone of the fcc lattice (the Wigner–Seitz cell of the reciprocal bcc lattice) is a truncated octahedron, as shown in Figure 9.29, with points of high symmetry labelled. Copper was the first material whose Fermi surface was determined experimentally, by Brian Pippard in 1955. This Fermi surface is shown in Figure 9.28 (right). Figure 9.30 shows a cross section of the Wigner–Seitz cell. The dark line in this diagram shows the section through the Fermi surface and for comparison the grey circle represents a section through the Fermi sphere of a free electron gas. Perhaps surprisingly, the Fermi sphere reaches just over 90% of the way

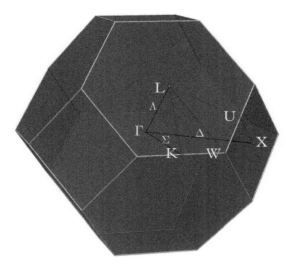

Fig. 9.29 The first Brillouin zone of the fcc lattice with the conventional labels for the points of high symmetry.

to the hexagonal faces of the Wigner–Seitz cell.[10] However, the free electron model fails for states whose wavelength is comparable to the spacing between planes of copper ions. These states have a stronger interaction with the ions, which reduces their energy relative to the corresponding free electron states. Their crystal momenta **k** are located in the necks that protrude from the Fermi sphere and intersect the hexagonal faces of the Wigner–Seitz cell. The energy of the states within the necks is lower than the Fermi energy so their crystal momenta lie within the Fermi surface.

The computer-generated band structure for copper is shown on the left of Figure 9.31. In the free electron model the s band has a parabolic shape, as described by equation (9.52). This can be seen in the figure, where the s band has an almost parabolic shape especially around Γ. Cutting through the s band are five rather complicated energy bands due to the overlapping d orbitals, which mix with the s orbitals. Above the d bands, the energy of the s band returns to its parabolic shape, as can be seen in the top half of the figure. The Fermi surface lies within the half-full s band. To the right of the figure the density of states is shown. As there is just one electron per atom in the broad s band and ten electrons per atom in the narrow d bands, the electron density peaks are in the region of the d bands.

We can see the relationship between the band structure and the Fermi surface by comparing Figure 9.31 to the section through the Fermi surface shown in Figure 9.30. The leftmost part of Figure 9.31 shows the band structure between points Γ and X. The dashed line at the Fermi level ε_F crosses the s band just before X. This corresponds to the point in Figure 9.30 where the path from Γ to X crosses the Fermi surface. Similarly, the s band

---

[10] The volume of the cubic cell of a bcc reciprocal lattice with vertices $(\pm 1, \pm 1, \pm 1)$ is 8. There are two lattice points per cell, so the volume of the Wigner–Seitz cell, which in this case is a truncated octahedron, is 4. With a single valence electron, the volume of the Fermi sphere is half the volume of the Wigner–Seitz cell, so the Fermi sphere has volume 2 and radius $\sqrt[3]{\frac{3}{2\pi}} \simeq 0.782$. The centre of a hexagonal face of the Wigner–Seitz cell is at $(\pm\frac{1}{2}, \pm\frac{1}{2}, \pm\frac{1}{2})$, so these points are a distance $\frac{\sqrt{3}}{2} \simeq 0.866$ from the origin.

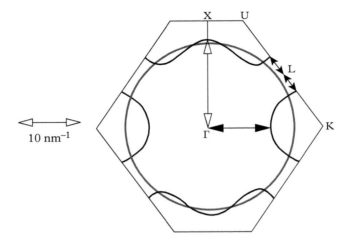

Fig. 9.30 Section through the Fermi surface of copper.

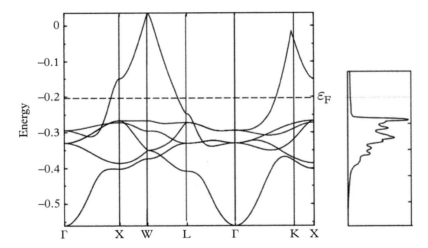

Fig. 9.31 Band structure in copper.

crosses the Fermi level between Γ and K in Figure 9.31, matching the point where the path from Γ to K crosses the Fermi surface in Figure 9.30. Note that there is no such crossing in either diagram along the line from Γ to L, as this path ends within a neck inside the Fermi surface.

9.6 Ferromagnetism

Thus far when considering the electronic structure of solids, we have adopted the independent electron approximation in which each molecular orbital provides two degenerate states that are available for two electrons of opposite spin. We have assumed that electrons fill the available states in order of increasing energy, and that electron spins are always paired to give no overall spin. This is true for core electrons, but not necessarily for the valence electrons. As we saw in section 9.2.1 when discussing the oxygen molecule, if the lowest available orbitals are degenerate there is a tendency for electrons to enter different orbitals and to *align* their spins. This may also happen in a solid if the exchange energy dominates the energy spacing of several partially filled valence bands, a situation that prevails in the transition metals with their partially filled d orbitals. The radii of the d orbitals in transition metal atoms are much smaller than the radii of the outer s orbitals. Their overlap with the orbitals on neighbouring atoms is therefore very small and the five d bands are both narrow and close in energy, as we have seen.

We can estimate when electron spins will align in a transition metal. The energy cost of two electrons occupying the same orbital with opposite spin is the exchange energy K, as discussed in section 9.2.1. If the total width of the d bands is W and we assume each of them has approximately the same width $\frac{1}{5}W$, then the cost of promoting one of the two electrons in the highest occupied d band to the next d band is $\frac{1}{5}W - K$. The condition for spin alignment to occur is therefore $K > \frac{1}{5}W$, and if this holds, the number of electrons with aligned spin will be maximized,[11] in accordance with Hund's first rule. For instance, the outer electronic structure of the iron ion Fe^{3+} is $3d^5$ with all five electrons in different d orbitals with aligned spins.

Aligned electron spins give an atom a magnetic moment, as we mentioned when discussing the oxygen molecule. In a solid, the effect may be dramatically magnified, as the magnetic moments associated with individual atoms may align throughout the crystal to produce a macroscopic magnetic field. This is known as *ferromagnetism*. It occurs in iron, cobalt and nickel below a transition temperature called the *Curie temperature*. The Curie temperature of iron is 1043 K. Alternatively, neighbouring atomic magnetic moments may align antiparallel to produce no overall magnetic moment. This long-range magnetic ordering is known as *antiferromagnetism*, and it occurs in many magnetic materials below a transition temperature known as the *Néel temperature*. For example, nickel oxide NiO is an antiferromagnet below 520 K. We will take a look at the theoretical analysis of spin alignment in section 10.12.

The first series of transition metals runs from scandium with outer electronic configuration $4s^2 3d^1$ to nickel with outer electronic configuration $4s^2 3d^8$. Each step through the series adds a proton to the atomic nucleus. The increasing nuclear charge reduces the radii of the d orbitals, so the overlap between d orbitals on adjacent atoms decreases and therefore W decreases. Simultaneously, the decrease in orbital radii forces electrons sharing the same d orbital closer together so the exchange energy K increases. These trends imply an increasing tendency for electron spins to align. We do indeed find that the elements in the second half of the first transition series show ferromagnetic and antiferromagnetic properties. The radii of the 4d and 5d orbitals in the next two transition series are larger, which leads to broader bands as well as smaller exchange energies, so magnetism is not a feature of these elements.

In the elements of the lanthanide series the 4f orbitals are partially filled. These orbitals have much smaller radii than the filled 5s, 5p and 6s orbitals, so they do not overlap

---

[11] Due to the antisymmetrization of the wavefunction, the total exchange energy is K times the number of pairs of electrons with aligned spins. For n such electrons this is $\frac{n(n-1)}{2}K$. The number of energy steps of size $\frac{1}{5}W$ through which the electrons are promoted in order to each occupy a different d orbital is $\sum_1^{n-1} m = \frac{n(n-1)}{2}$.

significantly and behave much like atomic orbitals. The condition for electron spin alignment is satisfied. The lanthanides have the largest magnetic moments of all atoms and these elements and their compounds have a range of interesting and useful magnetic properties. The strongest commercially available permanent magnets are known as NIB magnets. They are formed of an alloy of neodymium, iron and boron with chemical formula $Nd_2Fe_{14}B$. Neodymium is a lanthanide and has the outer electronic structure $5s^2 5p^6 6s^2 4f^4$.

9.7 Further Reading

For comprehensive discussions of atomic orbitals and molecular binding, see

C.S. McCaw, *Orbitals: with Applications in Atomic Spectra*, London: Imperial College Press, 2015.

A. Alavi, *Part II Chemistry, A4: Theoretical Techniques and Ideas*, Cambridge University, 2009. Available at: www-alavi.ch.cam.ac.uk/files/A4-notes.pdf

R.M. Nix, *An Introduction to Molecular Orbital Theory, Lecture Notes*, Queen Mary University of London, 2013. Available at: www.chem.qmul.ac.uk/software/download/mo/

For solid state physics, see

P.A. Cox, *The Electronic Structure and Chemistry of Solids*, Oxford: OUP, 1987.

H.P. Myers, *Introductory Solid State Physics (2nd ed.)*, London: Taylor and Francis, 1997.

N.W. Ashcroft and N.D. Mermin, *Solid State Physics*, Fort Worth TX: Harcourt, 1976.

S.L. Altmann, *Band Theory of Solids: An Introduction from the Point of View of Symmetry*, Oxford: OUP, 1991.

10
Thermodynamics

10.1 Introduction

Thermodynamics unifies our understanding of the physical world. It is relevant throughout the whole range of substances we call solids, liquids and gases, and is therefore important in physics, chemistry and even the biological sciences. Its immense significance may be judged by the fact that when applied to electromagnetic radiation it triggered the start of the quantum revolution. Thermodynamics describes the behaviour of elementary particles and it also plays a vital role in explaining the evolution of the entire cosmos, as this is largely the thermal history of matter and radiation in an expanding spacetime. Rather surprisingly, thermodynamics has important applications to black holes.

As demonstrated in earlier chapters, the dynamics of one or two bodies is readily accounted for mathematically. But as the number of bodies increases, the calculations rapidly become intractable. When we reach very large numbers, however, we arrive at another level of simplification and it is again possible to perform accurate statistical, or probabilistic calculations. This is the realm of thermodynamics and it is why physicists often refer to this branch of their subject as *statistical mechanics*. Whenever a physical system is composed of numerous subsystems which have some independence, but also some degree of contact, statistical mechanics is relevant. The classic example is a gas consisting of numerous molecules that collide from time to time.

Thermodynamics begins with our perceptions of hot and cold. We know from common experience that these properties coincide with readily observable changes in materials. For instance, solid iron changes colour from red to orange in a furnace. Gases expand when they get hotter, and so do liquids such as water and mercury. These changes may be used to calibrate a temperature scale, allowing us to construct devices to measure temperature. Traditionally, temperature has been measured using mercury thermometers. Today, thermometers often rely on steady changes in the electrical properties of materials that are correlated with temperature.

When two bodies are brought into contact their temperatures change, and given sufficient time they attain a steady condition called *equilibrium*, where their temperatures are equal. We always observe that heat flows from the body with the higher temperature to the body with the lower temperature, so the temperature of the hotter body decreases, and that of the cooler body increases. There appears to be no path towards equilibrium that violates this law. This raises the question of what exactly heat is.

The Physical World. Nicholas Manton and Nicholas Mee, Oxford University Press (2017).
© Nicholas Manton and Nicholas Mee. DOI 10.1093/oso/9780198795933.001.0001

10.1.1 What is heat?

The natural philosophers of the 18th century believed heat to be a fluid that they named caloric. The idea was that a hot body contains more caloric than a cold body, and that this fluid is transferred from a hot body to a cold body when they come into contact, thereby increasing the temperature of the cold body. Caloric was imagined to be a conserved fluid that could move around but never be created or destroyed. Clearly, this rather simplistic theory does not explain some rather obvious facts about heat. (Even so, vestiges of the theory have survived into the modern world. We talk of the flow of heat and some of us count our calories.) In 1798, Benjamin Thompson, also known as Count Rumford, published the results of his experiments on the generation of heat. Rumford measured changes in temperature resulting from mechanical work. Among the subjects of his enquiries was the boring of cannon, which required the application of a great deal of work by teams of horses. Rumford showed that the heat generated in this process could be used to bring large volumes of water to the boil and that as long as the boring continued, a seemingly inexhaustible supply of heat was available. The conclusion was clear. Heat could not be a fluid contained within the metal of the cannon, but was instead being transferred to the metal by the friction of the boring equipment, powered by the work of the horses. This indicated a close relationship between the heat, which produces the change in temperature, and the mechanical work, which is the energy transferred to the material through the operation of various forces. It was natural, therefore, to assume that heat is a form of energy.

The recognition that work can be converted into heat was critical in extending the concept of energy beyond the mechanical notions of kinetic and potential energy. Heat is often released in chemical reactions, such as the burning of fuel, so energy changes play an important role in chemistry. The chemical energy stored in various substances can be determined by measuring the heat produced in chemical reactions. This brings us full circle, as the energy generated by the horses observed by Rumford was derived from their food, whose chemical energy can be measured directly by burning it. The physical and chemical notions of energy are united here in the realm of biological processes and although the total energy must be conserved, heat by itself is not.

The need to understand the relationship between heat and mechanical energy became more urgent with the development of steam power during the Industrial Revolution. In engines, heat is partially converted to work. Thermodynamics grew out of a theoretical analysis of the interplay of heat and work in the boilers and pistons of steam engines. Indeed, the word thermodynamics consists of two halves—it is about the relationship between temperature and dynamics, that is, between heat, forces and motion.

10.1.2 The ideal gas law

In 1662, Robert Boyle published an account of his experiments on gases. He demonstrated that at a fixed temperature T, the pressure P and volume V of a fixed mass of gas are inversely proportional, a result since known as Boyle's law. Seventy years later the mathematician Daniel Bernouilli showed that Boyle's law could be explained by assuming that a gas is composed of small particles in constant irregular motion that are continually colliding with each other and with the walls of their container. This insight was not followed up for more than a century when it would become the cornerstone of the kinetic theory of gases.

In the kinetic theory, an *ideal gas* is modelled as a vast collection of point particles that undergo perfectly elastic collisions, but are non-interacting when separated. This model is a useful representation of many real gases at temperatures well above their boiling point. It works well when the gas has a much lower density than its liquid or solid phases and fails at high pressures when intermolecular forces and the finite size of the molecules become important. Helium offers the best physical realization of an ideal gas, but other gases formed of non-polar molecules, such as the nitrogen and oxygen found in dry air, also provide good examples. By contrast, water vapour does not, as water molecules are polar and interact strongly with each other. An ideal gas obeys the following simple relationship between its physical properties, which is a generalization of Boyle's law:

$$PV = A(T + T_0).$$ (10.1)

This is known as the *ideal gas law* and it is obeyed to varying degrees of accuracy by real gases. An important feature of the law is the inclusion of the added constant temperature T_0, whose value is about $273°C$ (centigrade). This constant is determined empirically and is found to be the same for all gases that offer good approximations to an ideal gas. The constant A is proportional to the mass of gas, and we shall be more precise about its value later.

In reality, the ideal gas law is not perfectly obeyed when the temperature is measured with a mercury thermometer. This is because a temperature scale based on the thermal expansion of a liquid such as mercury is not completely satisfactory. The same is true of a scale based on physical properties of a solid, such as electrical resistivity. Liquids and solids are materials formed of vast numbers of atoms that interact with each other in complicated ways. Any linear relationship between temperature and their physical properties may hold to a good approximation for a range of temperatures but it is unrealistic to expect such a relationship to be exact. A better way to define temperature is to use the properties of an ideal gas, in which case we can take equation (10.1) as the definition of temperature.

The shift by T_0 is very significant. It enables us to define $T_{abs} = T + T_0$ as the absolute temperature and to establish an absolute scale of temperature running from zero upwards. This is the Kelvin scale, named after William Thomson who was raised to the peerage as Lord Kelvin. T_{abs} is measured in degrees kelvin, which are denoted as K. On the Kelvin scale, the freezing point of water has the value of approximately 273.15 K, and the boiling point of water is approximately 373.15 K. Conversely, $-273.15°C$ is the absolute zero of temperature. At this temperature an ideal gas would exert no pressure. Alternatively, a finite pressure would compress an ideal gas to zero volume. From now on we will change notation and always use T to mean absolute temperature. So the ideal gas law becomes

$$PV = AT.$$ (10.2)

Of course, an ideal gas is a mathematical construction. We can only determine temperature in the real world by using the measured properties of real physical materials, but we know that at sufficiently low pressures, all gases behave like ideal gases. So we can define temperature as follows:

$$T = \frac{1}{A} \lim_{P \to 0} PV.$$ (10.3)

Experimental physicists have developed elaborate refrigeration techniques to reduce the temperature of samples of real substances. This becomes ever more difficult as the temperature approaches absolute zero and it proves impossible to reach a negative absolute temperature or even absolute zero itself.

10.1.3 The microscopic origin of heat

If a hot body contains more energy than a cold body, it is natural to ask what form this energy takes. The first person to offer a convincing answer was Henry Cavendish who realized that heating a solid increases the vibrations of the particles that form the solid. Cavendish died in 1810 and along with much of his work, his analysis of this question was left unpublished. By the middle of the 19th century, however, similar views were held by a number of researchers. Maxwell used the kinetic theory to analyse the problem statistically in the case of a gas. The interactions between particles are ignored in an ideal gas, so the energy of each particle consists solely of its kinetic energy $\frac{1}{2}mv^2$, where m is the mass of the particle and v is its speed.

Maxwell calculated that at room temperature the average speed of a nitrogen molecule in the air is around 500 m s^{-1}, which is more than twice the top speed of a passenger aircraft. Clearly the air as a whole is not moving at such a speed. Rather, it is the molecules within that are moving rapidly but in random directions, and they are continually bouncing off each other. One of the successes of the kinetic theory was the prediction of the speed of sound in a gas in terms of the speeds of its component particles.[1]

10.1.4 Iced tea

Heat always flows from hot objects to cold objects and never from cold objects to hot objects, but why is this the case? We know from our lifelong experience of interacting with the world around us that if we put ice in our tea, the ice will always cool the tea down and never warm it up. Nevertheless, the ice cubes that we take from our freezer contain plenty of heat; their temperature is much higher than a winter's day in Siberia, for instance. Perhaps some of this heat could be liberated from the ice and added to the tea, thereby warming the tea and cooling the ice further. Although this would conserve energy, we know that it never happens.

The impossibility of this process was explained in the middle years of the 19th century by inventing a new quantity, the *entropy S*. A change in entropy equals an amount of heat ΔE divided by the temperature T:

$$\Delta S = \frac{\Delta E}{T} . \tag{10.4}$$

For a given amount of heat, the entropy change is smaller at a high temperature than at a low temperature. If T_h represents a high temperature and T_l a low temperature, then

$$\frac{\Delta E}{T_h} < \frac{\Delta E}{T_l} , \quad \text{so} \quad \Delta S_h < \Delta S_l . \tag{10.5}$$

This means that when an amount of heat ΔE is transferred from a hot object to a cold object, the total entropy increases. If heat were transferred from a cold object to a hot object, then entropy would decrease. We are very familiar with the first of these processes, but the second process never happens. This observation is captured by the statement that

---

[1] At atmospheric pressure, the speed of sound in air is about 340 m s^{-1}, which is easily measured from the delay between seeing a gun fired and the time of arrival of the bang.

in any allowed process the total entropy of the universe must increase, a fact that was first recognized in the analysis of steam engines by the 19th-century engineer Sadi Carnot.

The purpose of this concept is highlighted by a comparison with the conservation of energy. If state A has the same energy as state B, then state A might evolve into state B or vice versa. The law of energy conservation allows either. By contrast, the law that entropy can increase but never decrease determines the direction in which processes occur. It limits the ways in which the universe can evolve and is somehow related to the passage of time and our perception of the direction of time. However, the origin of entropy was a mystery when it was introduced. The rule that entropy always increases might provide a succinct way to determine which processes are forbidden, but it does not tell us why they are not observed. They have simply been outlawed. We need to identify the microscopic variables that are responsible for the basic thermodynamic quantities, entropy and temperature.

Entropy is now recognized as the most fundamental quantity in thermodynamics. Maxwell showed that we can understand the overall effects of heat statistically without any detailed knowledge of the motion of each atom or molecule. The gross effect of the random motion is captured by the notion of temperature, but heat is a disordered kind of energy and entropy is the most direct measure of this disorder. Ludwig Boltzmann explained entropy as arising from the random velocities of molecules in terms of classical dynamics. However, the accuracy of classical dynamics is limited. As we have seen, when one tries to understand molecules and atoms in detail, or the properties of liquids and solids, quantum mechanics is required. The simplest definition of entropy is actually in terms of quantum states. This is what we will consider next.

10.2   Entropy and Temperature

A simple thermodynamic system is one that depends on two independent macroscopic variables, one of a thermal kind, and one of a dynamical kind. One such pair of variables is temperature T and volume V. Other pairs include energy E and pressure P. Theoretically, the most convenient pair is E and V. In this case, E is known as the *internal energy* and is often denoted by U. It is associated with the invisible thermal motion of the molecules, but excludes the kinetic energy of the system as a whole, if the system happens to be in motion.

The thermodynamic variables divide naturally into two types known as *intensive* and *extensive*. Intensive variables are physical properties of a system that do not depend on the size or amount of material in the system. Such variables include density, temperature and pressure. Extensive variables, on the other hand, are additive properties that are proportional to the amount of material in the system. Such variables include energy, mass and volume.

We assume that the system consists of a very large number of molecules, N, in a container of volume V. The molecules behave as a quantum system controlled by a Hamiltonian that may include terms describing interactions between the molecules. The wavefunction of each molecule must vanish at the walls of the container, so the probability of finding a particle outside the container is zero. This boundary condition implies that the energy levels depend on V but they do not depend in an important way on the shape of the container.

For the rest of this section, we will assume the volume V is fixed and just consider the effect of varying the system's energy. In section 10.3, we will consider the effect of varying both energy and volume.

The spacing between the energy levels of a single molecule decreases as the molecule's energy increases. This is also true for a system of N molecules, where the total density of energy levels increases staggeringly fast as the total energy E increases.

We define $\Omega(E)$ to be the number of independent quantum states of the N-particle system between energies $E - \Delta E$ and E. ΔE is taken to be a very small energy, less than the energy resolution that could be achieved experimentally, but its precise value does not matter very much. Loosely, one can say that Ω is the number of states of the system *at* energy E. More precisely,

$$\Omega(E) = g(E)\,\Delta E\,, \tag{10.6}$$

where $g(E)$ is the *density of states*,[2] the number of states per unit energy interval near E.

In macroscopic physical systems, the growth of Ω with energy is so rapid that one can also define $\Omega(E)$ to be the number of quantum states with energy less than E. Although this sounds rather different, it is practically the same because the number of states with energy significantly less than E is negligible by comparison with the number very close to E. There is an analogy here with the surface area of a sphere of unit radius in high dimensions. The area is essentially the same as the volume of the sphere, because almost all the volume is very close to the surface.

It is a fundamental postulate of statistical mechanics that there is nothing that significantly distinguishes the states of the system apart from their energy. So if the system has energy E, and is in equilibrium, then the system is equally likely to be in any one of the available states at this energy. Therefore, all states of energy E occur with the same probability $\frac{1}{\Omega(E)}$. This is the cornerstone of statistical mechanics. The best justification for the statement is that it expresses our total ignorance about any details of the microscopic state of the N particles other than their total energy. Consequently, the only important quantity, other than E, is the number of states $\Omega(E)$. This enables us to capture the notion of disorder in a thermal system.

Both Ω and g are truly vast; they depend exponentially on N. A tabletop system usually has more than $N = 10^{23}$ molecules. Any number $\exp cN$, where c is some modest coefficient and N is of this order, almost defies comprehension. This exponential dependence can be understood as follows. Any thermodynamic system can be regarded as made of subsystems in rather mild contact. For example, a sample of gas can be regarded as two smaller samples in contact along a common surface. For two systems with mild interactions the quantum states of the combined system can be expressed as $\Psi = \Psi_1 \Psi_2$, where Ψ_1 runs through all states available to system 1 and Ψ_2 runs through all states available to system 2. Therefore the number of states for the combined system is $\Omega = \Omega_1 \Omega_2$. This product rule is compatible with Ω depending exponentially on the number of particles, because if the systems are of the same type and $\Omega_1 = \exp cN_1$ and $\Omega_2 = \exp cN_2$, where N_1 and N_2 are the numbers of particles in the two systems, then $\Omega = \exp cN$ where $N = N_1 + N_2$ is the total number of particles.

We define the entropy S of the system to be

$$S(E) = \log \Omega(E)\,, \tag{10.7}$$

and although still incredibly large, this is much more manageable as it scales roughly with N. Entropy is an extensive, additive function. Taking the logarithm of equation (10.6), we

---

[2] Note that this is for the N-particle system, not just one particle as in section 8.2.1.

find that for practical purposes, $S(E) = \log \Omega(E) = \log g(E)$ because $\log \Delta E$ is fixed and not proportional to N. In terms of the entropy, the fundamental postulate is that each quantum state at energy E occurs with probability $\frac{1}{\Omega(E)} = e^{-S(E)}$. Entropy is often described as a measure of disorder. Disorder, when used in this sense, means the logarithm of the number of physical states, $\Omega(E)$, available to the system, and it represents our inability to distinguish between them due to our imprecise knowledge of the exact microscopic details of the system.[3]

How is entropy, which simply counts states, related to temperature? Let us consider two systems with fixed volumes and not necessarily of the same type. Suppose first of all that the systems are not in contact and that they have energies E_1 and E_2, and entropies $S_1(E_1)$ and $S_2(E_2)$. The total number of available states for the combined system is the product of the numbers of states available to the systems separately, $\Omega = \Omega_1 \Omega_2 = e^{S_1(E_1)} e^{S_2(E_2)}$. Taking logarithms, the total entropy is

$$S = S_1(E_1) + S_2(E_2). \tag{10.8}$$

Next, let the systems come into thermal contact by bringing them sufficiently close for energy to flow between them, but not so close that particles are exchanged. The total energy $E = E_1 + E_2$ is conserved, so we can write $E_2 = E - E_1$, and

$$S = S_1(E_1) + S_2(E - E_1). \tag{10.9}$$

The contact allows E_1 to vary.

Now, the function $\Omega = \Omega_1 \Omega_2$ has an incredibly high and narrow maximum at a particular value of E_1. This is because a small fractional change in Ω's large exponent has a colossal effect on Ω itself. Consequently, for a given total energy E, the vast majority of the states in the combined system have an energy split with this particular E_1. After the energy has flowed between the systems, the overwhelming likelihood is that the combined system will occupy one of these most probable states. After the initial flow of energy, E_1 will not change further, and we say that the combined system has then reached equilibrium.

S also has a maximum for this particular E_1, although the maximum is less dramatic than for Ω and not so sharp. To characterize this maximum thermodynamically, we differentiate with respect to E_1. The maximum occurs when

$$\frac{dS}{dE_1} = \frac{dS_1}{dE_1} + \frac{dS_2}{dE_2}\frac{dE_2}{dE_1} = 0. \tag{10.10}$$

As $E = E_1 + E_2$ is constant, $\frac{dE_2}{dE_1} = -1$, so the condition for equilibrium is

$$\frac{dS_1}{dE_1} = \frac{dS_2}{dE_2}. \tag{10.11}$$

This motivates the *thermodynamic definition of temperature*. For any system whose entropy as a function of energy (at a given volume V) is $S(E)$, one defines the system's temperature T as

$$\frac{1}{T} = \frac{dS}{dE}. \tag{10.12}$$

This definition of temperature is even more fundamental than the definition based on the ideal gas law. The thermodynamic temperature T of a system with energy E is simply the

---

[3] A pure quantum state of energy E will not by itself evolve into a disordered state, but if the system is weakly coupled to an external environment, referred to as a *heat bath*, then in time we can expect it to become disordered.

inverse of the slope of the graph of $S(E)$, where S is determined by the density of quantum states of the system. T is a function of E. Real physical systems all have the property that S increases as E increases, so T is positive. Disorder increases with energy. Moreover, the slope $\frac{dS}{dE}$ decreases with E, so temperature increases with energy. Equation (10.11) reflects the fact that two systems in equilibrium have the same temperature. This is the most basic property of temperature and is known as the *zeroth law of thermodynamics*.

The thermodynamic temperature T, defined by equation (10.12), may seem rather formal but it satisfies the key properties that were found using the phenomenological concepts of temperature and entropy. From now on it is the notion of temperature that we will use.

It is interesting to consider the direction of energy flow towards equilibrium. Differentiating equation (10.8) with respect to time, and remembering that $\frac{dE_2}{dt} = -\frac{dE_1}{dt}$, gives

$$\frac{dS}{dt} = \frac{dS_1}{dE_1}\frac{dE_1}{dt} + \frac{dS_2}{dE_2}\frac{dE_2}{dt} = \left(\frac{1}{T_1} - \frac{1}{T_2} \right) \frac{dE_1}{dt}\,. \qquad (10.13)$$

However, we have argued that in any process where two separate systems are brought together to form a single system, energy flows to move S towards its maximum, so entropy increases with time:

$$\frac{dS}{dt} \geq 0\,. \qquad (10.14)$$

The sign of $\frac{dE_1}{dt}$ is therefore positive if $T_2 > T_1$, and in this case E_1 increases. If $T_1 > T_2$, $\frac{dE_1}{dt}$ is negative, and E_1 decreases. In both cases the energy flows from the system of higher temperature to the system of lower temperature. This statement, as well as the more general equation (10.14), is known as the *second law of thermodynamics*.

It is not obvious that the thermodynamic and ideal gas scales of temperature agree. One temperature could be a rather complicated function of the other. Fortunately, this is not the case, because using our definitions of entropy and temperature, the pressure and entropy of an ideal gas can be derived from quantum mechanical first principles, as we will show in section 10.6. We will find that the ideal gas temperature is actually the same as the thermodynamic temperature.

S is dimensionless, so T has dimensions of energy. This is natural, but for historical reasons practical temperatures are measured in degrees kelvin (K) and a conversion factor k_B, known as Boltzmann's constant, is needed to relate a degree kelvin to energy in joules. We will set $k_B = 1$ and can then use energy units as units of temperature. In atomic physics the electron volt eV is a useful energy unit, and 1 eV corresponds to about 10^4 K.

It is convenient to work with infinitesimals and write equation (10.12) as

$$dE = T\,dS\,. \qquad (10.15)$$

When the energy of a system changes in a way that has nothing to do with a change in volume, this energy is regarded as heat. So $T\,dS$ is an infinitesimal quantity of heat. When heat is added to the system, the energy increases and as a result the temperature rises. The amount of heat needed to produce a unit change of temperature (when the volume V remains fixed) is called the *heat capacity* at constant volume and is denoted by C_V. More precisely, the heat capacity is $C_V = \frac{dE}{dT}$ and from equation (10.15) we obtain the relation

$$C_V = T\frac{dS}{dT}\,. \qquad (10.16)$$

The heat added to a system can be accurately determined, by using a resistive electric coil to supply the heat, for instance. The change in temperature is also readily determined, so C_V is easy to measure. Using equation (10.16) and integrating, one can then calculate the entropy S, using the formula

$$S(T) - S(\widetilde{T}) = \int_{\widetilde{T}}^{T} \frac{C_V}{T}\, dT\,, \tag{10.17}$$

where $\widetilde{T}$ is some fixed temperature. The disorder associated with the entropy cannot be measured directly, so this formula offers a practical method of calculating the entropy of the system. However, it leaves the additive constant $S(\widetilde{T})$ undetermined. It follows from quantum mechanics, however, that any quantum system has a *unique* ground state of minimum energy. At zero temperature this is the only state available, so $\Omega = 1$, and the entropy is $S(0) = \log \Omega = 0$, which fixes the constant. This proposition that the entropy of any system vanishes at absolute zero temperature is known as the *third law of thermodynamics*. Choosing $\widetilde{T} = 0$ in the integral (10.17) gives

$$S(T) = \int_{0}^{T} \frac{C_V}{T}\, dT\,, \tag{10.18}$$

which permits an experimental determination of absolute entropy from heat capacity measurements,[4] although in practice it may be difficult to measure or estimate C_V right down to absolute zero.

10.3   The First Law of Thermodynamics

The first law is a precise statement about energy and entropy in thermodynamics when both the energy E and volume V of the system are variable. The number of states at energy E is $\Omega(E, V)$, and the entropy is

$$S(E, V) = \log \Omega(E, V)\,. \tag{10.19}$$

The temperature is now defined through the partial derivative,

$$\left. \frac{\partial S}{\partial E} \right|_{V} = \frac{1}{T}\,, \tag{10.20}$$

where we have introduced the standard notation in which the subscript on the vertical line indicates the variables that are fixed when taking the partial derivative. (In this case the volume V is fixed.) As S always increases with E, it is possible to regard E as a function of the two independent variables S and V.

When the volume of a gas (or liquid) increases, work is done on the surrounding medium, and as a result some of the internal energy of the gas is lost. Suppose the pressure outside is marginally less than the pressure inside, so the expansion is steady and not explosive; then the system remains in equilibrium as it expands. Suppose further that no heat enters or leaves the system while it expands. An expansion like this is referred to as *adiabatic*, and a

---

[4] There are additional contributions from the *latent heat* at phase transitions below temperature T.

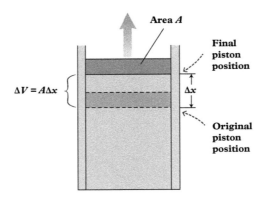

Fig. 10.1 Piston.

good example of such an expansion is where a pocket of air is swept up and over a mountain by the wind. The pressure decreases a little with height, so the pocket of air expands, loses energy and cools, but in the time this takes essentially no heat is transferred to or from the rest of the air.

The pressure P is the force per unit area that a gas exerts on a containing wall or on a surrounding region of gas (even when there is no dividing wall). The infinitesimal work done when the gas expands is $P\,dV$, where dV is the infinitesimal increase in volume. This is because dV is the area times the distance moved, as shown in Figure 10.1, so $P\,dV$ is pressure times area times distance. In turn, this is force times distance, which is the work done by the force.

In a process that does not involve heat, the work done must come from the internal energy E of the gas, so

$$dE = -P\,dV\,. \tag{10.21}$$

A general thermodynamic process involves both heat and work, so the general formula for a system's energy change is

$$dE = T\,dS - P\,dV\,. \tag{10.22}$$

This is the *first law of thermodynamics*.

The first law combines the statements that T is the derivative of E with respect to S, if V is fixed, and that $-P$ is the derivative of E with respect to V, if S is fixed. More precisely, T and P are the partial derivatives

$$T = \left.\frac{\partial E}{\partial S}\right|_V\,, \qquad P = -\left.\frac{\partial E}{\partial V}\right|_S\,, \tag{10.23}$$

and both are simultaneously functions of S and V. Two systems in contact via a thin, movable barrier are in equilibrium only if they have the same temperature and pressure.

There is an interesting mathematical consequence of equations (10.23). Mixed second partial derivatives are always symmetric, so in particular $\frac{\partial^2 E}{\partial V \partial S} = \frac{\partial^2 E}{\partial S \partial V}$. Therefore

$$\left.\frac{\partial T}{\partial V}\right|_S = -\left.\frac{\partial P}{\partial S}\right|_V\,. \tag{10.24}$$

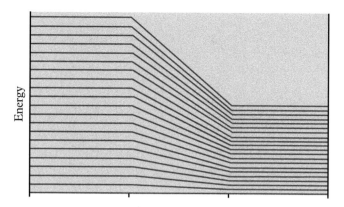

Fig. 10.2 Adiabatic change in energy levels.

This is an example of a *Maxwell relation*.

We defined an adiabatic change as one involving no heat. Such a change is one for which $dS = 0$, so S is constant. From our definition, however, S is the logarithm of the number of states available to the system. Are these ideas consistent? The answer is yes. When a gas expands, the volume increases and the quantum energy levels change. Typically, as the volume increases, the energy levels are lowered, but the number of quantum energy levels does not change, so S remains constant. The flow of energy levels is sketched in Figure 10.2. As the gas expands, the many available levels in the region of the initial energy all move together to become the same number of available levels close to the final lower energy. If the system occupies one of these levels, then, as the energy level changes, the system will continuously follow this level down in energy, because a quantum system does not jump during a smooth, slow change, if no heat is added to excite the system. So the entropy does not change.

10.3.1 New variables

We have stressed that a simple thermodynamic system is controlled by two independent variables, one thermal and one dynamical. In the first law of thermodynamics (10.22), the independent variables are taken to be S and V, but there are various alternative choices.

One convenient choice is to take the easily measured temperature T and volume V to be the independent variables. T is related to a derivative of S, as in equation (10.20), so this change of variables requires a bit of care. The standard procedure is to define a new energy function F, the *Helmholtz free energy* (or free energy, for short), named after Hermann von Helmholtz. The free energy is

$$F = E - TS \tag{10.25}$$

and is a function of S and V, but it can be converted to a function of T and V by expressing E and S in terms of T and V. An infinitesimal change in F is

$$
\begin{aligned}
dF &= dE - T\,dS - S\,dT \\
&= T\,dS - P\,dV - T\,dS - S\,dT \\
&= -S\,dT - P\,dV ,
\end{aligned}
\tag{10.26}
$$

where we have used the Leibniz rule $d(TS) = T\,dS + S\,dT$, then replaced dE using the first law of thermodynamics (10.22). The final expression is analogous to the first law, but for F rather than E. Notice that only the infinitesimal changes of the new independent variables, dT and dV, occur in this expression. So, regarding the free energy as a function $F(T, V)$, we see that

$$S = -\left.\frac{\partial F}{\partial T}\right|_V, \qquad P = -\left.\frac{\partial F}{\partial V}\right|_T, \tag{10.27}$$

and a new Maxwell relation $\left.\frac{\partial S}{\partial V}\right|_T = \left.\frac{\partial P}{\partial T}\right|_V$ follows from this.

$P\,dV$ is the work done by the system as it expands. It is equal to the decrease in free energy when the system expands with $dT = 0$, so free energy is the energy available for work in situations where temperature remains constant. Of course, to keep the system at constant temperature as it expands some heat must be supplied. The free energy F is a useful concept because it can be calculated in a way that is quite different from subtracting TS from E and changing variables, as we will see.

Another useful choice of independent variables is the pair S and P. Again it is convenient to define a modified energy function. This is the *enthalpy H*, defined as

$$H = E + PV. \tag{10.28}$$

An infinitesimal change of H is

$$\begin{aligned} dH &= T\,dS - P\,dV + P\,dV + V\,dP \\ &= T\,dS + V\,dP, \end{aligned} \tag{10.29}$$

so when H is regarded as a function of S and P, one has

$$T = \left.\frac{\partial H}{\partial S}\right|_P, \qquad V = \left.\frac{\partial H}{\partial P}\right|_S. \tag{10.30}$$

The enthalpy H is the most useful thermodynamic energy function in situations where pressure is constant. For example, it applies to gases and liquids under the influence of normal atmospheric pressure. H is therefore of particular interest to chemists as most chemical reactions, both in the laboratory and in industry, are carried out in systems at atmospheric pressure.

$T\,dS$ is the heat, as always, so the heat required to raise the temperature by one unit in a system at constant pressure is

$$C_P = T\left.\frac{\partial S}{\partial T}\right|_P = \left.\frac{\partial H}{\partial T}\right|_P. \tag{10.31}$$

C_P is the heat capacity at constant pressure.

The final modified energy function is the *Gibbs free energy G*, named after Josiah Willard Gibbs. (This is sometimes called the Gibbs potential and denoted by Φ). It combines the transformations that lead from E to F and from E to H. The Gibbs free energy is defined as

$$G = E - TS + PV. \tag{10.32}$$

Using the first law and the Leibniz rule as before one finds

$$dG = -S\,dT + V\,dP. \tag{10.33}$$

G is therefore naturally regarded as a function of the intensive variables T and P, neither of which depend on the size of the system, and

$$S = -\left.\frac{\partial G}{\partial T}\right|_P, \qquad V = \left.\frac{\partial G}{\partial P}\right|_T. \tag{10.34}$$

G itself (like E, F and H) is extensive, meaning that it is proportional to the number of molecules in the system N. Therefore G can be expressed as

$$G(T, P, N) = N\widetilde{G}(T, P) \tag{10.35}$$

where $\widetilde{G}(T, P)$ is the Gibbs free energy per molecule.[5]

We shall see later that $\widetilde{G}$ is the *chemical potential* of the thermodynamic system. This is a useful notion whenever particle numbers are variable, for example in chemical reactions, and also in the analysis of phase transitions.

10.4   Subsystems—The Gibbs Distribution

So far, we have considered a macroscopic system with given energy E. The entropy S is the logarithm of the number of available quantum states at that energy, and it is assumed that in equilibrium each state is occupied with equal probability. When the system is in thermal contact with another system, energy may be transferred between the two systems, but once the combined system has reached a new equilibrium and the temperatures are equal, further fluctuations of the energy are negligible.

Let us now consider a subsystem of the initial system, and assume the whole system has reached equilibrium. If the subsystem is macroscopic—for example, a 1% part of the initial system—then the subsystem will have a definite energy, with negligible fluctuations. In fact, the energy and entropy will be 1% of the total, and the temperature of the subsystem will be the same as that of the whole system. The situation is more interesting if the subsystem is microscopic.

An example of a microscopic subsystem is a single atom or molecule in a gas, or perhaps an impurity particle in the gas. Because the subsystem is in contact with the rest of the system, it will exchange energy in collisions with the rest of the system, so its energy will vary. At any particular instant, our incomplete knowledge of the state of the subsystem means that the best description we can expect is the probability of occupation of the subsystem's states. When the system is in equilibrium, this probability distribution will not change, even though the energy of the subsystem itself can fluctuate.

---

[5] The analogue for E would be a more complicated expression, $E = N\widetilde{E}(S/N, V/N)$.

Let us label the independent quantum states of the subsystem with an integer n, and denote the energy of the nth state by E_n. Some of these states may be degenerate in energy. Usually n will run from 0 to ∞, although for some spin systems the range is finite.

We shall now determine the probability that the subsystem is in the nth state. This probability depends on E_n and on the temperature T of the rest of the system. Suppose the complete system is isolated and has energy $E^{(0)}$. By conservation of energy, $E^{(0)}$ is constant. If the subsystem has energy E_n then the rest of the system has energy $E^{(0)} - E_n$. E_n is a microscopic fraction of $E^{(0)}$, but since the states of macroscopic systems are so numerous, and their number is so sensitive to energy, we must take this energy shift into account. The rest of the system, which may be considered as a heat bath, has entropy $S(E^{(0)} - E_n)$ when the subsystem has energy E_n. As E_n is small, we can use the Taylor expansion to first order,

$$S(E^{(0)} - E_n) \simeq S(E^{(0)}) - E_n \frac{dS}{dE} \tag{10.36}$$

$$= S(E^{(0)}) - \frac{E_n}{T} , \tag{10.37}$$

where the temperature T of the heat bath is evaluated at energy $E^{(0)}$. The number of states available to the heat bath is therefore

$$e^{S(E^{(0)} - E_n)} = e^{S(E^{(0)})} e^{-\frac{E_n}{T}} . \tag{10.38}$$

As the subsystem is in a definite state, the nth state, this is also the number of states of the whole system.

Now consider the total system again, without fixing the state of the subsystem. All of the states available to the total system are equally likely, and $e^{S(E^{(0)})}$ is a constant, so the relative probability that the subsystem is in the nth state is simply proportional to the number of states available when the subsystem is in this state. This probability is

$$P(E_n) \propto e^{-\frac{E_n}{T}} , \tag{10.39}$$

which is known as the *Gibbs distribution*, though also known to Ludwig Boltzmann in a more restricted context.

The constant of proportionality must be fixed so that the total probability is 1. We therefore define a quantity Z, known as the Gibbs sum or *partition function*, to be

$$Z = \sum_{n=0}^{\infty} e^{-\frac{E_n}{T}} . \tag{10.40}$$

The correctly normalized probability for a subsystem to be in a particular state with energy E_n, when it is part of a system at temperature T, is then

$$P(E_n) = \frac{1}{Z} e^{-\frac{E_n}{T}} . \tag{10.41}$$

The partition function is a remarkably useful quantity to consider.

It is important to note that although the subsystem must be in thermal contact with the heat bath, this contact should be sufficiently weak for the energy levels of the subsystem E_n

to be unaffected by the contact. In other words, the only role of the heat bath is to determine T. If the subsystem is more strongly coupled, for example a single atom in a solid, then it cannot be isolated and considered separately.

We can make a consistency check on the Gibbs distribution. The behaviour of a *macroscopic* subsystem should be the same as it would be if it was isolated and formed a complete system in its own right. Let its energy be E. Such a macroscopic subsystem has a partition function where the sum (10.40) can be replaced by an integral,

$$Z = \int_{E_{\min}}^{\infty} g(E) e^{-\frac{E}{T}} \, dE . \tag{10.42}$$

$g(E)$ is the density of states, which makes sense for a macroscopic subsystem. Writing g in terms of the subsystem's entropy gives

$$Z = \int_{E_{\min}}^{\infty} e^{S(E) - \frac{E}{T}} \, dE . \tag{10.43}$$

The exponent has a maximum with respect to E where

$$\frac{dS(E)}{dE} - \frac{1}{T} = 0 , \tag{10.44}$$

and this relates E to the temperature T in the way we expect for any thermodynamic system. Z itself is completely dominated by the contribution to the integral from the immediate neighbourhood of this maximum, so Z can be approximated by

$$Z = e^{S(E) - \frac{E}{T}} , \tag{10.45}$$

with S, E and T related by equation (10.44). Taking logarithms gives

$$-T \log Z = E - TS(E) , \tag{10.46}$$

but the right-hand side is exactly the definition of the free energy F for a macroscopic system at temperature T, so

$$F = -T \log Z , \tag{10.47}$$

or equivalently, $\frac{1}{Z} = e^{\frac{F}{T}}$. The normalized Gibbs distribution (10.41) for a macroscopic system in contact with a heat bath at temperature T is therefore

$$P(E) = e^{\frac{F - E}{T}} . \tag{10.48}$$

This is the probability that the system is in a particular microstate of energy E.

In fact, the overwhelming probability is to find the system with energy equal to the thermodynamic equilibrium value E. For this energy, $F = E - TS$. So the probability (10.48) becomes

$$P(E) = e^{\frac{E - TS - E}{T}} = e^{-S} . \tag{10.49}$$

However, as there are $e^{S(E)}$ states with energy E, the probability that a particular one is occupied should indeed be $e^{-S(E)}$. This completes the consistency check.

For a macroscopic system in contact with a heat bath, fluctuations away from the thermodynamic equilibrium energy E are possible, but significant fluctuations have negligible probability. They are suppressed by the entropy factor for smaller energies, and by the energy factor for larger energies. The magnitude of these energy fluctuations may be estimated through a more precise analysis and they are found to depend on the heat capacity of the system. The conclusion is that a macroscopic system at temperature T has the same thermodynamic properties, whether or not it is in contact with a heat bath.

Equation (10.47) shows the usefulness of the partition function $Z = \sum_n e^{-\frac{E_n}{T}}$. It gives the most direct route from knowledge of the quantum states of a macroscopic system to the system's thermodynamic properties. The free energy $F = -T \log Z$ is a function of T and V, as the energies E_n depends on the volume V. The derivative of $-F$ with respect to V is the pressure P, according to equation (10.27), and the expression for P in terms of T and V is called the *equation of state*. The derivative of $-F$ with respect to T is the entropy S, and the derivative of S with respect to T determines the heat capacity C_V, via equation (10.16). The partition function is rather easy to calculate for a system of N approximately non-interacting subsystems, for example a dilute gas of N atoms or molecules, as we shall see in section 10.6.

10.5   The Maxwell Velocity Distribution

Maxwell showed the importance of statistical mechanics when he derived the probability distribution for the velocities of particles in an ideal gas at temperature T. Following Maxwell, we can consider the molecules in a macroscopic container to behave classically to a good approximation. In this limit the discrete momenta predicted by quantum mechanics become quasi-continuous, and the usual classical formula for the kinetic energy of each molecule

$$\varepsilon = \frac{1}{2m}(p_x^2 + p_y^2 + p_z^2) = \frac{1}{2}m(v_x^2 + v_y^2 + v_z^2) \tag{10.50}$$

is valid, where $\mathbf{v} = (v_x, v_y, v_z)$ is its velocity and m is its mass. The potential energy depends on the relative positions of the molecules in the container, but this part separates out. If the inter-molecular forces are attractive, the molecules are more likely to be slightly clustered together, and if repulsive the molecules are more likely to be evenly spaced out, but the velocities of the molecules are unaffected by their spatial arrangement, and most importantly the velocity of each molecule may be treated independently.

Therefore, the probability density for the velocity of each molecule is

$$P(v_x, v_y, v_z) \propto e^{-\frac{m(v_x^2 + v_y^2 + v_z^2)}{2T}}. \tag{10.51}$$

This is the Maxwell distribution, which is a special case of the Gibbs distribution that applies when the available states simply correspond to the range of possible kinetic energies for each molecule. The correctly normalized probability density is

$$P(v_x, v_y, v_z) = \left(\frac{m}{2\pi T}\right)^{\frac{3}{2}} e^{-\frac{m(v_x^2 + v_y^2 + v_z^2)}{2T}}, \tag{10.52}$$

which satisfies $\int P(v_x, v_y, v_z)\, dv_x dv_y dv_z = 1$, as follows from the Gaussian integral (1.64), $\int_{-\infty}^{\infty} e^{-au^2}\, du = \left(\frac{\pi}{a}\right)^{\frac{1}{2}}$.

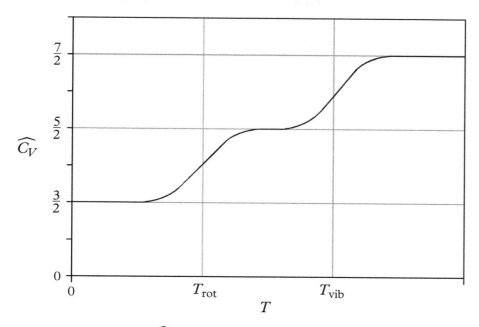

Fig. 10.3 At low temperatures, $\widehat{C_V}$, the heat capacity per molecule of a diatomic gas, is equal to $\frac{3}{2}$. At higher temperatures the molecules may be excited into rotational states and the heat capacity increases accordingly. At even higher temperatures the molecules are also excited into vibrational states.

The directions of motion are all equally likely, so the distribution of molecular speeds v has more significance than the distribution of the velocities. With $v^2 = v_x^2 + v_y^2 + v_z^2$, this probability distribution is

$$P(v) = \left(\frac{m}{2\pi T}\right)^{\frac{3}{2}} 4\pi v^2 e^{-\frac{mv^2}{2T}}, \tag{10.53}$$

and the mean speed and the variance of the distribution are easily found. In particular, the mean kinetic energy of each molecule is

$$\langle K \rangle = \left(\frac{m}{2\pi T}\right)^{\frac{3}{2}} \int_0^\infty \left(\frac{1}{2}mv^2\right) 4\pi v^2 e^{-\frac{mv^2}{2T}} \, dv = \frac{3}{2}T, \tag{10.54}$$

as follows from the Gaussian integral (1.66).

The total kinetic energy of the N molecules in the gas is $N\langle K \rangle = \frac{3}{2}NT$, and this is the total energy E if the gas consists of non-interacting structureless atoms. In this case the heat capacity is

$$C_V = \frac{dE}{dT} = \frac{3}{2}N. \tag{10.55}$$

It is straightforward to measure the heat capacity and this offers a way in which the number of atoms in a gas sample can be deduced. Historically, this was an important step in the development of the atomic picture of matter.

For most real gases, the total energy includes two other types of contribution. First, if the gas is not dilute and the distance between the molecules is comparable to the range of the inter-molecular potential, which is roughly the molecular size, then the potential energy must be included. Second, the internal structure of the molecules may also contribute, in which case there will be energy associated with molecular rotational and vibrational motion. (If either of these is very large, then the molecule will break up into atoms, but this only occurs at extremely high temperatures.) One must perform a Gibbs sum as in equation (10.40) to find the partition function of each molecule, and this requires a considerable knowledge of the quantum states of the molecule. However, at low temperatures, the ideal picture of a structureless molecule or atom holds. As the temperature increases, corrections arise, starting with the contribution of the excited states of lowest energy. Typically, the rotational states of a molecule have the lowest energies, vibrational states have higher energies, and electron excitations within atoms have higher energies still. The heat capacity of a molecular gas increases with temperature as these different quantum states become occupied, as shown in Figure 10.3.

10.6 Ideal Gases—Equation of State and Entropy

We can calculate further properties of an ideal gas, using the density of states in phase space that was mentioned in section 8.2.1. The partition function z for a single structureless molecule is

$$z = \int e^{-\frac{(p_x^2 + p_y^2 + p_z^2)}{2mT}} \frac{d^3x \, d^3p}{(2\pi\hbar)^3}. \tag{10.56}$$

In the classical limit, this integral over the molecule's position and momentum coordinates, with a normalization factor $2\pi\hbar$ for each dimension of space, replaces the sum over quantum states. The exponent is the classical expression for $-\frac{\varepsilon}{T}$ when only the kinetic energy ε contributes. The spatial integral gives the volume V, and the momentum integrals are Gaussian. Evaluating these gives

$$z = \frac{V}{(2\pi\hbar)^3}(2\pi mT)^{\frac{3}{2}} = V\left(\frac{mT}{2\pi\hbar^2}\right)^{\frac{3}{2}}. \tag{10.57}$$

For N molecules, the total partition function is

$$Z = \frac{1}{N!}z^N. \tag{10.58}$$

The factor z^N correctly accounts for the total energy, as the sum of the individual molecular energies is $E = \varepsilon_1 + \varepsilon_2 + \cdots + \varepsilon_N$, and the integral is over the positions and momenta of all the molecules. The combinatorial factor $N!$ compensates for the overcounting of states that are not physically distinct. According to quantum mechanics, identical particles cannot be labelled, so a permutation of the molecules has no physical effect, as discussed in section 8.7. (This is the case for both fermions and bosons.)

We are interested in the free energy $F = -T \log Z$ rather than Z itself. Substituting from equation (10.58), we find

$$F = -T \log\left(\frac{1}{N!}z^N\right) = -T(N \log z - \log N!). \tag{10.59}$$

Now the famous Stirling approximation for $N!$ comes in handy:

$$N! \simeq (2\pi N)^{\frac{1}{2}} N^N e^{-N}, \tag{10.60}$$

and a sufficiently good approximation for statistical mechanics is

$$\log N! \simeq N \log N - N. \tag{10.61}$$

(The next term is proportional to $\log N$ and can be neglected when N is of order 10^{23}.) Using this approximation and collecting the common factors N gives

$$F = -NT(\log z + 1 - \log N) = -NT \log \left(\frac{ze}{N} \right), \tag{10.62}$$

where $e = 2.718\ldots$ is Euler's constant, the base of natural logarithms. Substituting for z from equation (10.57), we find that the free energy of an ideal gas of N molecules in a volume V at temperature T is

$$F = -NT \log \left(\frac{Ve}{N} \left(\frac{mT}{2\pi\hbar^2} \right)^{\frac{3}{2}} \right). \tag{10.63}$$

Separating this into a term depending on the molecular density $\frac{N}{V}$ and a term depending only on temperature gives

$$F = -NT \log \left(\frac{Ve}{N} \right) - \frac{3}{2} NT \log \left(\frac{mT}{2\pi\hbar^2} \right). \tag{10.64}$$

The free energy is extensive, because it is proportional to N, and $\frac{F}{N}$ depends only on the intensive quantities $\frac{N}{V}$ and T. This would not have been the case without the contribution of the $N!$ factor.

We can now calculate the pressure P and the entropy S. The pressure is

$$P = -\left. \frac{\partial F}{\partial V} \right|_T = \frac{NT}{V}. \tag{10.65}$$

This is the equation of state for an ideal gas. It is the ideal gas law $PV = NT$, so the constant A in equation (10.2) is simply N, the number of molecules. (Its form is slightly different if Boltzmann's constant is included, or if the number of molecules is replaced by the number of moles of gas.) We have derived this equation from first principles and this implies that our rather abstract definition of temperature (10.12) is consistent with the ideal gas temperature scale defined in equation (10.3).

Similarly, the entropy is

$$S = -\left. \frac{\partial F}{\partial T} \right|_V = N \log \left(\frac{Ve}{N} \right) + \frac{3}{2} N \log \left(\frac{mT}{2\pi\hbar^2} \right) + \frac{3}{2} N. \tag{10.66}$$

Again, this is extensive, but with a rather complicated dependence on temperature and density. This expression for S is only valid at high temperatures, because it relies on the classical estimate of the partition function. It does not satisfy the third law of thermodynamics: $S = 0$ when $T = 0$. The expression for the heat capacity is simpler, and is

$$C_V = T \frac{\partial S}{\partial T}\bigg|_V = \frac{3}{2}N, \qquad (10.67)$$

as we found earlier. Other useful quantities, like the energy, the enthalpy, and the heat capacity at constant pressure C_P, are easily calculated.

If the molecules or atoms have internal structure but the gas remains dilute then the second term of the free energy has a different dependence on temperature, as this derives from the exponent in equation (10.56), which would include additional terms, so the entropy and heat capacity of the gas are different. However, the first term, which depends on the volume, is unchanged, and therefore the equation of state remains $PV = NT$. Two ideal gas samples in mechanical and thermal equilibrium (i.e. at the same pressure and temperature), and having equal volumes, therefore contain the same number of molecules, as Amedeo Avogadro first realized in the early 19th century. By weighing the two gas samples, one can find the ratio of their molecular weights. This enabled chemists to determine the atomic structure of simple molecules such as O_2, H_2O and CO_2.

10.7 Non-Ideal Gases

Molecular interactions are generally more complicated than atomic interactions, so let us just consider monatomic gases, such as the noble gases, whose molecules are single atoms. There are various ways in which even monoatomic gases deviate from being ideal. As the density increases, the chance of two atoms being close together rises and the interaction potential energy[6] has a greater effect. At high density and low temperature, it also becomes relevant whether the atoms are fermions or bosons. Previously we considered the single-atom states, with energies ε, and then assumed that the N atoms of the gas occupy these states independently. This is justified at high temperatures and low densities, because the probability that a particular state is occupied by a single atom is much less than 1. In this situation, the behaviour of fermions and bosons is similar. At low temperatures, however, the Gibbs distribution tells us that there is much greater probability of low energy states being occupied than high energy states. At most one fermionic atom can occupy a single-atom state, whereas an arbitrary number of bosonic atoms can occupy such a state, so now the combinatorial factor is not simply the constant $N!$. In either case, there is an effective interaction between the atoms, so the Gibbs distribution is modified and this affects the thermodynamic properties of the gas.

For gases that are close to ideal, there is a systematic way of writing the equation of state, as an expansion in the number density $\frac{N}{V}$. This is known as the *virial expansion*, and takes the form

$$P = \frac{NT}{V}\left(1 + \frac{NB(T)}{V} + \frac{N^2C(T)}{V^2} + \cdots\right). \qquad (10.68)$$

The leading term gives, of course, the ideal gas law. $B(T)$ and $C(T)$ are called the second and third virial coefficients. For a classical gas, $B(T)$ can be calculated as an integral involving

---

[6] There are very weak Van der Waals forces between noble gas atoms due to the electric dipoles produced by the fluctuating electron fields in the atoms.

the interaction potential of just one pair of atoms, although in general its dependence on T is not particularly simple. The simplest non-ideal gas may be modelled as a gas of hard sphere atoms, each of diameter l, so that the atom centres cannot approach each other closer than l. This represents a short-range hard repulsion between the atoms. The pressure is greater than for an ideal gas, due to the reduction in volume as space is taken up by the atoms themselves. For the hard sphere gas one finds that the second virial coefficient is independent of T and equal to $\frac{2}{3}\pi l^3$. $C(T)$ is more difficult to calculate, even for a hard sphere gas, because it is associated with three-atom interactions.

For a gas of fermions or bosons, the virial coefficients are non-zero even when there is no interaction potential and therefore no attractive or repulsive force between the atoms. One finds that

$$B(T) = \pm \frac{1}{2g}\left(\frac{\pi\hbar^2}{mT}\right)^{\frac{3}{2}}.$$

(10.69)

Here the upper sign corresponds to fermions and the lower sign to bosons. g is the number of independent spin states of the atom, which is one for a spin 0 boson and two for a spin $\frac{1}{2}$ fermion. For fermions the pressure is higher than that of a classical ideal gas, as a consequence of the Pauli exclusion principle. For bosons it is lower.

The effects we have been discussing occur at modest densities, but at high densities real gases usually liquefy, at least for sufficiently low temperatures. This discontinuous behaviour is a phase transition and is much more difficult to understand theoretically. We will say something about phase transitions in section 10.13.

10.8 The Chemical Potential

The chemical potential μ of a system, mentioned at the end of section 10.3.1, is an intensive quantity related to variations in the system's particle number N. There are many circumstances in which the chemical potential plays an important role. These include chemical reactions where the number of molecules of various chemical species may change. The chemical potential is also useful when considering a subvolume of a gaseous system, where particles may enter or leave. A third example is a phase transition, where particles are transformed from one phase into another.

The relationship between μ and N is rather similar to that between T and E. Recall that an isolated system has fixed energy E, and the temperature T is defined indirectly in terms of entropy. However, one can also consider a system that is in contact with a heat bath and therefore remains at a fixed temperature T, in which case the energy E adjusts to the imposed temperature. Similarly, an isolated system usually has a fixed number of particles N, but often it is useful to consider the system in contact via a porous membrane with a bath of particles at fixed chemical potential μ, and then N adjusts until this imposed value of μ is reached.

In the same way that we showed that two systems in thermal contact reach equilibrium and have maximum entropy when their temperatures are equal, so one can show that two systems that can freely exchange particles will reach an equilibrium state where their chemical potentials are equal. Let us now define chemical potential more precisely.

If we allow the particle number of a macroscopic system to vary, then the energy is a function of entropy, volume and particle number, $E(S, V, N)$. The chemical potential is then

$$\mu = \left.\frac{\partial E}{\partial N}\right|_{S,V} , \qquad (10.70)$$

a partial derivative with both entropy S and volume V fixed, and the first law generalizes to

$$dE = T\,dS - P\,dV + \mu\,dN . \qquad (10.71)$$

The term $\mu\,dN$ must be included in all the energy functions, including the Gibbs free energy G, so

$$dG = -S\,dT + V\,dP + \mu\,dN \qquad (10.72)$$

and $\mu = \left.\frac{\partial G}{\partial N}\right|_{T,P}$. Recall, however, that $G(T, P, N) = N\widetilde{G}(T, P)$ where $\widetilde{G}(T, P)$ is the Gibbs free energy per particle, so $\left.\frac{\partial G}{\partial N}\right|_{T,P} = \widetilde{G}(T, P)$. The chemical potential of a system is therefore not a completely new quantity, but equal to the Gibbs free energy per particle. The total Gibbs free energy is $G = \mu N$.

Consider now a system (not necessarily macroscopic) in contact with a much larger heat and particle bath, where the combined energy and particle number have the constant values $E^{(0)}$ and $N^{(0)}$. We are interested in the probability of finding the system in a particular quantum state where it has energy E and has N particles. For such a microstate, the bath will have energy $E^{(0)} - E$ and particle number $N^{(0)} - N$. As in the derivation of the Gibbs distribution, we can make a linear approximation to the entropy S of the bath,

$$S(E^{(0)} - E, N^{(0)} - N) = S(E^{(0)}, N^{(0)}) - \frac{E}{T} + \frac{\mu N}{T} , \qquad (10.73)$$

where we have assumed the volume to be fixed and used the first law in the form $dS = \frac{dE}{T} - \frac{\mu\,dN}{T}$. The number of states available to the combined system is therefore

$$e^{S(E^{(0)} - E, N^{(0)} - N)} = e^{S(E^{(0)}, N^{(0)})}\, e^{\frac{\mu N - E}{T}} , \qquad (10.74)$$

and the probability that the system occupies the microstate is

$$P(E, N) = \frac{1}{Z_G} e^{\frac{\mu N - E}{T}} . \qquad (10.75)$$

This is the analogue of the Gibbs distribution, taking into account the variable particle number. The normalization factor Z_G, ensuring the total probability sums to 1, is called the grand partition function. The probability distribution (10.75) determines the mean values of E and N in terms of T and μ. For a macroscopic system, these mean values are the thermodynamic values.

10.9 Fermion and Boson Gases at Low Temperature

Gases of fermions or bosons exhibit remarkable properties when the temperature is close to absolute zero. Bosons, in particular, can undergo a change of phase called Bose–Einstein condensation.

We assume the gas consists of N identical particles, which may be either fermions or bosons, in a box of volume V, and that there is no direct interaction between them. The only interaction is due to the quantum nature of the multi-particle wavefunction for identical particles. The lack of any direct interaction means that we can work with the energy levels for each individual particle, and as the box is macroscopic, the 1-particle states have a quasi-continuous spectrum of energies ε.

A 1-particle state can have an occupation number of either 0 or 1 if the particles are fermions, and 0 or any positive integer if they are bosons. In this situation, the chemical potential is useful as it avoids complicated combinatorial calculations. We can treat a 1-particle state with energy ε as a system in contact with a heat and particle bath formed of the rest of the gas, at temperature T and with chemical potential μ. The occupation number of the state is variable and when the occupation number is n, the system has energy $\varepsilon_n = n\varepsilon$, so $\mu n - \varepsilon_n = n(\mu - \varepsilon)$.

According to the distribution (10.75), with $N = n$ and $E = n\varepsilon$, the normalized probability that a 1-particle state with energy ε has occupation number n is

$$P(n) = \frac{1}{z} e^{\frac{n(\mu-\varepsilon)}{T}}, \tag{10.76}$$

where $z = \sum_n e^{\frac{n(\mu-\varepsilon)}{T}}$. For fermions $n = 0, 1$, so there are just two terms in the sum; for bosons $n = 0, 1, 2, \ldots$, so the sum is an infinite geometric series that converges and is easily summed provided $\mu < \varepsilon$. These 1-particle partition functions are

$$z_F = 1 + e^{\frac{\mu-\varepsilon}{T}} \qquad \text{for fermions}, \tag{10.77}$$

$$z_B = \frac{1}{1 - e^{\frac{\mu-\varepsilon}{T}}} \qquad \text{for bosons}. \tag{10.78}$$

10.9.1 The Fermi–Dirac function

More important than these probability distributions are the *mean occupation numbers* as a function of ε,

$$\overline{n}(\varepsilon) = \frac{1}{z} \sum_n n\, e^{\frac{n(\mu-\varepsilon)}{T}}. \tag{10.79}$$

For fermions, $z = z_F$ and n takes the values 0 and 1, so the mean is

$$\overline{n}(\varepsilon) = \frac{e^{\frac{\mu-\varepsilon}{T}}}{1 + e^{\frac{\mu-\varepsilon}{T}}} = \frac{1}{e^{\frac{\varepsilon-\mu}{T}} + 1} \equiv n_F(\varepsilon). \tag{10.80}$$

$n_F(\varepsilon)$ is called the *Fermi–Dirac function*. It is shown for various temperatures in Figure 10.4. For states with energy ε less than μ, the exponential term in the denominator is small so the mean occupation number is close to 1, whereas for states with energy greater than μ, the exponential term is large so the mean occupation number is close to 0. As the exponent is proportional to $\frac{1}{T}$, this transition is steep at low temperatures, and less so at high temperatures. The limiting case at zero temperature is the *degenerate Fermi gas*, where all states below μ are occupied, and all states above it are empty. In this case, μ is the energy of the highest occupied state and is the Fermi energy ε_F. At finite temperature the sharp discontinuity in the occupation of states at ε_F is broadened out.

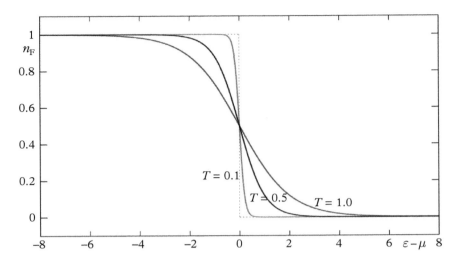

Fig. 10.4 The Fermi–Dirac function.

Suppose the density of 1-particle states in the box is $g(\varepsilon)$, and the possible energies run from a minimum value $\varepsilon_{\min}$ upwards. Then the total number of particles N equals the mean fermion occupation number $n_{\mathrm{F}}(\varepsilon)$ integrated over the entire range of available energies, weighted by the density of states,

$$N = \int_{\varepsilon_{\min}}^{\infty} \frac{g(\varepsilon)}{e^{\frac{\varepsilon - \mu}{T}} + 1} \, d\varepsilon . \tag{10.81}$$

This gives N as a function of μ and T. The value of μ as a function of N and T is determined by inverting this expression. The total energy of the gas is obtained from a similar integral, with an additional factor of ε in the integrand.

For the degenerate Fermi gas at $T = 0$, all the states are occupied up to the Fermi energy ε_{F}, so $\mu = \varepsilon_{\mathrm{F}}$ and ε_{F} is related to N by

$$N = \int_{\varepsilon_{\min}}^{\varepsilon_{\mathrm{F}}} g(\varepsilon) \, d\varepsilon . \tag{10.82}$$

For T small, μ differs only a little from ε_{F}, with the difference depending on the density of states $g(\varepsilon)$ close to ε_{F}.

10.9.2 Pressure of a degenerate electron gas

Consider N electrons in a box of volume V. Their electric charges are assumed to be screened by the positive charges of background ions, so the electrons can be treated as non-interacting. Suppose the temperature is so low that the electrons behave as a degenerate Fermi gas. The electron kinetic energies range from 0 upwards, and the density of states is $g(\varepsilon) = \frac{V}{2\pi^2} \left(\frac{2m}{\hbar^2} \right)^{\frac{3}{2}} \varepsilon^{\frac{1}{2}}$. Below the Fermi energy ε_{F} all states are occupied and above it, all states are empty. In section 8.7.1 we evaluated the Fermi energy of a degenerate electron gas, and we also evaluated the total energy of the N electrons, finding

$$E = \frac{3(3\pi^2)^{\frac{2}{3}}}{5} \frac{\hbar^2}{2m} \left(\frac{N}{V} \right)^{\frac{2}{3}} N . \tag{10.83}$$

The state of the degenerate electron gas is the one of lowest energy for N electrons, so it is the zero temperature state. As the state is unique, the entropy is zero.

Suppose now that the volume V varies. The entropy S remains zero, so $T dS$ is zero, and the first law of thermodynamics reduces to $dE = -P dV$. Therefore the pressure of the degenerate electron gas is

$$P = -\frac{dE}{dV} = \frac{2(3\pi^2)^{\frac{2}{3}}}{5} \frac{\hbar^2}{2m} \left(\frac{N}{V}\right)^{\frac{5}{3}}.\tag{10.84}$$

This substantial pressure, proportional to the density to the power $\frac{5}{3}$, is a result of the Pauli exclusion principle. It is called *electron degeneracy pressure*. A classical ideal gas with equation of state $PV = NT$ has zero pressure at zero temperature, but the exclusion principle requires almost all the electrons to have positive kinetic energy even at zero temperature, and the total energy increases as the volume decreases, which produces the pressure. This pressure plays a crucial role in the evolution of white dwarf stars, as we will see in section 13.7.1.

10.9.3 The heat capacity of an electron gas

In the Drude theory, the electrons in a metal are treated as a classical gas of free particles. Although this early theory had some successes, there was very poor experimental support for its prediction of the contribution of the electrons to the metal's heat capacity. If electrons really behaved as classical non-interacting particles, we would expect the energy spectrum to be the same as that for the ideal gas discussed in section 10.5. For N_e electrons, the heat capacity would then be $\frac{3}{2}N_e$, as given in equation (10.55) for the ideal gas. At normal room temperature, this is over 100 times the measured value.

For an accurate calculation of the heat capacity, quantum theory must be used. Electrons are fermions, so the mean occupation of a quantum state is given by the Fermi–Dirac function (10.80). Electrons in a metal have a high number density, so the Fermi energy ε_F of the electrons is much greater than room temperature T. The jump in the Fermi–Dirac function from 1 to 0 occurs over the relatively narrow energy range of order T around ε_F, so thermal excitations only affect a small fraction of the electrons – those with energies close to ε_F.

We will now perform a computation of the electronic heat capacity valid at temperatures low compared to ε_F. The thermodynamic energy of a Fermi gas of electrons is

$$E_e = \int_{\varepsilon_{\min}}^{\infty} \varepsilon g(\varepsilon) n_F(\varepsilon) \, d\varepsilon,\tag{10.85}$$

where $g(\varepsilon)$ is the density of states and $n_F(\varepsilon)$ is the Fermi–Dirac function. The density of states is not temperature dependent, but n_F is, so the heat capacity is

$$C_e = \frac{dE_e}{dT} = \int_{\varepsilon_{\min}}^{\infty} \varepsilon g(\varepsilon) \frac{\partial n_F}{\partial T} \, d\varepsilon.\tag{10.86}$$

At low temperature, the chemical potential μ is approximately ε_F and its temperature dependence can be ignored. Let $x = \frac{\varepsilon - \varepsilon_F}{T}$. The Fermi–Dirac function is then

$$n_F(x) = \frac{1}{e^x + 1}\tag{10.87}$$

and its derivative with respect to T is

$$\frac{\partial n_F}{\partial T} = \frac{dn_F}{dx}\frac{\partial x}{\partial T} = \frac{e^x}{(e^x+1)^2}\frac{\varepsilon - \varepsilon_F}{T^2} = \frac{1}{T}\frac{xe^x}{(e^x+1)^2}. \tag{10.88}$$

The last quantity can be re-expressed as

$$\frac{1}{T}\frac{x}{(e^{\frac{1}{2}x}+e^{-\frac{1}{2}x})^2}, \tag{10.89}$$

an odd function of x concentrated around $x = 0$, that is, around $\varepsilon = \varepsilon_F$. In the integral (10.86) we can therefore extend the range from $-\infty$ to ∞, treat $g(\varepsilon)$ as the constant $g(\varepsilon_F)$ and replace the energy factor ε by $(\varepsilon - \varepsilon_F) + \varepsilon_F = Tx + \varepsilon_F$. The last constant ε_F does not contribute to the integral because it multiplies an odd function. Then, replacing $d\varepsilon$ by $T dx$, we find

$$C_e = Tg(\varepsilon_F)\int_{-\infty}^{\infty}\frac{x^2}{(e^{\frac{1}{2}x}+e^{-\frac{1}{2}x})^2}\,dx. \tag{10.90}$$

This is a standard integral whose value is $\frac{\pi^2}{3}$. Therefore

$$C_e = \frac{\pi^2}{3}g(\varepsilon_F)\,T \tag{10.91}$$

at low temperatures.

In section 8.7.1 we calculated the density of states for free electrons to be

$$g(\varepsilon) = (2m^3)^{\frac{1}{2}}\frac{V}{\pi^2\hbar^3}\varepsilon^{\frac{1}{2}} = \frac{3}{2}\frac{N(\varepsilon)}{\varepsilon}, \tag{10.92}$$

where $N(\varepsilon)$ is the total number of electron states with energy ε or less. The density of states at the Fermi surface is therefore $g(\varepsilon_F) = \frac{3}{2}\frac{N_e}{\varepsilon_F}$, as $N(\varepsilon_F) = N_e$, and substituting into equation (10.91), we obtain

$$C_e \simeq \frac{\pi^2}{2}N_e\frac{T}{\varepsilon_F}. \tag{10.93}$$

As $T \ll \varepsilon_F$ for a metal at room temperature, this heat capacity is much less than the classical $\frac{3}{2}N_e$.

At low temperatures, the heat capacity of metals can be expressed as a sum of the electronic contribution, and a contribution due to lattice vibrations. Both the lattice contribution and the higher-order corrections to the electronic contribution are proportional to T^3. At low temperatures, the electronic contribution proportional to T is dominant. There is reasonable agreement between this simple model and measurements of the heat capacity in those metals for which the free electron model works well. These are the coinage metals and the alkali metals, where the predicted values of the electronic heat capacity are accurate to within $10\% - 30\%$.

10.9.4   The Bose–Einstein function

For bosonic particles, the mean occupation number of each 1-particle state can be calculated in a similar way as for fermionic particles. Using the probability distribution (10.76), with $z = z_B$ from equation (10.78), we find the mean occupation number is

$$
\begin{aligned}
\overline{n}(\varepsilon) &= \frac{1}{z_B} \sum_{n=0}^{\infty} n\, e^{\frac{n(\mu-\varepsilon)}{T}} = \left(1 - e^{\frac{\mu-\varepsilon}{T}}\right) \sum_{n=0}^{\infty} n e^{\frac{n(\mu-\varepsilon)}{T}} \\
&= 0 + e^{\frac{\mu-\varepsilon}{T}} + 2e^{\frac{2(\mu-\varepsilon)}{T}} + 3e^{\frac{3(\mu-\varepsilon)}{T}} + \cdots \\
&\quad - 0 - e^{\frac{2(\mu-\varepsilon)}{T}} - 2e^{\frac{3(\mu-\varepsilon)}{T}} - \cdots \\
&= \sum_{n=1}^{\infty} e^{\frac{n(\mu-\varepsilon)}{T}}.
\end{aligned}
\tag{10.94}
$$

For $\mu < \varepsilon$, this geometric series sums to

$$
\overline{n}(\varepsilon) = \frac{e^{\frac{\mu-\varepsilon}{T}}}{1 - e^{\frac{\mu-\varepsilon}{T}}} = \frac{1}{e^{\frac{\varepsilon-\mu}{T}} - 1} \equiv n_B(\varepsilon).
\tag{10.95}
$$

$n_B(\varepsilon)$ is called the *Bose–Einstein function*, and it differs from the Fermi–Dirac function $n_F(\varepsilon)$ only by replacing $+1$ by -1 in the denominator. Figure 10.5 shows the function n_B and also n_F for comparison.

The form of the denominator of $n_B(\varepsilon)$ means that the integral

$$
\int_0^{\infty} \frac{x^{n-1}}{e^x - 1}\, dx
\tag{10.96}
$$

often arises in the theory of Bose gases, for various n. Its value is

$$
\begin{aligned}
&\int_0^{\infty} \frac{x^{n-1}}{e^x - 1}\, dx \\
&= \int_0^{\infty} x^{n-1}\left(e^{-x} + e^{-2x} + e^{-3x} + \cdots\right) dx \\
&= \int_0^{\infty} x^{n-1} e^{-x}\, dx + \int_0^{\infty} \left(\frac{x'}{2}\right)^{n-1} e^{-x'} \frac{1}{2}\, dx' + \int_0^{\infty} \left(\frac{x''}{3}\right)^{n-1} e^{-x''} \frac{1}{3}\, dx'' + \cdots \\
&= \int_0^{\infty} x^{n-1} e^{-x}\, dx \left(1 + \frac{1}{2^n} + \frac{1}{3^n} + \cdots\right) \\
&= \Gamma(n)\zeta(n),
\end{aligned}
\tag{10.97}
$$

where $\Gamma(n) = \int_0^{\infty} x^{n-1} e^{-x}\, dx$ and $\zeta(n) = \sum_{k=1}^{\infty} \frac{1}{k^n}$. $\Gamma(n)$ is the Euler gamma function and $\zeta(n)$ is the Riemann zeta function. $\Gamma(n) = (n-1)!$ when n is a positive integer. Another useful value is $\Gamma(\frac{3}{2}) = \frac{1}{2}\pi^{\frac{1}{2}}$. The following values of the zeta function will also be of use to us: $\zeta(\frac{3}{2}) \simeq 2.612$, $\zeta(3) \simeq 1.202$, $\zeta(4) = \frac{\pi^4}{90}$.

The 1-particle states have energies that range from $\varepsilon_{\min}$ upwards. $n_B(\varepsilon)$ only makes sense if μ is less than every allowed energy ε, since mean occupation numbers cannot be negative,

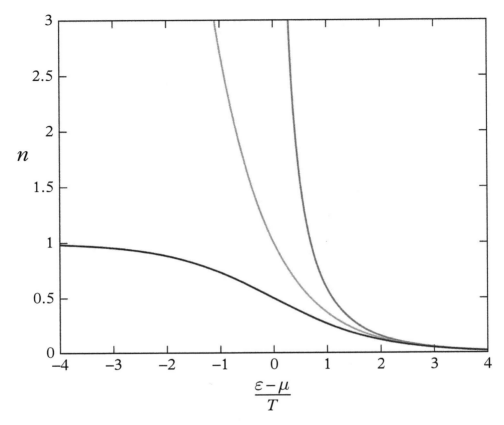

Fig. 10.5 Comparison of the Bose–Einstein and Fermi–Dirac functions (upper and lower curves). For $\varepsilon \gg \mu$ they are asymptotically the same, as the probability of multiple occupancy of a state is negligible. The asymptotic form is the Maxwell–Boltzmann function $e^{\frac{\mu - \varepsilon}{T}}$ (middle curve).

so $\mu < \varepsilon_{\min}$. The value of μ is determined by the requirement that the total number of particles is N, and this again leads to an integral constraint

$$N = \int_{\varepsilon_{\min}}^{\infty} \frac{g(\varepsilon)}{e^{\frac{\varepsilon - \mu}{T}} - 1} \, d\varepsilon . \tag{10.98}$$

For small T, the Bose–Einstein function decreases rapidly as ε increases, so if N is fixed, μ must approach $\varepsilon_{\min}$ from below as T decreases. Most particles then occupy a narrow range of excited states with energies at and just above $\varepsilon_{\min}$.

Remarkably, there is a critical temperature T_c below which only a finite fraction of the particles can be in states with $\varepsilon > \varepsilon_{\min}$. The remaining fraction are all in the ground state, with energy $\varepsilon_{\min}$. The ground state is then macroscopically occupied. This phenomenon is known as *Bose–Einstein condensation*.

To find T_c we will assume that the ground state is discrete and has energy $\varepsilon_{\min}$, but the 1-particle excited states have a quasi-continuous energy spectrum, with density of states

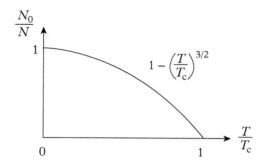

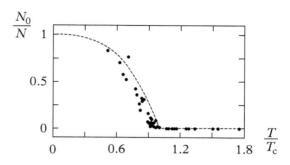

Fig. 10.6 Left: The predicted condensate fraction $\frac{N_0}{N}$ plotted against the normalized temperature $\frac{T}{T_c}$ for bosons in a uniform 3-dimensional box. Right: The observed condensate fraction plotted against the normalized temperature for an atomic Bose–Einstein condensate in a 3-dimensional harmonic trap. The dashed line is a plot of $\frac{N_0}{N} = 1 - (\frac{T}{T_c})^3$.

$g(\varepsilon)$. For non-relativistic spin 0 bosons of mass m, free to move in a box of volume V, we can set $\varepsilon_{\min} = 0$. The density of excited states is

$$g(\varepsilon) = \left(\frac{m^3}{2}\right)^{\frac{1}{2}} \frac{V}{\pi^2 \hbar^3} \varepsilon^{\frac{1}{2}}. \tag{10.99}$$

The critical temperature T_c is reached when $\mu = 0$. At this temperature, the fraction of particles in the ground state is still essentially zero, so the total number of particles is given by substituting $g(\varepsilon)$ into equation (10.98) and setting $\mu = 0$, to give

$$
\begin{aligned}
N &= \left(\frac{m^3}{2}\right)^{\frac{1}{2}} \frac{V}{\pi^2 \hbar^3} \int_0^\infty \frac{\varepsilon^{\frac{1}{2}}}{e^{\frac{\varepsilon}{T_c}} - 1} d\varepsilon \\
&= \left(\frac{mT_c}{2\pi\hbar^2}\right)^{\frac{3}{2}} V \frac{2}{\pi^{\frac{1}{2}}} \int_0^\infty \frac{x^{\frac{1}{2}}}{e^x - 1} dx \\
&= 2.612\, V \left(\frac{mT_c}{2\pi\hbar^2}\right)^{\frac{3}{2}}, \tag{10.100}
\end{aligned}
$$

where we substituted $x = \frac{\varepsilon}{T_c}$ to obtain the standard integral (10.96) with $n = \frac{3}{2}$ and then used the values $\Gamma(\frac{3}{2}) = \frac{1}{2}\pi^{\frac{1}{2}}$ and $\zeta(\frac{3}{2}) \simeq 2.612$. The critical temperature for Bose–Einstein condensation is therefore related to the number density by the expression

$$\frac{N}{V} = 2.612 \left(\frac{mT_c}{2\pi\hbar^2}\right)^{\frac{3}{2}}. \tag{10.101}$$

At lower temperatures, μ remains at 0, and the number of particles in the excited states is $2.612\, V \left(\frac{mT}{2\pi\hbar^2}\right)^{\frac{3}{2}}$, which is less than N. The rest are in the ground state. Denoting the number of these by $N_0(T)$, the total number of particles is now

$$N = N_0(T) + 2.612\, V \left(\frac{mT}{2\pi\hbar^2}\right)^{\frac{3}{2}}. \tag{10.102}$$

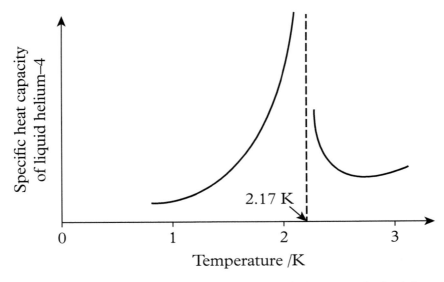

Fig. 10.7 The heat capacity of helium has a very sharp peak, known as the lambda peak due to its shape, at a temperature of 2.17 K.

Using expression (10.100) for N in terms of T_c, we find

$$N_0(T) = 2.612\, V \left(\frac{m}{2\pi\hbar^2} \right)^{\frac{3}{2}} \left(T_c^{\frac{3}{2}} - T^{\frac{3}{2}} \right), \tag{10.103}$$

and dividing by N gives the fraction of particles in the ground state,

$$\frac{N_0(T)}{N} = 1 - \left(\frac{T}{T_c} \right)^{\frac{3}{2}}. \tag{10.104}$$

On the left of Figure 10.6, this fraction is shown as a function of T.

There is a phase transition at the critical temperature T_c where Bose–Einstein condensation sets in, as the derivatives of the number of particles in the ground state $N_0(T)$ and the heat capacity C_V are discontinuous there.

Bose–Einstein condensation does not easily occur in physical systems as most naturally occurring gases of bosonic atoms liquefy at temperatures well above their critical temperature, and in the liquid phase the atomic interactions are no longer negligible. However, Bose–Einstein condensation is believed to occur in liquid helium, for atoms of the most abundant isotope ^{4}He. Helium remains a gas until it reaches about 4 K, where it liquefies. At the natural density of the liquid, the critical temperature T_c is predicted to be about 3 K. The fraction of atoms in the ground state is not directly measurable, but the heat capacity is. As shown in Figure 10.7, there is a very sharp peak in the heat capacity of helium at a temperature of 2.17 K. Below this temperature, helium is a superfluid and this phase transition is thought to be related to Bose–Einstein condensation.

There is indisputable evidence for Bose–Einstein condensation in systems of ultra-cold alkali metal atoms. Using lasers and an inhomogeneous magnetic field, researchers are able to

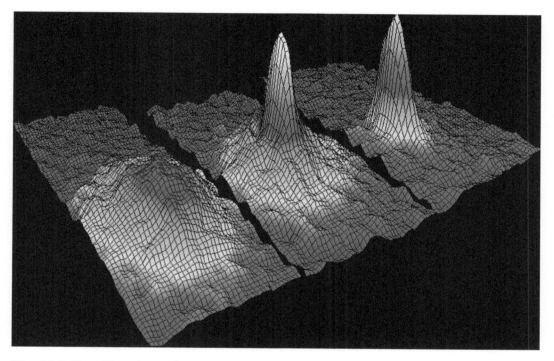

Fig. 10.8 Bose–Einstein condensate. By cooling rubidium atoms, which are bosons, to less than 1.7×10^{-7} K, Carl Wieman and Eric Cornell produced the first true Bose–Einstein condensate in 1995.

trap atoms such as rubidium-87 and sodium-23 in harmonic oscillator potentials. Typically, of the order of 10^4–10^7 atoms are trapped at a density that is much lower than that of liquid helium. The interactions of these atoms are therefore much weaker than in liquid helium and they conform to the assumption of ideal Bose gas theory much better. The low density means that the critical temperature T_c is in the region of a microkelvin. T_c can be calculated using a similar integral to (10.100), but the density of states in a harmonic trap[7] is proportional to ε^2 rather than $\varepsilon^{\frac{1}{2}}$. On changing variables to $x = \frac{\varepsilon}{T_c}$, $\varepsilon^2 \, d\varepsilon$ is replaced by $T_c^3 x^2 \, dx$, which results in the expression $\frac{N_0}{N} = 1 - (\frac{T}{T_c})^3$, as plotted on the right of Figure 10.6. Using these techniques, Carl Wieman and Eric Cornell were the first to create a Bose–Einstein condensate (BEC) in 1995 when they cooled a cloud of rubidium-87 atoms to ultra-low temperatures. These atoms are bosons, as they contain an even number of spin $\frac{1}{2}$ particles, namely 37 electrons, 37 protons and 50 neutrons. When cooled below 1.7×10^{-7} K, a large fraction of the rubidium atoms condense into the same state. The graphic in Figure 10.8 is a false-colour image of the spatial density distribution of a cloud of rubidium atoms at three different temperatures showing the atoms clustering into the same state as the temperature

---

[7] At level N of the 3-dimensional harmonic oscillator, the energy is $\varepsilon_N = (N + \frac{3}{2})\hbar\omega$ and the degeneracy is $g(\varepsilon_N) = \frac{1}{2}(N + 1)(N + 2)$, as we will show in section 11.3.2. Trading N for ε_N gives $g(\varepsilon_N) \propto (\varepsilon_N - \frac{1}{2}\hbar\omega)(\varepsilon_N + \frac{1}{2}\hbar\omega) = \varepsilon_N^2 - \frac{1}{4}\hbar^2\omega^2$. For large N, we can neglect the last term.

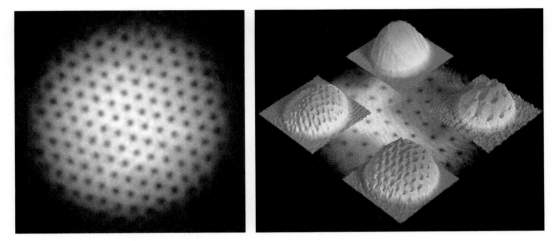

Fig. 10.9 Left: A regular lattice of vortices in a rotating condensate of sodium atoms. Lasers were used to set a condensate 60 micrometres in diameter and 250 micrometres in length in rotation. Right: The condensate was allowed to expand ballistically which resulted in a 20-fold magnification. The images represent 2-dimensional cuts through the density distribution and show the density minima due to the vortex cores. The examples shown contain 0, 16, 70 and 130 vortices.

is lowered. Red indicates low density, yellow and green indicate intermediate densities and high density is indicated by blue to white.

Lasers pulses can be used to spin Bose–Einstein condensates, and this results in the formation of an array of vortices within the condensate that carry the angular momentum. Figure 10.9 shows a rotating Bose–Einstein condensate.

10.10 Black Body Radiation

Thus far our discussions of Fermi and Bose gases have related to gases composed of massive atoms or electrons moving non-relativistically, but there are other collections of particles that one might not consider to be gases, whose properties can be analysed in a similar way. One of these is the gas of photons that we call *black body radiation*. This is the name given to the radiation produced by an idealized perfect emitter and absorber of electromagnetic radiation. The measured black body spectrum could not be explained by classical thermodynamics and this led to a crisis in physics at the end of the 19th century, which was only resolved with the advent of quantum physics.

Surprisingly, even an empty box containing no matter has thermodynamic properties. This is because the electromagnetic field within the box is thermally excited by the matter that forms the walls of the box. The box contains a gas of photons that are continually emitted and absorbed by the material in the walls. These photons rapidly reach thermal equilibrium, with a temperature T equal to that of the box.

The chemical potential μ represents the derivative of the energy with respect to particle number. μ is the energy cost of increasing the particle number by 1. Photons have zero rest mass, which means that the energy cost of emitting a photon can be arbitrarily small; the energy of a photon with an infinitely long wavelength is zero. As photon number is not conserved and photons are continually being emitted and absorbed, the photon chemical potential is therefore $\mu = 0$. We may consider the electromagnetic field throughout space to act as an infinite particle bath that maintains the photon chemical potential at zero.

Photons are massless spin 1 particles, so they are bosons and have two independent polarization states in the directions transverse to the direction of propagation. Regarded as relativistic particles, they obey the relation $E = |\mathbf{p}|$, the analogue of equation (4.27) for massless particles. Their interactions with each other are negligible, so they form an ideal Bose gas. In a finite box of volume V, with periodic boundary conditions, the allowed wavevectors $\mathbf{k}$ of the electromagnetic wave modes are discrete. The density of wave modes in $\mathbf{k}$-space is $\frac{2V}{(2\pi)^3}$, the result we derived in section 8.2.1, with an extra factor of 2 for the two polarization states. The density in the wavenumber k, the magnitude of the wavevector, is $\frac{8\pi k^2 V}{(2\pi)^3} = \frac{k^2 V}{\pi^2}$. It is more convenient to work with the frequency ω, but this equals k for an electromagnetic wave, so the density of modes in ω is

$$g(\omega) = \frac{\omega^2 V}{\pi^2} \,. \tag{10.105}$$

Each mode of electromagnetic radiation in the box can be occupied by any number of photons. If the mode has frequency ω, each photon has energy $\hbar\omega$, and if there are n photons, the total energy is $n\hbar\omega$. The mean number of photons in each mode is given by the Bose–Einstein function, with $\mu = 0$,

$$n_{\mathrm{B}}(\omega) = \frac{1}{e^{\frac{\hbar\omega}{T}} - 1} \,. \tag{10.106}$$

The number density of photons at frequency ω is the mean number of photons in a mode, $n_{\mathrm{B}}(\omega)$, times the density of modes $g(\omega)$,

$$N(\omega) = \frac{V}{\pi^2} \frac{\omega^2}{e^{\frac{\hbar\omega}{T}} - 1} \,. \tag{10.107}$$

From $N(\omega)$, the energy density of photons is obtained by multiplying by $\hbar\omega$, giving

$$E(\omega) = \frac{V\hbar}{\pi^2} \frac{\omega^3}{e^{\frac{\hbar\omega}{T}} - 1} \,. \tag{10.108}$$

This is known as the *Planck formula* and it is where Planck first introduced the constant h that bears his name. $E(\omega)$ is the spectral energy density of black body radiation. The energy density as a function of wavelength is plotted for various temperatures in Figure 10.10.

We will now determine how the peak radiance, the maximum of $E(\omega)$, changes with temperature. It is convenient to substitute $x = \frac{\hbar\omega}{T}$, so that $E(\omega) \propto \frac{x^3}{e^x - 1}$. Differentiating, we see that the maximum is where $3x^2(e^x - 1) - x^3 e^x = 0$, or equivalently where $\frac{x}{1 - e^{-x}} = 3$, which can be solved numerically to give $x \simeq 2.8214$. Therefore, the peak radiance of black body radiation is at

$$\hbar\omega_{\mathrm{peak}} \simeq 2.8214 \, T \,. \tag{10.109}$$

This linear increase with temperature is known as *Wien's displacement law*. It is often presented in the form $\lambda_{\mathrm{peak}} \propto \frac{1}{T}$, and this relationship between temperature and the peak wavelength can be seen in Figure 10.10. Whereas a hot cooking pot mainly emits infrared

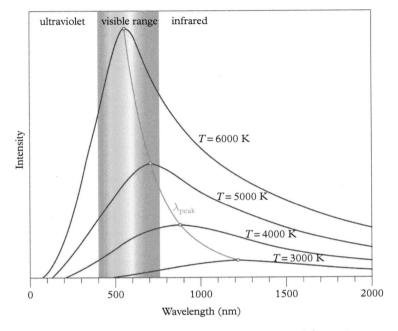

Fig. 10.10 The Planck formula for black body radiation plotted for various temperatures.

radiation, the Sun, whose surface temperature is about 6000 K, emits a substantial fraction of its radiation as visible and ultraviolet light. It is worth noting that a hotter body emits more radiation at all wavelengths, not just in the region of the peak. We will make use of Wien's displacement law when considering the physics of stars in Chapter 13.

The total number of photons in all modes is the integral of $N(\omega)$,

$$N = \frac{V}{\pi^2} \int_0^\infty \frac{\omega^2}{e^{\frac{\hbar\omega}{T}} - 1} \, d\omega = \frac{VT^3}{\pi^2\hbar^3} \int_0^\infty \frac{x^2}{e^x - 1} \, dx \simeq 2.404 \frac{VT^3}{\pi^2\hbar^3}, \tag{10.110}$$

where we substituted $x = \frac{\hbar\omega}{T}$ to obtain the integral (10.96) with $n = 3$ and then used the values $\Gamma(3) = 2! = 2$ and $\zeta(3) \simeq 1.202$.

The total energy of the black body radiation is the integral of $E(\omega)$,

$$E = \frac{V\hbar}{\pi^2} \int_0^\infty \frac{\omega^3}{e^{\frac{\hbar\omega}{T}} - 1} \, d\omega = \frac{VT^4}{\pi^2\hbar^3} \int_0^\infty \frac{x^3}{e^x - 1} \, dx = \frac{\pi^2}{15} \frac{VT^4}{\hbar^3}. \tag{10.111}$$

Here we used the integral (10.96) with $n = 4$ and the values $\Gamma(4) = 3! = 6$ and $\zeta(4) = \frac{\pi^4}{90}$. The result is usually presented as

$$E = 4\sigma VT^4, \tag{10.112}$$

where $\sigma = \frac{\pi^2}{60\hbar^3}$ is the *Stefan–Boltzmann constant*.

Classically, there would be infinitely many modes of radiation and each mode would carry an equal amount of thermal energy, so the total energy would be infinite. We now know that electromagnetic radiation is transmitted in discrete packets, or photons, and this suppresses the energy in extreme ultraviolet modes, leading to a finite energy in the quantum theory.

The total photon number N and total energy E of black body radiation are proportional to the volume and both are finite. This was the first success of the quantum ideas introduced by Planck.

Black body radiation has further thermodynamic properties. At fixed volume, the first law of thermodynamics states that $dE = T\,dS$, and from $E = 4\sigma V T^4$ it follows that $dE = 16\sigma V T^3\,dT$ so $dS = 16\sigma V T^2\,dT$. Integrating, we find the entropy of black body radiation is

$$S = \frac{16}{3}\sigma V T^3\,,\tag{10.113}$$

where the constant of integration is zero, as the entropy vanishes at zero temperature. Substituting for T from equation (10.112), we find the expression

$$E = \left(\frac{81}{1024\,\sigma}\right)^{\frac{1}{3}} V^{-\frac{1}{3}} S^{\frac{4}{3}}\,.\tag{10.114}$$

for E in terms of S and V. The radiation pressure due to the photons within a black body at temperature T is therefore

$$P = -\left.\frac{\partial E}{\partial V}\right|_S = \frac{1}{3V}E = \frac{4}{3}\sigma T^4\,.\tag{10.115}$$

This result will be useful in Chapter 13 when we consider the radiation pressure within a star.

Also important is the energy emitted at the surface of a black body. This can be calculated as follows. Suppose we have a small parcel of black body radiation at temperature T close to the surface of the body. The energy consists of photons travelling at unit speed (the speed of light) randomly in all directions. The energy density is $4\sigma T^4$, and if all this energy were travelling out of the body orthogonally to the surface, then the rate of energy emission per unit area would be $4\sigma T^4$, but only half the energy is travelling outwards from the surface, and for this half the component of the velocity orthogonal to the surface varies between 0 and 1. The average component of velocity orthogonal to the surface is $\frac{1}{2}$ (the average of $\cos\vartheta$ over the hemisphere $0 \le \vartheta \le \frac{\pi}{2}$). The rate of energy emission per unit area is therefore σT^4. This is known as the *Stefan–Boltzmann law* and it determines the luminosity of a star.

10.11 Lasers

In this section we investigate the interplay of black body radiation with photon absorption and emission in atoms, and discuss the technology of lasers that has emerged from the physics.

Bohr introduced the idea that numerous energy levels are available to electrons in atoms, and an electron in an excited state E_2 will spontaneously fall to a lower energy level E_1, with the emission of a photon of energy $E_2 - E_1 = \hbar\omega$, where ω is the frequency of the photon. Conversely, a photon with precisely this energy can promote an electron from energy level E_1 to E_2. Einstein realized that in the presence of photons of the appropriate energy, electrons may also be induced to fall to lower energy levels. Photons of energy $E_2 - E_1 = \hbar\omega$ will stimulate electrons in the excited state E_2 to emit further photons of the same energy and fall to energy level E_1.

Einstein's argument was based on a simple model of atoms described by a two-level system where $E_2 > E_1$, and the population of electrons at each level E_m is n_m. Einstein postulated that the rate at which population n_2 changes is

$$\frac{dn_2}{dt} = -n_2 A_{21} - n_2 B_{21} u(\omega) + n_1 B_{12} u(\omega) \,, \qquad (10.116)$$

where $u(\omega)$ is the spectral energy density of the photons per unit volume. Here, $-n_2 A_{21}$ is the rate of spontaneous emission, $-n_2 B_{21} u(\omega)$ is the rate of stimulated emission and $n_1 B_{12} u(\omega)$ is the rate of stimulated absorption that raises electrons from level E_1 to E_2. A_{21} is an intrinsic property of the energy level related to its half-life, whereas through $u(\omega)$ the terms containing B_{21} and B_{12} are dependent on the presence of other photons of frequency ω. For a gas of these atoms in thermal equilibrium with black body radiation, the populations of electrons in the two levels remain constant, i.e. $\frac{dn_1}{dt} = \frac{dn_2}{dt} = 0$, so from equation (10.116),

$$(n_1 B_{12} - n_2 B_{21}) u(\omega) = n_2 A_{21} \,, \qquad (10.117)$$

and rearranging gives

$$u(\omega) = \frac{A_{21}}{B_{12}} \frac{1}{\left(\frac{n_1}{n_2} - \frac{B_{21}}{B_{12}} \right)} \,. \qquad (10.118)$$

In thermodynamic equilibrium at temperature T, the ratio of the atom populations is given by the ratio of Gibbs factors

$$\frac{n_1}{n_2} = \frac{e^{-\frac{E_1}{T}}}{e^{-\frac{E_2}{T}}} = e^{\frac{\hbar\omega}{T}} \,, \qquad (10.119)$$

while $u(\omega) = \frac{1}{V} E(\omega)$ is given by the Planck formula (10.108), so we find

$$u(\omega) = \frac{\hbar\omega^3}{\pi^2} \frac{1}{e^{\frac{\hbar\omega}{T}} - 1} = \frac{A_{21}}{B_{12}} \frac{1}{\left(e^{\frac{\hbar\omega}{T}} - \frac{B_{21}}{B_{12}} \right)} \,. \qquad (10.120)$$

This equation can only be satisfied if $B_{12} = B_{21}$, and following this observation we derive the surprising relationship $\frac{A_{21}}{B_{21}} = \frac{\hbar\omega^3}{\pi^2}$.

Stimulated emission is a direct consequence of the bosonic nature of photons. Were it not for the stimulated emission term $B_{21} u(\omega)$ in equation (10.116), we would have obtained the result $u(\omega) = \frac{A_{21}}{B_{12}} e^{-\frac{\hbar\omega}{T}}$, appropriate to distinguishable particles. Speaking rather informally, bosons like to occupy the same state as other bosons.

We can evaluate the importance of stimulated emission by considering a couple of realistic examples. For red light of wavelength 632.8 nm, $\omega \simeq 3 \times 10^{15}$ s^{-1}. At room temperature, $T \simeq 0.025$ eV $= 4 \times 10^{-21}$ J and therefore $\frac{\hbar\omega}{T} \simeq \frac{10^{-34} \times 3 \times 10^{15}}{4 \times 10^{-21}} = 75$. From equation (10.120), and the equality of the B coefficients,

$$\frac{A_{21}}{B_{21} u(\omega)} = e^{\frac{\hbar\omega}{T}} - 1 \simeq e^{75} \,, \qquad (10.121)$$

so spontaneous emission of red light vastly exceeds stimulated emission at room temperature. However, for $\omega \simeq 10^{12}$ s^{-1}, corresponding to the high-frequency end of the microwave spectrum, the opposite is true.

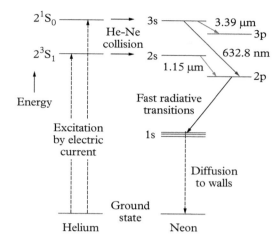

Fig. 10.11 Energy levels in helium atoms and neon atoms. The population inversion is between the 3s and 2p neon levels.

For such microwaves, $\frac{\hbar\omega}{T} \simeq 0.025$ and therefore

$$\frac{A_{21}}{B_{21}u(\omega)} = e^{0.025} - 1 \simeq 0.025 \,, \tag{10.122}$$

so stimulated emission dominates over spontaneous emission. In general, for radiation with much greater frequency than the peak of the black body spectrum, spontaneous emission will dominate, while for radiation with lower frequency than the black body peak, stimulated emission will dominate.

It might appear that the processes of spontaneous and stimulated emission are quite distinct, but this is not really the case. Even when the electromagnetic field does not contain any photonic excitations, each mode still has zero point energy $\frac{1}{2}\hbar\omega$ and spontaneous emission may be regarded as emission stimulated by these zero point oscillations.

The phenomenon of stimulated emission has given us a technology so important that its acronym *laser* (Light Amplification by the Stimulated Emission of Radiation) has entered our everyday language. To create a laser the system must be out of thermodynamic equilibrium, with energy supplied to maintain a population inversion in the electron energy levels. We will consider how this is possible shortly. In a two-level system, if we arrange for the majority of the electrons to occupy level E_2, then an electron that spontaneously falls to level E_1 will emit a photon of frequency ω that may stimulate the emission of further photons of the same frequency and trigger a cascade of photons. By pumping in energy to return the electrons to level E_2, we can obtain a continuous output of photons of frequency ω. Most significantly, each of the photons whose emission is stimulated not only has the same frequency but also the same polarization, and it is in phase with the stimulating wave and is emitted in the same direction.

The first practical lasers, demonstrated in 1960, were based on a gas of helium and neon atoms mixed in a ratio of 10:1. Helium atoms have excited states close to those of neon, as shown in Figure 10.11. Passing an electrical current through the gas mixture excites electrons in the helium atoms. The helium atoms then undergo collisions with the neon atoms and this

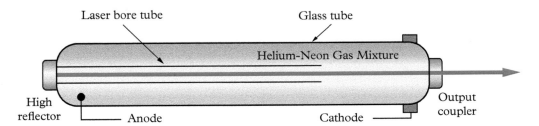

Fig. 10.12 Helium–neon laser. The bore tube is narrow so the neon atoms will diffuse rapidly to the walls, where collisions return electrons in the neon 1s excited state to the ground state.

raises the neon outer electrons into excited states. (The small energy difference is supplied by the kinetic energy of the helium atoms.) The neon 3s state is metastable and is therefore heavily populated, whereas the neon 2p state decays rapidly to the 1s state, so it has a very small population.[8] As long as a current is passed through the gas, this population inversion will persist.

To produce a laser, the gas mixture is held in a narrow chamber between two highly polished mirrors forming an optical resonant cavity, as shown in Figure 10.12. Spontaneous decays from the neon 3s to 2p levels produce photons that stimulate further such transitions. A beam of these photons reflects back and forth between the end mirrors. (Any photons that are emitted off the beam axis are lost.) The laser is essentially a very high frequency electromagnetic oscillator. One of the end mirrors is partially (99%) silvered, which allows a beam of photons to escape. The helium–neon laser emits red light of 632.8 nm wavelength. As a cheap and compact, but powerful monochromatic light source, it has found many applications including barcode scanners. Many other types of laser systems have now been developed based on a whole range of materials: gas, liquid, crystalline solids, semiconductors and insulators.

A laser beam is very different to light from a conventional source, such as an incandescent light bulb. When a tungsten filament is heated by an electrical current, it emits radiation with a thermal spectrum that satisfies the Planck formula to a very good approximation. The radiation is the result of random processes and is emitted in all directions, with all polarizations and a random distribution of phases, so the light is incoherent. By contrast, the photons in a laser are all in phase, and are emitted in the same direction in a narrow beam, with the same polarization. A standard lens will focus the coherent light emitted by a laser to a diffraction-limited spot whose size depends on the wavelength. Flux densities as high as 10^{17} W cm^{-2} can readily be achieved, which compares to 10^3 W cm^{-2} for an oxyacetylene torch. This has given rise to applications from welding to nuclear fusion research. The coherence of the laser beam is essential for many other applications including holography and gravitational wave interferometers. Other applications include data storage and retrieval from CDs and DVDs, eye surgery, guide stars for adaptive optics, laser printing and fibre optics. Much effort is being expended on developing photonics technology as an alternative to electronics. As photons travel at the speed of light and undergo negligible interactions with other photons, photonic devices promise significant advantages of smaller size and higher speed.

---

[8] We are using the Paschen notation for excited states here.

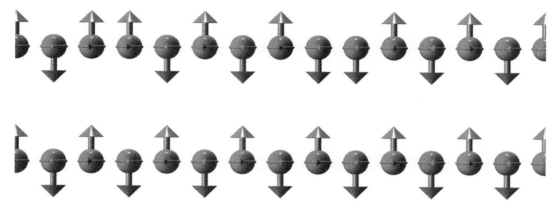

Fig. 10.13 Bottom: The ground state of the 1-dimensional antiferromagnet. Top: The 1-dimensional antiferromagnet with two defects.

10.12 Magnetization in Spin Systems

As we saw in Chapter 9, in a crystalline solid formed of a regular lattice of atoms, the atoms often have a net spin. As a consequence, each behaves like a microscopic magnet, with a magnetic moment that in some solids produces magnetic interactions between neighbouring atoms. There is a great deal of interest in explaining the thermodynamic properties and physical characteristics of such materials in terms of simple models based on a lattice system of spins. Let us consider atoms with spin $\frac{1}{2}$ so that there are only two independent quantum spin states, and the structure of the lattice forces each spin to be either up or down along one of the lattice axes, with spin projections $+\frac{1}{2}$ and $-\frac{1}{2}$. Superpositions of spin can be ignored. The magnetic moments will then point up or down, with a strength that is a constant times the spin projection.

Magnetic forces fall off rapidly with distance, so we will assume that the only contributions to the energy come from the interaction between spins at neighbouring lattice sites. In most common materials, the microscopic magnets behave in a manner that is familiar from pairs of bar magnets. Opposite poles prefer to be close together. In such materials, the lowest energy arrangement of atomic spins is with adjacent magnetic moments aligned in opposite directions. Any parallel magnetic moments increase the energy.

The simplest such model is a 1-dimensional chain consisting of a large number, $N+1$, of equally spaced spins. In the ground state the spins alternate, as shown in Figure 10.13 (bottom), and the state is said to be exactly antiferromagnetically ordered. Let us normalize the energy so that the ground state has zero energy. Excited states have defects where neighbouring spins are aligned in the same direction. The state shown in Figure 10.13 (top) has two defects.

We can study the thermodynamic properties of this system from first principles. A defect raises the energy by ε, say, and assuming only nearest neighbour spin interactions, the energy of n defects is $n\varepsilon$. There are N possible locations for the n defects, and the defects must all be at different locations. Assuming the spin at the left end of the chain is fixed to be up, the number of states Ω with n defects is the combinatorial factor

$$\binom{N}{n} = \frac{N!}{n!(N-n)!}, \tag{10.123}$$

which is the number of ways of choosing n locations from N. As N is large, let us write $n = \alpha N$, where α is the fractional density of defects. Then

$$\Omega = \binom{N}{\alpha N} = \frac{N!}{(\alpha N)!((1-\alpha)N)!}. \tag{10.124}$$

Using the approximation (10.61), $\log X! = X \log X - X$, we find the entropy as a function of α,

$$\begin{aligned}
S = \log \Omega &= N \log N - N - \alpha N \log(\alpha N) + \alpha N - (1-\alpha)N \log((1-\alpha)N) + (1-\alpha)N \\
&= N\{\log N - \alpha \log(\alpha N) - (1-\alpha)\log((1-\alpha)N)\} \\
&= N\{\log N - \alpha(\log \alpha + \log N) - (1-\alpha)(\log(1-\alpha) + \log N)\} \\
&= -N\{\alpha \log \alpha + (1-\alpha)\log(1-\alpha)\}, \tag{10.125}
\end{aligned}$$

and in terms of α, the energy is $E = \alpha N \varepsilon$.

Differentiating these expressions with respect to α, for fixed N, gives

$$\frac{dS}{d\alpha} = -N(\log \alpha + 1 - \log(1-\alpha) - 1) = N \log\left(\frac{1}{\alpha} - 1\right), \tag{10.126}$$

$$\frac{dE}{d\alpha} = N\varepsilon. \tag{10.127}$$

Therefore, for this system,

$$\frac{1}{T} = \frac{dS}{dE} = \frac{dS}{d\alpha}\frac{d\alpha}{dE} = \frac{1}{\varepsilon} \log\left(\frac{1}{\alpha} - 1\right). \tag{10.128}$$

If we invert this relation, we find

$$\alpha = \frac{1}{e^{\frac{\varepsilon}{T}} + 1}. \tag{10.129}$$

This is the fraction of neighbouring pairs of spins where the spins are aligned to form a defect in the antiferromagnet. At low temperatures $T \ll \varepsilon$, the fraction is $\alpha \simeq e^{-\frac{\varepsilon}{T}}$, which is exponentially small, so there are large blocks of spins with exact antiferromagnetic ordering. However, because of the small density of defects there is no long-range order as N increases to infinity. A spin that is a large distance from the left end of the chain is equally likely to be up or down. At high temperatures $T \gg \varepsilon$, the fraction α approaches $\frac{1}{2}$, so defects and non-defects are equally likely. The spins are completely randomized, with no correlation even between nearest neighbours as $T \to \infty$.

This is just the simplest model of lattice spins. There are countless others that have been devised. In one dimension there are ferromagnetic chains where neighbouring spins prefer to be aligned, as shown in Figure 10.14. Either type of chain can be subject to an external magnetic field, which affects the energy and tends to align the spins in one direction. The spin chain discussed above is a purely thermal system, but by including the external field it becomes a thermodynamic system with the external field replacing the pressure variable in a gas and the net magnetization replacing the volume.

It is also possible to consider lattices of spins in two or three dimensions, and to allow the spin projections to have more than two values to model atoms with spins greater than

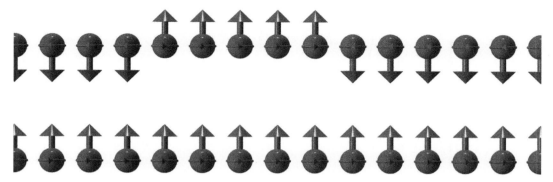

Fig. 10.14 Bottom: The ground state of the 1-dimensional ferromagnet. Top: The 1-dimensional ferromagnet with two defects.

$\frac{1}{2}$. For a cubic lattice, the couplings between neighbouring spins may differ depending on whether the neighbours are parallel to the spin projection or perpendicular.

There are also truly quantum spin lattices, where each atom has spin $\frac{1}{2}$, and has a spin operator **s**, but one cannot immediately assume each spin is either up or down. The quantum Hamiltonian for each neighbouring pair of spins, $\mathbf{s}^{(1)}$ and $\mathbf{s}^{(2)}$, may have the isotropic form $c\,\mathbf{s}^{(1)} \cdot \mathbf{s}^{(2)}$, or the more complicated form $c_1 s_x^{(1)} s_x^{(2)} + c_2 s_y^{(1)} s_y^{(2)} + c_3 s_z^{(1)} s_z^{(2)}$. The total Hamiltonian is a sum of such terms over all neighbouring pairs of spins.

Some of these lattice spin models have thermodynamic properties that can be calculated exactly, notably the *2-dimensional ferromagnetic Ising model*, where $c_1 = c_2 = 0$ and $c_3 < 0$. This was solved by Lars Onsager using more sophisticated combinatorial methods than we used for the 1-dimensional antiferromagnetic chain earlier. The most striking result for the Ising model is that at low temperatures there is infinite-range ferromagnetic order, even in the absence of any external magnetic field. This means that the majority of the system forms a connected region with the spins all pointing in the same direction. The defects occur as small islands of spins pointing in the opposite direction, and as the temperature increases the disorder increases. At a critical temperature T_{Curie}, called the *Curie temperature* in a ferromagnet, there is a phase transition, and the infinite-range order disappears. If a particular spin is up, then nearby spins are still more likely to be up than down, but the probability that a distant spin is up approaches $\frac{1}{2}$ as the distance increases.

10.13   A Little about Phase Transitions

The physical characteristics of many materials change dramatically at a precise temperature. The freezing and boiling of water are the most familiar examples of these transformations, which are known as phase transitions. A phase transition occurs when there is a discontinuity in the thermodynamic behaviour and often even the appearance of a system. At atmospheric pressure, water freezes into ice at 273 K (to the nearest degree) and steam turns into liquid water at 373 K. Ice is clearly different from liquid water or steam because it is a crystalline solid, and 334 joules, known as the *latent heat*, is required to melt 1 gram of ice. The difference between water and steam is not so easy to identify. There is certainly a discontinuity in the properties of water at 373 K because a substantial amount of latent heat is required to convert water to steam at this temperature. At atmospheric pressure, for 1 gram of water about 420 joules is required to raise the temperature from 273 to 373 K, and then 2270

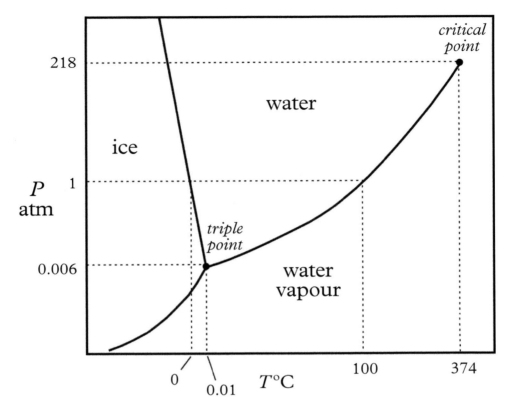

Fig. 10.15 Schematic T, P phase diagram for water. (The region around the triple point is enlarged, and the slope of the ice–water coexistence curve is exaggerated.)

joules to convert the water to steam. This is why boiling away all the water in a kettle takes a long time. The volume of steam is also much greater than the volume of water, and the entropy of steam is greater than the entropy of water.

We have met other types of phase transition, for example, the Bose–Einstein condensation of a gas of bosons, where below a critical temperature a finite fraction of the atoms are in the 1-particle ground state. There are also phase transitions in solid materials that are related to their electric and magnetic properties, such as the Curie temperature of a ferromagnet, below which a ferromagnetic material spontaneously gains a net magnetization due to a net alignment of the atomic spins. There are other phase transitions in which the crystalline structure of a solid is transformed. For instance, the crystal structure of iron changes from body centred cubic to face centred cubic at a critical temperature of 1044 K. (For descriptions of these packings, see Figure 9.20.) This is connected to the ferromagnetic transition in iron at its Curie temperature of 1043 K. Other phase transitions occur in mixtures and solutions of chemicals, such as liquid crystals. Another example is the phenomenon of superconductivity, which sets in suddenly when some materials are cooled below a critical temperature. Below this transition temperature, the superconductor will expel any magnetic field that passes through the material and its electrical resistance disappears, so any electric

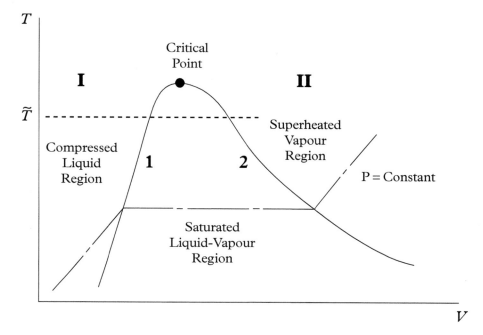

Fig. 10.16 V, T phase diagram.

current will persist indefinitely. One of the few model systems where a phase transition can be rigorously shown to occur is the Ising ferromagnetic system in two dimensions, as discussed previously.

A basic feature of distinct phases is that they can coexist in systems that are in contact (e.g., water in contact with steam), and still be in equilibrium. We know from general thermodynamic considerations that systems in equilibrium must have the same temperature, pressure and chemical potential. If not, they would exchange energy or particles, or the surface separating them would move. Let us suppose that the independent variables are temperature and pressure and that all other thermodynamic quantities are functions of these. In particular, the chemical potentials of two phases I and II are two distinct functions $\mu_I(T, P)$ and $\mu_{II}(T, P)$. The phases can then coexist in equilibrium on the phase transition curve in the (T, P)-plane where $\mu_I(T, P) = \mu_{II}(T, P)$. This is a single equation that relates P to T. (μ_I is a function defined in the region of phase I, but it is theoretically possible to extend its range some way beyond this curve into the region of phase II, despite phase I being unstable there. Similarly, μ_{II} can be extended into the region of phase I. This is related to the phenomenon of supercooling and superheating.)

Figure 10.15 is a typical phase diagram in the (T, P)-plane. Many systems have three distinct phases, or possibly more. Three phases can have equal chemical potentials only at isolated points in the (T, P)-plane. A point where three phases coexist is known as a triple point.

A phase diagram looks different if plotted in the (V, T)-plane. Here, by V we mean the total volume of a fixed mass of the substance, so the volume is not the same for the two phases. A typical liquid–gas phase diagram in the (V, T)-plane is shown in Figure 10.16.

Moving to the right along a fixed temperature line $T = \tilde{T}$, in region I the volume is increasing but small, and the substance is a pure liquid. The pressure is decreasing but this is not shown. On curve 1 the change of phase begins, and between curve 1 and curve 2 there are samples of liquid and gas coexisting and in contact. The volume increases between the curves, and the fraction of liquid decreases from 1 to 0, but the pressure and temperature are unchanging. At curve 2 there is pure gas, and in region II the volume continues to increase, with the pressure decreasing again.

There is an interesting relation between the latent heat and the slope of the phase separation curve in the (T, P)-plane, called the *Clausius–Clapeyron relation*. Adjacent points just to the left and right of the curve represent the system in phase I and phase II. T, P and μ all have equal values at two such points, and as particles are not created or destroyed in a phase transition, N also has equal values. Therefore the Gibbs free energy, $G = \mu N$, has equal values at the two points. The quantities that jump discontinuously as the system crosses the phase separation curve are the entropy S and volume V. Let us denote these quantities on the two sides of the curve by S_I, V_I and S_{II}, V_{II}. We also denote the Gibbs free energy on the two sides by G_I and G_{II} although these are equal.

Now consider an infinitesimal motion (dT, dP) along the curve to a neighbouring pair of adjacent points. The infinitesimal changes of G_I and G_{II} along the two sides of the curve are equal. Therefore, as $dG = -S\,dT + V\,dP + \mu\,dN$, and particle number is fixed,

$$-S_I\,dT + V_I\,dP = -S_{II}\,dT + V_{II}\,dP\,, \tag{10.130}$$

so

$$(S_{II} - S_I)\,dT = (V_{II} - V_I)\,dP\,, \tag{10.131}$$

which means that along the curve, the slope is

$$\frac{dP}{dT} = \frac{S_{II} - S_I}{V_{II} - V_I}\,. \tag{10.132}$$

Quite generally, $T\,dS$ is an infinitesimal amount of heat. At a phase transition, with T constant, we can integrate this and deduce that $T(S_{II} - S_I)$ is the latent heat L of the transition. Therefore equation (10.132) can be re-expressed as

$$\frac{dP}{dT} = \frac{L}{T(V_{II} - V_I)}\,, \tag{10.133}$$

and this is the Clausius–Clapeyron relation (Figure 10.17). The slope of the phase transition curve is proportional to the latent heat required for the transition. Both L and V are extensive, so the slope does not depend on the amount of material involved. In a liquid to gas transition, L and T are positive and $V_{II} \gg V_I$, so the slope is positive, which means that as the pressure increases, the boiling temperature increases. This is why a pressure cooker, in which steam is partially trapped, speeds up cooking, and also why on a high mountain, where atmospheric pressure is lower than at sea level, water boils at a temperature lower than 373 K.

In the melting transition of ice to water, the volume of liquid water is less than the volume of an equal mass of solid ice. This is a surprising consequence of the crystal structure of ice, as mentioned in section 9.4.1. As L and T are positive, the Clausius–Clapeyron relation

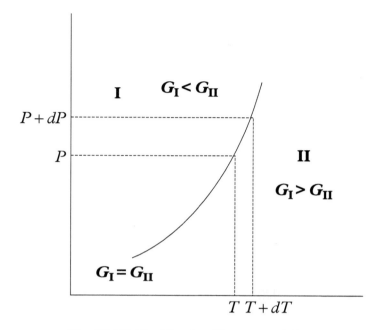

Fig. 10.17 The Clausius–Clapeyron relation.

implies that as the pressure decreases, the melting temperature increases, but the effect is very small because V_{II} is very close to V_I. At a pressure of around one-hundredth of atmospheric pressure, water has a triple point. At lower pressures, liquid water no longer exists, as shown in Figure 10.15.

10.14 Hawking Radiation

In section 6.11.1, we showed that the event horizon of a non-rotating black hole of mass M is a sphere of surface area

$$A = 4\pi r_S^2 = 16\pi G^2 M^2 \,, \tag{10.134}$$

where r_S is the Schwarzschild radius. Any material that falls into a black hole will increase its mass, and also its area. If two Schwarzschild black holes of mass M merge, the area of the resultant black hole of mass $2M$ is $64\pi G^2 M^2$ which is greater than the combined area $2 \times 16\pi G^2 M^2$ of the two original black holes. This is true in general, even if the black holes are rotating and charged. Indeed, starting from very general assumptions, Stephen Hawking proved in 1971 that in any process, the total area A of black hole event horizons in the universe must increase:

$$\frac{dA}{dt} \geq 0 \,. \tag{10.135}$$

This is known as the black hole area theorem.

At around the same time, Jacob Bekenstein was concerned that black holes appear to offer a sink down which the universe could lose some of its entropy. Any material that falls into a black hole is lost forever to the rest of the universe, along with the entropy that it contains. This seemed to be a way to reduce the entropy of the universe that would violate the second law of thermodynamics, so it was very puzzling. By 1972, Bekenstein realized that there might be a solution if it was possible to assign an entropy to the black hole. Bekenstein noted the similarity between Hawking's area theorem and the second law of thermodynamics and tentatively proposed that there might be a precise correspondence between the two, such that a black hole's area was actually a measure of its entropy. When material falls into a black hole, the area of its event horizon increases. If this was interpreted as an increase in the entropy of the black hole, then it might compensate for the loss of entropy to the rest of the universe. But how could this be? The area theorem of black holes is a geometrical result in general relativity, whereas the second law of thermodynamics is a statistical law about heat. Furthermore, black holes were believed to be essentially featureless, being characterized simply by their mass, angular momentum and electric charge. How could they have any statistical properties?

Hawking at first dismissed Bekenstein's idea. If an entropy is assigned to a black hole, then the black hole must behave like a body with a well defined temperature, so it must emit radiation. This contradicted everything that was believed about black holes. When a black hole is modelled in general relativity, its temperature must be zero because although radiation may fall into the black hole, nothing can ever come out. Hawking soon realized, however, that the situation is very different if quantum mechanics is taken into account. General relativity works incredibly well. It is the best theory of gravity that we have, but it is a classical theory. In reality the world is quantum mechanical, so the ultimate theory of gravity must be a quantum theory. Hawking showed that a quantum black hole does indeed emit radiation, now known as Hawking radiation, and therefore has a non-zero temperature.

Hawking's area theorem (10.135) is often called the second law of black hole mechanics, and following Bekenstein, it is regarded as the analogue of the second law of thermodynamics (10.14). Pursuing the analogy further, the first law of black hole mechanics, which is the analogue of the first law of thermodynamics (10.22), should relate changes in the mass of a black hole (or equivalently, its energy) to changes in the area of its event horizon and changes in its angular momentum J. From the area–mass relation (10.134), it follows that for a non-rotating black hole $dA = 32\pi G^2 M \, dM$, so $dM = \frac{\kappa}{8\pi G} dA$ where $\kappa = \frac{1}{4GM}$. This is the first law when $J = 0$. More generally, for a rotating black hole, the first law of black hole mechanics is

$$dM = \frac{\kappa}{8\pi G} dA + \Omega \, dJ \,, \tag{10.136}$$

where $\Omega = \frac{a}{2Mr_+}$ is the angular velocity of the event horizon, as defined in section 6.11.2. There is a further term in the first law if the black hole is electrically charged.

$\kappa = \frac{1}{4GM}$ is interpreted as the surface gravity at the event horizon, which is the relativistic generalization of the gravitational acceleration g experienced at the surface of an object.[9] The surface gravity κ must be constant over the event horizon of a black hole, which provides us with an analogue of the zeroth law of thermodynamics, with a multiple of κ playing the role of temperature. By studying photon emission from a black hole using quantum theory, Hawking calculated that a black hole has the *Hawking temperature*

$$T_{\mathrm{H}} = \frac{\hbar}{2\pi} \kappa = \frac{\hbar}{8\pi GM} \,. \tag{10.137}$$

---

[9] The derivation of the expression for κ requires a rigorous general relativistic treatment, but we can understand its form by considering the Newtonian acceleration at the Schwarzschild radius, which is $\frac{GM}{r_{\mathrm{S}}^2} = \frac{GM}{(2GM)^2} = \frac{1}{4GM}$.

It then follows from the first law of black hole mechanics (10.136) that $\frac{\kappa}{8\pi G}\,dA = \frac{T_\mathrm{H}}{4\hbar G}\,dA = T_\mathrm{H}\,dS_\mathrm{BH}$, so the entropy of the black hole is

$$S_\mathrm{BH} = \frac{A}{4\hbar G}. \tag{10.138}$$

The temperature of the black hole is the temperature that would be measured by someone well outside its event horizon, as deduced from the Hawking radiation that it emits. So despite the incredibly violent processes that might occur inside a black hole, its temperature can be very low. In fact, the temperature of a stellar-mass black hole is unmeasurably low.

The easiest way to see the connection between the mass of a black hole and its temperature is to consider the wavelength of the radiation that it emits. According to quantum mechanics, a black hole cannot confine radiation with a wavelength greater than its event horizon. The electromagnetic field within a black hole is continuously fluctuating. Any photons that are produced with a wavelength greater than the Schwarzschild radius may find themselves outside the black hole by quantum mechanical tunnelling and escape to a distant observer. If the typical wavelength of the Hawking radiation is $\lambda_\mathrm{H} \simeq r_\mathrm{S} = 2GM$, then its frequency is $\omega_\mathrm{H} = \frac{2\pi}{\lambda_\mathrm{H}} \simeq \frac{\pi}{GM}$, and from this we can roughly estimate the Hawking temperature. Using equation (10.109), relating the peak frequency of black body radiation to temperature, we obtain

$$T_\mathrm{H} \simeq \frac{\hbar\omega_\mathrm{H}}{2.8} \simeq \frac{\pi}{2.8}\frac{\hbar}{GM}. \tag{10.139}$$

Hawking's more precise calculation gives the temperature (10.137).

The Schwarzschild radius of a stellar-mass black hole is several kilometres, so the typical wavelength of its Hawking radiation is also several kilometres, corresponding to a temperature of less than 10^{-7} K. This is so low that such a black hole inevitably absorbs more radiation than it emits. The universe is bathed in radiation known as the cosmic microwave background produced shortly after the Big Bang. It has a spectrum corresponding to a black body temperature of around 2.7 K, far higher than the Hawking temperature of a stellar-mass black hole.

As the wavelength of Hawking radiation is comparable to the size of the event horizon, a black hole with a Schwarzschild radius of a few hundred nanometres will emit radiation in the visible spectrum and its temperature is therefore in the region of 10^3 K. This corresponds to a black hole of mass 10^{20} kg. Hawking speculated that such mini black holes might have formed immediately after the Big Bang when the universe was very dense. This would have occurred if the matter in the early universe was quite lumpy and some of the denser regions collapsed to form black holes. These hypothetical mini black holes are known as *primordial black holes*. They could, for instance, have a mass comparable to that of an asteroid packed into a region that is smaller than an atom.[10] Being hot, the mini black holes would emit large amounts of radiation and thereby gradually lose their mass. With this decrease in mass, their temperature would rise further, increasing the rate at which they radiate. In a runaway process the temperature of the mini black hole would rise dramatically in its final moments until it disappeared with a huge blast of radiation.

Primordial black holes with mountain-sized masses of around 10^{11} kg should currently be on the verge of exploding. Slightly larger and the black holes will continue emitting

[10] The mass of a large asteroid with a diameter of 1000 km is around 10^{20} kg.

X-rays and gamma rays for many aeons to come. Astronomers have searched for the gamma-ray blasts that would be produced by exploding mini black holes but none has ever been seen. Whether they exist or not, there is absolutely no doubt that the general principles of the theory are correct and that black holes do emit Hawking radiation. The combination of general relativity, quantum mechanics and thermodynamics mesh together so well that these ideas must play an important role in the fundamental structure of the universe. This was the first result that linked quantum mechanics to gravity, which is why it is among the most profound ideas of modern physics. The microscopic origin of black hole entropy is still not fully understood, but the entropy is believed to correspond to the logarithm of the number of microstates of quantum gravity, describing the quantum fluctuations of the spacetime metric.

10.15 Further Reading

K. Huang, *Introduction to Statistical Physics*, London: Taylor and Francis, 2001.

L.D. Landau and E.M. Lifschitz, *Statistical Physics (Part 1): Course of Theoretical Physics, Vol. 5 (3rd ed.)*, Oxford: Butterworth-Heinemann, 1980.

For a survey of phase transitions, see

J.M. Yeomans, *Statistical Mechanics of Phase Transitions*, Oxford: OUP, 1992.

For an account of Bose–Einstein condensation, see

C.J. Pethick and H. Smith, *Bose–Einstein Condensation in Dilute Gases*, Cambridge: CUP, 2002.

For a broad survey of light, optics, black body radiation and lasers, see

I.R. Kenyon, *The Light Fantastic*, Oxford: OUP, 2008.

11

Nuclear Physics

11.1 The Birth of Nuclear Physics

The existence of the nucleus was deduced by Rutherford in 1911 following the gold foil experiments of Hans Geiger and Ernest Marsden that he had initiated. The results from the scattering of alpha particles by the gold foil enabled Rutherford to calculate the radius of the nucleus of a gold atom.[1] The modern value is 7.3 fm, where fm is a femtometre or 10^{-15} m. By comparison the radius of a gold atom is 135 pm, where pm is a picometre or 10^{-12} m. This is greater by a factor of 18,500, so the nucleus occupies a tiny proportion of the atom. Rutherford went on to discover the first constituent of the nucleus later in the decade, which he named the proton. In 1920 Rutherford predicted that the atomic nucleus must also contain a neutral particle with a similar mass. During the 1920s Rutherford's team in Cambridge searched for this elusive particle, without success.

In 1930, Walther Bothe and Herbert Becker bombarded a beryllium sample with alpha particles from a radioactive polonium source and found that highly penetrating radiation was emitted. They assumed that the radiation consisted of high energy gamma ray photons. Soon afterwards Irène Curie and Frédéric Joliot found that this radiation would dislodge protons from a hydrogen-rich paraffin target. This meant that if the radiation really was composed of gamma rays, then they must have far higher energy than any previously discovered radiation, in excess of 52 MeV. When James Chadwick heard about the Joliot–Curie experiment, he realized that it might be the signature of Rutherford's neutral particle. In a series of meticulous experiments over a period of about six weeks in 1932, Chadwick proved that the bombardment of beryllium was indeed releasing a new particle, the neutron, the nuclear particle predicted by Rutherford. With a mass almost equal to that of the proton, a neutron requires far less kinetic energy than a gamma ray to eject a proton from paraffin. The discovery of the neutron transformed our understanding of the structure of matter. It was now clear that an atom consists of a positively charged nucleus surrounded by a cloud of orbiting electrons and that the nucleus itself is formed of a collection of two types of nucleons: protons and neutrons. Although the nucleus occupies a tiny region of an atom, it contains almost all the mass, so the density of the nucleus is vastly greater than the density of ordinary matter. The masses of the neutron and proton are 939.57 MeV and 938.27 MeV, which are 1838.7 and 1836.2 times the mass of the electron, respectively. As there are usually

---

[1] The alpha particles did not excite or disrupt the gold nuclei and obeyed Rutherford's elastic scattering formula, so the distance of closest approach to the nuclei could be determined by equating their kinetic energy to the Coulomb potential. This gave Rutherford a maximum radius for the gold nucleus, which is about three times the actual radius.

The Physical World. Nicholas Manton and Nicholas Mee, Oxford University Press (2017).
© Nicholas Manton and Nicholas Mee. DOI 10.1093/oso/9780198795933.001.0001

at least twice as many nucleons as electrons in an atom, the mass of the nucleus is around 4000 times the mass of the electrons.

Chadwick's discovery was the key to understanding and exploiting the atomic nucleus. The outbreak of war at the end of the decade led to the Manhattan project in the United States and an accelerated effort to develop nuclear technology. By the end of the war, just 13 years after the discovery of the neutron, the world had nuclear fission reactors and nuclear weapons had been used in war.

11.2   The Strong Force

The standard notation for a nucleus of element X is $^A_Z X_N$, where the atomic number Z is equal to the number of protons in the nucleus, and the nuclear mass number is $A = Z + N$, where N is the number of neutrons in the nucleus. Often this is simplified to $^A X$, as Z and N can easily be calculated from the name of the element and the mass number. For example, the alpha decay of uranium-238 is described by the nuclear transformation

$$^{238}_{92}U_{146} \quad \rightarrow \quad ^{234}_{90}Th_{144} \quad + \quad ^4_2He_2 \,. \tag{11.1}$$

Here $^{238}_{92}U_{146}$ and $^{234}_{90}Th_{144}$ represent uranium and thorium nuclei and $^4_2He_2$ is a helium nucleus or alpha particle. The Q value of a nuclear process is the energy released in the nuclear transformation; in this case Q is 4.27 MeV. This is shared between the kinetic energy of the alpha particle (4.198 MeV) and the kinetic energy of the recoiling thorium nucleus (0.070 MeV).

Many properties of the nucleus are readily deduced from a few basic experimental facts. A nucleus is composed of a tightly bound collection of protons and neutrons, with similar numbers of each. Protons carry a positive electric charge, whereas neutrons are electrically neutral, which immediately implies that there must be a force operating within the nucleus that is far stronger than the electromagnetic force to overcome the repulsion between the protons. We know this force as the strong force. Within a nucleus it is in the region of 100 times the strength of the electrostatic force between two protons. The electrostatic force is described by an inverse square law, which gives it an unlimited range. By contrast the strong force only operates within the nucleus. It has a range of about 1–3 fm. At distances shorter than 1 fm nucleons have a hard repulsive core.[2] If the strong force were not repulsive at these short distances, nuclear material would simply collapse.

The energies involved in chemical bonding are comparatively small. Although the mass of an atom decreases when it bonds to another atom, the difference in mass is too small to measure. This is in stark contrast to nuclear binding energies, which are best expressed in terms of the change in mass of the reacting nuclei. For instance, the simplest nuclear system is the deuteron, which is formed of one neutron and one proton. The deuteron mass m_d is significantly less than the sum of the neutron mass m_n and the proton mass m_p, as

$$m_d = 2.01355 \, u \tag{11.2}$$

and

$$m_n + m_p = 1.00866 \, u + 1.00728 \, u = 2.01594 \, u \,. \tag{11.3}$$

---

[2] The core radius of the proton is around 0.8 fm.

Here, u is the unified atomic mass unit, which is defined to be one-twelfth of the mass of an unbound neutral carbon atom in its ground state; u is equal to 931.4941 MeV or 1.660539×10^{-27} kg. The binding energy of the deuteron is therefore

$$\Delta m = (m_n + m_p) - m_d = 2.01594\,u - 2.01355\,u = 0.00239\,u\,, \tag{11.4}$$

which is 2.2 MeV. In general, the binding energy of a nucleus is given by

$$B(Z, A) = Zm_p + Nm_n + Zm_e - m(Z, A)\,, \tag{11.5}$$

where m_e is the mass of the electron and $m(Z, A)$ is the mass of the atom. (Atomic masses are usually quoted rather than nuclear masses, as they are easier to measure and the number of electrons remains constant in nuclear reactions so their masses cancel out.)

The most stable low-mass nuclei typically contain equal numbers of protons and neutrons, so $Z = N$. There are many such examples including ^{4}He, ^{12}C, ^{14}N and ^{16}O. This suggests that the strong force acts on protons and neutrons in the same way. Heavy nuclei contain more neutrons than protons; the most stable nuclei typically have $\frac{Z}{A} \simeq 0.4$ or $N \simeq 1.5\,Z$. This is readily understood when the electrostatic repulsion of the protons is taken into account. Figure 11.1 shows a plot of neutrons against protons in atomic nuclei. Stable nuclei contain neutrons and protons in just the right proportions to fall in the *valley of stability*, between the *radioactive nuclei* that contain an excess of either protons or neutrons. The value of Z determines the identity of each element. Nuclei with the same Z, but different N are referred to as different isotopes of the element. Many elements have more than one stable isotope. For example, carbon has two stable isotopes, ^{12}C and ^{13}C.

The distribution of nucleons within the nucleus has been explored in a wide range of scattering and other experiments.[3] The results show that the nucleon density remains approximately constant throughout the body of the nucleus, but falls off towards the edge, so the nucleus has a lower density skin. As shown in Figure 11.2, the skin depth is approximately the same for different nuclei and is around 2.3 fm. This is due to the limited range of the strong force.

Furthermore, apart from the very lightest nuclei, which are slightly less dense, the nucleon density is almost the same for all nuclei. As nucleons are added to a nucleus, the volume V of the nucleus grows in proportion. It is like gluing sticky balls together. The number of nucleons per unit nuclear volume is constant, so for a nucleus of mass number A,

$$V = \frac{4\pi}{3} R_0^3 A \quad \text{and} \quad R = R_0 A^{\frac{1}{3}}\,, \tag{11.6}$$

where R_0 is a length constant and R is the nuclear radius. This should be compared to the trends for atomic radii which are much less regular. For instance, the radius of an atom of element 11 sodium is 180 pm, whereas the radius of an atom of element 18 argon is just 70 pm.

---

[3] These include alpha decay analyses and the spectroscopy of pionic atoms (pion–nucleus bound states). Alpha decay is very sensitive to the nuclear potential, as we will discuss later. Pions are more massive and therefore more tightly bound to the nucleus than electrons. Their orbitals penetrate the nucleus and the nuclear structure has a significant effect on their energy levels.

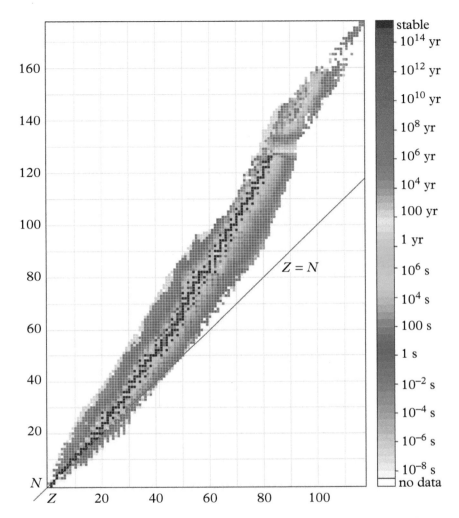

Fig. 11.1 Nuclear chart showing the valley of stability and the half-lives of nuclei.

A whole range of experiments give consistent values for R_0 in the region of 1.2 fm. Moreover, the protons and neutrons are distributed fairly uniformly throughout each nucleus, so nuclear density is constant with a value of

$$\rho_{\text{nuc}} = \frac{m_p A}{V} = \frac{3m_p}{4\pi(1.2)^3} \text{ fm}^{-3} \simeq 2.31 \times 10^{17} \text{ kg m}^{-3} . \tag{11.7}$$

In deriving this expression, we have given all the nucleons the same mass, as $m_n \simeq m_p$, and used equation (11.6).

11.2.1 The nuclear potential

It is a good approximation to treat the nucleons in a nucleus as moving quantum mechanically in an average potential produced by all the other nucleons. In accordance

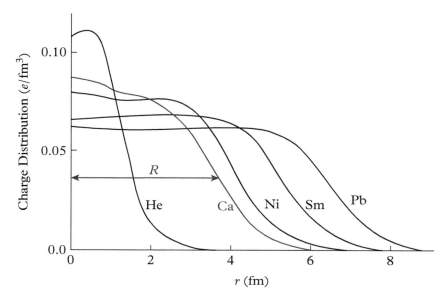

Fig. 11.2 Nuclear density profiles.

with this model, the nucleons are observed to have discrete energy levels. The potential seen by protons differs a little from that seen by neutrons, due to the Coulomb repulsion between protons. This raises the energy levels of the protons relative to those of the neutrons. These potentials are sketched in Figure 11.3. Their depth and radius depends on the nuclear mass number A.

Protons and neutrons are spin $\frac{1}{2}$ fermions, so they obey the Pauli exclusion principle and this is key to the structure of nuclei. It means that no two nucleons can occupy exactly the same state. In the nuclear ground state, the nucleons occupy the lowest energy states compatible with the Pauli principle. The nucleus can also exist in excited states where one or more protons or neutrons are promoted to higher energy levels. The nucleus may also have excited states that are due to a collective excitation of the entire nucleus. These include rotational and vibrational states. Collective excitations are not easy to understand in terms of the single nucleon energy levels.

Excited nuclei rapidly decay into their ground states by the emission of gamma ray photons. These photons typically have energies in the MeV range, which is around a million times the energy differences involved in electron transitions in atoms. Nuclei may also fall to lower energy states with the emission of other particles, such as alpha particles or neutrons or even by undergoing nuclear fission.

As the nuclear density is constant, a finite-depth spherical well of volume V is a good first approximation to the nuclear potential. This can be used to estimate the kinetic energy of the nucleons. In the ground state of the nucleus, the nucleons fill all states up to a maximum energy ε_F corresponding to a maximum momentum p_F, which is the radius of a sphere of occupied states in momentum space. ε_F is the Fermi energy and p_F the Fermi momentum, just as in solid state physics.

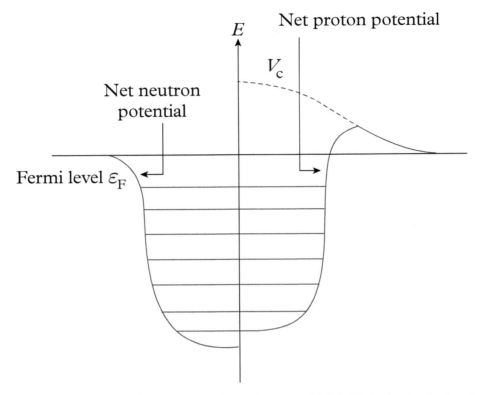

Fig. 11.3 Nuclear potential for neutrons (left) and protons (right). V_C is the Coulomb potential, which only affects protons.

The number of nucleon states in an infinitesimal region d^3p of momentum space is

$$dA = 2 \times 2 \times V \frac{d^3p}{(2\pi\hbar)^3} = 4V \frac{4\pi p^2 dp}{(2\pi\hbar)^3}, \tag{11.8}$$

where one factor of 2 is due to the two types of nucleon and the other factor of 2 represents the two spin states. This can be integrated to give the total number of nucleons in the nucleus

$$A = \frac{16\pi V}{8\pi^3\hbar^3} \int_0^{p_F} p^2 dp = \frac{2V}{3\pi^2\hbar^3} p_F^3, \tag{11.9}$$

so using equation (11.7),

$$p_F^2 = \left(\frac{3\pi^2\hbar^3 A}{2V}\right)^{\frac{2}{3}} = \left(\frac{3\pi^2\hbar^3 \rho_{\mathrm{nuc}}}{2m_p}\right)^{\frac{2}{3}}. \tag{11.10}$$

The kinetic energy of each of the highest energy nucleons is therefore

$$\varepsilon_F = \frac{p_F^2}{2m_p} = \frac{1}{2m_p}\left(\frac{3\pi^2\hbar^3 \rho_{\mathrm{nuc}}}{2m_p}\right)^{\frac{2}{3}} \simeq 35 \text{ MeV}. \tag{11.11}$$

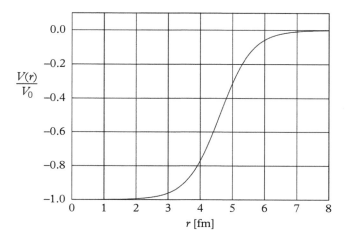

Fig. 11.4 Nuclei have a core with a constant density surrounded by a more diffuse skin. The density distribution is reflected in the nuclear potential, which is modelled by the Woods–Saxon potential, shown here. This potential is midway between a spherical potential well with a sharp edge and a 3-dimensional harmonic oscillator.

This is almost 4% of the rest mass of a nucleon, which gives the nucleon a velocity in excess of 25% of the speed of light. Even these highest energy nucleons are still bound, with a binding energy of around 8 MeV, so the spherical potential well of the nucleus must be at least 43 MeV deep. The radius of the well increases with A, but its depth is almost independent of A.

The nuclear potential may be modelled more accurately by the *Woods–Saxon potential* which represents the mean field due to the nucleon distribution. This potential takes into account the lower density skin of the nucleus, and has the form

$$V(r) = -\frac{V_0}{e^{\frac{r-R}{a}} + 1} \tag{11.12}$$

where V_0 is the approximately A-independent depth of the potential, a is the skin thickness and R is the nuclear radius. This potential is shown in Figure 11.4.

11.2.2 Nucleon pairing

In Chapter 9, we discussed the tendency for electrons to align their spins in accordance with Hund's first rule. The electrostatic force between two electrons is repulsive, so electrons preferentially occupy antisymmetric spatial wavefunctions in order to maximize their separation and thereby minimize the electrostatic potential energy. As electrons are fermions, 2-electron wavefunctions are antisymmetric, so a pair of electrons in an antisymmetric spatial wavefunction must have a symmetric spin wavefunction with aligned spins.

The corresponding situation for neutrons and protons in the nucleus is opposite because the strong force is attractive. It is energetically favourable for two identical nucleons to enter the same orbital as this minimizes their average separation and so maximizes their binding. The 2-nucleon spatial wavefunction is then symmetric, and the spin wavefunction must be antisymmetric. The binding energy of nucleons is therefore maximized by pairs of

The Incremental Binding Energies of Neutrons in the Isotopes of Tin

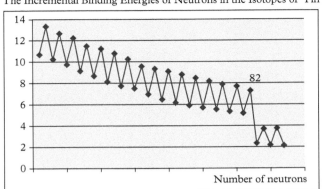

Fig. 11.5 Binding energies (in MeV) for the addition of successive neutrons in tin isotopes. The drop in binding energy for the 83rd neutron is explained later.

nucleons entering the same orbital with opposite spins and this rule is universally followed. The binding energies of neutrons in isotopes of tin are shown in Figure 11.5. The figure shows a see-saw pattern due to the increased binding of even numbers of neutrons, due to the pairing interaction. The pairing effect typically reduces energies by 1.0–1.5 MeV.

The binding energies for protons show a similar pattern and this has resulted in dramatic differences between the abundance of elements of even and odd atomic number Z, as shown in Figure 11.6. In the figure, the relative abundance of the elements in the solar system is shown on a logarithmic scale. For instance, element 14 silicon is 100 times as abundant as element 15 phosphorus, and element 16 sulphur is about 70 times as abundant as phosphorus.[4] Perhaps most remarkably, there are almost no stable nuclei where N and Z are both odd.

11.2.3 The liquid drop model

Figure 11.7 plots the experimentally determined nuclear binding energy per nucleon, $\frac{B}{A}$, for nuclei along the valley of stability. (This is the energy required for total breakup of the nucleus divided by the number of nucleons, not the energy required to remove just one nucleon.) $\frac{B}{A}$ increases with A until it reaches a broad peak in stability near $A = 60$ corresponding to the elements around iron, where $\frac{B}{A} \simeq 8.6$ MeV per nucleon. The most stable nucleus is $^{56}_{26}\text{Fe}_{30}$. This means that, in principle, it is possible to release energy by fusing light nuclei into nuclei of lower mass than ^{56}Fe, or by fissioning nuclei into fragments of higher mass than ^{56}Fe. Because of the plateau around iron, not much energy is released by fusing isotopes immediately below iron. Indeed, the available fusion energy comes mainly from the first step, fusing hydrogen to produce helium. Similarly, there is little energy available from fission until we reach elements that are much heavier than iron. Beyond $A = 60$ there is a gradual decline of $\frac{B}{A}$ to about 7.6 MeV per nucleon for the heaviest nuclei.

A very useful formula for the total nuclear binding energy $B(Z, A)$ as a function of Z and A can be constructed by considering the nucleus as analogous to a drop of liquid. We will refer to this as the *liquid drop formula*, but it is also known as the semi-empirical mass

---

[4] Phosphorus is a vital component of biological molecules such as DNA. It is far less abundant on the Earth than the other major atomic components of living organisms.

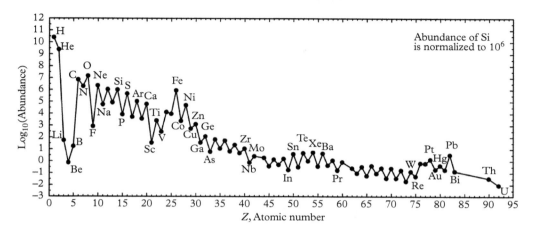

Fig. 11.6 Solar system nuclear abundances (logarithmic scale). The beryllium abundance is exceptionally low, because ^{8}Be is unstable and ^{9}Be, which is stable, is rapidly consumed by the fusion reactions in stars.

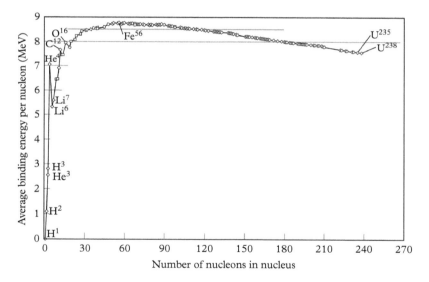

Fig. 11.7 Binding energy per nucleon.

formula or Bethe–Weizsäcker formula. The formula consists of five terms based on some simple observations about the structure of a nucleus, and is

$$B(Z, A) = a_V A - a_S A^{\frac{2}{3}} - a_C Z(Z-1) A^{-\frac{1}{3}} - a_A \frac{(A-2Z)^2}{A} + \delta(Z, A). \qquad (11.13)$$

The first term, $a_V A$, is proportional to the number of nucleons, and as the nuclear volume V scales with A, as described by equation (11.6), this term is proportional to the volume. This term dominates, so the binding energy per nucleon is approximately constant. The physical reason is that the range of the nuclear force is comparable to the size of a nucleon, so nucleons are only bound to their nearest neighbours. If there were an attraction between each nucleon and all the other nucleons in the nucleus, then the binding energy would scale as $A(A-1)$.

Close to the edge of the nucleus, nucleons are not completely surrounded by other nucleons and are therefore less tightly bound. The second term, $-a_S A^{\frac{2}{3}}$, is a surface energy term that compensates for the lower binding energy of nucleons when near the surface. It may be considered as a surface tension effect, proportional to the nuclear surface area. The surface area to volume ratio decreases as the size of the nucleus increases, so this term becomes less important with increasing A.

So far the formula treats all nucleons as identical, which they are with respect to the strong force. They are, however, very different with regard to the electromagnetic force. The third term is a correction due to the Coulomb repulsion of the protons. Each of the Z protons is repelled by the other $Z-1$ protons, so this term is proportional to $\frac{Z(Z-1)}{R}$. The nuclear radius R is proportional to $A^{\frac{1}{3}}$, so the Coulomb term takes the form $-a_C Z(Z-1)A^{-\frac{1}{3}}$.

If we could switch off the electromagnetic interaction, we would expect stable nuclei to contain equal numbers of neutrons and protons, to make the Fermi energies ε_F for neutrons and protons the same. The fourth term proportional to $\frac{(A-2Z)^2}{A}$, or equivalently $\frac{(N-Z)^2}{A}$, represents the energy penalty as nuclei move away from $N=Z$. This is known as the asymmetry term. Its coefficient is proportional to $\frac{1}{A}$ because higher energy levels are occupied by nucleons as A increases, and these are closer together.

Finally, as mentioned earlier, neutron pairs and proton pairs will form if at all possible. It is energetically favourable to form such pairs and this is reflected in the differences in the binding energies of nuclei with even numbers of neutrons and protons compared to nuclei with odd numbers of these particles. Indeed, this effect is so important for nuclear stability that there are only four stable odd–odd nuclei: ^{2}H, ^{6}Li, ^{10}B and ^{14}N. The pairing effect appears to be universally obeyed in nuclear ground states and the last term in the liquid drop formula quantifies it. It is represented by

$$\delta(Z,A) = \left\{ \begin{array}{ll} +\delta_0, & \text{for } N, Z \text{ even} \\ 0, & \text{for } A \text{ odd} \\ -\delta_0, & \text{for } N, Z \text{ odd} \end{array} \right.$$

where $\delta_0 = a_P A^{-\frac{3}{4}}$.

The constants a_V, a_S, a_C, a_A, a_P are determined by experiment and take the following values:

$$a_V \simeq 15.6 \text{ MeV}, \quad a_S \simeq 16.8 \text{ MeV}, \quad a_C \simeq 0.72 \text{ MeV},$$
$$a_A \simeq 23.3 \text{ MeV}, \quad a_P \simeq 34 \text{ MeV}. \tag{11.14}$$

The liquid drop formula implies that the binding energy per nucleon in MeV is

$$\frac{B(Z,A)}{A} = a_V - a_S A^{-\frac{1}{3}} - a_C Z(Z-1)A^{-\frac{4}{3}} - a_A \frac{(A-2Z)^2}{A^2} + \frac{\delta(A,Z)}{A}. \tag{11.15}$$

Figure 11.8 shows how the terms in this formula combine. The surface, Coulomb and asymmetry energies all contribute negatively, to lessen the binding energy. The binding energies derived from the liquid drop formula compare well with the measured binding energies shown in Figure 11.7.

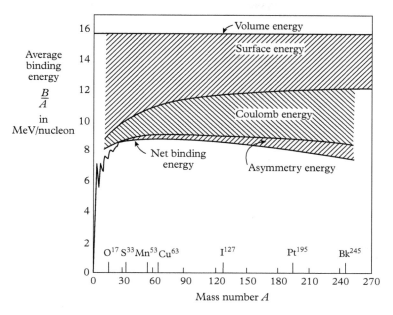

Fig. 11.8 Contributions to the binding energy per nucleon in the liquid drop model.

We can see the relative importance of each term by taking ^{238}U as an example. In this case the binding energy per nucleon is

$$\frac{B(92,238)}{A} = 15.6 - \frac{16.8}{(238)^{\frac{1}{3}}} - 0.72\frac{92 \times 91}{(238)^{\frac{4}{3}}} - 23.3\frac{(54)^2}{(238)^2} + \frac{34}{(238)^{\frac{7}{4}}}$$

$$= 15.6 - 2.7 - 4.1 - 1.2 + 0.002 = 7.6\,\text{MeV}. \qquad (11.16)$$

(Note the relative unimportance of the pairing term for heavy nuclei as it only corrects for the pairing interaction between the final couple of nucleons in the nucleus. This is why the pairing term does not appear in Figure 11.8.) The mass $m(92,238)$ of ^{238}U is $238.050788\,u = 238.050788 \times 931.4941$ MeV $= 221{,}742.9$ MeV. We can use equation (11.5) to work out the true binding energy of the nucleus

$$B(92,238) = 92m_p + 146m_n + 92m_e - m(92,238)$$

$$= 92 \times 938.2723 + 146 \times 939.5656 + 92 \times 0.5110 - 221{,}742.9$$

$$= 223{,}544.6 - 221{,}742.9$$

$$= 1801.7\,\text{MeV}, \qquad (11.17)$$

so the actual binding energy per nucleon is $\frac{1801.7}{238} = 7.57$ MeV. The liquid drop calculation agrees with the measured value to within the accuracy of its five parameters.

We can use the liquid drop formula to find the relationship between the proton number Z and the total number of nucleons A in stable nuclei. Differentiating equation (11.13) with respect to Z, while keeping A fixed, we find that the binding energy is maximized when

$$\frac{\partial B}{\partial Z} = -a_{\text{C}}(2Z-1)A^{-\frac{1}{3}} + 4a_{\text{A}}\frac{A - 2Z}{A} = 0. \qquad (11.18)$$

This can be rearranged to give

$$\frac{Z}{A} = \frac{1}{2}\left(\frac{4a_A + a_C A^{-\frac{1}{3}}}{4a_A + a_C A^{\frac{2}{3}}}\right) \simeq \frac{1}{2}\left(\frac{1}{1 + \frac{a_C}{4a_A}A^{\frac{2}{3}}}\right), \tag{11.19}$$

where we have used the fact that $a_C A^{-\frac{1}{3}}$ is always small. This expression gives the proportion of protons as $\frac{Z}{A} \simeq 0.5$ for small A, but as A increases $\frac{Z}{A}$ decreases. For $A = 238$, the formula predicts correctly that there are $0.39 \times 238 = 92$ protons in the nucleus.

The liquid drop model is very useful when considering alpha decay, fission and other types of radioactivity and even the structure of neutron stars.

11.3 The Nuclear Shell Model

The liquid drop formula matches the general trend of nuclear binding energies very well. However, there are nuclei with larger than expected binding energies, as shown in Figure 11.7. The binding energies of these nuclei exceed the values derived from the formula, and are observed to occur when the nuclei contain certain *magic numbers* of either protons or neutrons. The magic numbers are evident from a wide range of experiments. They are 2, 8, 20, 28, 50, 82 and 126. Nuclei with a magic number of neutrons or protons have an enhanced stability and are more abundant than their neighbours. They also exist in a larger number of stable isotopes or isotones (nuclei with equal numbers of neutrons, but different numbers of protons). For instance, the element with the greatest number of stable isotopes is tin, which is element number 50 with a magic number of protons. Tin has ten stable isotopes. By contrast, the neighbouring elements indium ($Z = 49$) and antimony ($Z = 51$) each have just one or two. Magic nuclei also have especially large first excitation energies, exceptionally long half-lives (if unstable) and very low neutron capture cross sections. Nuclei that are doubly magic, such as $^4_2\text{He}_2$, $^{16}_8\text{O}_8$ and $^{208}_{82}\text{Pb}_{126}$, are especially stable compared to their neighbours. Indeed, $^{208}_{82}\text{Pb}_{126}$ is the heaviest stable nucleus. ($^{209}_{83}\text{Bi}_{126}$ was recently shown to be quasi-stable with a half-life of 1.9×10^{19} years.)

11.3.1 The atomic shell analogy

So how can we explain the anomalous stability of these nuclei? There is an obvious analogy—the chemical stability of the inert gases (helium, neon, argon, krypton, xenon, radon). The properties of these atoms are readily understood in terms of their electronic structure, as discussed in section 9.1.2. Figure 9.10 shows that as atomic number changes, the energy required to remove an electron from an atom reaches a peak at each of the inert gases, which explains why they do not readily form chemical compounds and exist naturally as monatomic gases. The electron orbitals exist in shells that are separated in energy; an inert gas atom has a full shell configuration and is stable as it takes a lot of energy to excite an electron to the next empty energy level. The numbers of states in successive shells and the corresponding atomic magic numbers are set out in Table 9.1 and Figure 9.3. Similarly, it is believed that the nuclear magic numbers represent the numbers of protons or neutrons that completely fill shells of the nuclear potential.

If we pursue this analogy, we can understand why Figure 11.5 shows a steep drop in binding energy for the 83rd neutron and why this isotope also has a short lifetime compared to the neighbouring tin isotopes with fewer neutrons. It is because the 82nd neutron completes a shell, so the next neutron must go into a higher energy shell. The last neutron in the tin isotope $^{133}_{50}Sn_{83}$ is equivalent to the loosely bound outer electron in an alkali metal atom, such as potassium.

11.3.2 The harmonic oscillator

The nuclear magic numbers are different from the atomic magic numbers because the nuclear potential differs significantly from the Coulomb potential in atoms, so there are different numbers of states in each shell. The Schrödinger equation cannot be solved analytically for either the spherical potential well (a spherical box) or the Woods–Saxon potential (Figure 11.4). However, there is another potential with simple solutions that proves to be a good starting point for describing the nuclear potential—the 3-dimensional harmonic oscillator.[5]

As discussed in Chapter 7, the Schrödinger equation for the 1-dimensional harmonic oscillator,

$$\left(-\frac{\hbar^2}{2m}\frac{d^2}{dx^2} + \frac{1}{2}m\omega^2 x^2 \right) \chi(x) = E\chi(x) \,, \tag{11.20}$$

has (unnormalized) solutions

$$\chi_n(x) = H_n\left(\sqrt{\frac{m\omega}{\hbar}}x \right) e^{-\frac{m\omega}{2\hbar}x^2} \,, \tag{11.21}$$

where the functions H_n are the Hermite polynomials (7.37). The energy levels are

$$E_n = \left(n + \frac{1}{2} \right)\hbar\omega \,, \tag{11.22}$$

so the ground state has energy $\frac{1}{2}\hbar\omega$ and the excited states are separated by energy $\hbar\omega$.

The harmonic oscillator is readily generalized to any number of dimensions. In two dimensions, the potential is $V(x, y) = \frac{1}{2}m(\omega_x^2 x^2 + \omega_y^2 y^2)$. The solutions are products of the 1-dimensional solutions, and their energies equal the sum of 1-dimensional energies, $E_{p,q} = \left(p + \frac{1}{2} \right)\hbar\omega_x + \left(q + \frac{1}{2} \right)\hbar\omega_y$. In the isotropic case, where $\omega_x = \omega_y = \omega$, the energy levels are $E_n = (n + 1)\hbar\omega$, with $n = p + q$. The degeneracy of energy level E_n is $n + 1$, that being the number of ways in which n can be formed from p and q. (p may take any value from 0 to n, and then q is fixed.) Similarly in three dimensions, the isotropic harmonic oscillator has the potential $V(x, y, z) = \frac{1}{2}m\omega^2(x^2 + y^2 + z^2)$. The solutions are products of 1-dimensional solutions, labelled (p, q, r), with $N = p + q + r$. The energy levels are $E_N = (N + \frac{3}{2})\hbar\omega$ and have even greater degeneracy. At energy level E_N, $p + q$ may take any value n from 0 to N, and once p and q are fixed, so is r, so using the previous 2-dimensional result, the total degeneracy Δ_N is the sum

$$\Delta_N = \sum_{n=0}^{N}(n + 1) = \frac{1}{2}(N + 1)(N + 2) \,, \tag{11.23}$$

the $(N + 1)$th triangular number.[6]

---

[5] At first sight this might seem unsuitable, as we know that nucleons interact via the strong force, which is short-range, whereas the harmonic oscillator force is long-range and increases with distance. However, if we started with the spherical potential well, which is short-range and has a sharp cut-off, we would reach essentially the same final result.

[6] This is also the number of monomials in x, y, z of degree N.

In spherical polar coordinates, the Schrödinger equation for the 3-dimensional, isotropic harmonic oscillator is

$$\left(-\frac{\hbar^2}{2m}\nabla^2 + \frac{1}{2}m\omega^2 r^2\right)\chi = E\chi. \tag{11.24}$$

Separating into radial and angular coordinates, the solutions may be expressed as

$$\chi_{nlm}(r,\vartheta,\varphi) = \frac{1}{r}R_n(r)P_l^m(\vartheta,\varphi), \tag{11.25}$$

where n is now the radial quantum number, l is the orbital angular momentum and m is the z-component of the angular momentum. The angular momentum states are labelled as in atomic physics

$$l = 0, 1, 2, 3, 4, 5\ldots; \quad \text{s, p, d, f, g, h}\ldots. \tag{11.26}$$

The energy is

$$E_N = \left(2n + l - \frac{1}{2}\right)\hbar\omega, \tag{11.27}$$

so $N = 2n + l - 2$ and the energy increase due to one step in the radial quantum number n equals that due to two steps in the angular momentum l. For given N, the maximum value of l is N, and other allowed values differ by a multiple of 2.

The harmonic oscillator multiplets with degeneracy Δ_N can therefore be decomposed into angular momentum multiplets as follows:

$$\Delta_{2k} = 1 + 5 + 9 + \ldots + (4k + 1), \tag{11.28}$$

$$\Delta_{2k+1} = 3 + 7 + 11 + \ldots + (4k + 3). \tag{11.29}$$

For instance, the fifth excited level of the 3-dimensional harmonic oscillator with degeneracy Δ_5 decomposes as

$$\Delta_5 = 21 = 3 + 7 + 11 = 3\text{p} + 2\text{f} + 1\text{h}. \tag{11.30}$$

(The angular momentum multiplets are labelled sequentially in order of increasing energy. For instance, the *third* occurrence of p states is at level 5, so they are labelled 3p.) The low-lying states of the 3-dimensional harmonic oscillator are given in Table 11.1.

Taking into account the two spin states of a nucleon, this would give the magic numbers 2, 8, 20, 40, 70 and 112. Thus the lowest magic numbers are correctly reproduced, but the agreement disappears for the higher numbers. However, as Maria Goeppert-Mayer and Otto Haxel, Hans Jensen and Hans Suess independently showed, by modifying the harmonic oscillator the magic numbers problem can be solved. Their explanations were published in the same issue of *Physical Review* in 1949.

Following these authors, we postulate that the harmonic oscillator states are subject to a perturbation that is dependent on their angular momentum. At each oscillator level N, the energy of the low angular momentum states rises and the energy of the high angular momentum states falls. As a nucleon's mean radial position increases with its angular momentum, high angular momentum states tend to be nearer to the nuclear surface, and this has the effect of flattening the harmonic oscillator potential, bringing it closer to the Woods–Saxon potential.

Energy level N	States	Degeneracy (including spin) $2\Delta_N$	Magic numbers
0	1s	2	2
1	1p	6	8
2	1d, 2s	$10 + 2 = 12$	20
3	1f, 2p	$14 + 6 = 20$	40
4	1g, 2d, 3s	$18 + 10 + 2 = 30$	70
5	1h, 2f, 3p	$22 + 14 + 6 = 42$	112

Table 11.1 Table of states of the 3-dimensional harmonic oscillator.

If we simply add to the Hamiltonian a term proportional to $-\mathbf{l}^2$ this has the undesirable effect of reducing the gaps between the shells. To compensate for this we subtract $\langle \mathbf{l}^2 \rangle_N$, the average value of $\mathbf{l}^2$ at each oscillator level, so the term to be added becomes $-\beta(\mathbf{l}^2 - \langle \mathbf{l}^2 \rangle_N)$, where β is a positive constant determined by experiment. For example, consider the fifteen $N = 4$ oscillator states, which consist of the one 3s, five 2d and nine 1g orbitals that are degenerate in the isotropic harmonic oscillator potential. The eigenvalue of $\mathbf{l}^2$ is $l(l+1)$ and takes the values 0, 6 and 20 respectively for these orbitals, so the average value is $\langle \mathbf{l}^2 \rangle_4 = \frac{1}{15}(1 \times 0 + 5 \times 6 + 9 \times 20) = 14$. Therefore, the term that is added to the energy for the $N = 4$ states is $-\beta(l(l+1) - 14)$, and this breaks the degeneracy, increasing the energy of the 3s and 2d states and reducing the energy of the 1g states. In general, $\langle \mathbf{l}^2 \rangle_N = \frac{1}{2}N(N+3)$. The resulting sequence of energy levels for the modified 3-dimensional harmonic oscillator is shown to the left of Figure 11.10. Note that we could have started with the spherical potential well and added an angular momentum term with opposite sign to produce very similar results.

11.3.3 Spin–orbit coupling

Fig. 11.9 Within the nucleus the strong force aligns the orbital angular momentum and spin of the nucleons. The effect is to reduce the energy of nucleon states (left) with total angular momentum $j = l + \frac{1}{2}$ and increase the energy of nucleon states (right) with $j = l - \frac{1}{2}$.

The critical breakthrough made by Mayer and Haxel, Jensen and Suess was to include an additional *spin–orbit coupling*. This means there is a strong tendency for the spin of a nucleon to be parallel to its orbital angular momentum, so states must be labelled by their total angular momentum $j = l + \frac{1}{2}$ or $j = l - \frac{1}{2}$. For instance, the eighteen 1g states with

Energy level N	States	Degeneracy (including spin)	Magic number
0	$1s_{\frac{1}{2}}$	2	2
1	$1p_{\frac{3}{2}}, 1p_{\frac{1}{2}}$	$4 + 2 = 6$	8
2	$1d_{\frac{5}{2}}, 1d_{\frac{3}{2}}, 2s_{\frac{1}{2}}$	$6 + 4 + 2 = 12$	20
3	$1f_{\frac{7}{2}}$	8	28
4	$1f_{\frac{5}{2}}, 2p_{\frac{3}{2}}, 2p_{\frac{1}{2}}, 1g_{\frac{9}{2}}$	$6 + 4 + 2 + 10 = 22$	50
5	$1g_{\frac{7}{2}}, 2d_{\frac{5}{2}}, 2d_{\frac{3}{2}}, 3s_{\frac{1}{2}}, 1h_{\frac{11}{2}}$	$8 + 6 + 4 + 2 + 12 = 32$	82
6	$1h_{\frac{9}{2}}, 2f_{\frac{7}{2}}, 2f_{\frac{5}{2}}, 3p_{\frac{3}{2}}, 3p_{\frac{1}{2}}, 1i_{\frac{13}{2}}$	$10 + 8 + 6 + 4 + 2 + 14 = 44$	126

Table 11.2 Nuclear states grouped into complete shells.

$l = 4$ are split into ten $1g_{\frac{9}{2}}$ states and eight $1g_{\frac{7}{2}}$ states. The spin–orbit term reduces the energy of states with spin and angular momentum in the same direction and increases the energy of states with spin and angular momentum opposite, as illustrated in Figure 11.9. So the energy of the $1g_{\frac{9}{2}}$ states is reduced and the energy of the $1g_{\frac{7}{2}}$ states rises.

The spin–orbit coupling is a surface effect. The environment experienced by nucleons within the body of the nucleus is the same in all directions, so here the direction of orbital angular momentum loses meaning. For a nucleon in the skin, however, the radial and tangential directions are different, and orbital angular momentum is tangential. Spin–orbit coupling may be modelled in the Hamiltonian by a term $-2\alpha \, \mathbf{l} \cdot \mathbf{s}$, where α is positive. The effect of this term can be calculated by squaring the total angular momentum operator $\mathbf{j} = \mathbf{l} + \mathbf{s}$. This gives $\mathbf{j}^2 = (\mathbf{l} + \mathbf{s})^2 = \mathbf{l}^2 + \mathbf{s}^2 + 2\,\mathbf{l} \cdot \mathbf{s}$, so $2\,\mathbf{l} \cdot \mathbf{s} = \mathbf{j}^2 - \mathbf{l}^2 - \mathbf{s}^2$. In terms of the eigenvalues of these operators, we obtain $2\,\mathbf{l} \cdot \mathbf{s} = j(j+1) - l(l+1) - s(s+1)$. Plugging in values for j, and $s = \frac{1}{2}$, we find that for $j = l + \frac{1}{2}$ the spin–orbit term is $-\alpha l$, whereas for $j = l - \frac{1}{2}$ it is $\alpha(l+1)$, so overall for states with orbital angular momentum l the spin–orbit term produces an energy level splitting of $\alpha(2l + 1)$. This means that at the $N = 3$ level there is a bigger energy split between the $1f_{\frac{7}{2}}$ and $1f_{\frac{5}{2}}$ states than between the $2p_{\frac{3}{2}}$ and $2p_{\frac{1}{2}}$ states that are degenerate with them in the harmonic oscillator potential, and at the $N = 4$ level the $1g_{\frac{9}{2}}$ and $1g_{\frac{7}{2}}$ states are affected much more than the $2d_{\frac{5}{2}}$ and $2d_{\frac{3}{2}}$, or $3s_{\frac{1}{2}}$ states. In general, at each harmonic oscillator level N the spin–orbit term affects the states of high angular momentum far more than those of low angular momentum, and it reduces the energy of the highest angular momentum states sufficiently to lower them into the shell below, as shown on the right of Figure 11.10. For instance, the eight $1f_{\frac{7}{2}}$ states are lowered into a separate shell of their own and the ten $1g_{\frac{9}{2}}$ states are lowered one complete shell down. These large energy level shifts alter the number of states in the nuclear shells, and reproduce the observed nuclear magic numbers, as shown in Figure 11.10. The states in each shell are given in Table 11.2.

The spin–orbit coupling contribution to the strong force produces a very large effect, much larger than the electromagnetic spin–orbit coupling observed in atomic physics, which has no bearing on the composition of the atomic shells. The reason for the large size of the coupling strength α is not fully understood.

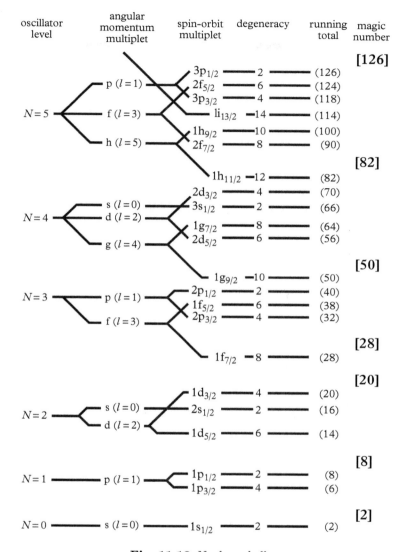

Fig. 11.10 Nuclear shells.

In the conventional parametrization of the Hamiltonian, α and β are expressed as $\alpha = \kappa\hbar\omega_0$ and $\beta = \mu\kappa\hbar\omega_0$, where ω_0 is the optimal oscillator frequency. The modified oscillator potential is

$$V_{\mathrm{MO}} = \frac{1}{2}m\omega_0^2 r^2 + V_l,\tag{11.31}$$

where

$$V_l = -\kappa\hbar\omega_0\left(\mu\left(\mathbf{l}^2 - \frac{1}{2}N(N+3)\right) + 2\,\mathbf{l}\cdot\mathbf{s}\right).\tag{11.32}$$

It depends on just three parameters: κ, μ and ω_0. The nuclear energy levels are reproduced very well with values of $\kappa \simeq 0.06$ and $\mu \simeq 0.4$. The oscillator frequency ω_0 varies from nucleus to nucleus; $\hbar\omega_0$ is approximately $41A^{-\frac{1}{3}}$ MeV. For $A = 125$ to 216, $A^{\frac{1}{3}} = 5$ to 6, therefore $41A^{-\frac{1}{3}}$ MeV $\simeq 7$ to 8 MeV, which gives an indication of the energy gaps between major shells for medium-heavy nuclei.

The shell model works very well for nuclei close to the magic numbers. Due to the pairing effect, all even–even nuclei have zero spin, as the nucleons pair up with opposite spins. The significance of shell structure is evident in nuclei such as $^{41}_{20}\text{Ca}_{21}$ which has a magic number of protons and one more than a magic number of neutrons. The extra unpaired neutron determines the overall spin of the nucleus. In its ground state, this nucleus has several filled shells of protons and neutrons plus one extra neutron which goes into a new shell. From Figure 11.10 we can see that the 21st neutron will find itself in a $1f_{\frac{7}{2}}$ state. We should expect this nucleus to have spin $J = \frac{7}{2}$, and indeed it does, as does the mirror nucleus $^{41}_{21}\text{Sc}_{20}$. Similarly, the nuclei $^{91}_{41}\text{Nb}_{50}$ and $^{91}_{40}\text{Zr}_{51}$ have spins $\frac{9}{2}$ and $\frac{5}{2}$ respectively, as can also be verified by examining Figure 11.10.

In nuclei with two nucleons outside a magic core, there is quite a strong *residual interaction* between the nucleons. This interaction is short-range and attractive, so the lowest energy states are those for which the overlap of the nucleon wavefunctions is largest. We can understand the consequence of this using an approximate classical picture. Recall that the nucleon spin is parallel to its orbital angular momentum because of the spin–orbit coupling. The nucleon orbits around an equator as in Figure 11.9 (left). The wavefunctions of two nucleons have large overlap when the classical orbits are along the same equator, either in the same or opposite directions. Their angular momentum vectors are then parallel or antiparallel.

For two neutrons or two protons, the Pauli principle forbids the particles to be in the same location and have the same spin state. This nullifies the short-range attraction if the angular momenta are parallel, and favours the angular momenta being antiparallel. In $^{42}_{20}\text{Ca}_{22}$ there are two neutrons in $1f_{\frac{7}{2}}$ states outside the magic $^{40}_{20}\text{Ca}_{20}$ core, and the total angular momentum can be $J = 0, 2, 4, 6$, consistently with the Pauli principle. The ground state has $J = 0$, because this is the state where the neutrons have antiparallel angular momenta, and as J increases the energy increases.

The nucleus $^{42}_{21}\text{Sc}_{21}$ is more interesting. Here there is a proton and a neutron outside the core, in $1f_{\frac{7}{2}}$ states. The total angular momentum J can take any value from 0 to 7, but the states distinct in structure from $^{42}_{20}\text{Ca}_{22}$ are those with odd J. The lowest of these states, with the proton and neutron angular momenta almost antiparallel, has $J = 1$, but remarkably, the state with $J = 7$ and the angular momenta parallel has only slightly higher energy. The states with $J = 3$ and $J = 5$ have considerably higher energy.

The $J = 7$ state cannot easily shed its excess energy, because of the large spin spacing and small energy gap to the $J = 1$ state. In fact, the $J = 7$ state of $^{42}_{21}\text{Sc}_{21}$ undergoes an inverse beta decay to $^{42}_{20}\text{Ca}_{22}$ (see section 11.3.4), and it has a half-life of more than one minute. It is unusual for a nuclear state above the ground state to have such a long half-life, and for this reason the $J = 7$ state of $^{42}_{21}\text{Sc}_{21}$ is known as an *isomer*, meaning a quasi-stable excited nucleus.

When there are three or four nucleons outside a magic core, the attractive residual interaction between them can be sufficient for these nucleons to form a cluster, for example, a tritium cluster ($^3_1$H$_2$) or an alpha particle. Such clustering is common and occurs, for example, in the nuclei $^{19}_9$F$_{10}$ and $^{20}_{10}$Ne$_{10}$, which have, respectively, tritium and alpha particle clusters outside an $^{16}_8$O$_8$ core.

11.3.4 Beta decay

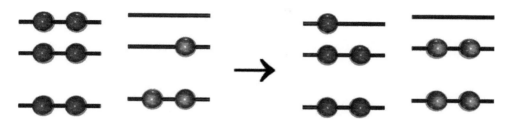

Fig. 11.11 Schematic representation of the energy levels of the neutrons and protons within a ^{9}Li nucleus (left) which undergoes beta decay with a half-life of 178 ms to form ^{9}Be (right).

Neutrons and protons fill up the nuclear energy levels separately, so the magic numbers apply to neutrons and protons separately. This is clear because although two neutrons cannot exist in the same state, a neutron and a proton can as they are different particles. Protons experience an electrostatic repulsion from all the other protons in the nucleus. This shifts their energy levels upward relative to the energy levels of the neutrons. In a stable nucleus, the highest occupied energy level—the Fermi level—must be the same for both neutrons and protons, otherwise energy can be released by converting a neutron into a proton or a proton into a neutron. This conversion is possible due to the weak force and is observed as radioactive *beta decay*. Nuclei with excess neutrons may undergo reactions such as

$$^9_3\text{Li}_6 \quad \rightarrow \quad ^9_4\text{Be}_5 \quad + \quad e^- \quad + \quad \bar{\nu}_e, \tag{11.33}$$

where e^- is an electron (or beta particle) and $\bar{\nu}_e$ denotes an antineutrino. The energy levels of the nucleons in these nuclei are shown in Figure 11.11. A free neutron will undergo beta decay with a half-life of around 10 minutes:

$$n \quad \rightarrow \quad p \quad + \quad e^- \quad + \quad \bar{\nu}_e. \tag{11.34}$$

Nuclei with excess protons may undergo reactions such as

$$^{23}_{12}\text{Mg}_{11} \quad \rightarrow \quad ^{23}_{11}\text{Na}_{12} \quad + \quad e^+ \quad + \quad \nu_e, \tag{11.35}$$

where e^+ is a positron (the antiparticle of the electron) and ν_e denotes a neutrino. This is known as inverse beta decay. It is also possible in certain heavier nuclei for a proton to convert into a neutron by capturing an inner atomic electron and emitting a neutrino. This process is known as *electron capture*.

These processes drive nuclei towards the valley of stability, shown in the nuclear chart of Figure 11.1. In the chart, nuclei to the left of the valley of stability undergo beta decay, whereas nuclei to the right undergo inverse beta decay. The proton energy levels are raised slightly in Figure 11.11 relative to the neutron energy levels. This shift increases markedly in heavier nuclei due to the Coulomb repulsion and this is why the valley of stability deviates from $Z = N$ in heavier nuclei.

Individual protons and neutrons are rarely spontaneously emitted from nuclei. This is because each proton and neutron has positive binding energy. However, if several neutrons are added to a nucleus to the left of the valley of stability, the binding energy of successive neutrons decreases until there is no binding energy for the addition of another neutron. At this point we have reached the *neutron dripline*. Neutron-rich nuclei beyond this point will shed neutrons on a timescale of around 10^{-23} s, which is the time for light to travel a distance equal to the neutron radius, so these nuclei fall apart as quickly as is causally possible. On the opposite side of the valley of stability we find the *proton dripline* where proton-rich nuclei rapidly shed protons.

11.3.5 The Nilsson model

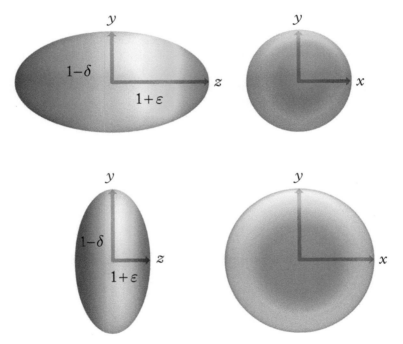

Fig. 11.12 Top: Prolate ellipsoid with $\varepsilon > 0$ shown with the z-axis horizontal (left) and with the z-axis coming out of the page (right). Bottom: Oblate ellipsoid with $\varepsilon < 0$ shown with the z-axis horizontal (left) and with the z-axis coming out of the page (right).

In the region of the magic numbers, nuclei are spherical and the spherical shell model described above works very well. When the nuclear shells are only partially filled, nuclei achieve lower-energy configurations by deforming into ellipsoids, thus breaking full rotational symmetry. This is fundamental to the analysis of these nuclei. They are described by a refinement of the shell model known as the Nilsson model after Sven Gösta Nilsson. These nuclei remain axisymmetric around an axis defined to be the z-axis, and have a volume equal to that of a spherical nucleus containing the same number of nucleons. The deformation is parametrized by a quantity ε. When ε is positive the nucleus is a *prolate* ellipsoid, and when ε is negative it is an *oblate* ellipsoid, as shown in Figure 11.12. Most large nuclei with partially filled outer shells of nucleons form prolate ellipsoids. Nuclei with a small number of vacant states (or holes) in their outer shell tend to be oblate.

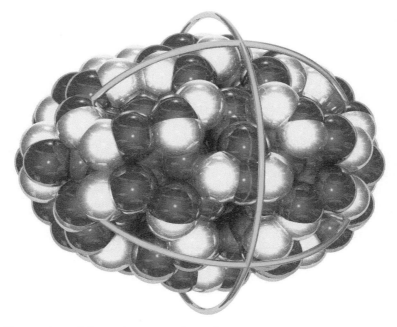

Fig. 11.13 If a nucleus deforms into a prolate ellipsoid, spherical symmetry is broken. Three orthogonal orbits of equal radius are shown. The orbit whose axis coincides with the remaining axis of symmetry is higher in energy, because most of the orbit lies outside the nucleus, so a nucleon in such an orbit will feel less of the attractive potential.

In the original spherical shell model, nucleon states with the same total angular momentum j are degenerate. For instance, all six $1\mathrm{d}\frac{5}{2}$ states have the same energy. However, a small prolate deformation produces a perturbation that splits this degeneracy; the energy now depends on the projection K of the angular momentum along the z-axis, the remaining symmetry axis. Figure 11.13 shows three hypothetical orbits with equal angular momentum around a prolate nucleus. One of the orbits lies in the (x, y)-plane, so its axis coincides with the z-axis, and $K = j$. The axes of the other two orbits are perpendicular to the z-axis. In these cases the angular momentum projections along z are $K = 0$. The $K = j$ orbit is largely outside the attractive potential well of the nucleus and is therefore less tightly bound and so higher in energy than the other two orbits, which are mostly within the nucleus. In general, for a prolate deformation, the energy of states whose orbits lie mainly within the nucleus fall, whereas those whose orbits lie largely outside the nucleus rise. The smaller the projection of the angular momentum of a state along z, the lower its energy. For instance, the sixfold degeneracy of the $1\mathrm{d}\frac{5}{2}$ states splits into three pairs of doubly degenerate states, which in order of increasing energy have $K = \pm\frac{1}{2}$, $K = \pm\frac{3}{2}$ and $K = \pm\frac{5}{2}$.

An ellipsoidal nucleus can be modelled using the Nilsson potential V_N, which is an anisotropic harmonic oscillator potential modified to include the same (or similar) angular momentum and spin–orbit terms V_l as those of the spherical shell model,

$$V_N = \frac{1}{2}m(\omega_x(x^2 + y^2) + \omega_z z^2) + V_l, \tag{11.36}$$

where $\omega_x^2 = \omega_y^2 = \omega_0^2(1 + \frac{1}{3}\varepsilon)$ and $\omega_z^2 = \omega_0^2(1 - \frac{2}{3}\varepsilon)$. Figure 11.14 shows the energy levels of the 1-nucleon states in this potential plotted against the deformation parameter ε. The centre line gives the energy levels for a spherical nucleus with $\varepsilon = 0$, and to the right ε is positive, producing a prolate nucleus. To the left, ε is negative and the nucleus is oblate.

For instance, for positive ε the six $1d_{\frac{5}{2}}$ states that are degenerate for $\varepsilon = 0$ split into three pairs of states labelled $\frac{1}{2}[220]$, $\frac{3}{2}[211]$ and $\frac{5}{2}[202]$ in order of increasing energy, with the energy splitting increasing with ε. For negative ε the energy ordering is reversed. We will now explain the labelling. The fraction that precedes the square bracket is $|K|$. Inside the square brackets, the first number gives the 3-dimensional (isotropic) harmonic oscillator level, the second gives the harmonic oscillator level in the z-direction and the third is the projection m of the orbital angular momentum in the z-direction, where $K = m \pm \frac{1}{2}$.

How does the prolate ellipsoidal deformation of the nucleus arise? The deformation is a self-consistent, collective effect of the nucleons to lower their total energy by increasing their mutual interactions. The combined density of all the wavefunctions in a j-multiplet is spherically symmetric, so all the filled lower shells produce a spherically symmetric core. In the outermost shell, however, the orbits with axes nearly perpendicular to the z-axis are occupied preferentially and this produces the prolate deformation along the z-axis, as well as lowering the energy of those orbits that contribute to the deformation. For instance, in a nucleus with half-filled $1h_{\frac{11}{2}}$ proton shell, corresponding to $Z = 76$, the deformation means that the protons may fill states with $K = \pm\frac{1}{2}, \pm\frac{3}{2}, \pm\frac{5}{2}$, all of which have lower energy than the equivalent, spherical shell model states, whereas those states whose energy has risen, with $K = \pm\frac{7}{2}, \pm\frac{9}{2}, \pm\frac{11}{2}$, are all vacant, so overall the deformation of the nucleus has lowered the total energy. This is why many osmium isotopes with $Z = 76$ are prolate.

There are further refinements of the Nilsson model. Ellipsoidal nuclei may be excited into collective rotational states in which they spin around an axis perpendicular to the z-axis. There is then an interaction between the angular momentum of each nucleon and the spin of the entire nucleus. By analogy with classical dynamics, this is referred to as the *Coriolis force*. This produces a further splitting of the nucleon energy levels. It lowers the energy of states with K in the same direction as the nuclear spin and raises the energy of those with opposite K.

11.4 Alpha Decay

Beyond the iron peak, as atomic number increases the nuclear binding energy per nucleon gradually decreases. This is shown in Figure 11.7 and modelled well by the liquid drop formula (11.15). So it is possible to release energy by breaking up a heavy nucleus. This is largely due to the increased Coulomb repulsion between the protons within the nucleus. However, the strong force binds the nucleons within a potential well and prevents immediate disintegration.

Heavy nuclei are unstable and decay at a random point in time characterized by a half-life for each possible decay process. The usual decay route for a heavy nucleus is by the emission

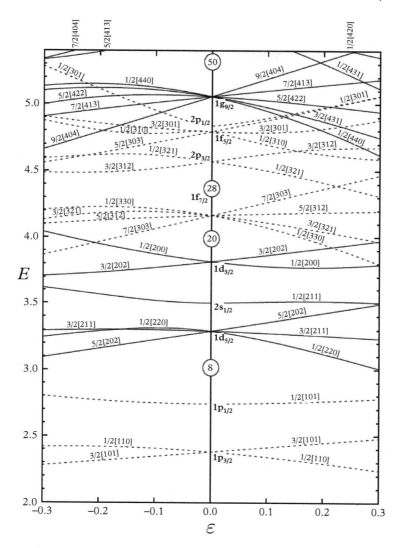

Fig. 11.14 Nilsson diagram.

of an alpha particle in a process such as (11.1). Most of the excess energy is carried away as the kinetic energy of the alpha particle. The half-life for alpha decay has a very strong dependence on the energy released, the Q_α value, which depends on the isotope that is decaying. The more energetic alpha particles are emitted with much shorter half-lives than those that are less energetic. For the isotopes of several radioactive elements that contain even numbers of both protons and neutrons, the half-lives are shown in Figure 11.15, and they fall on separate almost parallel lines. For each element, with fixed proton number but variable neutron number, the half-life $\tau_{\frac{1}{2}}$ has a logarithmic dependence on the energy

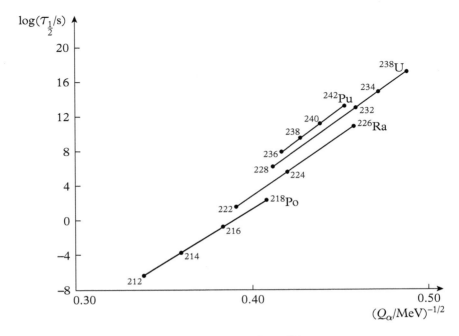

Fig. 11.15 Geiger–Nuttall law.

A	Q_α (MeV)	$1/\sqrt{Q_\alpha}$	$\tau_{\frac{1}{2}}$ (s)	$\log \tau_{\frac{1}{2}}$
218	9.85	0.319	10^{-7}	-7.00
220	8.95	0.334	10^{-5}	-5.00
222	8.13	0.351	2.8×10^{-3}	-2.55
224	7.31	0.370	1.04	0.017
226	6.45	0.394	1854	3.27
228	5.52	0.426	6.0×10^{7}	7.78
230	4.77	0.458	2.5×10^{12}	12.40
232	4.08	0.495	4.4×10^{17}	17.64

Table 11.3 Alpha decay energies and half-lives for thorium isotopes. An increase of 0.01 in $1/\sqrt{Q_\alpha}$ corresponds to an increase in $\log \tau_{\frac{1}{2}}$ of about 1.4.

released that can be expressed as

$$\log \tau_{\frac{1}{2}} = A + B \frac{1}{\sqrt{Q_\alpha}}. \tag{11.37}$$

This is known as the *Geiger–Nuttall law*. The results for isotopes of the element thorium, with 90 protons, are listed in Table 11.3.

Classically, an alpha particle within a nucleus would be trapped within the nuclear potential well and be unable to escape without the input of energy from elsewhere. As George Gamow pointed out in 1928, quantum mechanics allows the possibility of tunnelling

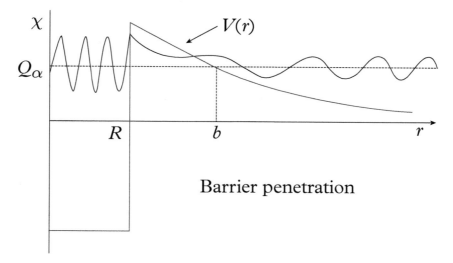

Fig. 11.16 Tunnelling through a potential barrier. The potential $V(r)$ and wavefunction $\chi(r)$ are shown.

through a potential barrier, so there is always a small probability that the alpha particle will find itself on the other side of the barrier. This insight led to an explanation of the Geiger–Nuttall law, an early triumph for quantum theory with implications not only for understanding alpha decay, but also for fission and fusion.

Following Gamow we can calculate the half-life for alpha decay. The alpha particle experiences a potential $V(r)$ that within the nucleus may be approximated as a finite-depth spherical well of radius R, and outside is a long-range repulsive Coulomb potential. This is shown in Figure 11.16. The zero energy line in the figure equals the total energy that the alpha particle and daughter nucleus would have if they came to rest after separating. The actual energy is $E = Q_\alpha$, which is positive but below the top of the potential barrier. So the alpha particle has to tunnel through the barrier. We will ignore any angular momentum dependence and treat the problem as 1-dimensional with a preformed alpha particle moving in the nuclear potential $V(r)$. Most of the wavefunction $\chi(r)$ of the alpha particle is inside the spherical well. The wavefunction decays exponentially within the potential barrier around the spherical well, and it has a small oscillatory tail extending beyond the barrier.

The Schrödinger equation for $r > R$, modelling the alpha particle outside the well, is

$$-\frac{\hbar^2}{2m_\alpha}\frac{d^2\chi}{dr^2} + (V(r) - Q_\alpha)\chi = 0\,, \tag{11.38}$$

where V is the Coulomb potential

$$V(r) = \frac{2Ze^2}{4\pi r}\,. \tag{11.39}$$

Ze is the charge of the daughter nucleus, $2e$ is the charge of the alpha particle, and m_α is the mass of the alpha particle.[7] The barrier lies in the region beyond R where $V(r) > Q_\alpha$.

---

[7] Strictly speaking this should be the reduced mass of the alpha particle.

Here the wavefunction is exponentially decaying and is approximately

$$\chi(r) \simeq \chi_0 \exp\left(-\sqrt{\frac{2m_\alpha}{\hbar^2}} \int_R^r \sqrt{V(r) - Q_\alpha}\, dr\right), \qquad (11.40)$$

where χ_0 is a normalization factor. The outer edge of the barrier is at radius b where the Coulomb potential $V(r)$ equals Q_α, so

$$b = \frac{2Ze^2}{4\pi Q_\alpha}. \qquad (11.41)$$

Beyond this radius, the alpha particle has positive kinetic energy and is repelled by the rest of the nucleus, accelerating until it gains kinetic energy Q_α. For example, in the alpha decay of thorium isotopes the charge on the daughter nucleus is $Z = 88$, and as $\frac{e^2}{4\pi} \simeq 1.440$ MeV fm we find

$$b = 1.440\frac{176}{Q_\alpha} \text{ fm} = 253.44\frac{1}{Q_\alpha} \text{ fm}, \qquad (11.42)$$

where Q_α is in MeV. The Q_α values in Table 11.3 give barrier radii b of 25.7–62.1 fm. (The corresponding radii of the daughter nuclei range from 7.2 to 7.3 fm.)

Using the formula (11.41) for b, the Coulomb potential can be re-expressed as

$$V(r) = Q_\alpha \frac{b}{r}. \qquad (11.43)$$

The rate of alpha decay $\Re_\alpha$ is proportional to the probability that the alpha particle penetrates the barrier. This tunnelling probability P_{tun} is the square of the amplitude of the wavefunction just outside the barrier, at $r = b$. From equations (11.40) and (11.43),

$$P_{\text{tun}} = \chi_0^2 \exp(-2G), \qquad (11.44)$$

where the Gamow factor G is

$$G = \sqrt{\frac{2m_\alpha}{\hbar^2}} \int_R^b \sqrt{V(r) - Q_\alpha}\, dr = \sqrt{\frac{2m_\alpha Q_\alpha}{\hbar^2}} \int_R^b \sqrt{\frac{b}{r} - 1}\, dr. \qquad (11.45)$$

The integral can be calculated exactly using the substitution $r = b\sin^2\theta$, but because $R \ll b$, a sufficiently good approximation is to set $R = 0$, giving

$$\int_R^b \sqrt{\frac{b}{r} - 1}\, dr \simeq \frac{b\pi}{2}. \qquad (11.46)$$

Using this approximation,

$$G \simeq \sqrt{\frac{2m_\alpha Q_\alpha}{\hbar^2}}\frac{b\pi}{2} = \sqrt{\frac{2m_\alpha}{\hbar^2 Q_\alpha}}\frac{Ze^2}{4}, \qquad (11.47)$$

where we have again used formula (11.41) for b. The half-life for alpha decay is the inverse of the decay rate $\mathfrak{R}_\alpha$; therefore

$$\tau_{\frac{1}{2}} = \frac{1}{\mathfrak{R}_\alpha} = a\exp(2G) \tag{11.48}$$

and

$$\log\tau_{\frac{1}{2}} = \log a + 2G = \log a + \sqrt{\frac{2m_\alpha}{\hbar^2}}\,\frac{e^2}{2}\,\frac{Z}{\sqrt{Q_\alpha}}, \tag{11.49}$$

where a is a constant. We have arrived at the Geiger–Nuttall law, equation (11.37).

Although Gamow's theory explains the alpha decay trends of even–even nuclei, it assumes the existence of a preformed alpha particle. There is no easy way to estimate the probability of such preformation, although for even–even nuclei we might expect that an alpha particle is readily formed from a neutron pair and a proton pair residing in the topmost occupied energy levels. The situation is quite different in a nucleus with A odd where the highest energy nucleon is unpaired. To form an alpha particle in such a nucleus, at least one of the nucleons must be derived from a lower energy level. This added complication means that we should not expect odd nuclei to show such simple alpha decay trends as the even–even nuclei.

The emission of alpha particles from heavy nuclei is well known, but what about the emission of other light nuclei? The energetics for the possible emission of ^{12}C nuclei are very favourable. For instance, $Q = 32$ MeV for ^{12}C emission from ^{220}Ra. If the appropriate modifications are made to the above calculation for the alpha decay rate, then one predicts a ^{12}C emission rate that is less by a factor of around 10^{-3}. However, ^{12}C emission is not observed. On the other hand, ^{223}Ra has a half-life of 11.2 days for alpha decay and the following decay has also been observed:

$$^{223}\text{Ra} \quad \rightarrow \quad ^{14}\text{C} \quad + \quad ^{209}\text{Pb}, \tag{11.50}$$

with a rate that is 10^{-9} times that of alpha decay. This suggests that the probability of forming a ^{14}C nucleus within a radium nucleus is around 10^{-6} times that of forming an alpha particle. It is interesting to note that although ^{12}C is the most stable carbon nucleus in free space, within the neutron-rich environment of a heavy nucleus, ^{14}C appears to be more stable, or at least more likely to form.

11.5   Fission

^{238}U is the heaviest naturally occurring nucleus on Earth. The uranium that is found in the Earth's crust is composed of its two longest-lived isotopes. It is 99.27% ^{238}U and 0.72% ^{235}U with traces of ^{234}U. The half-life for alpha decay of ^{238}U is 4.5×10^9 years, while the half-life of ^{235}U is 7.0×10^8 years, so the naturally occurring proportion of ^{235}U would have been greater in the distant past. The ratio of these isotopes has great technological significance, because the two nuclei have very different fission properties, as we will see.

The heaviest nuclei are unstable to fission, dissociating into two smaller, more tightly bound nuclei in reactions such as

$$^{238}_{92}\text{U}_{146} \quad \rightarrow \quad ^{143}_{55}\text{Cs}_{88} \quad + \quad ^{93}_{37}\text{Rb}_{56} \quad + \quad 2n. \tag{11.51}$$

The binding energy per nucleon of ^{238}U is about 7.6 MeV, whereas for nuclei half this size it is about 8.5 MeV. This means that the fission of a ^{238}U nucleus releases around

$238 \times 0.9 = 214$ MeV. (This is a rough estimate as there are many ways the nucleus may fission.) Despite this significant energy release, the probability of spontaneous fission is very low. The half-life for spontaneous fission of ^{238}U is around 10^{16} years, two million times longer than the half-life for alpha decay.

The reason is that the two fission fragments must pass over a barrier along the fission path that is higher in energy than the original nucleus. This barrier maintains the integrity of the nucleus. The potential energy profile along the fission path is shown in Figure 11.17.

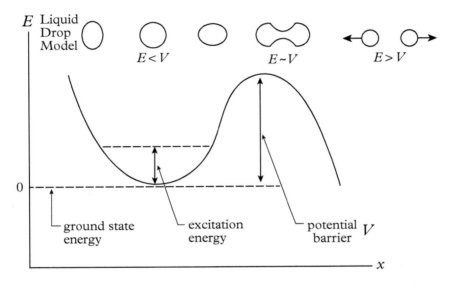

Fig. 11.17 The potential energy profile of two nuclear fragments plotted against the distance x separating their centres.

So fission, like alpha decay, involves tunnelling, but the probability of tunnelling is much less for a large nuclear fragment than for the relatively small alpha particle, as the Gamow factor (11.45) increases with the mass of the fragment. This explains the much greater half-life of ^{238}U for fission relative to alpha decay. The height of the potential barrier is known as the *activation energy*. For ^{238}U it is around 5.5 MeV.

In a fission event, such as (11.51), there is a high probability for additional neutron release, because more neutrons are required to stabilize heavy nuclei such as ^{238}U than to stabilize the two smaller fragments. For $A > 200$, $\frac{N}{Z} \simeq 1.5$, whereas for $70 < A < 160$, $\frac{N}{Z} \simeq 1.3$–1.4. So after fission there are surplus neutrons. In order to reach the valley of nuclear stability these neutrons are shed in various ways. Some neutrons are released immediately upon fission. These are known as *prompt neutrons*. Others are emitted by the highly excited fission products in reactions such as

$$^{90}_{36}\text{Kr}^*_{54} \quad \rightarrow \quad ^{89}_{36}\text{Kr}_{53} \quad + \quad n, \tag{11.52}$$

where the $*$ denotes an excited state. The time lag provided by these *delayed neutrons* is vital for the control of fission reactors. Even if neutrons are not released, many fission products are highly radioactive and undergo beta decay converting neutrons to protons, but this can take many years.

Fig. 11.18 Neutron-induced nuclear fission.

As there is no Coulomb barrier, the released neutrons are readily absorbed by other uranium nuclei. The absorption of a neutron by a uranium nucleus releases energy, so the new uranium isotope is born in an excited state. The excitation energy can be sufficient to push the nucleus over the fission potential barrier, as illustrated in Figure 11.18. For instance, ^{235}U contains an odd number of neutrons, so the binding of another neutron to form ^{236}U in the reaction

$$^{235}_{92}\text{U}_{143} \quad + \quad n \quad \rightarrow \quad ^{236}_{92}\text{U}^*_{144} \tag{11.53}$$

is favoured, and results in the creation of a ^{236}U nucleus in a highly excited state with energy

$$
\begin{aligned}
m(92, 236)^* &= m(92, 235) + m_n = (235.043924\,u + 1.0086665\,u) \\
&= 236.052589\,u.
\end{aligned} \tag{11.54}
$$

The excitation energy is

$$
\begin{aligned}
Q_{\text{exc}} &= m(92, 236)^* - m(92, 236) = 236.052589\,u - 236.045563\,u \\
&= 6.5\,\text{MeV},
\end{aligned} \tag{11.55}
$$

whereas the activation energy of ^{236}U is 6.2 MeV, so the newly formed nucleus has sufficient energy to easily pass over the fission barrier. Each neutron-induced fission releases further neutrons and therefore a chain reaction is possible in ^{235}U, which, if uncontrolled, may lead to an explosion.

By contrast, ^{238}U nuclei contain an even number of neutrons, so the binding energy of an additional neutron is much lower. A similar calculation to the one above gives an excitation energy of 4.8 MeV, lower than the fission activation energy of 5.5 MeV. Therefore a chain reaction is not possible in ^{238}U unless the neutrons are absorbed with sufficient kinetic energy (0.7 MeV) to make up the difference. The properties of these uranium isotopes determine the possible designs for functional nuclear reactors.

11.6   Fusion

Fusion reactions were vital for the cosmic synthesis of all the elements beyond hydrogen, and are therefore critical to our existence. They are also responsible for the energy generation in stars, as will be described in Chapter 13. It has long been a dream of physicists to generate cheap and plentiful energy using nuclear fusion. In pursuit of this goal various experimental fusion reactors have been built in the past fifty or so years and the ITER (International Thermonuclear Experimental Reactor) is currently under construction at Cadarache in France (Figure 11.19).

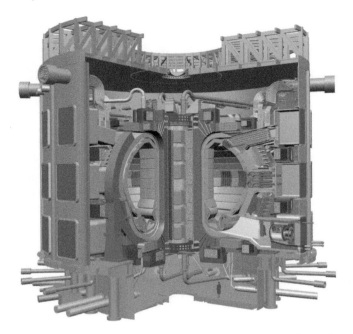

Fig. 11.19 Cut-away diagram of the ITER fusion reactor currently under construction.

For fusion to occur, two nuclei must find themselves close enough for the strong force to come into play. All nuclei are positively charged, so the Coulomb barrier prevents them approaching closely enough at ordinary temperatures and pressures. The barrier is much lower for nuclei with small values of Z, so the temperatures required for fusion are least for the lightest nuclei.

Take a process such as

$$B \quad + \quad X \quad \rightarrow \quad Y \tag{11.56}$$

in the rest frame of particle X. The collision cross section σ_{coll} of a non-relativistic particle B with momentum p is the area of a disc with radius equal to the de Broglie wavelength $\frac{2\pi\hbar}{p}$. Therefore the collision cross section is inversely proportional to the particle's kinetic energy E,

$$\sigma_{\text{coll}} = \pi \frac{(2\pi\hbar)^2}{p^2} \propto \frac{1}{E} . \tag{11.57}$$

In order to fuse, two nuclei must collide and then tunnel through the potential barrier, so the cross section for fusion σ_{fus} is the product of the collision cross section and the tunnelling probability. This is like the reverse of alpha decay; the higher the kinetic energy of B, the narrower the potential barrier, and the higher the probability of tunnelling.

We can therefore use the results from section 11.4. The tunnelling probability is proportional to $\exp(-2G(E))$, where $G(E)$ is the Gamow factor, so the fusion cross section is

$$\sigma_{\text{fus}}(E) = \frac{S(E)}{E} \exp(-2G(E)), \tag{11.58}$$

where $S(E)$ is a slowly varying nuclear structure function that may be determined by experiment or estimated semi-empirically. The Gamow factor can be adapted from equation (11.45),

$$G(E) = \sqrt{\frac{2\mu}{\hbar^2}} \int_R^b \sqrt{V(r) - E}\, dr = \sqrt{\frac{2\mu E}{\hbar^2}} \int_R^b \sqrt{\frac{b}{r} - 1}\, dr, \tag{11.59}$$

where μ is the reduced mass of B and X, and b is the distance at which the kinetic energy of B equals the Coulomb potential energy, so $E = V(b) = \frac{Z_B Z_X e^2}{4\pi b}$. $G(E)$ may be evaluated as before, to give a modified version of equation (11.47),

$$2G(E) \simeq 2\sqrt{\frac{2\mu}{\hbar^2 E}}\frac{Z_B Z_X e^2}{8} = \sqrt{2\pi^2 \alpha^2 m_p}\sqrt{\frac{\mu}{m_p}}\frac{Z_B Z_X}{\sqrt{E}} = \sqrt{\frac{E_G}{E}}, \tag{11.60}$$

where $\alpha = \frac{e^2}{4\pi\hbar}$ is the *fine structure constant*, and we have collected together the constant terms to define the Gamow energy

$$\begin{aligned}
E_G &= 2\pi^2 \alpha^2 m_p \frac{\mu}{m_p}(Z_B Z_X)^2 = 2\pi^2 \alpha^2 \times (938\ \text{MeV}) \left(\frac{\mu}{m_p}\right)(Z_B Z_X)^2 \\
&= \left(\frac{\mu}{m_p}\right)(Z_B Z_X)^2 \times 987\ \text{keV}.
\end{aligned} \tag{11.61}$$

(To obtain the final expression for the Gamow energy, we inserted the rest mass of the proton $m_p = 938$ MeV and used the well known approximation $\alpha \simeq \frac{1}{137}$.)

When a mono-energetic beam of particles B with number density n_B and velocity v impinges on a target composed of the nucleus X with number density n_X, such as may occur in a particle accelerator, the total number of fusion reactions in volume dV and time dt is

$$dN_{\text{fus}} = \frac{n_B n_X}{1 + \delta_{BX}}\sigma_{\text{fus}}(E)v\, dt\, dV, \tag{11.62}$$

with $\sigma_{\text{fus}}(E)$ given by formula (11.58). Here $E = \frac{1}{2}\mu v^2$ and to avoid double counting when identical nuclei fuse, $\delta_{BX} = 1$ when B and X are identical and is zero otherwise.

11.6.1 Thermonuclear fusion

Within a hot ionized plasma, such as in a star, nuclei have a random distribution of velocities and the probability of fusion depends on the relative velocity of any two incident particles. This probability increases dramatically as the velocity increases, and fusion is only possible at very high temperatures, T. It is therefore known as *thermonuclear fusion*. In thermal equilibrium the velocity distribution of the nuclei is the Maxwell distribution (10.53). To find the fusion rate we must integrate the cross section over this distribution. From equation (11.62), the integrated reaction rate per unit volume is

$$\Re_{\text{fus}} = \frac{dN_{\text{fus}}}{dV\,dt} = \frac{n_B n_X}{1 + \delta_{BX}} \langle \sigma_{\text{fus}}(E)v \rangle, \tag{11.63}$$

where

$$\langle \sigma_{\text{fus}}(E)v \rangle = \int_0^\infty \left(\frac{\mu}{2\pi T} \right)^{\frac{3}{2}} \exp\left(-\frac{E}{T} \right) \sigma_{\text{fus}}(E) \, v \, 4\pi v^2 \, dv$$

$$= \left(\frac{2}{\pi \mu T^3} \right)^{\frac{1}{2}} \int_0^\infty \exp\left(-\frac{E}{T} \right) \sigma_{\text{fus}}(E) E \, dE, \tag{11.64}$$

and the factors involving T come from the Maxwell distribution. Substituting from equation (11.58), $\sigma_{\text{fus}} = \frac{S(E)}{E} \exp(-2G(E))$, and inserting the Gamow factor $2G(E) = \sqrt{\frac{E_G}{E}}$ from equation (11.60) gives

$$\langle \sigma_{\text{fus}}(E)v \rangle = \left(\frac{2}{\pi \mu T^3} \right)^{\frac{1}{2}} \int_0^\infty S(E) \exp\left(-\frac{E}{T} - \sqrt{\frac{E_G}{E}} \right) dE. \tag{11.65}$$

Even within stars, it is only the exceptionally energetic particles in the Maxwell tail that have sufficient kinetic energy to contribute to fusion. The factor $\exp(-\frac{E}{T})$ decreases rapidly with energy E and the tunnelling probability $\exp\left(-\sqrt{\frac{E_G}{E}} \right)$ increases rapidly. Only in the region where these two exponential functions overlap significantly, known as the *Gamow peak*, is fusion possible. The relatively narrow range around the maximum of the Gamow peak is known as the *optimum bombarding energy* E_o. As T rises, the Maxwell distribution shifts to higher energies and this increases both the height and width of the Gamow peak, as shown in Figure 11.20, so the fusion rate increases dramatically.

The integral in equation (11.65) cannot be calculated analytically, but the exponent $f(E) = -\frac{E}{T} - \sqrt{\frac{E_G}{E}}$ has a sharp maximum, so the integral may be evaluated approximately using the method of steepest descents, which was described in section 1.4.4. It gives a good estimate of the fusion rate if there are no nuclear resonances close to the optimal bombarding energy. Differentiating the exponent, we find that its maximum occurs when

$$f'(E) = -\frac{1}{T} + \frac{1}{2E}\sqrt{\frac{E_G}{E}} = 0. \tag{11.66}$$

Solving this gives the optimal bombarding energy and the height of the Gamow peak,

$$E_o = \left(\frac{E_G T^2}{4} \right)^{\frac{1}{3}} \quad \text{and} \quad f(E_o) = -3 \left(\frac{E_G}{4T} \right)^{\frac{1}{3}}. \tag{11.67}$$

Calculating the second derivative and substituting for E_o, we obtain

$$f''(E) = -\frac{3}{4E^2}\sqrt{\frac{E_G}{E}}, \quad \text{so} \quad \frac{1}{\sqrt{|f''(E_o)|}} = \frac{1}{\sqrt{3}}(2E_G T^5)^{\frac{1}{6}}, \tag{11.68}$$

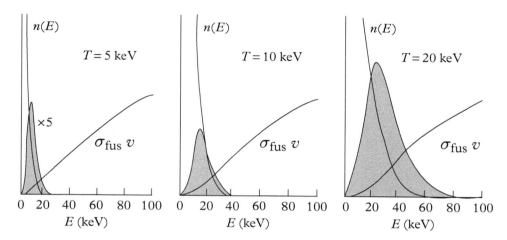

Fig. 11.20 As temperature T increases, the Boltzmann factor $n(E) = \exp(-\frac{E}{T})$ moves to the right. This increases the overlap with the tunnelling probability $\sigma_{\text{fus}}(E)v$, which dramatically increases the size of the Gamow peak (shaded) and results in a strong temperature dependence for the fusion rate.

where $\dfrac{1}{\sqrt{|f''(E_o)|}}$ is the width of the Gamow peak. The steepest descents formula (1.77) then gives

$$\langle \sigma_{\text{fus}}(E)v \rangle \simeq \left(\frac{2}{\pi \mu T^3} \right)^{\frac{1}{2}} S(E_o) \exp(f(E_o)) \sqrt{\frac{2\pi}{|f''(E_o)|}}$$

$$= \left(\frac{4}{\mu T^3} \right)^{\frac{1}{2}} \frac{1}{\sqrt{3}} (2 E_G T^5)^{\frac{1}{6}} S(E_o) \exp \left\{ -3 \left(\frac{E_G}{4T} \right)^{\frac{1}{3}} \right\}. \qquad (11.69)$$

Substituting into equation (11.63), we find the fusion rate

$$\Re_{\text{fus}} \simeq \frac{n_B n_X}{1 + \delta_{BX}} \left(\frac{2}{\sqrt{3\mu}} \right) \frac{S(E_o)(2 E_G)^{\frac{1}{6}}}{T^{\frac{2}{3}}} \exp \left\{ -3 \left(\frac{E_G}{4T} \right)^{\frac{1}{3}} \right\}. \qquad (11.70)$$

Astrophysicists often quote the temperature dependence of fusion rates as $\Re_{\text{fus}} \propto T^n$, where the exponent n is obtained by taking the logarithmic derivative of $\Re_{\text{fus}}$, i.e. $n = \frac{T}{\Re_{\text{fus}}} \frac{d\Re_{\text{fus}}}{dT}$. Although the exponent n varies with temperature, such formulae still prove useful when fusion reactions are occurring at a set temperature within a star. Taking the logarithm of equation (11.70) gives

$$\ln \Re_{\text{fus}} = -\frac{2}{3} \ln T - 3 \left(\frac{E_G}{4T} \right)^{\frac{1}{3}} + \text{constant}, \qquad (11.71)$$

and the logarithmic derivative is

$$n = \frac{T}{\Re_{\text{fus}}} \frac{d\Re_{\text{fus}}}{dT} = T \frac{d \ln \Re_{\text{fus}}}{dT} = -\frac{2}{3} + \left(\frac{E_G}{4T} \right)^{\frac{1}{3}}. \qquad (11.72)$$

Substituting for E_G from equation (11.61), we find that the optimal bombarding energy and width of the Gamow peak are

$$E_o \simeq 1220 \times \left(\left(\frac{\mu}{m_p} \right) (Z_B Z_X)^2 (T_6)^2 \right)^{\frac{1}{3}} \text{eV}, \tag{11.73}$$

$$\frac{1}{\sqrt{|f''(E_o)|}} \simeq 265 \times \left(\left(\frac{\mu}{m_p} \right) (Z_B Z_X)^2 \right)^{\frac{1}{6}} (T_6)^{\frac{5}{6}} \text{eV}, \tag{11.74}$$

where $T_6 = \frac{T}{10^6 \text{K}}$ is a convenient, dimensionless notation for stellar temperature in units of a million kelvin. As $1 \text{ K} = 8.62 \times 10^{-5}$ eV, a temperature T_6 corresponds to $86.2 \, T_6$ eV. Using this notation, we also obtain

$$n = -\frac{2}{3} + \left(\frac{2860 \left(\frac{\mu}{m_p} \right) (Z_B Z_X)^2}{T_6} \right)^{\frac{1}{3}}. \tag{11.75}$$

Most of the Sun's energy is generated in a process known as the *proton–proton* chain, or *pp* chain, involving the fusion of two protons, as we will discuss in section 13.5.1. Both B and X are protons, so $Z_B = Z_X = 1$, and the reduced mass of the fusing protons is $\mu = \frac{1}{2} m_p$. Fusion takes place in the solar core at around $T = 1.6 \times 10^7$ K, so $T_6 = 16$ and therefore

$$E_o(pp \text{ chain}) \simeq 1220 \times \left(\left(\frac{1}{2} \right) (16)^2 \right)^{\frac{1}{3}} \text{eV} \simeq 6.2 \text{ keV},$$

$$\frac{1}{\sqrt{|f''(E_o)|}} (pp \text{ chain}) \simeq 265 \times \left(\frac{1}{2} \right)^{\frac{1}{6}} (16)^{\frac{5}{6}} \text{eV} \simeq 2.4 \text{ keV}. \tag{11.76}$$

At this temperature, the typical energy of a proton is 86.2×16 eV $\simeq 1.4$ keV. As the optimal bombarding energy E_o is 6.2 keV, most of the fusion energy produced in the Sun is due to collisions at over four times this typical energy.

Stars that are more massive than the Sun generate the bulk of their energy through the CNO cycle, a fusion process catalysed by carbon, nitrogen and oxygen nuclei, as we will see in section 13.5.2. The core temperature of such stars is about $T = 2.0 \times 10^7$ K, so $T_6 = 20$. The bottleneck of the CNO cycle is the reaction

$$p + {}^{14}\text{N} \rightarrow {}^{15}\text{O}. \tag{11.77}$$

In this case $Z_B = 1$, $Z_X = 7$ and the reduced mass of the colliding proton and nitrogen nucleus is $\mu = \frac{14}{15} m_p \simeq m_p$. Substituting these values into equations (11.73) and (11.74) gives

$$E_o(\text{CNO cycle}) \simeq 1220 \times \left((7)^2 (20)^2 \right)^{\frac{1}{3}} \text{eV} \simeq 33 \text{ keV},$$

$$\frac{1}{\sqrt{|f''(E_o)|}} (\text{CNO cycle}) \simeq 265 \times (7)^{\frac{1}{3}} (20)^{\frac{5}{6}} \text{eV} \simeq 6.2 \text{ keV}. \tag{11.78}$$

The typical proton energy is now around 1.7 keV, so most of the energy produced in the CNO cycle is due to a tiny fraction of protons with exceptionally high energies.

At the solar core temperature $T_6 = 16$, the exponent n determining the temperature dependence of the pp chain fusion rate is

$$n(pp \text{ chain}) \simeq -\frac{2}{3} + \left(\frac{2860}{2 \times 16}\right)^{\frac{1}{3}} = -0.67 + 4.47 \simeq 3.8 \,, \qquad (11.79)$$

whereas in the case of the CNO cycle at temperature $T_6 = 20$,

$$n(\text{CNO cycle}) \simeq -\frac{2}{3} + \left(\frac{2860 \times (7)^2}{20}\right)^{\frac{1}{3}} \simeq -0.67 + 19.1 \simeq 18 \,, \qquad (11.80)$$

so the rate of fusion in the CNO cycle is extremely sensitive to temperature. We will use these results in Chapter 13.

11.6.2 Controlled nuclear fusion

Returning to the prospects for controlled fusion, hydrogen isotopes are the obvious candidates for the nuclear fuel. The isotopes of hydrogen are deuterium ^{2}H, whose nuclei consist of a proton and one neutron, and tritium ^{3}H, whose nuclei consist of a proton and two neutrons. These nuclei are often called deuterons and tritons. Fusing deuterons into a ^{4}He nucleus is not a very effective process, as the energy released is 23.8 MeV, which is sufficient to dissociate a neutron or a proton. The following reactions, which occur with equal probability, are much more likely:

$$^2\text{H} \;+\; ^2\text{H} \;\rightarrow\; ^3\text{He} \;+\; n \;+\; 3.3 \text{ MeV} \qquad (11.81)$$
$$^2\text{H} \;+\; ^2\text{H} \;\rightarrow\; ^3\text{H} \;+\; p \;+\; 4.0 \text{ MeV} \,. \qquad (11.82)$$

ITER will generate energy through the much more effective fusion of deuterons and tritons:

$$^2\text{H} \;+\; ^3\text{H} \;\rightarrow\; ^4\text{He} \;+\; n \;+\; 17.6 \text{ MeV} \qquad (11.83)$$
$$^3\text{H} \;+\; ^3\text{H} \;\rightarrow\; ^4\text{He} \;+\; 2n \;+\; 11.3 \text{ MeV} \,. \qquad (11.84)$$

The deuterium–tritium plasma will be confined within a toroidal *tokamak* where its temperature will be raised to 10^8 K in order to achieve fusion. The goal is to generate ten times the energy required to run the reactor. If all goes to plan, fusion will be sustained for up to ten minutes at a time, while generating 500 MW of power.

Tritium has a half-life of 12.3 years and is not readily available in nature. The core of ITER will be surrounded by a lithium blanket, so that the reactor can generate its own tritium fuel through the following processes:

$$n \;+\; ^6\text{Li} \;\rightarrow\; ^4\text{He} \;+\; ^3\text{H} \qquad (11.85)$$
$$n \;+\; ^7\text{Li} \;\rightarrow\; n \;+\; ^4\text{He} \;+\; ^3\text{H} \,. \qquad (11.86)$$

Lithium is available in salt deposits and in low concentrations in sea water.

11.7 The Island of Stability

The heaviest naturally occurring nucleus is ^{238}U, as the half-lives of all heavier nuclei are much shorter than the age of the Earth. Since 1940 many of these *transuranic elements* have been artificially created and studied in the laboratory. Nuclei with $Z = 93$–100 were first produced by Glenn Seaborg and his team at the University of Berkeley, California by prolonged exposure of uranium to intense neutron fluxes from nuclear reactors.[8] Transuranic elements are formed following beta decay of the new neutron-rich nuclei. These elements can then be chemically isolated and purified. There is a limitation to the creation of new elements in this way. The nuclear half-lives decrease dramatically as atomic number Z increases. The longest-lived isotope of plutonium is $^{244}_{94}\text{Pu}_{150}$ with a half-life of 8×10^7 years, while the longest half-life of a californium isotope is 898 years for $^{251}_{98}\text{Cf}_{153}$, and just 100.5 days for the fermium isotope $^{257}_{100}\text{Fm}_{157}$. The advance to even heavier nuclei is blocked by the next isotope $^{258}_{100}\text{Fm}_{158}$, which has a half-life of just 0.3 ms. To reach the nuclei beyond this, it is necessary to accelerate light nuclei and fire them at a target formed from a heavy nucleus. For instance, element 104 rutherfordium was created in 1969 at Berkeley by firing ^{12}C nuclei at californium nuclei, with the result

$$^{12}_{6}\text{C}_6 \quad + \quad ^{249}_{98}\text{Cf}_{151} \quad \rightarrow \quad ^{257}_{104}\text{Rf}_{153} \quad + \quad 4n \,. \tag{11.87}$$

High energy heavy ion accelerators are required for this research as there is a large Coulomb barrier proportional to $Z_1 Z_2$ to overcome in a reaction between ions with proton numbers Z_1 and Z_2. In the above case, $Z_1 Z_2 = 6 \times 98 = 588$.

The filling of nuclear shells confers added stability to nuclei. In the late 1960s, Seaborg suggested that the instability trend should reverse as Z and N approach the next magic numbers. This was supported by calculations of the expected rates of alpha and beta decay and fission half-lives by Nilsson and his collaborators. These new magic numbers are not known precisely, but are expected to be $Z = 114, 118$ or 126 and somewhere around $N = 184$. The relatively long-lived nuclei that might exist close to these values are referred to as the *island of stability*. Calculations suggest that the half-life of $^{294}_{110}\text{Ds}_{184}$, darmstadtium-294, could be as long as 10^6 years. Reaching the island of stability is beyond current technology, but only just.

0.2% of naturally occurring calcium is formed of the doubly magic neutron-rich nucleus $^{48}_{20}\text{Ca}_{28}$. Its large proportion of neutrons makes it an ideal projectile for producing neutron-rich heavy nuclei. Researchers at the Flerov Laboratory of Nuclear Reactions in Dubna, Russia and the GSI Helmholtz Centre for Heavy Ion Research in Darmstadt, Germany have produced nuclei of all the elements up to atomic number 118 by firing beams of ^{48}Ca nuclei at heavy targets formed of elements such as Pu, Am, Cm, Bk and Cf. For example, atoms of the recently named element number 117 tennessine were created in 2014 in Darmstadt by firing $^{48}_{20}\text{Ca}_{28}$ ions at a berkelium target. Around 13 mg of $^{249}_{97}\text{Bk}_{152}$, which has a half-life of just 330 days, was produced specifically for the experiment by the Oak Ridge National Laboratory in the United States. Tennessine nuclei were then created in the following reaction:

$$^{48}_{20}\text{Ca}_{28} \quad + \quad ^{249}_{97}\text{Bk}_{152} \quad \rightarrow \quad ^{293}_{117}\text{Ts}_{176} \quad + \quad 4n \,. \tag{11.88}$$

In this reaction, $Z_1 Z_2 = 20 \times 97 = 1940$, which gives some idea of the increasing beam energies required to synthesize the superheavy nuclei.

The half-lives of the known isotopes of the heaviest nuclei increase with increasing neutron number and are generally in good agreement with predictions, so we may be within sight of the island of stability.

---

[8] This is equivalent to the r-process known to astrophysicists.

11.8 Exotic Nuclei

Most of the nucleons in the very lightest nuclei are on the nuclear surface, so their bonds are not saturated. These nuclei do not fit the liquid drop formula and are less dense than might be expected from the density trend described by equation (11.6). There are other nuclei that do not match the predictions of the shell model too well. For instance, Figure 11.7 shows that ^{12}C has a relatively high binding energy per nucleon, even though it does not have a magic number of neutrons or protons. There are a number of competing models that describe light nuclei, each capturing certain aspects of their structure.

The lightest composite nucleus is the deuteron, formed of one neutron and one proton. It has a binding energy of just 2.2 MeV, which is much less than the binding energy per nucleon of large nuclei, and it has no bound excited states. Consequently the mean radius of the deuteron, which is 2.14 fm, is significantly larger than might be expected from equation (11.6). The most stable of the light nuclei is ^{4}He. So much so that adding an extra neutron or proton to form ^{5}He or ^{5}Li, or fusing two ^{4}He nuclei to form ^{8}Be, proves impossible. This presents a significant barrier to the creation of elements heavier than helium. Although the addition of one extra particle does not produce a stable nucleus, the addition of two extra nuclear components to ^{4}He does lead to stable nuclei such as ^{6}Li, ^{9}Be and ^{12}C.

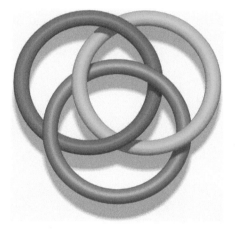

Fig. 11.21 The Borromean rings.

These latter nuclei are known as *Borromean nuclei*, because they have the curious property that if one component is removed they fall into three parts. This is reminiscent of the Borromean rings on the coat of arms of the Italian Borromeo family, shown in Figure 11.21. Although the three rings cannot be separated, no two rings are linked, so if one ring is removed the other two fall apart. Other examples of Borromean nuclei are ^{6}He, ^{11}Li, ^{14}Be and ^{22}C. They are not readily understood in terms of the shell model, but seem to have an almost molecular structure.

For example, ^{9}Be appears to be composed of two alpha particles plus a neutron, as shown in Figure 11.22 (left), and the nucleus is readily broken down into these components,

$$^9\text{Be} \quad \rightarrow \quad ^4\text{He} \quad + \quad ^4\text{He} \quad + \quad n. \tag{11.89}$$

Fig. 11.22 Left: ^{9}Be nucleus composed of two alpha particles plus a neutron. Right: ^{12}C composed of three alpha particles.

When beryllium is subjected to alpha particle bombardment, neutrons are released, which was key to the discovery of the neutron, as we saw in section 11.1.

There is also evidence that nuclei such as ^{12}C, ^{16}O, ^{20}Ne and ^{24}Mg may be viewed as clusters of alpha particles. The ground state of ^{12}C appears to be shaped like an equilateral triangle formed of three alpha particles, as shown in Figure 11.22 (right). Fred Hoyle predicted the existence of an excited state of ^{12}C that plays a very important role in nucleosynthesis, to be discussed in Chapter 13. This excited state is thought to correspond to a linear (or perhaps bent) chain of three alpha particles.

Fig. 11.23 Left: The halo nucleus ^{11}Li. Right: The nucleus ^{208}Pb, which is comparable in size to ^{11}Li.

Since the 1980s there has been a great deal of interest in neutron-rich, low-mass nuclei close to the neutron dripline. These nuclei are much larger than would be expected from the relationship given in equation (11.6). They are known as *halo nuclei*, as they contain neutrons that only weakly bound to the nuclear core. The loosely bound halo nucleons spend less than half their time within the core. One example is ^{6}He, which contains two neutrons weakly bound to a ^{4}He core. A second much-studied halo nucleus is ^{11}Li which contains two halo neutrons weakly bound to a ^{9}Li core. This nucleus is comparable in size to ^{208}Pb, as illustrated in Figure 11.23. ^{11}Li has a half-life of 9 ms. It is a Borromean nucleus and just 0.3 MeV is required to remove both its halo neutrons.

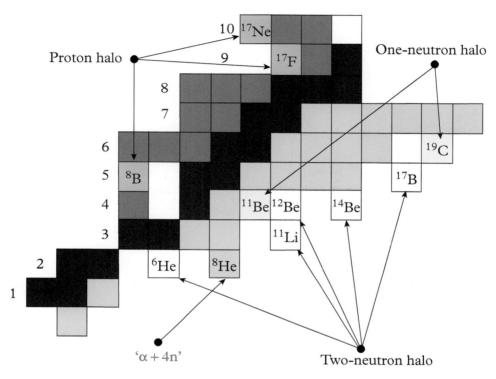

Fig. 11.24 Halo nuclei. Z is plotted vertically.

Other halo nuclei are shown in Figure 11.24. They include ^{8}He, ^{11}Be, ^{14}Be, ^{17}B and ^{19}C. There are also nuclei such as ^{8}B and ^{17}Ne that appear to contain loosely bound halo protons.

11.9 Pions, Yukawa Theory and QCD

So far, we have used phenomenological models like the shell model and liquid drop model to study the nucleus. With more sophisticated models of the nucleon–nucleon interaction, and high-performance computers, it is now possible to accurately calculate the masses of many of the smaller nuclei, and the energies of their excited states. The nucleus needs to be treated as a multi-nucleon quantum mechanical system, with fundamental 2-nucleon and 3-nucleon potentials. To obtain a deeper understanding, one should derive these nuclear potentials from more fundamental interactions between the nucleons. In 1935 Hideki Yukawa attempted to do this. He proposed a theory that would explain the short range of the strong force and the diminishing nuclear density in the skin of the nucleus. His idea was that the strong force between nucleons is due to the exchange of a new particle. Unlike the photon which is exchanged to produce the electromagnetic force, Yukawa's particle has a non-zero mass.

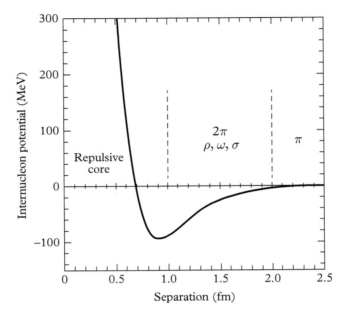

Fig. 11.25 Meson exchange potential.

The force produced by the exchange of such a particle is described by the Yukawa potential

$$V(r) = -\frac{\lambda^2}{4\pi r} \exp\left(-\frac{r}{a}\right) , \tag{11.90}$$

where λ^2 determines the strength of the interaction. The range of the force is $a = \frac{\hbar}{m_\pi}$, where m_π is the mass of the exchanged particle. Note that with zero-mass exchange particles, the Yukawa potential reduces to the infinite-range Coulomb potential. To match the known range of the strong force, Yukawa predicted that the particle would have a mass at least 200 times the mass of the electron, in the region of 130 MeV. A triplet of particles, π^-, π^0, π^+, with the desired properties was discovered by Cecil Powell, César Lattes and Giuseppe Occhialini in cosmic rays in 1947. They are known as *pions* and are the lightest of the family of particles known as *mesons*. The mass of the electrically neutral π^0 is 135.0 MeV, and the mass of the charged π^- and π^+ is 139.6 MeV.

The nucleon–nucleon potential shown in Figure 11.25 has been determined in scattering experiments, but is partly explained in terms of pion exchange. Following Yukawa's work, other Japanese theorists developed more sophisticated models of the nucleus. In particular, Mituo Taketani and collaborators proposed in the 1950s that the nuclear potential may be understood in three layers. Single pion exchange, as described by the Yukawa potential, accounts for the outermost region beyond 2.0 fm. Between about 1.0 fm and 2.0 fm the potential is best explained by two-pion exchange, which produces the following version of a Van der Waals-type potential,

$$V(r) \simeq -\frac{P(m_\pi r)}{r^6} \exp\left(-\frac{2r}{a}\right) , \tag{11.91}$$

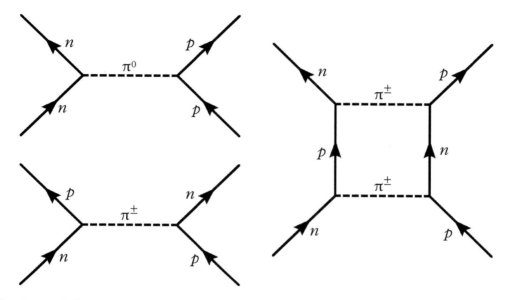

Fig. 11.26 Left: Two examples of single pion exchange. Right: An example of two-pion exchange.

where P is a polynomial in $m_\pi r$. Diagrams representing these pion exchange interactions are shown in Figure 11.26. The exchange of other heavier mesons is also important at these shorter distances. These include the $\eta(549)$, $\rho(770)$ and $\omega(782)$ mesons,[9] where the numbers in brackets are their masses in MeV. In the inner region, at distances less than 1.0 fm, there is a sharp repulsion between nucleons which gives them a hard core and prevents their merging.

In reality, the physics of the strong force is more complicated than this. The potential is not the single function shown in Figure 11.25, because it depends on the spins of the nucleons, and also on whether or not the nucleons are of the same type. The best available 2-nucleon and 3-nucleon potentials are still phenomenological, as they are adjusted to match scattering data and the properties of a few bound states, but their features can be understood and to some extent predicted in terms of an effective field theory of strongly interacting pions.

We now know that neither the proton nor the neutron is actually a fundamental particle. The proton contains two up quarks and a down quark, whereas the neutron contains one up quark and two down quarks. The quarks are bound together by the colour force as described by quantum chromodynamics (QCD). We will have more to say about this in Chapter 12. The term meson was originally used for particles intermediate in mass between electrons and nucleons. This is no longer tenable. The modern meaning of meson is a particle formed of a quark–antiquark pair. The pions, the lightest such particles, are formed of pairs of up and down quarks and antiquarks.

QCD is now a very well established theory with huge amounts of experimental support. It is ironic that we have a deeper understanding of the force that binds quarks together into nucleons than we have of the force between nucleons that determines the structure of the nucleus. In principle it should be possible to derive the nucleon–nucleon potential and other

---

[9] There is also the σ meson, whose structure is not yet fully understood.

properties of nuclei from QCD, but this is an exceedingly difficult problem. For instance, the spin–orbit coupling between a nucleon and a nucleus is still not well understood. With the use of a great deal of computing power, it has recently become possible to derive the masses of strongly interacting particles, including the proton and neutron and the various mesons, directly from QCD with around 5% precision. This is a great achievement. Now, the interactions between nucleons are also being explored on the computer and the early results appear to show that QCD does indeed produce a nucleon–nucleon potential that changes sharply from an attractive force at longer distances to a hard repulsive force at distances below 1.0 fm, matching the observed potential shown in Figure 11.25.

11.10   Further Reading

K. Heyde, *Basic Ideas and Concepts in Nuclear Physics: An Introductory Approach (3rd ed.)*, Bristol: IOP, 2004.

K.S. Krane, *Introductory Nuclear Physics*, New York: Wiley, 1988.

R.F. Casten, *Nuclear Structure from a Simple Perspective (2nd ed.)*, Oxford: OUP, 2000.

The shell model, including the effects of nuclear deformation, is surveyed in

S.G. Nilsson and I. Ragnarsson, *Shapes and Shells in Nuclear Structure*, Cambridge: CUP, 1995.

For a review of the production and properties of superheavy nuclei, see

Y. Oganessian, *Synthesis and Decay Properties of Superheavy Elements*, Pure Appl. Chem. **78** (2006) 889-904.

For a review of halo nuclei, see

J. Al-Khalili, *An Introduction to Halo Nuclei* in *The Euroschool Lectures on Physics with Exotic Beams, Vol. 1*, eds. J.S. Al-Khalili and E. Roeckl, pp 77-112, Berlin, Heidelberg: Springer, 2004.

For a survey of Yukawa's pion theory, see

W. Weise, *Yukawa's Pion, Low-Energy QCD and Nuclear Chiral Dynamics*, Prog. Theor. Phys. Suppl. **170** (2007) 161-184.

12
Particle Physics

12.1 The Standard Model

In this chapter we will investigate the fundamental components of matter. We have so far considered the structure of the universe and the matter that it contains in terms of four distinct forces: gravity and electromagnetism plus the two nuclear forces, the strong force and the weak force. Gravity determines the large-scale structure of the universe, but the intrinsic weakness of gravity means that it has an utterly negligible effect on the interactions between pairs of fundamental particles, such as the electrons in an atom. By contrast, all three of the other forces play important roles in particle physics. The electromagnetic force holds atoms together. The strong force binds protons and neutrons in the nucleus of an atom. It is also important in alpha decay and nuclear fission. The weak force is responsible for beta decay of nuclei and plays a major role in the synthesis of the elements in the stars. When the weak force acts, particles may change their identity. For instance, in beta decay a neutron is converted into a proton and, simultaneously, an electron and an electron antineutrino are created. Today these three forces are understood within a single structure that includes a unified theory of the electromagnetic and weak forces, combined with a rather similar theory of the strong force. This incredibly successful theory is one of the triumphs of modern physics. It is known as the *Standard Model*.

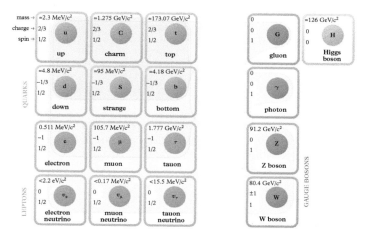

Fig. 12.1 Standard Model particle table. The first three columns show the three generations of fermions. The final columns show the bosons.

The Physical World. Nicholas Manton and Nicholas Mee, Oxford University Press (2017).
© Nicholas Manton and Nicholas Mee. DOI 10.1093/oso/9780198795933.001.0001

12.1.1 Fundamental particles

All known particles and forces can be reduced to the interplay of a small number of fundamental particles, as shown in Figure 12.1. Each type of particle is completely defined by its rest mass, its spin and the various charges that determine its interactions. Particles may have numerous other properties, such as their magnetic moments, and the rates at which they decay into various other types of particle. These can all be calculated from theory. Quantum mechanics treats particles of the same type as indistinguishable. They divide naturally into two classes: bosons which have integer spin and obey Bose–Einstein statistics, and fermions which have half-integer spin, and obey Fermi–Dirac statistics and the exclusion principle. There are just 12 fundamental fermions, all of spin $\frac{1}{2}$, along with their antiparticles, as listed in Table 12.1. They are divided into two categories, *quarks* and *leptons*, depending on whether they are affected by the strong force. Quarks interact through the strong, weak and electromagnetic forces. The charged leptons, namely the electron, the muon and the tauon, interact via the electromagnetic and weak forces, but not the strong force. The uncharged leptons are the neutrinos.[1] They only interact via the weak force.

These 12 fermions form three *generations* of four particles. The first generation consists of the *up* and *down* quarks, the *electron* and the *electron neutrino*. Ordinary matter is formed from the first three of these particles. The proton is composed of two up quarks and a down quark bound together. The neutron is composed of two down quarks and one up quark. The two quarks in the second generation are known as *charm* and *strange*, and the two leptons are the *muon* and the *muon neutrino*. The third generation consists of the *top* and *bottom* quarks, the *tauon* and the *tauon neutrino*. Each particle in the second and third generations carries the same charges as the corresponding particle of the first generation and seems to be just a heavier replica of it.

Most particles are unstable and decay via processes that are controlled by one or other of the fundamental forces. Heavier particles decay to two or more lighter particles, releasing energy that is carried away as kinetic energy of the decay products. If such particles have a mean lifetime T (i.e. half-life $(\log 2)T$) and there are $N(0)$ of them at time $t = 0$, then the number of particles at a later time is

$$N(t) = N(0)\exp(-\Re t) \tag{12.1}$$

where the decay rate is $\Re = \frac{1}{T}$. Typically, it is the decay width

$$\Gamma = \hbar\Re = \frac{\hbar}{T} \tag{12.2}$$

that is measured (in units of energy) in collider experiments. The rest mass (or energy) of short-lived particles is not precisely defined due to the quantum mechanical uncertainty relations. Γ is the width of the rest mass distribution, as shown, for example, in Figure 12.26. If the particle decays in several ways, then

$$\Gamma_{\text{total}} = \sum_i \Gamma_i \quad \text{and} \quad \text{Br}_i = \frac{\Gamma_i}{\Gamma_{\text{total}}}, \tag{12.3}$$

where Γ_i is the partial decay width for decay mode i and Br_i is the branching fraction into this mode.

---

[1] Figure 12.1 gives current limits on the neutrino masses obtained directly from particle decays. Table 12.1 gives the much more stringent limits obtained from cosmology and neutrino oscillation experiments.

Generation	Leptons	q	Mass (GeV)	Quarks	q	Mass (GeV)
I	electron (e^-)	-1	0.0005	up (u)	$\frac{2}{3}$	0.002
I	neutrino (ν_e)	0	$< 10^{-9}$	down (d)	$-\frac{1}{3}$	0.005
II	muon (μ^-)	-1	0.106	charm (c)	$\frac{2}{3}$	1.3
II	neutrino (ν_μ)	0	$< 10^{-9}$	strange (s)	$-\frac{1}{3}$	0.1
III	tauon (τ^-)	-1	1.78	top (t)	$\frac{2}{3}$	173
III	neutrino (ν_τ)	0	$< 10^{-9}$	bottom (b)	$-\frac{1}{3}$	4.2

Table 12.1 Fundamental spin $\frac{1}{2}$ fermions of the Standard Model. q is electric charge in units of the proton charge. (Each particle has an antiparticle of the same mass and opposite charge.)

Bosons	q	Mass (GeV)	Spin	Role
gluon (G)	0	0	1	QCD exchange boson
photon (γ)	0	0	1	QED exchange boson
$W^\pm$ boson	± 1	80.4	1	weak exchange boson
Z boson	0	91.2	1	weak exchange boson
Higgs boson (H)	0	125	0	Higgs mechanism

Table 12.2 The fundamental bosons of the Standard Model.

The lifetimes seen in particle decays are typically: strong force, $10^{-24} - 10^{-20}$ s; electromagnetic force, $10^{-19} - 10^{-16}$ s; weak force, $10^{-12} - 10^{-6}$ s. The quarks and charged leptons of the second and third generations are unstable and rapidly decay, principally via the weak force.

Two conservation laws seem to be universally valid in all particle collisions and decays. These are lepton number conservation, where each lepton contributes $+1$ and each antilepton -1, and baryon number conservation, where each quark contributes $\frac{1}{3}$ and each antiquark $-\frac{1}{3}$. The factor of $\frac{1}{3}$ occurs because baryons, such as protons and neutrons, are composed of three quarks, and antibaryons of three antiquarks.

The fundamental bosons are shown in Table 12.2. The exchange of the small number of spin 1 bosons listed here produces the strong, electromagnetic and weak forces, as we will discuss later. The Higgs boson is the only known fundamental spin 0 or scalar particle. It has a unique role in the Standard Model of breaking the symmetry of the electroweak force by giving mass to the W and Z bosons and the fundamental fermions, but not to the photon. The masses of the fundamental particles, both fermions and bosons, vary greatly.

The behaviour of all the particles and forces of the Standard Model can only be explained by theories that incorporate both quantum mechanics and relativity. The appropriate language for this union is *quantum field theory*, to which we now turn.

12.2 Quantum Field Theory

Einstein first invoked quantum mechanics to explain the photoelectric effect, treating electromagnetic waves as composed of photons, which are massless particles that travel at the speed of light. Yet standard quantum mechanics is clearly a non-relativistic theory. The Schrödinger equation includes a single derivative with respect to time, but double derivatives with respect to space. Furthermore, the relation of energy to momentum for a quantum mechanical particle of mass m, not subject to a potential, is $E = \frac{\mathbf{p}^2}{2m}$, rather than the relativistic relation $E^2 = \mathbf{p}^2 + m^2$. To describe photons and other particles travelling at close to the speed of light it is necessary to find a version of quantum mechanics that is compatible with special relativity.

The earliest approach was to seek a relativistically invariant equivalent of the Schrödinger equation with the expectation that its solutions would be wavefunctions of a relativistic particle. Two equations initially looked promising. One was the Klein–Gordon equation, which we mentioned in section 3.2. In this equation, the single time derivative is removed, and the negative Laplacian $-\nabla^2$ is replaced by the wave operator $\frac{\partial^2}{\partial t^2} - \nabla^2$. The second more radical option was the Dirac equation. Dirac recognized the importance of retaining a single time derivative in the quantum mechanical wave equation, so that the evolution of a wavefunction into the future only depends on the wavefunction itself at an initial time and not on its first time derivative. This is consistent with the postulate of wavefunction collapse, which implies that the wavefunction is determined by the outcome of a measurement, and its subsequent evolution is determined by the wave equation. The Dirac equation combines a single time derivative with single space derivatives. We will describe the novel way in which this is achieved in section 12.3.

However, neither the Klein–Gordon equation nor the Dirac equation can be considered as true, relativistic single-particle analogues of the Schrödinger equation. The drawback with using the Klein–Gordon equation in this way is that a suitable probability density for the position of a particle cannot be defined. The obstacle for the Dirac equation is that, in addition to the usual positive energy states, its solutions include particle states with negative energy of arbitrarily large magnitude.

The underlying reason for these problems is that particle interactions at relativistic velocities may involve sufficient energy to create new particles. In standard quantum mechanics it is possible to define a wavefunction for several interacting particles, as we did in section 8.7, but the number of particles does not change with time. When a particle is travelling close to the speed of light, however, its total energy may be several times its rest mass. In a high energy particle collision, energy is readily converted into new particles, so in relativistic physics, particle numbers often do change. This means that relativistic quantum mechanics is necessarily a many-particle theory, and particle number is not fixed. Historically, the progress in understanding multi-particle quantum mechanics with variable numbers of particles was rather tortuous, but the end result is *quantum field theory*. The Klein–Gordon and Dirac equations reappear here, but as relativistic field equations with an interpretation that is closer to that of the classical Maxwell equations. Our discussion of the subject will be largely descriptive, as performing calculations in quantum field theory involves numerous technical details and complicated mathematical machinery.

In classical physics there is a clear difference between particles of matter, and fields. Particles are point-like or at least highly localized, whereas fields extend throughout space and time; fields are also responsible for the forces between particles. Quantum field theory largely dissolves this difference. For example, there is an electron field associated with electrons, just as there is an electromagnetic field associated with photons. Some difference remains, however, because matter consists mostly of fermionic particles, and fermions with half-integer spin (like the electron) and bosons with integer spin (like the photon) are associated with different types of quantized field.

A powerful feature of quantum field theory is that it accounts for all the ways that particles interact, gravitational interactions excepted. So the theory of particle bound states and particle scattering, normally described in terms of attractive or repulsive forces, is unified with the theory of particle production and decay. That is why we speak of neutron decay, or Z boson production, as being due to the *weak force*.

12.2.1 Quantizing the electromagnetic field

It is currently believed that there is a field for every type of fundamental particle, and these fields are part of the fabric of spacetime. A field has a dynamical degree of freedom at each point of space, so altogether, a field has infinitely many degrees of freedom. The field values at different points are coupled together, with the result that in the simplest cases, the basic dynamical solutions of the field equation are wave modes, with definite wavevectors $\mathbf{k}$ and frequencies ω. The field equation determines ω in terms of $\mathbf{k}$, but there are still infinitely many solutions, because $\mathbf{k}$ can be any 3-vector. Classically, the magnitude $A(\mathbf{k})$ of each wave mode oscillates independently.

In the quantum field theory, each of these magnitudes is treated as a quantum harmonic oscillator, and there is one oscillator for each $\mathbf{k}$. (The analogy is between the variable x of a harmonic oscillator and the variable $A(\mathbf{k})$ of the wave mode. x is usually a spatial position but this is unimportant here.) Particles are the excited states of these quantized wave modes. They are not spatially localized, but they have a definite energy and momentum, related to ω and $\mathbf{k}$.

Spin 0 particles are excitations of fields that obey the Klein–Gordon equation or a variant of this. Spin $\frac{1}{2}$ particles are excitations of fields that obey the Dirac equation. We will describe these particles in due course but first we will consider photons, which have spin 1. They are quantized excitations of the electromagnetic field, which is more familiar. We have already examined the classical theory of the electromagnetic field in Chapter 3, and photons featured in our discussion of black body radiation in section 10.10.

When constructing a quantum field theory for electromagnetism, we begin with the classical electromagnetic field obeying the source-free, relativistic Maxwell equations. The field has an infinite number of wave modes, with two independent polarization directions for each non-zero wavevector $\mathbf{k}$. (The mode with zero wavevector is unphysical, as it can be removed by a gauge transformation.) Each wave mode has a magnitude of oscillation, described by the strength of the 4-vector potential $\mathcal{A}$, or equivalently, the correlated strengths of the electric and magnetic fields. The Maxwell equations imply that the frequency of a wave mode with wavevector $\mathbf{k}$ is $\omega = |\mathbf{k}|$. Quantizing the electromagnetic field means quantizing all these wave modes. More precisely, it means constructing a quantum Hamiltonian operator for an infinite-dimensional set of harmonic oscillators. This is an infinite extension of the 3-dimensional oscillator that appeared, for example, in the Nilsson model of nuclei in section 11.3.5. Implicitly, there is a Schrödinger equation for all the oscillators, whose combined solution is known as a *wavefunctional* of the field.

For the mode with wavevector $\mathbf{k}$ and frequency ω, and a choice of polarization, the energy levels of the oscillator are $E_n = \left(n + \frac{1}{2}\right)\hbar\omega$, $n = 0, 1, 2, \ldots$. The ground state is interpreted as the state with no photons. The first excited state has energy $\hbar\omega$ greater than the ground state, and is interpreted as a single photon of wavevector $\mathbf{k}$ and frequency $\omega = |\mathbf{k}|$, which agrees with Einstein's hypothesis that electromagnetic waves of frequency ω consist of photons of energy $\hbar\omega$. The nth excited state is a state of n photons, each with wavevector **k** and energy $\hbar\omega$, and this n-photon state is unique, so a permutation of the photons has no effect. Also, there is no constraint on n, so the theory correctly describes photons as bosons. Stationary states of the quantum field theory are best described in terms of occupation numbers—the number of photons with various wavevectors **k**. A general state is a superposition of these.

If every mode is in its ground state, then the entire electromagnetic field is in its ground state. This state is known as the *vacuum* and has no photons at all. The ground state energy of the harmonic oscillator of frequency ω is $\frac{1}{2}\hbar\omega$, so the sum of the ground state energies for all wave modes, which are infinite in number, appears to be an infinite total energy. However, this energy is physically undetectable, so it is simply dropped and the ground state energy is defined to be zero.[2] This can be achieved by shifting the quantum Hamiltonian of each oscillator by a constant so that its ground state energy is zero rather than $\frac{1}{2}\hbar\omega$.

The classical electromagnetic field carries momentum as well as energy. There is an expression for the momentum in terms of the strengths of the wave modes and their wavevectors **k**, and from this one can derive a momentum operator in the quantum field theory. It is a sum of terms for each wave mode individually. For a wave mode of wavevector **k**, the momentum operator is similar to the Hamiltonian, but with ω replaced by **k**. Therefore a 1-photon state of this mode simultaneously has momentum $\mathbf{p} = \hbar\mathbf{k}$ and energy $E = \hbar\omega$. As $\omega = |\mathbf{k}|$, this implies that a photon obeys the relativistic energy–momentum relation $E = |\mathbf{p}|$. So a photon is *massless*.

Photons are spin 1 (vector) particles. A massive spin 1 particle would have three independent polarization states because in its rest frame all three orthogonal spatial directions are available, but as the photon is massless and cannot be at rest (i.e. cannot have zero momentum), it is consistent that it has only two independent polarization states, with no polarization state along the direction of **k**. This is a direct consequence of the fact that the longitudinal part of the classical vector potential, parallel to **k**, can be removed by a gauge transformation, as described in section 3.7. However, a photon is still a vector particle, as its polarization rotates like a vector under a spatial rotation around the **k**-axis.

Quantization of the electromagnetic field leads to a complete theory of photons as massless, spin 1 particles. The success of this approach suggests that other particles might be understood by quantizing different types of field, satisfying appropriate field equations.

12.2.2 The quantized scalar Klein–Gordon field

Often the most convenient starting point for a quantum field theory is a classical Lagrangian density for the field. From this, using the principle of least action, one can derive the classical dynamical field equation, as described in section 2.3.

Let us consider the following Lagrangian density for a real scalar field $\phi(\mathbf{x}, t)$:

$$\mathcal{L} = \frac{1}{2}\partial\phi \cdot \partial\phi - \frac{1}{2}m_0^2\phi^2 = \frac{1}{2}\left(\frac{\partial\phi}{\partial t}\right)^2 - \frac{1}{2}\nabla\phi \cdot \nabla\phi - \frac{1}{2}m_0^2\phi^2\,, \tag{12.4}$$

where m_0 is a positive mass parameter. The total action is the integral of $\mathcal{L}$ over spacetime,

$$S = \int \mathcal{L}\,d^4x\,, \tag{12.5}$$

---

[2] This procedure does not introduce any problem in the quantum theory of the electromagnetic field or the other fields that we will consider, but it would be an issue for a quantum theory of gravity, as all energy is a source for the gravitational field.

and the field equation derived from this action is the Klein–Gordon equation

$$\frac{\partial^2 \phi}{\partial t^2} - \nabla^2 \phi + m_0^2 \phi = 0. \tag{12.6}$$

This is our equation (3.18), but as it is now relativistic, we have set $c = 1$. (We have also replaced μ by m_0.)

The Klein–Gordon equation is simpler than the Maxwell equations, as ϕ only has a single magnitude, and not a vector polarization, and there are no gauge transformations to consider. It is linear in ϕ, so its independent solutions are again wave modes, with wavevector $\mathbf{k}$ and frequency ω. Substituting the wave form

$$\phi(\mathbf{x}, t) = A e^{i(\mathbf{k} \cdot \mathbf{x} - \omega t)} \tag{12.7}$$

into equation (12.6), we find the relation

$$\omega^2 = \mathbf{k}^2 + m_0^2. \tag{12.8}$$

This particular wave is complex, but as in other problems involving oscillations, real solutions can be found using linear combinations of complex solutions. The field then oscillates harmonically with positive frequency $\omega = \sqrt{\mathbf{k}^2 + m_0^2}$.

Quantization proceeds by treating each wave mode as a quantum harmonic oscillator, just as in the electromagnetic case. For each mode, one quantizes the oscillator amplitude A. The ground state of a mode with frequency ω has energy $\frac{1}{2}\hbar\omega$, and there are excited states separated by energy gaps $\hbar\omega$. In the vacuum, all modes are in their ground state. The total ground state energy contributed by the infinitely many modes is again discarded, and the vacuum defined to have zero energy. The first excited state of the mode with wavevector $\mathbf{k}$ is interpreted as a 1-particle state, similar to a single photon. The particle has energy $E = \hbar\omega = \hbar\sqrt{\mathbf{k}^2 + m_0^2}$, and one can again show, by finding the quantum operator representing the total field momentum, that the particle has momentum $\mathbf{p} = \hbar\mathbf{k}$. Multiplying equation (12.8) by $\hbar^2$ shows that the particle's energy and momentum are related by

$$E^2 = \mathbf{p}^2 + (\hbar m_0)^2, \tag{12.9}$$

with E positive. This is the energy–momentum relation for a relativistic particle of mass $m = \hbar m_0$. So the quantized, real Klein–Gordon field theory describes a single type of particle of mass m. If a given oscillator mode is in its nth excited state, then this state represents n identical particles, each with the same momentum and energy. As for photons, we have a theory of bosons. The Klein–Gordon particles have spin 0, because a 1-particle state is completely determined by its momentum, and there is no polarization vector. One calls the particles *scalar bosons*.

Note that the particle mass $m = \hbar m_0$ is a quantum phenomenon, involving Planck's constant. This is quite surprising. m_0 is the mass parameter of the field, sometimes loosely called the mass of the field, but by itself it has the wrong dimensions to be a particle mass.

Wave modes are not localized in space, so particles that arise as excitations of the Klein–Gordon field do not have an obvious position. The relation between the momentum and the wavevector is the de Broglie relation, $\mathbf{p} = \hbar\mathbf{k}$, so in the non-relativistic limit there is a rather close connection between the Klein–Gordon field and the wavefunction of a single, quantum mechanical particle. A solution of the Klein–Gordon field equation, with wavevectors that are small compared with m_0, behaves like the quantum state of a non-relativistic particle. This is because for small momentum the energy–momentum relation (12.9) becomes $E \simeq m + \frac{\mathbf{p}^2}{2m}$. The constant term m simply produces a universal, time-dependent exponential factor $e^{-im_0 t}$ in the field. After extracting this, what remains is a field satisfying the Schrödinger equation for a free particle of mass m. By combining modes with different momenta, one can create a spatially localized 1-particle state, as in non-relativistic quantum mechanics. However, the momenta must be small, and this limits the extent of the spatial localization. Any attempt to localize one particle too much creates what is really a multi-particle state, so there is no exact truncation to a single-particle quantum theory that is consistent with relativity.

12.3   The Dirac Field

The Dirac field ψ satisfies a relativistic equation that is first order in time and space derivatives. Such an equation is only possible if it is constructed using four 4×4 matrices γ, now called Dirac matrices or *gamma matrices*, so ψ must have four complex components arranged in a column. In section 8.5 we introduced the idea that the wavefunction of an electron with spin $\frac{1}{2}$ is a 2-component spinor. The Dirac field ψ is a modification of this, and is called a 4-component Dirac spinor.[3] When quantized, it describes the states of two related particles, both with spin $\frac{1}{2}$. We will explain this more carefully in the following, but to anticipate: if one of these particles is an electron, then the other particle is its antiparticle, the positron, an example of *antimatter*. Quantization of the Dirac field gives a multi-particle theory, but as in the Klein–Gordon case, the 1-particle states are the basic ones.

12.3.1   The Dirac equation

In the shorthand 4-vector notation we used in Chapters 4 and 6, the Dirac equation is

$$(i\gamma \cdot \partial - m_0)\psi = 0, \tag{12.10}$$

where m_0 is a mass parameter. Expanding out the 4-vector dot product $\gamma \cdot \partial$, the equation becomes $i\gamma^\mu \frac{\partial \psi}{\partial x^\mu} - m_0 \psi = 0$, or in full,

$$i\gamma^0 \frac{\partial \psi}{\partial x^0} + i\gamma^1 \frac{\partial \psi}{\partial x^1} + i\gamma^2 \frac{\partial \psi}{\partial x^2} + i\gamma^3 \frac{\partial \psi}{\partial x^3} - m_0 \psi = 0. \tag{12.11}$$

We see that it involves single time ($x^0 = t$) and space (x^1, x^2, x^3) derivatives in a symmetrical way, and certainly looks relativistic. At first sight, it appears that γ is some constant, universal 4-vector, but this cannot be correct, as choosing a particular 4-vector breaks the Lorentz transformation symmetry that is essential for a relativistic theory. Different observers, moving relative to one another, would require γ to be different 4-vectors.

Dirac found a resolution of this problem. Instead of having ordinary numbers for the components, he constructed the 4-vector out of four constant square matrices $\gamma = (\gamma^0, \gamma^1, \gamma^2, \gamma^3)$ that collectively have the desired Lorentz transformation properties. These four gamma matrices need to satisfy certain algebraic relations, designed to ensure consistency with special relativity. Their consequence is that each component of ψ satisfies the relativistic Klein–Gordon equation. Wave mode solutions then satisfy the relativistic relation between frequency and wavevector, $\omega^2 = \mathbf{k}^2 + m_0^2$.

---

[3] It is not a 4-vector, because spinors and vectors transform differently under Lorentz transformations.

To find these algebraic relations, we assume that ψ satisfies the Dirac equation (12.10) and then act from the left with the operator $i\gamma \cdot \partial + m_0$, obtaining

$$(i\gamma \cdot \partial + m_0)(i\gamma \cdot \partial - m_0)\psi = 0 \tag{12.12}$$

or, more concretely,

$$\left(i\gamma^\nu \frac{\partial}{\partial x^\nu} + m_0\right)\left(i\gamma^\mu \frac{\partial}{\partial x^\mu} - m_0\right)\psi = 0. \tag{12.13}$$

Expanding this equation out gives

$$\gamma^\nu \gamma^\mu \frac{\partial^2 \psi}{\partial x^\nu \partial x^\mu} + m_0^2 \psi = 0. \tag{12.14}$$

As the double partial derivative is symmetric under the exchange of μ and ν, the only part of $\gamma^\nu \gamma^\mu$ that contributes is the symmetrical combination $\frac{1}{2}(\gamma^\mu \gamma^\nu + \gamma^\nu \gamma^\mu)$. For each component of ψ, equation (12.14) reduces to the Klein–Gordon equation

$$\frac{\partial^2 \psi}{\partial t^2} - \nabla^2 \psi + m_0^2 \psi = 0, \tag{12.15}$$

provided that

$$\gamma^0 \gamma^0 = 1_n, \quad \gamma^1 \gamma^1 = \gamma^2 \gamma^2 = \gamma^3 \gamma^3 = -1_n, \tag{12.16}$$

where 1_n is the unit $n \times n$ matrix for some n, and

$$\gamma^\mu \gamma^\nu + \gamma^\nu \gamma^\mu = 0 \tag{12.17}$$

whenever μ and ν are different. A more compact way of writing the algebraic relations (12.16) and (12.17) is

$$\gamma^\mu \gamma^\nu + \gamma^\nu \gamma^\mu = 2\eta^{\mu\nu} 1_n, \tag{12.18}$$

where $\eta^{\mu\nu} = \mathrm{diag}(1, -1, -1, -1)$ is the (inverse) Minkowski metric tensor, as defined by equation (6.11).

The relations (12.18) are called the *Dirac algebra*, or alternatively the *gamma matrix anticommutation relations* (anti- because of the plus sign between the two terms on the left-hand side). These can indeed be satisfied and the basic solution is in terms of 4×4 matrices. There are no smaller matrices that work. For instance, 1×1 matrices are just numbers and although equations (12.16) could be solved using ± 1 and $\pm i$, equation (12.17) could not be simultaneously satisfied. One way of presenting the solution is to write the 4×4 matrices in 2×2 blocks, with the blocks being either zero matrices, unit matrices 1_2, or Pauli matrices $\sigma_1, \sigma_2, \sigma_3$, as defined in section 8.5. The gamma matrices are then

$$\gamma^0 = \begin{pmatrix} 1_2 & 0 \\ 0 & -1_2 \end{pmatrix}, \quad \gamma^i = \begin{pmatrix} 0 & \sigma_i \\ -\sigma_i & 0 \end{pmatrix}, \quad i = 1, 2, 3, \tag{12.19}$$

and these work because $\sigma_i \sigma_j + \sigma_j \sigma_i = 2\delta_{ij} 1_2$. Although this solution is not unique, variant solutions using 4×4 matrices only differ by a change of basis in the space of Dirac spinors, and this makes no physical difference. There are solutions with larger matrices, but these are just built from several copies of the 4×4 matrix solution above, and correspond to having several spinors representing multiple Dirac fields. So the solution we have given is essentially unique.

There is a version of the Dirac algebra in any spacetime dimension, and each of the spacetime coordinates can be either time-like or space-like. In particular, there is a version in Euclidean space of any dimension. Whenever the dimension increases by two, the size of the Dirac matrices doubles. So in ten spacetime dimensions, the matrices are 32×32, but they can still be built up from 1_2 and the Pauli matrices.

Now let us return to four spacetime dimensions, and seek solutions of the Dirac equation. As the gamma matrices are written in terms of 2×2 blocks, it is convenient to split ψ into a pair of 2-component spinors. The Dirac equation contains no explicit functions of space and time, so it is natural to seek wave mode solutions of the form

$$\psi(\mathbf{x}, t) = e^{i(\mathbf{k} \cdot \mathbf{x} - \omega t)} \begin{pmatrix} \chi \\ \xi \end{pmatrix}, \tag{12.20}$$

where χ and ξ are constant 2-component spinors. Substituting this into the Dirac equation (12.10), we obtain the coupled equations

$$\begin{aligned} (\omega - m_0)\chi - \mathbf{k} \cdot \boldsymbol{\sigma}\, \xi &= 0 \\ \mathbf{k} \cdot \boldsymbol{\sigma}\, \chi - (\omega + m_0)\xi &= 0, \end{aligned} \tag{12.21}$$

where

$$\mathbf{k} \cdot \boldsymbol{\sigma} = k_1 \sigma_1 + k_2 \sigma_2 + k_3 \sigma_3 = \begin{pmatrix} k_3 & k_1 - ik_2 \\ k_1 + ik_2 & -k_3 \end{pmatrix} \tag{12.22}$$

is a 2×2 matrix. If we take χ to be an arbitrary constant 2-spinor, then the second equation determines ξ to be

$$\xi = \frac{\mathbf{k} \cdot \boldsymbol{\sigma}\, \chi}{\omega + m_0}. \tag{12.23}$$

Substituting in the first equation and using the identity $(\mathbf{k} \cdot \boldsymbol{\sigma})^2 = \mathbf{k}^2\, 1_2$, we find

$$(\omega - m_0)\chi - \mathbf{k}^2 \frac{\chi}{\omega + m_0} = 0. \tag{12.24}$$

A non-trivial solution generally has χ non-zero, which requires $\omega^2 - m_0^2 - \mathbf{k}^2 = 0$ or equivalently

$$\omega^2 = \mathbf{k}^2 + m_0^2. \tag{12.25}$$

This condition ensures that the Dirac equation is satisfied, and has the consequence that each component of the Dirac spinor ψ satisfies the Klein–Gordon equation.

There is no reason here to fix the sign of ω. For given $\mathbf{k}$, the frequency ω is the square root of $\mathbf{k}^2 + m_0^2$ of either sign. If m_0 is fixed as positive, then $\omega \geq m_0$ if ω is positive, and $\omega \leq -m_0$ if ω is negative. (If $\mathbf{k}$ is zero and $\omega = -m_0$, then χ is zero and ξ is an arbitrary constant 2-spinor.) The Dirac equation is linear, so the various wave mode solutions are independent and can be superposed.

Since each component of ψ obeys the Klein–Gordon equation, it is now plausible that the Dirac equation is relativistically consistent. However the four components of the spinor field ψ are not independent, and finding the appropriate Lorentz transformation of ψ requires more algebraic work. One can in fact verify that the Dirac equation allows for Lorentz transformations. A boost mixes the upper and lower 2-spinors χ and ξ, but a spatial rotation acts more simply. The 2-spinors each rotate in the same way that a non-relativistic 2-spinor rotates. This is exactly what is required. It means that χ transforms like the spin state of a spin $\frac{1}{2}$ particle, with spin aligned in any direction. With χ fixed, ξ is determined by the formula (12.23) and transforms the same way.

12.3.2 Quantizing the Dirac field—particles and antiparticles

We could try quantizing the Dirac field by treating each wave mode as a harmonic oscillator, just as we did for other types of field. In the naïve vacuum, all modes would be unexcited. However there is a serious problem with this procedure. The solutions with negative frequency ω correspond to particle states of negative energy. Every excitation of a negative frequency mode decreases the total energy, and this would produce an unstoppable collapse in energy. An interaction between the Dirac field and another field would excite positive and negative frequency modes, leading to an unlimited number of negative energy particles. In short, the theory would be unstable.

Now, the Dirac field was invented to describe electrons, which are fermions. Dirac realized that the problem of negative energy states is closely related to the Pauli exclusion principle and proposed the following explicitly fermionic quantization of each mode. Each wave mode has just two possible quantum states. It is either unexcited, in which case no particle is present, or excited, in which case exactly one particle is present. In other words, the wave mode is either unoccupied, or occupied once. No two particles can be in the same state of momentum, energy and spin. In the vacuum state, the positive frequency modes are all unoccupied. An occupied state for a mode of positive frequency ω has energy $\hbar\omega$ greater than the empty state, and is interpreted as a positive energy particle. If the wavevector is **k**, the particle has momentum $\hbar\mathbf{k}$ and energy $\sqrt{\hbar^2\mathbf{k}^2 + \hbar^2 m_0^2}$, so as in the Klein–Gordon theory, the particle mass is $m = \hbar m_0$.

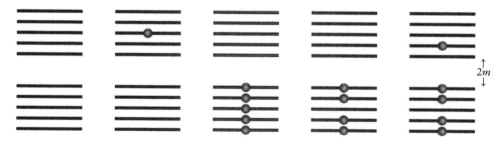

Fig. 12.2 From left to right: The naïve vacuum; a naïve 1-particle state; the filled Dirac sea, the true vacuum state; a 1-hole Dirac state representing an antiparticle; a state with one particle and one antiparticle.

What about the negative frequency modes? Dirac postulated that in the true vacuum state all these modes are *occupied*. This Dirac vacuum state is also referred to as the *filled Dirac sea*. It is illustrated in the middle of Figure 12.2. The Dirac vacuum gives the theory a remarkable symmetry that is different from the symmetry of the naïve vacuum. It is a symmetry that exchanges positive and negative energies, and at the same time trades occupied for unoccupied states.

Consider now what happens when a negative frequency wave mode changes from the occupied to the unoccupied state. As ω is negative, the energy *decreases by a negative amount*; in other words, the energy increases. The unoccupied state has energy $\hbar|\omega|$ greater than the occupied state. We can regard this positive energy as the energy of a new type of particle, an *antiparticle*. (The antiparticle was originally called a *hole*, as it is a hole in the Dirac sea.)

The theory describes particles and antiparticles, both of which may exist in states with momentum $\mathbf{p} = \hbar\mathbf{k}$ and spin either up or down. A particle of momentum $\mathbf{p} = \hbar\mathbf{k}$ is an excitation of a wave mode with wavevector $\mathbf{k}$. An antiparticle of momentum $\mathbf{p} = \hbar\mathbf{k}$ is a de-excitation of a wave mode with wavevector $-\mathbf{k}$. Both particle and antiparticle have the same positive mass $m = \hbar m_0$, and both have positive energy $\sqrt{\mathbf{p}^2 + m^2}$. What then distinguishes particle from antiparticle? In general, they will undergo different interactions. In many cases, we can distinguish them by their electric charge.[4] When coupled to the electromagnetic field, the Dirac field has a definite charge that is the same for all wave modes. When an unoccupied state changes to an occupied state, the charge changes by a fixed amount q. So a particle has charge q. But the antiparticle arises when an occupied state changes to an unoccupied state. This changes the charge by $-q$, so the antiparticle has charge $-q$.

When the Dirac field is coupled to a quantized electromagnetic field, it is possible for a photon to raise an occupied negative energy state (a negative energy particle) from the Dirac sea into a positive energy state. This produces both a hole in the Dirac sea and a positive energy particle. It is interpreted as the creation of a particle–antiparticle pair, as shown in Figure 12.3. Charge is conserved because both the photon and the particle–antiparticle pair have zero net charge. The energy gap between negative and positive energy states is $2m$, so the process can only occur for a photon with at least this energy.

If q is the charge on the electron $-e$, and m the electron mass m_e, then the Dirac theory describes relativistic electrons (e^-) and also predicts the existence of the antiparticles of electrons, with charge e. These are now called *positrons* (e^+), as they are positively charged. This was the first prediction of antimatter. In 1932, shortly after Dirac developed the theory, the positron was discovered by Carl Anderson, who was studying cosmic rays with a cloud chamber. Figure 12.4 shows the first published photograph of a track of a positron. As predicted, positrons have the same mass as electrons but opposite electric charge. Their discovery was a triumph for the theory.

Historically, the idea of a filled Dirac sea was very important as it led to the discovery of antiparticles, but it raises many issues about the vacuum state. This state must have zero net energy, momentum and electric charge, yet it is supposed to be filled by an infinite sea of negative energy particles, with potentially infinite negative total energy and charge. Although antiparticles were originally interpreted as holes in the Dirac sea in the way we have described, today's particle physicists regard the filled Dirac sea as an unnecessary crutch. It is usual today to simply discuss particles and antiparticles as independent excitations of the Dirac field. This avoids having to argue away an infinite sea of negative energy particles.

---

[4] For Dirac particles that are not electrically charged, such as neutrinos, we could define a new charge called ψ-particle number and define the ψ particle to have charge $+1$ and the ψ antiparticle to have charge -1.

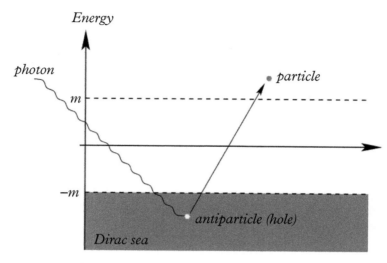

Fig. 12.3 Pair creation from the Dirac sea.

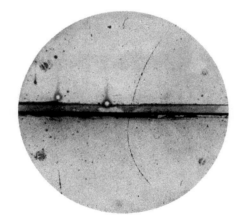

Fig. 12.4 The first photograph of a positron. Its path is curved by a magnetic field in the detector. The positron arrives from the top right and slows down as it passes through the lead plate across the centre of the detector.

The quantized Dirac field does not just apply to electrons and positrons; it is appropriate for all spin $\frac{1}{2}$ particles. In particular, there is a separate Dirac field for each of the other massive leptons, the muon and tauon together with their antiparticles, and also for each type of quark and antiquark. These particles are distinguished by their masses and charges, and by how their fields interact with other fields.

Neutrinos are more complicated and still not fully understood. They also have spin $\frac{1}{2}$, but the most appropriate way to describe them in quantum field theory is yet to be determined. Until the 1990s, neutrinos were usually described by a variant of the Dirac field with zero mass parameter. We now know that neutrinos have a tiny, but non-zero, rest mass. Rather remarkably, they can convert from one type to another in empty space and it is even possible that neutrinos are identical to their antiparticles. We will describe the ongoing research into their properties in section 12.9.

12.4   Actions and Interactions

So far, we have been considering the quantization of fields obeying linear field equations, whose excitations represent non-interacting, free particles. An application is to a box of photons, modelling black body radiation. Large numbers of photons may be trapped inside the box, but they do not scatter off each other to any significant extent.

Interacting particles scatter off each other in high energy collisions, and their kinetic energy may be converted into new particles. This is what makes particle physics an exciting experimental subject. After particles have collided, further interactions within detectors are essential if we are to see an outgoing particle's track and measure its energy and momentum. Interactions also lead to particle decay, in which an unstable particle typically decays into two or more lighter particles.

To describe interacting particles we must consider interacting fields, and that requires nonlinear terms in the field equations. One field is then a source for another type of field. In a classical nonlinear theory, the oscillation of a wave mode of one field can stimulate oscillations in the wave modes of another field. In the quantum field theory, this corresponds to particle production and decay. Even with a single field, nonlinear terms can couple wave modes of different frequencies and wavevectors. This is interpreted quantum mechanically as particle scattering, in which energy is transferred in particle collisions from motion in one direction to motion in another direction.

A field Lagrangian offers the most concise way to encode field and particle interactions. A quadratic Lagrangian leads, via the principle of least action, to linear field equations, and the quantized field theory has no particle interactions. A Lagrangian including higher powers of the fields leads to nonlinear field equations, and particle interactions. Other than in a few exceptional cases, it is not possible to exactly solve interacting quantum field theories. The usual strategy is to assume that the coefficients of any non-quadratic terms are small, so that the interactions produce small corrections to the free-particle theory. These coefficients are called coupling constants. The amplitudes for any measurable quantities predicted by the theory can then be calculated as a series expansion in the coupling constants, a procedure known as perturbation theory. It is this approach that leads to Feynman diagrams.

12.4.1   Quantum electrodynamics

The perturbative approach to particle interactions was first developed by physicists modelling the interactions of photons and charged particles. The culmination of these efforts is one of the most successful theories ever devised, quantum electrodynamics or QED. In this theory the electromagnetic force derives from the exchange of photons between electrically charged particles such as protons and electrons or, at a more fundamental level, between quarks and charged leptons. QED predictions have been tested in the laboratory and agree with the experimental measurements to an incredible precision, in some cases close to one part in a trillion (10^{12}).

A very important aspect of electromagnetism is the freedom that exists to redefine the potentials by performing a gauge transformation, as discussed in Chapter 3. A gauge transformation leaves the electromagnetic field $\mathcal{F}$ unchanged and therefore has no effect on any physically measurable quantity. It is simply a reflection of the redundancy in our description of electromagnetism, but it is an essential feature of both the classical theory and the quantum field theory. Any Lagrangian density that includes electromagnetism must be unchanged by a gauge transformation. In other words, it must be *gauge invariant*.

The way to achieve gauge invariance, when constructing the Lagrangian, is to use the following modification of the derivative term acting on any field that carries an electric charge q:

$$\partial \quad \longrightarrow \quad D = \partial - iq\mathcal{A}\,, \tag{12.26}$$

where $\mathcal{A}$ is the electromagnetic 4-vector potential. This introduces interactions while preserving gauge invariance. The QED Lagrangian density for charged spin $\frac{1}{2}$ particles interacting via the electromagnetic field is[5]

$$\mathcal{L} = -\frac{1}{4}\mathcal{F} \cdot \mathcal{F} + i\overline{\psi}\gamma \cdot D\psi - m_0\overline{\psi}\psi\,. \tag{12.27}$$

The fields here are the 4-vector potential $\mathcal{A}$, with its field strength

$$\mathcal{F} = \partial\mathcal{A} - (\partial\mathcal{A})^T\,, \tag{12.28}$$

and the Dirac field ψ, with mass parameter m_0 and charge q. $\overline{\psi}$ is the Dirac congugate of ψ, a row 4-spinor constructed from the complex conjugates of the components of ψ (with the third and fourth components having reversed signs, to achieve Lorentz invariance). $D\psi$ denotes $\partial\psi - iq\mathcal{A}\psi$.

A gauge transformation acts on the fields as follows:

$$\mathcal{A} \to \mathcal{A} - \partial\lambda\,, \quad \psi \to e^{-iq\lambda}\psi\,, \tag{12.29}$$

where λ is an arbitrary function of space and time. The transformation of $\mathcal{A}$ puts into 4-vector form the gauge transformation given in equation (3.58). ψ changes by a phase factor that depends on λ and also the charge q.

Let us check the gauge invariance of the QED Lagrangian density (12.27). Under the gauge transformation (12.29), $\partial\mathcal{A}$ acquires an additional term $-\partial\partial\lambda$ and $(\partial\mathcal{A})^T$ acquires the transpose of this, which reverses the order of the partial derivatives. By the symmetry of mixed partial derivatives, these additional terms cancel in $\mathcal{F}$, so the first term of $\mathcal{L}$ is gauge invariant. The final, mass term for the Dirac field is invariant because $\overline{\psi}$ involves the complex conjugate of ψ, and this transforms with a phase factor $e^{iq\lambda}$ which cancels against the phase factor multiplying ψ. The middle term is the most interesting. The modified derivative $D\psi = \partial\psi - iq\mathcal{A}\psi$ gauge transforms to

$$\begin{aligned}
\partial\psi - iq\mathcal{A}\psi \quad &\to \quad \partial\left(e^{-iq\lambda}\psi\right) - iq(\mathcal{A} - \partial\lambda)e^{-iq\lambda}\psi \\
&= \quad e^{-iq\lambda}\left(\partial\psi - iq(\partial\lambda)\psi - iq\mathcal{A}\psi + iq(\partial\lambda)\psi\right) \\
&= \quad e^{-iq\lambda}(\partial\psi - iq\mathcal{A}\psi)\,. \tag{12.30}
\end{aligned}$$

In other words, $D\psi$ transforms to $e^{-iq\lambda}D\psi$, simply picking up the same phase factor as ψ itself, and for this reason $D\psi$ is called the *gauge covariant derivative* of ψ. The middle term

---

[5] The γ shown in this and subsequent equations represents the 4-vector of gamma matrices and should not be confused with the photon.

of $\mathcal{L}$ has a product of $\overline{\psi}$ and $D\psi$ which pick up cancelling phase factors, so this term is gauge invariant too.

Gauge invariance is important for the following reasons. It ensures that the photon has no physical states with longitudinal polarization. This is consistent with simple experiments involving polarizers and a beam of light. Two orthogonal polarizers cut both types of transverse polarization and stop the beam totally. Also, if light had a longitudinal polarization state, the formulae for the energy and entropy of black body radiation would be different, disagreeing with measurements of radiation pressure. Finally, and perhaps most importantly, without gauge invariance photons could acquire mass through their interactions. Then light would not travel in a vacuum at the fixed 'speed of light', undermining the many successes of relativity in all areas of physics.

In the QED Lagrangian density all terms are quadratic except the term

$$\mathcal{L}_{\text{int}} = q(\overline{\psi}\gamma\psi) \cdot \mathcal{A} \tag{12.31}$$

that comes from the second part of the gauge covariant derivative. This is the term responsible for interactions between the charged particles, their antiparticles, and photons. We will now look in more detail at the physics that arises from these particle interactions.

12.4.2 Feynman diagrams

Feynman devised a very useful diagrammatic approach to visualizing the interactions of particles. For electrons, positrons and photons, the interaction term in the QED Lagrangian density is

$$\mathcal{L}_{\text{int}} = -e(\overline{\psi}\gamma\psi) \cdot \mathcal{A}, \tag{12.32}$$

and it can be represented as a simple diagram known as a vertex, which is shown in Figure 12.5 (left) and (centre).

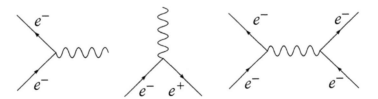

Fig. 12.5 Feynman diagrams representing photon emission or absorption from an electron, photon production in electron–positron annihilation, and the exchange of a photon between two scattering electrons.

Time passes upwards in the diagrams. The solid lines represent electrons and positrons, and the wavy lines represent photons. A forward arrow indicates an electron and a backwards arrow indicates a positron. The same vertex can represent different processes depending on its orientation. The vertex on the left shows an electron emitting or absorbing a photon. The diagram in the centre represents the annihilation of an electron and a positron with the creation of a photon. The strength of the interaction is determined by the coupling constant $-e$. At each vertex, all the conservation laws pertinent to the theory, such as the conservation of electric charge and the conservation of energy and momentum, are obeyed.

This is achieved automatically when the full machinery of quantum field theory is used to construct the diagrams.

The vertices can be combined to produce Feynman diagrams that represent particle scattering processes. The simplest diagram representing the scattering of two electrons is shown in Figure 12.5 (right). In this diagram a single photon is exchanged between two electrons. As the photon carries energy and momentum, its exchange transfers energy and momentum between the electrons and thereby alters their trajectories. This diagram can be evaluated to calculate the amplitude for the scattering of two electrons to lowest order in perturbation theory. The amplitude is proportional to e^2, as there is one factor of the charge for each vertex. Physically measurable quantities, such as cross sections, depend on the scattering probability, which is calculated by taking the modulus squared of the amplitude, so the scattering cross section is proportional to e^4.

The full result is a relativistic extension of the result obtained quantum mechanically for two electrons scattering through the repulsive $\frac{e^2}{4\pi r}$ Coulomb potential. This is a significant achievement, because in non-relativistic quantum mechanics the Coulomb potential is simply chosen from a range of many possibilities. The construction of viable quantum field theories is much more difficult. Only those with the simplest interaction vertices are consistent, so there is no freedom to select the potential. We were led to our lowest-order result by defining the simplest possible vertex for electron–photon interactions in a gauge invariant Lagrangian. There is essentially no other way to couple a charged fermion to the electromagnetic field, and the only freedom is to change the value of the coupling by changing the charge q. So quantum field theory provides a much deeper explanation of the Coulomb force.

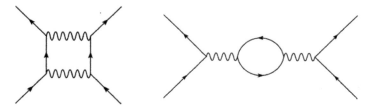

Fig. 12.6 One-loop QED Feynman diagrams.

We can construct Feynman diagrams that represent electron–electron scattering at higher order in perturbation theory, such as those shown in Figure 12.6. The inclusion of these higher-order diagrams will give a series of quantum corrections to the basic result. Feynman diagrams are the easiest way to keep track of these higher-order terms. Each can be evaluated as a contribution to the quantum amplitude for scattering. If the initial and final states are the same for different diagrams, then there may be quantum interference between these contributions, as the amplitudes corresponding to the diagrams must be added before the final scattering cross section is worked out. To compute the scattering amplitude up to order e^4, we must include all possible diagrams with at most four vertices, and with the external lines representing electrons. Internal lines represent particles with only a fleeting existence, called *virtual particles*. They can be photons, electrons or positrons. The diagrams appear to give a spacetime picture of a physical process, even though each diagram with an internal loop actually corresponds to a rather complicated integral over the possible energies and momenta of the virtual particles. As long as the coupling constant is small, the exchange of a single photon, as shown in Figure 12.5 (right), gives the dominant contribution to the scattering amplitude. Each additional vertex in a diagram has an additional factor of e, which reduces the size of its contribution. The sequence of diagrams represents a perturbation expansion that can in principle be used to evaluate the scattering amplitude to an arbitrary degree of accuracy.

Evaluating the diagrams is not at all easy, especially when they contain a number of internal loops, because of the multiple 4-momentum integrals that they encode. It relies on an elaborate and technical procedure known as *renormalization*. One feature of renormalization is that we have to accept that the masses and coupling parameters of the particles are not determined by the theory and must be measured and treated as experimental inputs. We can then calculate scattering cross sections and other measurable quantities with high precision and up to arbitrarily high energy. This gives quantum field theory a powerful predictive character.

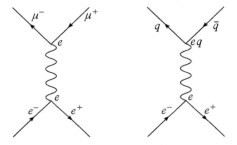

Fig. 12.7 Feynman diagrams representing electron–positron annihilation with the creation of a muon–antimuon pair (left), or quark–antiquark pair (right).

There are many types of charged particle. In the full theory of QED we must include a Dirac field for each of the fundamental fermions. We then obtain a theory that describes electromagnetic interactions between all these different charged particles. One example of such a QED process is the conversion of an electron–positron pair into a muon–antimuon pair. This is mediated by a virtual photon that briefly carries all the energy and momentum. The lowest-order prediction for the total cross section σ for this process, as represented by the Feynman diagram shown in Figure 12.7 (left), is

$$\sigma(e^+e^- \to \mu^+\mu^-) = \frac{4\pi\alpha^2}{3E^2}, \tag{12.33}$$

where $\alpha = \frac{e^2}{4\pi\hbar} \simeq \frac{1}{137}$ is the *fine structure constant* and E is the centre of mass energy. The dimensionless α is the true expansion parameter of the perturbation series of QED.

12.5   The Strong Force

Yukawa proposed in 1935 that the strong force between protons p and neutrons n, holding the atomic nucleus together, could be explained by the exchange of three spin 0 particles now known as the pions π^+, π^- and π^0. This was described in section 11.9, but quantum field theory gives deeper insights. In the actual theory devised by Yukawa, there is a doublet of Dirac fields representing the nucleons $N = (p, n)$. The two members of the doublet are distinguished by their isospin;[6] the proton p has isospin $\frac{1}{2}$ and the neutron n has isospin $-\frac{1}{2}$. The scalar pion fields (π^+, π^0, π^-) have isospins $1, 0$ and -1, respectively.

A simplified version of Yukawa theory has just one Dirac field ψ with mass parameter M_0 interacting with one scalar Klein–Gordon field ϕ with mass parameter m_0 The Lagrangian density is

$$\mathcal{L} = \frac{1}{2}\partial\phi \cdot \partial\phi - \frac{1}{2}m_0^2\phi^2 + i\overline{\psi}\gamma \cdot \partial\psi - M_0\overline{\psi}\psi + \lambda\overline{\psi}\psi\phi \,. \tag{12.34}$$

The first four terms describe the free Klein–Gordon field and the free Dirac field, and only the last, non-quadratic term has a coupling $\lambda\overline{\psi}\psi\phi$ between these fields, corresponding to the vertex shown in Figure 12.8. This is known as a Yukawa coupling. The quantized theory has a spin $\frac{1}{2}$ Dirac particle, which we also call ψ, and its antiparticle, antiψ, together with a spin 0 scalar particle, ϕ. In the Feynman diagrams, the solid lines represent ψ and antiψ particles, and the dashed lines represent ϕ particles. The scattering of two ψ particles is represented by the Feynman diagrams shown in Figure 12.9. These are examples of *tree diagrams*—diagrams without loops.

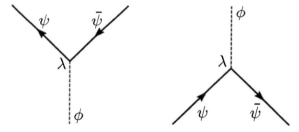

Fig. 12.8 Yukawa vertices; λ is the coupling constant at the vertex. Left: A ϕ particle converts into a ψ particle and an antiψ particle. Right: A ψ particle and an antiψ particle annihilate to form a ϕ particle.

The free ϕ field satisfies the Klein–Gordon equation (12.6),

$$\frac{\partial^2\phi}{\partial t^2} - \nabla^2\phi + m_0^2\phi = 0 \,, \tag{12.35}$$

or for static fields

$$\nabla^2\phi(\mathbf{x}) = m_0^2\phi(\mathbf{x}) \,. \tag{12.36}$$

The solution of this equation, describing the interaction of a ψ particle at the origin with a ψ particle at distance r away, is the Yukawa potential

$$V(r) = -\frac{\lambda^2}{4\pi r}\exp(-m_0 r) \,, \tag{12.37}$$

whose range is $\frac{1}{m_0}$. This potential also emerges from the scattering amplitude calculated using the diagrams in Figure 12.9. Recall that the mass of the ψ particle is $M = \hbar M_0$. If this is much greater than the kinetic energies available, then pair creation of the ψ particles

---

[6] Isospin has some similarity to spin and is used to classify strongly interacting particles. For our purposes we will only need to consider one component of isospin, analogous to the s_3 component of spin.

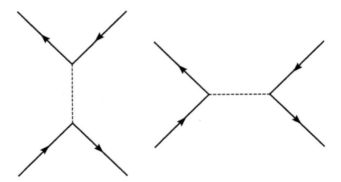

Fig. 12.9 Yukawa tree diagrams for ψ–antiψ scattering. Left: A ψ and an antiψ annihilate to form a virtual ϕ, which then converts back into a ψ and an antiψ. Right: A ψ and an antiψ exchange a ϕ.

is impossible. However, slowly moving ψ particles interact quantum mechanically through the Yukawa potential. This is the basis for the Yukawa theory of nucleons and pions as a model of nucleon–nucleon forces.

In the more complete Yukawa theory, with its doublet of nucleons and triplet of pions, the fields in the doublet mix in the interactions and so do the fields in the triplet. This means that the identity of a particle may change following an interaction. Neutrons may become protons and vice versa, as shown in the diagrams in Figure 11.26. The isospin is conserved at each vertex and this constrains these transformations. For instance, in Figure 11.26, the top-left diagram shows a process mediated by the exchange of a π^0, in which the isospins of the nucleons do not change, but in the bottom-left diagram, at the left-hand vertex a neutron emits a π^- with isospin -1 and becomes a proton. The proton has one unit of isospin more than the neutron, so isospin is conserved at the vertex. At the right-hand vertex the π^- is absorbed by a proton, which is transformed into a neutron, and again isospin is conserved. These processes continually occur to the neutrons and protons in a nucleus. In these examples, isospin conservation might simply appear to be a way of ensuring electric charge conservation, but isospin is actually an internal rotational symmetry of the strong interactions that implies more. For example, it relates the strength of the π^0 couplings to the strength of the $\pi^\pm$ couplings.

To explain the range of the strong force, Yukawa predicted that the mass of the pions m_π should be in the region of 130 MeV. The observed rest mass m_π of the charged pions π^+ and π^- is 139.6 MeV. Inside a nucleus, pions do not usually decay, but as free particles they decay with a mean lifetime of 2.6×10^{-8} s via the weak force as follows:

$$\pi^- \to \mu^- + \bar{\nu}_\mu, \qquad \pi^+ \to \mu^+ + \nu_\mu, \qquad (12.38)$$

where μ^- is the muon and μ^+ the antimuon. The mass of the neutral pion π^0 is 135.0 MeV. It decays via the electromagnetic force which gives it a much shorter half-life of 8.4×10^{-17} s. The chief decay modes (with γ denoting a photon) are

$$\pi^0 \to 2\gamma \qquad (\text{Br} = 0.988)\,,$$
$$\pi^0 \to \gamma + e^- + e^+ \qquad (\text{Br} = 0.012)\,, \qquad (12.39)$$

where the branching fraction Br is the proportion of decays of each kind.

There was a substantial effort to develop Yukawa theory in the 1940s and 1950s. Interaction terms between the pion fields must be included to produce a realistic Lagrangian and these are quite complicated and difficult to determine experimentally. However, in principle, by following this approach, all nuclear forces and the properties of nuclei would be predictable in terms of a small number of coupling constants. In practice, calculations cannot be reliably carried out because the coupling constant λ is large. That is why the strong interactions are called *strong*. Higher-order Feynman diagrams, analogous to the loop diagrams in Figure 12.6, produce large effects especially at short range, so although the Yukawa potential gives a good description of the strong nucleon–nucleon force at medium range (1–5 fm), it doesn't have predictive power at shorter range.

A further complication is that pions interact sufficiently strongly among themselves that there are particles that can be interpreted as short-lived bound states of two pions or three pions. These are called the ρ and ω mesons. In principle, their effect is already fully contained in the Yukawa theory, but it is often simpler to include them as independent particles that couple to nucleons.

By around 1960, high energy collisions of nucleons, and of pions and nucleons, had led to the discovery of yet more particles, and the theory of the strong interactions of all these particles became extremely complicated and unsatisfactory. It seemed that all efforts to produce a quantum field theory of the strong force were doomed to fail. But then a remarkable breakthrough occurred, paving the way for the current theory of strong interactions.

12.5.1 Quarks

Particles such as protons, neutrons and pions that interact via the strong force are known as *hadrons*. (This is the origin of the name of the Large Hadron Collider (LHC) at CERN, Geneva, which is a proton collider.) Hadrons come in two types, the *mesons*, such as the pions, and the *baryons*, such as the proton and the neutron. The particle accelerators of the 1950s and 1960s discovered many mesons and baryons. We now understand the properties of these particles in terms of their substructure. Murray Gell-Mann realized that they could be explained by the existence of spin $\frac{1}{2}$ component particles that he named *quarks*, and this suggests that the Yukawa force between protons and neutrons is the result of a much deeper interaction between quarks that is ultimately the basis of the strong force.

Originally, three types of quark were postulated: the up quark u, the down quark d and the strange quark s. The u and d quarks together make up an isospin doublet, and the s quark is an isospin singlet.[7] An important part of Gell-Mann's scheme is that these three quarks are unified into a bigger symmetry structure. We now know of three more quark types named charm c, bottom b and top t, which are heavier. The six quark types are called the six *flavours* of quarks. Gell-Mann's idea was that mesons, such as pions and kaons (K), are formed from a quark and an antiquark bound together. For instance, the positively charged pion (π^+) is composed of an up quark and an antidown quark, $u\bar{d}$. Figure 12.10 shows the quark constituents of the lightest mesons. All particles in this *meson octet* have spin 0, with the spins of the constituent quark and antiquark aligned antiparallel. There is a similar collection of higher-mass spin 1 mesons, including the ρ and ω mesons, where the spins of

---

[7] The s quark is also said to carry one (negative) unit of strangeness.

the quark and antiquark are parallel. There are further mesons with even higher spin, in which the quark constituents carry some orbital angular momentum. These mesons have correspondingly greater mass.

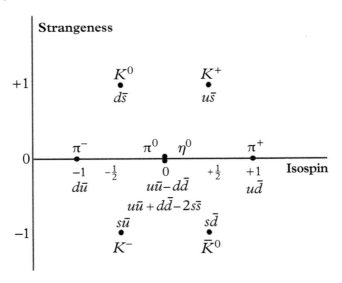

Fig. 12.10 Spin 0 mesons and their quark constituents u, d and s. A bar over the letter indicates an antiquark. The uncharged mesons π^0 and η^0 are composed of orthogonal superpositions of the quarks and antiquarks.

Crucially, there is another way in which particles are formed out of quarks in Gell-Mann's model. Three quarks may bind together to form a baryon. For instance, the proton is formed of one down quark and two up quarks, duu; the neutron is composed of two down quarks and one up quark, ddu. Neutrons and protons have spin $\frac{1}{2}$. This is because the spin of one quark is aligned opposite to the spin of the other two. The quark content of each particle in the *baryon octet* of spin $\frac{1}{2}$ baryons is shown in Figure 12.11 (left). Just as with the mesons, there are higher mass baryons with different configurations of quark spin and orbital angular momentum. There are ten spin $\frac{3}{2}$ baryons formed from the u, d and s quarks where the quark spins are all aligned parallel. Gell-Mann named this collection of baryons the decimet, but they are now usually referred to as the *baryon decuplet*. The quark content of each of these particles is shown in Figure 12.11 (right). The hexagonal and triangular structures of these collections of particles is a successful prediction of Gell-Mann's symmetry of quark flavours.

Other than the proton, all these particles are unstable. The Deltas (Δ) decay by the strong force into nucleons and pions; they decay so rapidly that they are not seen directly, and the primary evidence for their existence is that cross sections are strongly enhanced when a pion interacts with a nucleon at a centre of mass energy around 1230 MeV. Free neutrons decay via the weak force with a half-life of about 10 minutes. All the other particles contain at least one strange quark and also decay via the weak force, for example,

$$\Sigma^+ \to p + \pi^0 \qquad (\mathrm{Br} = 0.52)\,,$$
$$\Sigma^+ \to n + \pi^+ \qquad (\mathrm{Br} = 0.48)\,, \tag{12.40}$$

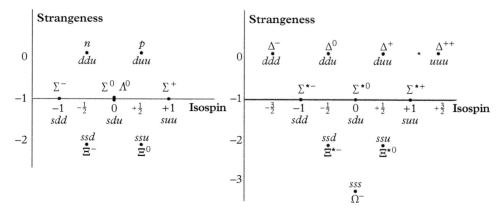

Fig. 12.11 Left: The quark content of the spin $\frac{1}{2}$ baryons. The Σ^0 and the Λ^0 are composed of orthogonal superpositions of u, d and s quarks. Right: The quark content of the decuplet of spin $\frac{3}{2}$ baryons.

and they leave tracks in particle detectors before decaying.

12.5.2 Confinement

Although Gell-Mann's quark hypothesis provides a neat explanation of the properties of the hundreds of hadrons observed in particle accelerators, there is an apparent problem. No quarks have ever been observed. To match the charges of the proton, neutron and other hadrons, the charge of an up quark must be $q_u = \frac{2}{3}$ and the charge of a down quark must be $q_d = -\frac{1}{3}$ (in units of the proton charge e). A fractionally charged particle would be easy to distinguish from any other particle. The width of the particle's track in a bubble chamber photograph would be much narrower, for instance. However, in over forty years of searching, no evidence for the existence of free quarks has ever been found.

Fortunately, it is possible to probe protons and neutrons and demonstrate that they contain these point-like constituents even without extracting and isolating individual quarks. A series of experiments was performed at the Stanford Linear Accelerator Center (SLAC) in California in the late 1960s and early 1970s that aimed to find any hidden substructure within the proton. The method was very similar to Rutherford's alpha particle experiments, but on an altogether grander scale. A liquid hydrogen target was bombarded by electron beams with energies between 5 GeV and 20 GeV. The results showed that protons do indeed contain tiny, hard, spin $\frac{1}{2}$ components that were scattering the electrons. It is natural to infer that these nuggets within the proton are Gell-Mann's quarks.

The force between quarks is so strong that it is impossible for a quark to escape imprisonment and exist as an independent free particle. They are always found caged within composite particles such as pions, protons or neutrons. This surprising property is called *confinement*. It led to a great deal of consternation in the early years of the quark model, and it was only when the force binding quarks together was understood better that their real physical existence became accepted.

The electric charges assigned to the quarks can be tested in electron–positron annihilation experiments. When an electron and a positron annihilate, there are several possible outcomes. One is the production of a muon and an antimuon, as depicted in Figure 12.7

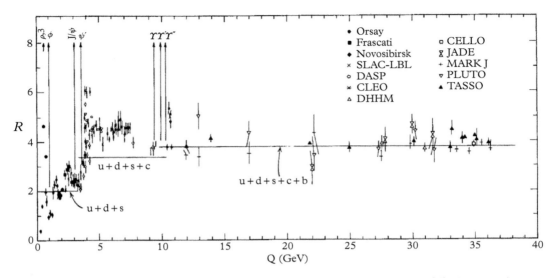

Fig. 12.12 The proportion of hadronic debris found in electron–positron annihilation experiments, as measured through R, the cross section ratio. Q is the collision energy.

(left). The cross section for such an event is given by equation (12.33). Alternatively, a quark and an antiquark may be produced, as shown in Figure 12.7 (right). At low energies, there are three possible quark–antiquark pairs that can be created: a $u\bar{u}$ pair, a $d\bar{d}$ pair or an $s\bar{s}$ pair. These quarks are not seen, as the strong force comes into play as soon as they are created, and what the experiments observe is a collection of hadrons emanating from the interaction point. This process is known as *hadronization*. The hadrons tend to be highly collimated into *jets* of particles that emanate from the point of the electron–positron impact.

The annihilation of the electron and positron is a purely electromagnetic interaction, dominated by the Feynman diagram in Figure 12.7 (right) with two vertices. The coupling at the first vertex is e and the coupling at the second is the quark charge eq, so the amplitude for the production of each type of quark–antiquark pair is proportional to $e^2 q$. The cross section σ is therefore proportional to $e^4 q^2$, but is otherwise exactly the same as for muon–antimuon production. This gives the following ratio in the case of strange quarks, for instance:

$$\frac{\sigma(e^+e^- \to s\bar{s})}{\sigma(e^+e^- \to \mu^+\mu^-)} = \frac{e^4 q_s^2}{e^4} = \frac{1}{9}. \tag{12.41}$$

We can estimate the proportion of hadronic debris produced in electron–positron collisions as the ratio R of the total cross section for quark production to the cross section for muon production:

$$R = \frac{\sigma(e^+e^- \to \text{hadrons})}{\sigma(e^+e^- \to \mu^+\mu^-)} = \frac{\sum \sigma(e^+e^- \to \text{quark antiquark})}{\sigma(e^+e^- \to \mu^+\mu^-)} = \sum_{\text{flavours}} q^2. \tag{12.42}$$

With Gell-Mann's three quark flavours u, d and s, this ratio is

$$R^{u,d,s} = \frac{4}{9} + \frac{1}{9} + \frac{1}{9} = \frac{2}{3} \tag{12.43}$$

in the simple quark model. However, in the experiments, the ratio is found to be close to 2, so there is three times as much hadronic debris as expected. At higher collision energies it is possible to generate the more massive quarks and more hadronic debris is produced. The amount of hadronic debris increases when the collision energy Q passes the threshold for creating charm quarks and antiquarks, which is around $2m_c \simeq 3.0$ GeV. Above this threshold, $R^{u,d,s,c} = \frac{4}{9} + \frac{1}{9} + \frac{1}{9} + \frac{4}{9} = \frac{10}{9}$. There is a further increase above the bottom–antibottom threshold, $2m_b \simeq 10$ GeV. Now, $R^{u,d,s,c,b} = \frac{4}{9} + \frac{1}{9} + \frac{1}{9} + \frac{4}{9} + \frac{1}{9} = \frac{11}{9}$. But the experiments show that there is always three times as much hadronic debris as expected from the simple quark model. The results are presented in Figure 12.12.

To explain these results, it is hypothesized that each flavour of quark comes in three *colours*: red r, blue b and green g, so there are three times as many different types of quark. This apparently ad hoc proposal leads to the theory known as *quantum chromodynamics*, and is the key to understanding the force between quarks.

12.6 QCD

Quantum electrodynamics (QED) would eventually be developed into successful theories of both the weak and strong forces. In 1954, Chen Ning Yang and Robert Mills devised a way to generalize the gauge invariance of electromagnetism to construct theories that resemble QED but based around bigger gauge symmetries. Electromagnetism is mediated by the photon, which is a massless spin 1 boson. In Yang–Mills theories the interactions are generated by a matrix of fields that form a set of closely related massless spin 1 bosons. In the early 1970s physicists realized that there was a Yang–Mills theory that offered the perfect way to explain the force between quarks. This force is known as the colour force or more formally quantum chromodynamics (QCD). It binds three quarks to form a proton or any other type of baryon. The term *colour* is used by analogy with the mixing of red, blue and green light to form white light, which is how white light is produced on a television screen or computer monitor. Just as the electromagnetic force can be the same for different particles of the same charge, for example protons and positrons, so the colour force is the same for the six different flavours of quarks.

QCD is similar to QED, but with a number of very important differences. Whereas interactions in QED depend on a single electric charge, in chromodynamics there are three different charges: red, blue and green. Combining three quarks, one with each of these colour charges, produces a particle, such as a proton, that is neutral with respect to the colour charges. In other words, the sum of a red charge, a blue charge and a green charge is no overall colour charge, hence the colour analogy.

Each colour charge has a negative. They are known as antired, antiblue and antigreen. This offers another way to produce a colour-neutral particle. A quark and an antiquark can be bound together to form a meson if the quark carries one of the three colour charges and the antiquark carries the corresponding anticolour charge, so that the charges cancel and together there is no overall colour. For example, the quark might carry the red charge and the antiquark the antired charge, $r\bar{r}$. Alternatively, the quark might be carrying the blue charge and the antiquark the antiblue charge, $b\bar{b}$.

12.6.1 Gluons

The colour force is due to the exchange of the Yang–Mills particles known as gluons, so-named because they provide the glue that sticks quarks together. We denote the gluons by G.

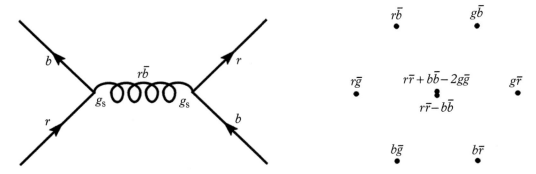

Fig. 12.13 Left: Feynman diagram showing quark scattering due to gluon exchange. Right: The colour charges of the gluons in the gluon octet.

Figure 12.13 (left) shows a QCD Feynman diagram representing an interaction between two quarks that scatter off each other. Here, the lines with arrows on them represent quarks. The curly line represents a gluon that is being exchanged between the two quarks. The strength of the coupling is g_s, where the subscript refers to the strong force.

There are eight types of gluon that are exchanged between the quarks. Each gluon carries two charges, both a colour charge and an anticolour charge. (This is due to the gluon field being a matrix.) The eight QCD gluons fit together into a colour-symmetry octet, as shown in Figure 12.13 (right). The octet diagram shows the colour charge and the anticolour charge of each of the eight gluons. For instance, the gluon represented by the point at the top left of the diagram carries a red charge and an antiblue charge. This particular gluon will participate in interactions such as that shown in the Feynman diagram. To the left of the diagram, the incoming red quark emits the (red, antiblue) gluon and is thereby transformed into a blue quark. This interaction conserves red charge because the red charge is transferred to the gluon. It also conserves blue charge, because the blue charge on the quark and the antiblue charge on the gluon have been created simultaneously. Then, on the right, the (red, antiblue) gluon interacts with a blue quark. The antiblue charge on the gluon cancels the blue charge on the quark and the red charge on the gluon is transferred to the quark. The overall effect of the exchange of the gluon is to transfer energy and momentum between the quarks and to swap over their colour charges. The same effect would be produced if the blue quark emitted a (blue, antired) gluon that was absorbed by the red quark; a single Feynman diagram represents both these processes. Taking into account all the gluon types, the colour state of a meson is not simply $r\bar{r}$ or $b\bar{b}$ as suggested earlier. It is a colour-neutral, symmetric superposition, $r\bar{r} + b\bar{b} + g\bar{g}$. At the same time, there is no $r\bar{r} + b\bar{b} + g\bar{g}$ gluon, because it would have no colour coupling strength.

Photons are electrically neutral, so they do not feel the electromagnetic force themselves and do not interact directly with other photons. Gluons, on the other hand, carry the colour charges. So gluons feel the colour force themselves, and interact with other gluons. The QCD Lagrangian includes both cubic $g_s G^3$ and quartic $g_s^2 G^4$ gluon self-couplings, where g_s is the strong-force coupling constant, so a gluon can split into two or three gluons and two or three gluons can combine to form a single gluon. This greatly complicates the colour interactions and makes the colour force very different to electromagnetism.

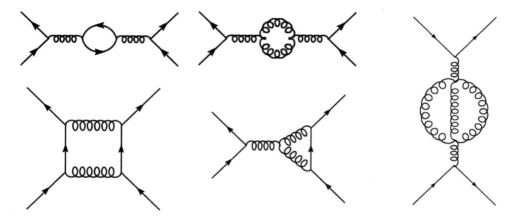

Fig. 12.14 Higher-order QCD Feynman diagrams.

Figure 12.14 shows several higher-order QCD diagrams. Again, the solid lines represent quarks and the curly lines represent gluons. The two diagrams on the left are familiar from QED, and replacing quarks with electrons and gluons with photons would give equivalent QED diagrams. The other three QCD diagrams involve gluons interacting with other gluons, so they have no QED equivalents. For instance, in the centre-top diagram, a quark emits a gluon, and this gluon dissociates into a pair of gluons which then recombine to form a single gluon, and this gluon is absorbed by the second quark. As well as making QCD calculations even more formidable than QED calculations, the additional diagrams indicate that the force behaves in a completely different way.

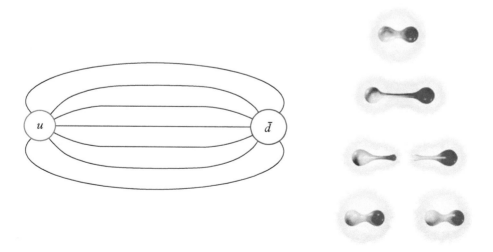

Fig. 12.15 Left: The colour field lines between two quarks forms a flux tube. Right: As two quarks separate, the energy in the colour field between them is converted into new quarks.

The colour force is actually quite feeble at very short distances, becoming strong only at longer distances. In this respect it is very different to the electromagnetic force. This is one of the great successes of QCD because it seems to correctly describe the operation of the strong force as observed in collider experiments. At longer distances, we can imagine the force between two quarks as the net result of a horrendously complicated exchange of multitudes of gluons that are simultaneously interacting with one another. This mass of tangled gluons effectively forms a tube of colour flux that behaves a bit like an elastic band between the quarks, as illustrated in Figure 12.15. It means that the colour force between two quarks is independent of their separation because the energy in the colour field increases approximately linearly with separation. This suggests that there is never enough energy to pull quarks completely apart. In fact, when the separation approaches the typical size of a hadron, which is around 10^{-15} m, the energy in the colour field is sufficient for new particles to form, as shown in Figure 12.15 (right). This is the practical manifestation of confinement. Individual quarks are not seen, because of the quark–antiquark pair creation as quarks separate.

As discussed earlier, the head-on collision of an electron and a positron results in their complete annihilation, and from the energy that is released a quark and an antiquark may be produced with a large amount of kinetic energy. As the quark and antiquark recede from one another, the energy in the colour field between them is converted into a shower of other quarks and antiquarks. All the quarks and antiquarks rapidly *hadronize*, so that their naked colour charges become hidden within colour-neutral particles, which are the particles seen in the detector. Such an event appears as two narrow jets of particles emitted in opposite directions from the point of electron–positron impact.

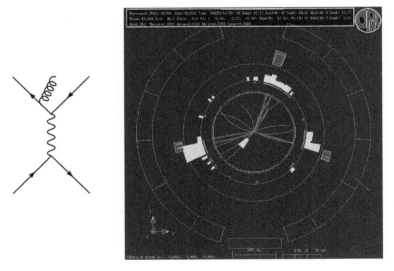

Fig. 12.16 Left: Feynman diagram showing an electron–positron annihilation event in which a quark–antiquark pair is produced and the quark emits a gluon. This would be seen as a three-jet event. Right: Three-jet event as seen in the OPAL detector at CERN.

Sometimes three jets of particles are produced, as shown in Figure 12.16 (right), and QCD can account for this. Occasionally, either the quark or the antiquark created in the electron–positron impact emits a gluon at the moment it comes into existence, as described by the Feynman diagram shown in Figure 12.16 (left). The emission of the gluon produces a third jet of hadrons. Sometimes both quark and antiquark emit a gluon, in which case four jets are seen. The amplitude for events in which a single gluon is emitted includes an additional factor of g_s, so the rate for the emission of a single gluon is proportional to $\alpha_s = \frac{g_s^2}{4\pi\hbar}$ times the rate for quark–antiquark production, where α_s is the strong force equivalent of the electromagnetic fine structure constant. This produces a second-order correction to the cross section for the production of hadrons that is proportional to $\frac{\alpha_s}{\pi}$, which gives

$$R_{\mathrm{QCD}} = \frac{\sigma(e^+e^- \to \text{hadrons})}{\sigma(e^+e^- \to \mu^+\mu^-)} = 3 \times \left(1 + \frac{\alpha_s}{\pi}\right) \sum_{\text{flavours}} q^2, \qquad (12.44)$$

where the factor of 3 is due to the sum over colours. For $\alpha_s \simeq 0.15$, the additional term produces a 5% correction to the expression for R, and this further improves the agreement between theory and experiment.

By analysing the distribution of all the particles within a jet it is possible to distinguish between a jet that has formed from a quark and one that has formed from a gluon. It is also possible to determine the spin of the gluon from the angular distribution of the jets. This confirms that gluons are not scalar particles. They have spin 1, as they must if they are the mediators of a Yang–Mills force.

The deep inelastic scattering experiments at SLAC and later accelerators have demonstrated that the structure of hadrons is rather more complicated than originally believed. Gell-Mann proposed that the composition of the proton is duu. These three quarks are known as the *valence quarks*. There are also quark–antiquark pairs within the proton that fluctuate in and out of existence and these may also scatter incident particles. They are known as *sea quarks*. Furthermore, these experiments suggest that typically just half the momentum of a proton is carried by its quark components, and the rest is carried by the gluons that are flying around within the proton and holding the quarks together.

12.6.2 Lattice QCD

The world's leading accelerators continue to confirm the predictions of QCD, but some QCD calculations are too difficult to perform manually, so physicists must turn to the supercomputer. QCD is modelled on a discrete grid representing space and time, an approach that is known as lattice QCD. One question that physicists would like to answer is the relationship between QCD and confinement. All the evidence suggests that QCD implies confinement and lattice QCD has given support to this view, but a definitive proof is still lacking. Another aim of lattice QCD is to predict the masses of the particles formed out of quarks from first principles. This is similar to working out the energy levels in an atom but vastly more complicated. When the computed masses of the various mesons and baryons are set against the values that are measured in particle accelerators, the agreement is very good; typically, QCD predictions match experimental results to better than 4% accuracy and only for the lightest mesons, the pions, are the results less convincing. Although this is not comparable to the incredible precision of the predictions of quantum electrodynamics or the equivalent results in atomic physics, it is still very impressive. The accuracy of the calculations will increase as greater computing power becomes available and as the techniques for performing the calculations are further refined.

12.6.3 Heavy quarks and exotic hadrons

In 1974, the discovery of a new meson with a mass of 3.1 GeV was simultaneously announced by a team led by Samuel Ting at Brookhaven National Laboratory, who named it J, and a team led by Burton Richter at SLAC who named it Psi (Ψ). Ever since, it has been known as the J/Ψ. The significance of the new meson was that it was the first appearance of the fourth quark flavour, charm. The composition of the J/Ψ meson is $c\bar{c}$. Just three years later a team at Fermilab led by Leon Lederman discovered the upsilon meson (Υ), the first particle containing the fifth quark, bottom. The upsilon is the lightest meson with quark composition $b\bar{b}$. A sixth quark is required to complete the third generation of fermions and physicists searched for it for many years without success.

The top quark was eventually discovered at Fermilab in 1995 almost two decades after the discovery of the bottom quark. This discovery was made with the Tevatron, a proton–antiproton collider operating at energies of up to 0.98 TeV per beam, through the process

$$p + \bar{p} \rightarrow t + \bar{t} + X^0 \,, \tag{12.45}$$

where X^0 represents other hadrons. The mass of the top quark is 173 GeV, which is about the mass of a gold atom and almost 40 times the mass of the bottom quark. The lifetime of the top quark is extremely short, around 4×10^{-25} s. This means that unlike the other quarks, the top decays before any hadron can form. (The bottom and charm quarks have lifetimes of around 10^{-12} s.)

In the half-century following the publication of Gell-Mann's quark model, hundreds of hadrons have been discovered, all of which can be categorized as mesons with quark content $q\bar{q}$, baryons with quark content qqq or antibaryons $\bar{q}\bar{q}\bar{q}$. However, QCD does not rule out the existence of *exotic* hadrons with other colour-neutral quark compositions such as the *dimeson* or *tetraquark* $qq\bar{q}\bar{q}$, the *pentaquark* $qqqq\bar{q}$, and even the combination $q\bar{q}G$, where G is a gluon. After many fruitless searches for these exotics, in 2014 the LHC established the existence of a hadronic resonance named $Z(4430)$ that appears to be a dimeson composed of two quarks and two antiquarks with mass 4430 MeV. The following year CERN announced evidence for pentaquark states. These have been named $P_c(4380)^+$ and $P_c(4450)^+$.

12.7 The Weak Force

The weak force was first observed in radioactive beta decay, as described in section 11.3.4. This occurs in certain atomic nuclei when a neutron changes into a proton with the emission of an electron. When a nucleus undergoes alpha decay, the alpha particle is emitted with a sharply defined energy. By contrast the electrons emitted in beta decay have a broad range of energies. Enrico Fermi realized that another particle was being emitted with the electron and taking away some of the energy and momentum released in beta decay. We now know this particle as the electron antineutrino ($\bar{\nu}_e$). So the beta decay of a neutron is

$$n \rightarrow p + e^- + \bar{\nu}_e \,. \tag{12.46}$$

The particle emitted along with the electron is defined as the antineutrino for the following reason. Leptons are always observed to be created or destroyed in pairs. This fact is formalized by introducing the charge known as *lepton number* that is conserved in all reactions including weak interactions. Hadrons have lepton number 0, so if the electron has lepton number $+1$ and lepton number is conserved, then the other lepton released in beta decay must have lepton number -1 and is therefore the antineutrino. Neutrinos are not electrically charged; they only interact via the weak interaction. It is a staggering fact that neutrinos would typically pass through solid material with a thickness of several light years without interacting.

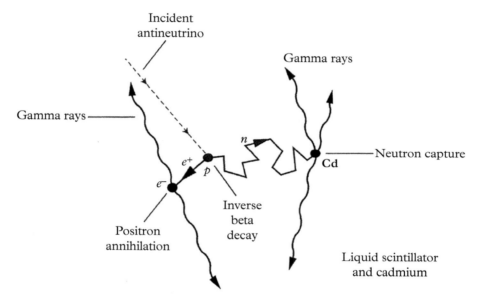

Fig. 12.17 The signal that an antineutrino had been detected in the Reines–Cowan experiment was the coincidence of a positron annihilation closely followed by the gamma decay of a cadmium nucleus.

Fermi's prediction of the neutrino was not verified until an extremely intense source of these particles became available with the construction of nuclear fission reactors. In 1956 Frederick Reines and Clyde Cowan detected antineutrinos emitted from the Savannah River reactor in the United States. The detector consisted of 300 litres of a solution containing cadmium chloride. Antineutrinos from the nuclear reactor are detected through the inverse beta decay reaction with protons in the solution,

$$\bar{\nu}_e + p \rightarrow n + e^+. \tag{12.47}$$

The positron rapidly annihilates with an electron to produce two oppositely directed 0.511 MeV gamma-ray photons and the neutron is captured by a cadmium nucleus, which has a large neutron capture cross section:

$$n + {}^{108}\text{Cd} \rightarrow {}^{109}\text{Cd}^* \rightarrow {}^{109}\text{Cd} + \gamma. \tag{12.48}$$

The ${}^{109}\text{Cd}^*$ nucleus is formed in an excited state and undergoes gamma decay within a few microseconds. The delayed coincidence of the positron annihilation closely followed by the gamma decay of the cadmium nucleus is the signal that an antineutrino has been detected, as depicted in Figure 12.17.

Prior to the confirmation of neutrinos another lepton had already been discovered. In 1937, the muon (μ^-) was revealed in the cosmic ray research of Carl Anderson and Seth Neddermeyer, although its identity was not definitively pinned down for another decade. The muon looks like a heavy replica of the electron and was a completely unexpected discovery. The mass of the muon is 105.7 MeV, which is 207 times the mass of the electron. The muon is unstable with a mean lifetime of 2.2×10^{-6} s and decays as follows:

$$\mu^- \rightarrow e^- + \bar{\nu}_e + \nu_\mu. \tag{12.49}$$

As shown here, associated with the muon there is a second neutrino known as the muon neutrino and designated ν_μ. The muon neutrino is distinct from the electron antineutrino emitted in beta decay, as was first explicitly demonstrated in 1962 by Lederman, Melvin Schwartz and Jack Steinberger.

In the mid-1970s Martin Perl discovered the third charged lepton, the tauon (τ^-). The mass of the tauon is 1777 MeV, almost 3500 times the mass of the electron, and its lifetime is 2.9×10^{-13} s. It has numerous decay routes. The most common are

$$\begin{aligned}
\tau^- &\rightarrow \pi^- + \pi^0 + \nu_\tau &&(\text{Br} = 0.255)\,, \\
\tau^- &\rightarrow e^- + \bar{\nu}_e + \nu_\tau &&(\text{Br} = 0.178)\,, \\
\tau^- &\rightarrow \mu^- + \bar{\nu}_\mu + \nu_\tau &&(\text{Br} = 0.174)\,.
\end{aligned} \tag{12.50}$$

The existence of the third, distinct neutrino, the tauon neutrino ν_τ, was confirmed at Fermilab in 2000.

The existence of these particles and decay processes, most of which involve neutrinos, requires a detailed theory of weak interactions. This theory was developed slowly, with contributions from several physicists. The surprising violation of the *discrete symmetries* of parity, charge conjugation, and time reversal in weak interactions provided some of the important clues.

12.7.1 Parity violation

It was recognized in the early days of quantum mechanics that atomic wavefunctions can be classified as even or odd under a spatial inversion, $\mathbf{x} \rightarrow -\mathbf{x}$. Spatial inversion is denoted by the parity operator P, which transforms a wavefunction $\Psi(\mathbf{x}, t)$ to $\Psi'(\mathbf{x}, t)$, where

$$\Psi'(\mathbf{x}, t) = P\Psi(\mathbf{x}, t) = \Psi(-\mathbf{x}, t)\,. \tag{12.51}$$

The eigenvalue of P is called the *parity*. If $\Psi(-\mathbf{x}, t) = \Psi(\mathbf{x}, t)$ the parity is positive, and if $\Psi(-\mathbf{x}, t) = -\Psi(\mathbf{x}, t)$ the parity is negative.[8] Spatial inversion is important, because in most areas of physics it is a symmetry. For example, if $\Psi(\mathbf{x}, t)$ is the wavefunction of an electron orbiting a nucleus at the origin, then $\Psi(-\mathbf{x}, t)$ is a related wavefunction that has the same energy if it is a stationary state, and evolves similarly if it is not. Sometimes $\Psi(\mathbf{x}, t)$ and $\Psi(-\mathbf{x}, t)$ are physically distinct, but often they are equal, or just differ by a sign; in other words, the wavefunction has definite parity.

---

[8] These parities are also referred to as even and odd.

Similarly, we can define the time reversal operator T, where

$$\Psi'(\mathbf{x}, t) = T\Psi(\mathbf{x}, t) = \Psi(\mathbf{x}, -t), \tag{12.52}$$

and the charge conjugation operator C, which transforms particles into antiparticles and vice versa,

$$\Psi'(\mathbf{x}, t) = C\Psi(\mathbf{x}, t) = \overline{\Psi}(\mathbf{x}, t), \tag{12.53}$$

where $\overline{\Psi}$ is the complex conjugate of Ψ.

When each of these discrete operations is applied twice, the original wavefunction is recovered. For instance, the parity operator applied twice gives

$$PP\Psi(\mathbf{x}, t) = P\Psi(-\mathbf{x}, t) = \Psi(\mathbf{x}, t). \tag{12.54}$$

The eigenvalues of P are therefore ± 1, and similarly the eigenvalues of T and C must also be ± 1. Under very general assumptions it can be shown that if physics can be described by quantum field theories, then it must be invariant under the simultaneous application of all three of these operators, so PCT is necessarily a symmetry of the theory. This is known as the PCT theorem. It means that if $\Psi(\mathbf{x}, t)$ is a physical state then so is $PCT(\Psi(\mathbf{x}, t)) = \overline{\Psi}(-\mathbf{x}, -t)$. We might imagine that physics must be invariant under each of these transformations when applied separately, but the reality is not so straightforward.

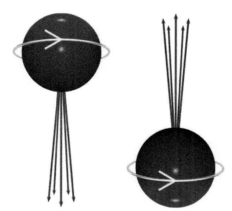

Fig. 12.18 Left: When ^{60}Co nuclei undergo beta decay, the electrons are preferentially emitted antiparallel to the nuclear spin. Right: A parity transformation inverts the relationship between the ^{60}Co spin and the direction in which the electrons travel. But this picture is unphysical. Far fewer electrons are emitted parallel to the ^{60}Co spin, so parity symmetry is violated.

If spatial inversion is a true symmetry of fundamental physics, then for every process that is observed there would be an equally probable mirror image process. Quantum mechanically, parity would be conserved. In 1956, Tsung Dao Lee and Chen Ning Yang questioned this assumption. They spent several weeks reviewing past experiments and came to the conclusion that there were many experiments that confirmed parity conservation in electromagnetic and strong interactions, but none of the experiments had any bearing on whether parity is conserved in weak interactions. They proposed several experiments where this could be tested, one of which was taken up by Chien Shiung Wu. She set out to study beta decay in cobalt-60, a nucleus with a high spin value of 5.

Cobalt-60 nuclei undergo the beta decay

$$^{60}_{27}\text{Co} \rightarrow ^{60}_{28}\text{Ni} + e^- + \bar{\nu}_e. \tag{12.55}$$

Wu cooled a sample of ^{60}Co to 0.01 K and placed it in an intense magnetic field to align the spin axes of the ^{60}Co nuclei. Each nucleus is initially in a state of definite parity. If parity is conserved, then there should be no correlation between the direction in which the electrons are emitted and the spin axes of the nuclei. We can see this as follows. Spatial inversion is the operation of simultaneous inversion in all three spatial directions. Inversion in the x-axis ($x \rightarrow -x$) followed by inversion in the y-axis ($y \rightarrow -y$) is equivalent to a rotation by 180° around the z-axis. If we position a ^{60}Co nucleus at the origin and take its spin axis to be the z-axis, then the spin is unchanged by this rotation. It is also unchanged by inversion in the z-direction ($z \rightarrow -z$). If an electron is emitted in the direction opposite to the nuclear spin, its direction of flight is also unaltered by the rotation around the z-axis, but its direction is reversed by the inversion in the z-direction. If parity is conserved, then all processes must occur at the same rate as the mirror image processes, so conservation of parity implies that equal numbers of electrons must be emitted parallel and antiparallel to the spin of the ^{60}Co nuclei. But this is not what Wu found. She showed that the electrons are preferentially emitted in the direction opposite to the cobalt-60 spin, as illustrated in Figure 12.18 (left), and established that parity symmetry is violated in weak interactions.

Following the discovery of parity violation, it was assumed that the combination of parity and charge conjugation CP would be conserved. However, in 1964 it was discovered that the weak interaction also violates CP conservation. Unlike the violation of parity, the violation of CP is a very small quantum effect. The violation of CP can be understood if there are three or more generations of fundamental fermions. (By the PCT theorem, violation of CP is equivalent to violation of time reversal symmetry T.) P and CP violation are now incorporated in the unified theory of the electromagnetic and weak forces, which forms a large part of the Standard Model.

12.8   The Theory of the Electroweak Force

QED explains the electromagnetic force as due to the exchange of virtual photons between charged particles, and QCD accounts for the strong force as due to the exchange of gluons between quarks, antiquarks and gluons. This begs the question of whether a similar, gauge invariant Yang–Mills theory is possible in the case of the weak force.

In the 1930s, Fermi proposed an early theory of the weak force, in which the beta decay event (12.46) is due to a single interaction vertex at which four particles couple with a strength determined by the Fermi coupling constant G_F, as shown in Figure 12.19 (left), where $\frac{G_\text{F}}{\hbar^3} \simeq 1.17 \times 10^{-5}$ GeV^{-2}. The observed weakness of the weak force can be accounted for if this interaction is due to the exchange of heavy W bosons with mass M_W, as shown in Figure 12.19 (centre). This diagram has two vertices whose strength g_w is the weak coupling constant. For low-energy interactions with $E^2 \ll M_W^2$, this leads to a relationship $\frac{G_\text{F}}{\sqrt{2}} = \frac{g_\text{w}^2 \hbar^2}{M_W^2}$. (The factor of $\sqrt{2}$ is due to the historical definition of G_F.) The dimensionless weak equivalent of the fine structure constant is $\alpha_\text{w} = \frac{g_\text{w}^2}{4\pi\hbar}$, and if this is assumed to be

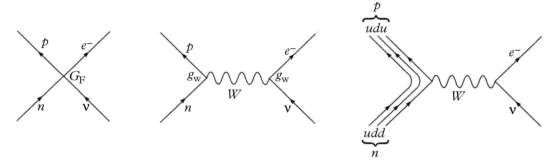

Fig. 12.19 Left: Four-particle vertex in Fermi theory. Centre: W boson exchange. Right: Beta decay in terms of quarks.

comparable in strength to the electromagnetic fine structure constant then $\alpha_{\rm w} \simeq \alpha \simeq \frac{1}{137}$. This gives us

$$M_W^2 = \frac{4\pi \hbar^3 \alpha_{\rm w} \sqrt{2}}{G_{\rm F}} \simeq \frac{4\pi \sqrt{2}}{137 \times 1.17 \times 10^{-5}} {\rm GeV}^2 , \tag{12.56}$$

which suggests that the mass of the exchange boson must be in the region of 100 GeV. Weak interactions are much weaker than electromagnetic interactions when the available energy is much less than 100 GeV, as for example in neutron beta decay, but they become comparable at higher energies.

Replacing Fermi's interaction vertex with a W boson exchange has some theoretical advantages, which were recognized before the W boson was experimentally discovered. It hints at a unification of the electromagnetic and weak interactions. However, there is a problem. Adding a mass term $\frac{1}{2}M_W^2 W \cdot W$ for the exchange bosons to the Yang–Mills Lagrangian ruins the gauge invariance of the theory, and renders it mathematically inconsistent. Initially this was seen as a serious stumbling block for such a model. In 1964, a resolution to this issue was independently found by several theorists: Peter Higgs; Robert Brout and François Englert; Gerald Guralnik, Carl Hagen and Tom Kibble. It is known as the *Higgs mechanism* and we will describe how it works in section 12.8.1. This is the key to the unified theory of electromagnetic and weak interactions known as the GWS theory after Sheldon Glashow, Steven Weinberg and Abdus Salam, that was developed in the late 1960s. The GWS electroweak theory is a Yang–Mills gauge theory coupled in a special way to a scalar Higgs field, and to quarks and leptons.

According to the GWS theory, the weak force is produced by the exchange of three massive spin 1 bosons: the W^-, W^+ and Z bosons. In terms of quarks and leptons, beta decay is explained by the exchange of a W^-, as shown in Figure 12.19 (right). One of the down quarks in the neutron emits a virtual W^- particle and is thereby transformed into an up quark. This changes the neutron into a proton. The emitted virtual W^- particle then immediately decays into the electron and electron antineutrino that are released from the nucleus. Similarly, inverse beta decay is accounted for by W^+ exchange.

12.8.1 The Higgs mechanism

We will describe the original version of the Higgs mechanism presented by Higgs in an illustrative model based on electromagnetism. Extending the mechanism to the physically

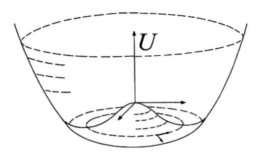

Fig. 12.20 Mexican hat potential.

important case of the electroweak theory is algebraically more complicated, but the principles are the same. Higgs proposed the Lagrangian density

$$\mathcal{L} = -\frac{1}{4}\mathcal{F} \cdot \mathcal{F} + \frac{1}{2}\overline{D\Phi} \cdot D\Phi + \frac{1}{2}\mu^2|\Phi|^2 - \frac{1}{4}\lambda|\Phi|^4 \qquad (12.57)$$

to describe a complex scalar field Φ of unit charge coupled to the electromagnetic 4-vector potential $\mathcal{A}$, with electromagnetic field strength $\mathcal{F}$. $D\Phi = \partial\Phi - i\mathcal{A}\Phi$ is the covariant derivative of the scalar field, and the Higgs potential is

$$U(|\Phi|) = -\frac{1}{2}\mu^2|\Phi|^2 + \frac{1}{4}\lambda|\Phi|^4 \,, \qquad (12.58)$$

where λ and μ are positive constants and $|\Phi|^2 = \overline{\Phi}\Phi$. U is known as the *Mexican hat potential* and is shown in Figure 12.20. It is unchanged under a gauge transformation, as U is independent of the phase of Φ. The terms $-\frac{1}{4}\mathcal{F} \cdot \mathcal{F} + \frac{1}{2}\overline{D\Phi} \cdot D\Phi$ are also gauge invariant, in essentially the same way as we discussed for the QED Lagrangian density in section 12.4.1. For convenience, we will shift the Higgs potential by the constant $\frac{1}{4}\lambda v^4$, where $\mu^2 = \lambda v^2$. This has no effect on the field equations, but now

$$U(|\Phi|) = \frac{1}{4}\lambda(|\Phi|^2 - v^2)^2 \,. \qquad (12.59)$$

The physical particles in a quantum field theory are the quantized excitations around the vacuum configuration, as we discussed in section 12.2. In the Higgs model, the vacuum of the quantum field theory is located at the minimum of the potential U. Previously we have assumed that in the vacuum state $\Phi = 0$, so the field vanishes. However, the Higgs potential has been constructed so that this is not the case; it is minimized by any field that satisfies $|\Phi| = v$. Quantum field theories must have a unique vacuum state, but here the vacuum appears to be degenerate, and this is critically important to the theory. Mathematically, there is a circle of possible vacuum states given by $|\Phi| = v$, but these vacuum states are physically indistinguishable, because they only differ by a gauge transformation, which changes the phase of Φ. We assume that in the very early evolution of the universe a unique vacuum state arises following random quantum fluctuations of the system. For simplicity, we will choose this to be $\Phi = v$. (We are free to do this by a convenient choice of gauge fixing.) We now have a non-zero field $\Phi = v$ throughout empty space even when the Φ field is unexcited and there are no Φ particles present, and this is due to the nonlinear self-coupling of the Φ field.

It is worth noting that this is only possible for a scalar field, because the existence of a non-zero background vector field would define a special direction in space and therefore break Lorentz invariance. In the vacuum, the electromagnetic potential must be $\mathcal{A} = 0$.

In choosing a unique vacuum state, the system seems to have lost the original symmetry of the Higgs potential. This is known as *spontaneous symmetry breaking*. Often the symmetry is described as hidden rather than broken, as the theory retains the original underlying gauge symmetry, but it is manifested in a more complicated nonlinear way. The knock-on effect is that the gauge symmetry of the electromagnetic field is spontaneously broken and the photon becomes a massive particle. To show this, we make the expansion

$$\Phi(\mathbf{x}, t) = v + \eta(\mathbf{x}, t) \tag{12.60}$$

and substitute back into the Lagrangian. The term $\frac{1}{2}\overline{D\Phi} \cdot D\Phi$ includes the piece $\frac{1}{2}(\overline{-i\mathcal{A}\Phi}) \cdot (-i\mathcal{A}\Phi) = \frac{1}{2}\mathcal{A} \cdot \mathcal{A}|\Phi|^2$. Close to the vacuum $\Phi = v$, the leading part of this is $\frac{1}{2}v^2 \mathcal{A} \cdot \mathcal{A}$, which is a mass term for the field $\mathcal{A}$, so the theory now describes a massive vector boson with mass parameter $M = v$.

After rewriting the rest of the Lagrangian in terms of $\Phi = v + \eta$, the terms involving η are

$$\mathcal{L}_\eta = \frac{1}{2}\partial\eta \cdot \partial\eta - \frac{1}{4}\lambda((v+\eta)^2 - v^2)^2 = \frac{1}{2}\partial\eta \cdot \partial\eta - \lambda v^2 \eta^2 + \dots . \tag{12.61}$$

η is a real dynamical field, the deviation of Φ from the vacuum v, and the particle that arises by quantizing this field is known as the Higgs boson. The coefficient of the η^2 term is λv^2, so the mass parameter of the Higgs boson is $m_\eta = \sqrt{2\lambda}v$.

A massless spin 1 particle has two polarizations, whereas massive spin 1 particles (vector bosons) have three polarizations. How does the extra polarization state arise? The Φ field is complex, so it has two degrees of freedom. One is the field η, and the other is the angular variable of the Mexican hat potential that connects the degenerate vacua. This second degree of freedom becomes the longitudinal polarization of the vector boson. An alternative point-of-view is that in the gauge where Φ is real, one cannot also impose a gauge condition on $\mathcal{A}$ forcing its longitudinal component to vanish. So again, what was the massless photon has acquired an additional polarization state to become a massive vector boson.

The Lagrangian of the GWS theory is more complicated. It starts with four massless spin 1 bosons that mediate the electroweak force. It also includes a scalar field Φ with self-interactions described by a Higgs potential. In this case, the field Φ is a complex doublet, so it has four real degrees of freedom. The symmetry of the Higgs potential is spontaneously broken with the result that three of the four spin 1 bosons become the massive W^+, W^- and Z bosons that mediate the weak force. Three of the degrees of freedom of the Φ field become the longitudinal polarizations of these particles. The remaining degree of freedom, analogous to the field η, is independent of these, and after quantization it gives rise to a spin 0 scalar particle, the Higgs boson H. The fourth spin 1 boson does not interact with the Φ field and remains massless. This is the physical photon. The original electroweak force has been divided into two apparently quite different forces: the powerful and long-range electromagnetic force and the feeble short-range weak force.

12.8.2 Fermion masses

We now know that parity is maximally violated by the weak interaction and this is because, rather surprisingly, the W bosons only couple to left-handed leptons and quarks and right-handed antileptons and antiquarks. A left-handed massless particle is one whose spin axis is opposite to its momentum. W bosons do not see right-handed leptons and quarks or left-handed antileptons and antiquarks.

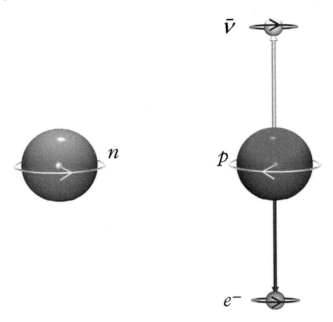

Fig. 12.21 Left: A neutron in a ^{60}Co nucleus with its spin oriented upwards prior to beta decay. Right: Following beta decay the proton has spin oriented downwards. The antineutrino emitted in beta decay is always right-handed, so to conserve angular momentum it must be emitted upwards. To conserve linear momentum the electron must be emitted downwards.

We can now understand the observed parity violation in ^{60}Co. When the neutron undergoes beta decay, the ^{60}Co nucleus loses one unit of spin. If the spin of the original neutron is aligned in the positive z-direction (by the magnetic field), then following beta decay the spin of the resulting proton is aligned in the negative z-direction. Angular momentum is conserved, so the antineutrino and electron must have a total spin of 1 aligned opposite to that of the proton, so both the antineutrino and electron spins must be aligned in the positive z-direction. Linear momentum is also conserved and as the nucleus has essentially zero momentum before and after beta decay, this means that the antineutrino and the electron must be emitted in opposite directions. The antineutrino has a tiny mass, less than 1 eV, so it recedes at an ultra-relativistic velocity. The virtual W boson emitted by the decaying neutron only couples to the right-handed antineutrino, so the spin axis of the antineutrino must be aligned with its momentum vector. The antineutrino is therefore emitted in the positive z-direction and the electron is emitted in the negative z-direction, as shown in Figure 12.21.

Parity violation has a number of consequences for the electroweak theory, including the question of the origin of the fermion masses. Fermion masses are an issue because of the following argument. Consider an electron travelling at velocity **v** in the positive

z-direction with its spin vector aligned in the same direction. If we boost to a frame moving at velocity **u** in the z-direction, where $|\mathbf{u}| > |\mathbf{v}|$, then in the new frame the electron will be travelling in the negative z-direction, but its spin will still be aligned in the positive z-direction. We have transformed a right-handed electron into a left-handed electron simply by changing frames. So a massive fermion like an electron needs both left and right parts in its field, which a Dirac 4-spinor ψ indeed has, and it can then have a Dirac mass term in the Lagrangian density of the type $M_0\overline{\psi}\psi$. However, the electroweak force treats left-handed and right-handed particles differently, and it is not possible to include a Dirac mass term for the fermions in the Standard Model Lagrangian, compatible with the electroweak gauge symmetry. Yet most of the quarks and leptons do have mass. This paradox is resolved by starting with massless fermions and enabling them to acquire their rest mass dynamically through the Higgs mechanism.

Recall that Yukawa interactions couple fermions, antifermions and scalars. In the GWS model the fermions are added to the Lagrangian as massless particles with a Yukawa coupling to the Higgs field. Only the left-handed fermions undergo weak interactions, so the left-handed fermion fields are combined into doublets that couple to the W and Z gauge bosons. There is a u and d quark doublet, an e and ν_e lepton doublet, and similarly for the other two generations. The right-handed fields do not couple to the W and Z, so they are singlets and do not carry any weak interaction charges. Mass terms in the Lagrangian must involve both left- and right-handed fields, and the only way to couple a left-handed doublet to a right-handed singlet in a gauge invariant way is to involve the complex doublet Higgs field Φ. With the Higgs field, one can have a Yukawa term of the form

$$\mathcal{L}_{\mathrm{Yuk}} = g_f\overline{\psi}_L\Phi\psi_R \tag{12.62}$$

in the Lagrangian. The product of the left-handed fermion doublet $\overline{\psi}_L$ with the doublet Higgs Φ is gauge invariant, and the right-handed fermion singlet ψ_R is too. So the Yukawa term is gauge invariant.

Now the Higgs mechanism comes in. Φ has a non-zero vacuum value. If we expand around the vacuum by substituting $\Phi = v + H$ into the Yukawa term, we obtain two terms, $g_f v\overline{\psi}\psi$ and $g_f\overline{\psi}\psi H$. The first is a mass term for the fermion and the second is a coupling to the Higgs boson H. In the GWS model, the Yukawa coupling constants g_f are determined by the requirement that the Higgs mechanism is responsible for giving each fermion *all* of its rest mass. The mass is essentially $g_f v$. v is known from the physics of the W and Z particles, so a measurement of the mass of each fermion f determines the coupling g_f. There is unfortunately, as yet, no independent understanding of the fermion masses by which the couplings g_f could be predicted. However, with g_f known to be proportional to the fermion mass, it follows that the Yukawa coupling $g_f\overline{\psi}\psi H$ between the Higgs boson H and each of the fermions of the Standard Model is proportional to the fermion mass and therefore different for each one. The Higgs couplings to the heavy quarks, t and b, and to the heavy tauon, are much greater than the couplings to the light quarks and lighter leptons, so the Higgs boson decays preferentially to the heavier particles. The branching fractions for the Higgs boson decay channels are currently being measured. The latest results appear to confirm the predictions of the GWS model.

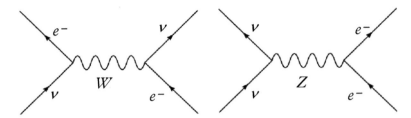

Fig. 12.22 Left: The exchange of a W boson between a neutrino and an electron. Right: The exchange of a Z boson between a neutrino and an electron.

12.8.3 Discovering the W and Z bosons and the Higgs boson

In the early 1970s no effect had ever been seen that could be attributed to the exchange of the electrically neutral Z boson, but as the Z couples with the same strength as the Ws, such effects are very small and difficult to detect. These weak neutral current effects, as they became known, include neutrino scattering processes resulting from the exchange of a Z, shown in Figure 12.22 (right), and were first observed in 1974 in the Gargamelle bubble chamber at CERN. Their discovery was critical in gaining acceptance for the GWS electroweak theory.

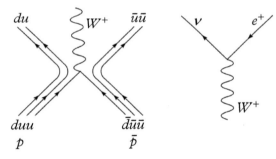

Fig. 12.23 Left: production of a W^+ boson in a proton–antiproton collision. Right: the decay of a W^+ boson into a positron and a neutrino.

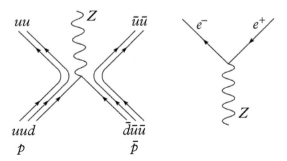

Fig. 12.24 Left: Production of a Z boson in a proton–antiproton collision. Right: The decay of a Z boson into an electron and a positron.

Following the discovery of weak neutral currents, Carlo Rubbia, Peter McIntyre and David Cline convinced CERN to convert the new Super Proton Synchrotron (SPS) machine into a proton–antiproton collider and to construct two new detectors known as UA1 and UA2. The aim was to hunt for the W and Z bosons. The first collisions were recorded in these detectors in 1981. Figure 12.23 (left) shows one of the modes for the production of a W^+ boson in the proton–antiproton collider. The W^+ then decays into a charged lepton, and a neutrino that is not detected, as shown in Figure 12.23 (right). The signal of a W^+ is therefore the detection of a high energy lepton with an energy equal to half the rest mass of the W^+. Figure 12.24 (left) shows one of the modes for the production of a Z boson. The Z has various decay modes. The most distinctive are the decays into charged lepton–antilepton pairs, as shown in Figure 12.24 (right). The signal of a Z is the detection of a high energy lepton and a high energy antilepton in the opposite direction, both with an energy equal to half the mass of the Z in the Z-boson rest frame. The discovery of the W bosons was announced in January 1983 and the discovery of the Z boson was announced later the same year. The mass of the W is 80.4 GeV. The mass of the Z is 91.2 GeV.

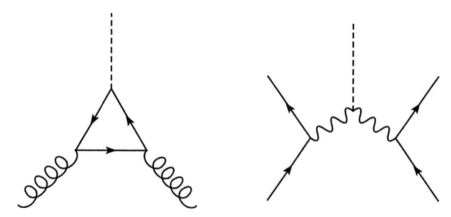

Fig. 12.25 Two ways in which a Higgs boson may be produced in a proton–proton collision. Left: gluon fusion via a loop of top quarks. Right: W or Z boson exchange.

The first task of the LHC when it entered service in 2008 was to complete the Standard Model by finding the final missing particle, the Higgs boson H. As discussed earlier, the role of the Standard Model Higgs mechanism is to give rest masses to the W and Z bosons and to the fundamental fermions, from which it follows that the strength of the interactions between these particles and the H is proportional to their masses. This is important for both the production and decay of the H. Protons contain quark and gluon subcomponents and this offers a number of ways in which Higgs bosons might be produced in proton–proton collisions. The mass of the top quark is much greater than that of the other quarks and leptons, so it has by far the greatest coupling to the H. Figure 12.25 (left) shows how an H might be produced through gluon fusion via a loop of top quarks. This is the most important channel for producing a Higgs boson at the LHC. Figure 12.25 (right) shows H production via the exchange of Ws from two quarks.

In July 2012, CERN announced that the LHC had discovered the Higgs boson with a mass of 125 GeV. The decays of the H are now being studied intensely to see whether they match expectations based on the masses of the particles of the Standard Model. So far, the coupling of the Higgs boson to the W and Z and the various quarks and leptons agrees well with theory. The spin of the Higgs boson has also been measured and confirmed to be zero. Any deviations from these patterns will be an indication of new physics beyond the Standard Model.

12.8.4 Quark mixing

In a simple version of the GWS model, weak forces act independently within each generation, with the fermionic part of the GWS Lagrangian constructed from the left-handed doublets and right-handed singlets of each generation, as described earlier. However, if this were all, then the lightest quarks in the second and third generations would be completely stable, and could not be transformed into the quarks of the first generation. This would mean that once spatially separated, $K^-(s\bar{u})$ mesons created in strong interaction processes would not decay. In reality, however, the lifetime of a K^- meson is quite short due to the decay of the s quark into a u quark mediated by a W^-. To construct the GWS model correctly we must also take into account a few other experimental results. Although the weak force coupling to each of the charged leptons has the same strength, the coupling at the weak interaction quark vertex udW is 5% smaller than at the $\mu\nu_\mu W$ vertex. Also, the measured decay rate of $K^-(s\bar{u}) \to \mu^-\bar{\nu}_\mu$ compared to $\pi^-(d\bar{u}) \to \mu^-\bar{\nu}_\mu$ is around one-twentieth of what would be expected if the weak coupling were the same for all quarks.

The resolution of these issues is to introduce some quark mixing into the theory. The GWS Lagrangian can only be constructed in terms of left-handed quark doublets (and right-handed singlets), so we must combine the six quark flavours into three doublets. By convention, the upper quarks of the three doublets are simply (u, c, t). The lower quarks of the doublets are not the standard ones (d, s, b) from the fermion table, but mixtures of these, denoted by (d', s', b'). (If electric charge is to be conserved, it is only possible to mix quarks that have the same electric charge.)

Each flavour of quark in the fermion table is a mass eigenstate, and this is what is seen in a meson whose mass is measured. We have no reason, however, to suppose that the quark states as seen by the weak interaction are the same as the mass eigenstates. Ignoring the third generation of quarks for the moment, the quark mixing is parametrized by one angle, the *Cabibbo angle* θ_C. The relation of the weak eigenstates (d', s') to (d, s) takes the form

$$\begin{pmatrix} d' \\ s' \end{pmatrix} = \begin{pmatrix} \cos\theta_C & \sin\theta_C \\ -\sin\theta_C & \cos\theta_C \end{pmatrix} \begin{pmatrix} d \\ s \end{pmatrix} \tag{12.63}$$

The weak interaction couples the up quark u directly to d', and the charm quark c to s'. Therefore the weak coupling of the u quark to the s quark is proportional to $g_w^2 \sin^2\theta_C$, while the coupling of the u quark to the d quark is proportional to $g_w^2 \cos^2\theta_C$, and their ratio is $\tan^2\theta_C$. To account for the observed decay rates of the kaons and pions, and other weak interaction rates, $\tan^2\theta_C \simeq 0.05$. The best current value for the Cabibbo angle is $\theta_C \simeq 0.23$.

This scheme, developed in the 1960s, accounts for the decay of mesons containing strange quarks, but it also suggests that another quark c is required to pair up with the s quark. The confirmation of this prediction, with the discovery of the fourth quark charm in November

1974 played a critical role in gaining acceptance for the existence of quarks and brought the Standard Model to the forefront of particle physics.

We now know that there are three generations of quarks. The full mixing of the flavours in weak interactions is described by the Cabibbo–Kobayashi–Maskawa (CKM) matrix

$$
\begin{pmatrix} d' \\ s' \\ b' \end{pmatrix} = \begin{pmatrix} V_{ud} & V_{us} & V_{ub} \\ V_{cd} & V_{cs} & V_{cb} \\ V_{td} & V_{ts} & V_{tb} \end{pmatrix} \begin{pmatrix} d \\ s \\ b \end{pmatrix}.
\tag{12.64}
$$

The $q = \frac{2}{3}$ quarks (u, c, t) couple directly to the $q = -\frac{1}{3}$ quarks (d', s', b') in electroweak interactions, and via the CKM matrix to (d, s, b). Without the CKM matrix, the heavier generations of quarks could not decay to lighter generations. Constraints on the matrix elements in the CKM matrix reduce its elements to three independent rotation angles and a complex phase. The complex phase is non-zero and this leads to the violation of CP symmetry in weak interactions. The two-generation model of Cabbibo cannot naturally accommodate the CP-violating phase, so there is some inevitability about a theory with three generations of quarks. However, there is as yet no deeper understanding that explains the values of the three angles and complex phase.

12.8.5 How many generations?

It is a remarkable fact that with an arbitrary collection of fundamental particles the Standard Model would be mathematically inconsistent. These potential inconsistencies are known as *anomalies*, and they arise because of the difference in the weak couplings to the left-handed and right-handed fermions. For the theory to be consistent there must be cancellations between the anomalies due to different fundamental particles and these cancellations imply relationships between the charges carried by the particles. It turns out that the anomalies cancel between the four particles in each generation of fermions. This means that the Standard Model is consistent for complete generations. For this reason, when the tauon was discovered in the mid-1970s physicists could confidently predict the existence of the other three members of the third generation, even though it would be twenty years before the top quark was observed and another five years before the final member of the generation, the tauon neutrino, was observed. So there are at least three generations of fundamental fermions, but how many generations are there in total? Remarkably, we now have a definitive answer to this question.

From 1989 to 2000 CERN operated the Large Electron–Positron collider (LEP). The energies of the electron and positron beams were tuned to generate the Z boson in vast quantities, so that its decay properties could be determined. There are several ways in which a Z can decay. It can decay into a quark–antiquark pair, where the quarks can be any of the five lightest quarks, in which case the end products are hadrons. It can decay into each of the three charged leptons (l^-) and their antiparticles (l^+). It can also decay into neutrino–antineutrino pairs, for example, $Z \to \nu_e \bar{\nu}_e$. Crucially, assuming that neutrinos are much lighter than the Z boson, the number of ways in which this is possible equals the number of generations. Each of these decay modes makes a contribution to the decay rate of the Z. At LEP a total of 17 million Z decays were observed. This has enabled physicists to establish the number of neutrino types and therefore the number of generations. Following the Z boson decays to neutrinos, the neutrinos escape undetected. Nonetheless, the partial

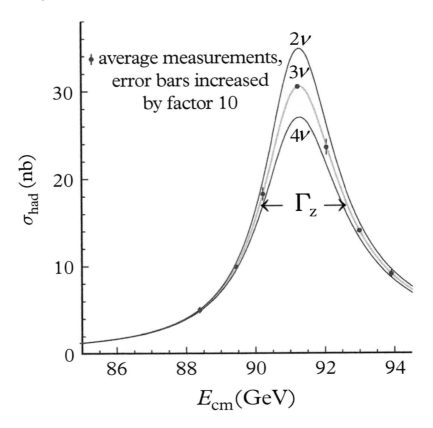

Fig. 12.26 Comparison between the measured decay width Γ_Z of the Z boson and the calculated total decay width for two, three and four types of neutrino. The measurements at LEP match the predicted decay width for three neutrino types.

decay width into neutrinos $\Gamma(Z \to \nu_l \bar{\nu}_l)$ can be deduced from the formula for the total width

$$\Gamma_Z = \Gamma(Z \to \text{hadrons}) + 3\,\Gamma(Z \to l^+ l^-) + N(\nu)\,\Gamma(Z \to \nu_l \bar{\nu}_l)\,, \qquad (12.65)$$

where $N(\nu)$ is the number of neutrino types. The total Z boson decay width is measured to be $\Gamma_Z = 2.490 \pm 0.007$ GeV. The partial decay width to hadrons is measured to be $\Gamma(Z \to \text{hadrons}) = 1.741 \pm 0.006$ GeV and the partial decay width to each type of charged lepton is $\Gamma(Z \to l^+ l^-) = 0.0838 \pm 0.0003$ GeV. These measurements agree with the Standard Model calculations. The calculated decay width to each neutrino type is $\Gamma(Z \to \nu_l \bar{\nu}_l) = 0.166$ GeV. If the figures quoted above are inserted into equation (12.65), we find $N(\nu) = 2.9840 \pm 0.0082$, which demonstrates very clearly that there are three neutrino types and no more. The Standard Model is only consistent if the fermions belong to complete generations, so this result shows that there are exactly three generations of fundamental fermions.

12.9 Neutrino Oscillations

Stars emit vast quantities of neutrinos due to the fusion reactions in their cores. In the late 1960s Ray Davis designed and built a neutrino detector 1.5 km underground in the Homestake Gold Mine in South Dakota to investigate the neutrinos emitted by the Sun. On average, 0.48 solar neutrinos were detected each day, compared to an expected rate of around 1.5 per day according to calculations of the solar neutrino flux by Davis' collaborator John Bahcall. This discrepancy was initially dismissed by most physicists, as the detector was relatively unsophisticated and calculations of the neutrino flux depended on complicated stellar models that were difficult to test. In the years since Davis' groundbreaking experiment it has become clear that the discrepancy is real and that it can be explained by the rather surprising behaviour of neutrinos.

In recent decades, stellar theory has received support from the new science of *helioseismology*. Turbulence in the layers close to the surface of the Sun generate pressure waves that produce Doppler shifts in the absorption lines of the solar spectrum. These waves have been monitored continuously since 2000 by the SOHO (Solar and Heliospheric Observatory) space probe which is situated at the Earth–Sun L_1 Lagrange point. Just as the seismic waves produced by earthquakes can be used to probe the internal structure of the Earth, so the solar pressure waves are a valuable source of information about the structure of the Sun. This is why their study is known as helioseismology. Their analysis has enabled astrophysicists to accurately determine the vital characteristics of the Sun, such as its density profile, core temperature and core composition. These measurements have provided a crucial vindication of the stellar models that we will discuss in Chapter 13. Indeed, the observations confirm Bahcall's Standard Solar Model to an accuracy of around 0.5%, which leaves no room to explain the neutrino deficit by refining the description of fusion reactions in the Sun.

Alongside these developments in solar physics, there have been major advances in the construction of neutrino detectors around the world. The largest is the Super-Kamiokande detector in Japan, which contains 50,000 tonnes of ultra-pure water surrounded by photo-multiplier tubes that can detect individual photons. Occasionally a neutrino scatters off an electron in the water. This kick gives the electron a relativistic velocity that is closely correlated with the direction of the incoming neutrino. As the electron races through the water it emits Čerenkov radiation[9] that is detected by the photo-multiplier tubes, and this enables the detector to determine the direction from which the neutrino originated. Super-Kamiokande confirms that the neutrinos detected by the Homestake experiment really are coming from the Sun. It has also been possible to analyse neutrinos produced by cosmic ray interactions in the atmosphere and compare the number that arrive from above the detector with the number that have travelled through the Earth after being created in the atmosphere on the opposite side of the globe.

Another sophisticated neutrino facility is the Sudbury Neutrino Observatory (SNO) in Ontario, Canada. This detector is composed of 1000 tonnes of heavy water and can detect neutrinos in three ways. The charged current channel determines the flux of electron neutrinos $\Phi(\nu_e)$ via the following interaction with a deuterium nucleus due to the exchange

---

[9] A cone of Čerenkov radiation is emitted by a charged particle travelling through a medium at a speed greater than the speed of light in the medium. This is a shock wave analogous to the sonic boom generated by an object travelling faster than the speed of sound.

of a virtual W boson:

$$\nu_e \quad + \quad {}^2\text{H}(p,n) \quad \rightarrow \quad p \quad + \quad p \quad + \quad e^-. \tag{12.66}$$

Next there is the neutral current channel in which the deuterium nucleus is dissociated due to the exchange of a virtual Z boson:

$$\nu \quad + \quad {}^2\text{H}(p,n) \quad \rightarrow \quad p \quad + \quad n \quad + \quad \nu. \tag{12.67}$$

This interaction is available to all three types of neutrino and so determines the total flux of neutrinos $\Phi(\nu_e) + \Phi(\nu_\mu) + \Phi(\nu_\tau)$. All three types of neutrino may also undergo elastic scattering with electrons. This is known as the elastic scattering channel:

$$\nu \quad + \quad e^- \quad \rightarrow \quad \nu \quad + \quad e^-. \tag{12.68}$$

Electron neutrinos may scatter off electrons due to the exchange of either W bosons or Z bosons, as shown in Figure 12.22, whereas muon neutrinos and tauon neutrinos can only scatter due to the exchange of Z bosons, so the rate is different for electron neutrinos. The flux combination determined by this channel is calculated to be $\Phi(\nu_e) + 0.15(\Phi(\nu_\mu) + \Phi(\nu_\tau))$. The results from SNO give the following values for the neutrino fluxes in units of 10^{-8} cm^{-2}s^{-1}:

$$\Phi(\nu_e) \quad = \quad 1.76 \pm 0.01,$$
$$\Phi(\nu_e) + \Phi(\nu_\mu) + \Phi(\nu_\tau) \quad = \quad 5.09 \pm 0.63, \tag{12.69}$$

so the total neutrino flux is close to three times the electron neutrino flux. Furthermore, in the same units, Bahcall's Standard Solar Model predicts that the flux of electron neutrinos generated by the solar core, with sufficient energy (> 2 MeV) to dissociate a deuterium nucleus, is

$$\Phi_{\text{BSSM}}(\nu_e) = 5.05 \pm 1.01. \tag{12.70}$$

This confirms the original findings from the Homestake neutrino detector and suggests very strongly that neutrinos created within the Sun, which all start out as electron neutrinos,[10] have somehow been transformed into muon and tauon neutrinos by the time they reach detectors on Earth. This interpretation has been strengthened by a variety of other neutrino experiments, including studies of the neutrinos produced by nuclear power stations and by cosmic rays hitting the Earth's atmosphere, as well as experiments with neutrino beams generated by particle accelerators.

Transmutation between neutrino species can be understood if the three types of neutrino each have a very small, but different, mass and furthermore if the weak eigenstates are not the same as the mass eigenstates, just as we found when considering the quarks. We can be sure that the neutrino that entered our detector was an electron neutrino if it interacts via the charged current channel and produces an electron. Similarly, when a nucleus undergoes inverse beta decay we know that the neutrino emitted along with the positron is an electron neutrino. By definition, the neutrino that is transformed into an electron when it couples to

---

[10] Muon neutrinos are only produced in processes that involve muons, and the temperature at the core of the Sun is far too low for the creation of muons.

a W boson is an electron neutrino and similarly for the muon and tauon neutrinos. These types of neutrino ν_e, ν_μ, ν_τ are known as the weak eigenstates. We can determine the weak eigenstate of a particular neutrino when it interacts, but we have no way to determine the mass eigenstate, so we have no reason to suppose that the weak eigenstates are the same as the mass eigenstates. In accordance with quantum mechanics we can only assume that when an electron neutrino is created it is in a superposition of the three mass eigenstates. We can represent this as

$$\Psi(\nu_e) = U_{e1}\Psi(\nu_1) + U_{e2}\Psi(\nu_2) + U_{e3}\Psi(\nu_3), \tag{12.71}$$

where ν_1, ν_2, ν_3 are the three mass eigenstates and U_{e1}, U_{e2}, U_{e3} are the components of a 3×3 matrix analogous to the CKM matrix. More generally

$$\begin{pmatrix} \Psi(\nu_e) \\ \Psi(\nu_\mu) \\ \Psi(\nu_\tau) \end{pmatrix} = \begin{pmatrix} U_{e1} & U_{e2} & U_{e3} \\ U_{\mu1} & U_{\mu2} & U_{\mu3} \\ U_{\tau1} & U_{\tau2} & U_{\tau3} \end{pmatrix} \begin{pmatrix} \Psi(\nu_1) \\ \Psi(\nu_2) \\ \Psi(\nu_3) \end{pmatrix}. \tag{12.72}$$

For simplicity we will consider a two-generation model. We can parametrize the mixing of the mass eigenstates by an angle θ, thus

$$\begin{pmatrix} \Psi(\nu_e) \\ \Psi(\nu_\mu) \end{pmatrix} = \begin{pmatrix} \cos\theta & \sin\theta \\ -\sin\theta & \cos\theta \end{pmatrix} \begin{pmatrix} \Psi(\nu_1) \\ \Psi(\nu_2) \end{pmatrix}. \tag{12.73}$$

An electron neutrino created at time $t = 0$ may be represented as a quantum state with wavefunction

$$\Psi(0) = \Psi(\nu_e) = \cos\theta\,\Psi(\nu_1) + \sin\theta\,\Psi(\nu_2). \tag{12.74}$$

The evolution of the neutrino wavefunction is described by the time-dependent free Schrödinger equation. At time t and distance z from the source, the wavefunction will be

$$\Psi(z,t) = \cos\theta\,\Psi(\nu_1)e^{i\phi_1} + \sin\theta\,\Psi(\nu_2)e^{i\phi_2}, \tag{12.75}$$

where $\phi_i = \frac{1}{\hbar}(p_i z - E_i t)$ and $E_i^2 - p_i^2 = m_i^2$. This is how the masses come in, and why $\Psi(\nu_i)$ is called a mass eigenstate. If the masses m_1 and m_2 of the two eigenstates are different, then the relative phase of the two components of the neutrino wavefunction will change. The inverse of the mixing matrix can be used to decompose the mass eigenstates back into weak eigenstates,

$$\begin{aligned} \Psi(\nu_1) &= \cos\theta\,\Psi(\nu_e) - \sin\theta\,\Psi(\nu_\mu) \\ \Psi(\nu_2) &= \sin\theta\,\Psi(\nu_e) + \cos\theta\,\Psi(\nu_\mu). \end{aligned} \tag{12.76}$$

Substituting these expressions into $\Psi(z,t)$, we find

$$\begin{aligned} \Psi(z,t) &= \cos\theta(\cos\theta\,\Psi(\nu_e) - \sin\theta\,\Psi(\nu_\mu))e^{i\phi_1} \\ &\quad + \sin\theta(\sin\theta\,\Psi(\nu_e) + \cos\theta\,\Psi(\nu_\mu))e^{i\phi_2} \\ &= (e^{i\phi_1}\cos^2\theta + e^{i\phi_2}\sin^2\theta)\Psi(\nu_e) - (e^{i\phi_1} - e^{i\phi_2})\sin\theta\cos\theta\,\Psi(\nu_\mu) \\ &= e^{i\phi_1}\{(\cos^2\theta + e^{i\Delta\phi}\sin^2\theta)\Psi(\nu_e) - (1 - e^{i\Delta\phi})\sin\theta\cos\theta\,\Psi(\nu_\mu)\} \\ &= c_e\Psi(\nu_e) + c_\mu\Psi(\nu_\mu), \end{aligned} \tag{12.77}$$

where the coefficients are $c_e = e^{i\phi_1}(\cos^2\theta + e^{i\Delta\phi}\sin^2\theta)$ and $c_\mu = -e^{i\phi_1}(1 - e^{i\Delta\phi})\sin\theta\cos\theta$, and we have defined the phase difference between the two mass eigenstates $\Delta\phi = \phi_2 - \phi_1 = \frac{1}{\hbar}((p_2 - p_1)z - (E_2 - E_1)t)$.

If $\Delta\phi = 0$, then $|c_e| = 1$ and $c_\mu = 0$, in which case an electron neutrino remains an electron neutrino. However, if $\Delta\phi \neq 0$, then $c_\mu \neq 0$ and the probability of detecting a muon neutrino in a beam of what were initially electron neutrinos is

$$
\begin{aligned}
P_\mu &= |c_\mu|^2 \\
&= \left(1 - e^{i\Delta\phi}\right)\left(1 - e^{-i\Delta\phi}\right)\sin^2\theta\cos^2\theta \\
&= (2 - 2\cos\Delta\phi)\sin^2\theta\cos^2\theta \\
&= \sin^2\left(\frac{\Delta\phi}{2}\right)\sin^2(2\theta).
\end{aligned}
\tag{12.78}
$$

$\Delta\phi$ is a function of time and position, so this probability will oscillate. The factor of $\sin^2(2\theta)$ gives the strength of the mixing. The effects of $\Delta\phi$ and θ are separated by measuring P_μ at various distances from the electron neutrino source (see Figure 12.27). Maximal mixing requires $\sin^2(2\theta) = 1$, in which case a beam of electron neutrinos would periodically transform completely into a beam of muon neutrinos and back again. (The observed value of θ for electron and muon neutrinos is close to satisfying this.)

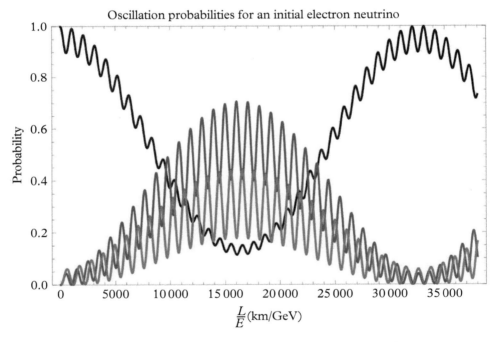

Fig. 12.27 Starting with a beam of electron neutrinos of energy E (in GeV), the black line shows the probability that a neutrino is detected as an electron neutrino at a distance of L km. The blue and red lines show, respectively, the probability that the neutrino is a muon or tauon neutrino.

The three-generation neutrino mixing matrix is known as the PMNS matrix, after Bruno Pontecorvo, Ziro Maki, Masami Nakagawa and Shoichi Sakata. Like the CKM matrix for quarks, it can be reduced to three independent angles and a phase δ. The current best values for the mixing angles are

$$\sin^2(2\theta_{12}) = 0.87 \pm 0.04\,, \quad \sin^2(2\theta_{23}) > 0.92\,, \quad \sin^2(2\theta_{13}) \simeq 0.10 \pm 0.01\,, \quad (12.79)$$

where the subscript pair denotes the generations involved. At the moment, the size of the phase δ is unknown. If it is non-zero, then neutrino oscillations will violate CP conservation.

The neutrino masses are tiny compared to those of the quarks and charged leptons. They can be independently determined from the momenta and energies of observed particles in processes involving an unobserved neutrino. As yet, they are not known precisely, but they are certainly less than 1 eV. (By comparison, the electron mass is 511 keV.) Nonetheless, the neutrino mass *differences* have been determined to better than 5% accuracy in neutrino oscillation experiments. The results are

$$\begin{aligned} \Delta m_{21}^2 &= m_2^2 - m_1^2 &\simeq& \quad (7.6 \pm 0.2) \times 10^{-5} \text{ eV}^2\,, \\ |\Delta m_{32}^2| &= |m_3^2 - m_2^2| &\simeq& \quad (2.3 \pm 0.1) \times 10^{-3} \text{ eV}^2\,, \end{aligned} \quad (12.80)$$

and therefore $|\Delta m_{31}^2| \simeq |\Delta m_{32}^2|$.

These results were found as follows. For a neutrino with definite energy much greater than the neutrino mass, $E_1 = E_2 = E_\nu$ and $p_i = (E_i^2 - m_i^2)^{\frac{1}{2}} \simeq E_\nu - \frac{m_i^2}{2E_\nu}$. The difference $p_2 - p_1$ is therefore proportional to $\Delta m^2 = m_2^2 - m_1^2$, and the variation of phase with distance z is $\Delta\phi = \frac{1}{\hbar}((p_2 - p_1)z - (E_2 - E_1)t) \simeq -\frac{(\Delta m^2)z}{2E_\nu \hbar}$ for each type of neutrino oscillation. For a beam of neutrinos with energy $E_\nu = 1$ GeV, we find

$$\frac{|\Delta\phi|}{2} = \frac{(\Delta m^2)z}{4 \times 10^9 \times 1.97 \times 10^{-7}} \text{ eV}^{-2}\text{m}^{-1} = 1.27 \times 10^{-3}(\Delta m^2)z\,\text{eV}^{-2}\text{m}^{-1}\,, \quad (12.81)$$

where we have used the conversion factor $\hbar = 1.97 \times 10^{-7}$ eV m. Between adjacent peaks of the neutrino oscillation, the argument of the function $\sin^2\left(\frac{\Delta\phi}{2}\right)$ changes by π. Setting $\frac{|\Delta\phi|}{2} = \pi$ and using the value of $\Delta m_{21}^2 = 7.6 \times 10^{-5}$ eV^2 in equation (12.81) gives the wavelength $z = \frac{\pi}{1.27 \times 7.6} \times 10^8\,\text{m} \simeq 33{,}000$ km for electron–muon neutrino oscillations, matching the observed oscillations shown in Figure 12.27. The value of $\Delta m_{32}^2 = 2.3 \times 10^{-3}$ eV^2 produces muon–tauon neutrino oscillations with a wavelength of $z = \frac{\pi}{1.27 \times 2.3} \times 10^6\,\text{m} \simeq 1100$ km.

Currently, there is a large experimental effort to measure these wavelengths more precisely. This includes long-baseline measurements, where neutrinos produced at one laboratory (J-PARC, CERN, Fermilab) are detected hundreds of kilometers away at another laboratory (Super-Kamiokande, Gran Sasso, Soudan Mine). The receiving laboratories are all underground, to limit the background of particles other than neutrinos due to cosmic rays.

12.10 Further Reading

For an overview of Feynman's approach to quantum theory and especially the interactions of photons with matter, see

R.P. Feynman, *QED: The Strange Theory of Matter and Light*, London: Penguin, 1985.

For historical introductions to particle physics, especially quarks, electroweak theory and the Higgs mechanism, see

N.J. Mee, *Higgs Force: Cosmic Symmetry Shattered*, London: Quantum Wave, 2012.

A. Watson, *The Quantum Quark*, Cambridge: CUP, 2004.

For comprehensive coverage of particle physics and quantum field theory, see

M. Thomson, *Modern Particle Physics*, Cambridge: CUP, 2013.

M.D. Schwartz, *Quantum Field Theory and the Standard Model*, Cambridge: CUP, 2014.

A. Zee, *Quantum Field Theory in a Nutshell*, Princeton: PUP, 2003.

For recent surveys of neutrino physics, see

K. Zuber, *Neutrino Physics (2nd ed.)*, Boca Raton FL: CRC Press, 2012.

S. Boyd, *Neutrino Physics Lecture Notes—Neutrino Oscillations: Theory and Experiment*, Warwick University, 2015.

13

Stars

We live in a galaxy of several hundred billion stars, in a universe of perhaps a trillion galaxies. The stars lie at vast distances, but their significance for our existence cannot be overstated. We are composed of atoms forged in previous generations of stars and we depend on the nearest star, the Sun, for the essentials of life—warmth, light and food. In this chapter we will reach for the stars.

13.1 The Sun

Astronomers first established the dimensions of the solar system in the 18th century. The transit of Venus in 1761 was viewed from a number of well separated locations, following a suggestion by Edmond Halley, so that the apparent shift in the position of Venus due to *parallax* could be measured. When combined with some simple geometry, this gives the distance from the Earth to the Sun,

$$d_\odot = 1.50 \times 10^{11} \text{ m}. \tag{13.1}$$

The solar radius is then easily calculated from the observed size of the Sun's disc. The modern figure is

$$R_\odot = 6.96 \times 10^8 \text{ m}. \tag{13.2}$$

Given $d_\odot$ and Newton's constant G, as measured by the Cavendish experiment, the Sun's mass can be determined by applying Kepler's third law (2.100) to the Earth's orbit. Setting T to be the Earth's year one finds

$$M_\odot = 1.99 \times 10^{30} \text{ kg}. \tag{13.3}$$

The average density of the Sun $\bar{\rho}_\odot$ is surprisingly low. From the figures just quoted, we calculate

$$\bar{\rho}_\odot = \frac{3M_\odot}{4\pi R_\odot^3} = 1.41 \times 10^3 \text{ kg m}^{-3}, \tag{13.4}$$

which is just 1.4 times the density of water, but at the centre of the Sun the density is far higher.

The Sun is much closer than any other star and we know considerably more about it, so it is a good starting point for modelling other stars. The Sun seems to be fairly typical with a mass somewhere in the middle of the range possible for stars, but around 85% of stars are less massive than the Sun. The solar mass $M_\odot$ is the standard by which these other stars are ranked.

The Physical World. Nicholas Manton and Nicholas Mee, Oxford University Press (2017).
© Nicholas Manton and Nicholas Mee. DOI 10.1093/oso/9780198795933.001.0001

To a very good approximation the Sun is spherical. It is spinning rather sedately, taking almost one month to complete a full rotation. The Sun is formed of a writhing mass of plasma. Any pulsations of this material must be quite gentle as otherwise the Sun's luminosity would be variable.

The Sun is subject to resonant oscillations that can be monitored through the Doppler shifts that they produce in the spectral lines. Although these oscillations do not significantly affect the stellar models that we will discuss, they have offered astronomers a window on the interior of the Sun. The analysis of these waves has enabled astronomers to measure the density, temperature and pressure profiles within the Sun and demonstrated the validity of stellar models that were developed long before any of this information became available.

13.2 The Herzsprung–Russell Diagram

Fig. 13.1 The distances to nearby stars have been determined by measuring the small shift in their positions as the Earth moves around the Sun.

Distances to nearby stars may be determined by measuring the small shift in their positions in the sky as the Earth moves around the Sun, as shown in Figure 13.1. Between 1989 and 1993 the European Space Agency's satellite Hipparcos determined the distances to almost 120,000 of our closest stellar neighbours. This precision astrometry forms the first step in the distance ladder that ultimately enables astronomers to deduce the distances to objects as remote as the most distant galaxies. It also underpins our understanding of the astrophysics of stars as it offers a wide sample of star types at known distances whose intrinsic luminosity can be determined. Figure 13.2 shows the results of a series of Hipparcos measurements of the position of a star.

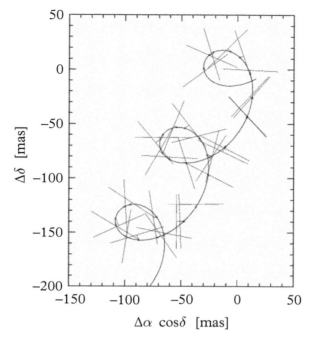

Fig. 13.2 The path on the sky of one of the Hipparcos Catalogue objects, over a period of three years. Each straight line indicates the observed position of the star at a particular epoch. The curve is the modelled stellar path fitted to all the measurements. The inferred position at each epoch is indicated by a dot. The amplitude of the oscillatory motion gives the star's parallax, with the linear component representing the star's proper motion. δ and α are angular celestial coordinates in units of milli-arcseconds (mas).

The observed or apparent luminosity of a star is the rate at which light energy is received at the Earth's surface, per unit area orthogonal to the line of sight. Radiation intensity falls off with the inverse square of distance, so if we know the distance d to a star, we can calculate its intrinsic or absolute luminosity L from its observed luminosity I, using the formula

$$I = \frac{L}{4\pi d^2}. \tag{13.5}$$

A second fundamental feature of a star is its surface temperature T_{surf}, which is obtained through Wien's law of black body radiation (10.109), $\hbar\omega_{\mathrm{peak}} = 2.8214\,T_{\mathrm{surf}}$. This converts to

$$T_{\mathrm{surf}} = \frac{2.898 \times 10^6}{\lambda_{\mathrm{peak}}}, \tag{13.6}$$

with T_{surf} in kelvin (K) and λ_{peak} the wavelength in nanometres (nm) corresponding to the peak intensity of the emitted radiation. For instance, the peak of the Sun's radiation is in the green part of the spectrum with a wavelength around 500 nm. This corresponds to a surface temperature of $T_{\odot\,\mathrm{surf}} \simeq 5800\,\mathrm{K}$. The peak wavelength of the radiation emitted by many stars in the night sky is between red and blue, so the Sun is rather typical.

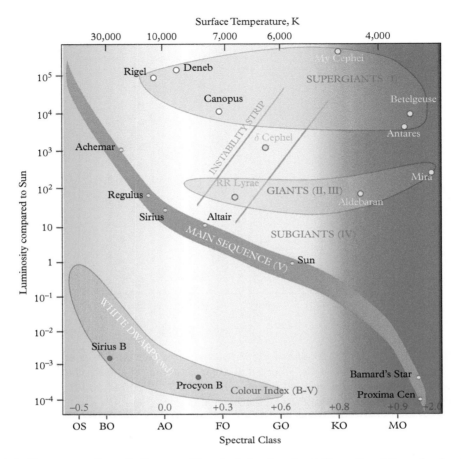

Fig. 13.3 Herzsprung–Russell diagram. The star's luminosity L (in units of the solar luminosity $L_\odot$) is plotted using a logarithmic scale against the logarithm of its surface temperature T_{surf}, as deduced from the peak wavelength of the radiation it emits.

These two characteristics of stars, L and T_{surf}, are the key to many of their other properties. Since around 1910, they have been plotted on diagrams known as *Herzsprung–Russell (HR) diagrams* after Ejnar Herzsprung and Henry Norris Russell, as shown in Figure 13.3. The vertical scale is the logarithm of intrinsic luminosity L, and the horizontal scale is the logarithm of the surface temperature T_{surf}. It is a quirk of HR diagrams that by convention temperature increases to the left along the horizontal axis. One of the challenges for the astrophysicist is to explain the patterns seen in the HR diagram.

Not all regions of the HR diagram are equally populated. The vast majority of stars occupy a diagonal band crossing the diagram from bottom right to top left. At the bottom right are cool faint stars and at the top left are hot bright stars. This band is known as the *main sequence*. The Sun is on the main sequence, as are most of our stellar neighbours such as Sirius and Vega. The reason that we find so many stars on the main sequence is that this is where a star spends the majority of its lifetime.

The HR diagram also contains a smattering of stars outside the main sequence. There are some very bright, but relatively cool stars at the top right. These stars have swelled to enormous dimensions, which gives them their great luminosity, but their outer layers are cool. They are known as *red giants*. Above them are the even larger and more luminous *supergiants*. Giants and supergiants are rare, but they are over-represented in our night sky because they are so bright. Stars such as Aldebaran, Antares and Betelgeuse reside in this region of the HR diagram. The enormous size of the supergiants was confirmed as long ago as 1920 by Albert Michelson and Francis Pease, who constructed an interferometer at the Mount Wilson Observatory in California and measured the diameter of Betelgeuse. More recent measurements suggest its diameter is about 1000 times that of the Sun, but this is not precise because Betelgeuse varies in size and shape and it does not have a sharp edge.

Towards the bottom left of the HR diagram there is a band of hot but very faint stars. These are the *white dwarfs*. A white dwarf is the highly compressed core of a star that has exhausted its nuclear fuel. It cannot generate energy, so it gradually cools as it radiates heat into space. No white dwarfs are visible to the naked eye. The closest example is Sirius B, the binary companion of Sirius whose orbit was discussed in section 2.10.1.

13.3   The Birth of Stars

Stars form through the gravitational collapse of gas clouds composed mainly of hydrogen and helium. As a cloud collapses gravitational energy is released, the cloud's temperature rises and the resulting thermal pressure resists further collapse. Some of the released gravitational energy must be radiated into space for continued collapse to occur. But the *protostar*, as it is known, is opaque, so it takes considerable time for radiation to diffuse to its surface. Consequently this collapse phase may last for ten million years or more. Eventually, the central region of the protostar reaches sufficiently high temperatures for nuclear fusion reactions to begin, and the thermal pressure that is generated halts further collapse. The protostar has become a main sequence star, a phase that may last for billions of years.

13.3.1   Stellar composition

The chemical composition of the stars was first determined in the 1920s by Cecilia Payne. With the aid of spectroscopy, she discovered that stars are composed almost entirely of hydrogen and helium, which was completely unexpected at the time. We now know that stars condense from gas clouds that are around three-quarters hydrogen (^{1}H) and one-quarter helium-4 (^{4}He) by mass. Almost all this helium was produced in the *primordial nucleosynthesis* immediately following the Big Bang. The density of primordial matter was insufficient for fusion reactions to create any heavier elements except traces of a few other light isotopes, such as deuterium (^{2}H), helium-3 (^{3}He) and lithium-7 (^{7}Li). In all stars other than the very first generation that formed shortly after the Big Bang, there is also a small admixture of heavier elements synthesized in previous generations of stars. This amounts to 1.69% by mass of the material that formed the Sun, a small but significant fraction.

The temperature within a star or protostar is far too high for atoms of hydrogen and helium to exist, so stars consist of dissociated electrons, protons and helium nuclei plus a small proportion of heavier ions. High temperature ionized matter such as this is known as a *plasma*. It can be treated as an ideal gas formed of electrons and ions. The appropriate equation of state is therefore the ideal gas law (10.65). Both the electrons and the ions contribute to the gas pressure P and for a non-interacting gas the partial pressures are additive,[1] so

$$P = \frac{(N_e + \sum_i N_i)}{V} T = \left(n_e + \sum_i n_i\right) T \tag{13.7}$$

where N_e and N_i are the numbers of electrons and various species of ions in a volume V of the star, and $n_e = \frac{N_e}{V}$ and $n_i = \frac{N_i}{V}$ are the electron and ion number densities. It is more convenient to express P in terms of the mass density $\rho = n_e m_e + \sum_i n_i m_i$. The electron mass m_e is negligible compared to the ion masses m_i, so to a very good approximation $\rho = \sum_i n_i m_i$. Therefore

$$P = \frac{n_e + \sum_i n_i}{\sum_i n_i m_i}\left(\sum_i n_i m_i\right) T = \frac{1}{\mu m_p}\rho T\,, \tag{13.8}$$

where

$$\mu m_p = \frac{\sum_i n_i m_i}{n_e + \sum_i n_i} \tag{13.9}$$

is the average atomic mass of the particles in the plasma, written as a multiple of the proton mass.

Clearly, μ depends on the composition of the plasma. If the plasma is formed of hydrogen, then it consists of equal numbers of protons and electrons, so $\mu = \frac{1}{2}$. If the plasma is pure ^{4}He then there are two electrons for every nucleus and the mass of each nucleus is to a good approximation $4m_p$, so $\mu = \frac{4}{3}$. Therefore, as hydrogen is converted into helium in the core of a star, μ increases. When the Sun formed, it was around three-quarters hydrogen ions and one-quarter helium ions by mass, with one electron for each hydrogen ion and two for each helium ion. The ratio $\frac{3}{4}$ to $\frac{1}{4}$ by mass corresponds to a ratio of $\frac{3}{4}$ to $\frac{1}{16}$ by number, so originally

$$\mu_{\text{primordial}} \simeq \frac{(\frac{3}{4} \times 1 + \frac{1}{16} \times 4)}{(\frac{3}{4} + 2 \times \frac{1}{16} + \frac{3}{4} + \frac{1}{16})} = \frac{16}{27} \simeq 0.59\,. \tag{13.10}$$

It is estimated that for the current composition of the Sun's core, $\mu_\odot \simeq 0.62$.

13.3.2 The virial theorem

We will model a typical star as a perfectly spherical ball of gas of a uniform composition that is in thermal equilibrium. Stars evolve slowly and can be treated as quasi-static. This is reasonable; we know that the Sun has been very stable for billions of years. Even a relatively small change in luminosity during this vast period of time would have extinguished life on Earth. So we assume the star is not pulsating and also ignore its rotation.

Let M be the total mass of the model star and R its radius. The mass is not evenly distributed throughout the star, but is concentrated towards its centre. We define a radial mass function $m(r)$ as the amount of mass within distance r of the centre. $m(r)$ is in the range $0 \leq m(r) \leq M$, and $m(0) = 0$, $m(R) = M$. $m(r)$ is related to the density $\rho(r)$ by

$$\frac{dm}{dr} = 4\pi r^2 \rho\,. \tag{13.11}$$

The assumption of spherical symmetry implies that the pressure, temperature and outward energy flux, $P(r)$, $T(r)$ and $F(r)$, have no angular dependence. $F(R)$, the outward energy

---

[1] This is known as *Dalton's law of partial pressures.*

flux at the surface, is the luminosity L. (For brevity we will drop the explicit dependence of these variables on r in much of what follows.)

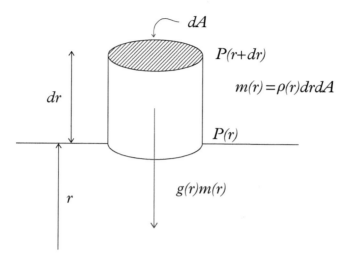

Fig. 13.4 Hydrostatic balance.

Throughout the galaxy there are gas clouds, and these are the nurseries where new stars form. The collision of two gas clouds or the shock wave from a supernova explosion may cause regions of gas to collapse under gravity. As the gas contracts, gravitational energy is released which heats the gas and as the temperature rises the thermal radiation emitted into space increases.

There is a very simple relationship known as the *virial theorem* that connects the gravitational energy of a star to its thermal or internal energy. In a stable star the tendency to collapse under gravity is balanced by the thermal pressure exerted within the plasma. Consider a small volume of plasma $dV = dr\,dA$, situated between r and $r + dr$ within the star, with mass $\rho(r)\,dr\,dA$, as shown in Figure 13.4. The downward force on this mass due to gravity is $g(r)\rho(r)\,dr\,dA$, where $g(r) = \frac{Gm(r)}{r^2}$. The upward force due to the thermal pressure of the hot plasma is $-\frac{dP}{dr}\,dr\,dA$, where $-\frac{dP}{dr}\,dr$ is the difference in pressure between the bottom and top surfaces. Equating these forces gives us the *equation of hydrostatic balance*

$$\frac{dP}{dr} = -\frac{Gm\rho}{r^2}, \tag{13.12}$$

which can be assumed to hold throughout the star.

The volume of the star within radius r is $V(r) = \frac{4\pi}{3}r^3$. Multiplying equation (13.12) on the left by $V(r)\,dr$ and on the right by $\frac{4\pi}{3}r^3\,dr$ gives

$$V\,dP = -\frac{4\pi G}{3}m\rho r\,dr. \tag{13.13}$$

The mass in a spherical shell is $dm = 4\pi r^2\rho\,dr$, so eliminating dr in favour of dm gives

$$3V\,dP = -\frac{Gm}{r}\,dm. \tag{13.14}$$

Integrating from the centre to the surface of the star we obtain

$$3 \int_{P_{\mathrm{cen}}}^{P_{\mathrm{surf}}} V \, dP = - \int_0^M \frac{Gm}{r} \, dm \,, \tag{13.15}$$

where P_{cen} and P_{surf} are the pressures at the centre and surface of the star, respectively. Integration by parts of the left-hand side gives

$$\begin{aligned}
\int_{P_{\mathrm{cen}}}^{P_{\mathrm{surf}}} V \, dP &= \left[PV \right]_{\mathrm{cen}}^{\mathrm{surf}} - \int_0^{V_{\mathrm{surf}}} P \, dV \\
&= - \int_0^{V_{\mathrm{surf}}} P \, dV \,, \\
&= - \int_0^M \frac{P}{\rho} \, dm \,, \tag{13.16}
\end{aligned}$$

where we have used the simple observations that the volume is zero at the centre of the star and the pressure is zero at the surface, and in the last step substituted $dV = \frac{1}{\rho} \, dm$. Using this result in equation (13.15) gives us the *virial theorem*

$$3 \int_0^M \frac{P}{\rho} \, dm - \int_0^M \frac{Gm}{r} \, dm = 0 \,. \tag{13.17}$$

As $-\frac{Gm}{r}$ is the gravitational potential at radius r, the second integral is the total gravitational potential energy of the star, Ω. This is a negative quantity. $|\Omega|$ is the amount of energy released by gravitationally binding the mass of the star. (Alternatively, it is the amount of energy required to remove all the particles forming the star far from their mutual gravitational attraction.)

If we assume the ideal gas law is valid within the star, then the total thermal energy of the particles of the plasma, per unit volume, is $\frac{3}{2} \left(n_e + \sum_i n_i \right) T$ in accordance with equation (10.54). The mass per unit volume is $\sum_i n_i m_i$, so the thermal energy per unit mass is

$$u = \frac{3}{2} \frac{n_e + \sum_i n_i}{\sum_i n_i m_i} T \,, \tag{13.18}$$

where n_e and n_i are the number densities of electrons and ions. The ideal gas law as given by the combination of equations (13.8) and (13.9) is

$$\frac{P}{\rho} = \frac{n_e + \sum_i n_i}{\sum_i n_i m_i} T \,, \tag{13.19}$$

so

$$\frac{P}{\rho} = \frac{2}{3} u \,, \tag{13.20}$$

a simple relation that is valid throughout the star. Integrating from the centre to the surface of the star, we find

$$3 \int_0^M \frac{P}{\rho} \, dm = 2 \int_0^M u \, dm = 2U \,, \tag{13.21}$$

where U is the total thermal energy of the star. The virial theorem (13.17) now reduces to

$$2U + \Omega = 0, \quad \text{or} \quad U = \frac{1}{2}|\Omega|. \tag{13.22}$$

It relates the thermal energy of the star to its gravitational energy.

We have arrived at this result, assuming that the ideal gas law holds. This is a good approximation if the plasma particles are small compared to the inter-particle spacing and if the particles can be treated as free, so their energy consists of kinetic energy alone and there is no electromagnetic potential energy due to interactions between the particles. In a plasma, the largest particles are nuclei with dimensions that are much smaller than an atom, so these conditions hold until pressures reach extremely high values. For most of the life of a star we can assume the ideal gas law describes the plasma very well. (Note that white dwarf stars are very dense and are supported by electron degeneracy pressure, where the fermionic nature of the electrons matters, so the ideal gas law does not hold and the following discussion will not apply.)

In the next section we will take a look at the implications of the virial theorem for star formation. We are familiar with everyday solid objects that cool down without any noticeable effect on their physical structure. Stars obey the ideal gas law and this means that they behave quite differently.

13.3.3 Star formation

As a gas cloud or protostar contracts under its own gravity, its gravitational binding energy Ω becomes more negative, and the virial theorem implies that the thermal energy of the protostar U must inevitably increase. This means that the temperature of the protostar will rise, leading to an increased emission of radiation. The protostar will therefore lose energy and, as it is an ideal gas, the pressure within the protostar will fall, leading to further contraction and a further release of gravitational binding energy raising the thermal energy again. So as a protostar loses energy it heats up. This is a general feature of gravitating systems. It is sometimes described as a negative heat capacity.

The total energy must be conserved, so equation (13.22) implies that half the gravitational binding energy released as the star forms heats the star and becomes the thermal energy of the star, while the other half of the binding energy is radiated into space. In fact, the gas cloud cannot contract to form a star unless it is able to lose half of the binding energy in this way, but, unlike a hot stone (or a white dwarf), an ideal gas cannot radiate and cool without a significant decrease of pressure. Fortunately, it is quite difficult for protostars and stars to lose energy by radiating photons, otherwise they would rapidly collapse. As stars are composed of a plasma of charged particles, they are opaque, so photons cannot leave the star without undergoing countless interactions with electrons and ions.

Opacity is a measure of the distance that a photon can travel before interacting. The mean free path of a photon is $\bar{l} = \frac{1}{\kappa\rho}$, where κ is the opacity per unit mass. Opacity is low at both low temperatures and very high temperatures. At high temperatures, such as those found in the core of a star, most photons have very high energies and are not easily absorbed, and the main cause of opacity is the scattering of photons by free electrons. The central density of the Sun is around 10^5 kg m^{-3} and the opacity per unit mass is about 0.1 m^2 kg^{-1}, so the distance travelled by a photon before scattering off an electron is just $\bar{l} = \frac{1}{\kappa\rho} \simeq 10^{-4}$ m. For a given plasma composition, at the temperatures and pressures found in a stellar core, the opacity is constant to a first approximation. As we move outwards through the star, the opacity increases as the temperature falls, but crucially the opacity maintains the plasma and radiation in thermal equilibrium. At much lower temperatures, such as may occur in the outer envelope of a red giant star, atoms form and this dramatically reduces the opacity, as most photons have insufficient energy to ionize an atom and there are few ions and free electrons available to scatter the photons.

For a contracting protostar to reach a stable density and stop contracting, a power source within the protostar must be triggered. When stars were first modelled, this power source was a mystery. We now know that the protostar contracts until its core reaches a sufficiently high temperature for nuclear fusion to begin. The energy that is released provides the thermal pressure that prevents further contraction. The protostar has now become a *star*. Half its gravitational binding energy has been radiated into space and the other half forms the initial thermal energy of the star. As long as the nuclear fuel continues to burn, the star remains stable and the gravitational binding energy of the star remains fixed. Similarly, the thermal energy of the star and its temperature profile remain fixed. This means that as long as the hydrostatic balance is maintained, the rate at which energy is radiated from the surface of the star must equal the rate at which energy is generated by fusion in the core of the star.

So how long does it take for a star to form? It is the length of time taken for the star to radiate half its gravitational binding energy away. This is known as the *thermal timescale* of the star. It can be estimated as

$$\tau_{\text{th}} \simeq \frac{|\Omega|}{2L} \simeq \frac{GM^2}{RL}, \tag{13.23}$$

where L is the luminosity of the star and $\frac{GM^2}{R}$ is an estimate of the binding energy, accurate to within a numerical factor close to 1. The luminosity of the Sun is $L_\odot = 3.846 \times 10^{26}$ W. If we put the mass $M_\odot$ and radius $R_\odot$ of the Sun into equation (13.23) we find that $\tau_{\odot\,\text{th}} \simeq 1.6 \times 10^7$ years. This is a rough estimate of the time that it would take for a gas cloud to contract and form a star with the mass of the Sun. Physicists of the 19th century attempted to determine the age of the Sun in this way, in the mistaken belief that its luminosity was solely due to the energy released by its gravitational contraction. On this basis, Kelvin and Helmholtz estimated the total lifespan of the Sun to be not much more than the thermal timescale, but this was in conflict with the age of the Earth as deduced by geologists and biologists. The physicists proved to be wrong and the resolution came with the discovery of nuclear fusion energy.

The thermal timescale also represents the time taken for energy generated within the core of the Sun to diffuse to the surface. If a photon created at the centre of the Sun did not interact before exiting the Sun, it would reach the surface in a couple of seconds, but within the Sun photons are constantly being scattered, absorbed and emitted by the electrons and other charged particles in the plasma before escaping into space.

The Sun also loses energy through its neutrino (ν) flux, which has a luminosity

$$L_{\nu\odot} = 0.023 L_\odot \tag{13.24}$$

as measured by neutrino detectors such as Super-Kamiokande in Japan. The mean free path of a neutrino is vastly greater than the radius of a star, so the neutrino radiation represents an instant loss of energy from the core. Neutrino emission means that the thermal energy yield of the fusion reactions is lower than would otherwise be the case, as the neutrino energy is not trapped within the star and cannot contribute to the thermal pressure supporting the star. In order to maintain hydrostatic balance the temperature and pressure within the core must be higher than they would otherwise be, requiring that the nuclear fuel burns faster and releases more energy. This is a relatively small effect in the Sun, but the loss of energy due to the emission of neutrinos greatly increases the rate of fuel burning in the later stages of very massive stars thereby shortening these stages dramatically.

13.4 Stellar Structure

Modelling stars is a complex problem involving thermodynamics, hydrodynamics and nuclear physics. There are now very good computer models that accurately describe the structure and evolution of all varieties of stars. Computer-based calculations play a very important role but do not always offer much insight into the underlying physics. Fortunately, many of the essential details of stellar structure can be gleaned by studying simplified models, so this is where we will concentrate our attention. This insight can then be sharpened by taking into account the precise results obtained numerically.

We will assume that stars are static and in thermal equilibrium with a source of energy within their core and that they have a uniform composition, but not uniform density. This means neglecting any evolution of the star with time, such as the depletion of its nuclear fuel. These assumptions are appropriate for young stars that have recently condensed out of a gas cloud and initiated fusion reactions within their core. As we will see, they provide a starting point for understanding stars in the main sequence. Much can be deduced about stars without any knowledge of the power source within their cores. Indeed, many of the fundamentals of stellar structure were worked out by Eddington before fusion energy was understood.

The vast majority of stars are found on the main sequence, which forms a diagonal line across the HR diagram. As $\log L$ is plotted against $\log T_{\mathrm{surf}}$ in the diagram, this implies a relationship between luminosity and surface temperature of the form

$$L \propto T_{\mathrm{surf}}^{a} , \tag{13.25}$$

where a is the slope of the line. In fact, the slope of the main sequence is greater for the brightest stars than for stars of average brightness. This can be explained by assuming that main sequence stars are powered by hydrogen fusion reactions in their core. We will deduce from this assumption luminosity–temperature relationships that describe the slope in both parts of the main sequence.

The rate of energy generation per unit mass at radius r, denoted by $q(r)$, may be expressed approximately as

$$q = q_0 \rho^b T^n , \tag{13.26}$$

where the powers b and n depend on the fusion process. Most fusion reactions involve the collisions of two particles. The rate of such reactions is proportional to the square of the density, ρ^2, so the rate of energy generation per unit mass q is proportional to ρ and therefore $b = 1$. For a three-particle process, the rate of collisions is proportional to ρ^3 and therefore q is proportional to ρ^2, so $b = 2$. The energy flux satisfies

$$\frac{dF}{dr} = 4\pi r^2 \rho q . \tag{13.27}$$

Within the core, where energy is being generated, $q > 0$ so the energy flux F through a spherical shell increases with r. Beyond the edge of the core $q = 0$, so F remains constant.

We also need an equation that determines how energy is transmitted through the star. Photons are continually scattered, absorbed and emitted by the electrons and ions in the plasma, bringing the radiation and the plasma into thermal equilibrium. The radiation is therefore isotropic with a black body spectrum, and heat steadily diffuses to the surface only because there is a very gradual temperature gradient from the core to the surface of the star. In the case of the Sun the average gradient is just $\frac{T_{\odot \, \text{cen}}}{R_\odot} = \frac{1.6 \times 10^7}{7.0 \times 10^8}$ K m$^{-1} \simeq 0.023$ K m^{-1}. Eddington found a relationship between the temperature gradient and the energy flux by considering the rate at which a slab of material situated between radial distance r and $r + dr$ would absorb momentum. The energy flux per unit area is $\frac{F}{4\pi r^2}$, so the energy absorbed by a slab of material of unit area is $\frac{F\kappa\rho}{4\pi r^2} dr$, where κ is the opacity per unit mass.

For photons $p = E$, and therefore the momentum absorbed equals the energy absorbed. The momentum absorbed results in a radiation pressure gradient, so

$$\frac{F\kappa\rho}{4\pi r^2} = -\frac{dP_{\text{rad}}}{dr} \, . \tag{13.28}$$

In Chapter 10, we showed that black body radiation pressure is given by equation (10.115),

$$P_{\text{rad}} = \frac{4}{3}\sigma T^4 \, , \tag{13.29}$$

where $\sigma = \frac{\pi^2}{60\hbar^3}$ is the Stefan–Boltzmann constant. This implies

$$\frac{dP_{\text{rad}}}{dr} = \frac{16}{3}\sigma T^3 \frac{dT}{dr} \, . \tag{13.30}$$

Combining equations (13.28) and (13.30), we obtain Eddington's relationship

$$\frac{dT}{dr} = -\frac{3}{64\pi} \frac{\kappa\rho}{\sigma T^3 r^2} F \, . \tag{13.31}$$

The gradual diffusion of photons is the main mechanism for the transmission of energy through the star. The radiation pressure combined with the opacity of the plasma is responsible for maintaining the temperature gradient within the star. However, it is the thermal plasma pressure, which is largely due to electrons, that is responsible for supporting a star against gravitational collapse, and not the radiation pressure. Nevertheless, radiation pressure becomes increasingly important for stars of greater mass. Stars of significantly higher mass than the Sun gradually lose most of their outer envelope as radiation pressure propels particles from the envelope outwards into space. Furthermore, stars of extremely high mass are unstable due to the great intensity of the radiation that they generate. Radiation pressure produces an upper limit on the mass of a stable star that is thought to be around 120 $M_\odot$.

13.4.1 The structure functions

It is convenient to express all the stellar variables in terms of the radial mass function $m(r)$ rather than the radial position r using equation (13.11), which may be expressed as

$$\frac{dr}{dm} = \frac{1}{4\pi r^2 \rho} \,. \tag{13.32}$$

This can be used to transform the equation of hydrostatic balance (13.12) into

$$\frac{dP}{dm} = \frac{dP}{dr}\frac{dr}{dm} = -\frac{Gm}{4\pi r^4} \,. \tag{13.33}$$

Similarly, from equations (13.27) and (13.26) we obtain

$$\frac{dF}{dm} = \frac{dF}{dr}\frac{dr}{dm} = q_0 \rho^b T^n \,, \tag{13.34}$$

and from equation (13.31),

$$\frac{dT}{dm} = \frac{dT}{dr}\frac{dr}{dm} = -\frac{3}{16}\frac{\kappa F}{\sigma T^3 (4\pi r^2)^2} \,. \tag{13.35}$$

With the addition of the ideal gas law (13.8),

$$P = \frac{1}{\mu m_p}\rho T \,, \tag{13.36}$$

we have a set of five coupled nonlinear differential equations (13.32)–(13.36) for r, ρ, P, F and T.

We can deduce much about the structure of main sequence stars through dimensional analysis of these equations. If we define the fractional mass as

$$x(r) = \frac{m(r)}{M} \,, \tag{13.37}$$

this will help in the comparison of stars of different masses. We can then replace $r(m)$, $P(m)$, $\rho(m)$, $T(m)$ and $F(m)$ by dimensionless functions of x, as follows:

$$r = f_1(x)R_* \,, \quad P = f_2(x)P_* \,, \quad \rho = f_3(x)\rho_* \,, \quad T = f_4(x)T_* \,, \quad F = f_5(x)F_* \,. \tag{13.38}$$

For any particular star, R_*, P_*, ρ_*, T_* and F_* are dimensional constants. They vary from star to star depending on the total mass M, and are called the stellar variables. We will determine the relationship between these quantities and M in the following. $f_i(x)$ are dimensionless structure functions that encode the profile of the thermodynamic variables from the centre to the surface of the star, as x runs from 0 to 1. The structure functions are plotted in Figure 13.5. The equations need only be solved for a standard star such as the Sun and the same structure functions will apply to all other stars that satisfy our assumptions, simply by scaling according to the mass of the star. Stars described by this simple model are known as homologous. The model works well for main sequence stars.

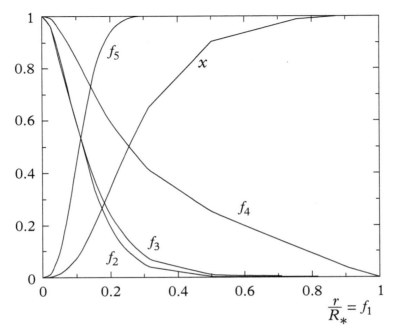

Fig. 13.5 The profiles of the thermodynamic variables from the centre to the surface of a star. $x = \frac{m}{M_*}$, $f_2 = \frac{P}{P_*}$, $f_3 = \frac{\rho}{\rho_*}$, $f_4 = \frac{T}{T_*}$ and $f_5 = \frac{F}{F_*}$ are plotted as functions of $f_1 = \frac{r}{R_*}$.

For instance, the temperature at the centre of a star is $T_{\text{cen}} = f_4(0)T_*$ and the temperature at the surface is $T_{\text{surf}} = f_4(1)T_*$. Given the surface temperature from observation, we can use f_4 to work out the temperature at any other point in the star. We also know how the surface temperature scales with the mass of the star, so from observations of the surface temperature of a main sequence star we can work out its mass, as we will see shortly.

A great deal of information can be extracted from these equations even without explicitly finding the structure functions. First, we can separate the variables in the structure equations. For example, the left-hand side of equation (13.33) can be rewritten as

$$\frac{dP}{dm} = \frac{dP}{dx}\frac{dx}{dm} = \frac{df_2}{dx}\frac{P_*}{M}, \tag{13.39}$$

where we have used the second of equations (13.38) and equation (13.37). When combined with the right-hand side, and substituting $m = Mx$ and $r = f_1 R_*$, this gives

$$\frac{df_2}{dx}\frac{P_*}{M} = -\frac{GMx}{4\pi f_1^4 R_*^4}. \tag{13.40}$$

We can separate this into two parts. One is a relationship between the universal structure functions,

$$\frac{df_2}{dx} = -\frac{x}{4\pi f_1^4}, \tag{13.41}$$

and the other relates the stellar variables,

$$P_* = \frac{GM^2}{R_*^4}. \tag{13.42}$$

Any constant of proportionality between these two sub-equations can be absorbed into the structure functions.

Proceeding in the same manner for the other structure equations gives

$$\frac{df_1}{dx} = \frac{1}{4\pi f_1^2 f_3} \quad , \qquad\qquad \rho_* = \frac{M}{R_*^3}, \tag{13.43}$$

$$f_2 = f_3 f_4 \quad , \qquad\qquad T_* = \frac{\mu m_p P_*}{\rho_*}, \tag{13.44}$$

$$\frac{df_4}{dx} = -\frac{3 f_5}{16 f_4^3 (4\pi f_1^2)^2} \quad , \qquad\qquad F_* = \frac{\sigma}{\kappa} \frac{T_*^4 R_*^4}{M}, \tag{13.45}$$

$$\frac{df_5}{dx} = f_3^b f_4^n \quad , \qquad\qquad F_* = q_0 \rho_*^b T_*^n M. \tag{13.46}$$

Equation (13.41) and the equations on the left in (13.43)–(13.46) form a closed set of differential equations for the structure functions f_i, so a unique set of solutions can be found numerically.

13.4.2 The mass–luminosity relationship

We can now deduce various simple relationships between the stellar variables. The key feature of a star is its mass M. From this almost everything else about the star follows. Substituting for P_* and ρ_* from equations (13.42) and (13.43) into equation (13.44) gives

$$T_* = \mu m_p \left(\frac{GM^2}{R_*^4} \right) \left(\frac{R_*^3}{M} \right) = G\mu m_p \frac{M}{R_*}. \tag{13.47}$$

We can use this relation to substitute for $T_* R_*$ in equation (13.45), obtaining

$$F_* = \frac{\sigma}{\kappa} (G\mu m_p)^4 M^3. \tag{13.48}$$

Therefore, the energy flux F is proportional to M^3, and as $L = F(1)$, this gives us an important result relating stellar luminosity to mass, $L \propto M^3$. For example, a main sequence star of 10 solar masses has 1000 times the luminosity of the Sun. The amount of nuclear fuel available to a star is proportional to its mass, so this immediately translates into an estimate of the main sequence lifetime of stars,

$$\tau_{\mathrm{MS}} \propto \frac{M}{L} \propto \frac{1}{M^2}. \tag{13.49}$$

This relationship is readily understood. Stars of greater mass have higher temperatures in their core, so the nuclear reactions proceed faster. They burn their nuclear fuel at an increased rate and race through their life much quicker than lower-mass stars, and this is one reason why they are rare. Later we will estimate the Sun's lifetime as a main sequence star to be around 10^{10} years. We can expect a star of 10 solar masses to have a main sequence lifetime just one-hundredth of this, around 10^8 years.

Also from equation (13.48) we see that the luminosity of a main sequence star is proportional to μ^4. μ is a function of the composition of the star, as defined in section 13.3.1. It increases as nuclear fusion proceeds, which means that the luminosity of a star increases as it burns its nuclear fuel. The luminosity of the Sun is believed to be about 30% greater now than when it formed around 4.6×10^9 years ago.

13.4.3 The density–temperature relationship

If we cube equation (13.42) and substitute for R_*^3 from equation (13.43) we obtain

$$P_*^3 = \frac{G^3 M^6}{R_*^{12}} = G^3 M^2 \rho_*^4 . \tag{13.50}$$

We can now substitute for P_* from equation (13.44) to give

$$\frac{\rho_*^3 T_*^3}{\mu^3 m_p^3} = G^3 M^2 \rho_*^4 . \tag{13.51}$$

Dividing by ρ_*^3 and rearranging produces

$$\rho_* = \frac{1}{(G\mu m_p)^3} \frac{T_*^3}{M^2} , \tag{13.52}$$

an M-dependent relationship between ρ_* and T_*. Using the structure functions, we obtain a similar relationship between the density and temperature that holds at any point within the star. It indicates that the core density of a more massive star is lower for a given temperature.

When a star has exhausted its nuclear fuel, its core contracts and it may reach a density where it is supported against collapse by electron degeneracy pressure. However, this only occurs at extremely high densities. Relationship (13.52) implies that more massive stars will need to attain higher temperatures before reaching the density where electron degeneracy pressure becomes important. As nuclear fusion reactions are very temperature dependent, this means that higher-mass stars may undergo several rounds of nuclear fusion that are not accessible to lower-mass stars. We will now take a closer look at nuclear fusion reactions and nucleosynthesis in stars.

13.5 Nucleosynthesis

Eddington first suggested in 1920 that the fusion of hydrogen nuclei into helium nuclei might provide the energy that keeps the Sun and other stars shining. The nucleus of a hydrogen atom is a single proton whereas the nucleus of a helium atom consists of two protons and two neutrons. Eddington realized that if a helium nucleus could be forged from four protons, then around 26 MeV would be released. This is the difference between the mass of four protons, which is 4×938.3 MeV $= 3753$ MeV, and the mass of the helium nucleus, which is 3727 MeV. Thus, about 0.7% of the mass of the protons would be converted into energy.

The strong nuclear force is very short range. In order for fusion reactions to occur, nuclei must approach each other to within about a femtometre (10^{-15} m). However, as nuclei are all positively charged, there is a large Coulomb barrier for them to overcome. In the early

20th century many physicists believed that the temperature of 1.6×10^7 K at the centre of the Sun was not hot enough for fusion reactions to occur. However, as we saw in Chapter 11, there are two factors that enable fusion reactions to proceed at these lower temperatures. One is that the Maxwell distribution of thermal kinetic energies has a long tail, so there are always a small number of nuclei with much greater energy than average. The second is that quantum tunnelling enables nuclei to pass through the Coulomb barrier even if they have insufficient energy to reach the top of the barrier. The long tail means that the fusion of hydrogen into helium is more of a slow fizzle than a bang, but this is sufficient to maintain the thermal pressure and support the star against gravitational collapse. As thermal energy cannot easily escape the star, a slow fizzle is enough. It is an amusing fact that the rate of energy generation per unit mass in a human body is greater than that in the Sun.

From the luminosity of the Sun, we can calculate that the Sun's total mass of hydrogen, $M_{\text{H}\odot}$, is being consumed by fusion reactions at a rate

$$\left| \frac{dM_{\text{H}\odot}}{dt} \right| = \frac{L_\odot + L_{\nu\odot}}{0.007c^2} = \frac{1.023 \times 3.846 \times 10^{26}}{0.007 \times 9 \times 10^{16}} \text{ kg s}^{-1} = 6.25 \times 10^{11} \text{ kg s}^{-1}, \qquad (13.53)$$

where we have included the neutrino flux given in equation (13.24) and used the fact that in fusing hydrogen to helium, about 0.7% of the mass is released as energy. The factor c^2 is needed to convert the rate of energy emission in watts to mass consumption in kg s^{-1}. Although the Sun loses 6.25×10^{11} kg of hydrogen per second, this does not put much of a dent in its total mass, which is $M_\odot = 2 \times 10^{30}$ kg. Hydrogen burning only occurs in the core of the Sun, so most of its hydrogen will never be burnt. Assuming that the luminosity of the Sun is constant and that during its main sequence lifetime around 15% of its hydrogen will be converted into helium, we can estimate the time that the Sun will spend on the main sequence as

$$\tau_\odot = 0.15 \times 0.75 \times \frac{M_\odot}{\left| \frac{dM_{\text{H}\odot}}{dt} \right|} = \frac{2.25 \times 10^{29}}{6.25 \times 10^{11}} = 3.6 \times 10^{17}\text{s} = 1.1 \times 10^{10} \text{ years}, \qquad (13.54)$$

where we have assumed that initially the Sun was 75% hydrogen by mass. Detailed models give a figure closer to 1.0×10^{10} years for the Sun's main sequence lifetime, so we are approaching the half-way point.

13.5.1   The proton–proton chain

The process by which hydrogen is fused into helium in low mass stars is known as the proton–proton chain. This mechanism was worked out by Hans Bethe and Charles Critchfield in 1938. Crucially, although the strong force will bind a single proton and a single neutron into a deuterium nucleus, it is not quite strong enough to form a nucleus from just two protons or just two neutrons. This means that in the first step towards a helium nucleus, two protons must collide and tunnel through the Coulomb barrier, and at the exact moment of impact one of the protons must undergo inverse beta decay. This proton is thereby converted into a neutron, with the emission of a positron and a neutrino. The other proton and the newly formed neutron then bind together to form a deuterium nucleus,

$$^1\text{H} + {}^1\text{H} \rightarrow {}^2\text{H} + e^+ + \nu_e. \qquad (13.55)$$

The positron e^+ quickly annihilates with an electron in the plasma to produce photons, and the electron neutrino ν_e escapes. The extreme weakness of the weak force, which is responsible for beta decay, makes this critical first step exceedingly slow. Just one collision in 10^{22} between two protons in a star such as the Sun will result in the production of a deuterium nucleus. Typically, a proton will ricochet off other protons for about ten billion years before it undergoes this reaction, so this is the bottleneck that determines the overall rate for the conversion of hydrogen into helium.

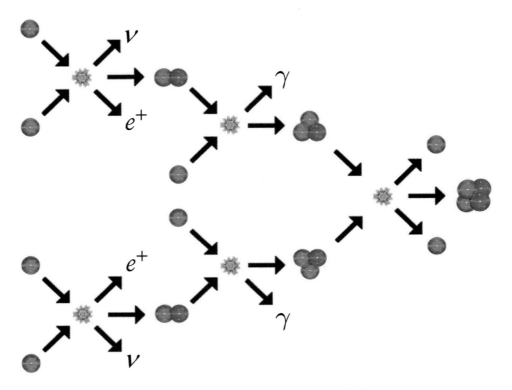

Fig. 13.6 The proton–proton chain (*pp*I).

The next step happens almost immediately. Within one second, the deuterium nucleus captures another proton to form a nucleus of helium-3, with the binding energy emitted as a photon,

$$^{2}\text{H} + {}^{1}\text{H} \rightarrow {}^{3}\text{He} + \gamma. \tag{13.56}$$

On average it takes another million years for the helium-3 nucleus to meet another helium-3 nucleus in the core of the Sun and undergo the reaction

$$^{3}\text{He} + {}^{3}\text{He} \rightarrow {}^{4}\text{He} + {}^{1}\text{H} + {}^{1}\text{H} \tag{13.57}$$

that results in the production of helium-4, with two protons being released back into the plasma. Overall, the result of these reactions is the conversion of four protons into a helium-4 nucleus (see Figure 13.6). (Electrical neutrality is maintained because two electrons annihilate with the two positrons emitted in the first step.) The 26 MeV of released energy is mostly in the form of photons but partly carried away by two neutrinos.

This process is often known as *pp*I. Fusion also proceeds by the following alternative route known as *pp*II, in which the ^{3}He nucleus fuses with a ^{4}He nucleus:

$$\begin{aligned}
^3\text{He} + {}^4\text{He} &\rightarrow {}^7\text{Be} + \gamma \\
^7\text{Be} + e^- &\rightarrow {}^7\text{Li} + \nu_e \\
^7\text{Li} + {}^1\text{H} &\rightarrow {}^4\text{He} + {}^4\text{He}\,.
\end{aligned} \tag{13.58}$$

The proton–proton chain processes in the Sun are 86% *pp*I reactions and 14% *pp*II reactions.

The rate at which energy is released by these proton–proton chain processes is determined by the first step (13.55), which involves an encounter between two protons. It is therefore proportional to the square of the density ρ^2, so the rate of energy generation per unit mass q_{pp} is proportional to ρ. In section 11.6.1 we estimated the temperature dependence of the rate of proton–proton fusion and obtained the exponent $n = 3.8$ at temperatures in the region of 1.6×10^7 K, as found in the core of the Sun. To simplify some of the formulae that follow, we will round this up to $n = 4$ and approximate the rate of energy generation by

$$q_{pp} \propto \rho T^4\,. \tag{13.59}$$

13.5.2 The CNO cycle

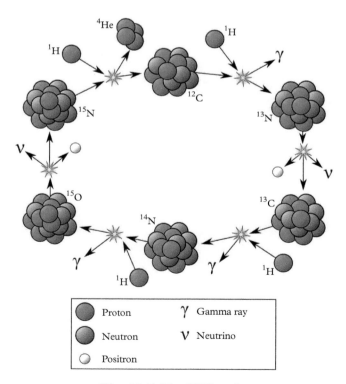

Fig. 13.7 The CNO cycle.

There is another process by which the Sun converts hydrogen to helium, accounting for around 5% of its energy production. This is known as the CNO or CNOF cycle, because it is catalysed by nuclei of carbon, nitrogen, oxygen and fluorine. The greater charge on the catalysing nuclei means that there is a bigger Coulomb barrier to overcome and therefore higher temperatures are required than for the proton–proton chain. The CNO cycle becomes the dominant hydrogen fusion process when core temperatures exceed about 2×10^7 K, as found in main sequence stars of mass more than 1.4 $M_\odot$. The six steps in the cycle can be represented as

$$
\begin{aligned}
^{12}\mathrm{C} + {}^1\mathrm{H} &\rightarrow {}^{13}\mathrm{N} + \gamma \\
^{13}\mathrm{N} &\rightarrow {}^{13}\mathrm{C} + e^+ + \nu_e \\
^{13}\mathrm{C} + {}^1\mathrm{H} &\rightarrow {}^{14}\mathrm{N} + \gamma \\
^{14}\mathrm{N} + {}^1\mathrm{H} &\rightarrow {}^{15}\mathrm{O} + \gamma \\
^{15}\mathrm{O} &\rightarrow {}^{15}\mathrm{N} + e^+ + \nu_e \\
^{15}\mathrm{N} + {}^1\mathrm{H} &\rightarrow {}^{16}\mathrm{O}^* \rightarrow {}^{12}\mathrm{C} + {}^4\mathrm{He},
\end{aligned}
\tag{13.60}
$$

as depicted in Figure 13.7. In the final step, a proton fuses with a nitrogen-15 nucleus to form an oxygen-16 nucleus in an excited state, denoted by $^{16}\mathrm{O}^*$, which almost immediately dissociates into a carbon nucleus and a helium nucleus.

There is a second possibility, the CNOF cycle, in which the excited oxygen nucleus emits a gamma ray photon and falls to a more stable lower energy state, as shown in the third step below. In this case, starting with $^{14}\mathrm{N}$ the cycle takes the following form:

$$
\begin{aligned}
^{14}\mathrm{N} + {}^1\mathrm{H} &\rightarrow {}^{15}\mathrm{O} + \gamma \\
^{15}\mathrm{O} &\rightarrow {}^{15}\mathrm{N} + e^+ + \nu_e \\
^{15}\mathrm{N} + {}^1\mathrm{H} &\rightarrow {}^{16}\mathrm{O}^* \rightarrow {}^{16}\mathrm{O} + \gamma \\
^{16}\mathrm{O} + {}^1\mathrm{H} &\rightarrow {}^{17}\mathrm{F} + \gamma \\
^{17}\mathrm{F} &\rightarrow {}^{17}\mathrm{O} + e^+ + \nu_e \\
^{17}\mathrm{O} + {}^1\mathrm{H} &\rightarrow {}^{14}\mathrm{N} + {}^4\mathrm{He}.
\end{aligned}
\tag{13.61}
$$

The result of both these cycles is that four protons are converted into a helium-4 nucleus with the release of the 26 MeV binding energy, and the carbon, nitrogen, oxygen and fluorine nuclei are recycled. Two of the steps are controlled by the weak interaction and include the emission of a neutrino. These neutrinos leave the star without interacting further and take around 1 MeV with them, so about 25 MeV remains in the star for each turn of the cycle. The CNO cycle would not have been possible in the first generation of stars as there were no nuclei as heavy as carbon in the immediate aftermath of the Big Bang.

The energy generation rate q_{CNO} is very sensitive to temperature and is usually approximated by the power law

$$
q_{\mathrm{CNO}} \propto \rho T^n,
\tag{13.62}
$$

with n quoted as somewhere between 16 and 20. We will use the figure $n = 18$ that we calculated in section 11.6.1. The exact temperature dependence is much less important than the fact that once a critical temperature is reached the rate of energy generation increases extremely rapidly with temperature. This is also true of all subsequent stages of stellar fusion.

13.5.3   The mass–radius relationship

Assuming the energy generation of main sequence stars is due to the fusion of hydrogen we can derive a relationship between stellar radius and mass. Combining equations (13.46) and (13.48) gives

$$q_0 \rho_*^b T_*^n M = \frac{\sigma}{\kappa}(G\mu m_p)^4 M^3 \,. \tag{13.63}$$

For hydrogen fusion reactions $b = 1$, so

$$\rho_* T_*^n \propto M^2 \,, \tag{13.64}$$

and after substituting for T_* from equation (13.44) we find

$$\frac{P_*^n}{\rho_*^{n-1}} \propto M^2 \,. \tag{13.65}$$

Substituting for P_* and ρ_* from equations (13.42) and (13.43) then gives

$$\left(\frac{M^2}{R_*^4}\right)^n \left(\frac{M}{R_*^3}\right)^{1-n} \propto M^2 \,, \tag{13.66}$$

and after simplifying the indices,

$$R_* \propto M^{\frac{n-1}{n+3}} \,. \tag{13.67}$$

For lower-mass stars burning hydrogen through the proton–proton chain the rate of energy generation is described by equation (13.59), so $n = 4$, which gives $R_* \propto M^{\frac{3}{7}}$. On the other hand, for higher-mass stars burning hydrogen through the CNO cycle, $n = 18$, which gives $R_* \propto M^{\frac{17}{21}} \simeq M^{0.81}$, so the radius of the more massive main sequence stars is almost proportional to their mass.

13.5.4   The mass–temperature relationship

From equation (13.47) we see that $T_* \propto \frac{M}{R_*}$. For stars fusing hydrogen in the proton–proton chain, $R_* \propto M^{\frac{3}{7}}$ so

$$T_* \propto M^{\frac{4}{7}} \,, \tag{13.68}$$

while for stars fusing hydrogen through the CNO cycle, $R_* \propto M^{\frac{17}{21}}$ so

$$T_* \propto M^{\frac{4}{21}} \simeq M^{0.19} \,. \tag{13.69}$$

These relationships are important, because measuring the surface temperature of a star is straightforward using Wien's law (13.6), as we discussed earlier. From this observational input, we can deduce the mass of the star, and this is the key to the other properties of the star.

13.5.5   Minimum mass of main sequence stars

We can use the mass–temperature relationship (13.68) to estimate the minimum mass M_{min} of a main sequence star burning fuel through the proton–proton chain. The core temperature of the Sun is 1.6×10^7 K and the minimum temperature at which the proton–proton chain will occur is estimated to be 4×10^6 K, so

$$\frac{M_{\min}}{M_\odot} = \left(\frac{T_{\min}}{T_\odot}\right)^{\frac{7}{4}} \simeq \left(\frac{4}{16}\right)^{\frac{7}{4}} \simeq 0.1 . \tag{13.70}$$

A more precise analysis suggests that $M_{\min}$ is about $0.08\ M_\odot$, which is approximately $80\ M_J$, where M_J is the mass of Jupiter.

Low-mass main sequence stars are known as red dwarfs. Even lower in mass are very faint objects known as brown dwarfs. The centre of a brown dwarf is too cool for hydrogen to be converted into helium via the proton–proton chain, but there are fusion reactions that are still possible.

Brown dwarfs with a greater mass than $65\ M_J$ fuse lithium which was produced in small quantities in the Big Bang. A lithium nucleus can absorb a proton to form ^{8}Be, which is unstable and immediately splits into two ^{4}He nuclei,

$$^7\text{Li} + {}^1\text{H} \quad \rightarrow \quad {}^8\text{Be} \quad \rightarrow \quad {}^4\text{He} + {}^4\text{He} + \gamma . \tag{13.71}$$

Brown dwarfs with masses greater than $13\ M_J$ can sustain fusion reactions with deuterium, which was also produced in small quantities in the Big Bang,

$$^2\text{H} + {}^1\text{H} \quad \rightarrow \quad {}^3\text{He} + \gamma . \tag{13.72}$$

For masses below $13\ M_J$ no fusion reactions are possible at all. This is recognized as the boundary between stars and planets.

13.5.6 The temperature–luminosity relationship

The Stefan–Boltzmann law implies that the luminosity of a star is

$$L = 4\pi R^2 \sigma T_{\text{surf}}^4 , \tag{13.73}$$

where R is the stellar radius and T_{surf} is the surface temperature. R is proportional to R_*, given by equation (13.67), so

$$L \propto M^{\frac{2(n-1)}{n+3}} T_{\text{surf}}^4 . \tag{13.74}$$

In subsection 13.4.2 we showed that $L \propto M^3$; therefore,

$$L \propto L^{\frac{2(n-1)}{3(n+3)}} T_{\text{surf}}^4 . \tag{13.75}$$

Reorganizing the powers, this implies that

$$L^{\frac{n+11}{3(n+3)}} \propto T_{\text{surf}}^4 \quad \text{or} \quad L \propto T_{\text{surf}}^{\frac{12(n+3)}{n+11}} . \tag{13.76}$$

For lower-mass main sequence stars powered by the proton–proton chain, where $n = 4$,

$$L \propto T_{\text{surf}}^{\frac{28}{5}} = T_{\text{surf}}^{5.6} . \tag{13.77}$$

For higher-mass main sequence stars on the CNO cycle, where $n = 18$,

$$L \propto T_{\text{surf}}^{\frac{252}{29}} \simeq T_{\text{surf}}^{8.7} . \tag{13.78}$$

The HR diagram, Figure 13.3, is a log–log plot, so these relationships imply that the slope of the lower section of the main sequence should be around 5.6, while the upper section should have a steeper slope of around 8.7. This is in reasonable agreement with observations.

In constructing this model of main sequence stars we assumed that the star had a uniform composition and that the energy generated in the core of the star diffuses out as thermal radiation. We also assumed a constant opacity per unit mass. These assumptions apply well, especially in the earlier stages of hydrogen burning. But they hold less well for the later evolution of stars. As time passes, the composition of the core changes, so it no longer matches the composition of the envelope of the star. Convection may also become important. This will mix the constituents and enable the star to consume more of its nuclear fuel. It will also affect the energy diffusion through the star. Convection is not easy to model, even numerically. It is also difficult to model the amount of mass lost through radiation pressure, although it is expected to be considerable. In the later stages of a star's life, the star's radiation pressure expels streams of particles from its outer layers into space as a stellar wind that is much stronger than the currently observed solar wind.

13.6 Giant Stars Beyond the Main Sequence

When a star has used up a large proportion of the nuclear fuel in its core, its energy generation declines. The thermal pressure is now insufficient to balance the gravitational pressure, so the core contracts until the thermal pressure can be restored. The density and temperature of the core increase until conditions becomes sufficiently extreme for a new phase of nuclear fusion reactions to begin.

Thermal radiation takes a long time to reach the surface of a star. Following the exhaustion of the hydrogen fuel, core contraction occurs on a much shorter timescale than the thermal timescale that we introduced in section 13.3.3, so the star cannot readily lose the energy released by the contraction. We have so far treated the opacity within a star as constant, but it is actually a function of temperature. As one moves outwards from the core, the temperature and pressure decrease and the opacity rises. When a new round of energy generation is triggered, there is an increase in the thermal radiation and, as it cannot readily escape, the additional radiation pressure forces the envelope to inflate. Through this expansion the temperature and pressure of the envelope fall, which increases the opacity and hinders the escape of thermal radiation, leading to further inflation of the envelope. The effect of this positive feedback is a huge expansion and cooling of the envelope of the star. Eventually the envelope becomes cool enough for helium and hydrogen atoms to form. This suddenly reduces the opacity allowing the radiation to escape the star. The star has left the main sequence and has been transformed into a *red giant*.

Many red giants are unstable to pulsations, which leads to variability in their luminosity. *Cepheid variables*, for example, expand until their outer envelope cools sufficiently for neutral helium atoms to form. At this point the opacity drops dramatically and the radiation trapped in the star escapes. After losing this heat the envelope contracts under gravity, the temperature rises, the helium atoms become singly ionized, and the opacity of the envelope again rises sharply. The contraction continues until sufficient energy is trapped to stop the contraction and the expanding phase begins again. The period of the cycle depends on the mass of the star and as the mass also determines the luminosity of the star, there is a relationship between the intrinsic luminosity L of a Cepheid variable and its cycle period that can be used, together with the relation (13.5), to determine the distance d to the star. For this reason, Cepheid variables have proved to be very important standard candles that enable astronomers to calculate the distances to the nearest galaxies, as we will discuss in section 14.2.

13.6.1 The triple alpha process

Red giants have consumed all the hydrogen in their core. They can only generate further energy by fusing helium. However, helium nuclei are very stable, as can be deduced from the nuclear binding energy plot (Figure 11.7 in Chapter 11). This is why they are emitted as alpha particles from radioactive heavy elements and why so much energy is released in their synthesis in stars. Their stability makes further fusion reactions difficult to achieve. Combining two helium nuclei to forge element 4, beryllium, proves to be impossible. The resulting ^{8}Be nucleus immediately falls apart again into two ^{4}He nuclei.

Edwin Salpeter suggested in 1951 that carbon (^{12}C) might be synthesized in a near simultaneous collision of three helium nuclei, but Fred Hoyle pointed out that such triple collisions are extremely unlikely events because the intermediate ^{8}Be nucleus survives for a mere 10^{-16} s. His calculations showed that the process would occur at a tiny fraction of the rate required to explain the abundance of carbon and heavier elements in the cosmos. Hoyle then proposed that the presence of carbon in stars could only be explained by the existence of an excited state of the ^{12}C nucleus, a resonance, with just the right energy to enhance its production in the triple alpha reaction. Despite their great scepticism about Hoyle's reasoning, a team of nuclear physicists at Caltech looked for the resonance and to their amazement discovered it at precisely the energy predicted by Hoyle. Hoyle's resonance is the second excited state of the ^{12}C nucleus, 7.65 MeV above the ground state, and just 0.25 MeV above the energy of three separate helium nuclei. Its existence increases the rate of carbon production in stars by a factor of around 10^8.

The Coulomb barrier faced by colliding helium nuclei is four times that for colliding protons, so the triple alpha process only occurs at around 10^8 K. The cores of giant stars that have consumed their hydrogen fuel must contract until they reach these temperatures, and then helium fusion can begin. The triple alpha reaction can be represented as

$$^4\text{He} + {}^4\text{He} + {}^4\text{He} \rightarrow {}^{12}\text{C} + \gamma. \tag{13.79}$$

This process involves the collision of three nuclei, so the rate of energy generation per unit mass is proportional to ρ^2, as discussed in section 13.4, and it is extremely sensitive to temperature. The triple alpha rate $q_{3\alpha}$ may be approximated by

$$q_{3\alpha} \propto \rho^2 T^{40}. \tag{13.80}$$

At helium-burning temperatures, the ^{12}C nucleus will also undergo the next step in nuclear burning and sometimes fuse with another ^{4}He nucleus to form the oxygen nucleus ^{16}O. Fortunately, unlike the ^{12}C nucleus, the ^{16}O nucleus does not have a resonance with an energy that would enhance its synthesis. If it did, then any ^{12}C would immediately be converted to ^{16}O and the universe would contain very little carbon.

Helium burning generates less energy than the burning of hydrogen to form helium, so this stage of the star's life is correspondingly shorter. 7.4 MeV is released for each ^{12}C nucleus that is synthesized, compared with the 26 MeV released in the production of each of the three helium nuclei from which the carbon nucleus is formed. Altogether, $3 \times 26 = 78$ MeV is generated when the three helium nuclei were originally produced, ten times the energy released when they fuse to form carbon. For this reason, the fusion of helium to form carbon continues for less than one-tenth of the time it takes for the helium to be forged from hydrogen. In a star like the Sun, the helium-burning phase will last for about a billion years.

The supergiant Betelgeuse has already consumed the hydrogen in its core and entered the helium-burning phase. With a mass around twenty times that of the Sun, it will burn its helium rapidly. Twenty times as much helium fuel will be burned in just a couple of million years.

13.7 Late Evolution

Fig. 13.8 Examples of planetary nebulae.

Eventually, the helium fuel of a star runs out. The core is now a mixture of carbon and oxygen. With the energy supply diminished, the core contracts and the temperature rises again. What happens next depends on the mass of the star.

13.7.1 White dwarfs

The temperatures within the cores of stars of relatively low mass ($M \leq 1.5 M_\odot$) do not rise sufficiently to trigger a new round of nuclear fusion. Their density will reach the point where the electrons within the core become a degenerate electron gas and resist further compression, thereby supporting the core against further contraction. When this happens, the outer layers of the star's envelope may disperse into space to form a *planetary nebula*. (Several examples are shown in Figure 13.8.) These objects were so-named by William Herschel shortly after his discovery of the planet Uranus, due to their disc-like appearance through a telescope. They are actually luminous clouds of gas and have nothing to do with planets. Within the planetary nebula the naked shrunken core remains, radiating into space at a temperature around 10^5 K. Such an object is known as a *white dwarf*. White dwarfs are thought to have a core composed of carbon and oxygen that may be surrounded by a shell of helium and hydrogen. Within 10,000 years or so, the planetary nebula disperses into the background interstellar gas leaving the tiny white dwarf blazing away.

We can estimate the size of a white dwarf based on the fact that it is supported by electron degeneracy pressure. We have seen in section 10.9.2 that for N_e electrons of mass m_e in a volume V, the electron degeneracy pressure is

$$P = \frac{2}{5}(3\pi^2)^{\frac{2}{3}}\frac{\hbar^2}{2m_e}\left(\frac{N_e}{V}\right)^{\frac{5}{3}}. \tag{13.81}$$

This is the appropriate expression for the pressure not just at zero temperature, but at all temperatures much lower than the Fermi energy. It is valid for white dwarfs, because the electron density and hence the Fermi energy is so high.

It is more convenient to express the electron degeneracy pressure in terms of the overall mass density $\rho = \frac{M}{V}$, which is largely due to the nucleons. So let N_N be the number of nucleons in the volume V. The total mass is approximately $M = N_N m_p$ where m_p is the proton mass. Let ξ be the average number of electrons per nucleon. (For hydrogen $\xi = 1$; for helium, carbon, oxygen and other light elements $\xi \simeq 0.5$.) Then $N_e = \xi N_N = \frac{\xi M}{m_p}$ so

$$\frac{N_e}{V} = \frac{\xi M}{m_p V} = \frac{\xi \rho}{m_p}. \tag{13.82}$$

Therefore, the pressure of the degenerate electron gas can be re-expressed as

$$P = \frac{2}{5}(3\pi^2)^{\frac{2}{3}}\frac{\hbar^2}{2m_e}\left(\frac{\xi\rho}{m_p}\right)^{\frac{5}{3}}, \tag{13.83}$$

or more succinctly

$$P = K_1\rho^{\frac{5}{3}}, \tag{13.84}$$

where

$$K_1 = \frac{2}{5}(3\pi^2)^{\frac{2}{3}}\frac{\hbar^2}{2m_e}\left(\frac{\xi}{m_p}\right)^{\frac{5}{3}}. \tag{13.85}$$

An expression for the gravitational pressure within a star in terms of its density may be obtained by combining equations (13.42) and (13.43) to give

$$P = GM^{\frac{2}{3}}\rho^{\frac{4}{3}}, \tag{13.86}$$

where we have identified the pressure and density with the stellar variables that characterize the star overall. In a white dwarf the electron degeneracy pressure balances this gravitational pressure, so

$$K_1\rho^{\frac{5}{3}} = GM^{\frac{2}{3}}\rho^{\frac{4}{3}}, \tag{13.87}$$

and therefore

$$\rho = \frac{G^3 M^2}{K_1^3}. \tag{13.88}$$

As $\rho = \frac{M}{V}$, the volume of the white dwarf is

$$V = \frac{K_1^3}{G^3 M}, \tag{13.89}$$

inversely proportional to its mass. This is unlike main sequence stars whose radius increases with mass, as we saw in section 13.5.3. Knowing its volume, we can estimate the radius of a white dwarf of solar mass $M_\odot$ to be

$$R_{\mathrm{WD}} = \frac{K_1}{G} \left(\frac{3}{4\pi M_\odot}\right)^{\frac{1}{3}} = \frac{3\pi\hbar^2}{5m_e G} \left(\frac{\xi}{m_p}\right)^{\frac{5}{3}} \left(\frac{1}{4M_\odot}\right)^{\frac{1}{3}}. \tag{13.90}$$

Plugging in the numbers, including $\xi = 0.5$, we would expect white dwarfs to have a radius of a few thousand kilometres. Astronomers have determined the radius of Sirius B to be $0.0084\,R_\odot = 5800$ km. By comparison, the radius of the Earth is 6400 km, so Sirius B, whose mass is almost that of the Sun, is packed into a volume smaller than the Earth. Being so small, white dwarfs are very faint compared to normal stars. Fusion reactions have ceased so they cool down as they radiate, but as they are supported by electron degeneracy pressure (which is almost independent of temperature) and not thermal gas pressure, their size remains fixed. The Stefan–Boltzmann law (13.73) implies that the luminosity of a white dwarf is proportional to the fourth power of its temperature. The HR diagram is a log–log plot of luminosity against temperature, so over time, as it cools, a white dwarf will gradually follow a path near the bottom left of the HR diagram with a slope of 4. When first formed, a white dwarf consists of a very hot liquid of carbon and oxygen nuclei within a sea of electrons. As the white dwarf cools, its carbon–oxygen core is thought to crystallize into an extremely dense diamond-like structure.

There is a maximum mass for which a white dwarf is stable, so not all stars end their evolution as white dwarfs. As the mass of a white dwarf rises, its volume decreases. This increases the momentum gaps between electron states, and because electrons are subject to the exclusion principle, they are forced into higher momentum states. In a white dwarf of sufficiently high mass, most of the electrons will attain relativistic velocities and this dramatically changes the equation of state.

The density of states in momentum space is still given by equation (8.12). Allowing for the two spin states of the electron, the density in p is

$$\widetilde{g}(p) = \frac{Vp^2}{\pi^2\hbar^3}. \tag{13.91}$$

When electrons are highly relativistic, their energy is $\varepsilon = p$, so the density of states in ε is

$$g(\varepsilon) = \frac{V\varepsilon^2}{\pi^2\hbar^3}. \tag{13.92}$$

Integrating this density up to the Fermi energy ε_{F}, we find the following relation between ε_{F} and N_e, the total number of electrons in volume V,

$$N_e = \frac{V}{\pi^2\hbar^3} \int_0^{\varepsilon_{\mathrm{F}}} \varepsilon^2 \, d\varepsilon = \frac{V}{3\pi^2\hbar^3} \varepsilon_{\mathrm{F}}^3. \tag{13.93}$$

It follows that $\varepsilon_{\mathrm{F}} = \left(3\pi^2\hbar^3 \frac{N_e}{V}\right)^{\frac{1}{3}}$, and the total energy of the electrons is

$$E = \frac{V}{\pi^2\hbar^3} \int_0^{\varepsilon_{\mathrm{F}}} \varepsilon^3 \, d\varepsilon = \frac{V}{4\pi^2\hbar^3} \varepsilon_{\mathrm{F}}^4 = \frac{3}{4}(3\pi^2)^{\frac{1}{3}} \hbar N_e^{\frac{4}{3}} V^{-\frac{1}{3}}, \tag{13.94}$$

so for the relativistic degenerate electron gas the pressure is

$$P = -\frac{dE}{dV} = \frac{1}{4}(3\pi^2)^{\frac{1}{3}}\hbar\left(\frac{N_e}{V}\right)^{\frac{4}{3}}. \tag{13.95}$$

Substituting for the number density from the relation (13.82), $\frac{N_e}{V} = \frac{\xi\rho}{m_p}$, we obtain

$$P = \frac{1}{4}(3\pi^2)^{\frac{1}{3}}\hbar\left(\frac{\xi\rho}{m_p}\right)^{\frac{4}{3}}, \tag{13.96}$$

or more succinctly

$$P = K_2\rho^{\frac{4}{3}}, \tag{13.97}$$

where $K_2 = \frac{1}{4}(3\pi^2)^{\frac{1}{3}}\hbar\left(\frac{\xi}{m_p}\right)^{\frac{4}{3}}$.

If we now equate the electron degeneracy pressure and the gravitational pressure, as given in equation (13.86), we obtain

$$K_2\rho^{\frac{4}{3}} = GM^{\frac{2}{3}}\rho^{\frac{4}{3}}. \tag{13.98}$$

The density cancels out of this equation, so we must conclude that electron degeneracy pressure cannot balance the gravitational pressure when M is too large. No stable density exists for a white dwarf if its electrons have relativistic velocities.

The maximum mass of a white dwarf is known as the *Chandrasekhar limit* after the astrophysicist Subrahmanyan Chandrasekhar who first calculated it. From equation (13.98) we can estimate the maximum mass as

$$M_{\text{Ch}} = \left(\frac{K_2}{G}\right)^{\frac{3}{2}} \simeq \left(\frac{\hbar}{G}\right)^{\frac{3}{2}}\left(\frac{\xi}{m_p}\right)^2 \simeq 4\,\xi^2\,M_\odot. \tag{13.99}$$

The Chandrasekhar limit depends on the composition of the white dwarf. More precise analyses estimate the limit to be

$$M_{\text{Ch}} = 5.83\,\xi^2\,M_\odot. \tag{13.100}$$

For a white dwarf composed of helium, carbon or oxygen, $\xi = 0.5$ so $M_{\text{Ch}} = 1.46M_\odot$. A white dwarf with a mass exceeding this limit must collapse further. Astronomers have never found a white dwarf with a mass exceeding the theoretical Chandrasekhar limit. (In the later stages of a star's life, large quantities of material may be lost due to radiation pressure, so the upper mass for a star that will evolve into a white dwarf is not known with certainty and could be between $6M_\odot$ and $8M_\odot$.)

13.7.2 Gravitational collapse of massive stars

As we have seen, when the fuel for one round of nuclear fusion is exhausted, a star's core will contract. For sufficiently massive stars the temperature will rise until the next stage of fusion is triggered. In the most massive stars, greater than about eight solar masses, there are six main stages of nuclear fusion: hydrogen burning, helium burning, carbon burning, neon burning, oxygen burning and silicon burning, where the ash from one round of nuclear burning becomes the fuel for the next round. Heavier elements have greater nuclear charge Z and therefore greater Coulomb barriers to overcome if fusion reactions are to occur. As Z increases so does the temperature required for nuclear fusion. Most of the elements heavier than helium in gas clouds where new stars form were forged within stars with masses in excess of 25 solar masses. The definitive proof that the creation of the elements must be an ongoing process came in 1952 when the spectral lines of element $Z = 43$ technetium were detected in a star. Technetium has no stable isotopes. The longest lived isotope is ^{98}Tc, with a half-life of just 4.2 million years.

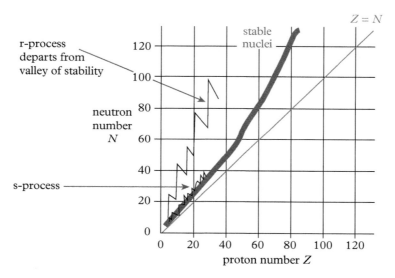

Fig. 13.9 Heavy nuclei form through the continued absorption of neutrons and subsequent beta decay. This occurs via the s-process in the cores of giant stars, which generates isotopes close to the valley of stability, and via the r-process in supernova explosions, which generates neutron-rich isotopes away from the valley of stability.

The temperature required for nuclear fusion is roughly proportional to the product of the charges on the fusing nuclei, although the precise details are determined by the nuclear energy levels. Hydrogen fusion occurs at around 1.6×10^7 K, helium fusion around 1.0×10^8 K and carbon fusion occurs in giant stars whose cores reach temperatures of 0.6–1.0×10^9 K. The main carbon fusion processes are

$$^{12}\text{C} + {}^{12}\text{C} \rightarrow {}^{20}\text{Ne} + {}^{4}\text{He}$$
$$^{12}\text{C} + {}^{12}\text{C} \rightarrow {}^{23}\text{Na} + {}^{1}\text{H}$$
$$^{12}\text{C} + {}^{12}\text{C} \rightarrow {}^{23}\text{Mg} + n \,. \tag{13.101}$$

The first two of these processes release energy. The third is endothermic, but will still occur at these extremely high temperatures and it is important because it produces free neutrons. These neutrons readily penetrate any nearby nuclei, as there is no Coulomb barrier to prevent their absorption. In this way new isotopes are created. A number of neutrons may be absorbed until an unstable nucleus with a surplus of neutrons is reached. This nucleus will then undergo beta decay and one of the neutrons will convert into a proton. Many isotopes of the lighter elements are formed like this. It is especially important for the synthesis of the elements with odd proton number Z. This process is known as the *slow* or *s-process*, as typically the interval between the absorption of two neutrons is longer than the half-lives for beta decay of any unstable nuclei that are created. The new nuclei that are created in the s-process will therefore lie close to the valley of stability, as shown in Figure 13.9.

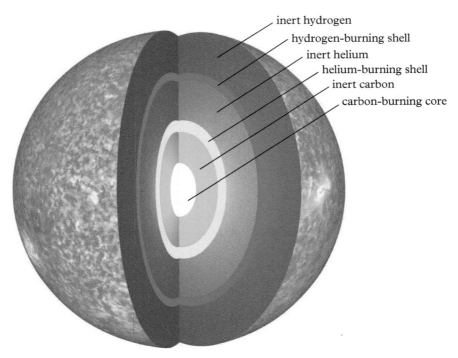

inert hydrogen
hydrogen-burning shell
inert helium
helium-burning shell
inert carbon
carbon-burning core

Fig. 13.10 In its later stages, a giant star that is fusing carbon in its core may also contain a shell of fusing helium and a shell of fusing hydrogen, as shown here. The relative sizes of the regions within the star are not to scale.

As the temperature increases steadily in the core, conditions may become suitable for the burning of material in shells around the core. For instance, take a star burning carbon in its core. Surrounding the core is a shell of carbon that is not quite hot enough for fusion to occur. This inert layer of carbon is surrounded by a shell of burning helium. As this helium is depleted, the burning shell will gradually move outwards and the carbon *ash* that it produces will accumulate around the inert carbon shell and thereby increase its radius. Outside the shell of burning helium will be a shell of helium that is too cool to fuse. Further out still will be a shell of burning hydrogen that is moving outwards as it fuses and adds to the helium below. Beyond the shell of burning hydrogen will be the hydrogen envelope of the star. A schematic diagram of the shell structure of such a star is shown in Figure 13.10.

With the exhaustion of each stage of nuclear burning, the temperature of the nuclear furnace rises and an ever wider range of fusion reactions becomes possible. The analysis of these reactions becomes increasingly complicated; nuclei may be built up in fusion reactions and dissociated by high energy gamma rays. During each round of nuclear burning, the concentrations of each isotope tend towards a state of nuclear statistical equilibrium. Carbon burning produces a core composed of oxygen, neon and magnesium. This is followed at temperatures in the region of 1.5×10^9 K by neon burning in the reaction

$$^{20}\text{Ne} + {^4}\text{He} \rightarrow {^{24}}\text{Mg} + \gamma \,. \tag{13.102}$$

This is swiftly followed by oxygen burning at 2.0×10^9 K, which mainly involves the reactions

$$^{16}\text{O} + {^{16}}\text{O} \rightarrow {^{28}}\text{Si} + {^4}\text{He}$$
$$^{16}\text{O} + {^{16}}\text{O} \rightarrow {^{31}}\text{P} + {^1}\text{H} \,. \tag{13.103}$$

The later stages of nuclear fusion in high mass stars generate significantly less energy and a large proportion of this energy is lost in neutrino and antineutrino emission, so these stages last for correspondingly shorter periods of time. Carbon burning lasts for a few centuries, while neon burning takes just a few years and oxygen burning lasts for around eight months to one year. The final stage of silicon burning occurs at a temperature of 3.5×10^9 K and lasts just a few days, in the process of which most of the core is converted into ^{56}Ni. This is almost the natural endpoint for stellar nuclear reactions.

The nucleus ^{56}Ni is unstable. Given sufficient time, it undergoes two stages of inverse beta decay to reduce its proton number Z, decaying into ^{56}Co with a half-life of six days, which subsequently decays into ^{56}Fe with a half-life of 77 days. The iron nucleus ^{56}Fe is the densest and therefore the most stable of all atomic nuclei, as discussed in Chapter 11, so no further energy is released in fusion reactions involving the iron nucleus or its neighbours. Once this stage has been reached, the star cannot generate any further energy by nuclear fusion, and without the continued release of energy, the outward pressure preventing the gravitational collapse of the star cannot be maintained.

At this point, the final core collapse begins. The temperature rises until at around 10^{11}K the thermal radiation consists of gamma rays with energies sufficient to break down the nickel, iron and other dense nuclei in the core into free nucleons. The core reaches nuclear density, at which point the nuclear material resists further compression and the collapsing core undergoes a bounce. So much gravitational binding energy has now been released that the star blasts itself apart in a *supernova* explosion which may be as bright as an entire galaxy of 100 billion stars.

As the supernova explodes, the nuclei within the fireball are bathed in neutrons from nuclei that have been dissociated by the gamma radiation. These neutrons build up the heavier elements in the Periodic Table in what is known as the *rapid* or *r-process*, as numerous neutrons may be absorbed even by unstable nuclei before they decay. This produces neutron-rich isotopes well away from the valley of stability, as shown in Figure 13.9.

Many elements, such as gold and uranium, that have great significance for us, can only be created in supernova explosions. There are several hundred naturally occurring isotopes of such elements. Astrophysicists now understand the detailed nuclear processes by which each isotope formed, and can quantitatively account for its cosmic abundance. The principal element formation sites are indicated in Figure 13.11. Recall also Figure 11.6, giving the solar system abundance of all the elements.

Following a supernova blast, the core of the star may have become transformed into an object with a radius of around 15 kilometres, about the size of a major city, but with the

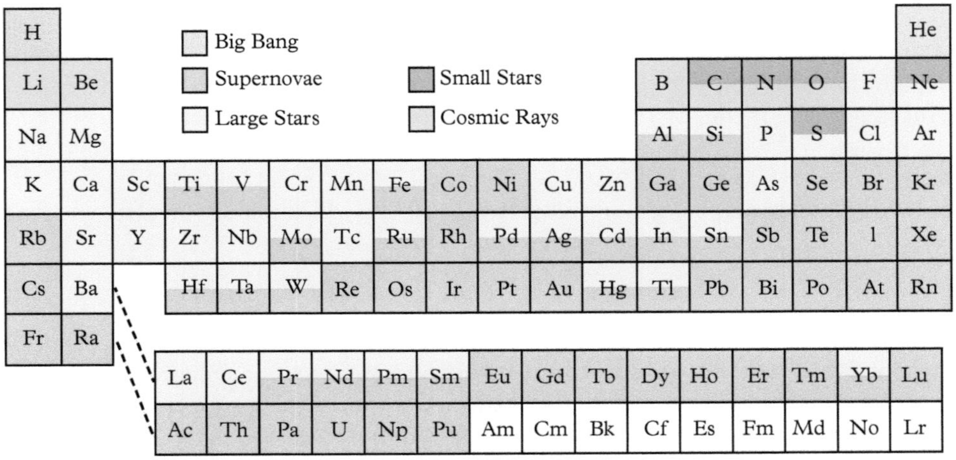

Fig. 13.11 The Periodic Table with each element colour-coded to indicate its origin. (The main source of the elements Li, Be and B is the breakdown of heavier nuclei such as C, N and O through the impact of cosmic rays, a process known as spallation.)

density of an atomic nucleus. This remarkable object is known as a *neutron star*. If the mass of the collapsing core is more than two or three times the mass of the Sun, its collapse cannot be stopped at all. The result is a black hole.

13.8   Neutron Stars

We saw in section 13.7.1 that above a certain mass a white dwarf is unstable. A more massive star continues its gravitational collapse and eventually explodes as a supernova. In the final milliseconds of collapse, the Fermi energy of the electrons becomes so high that the electrons and protons interact via the weak force and undergo inverse beta decay. The electrons and protons merge into neutrons, and neutrinos are emitted:

$$e^- + p \rightarrow n + \nu_e. \tag{13.104}$$

At extremely high pressures and temperatures, neutrons collectively have lower energy than protons, as they do not interact electrostatically. The result is that the core of the star is transformed into a body that is composed almost entirely of neutrons. This is the extraordinary object that will remain after the outer layers of the star have been blasted deep into space by the supernova. The density of the resulting neutron star is comparable to the density of an atomic nucleus. The star has, in effect, turned itself into a giant atomic nucleus. Just one teaspoonful of neutron star material weighs about two billion tonnes. At least, this would be the weight of an average teaspoonful of neutron star. Neutron stars are compressed to much higher densities towards their centres than in their crusts.

By analogy with the maximum mass of a white dwarf, there is also a maximum mass for a neutron star, but as the physics of these objects is so exotic this limit is not known with the same certainty. However, it must be in the range of 2–$3M_\odot$. A neutron star with a mass exceeding this value would inevitably collapse to form a black hole. Currently, the highest, accurately determined mass is $2.01 \pm 0.04 \ M_\odot$, for the neutron star J0348+0432.

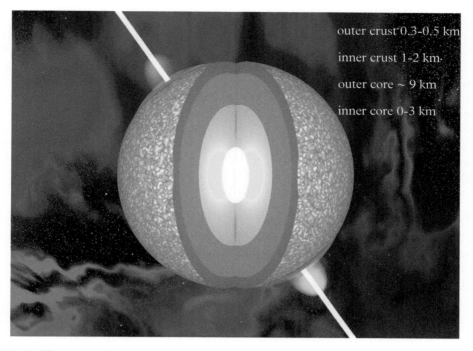

outer crust 0.3–0.5 km

inner crust 1–2 km

outer core ~ 9 km

inner core 0–3 km

Fig. 13.12 The internal structure of a neutron star. The outer crust consists of a lattice of white-dwarf-like matter and is supported by electron degeneracy pressure. The inner crust is composed of a lattice of heavy nuclei and a superfluid of free neutrons. The outer core consists of a superfluid of neutrons plus a small number of superconducting protons and is supported by neutron degeneracy pressure. The composition of the inner core is uncertain, but could be some sort of quark–gluon plasma.

We might naïvely expect that neutron stars are supported at nuclear densities by neutron degeneracy pressure and therefore that their radius can be determined by the same method as used for white dwarfs, simply by exchanging neutrons for electrons in the calculation. As the radius of a white dwarf R_{WD} is inversely proportional to m_e, as given by equation (13.90), this argument would imply that the neutron star radius is

$$R_{\mathrm{NS}} \simeq \frac{m_e}{m_n} R_{\mathrm{WD}}\,, \tag{13.105}$$

substantially less than the radius of a white dwarf, as $m_n \simeq 1838 m_e$. However, a radius of just 3 km, as indicated by this calculation, would put the neutron star within its Schwarzschild radius $2GM$, which is around 4.5 km for a neutron star with the Chandrasekhar mass. A neutron star could not exist with this radius, as it would immediately collapse into a black hole. The resolution appears to be that the strong force produces a strong repulsion between neutrons when they are compressed beyond nuclear densities and it is this that supports a neutron star against gravitational collapse. Neutron stars are thought to have radii of around 15 km, which is about twice the Schwarzschild radius for a neutron star of mass $2.5 M_\odot$.

The structure of a neutron star is represented in Figure 13.12. These weird objects are believed to have a hot plasma atmosphere that is a few centimetres thick surrounding an outer crust of white-dwarf-like matter, with a density of around 10^9 kg m^{-3}, consisting of heavy nuclei in a degenerate sea of electrons. Moving inwards we reach the inner crust at the neutron drip density. The proportion of neutrons in the nuclei dramatically increases along with the density of free neutrons until we reach a transition density of around 1.7×10^{17} kg m^{-3}. We now enter the outer core, which consists almost exclusively of neutrons plus a small quantity of protons, electrons and muons. The physical structure of the material that forms the inner core at the heart of a neutron star is open to speculation. There have been a number of suggestions including densely packed strange baryons and Bose–Einstein condensates of pions and kaons. Another possibility is that it consists of some sort of quark–gluon plasma.

13.8.1 Pulsars

A star's rate of rotation must increase dramatically as it collapses. We can estimate the angular velocity ω of a newly formed neutron star by imagining the Sun collapsing into a star with a 15 km radius. If no mass is lost, the mass of the neutron star will be $M_{NS} = M_\odot$. The angular momentum of the Sun is

$$J_\odot \propto M_\odot R_\odot^2 \omega_\odot .$$ (13.106)

As angular momentum is conserved, the angular momentum of the neutron star J_{NS} equals $J_\odot$ and the rotational period of the neutron star τ_{NS} will therefore be

$$\tau_{NS} = \left(\frac{R_{NS}}{R_\odot}\right)^2 \tau_\odot = \left(\frac{15}{7 \times 10^5}\right)^2 (2.1 \times 10^6) \text{ s} \simeq 10^{-3} \text{ s} ,$$ (13.107)

where we have used the fact that the rotational period of the Sun is 24.5 days, which is 2.1×10^6 seconds. This seems to be a reasonable estimate for the rate of rotation of newly born neutron stars. The magnetic field strength of a neutron star is enormous, around a trillion times that of the Earth. As on the Earth, the magnetic poles need not coincide with the poles of rotation, so the magnetic dipole field spins around the neutron star. This time-dependent magnetic field generates an electric field that accelerates electrons and other charged particles outwards from the magnetic poles of the neutron star, producing two intense beams of radiation that are blasted into space from the poles. These beams are known as *pulsars*. They sweep around the heavens like a cosmic lighthouse, as illustrated in Figure 13.13. On Earth, radio astronomers detect a pulse of radio waves once every rotation when the pulsar beam points in our direction, which may be numerous times a second.

The spinning magnetic field acts as a brake on the rotation of the neutron star, and the pulsar that it generates transmits the lost rotational energy to the surrounding nebula. Gradually the angular velocity of the neutron star decreases. The neutron star within the Crab Nebula was formed around 1000 years ago and now spins about thirty times a second. It has been calculated that the amount of rotational energy lost by the spinning neutron star in the Crab Nebula matches the energy required to illuminate the nebula. Pulsars have a limited lifespan. In around one million years the period of rotation of the neutron star will have increased to about one second, at which point there will be insufficient energy to power the pulsar and it will disappear from view.

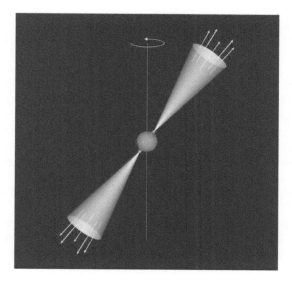

Fig. 13.13 A pulsar is a beam of radiation produced by a rapidly rotating neutron star.

A neutron star in a binary system with a normal star may accrete material from its companion. This could happen if an aging companion sheds its outer layers or as the companion inflates into a giant star. When this happens, an accretion disc may form around the neutron star. As material spirals on to its surface from the inner edge of the accretion disc, the neutron star will spin up, reviving its long-dead pulsar. This is thought to be the origin of *millisecond pulsars*, which are observed to have extremely short periods. Currently, the pulsar with the shortest known period of rotation is PSR J1748-2446ad, which spins an incredible 716 times a second.

13.9 Supernovae

A star that suddenly appears, as if from nowhere, is known as a *nova*, meaning new star. These events are seen quite regularly by astronomers and have been observed for thousands of years. Gradually the nova fades, and it eventually disappears. In the early 1930s, stellar explosions of a much brighter type were recognized by Walter Baade and Fritz Zwicky who dubbed them *supernovae*. They identified several examples in other galaxies and instigated systematic searches for more. Until quite recently, astronomers had few examples to work on. Now, hundreds are discovered and studied in detail every year. The sample size has been transformed in recent decades by the deployment of automatic searches, such as the Palomar Transient Factory (PTF), as well as systematic searches by teams of amateurs. The computer modelling of supernova explosions is also advancing with increases in processing power. As a consequence, the physics of supernovae is a rapidly developing field. In the 1950s, a broad classification was introduced based on the spectral lines displayed by supernovae. With so many more examples to analyse, the classification system has become stretched to include numerous unusual cases that do not fit the standard patterns.

We now know that there are two principal mechanisms that result in supernovae. Type Ia supernovae are produced by a runaway thermonuclear explosion of a white dwarf star close to the Chandrasekhar limit. Type II supernovae are produced by the core collapse of a star that has exhausted its nuclear fuel, as described in section 13.7.2. The ultimate power source of Type II supernovae is the release of gravitational binding energy.

A large proportion of stars live in binary or multiple star systems. The interactions between such stars can be varied and complicated especially in the later supergiant stages of evolution of very massive stars. These interactions have a significant effect on the environment in which a supernova might occur. It is becoming clear that many of the subdivisions in the classification system are simply due to environmental factors. For instance, a star that undergoes core collapse might have an envelope that is hydrogen-rich or helium-rich or it might have completely lost its envelope due to interactions with a partner star, and this will, of course, affect the spectral lines that appear in the light from the supernova. The appearance of the supernova is also modified by any material in the region around the exploding star, such as might have been ejected by the star in earlier, lesser eruptions.

Interactions within a binary system are also considered to be critial for the origin of Type Ia supernovae, although there is still some debate about the precise mechanism. The more massive star in a binary system may evolve into a white dwarf, while some time later its companion inflates to become a red giant. The white dwarf is a dead star that is no longer undergoing nuclear fusion reactions, but as it travels around the red giant it may accumulate material from the red giant's outer layers. This material is drawn to the white dwarf and compressed by its intense gravity to form a shell around its surface. Eventually, a critical density is reached and the shell erupts in a huge nuclear fusion explosion that may be visible from the other side of the galaxy. We see such an event as a nova; about ten appear in the Milky Way galaxy each year. (Another 30 or so are thought to be obscured from view by dust and gas clouds.) The process leading to the nova will repeat as the white dwarf continues to attract material from its companion star. The period between eruptions is typically several thousand years, but it may be as short as a decade or two. For example, the star RS Ophiuchi erupted in 1898, 1933, 1958, 1967, 1985 and 2006. Astrophysicists are unsure whether the white dwarf mass steadily increases in this process or decreases due to ablation in each blast.

When on the main sequence, any temperature rise within a star increases the thermal pressure and this acts like a valve to regulate the temperature and fusion rate. But this mechanism cannot occur in a white dwarf. When the mass of a white dwarf approaches the Chandrasekhar limit, the detonation of a shell of material accreted to its surface may trigger carbon fusion in its core. The white dwarf's intense gravity squeezes the core. It is highly degenerate and supported by electron degeneracy pressure, which is essentially independent of temperature, so any energy released by carbon fusion produces a dramatic rise in temperature and this in turn produces a dramatic increase in the fusion rate. The resulting runaway thermonuclear explosion obliterates the white dwarf, and it lights up as a Type Ia supernova, which may be as much as 100,000 times brighter than the earlier novae. All Type Ia supernovae are believed to have similar intrinsic luminosities, and since the 1990s they have been used as standard candles[2] to determine the distances to far-flung galaxies. Since 1998, studies of distant Type Ia supernova explosions have been used to show that the expansion of the universe is accelerating. Recently, however, some doubt has been

---

[2] Strictly speaking they are *standardizable*. Their intrinsic luminosities may vary by up to a factor of 10, but their peak intrinsic luminosity can be deduced from their light curve, which plots their decreasing luminosity against time.

cast on the reliability of this method of distance determination. Some supernovae classed as Type Ia are thought to be due to the merger of a binary system formed of two white dwarfs, in which case the total mass and the merger process will be different for each binary system and the luminosities of the resulting supernovae will vary. If correct, this will have implications for distance determination on the very longest scales.

In 1987 astronomers witnessed the closest visible supernova since the dawn of the telescope age. This object was a Type II supernova and is referred to as SN1987A. It is located in a dwarf galaxy known as the Large Magellanic Cloud, that is gravitationally bound to our own galaxy at a distance of around 168,000 light years. Even at this distance the supernova appeared as a moderately bright star in the skies of the southern hemisphere. The appearance of a relatively close supernova offered astronomers the opportunity to test their ideas about these cosmic cataclysms. Within days, the progenitor star of SN1987A had been found in photographic records. This star, designated Sanduleak $-69°$ 202, was a blue supergiant, so the supernova is the result of core collapse, as expected for a Type II supernova. Nevertheless it has some atypical features.

Core collapse generates an intense flux of neutrinos and antineutrinos. At the incredibly high pressures within the core, it is energetically favourable for protons and electrons to combine to form neutrons by inverse beta decay, thus releasing neutrinos:

$$p + e^- \rightarrow n + \nu_e. \tag{13.108}$$

Convection within the core may bring extremely neutron-rich nuclei into regions at slightly lower pressures where they rapidly undergo beta decay and release antineutrinos:

$$n \rightarrow p + e^- + \overline{\nu_e}. \tag{13.109}$$

However, another process is even more significant. Gamma rays in the collapsing core have sufficient energy (> 1.02 MeV) to form electron–positron pairs when they scatter off ions in the plasma. Some of these electrons and positrons then annihilate to produce neutrino–antineutrino pairs (of all three types):

$$e^- + e^+ \rightarrow \nu + \overline{\nu}. \tag{13.110}$$

Within a matter of seconds during core collapse, around 10^{58} neutrinos and antineutrinos are created. When supernovae are modelled, the initial explosion always stalls and this vast surge of neutrinos is required to reignite the supernova blast that tears the dying star apart. For an instant, the neutrinos are trapped and form a degenerate gas, filling all available states, like electrons in a metal.

Two to three hours prior to the appearance of SN1987A, neutrino observatories in Japan, the United States and Russia detected a total of 24 antineutrinos in just ten seconds. This was the first ever detection of neutrinos from outside the solar system. It has been estimated that there should be a supernova explosion in the Milky Way galaxy every 30 or so years, but most of these are hidden from view by intervening dust clouds. The last one seen by astronomers was in 1604. The next occurence should easily be picked up by the more sophisticated neutrino observatories of today. The Super-Kamiokande detector alone would expect to see a pulse of around 10,000 neutrinos and antineutrinos from a supernova in our galaxy.

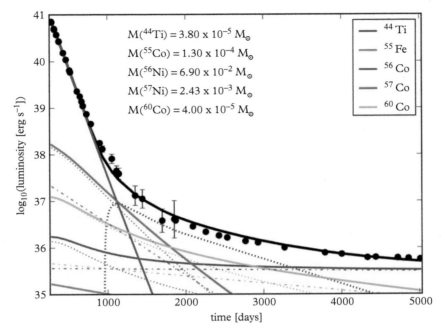

Fig. 13.14 The light curve of SN1987A (dots) matched to the combined radiation emissions from the long-lived radioactive nuclei ^{44}Ti, ^{55}Fe, ^{56}Co, ^{57}Co and ^{60}Co.

The light from SN1987A gradually faded in the months following its appearance. Its decline in luminosity fits our theoretical understanding very well, as it matches the decay half-life for ^{56}Ni $\rightarrow$ ^{56}Co, which is six days, followed by the half-life for ^{56}Co $\rightarrow$ ^{56}Fe, which is 77.3 days. Figure 13.14 compares the decrease in luminosity of SN1987A to that expected due to radioactive decay of the various isotopes created by the supernova. Up to about 1000 days, the light curve matches the decay of ^{56}Co. Most of the light that is currently received from the supernova remnant is thought to be due to the decay of the titanium isotope ^{44}Ti, which has a half-life of sixty years.

Astrophysicists were surprised to discover that the progenitor of SN1987A was a blue supergiant with a mass of $20M_{\odot}$ and a radius of $40R_{\odot}$, rather than a red supergiant with a radius of around $1000R_{\odot}$, as expected from stellar evolution theory. In 1990, the supernova remnant was discovered to lie within an odd-looking triple ring system, as shown in Figure 13.15. These and other unusual features of SN1987A now have a convincing explanation. Philipp Podsiadlowski, Thomas Morris and Natasha Ivanova have proposed that the progenitor star began as a binary system containing stars with approximate masses of 15–20$M_{\odot}$ and $5M_{\odot}$ orbiting each other with a period of at least ten years. When the greater star entered its red supergiant phase, it engulfed the companion star in its outer envelope. Friction due to the diffuse envelope caused the stars to spiral together until the companion star collided with the core of the red supergiant around 20,000 years ago. The collision mixed material from the core into the envelope of the red supergiant, which along with the added mass transformed the star from a red to a blue supergiant. Computer models of the hydrodynamics of the merger show that as the companion star

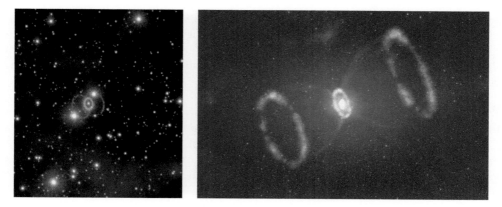

Fig. 13.15 Left: Hubble Space Telescope (HST) image of SN1987A showing three rings around the central supernova. Right: An artist's impression of the triple-ring system from a different angle showing its 3D structure.

spirals within the red supergiant, its envelope is spun up to form a disc around the bloated host. Also, the inspiral of the companion heats up the envelope and causes around $0.5M_\odot$ of material to be ejected. The escape of this material is hindered by the presence of the equatorial disc, with the greatest concentrations of material leaving at angles of $\pm 45°$ to the disc. This forms outflowing rings above and below the plane of the disc. After the merger, the temperature within the red supergiant rises and it shrinks to become a blue supergiant, leaving the disc to become a third outflowing ring of several solar masses. The end result is a system of three rings, as seen in the Hubble Space Telescope image in Figure 13.15.

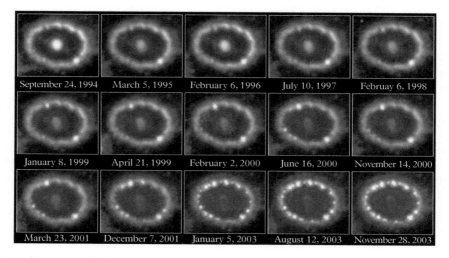

Fig. 13.16 Sequence of HST images of SN1987A. The shock wave from the explosion is catching up with and exciting the dense central ring of material that was ejected from the star 20,000 years prior to the supernova explosion.

Figure 13.16 shows the evolving appearance of SN1987A. The material expelled in the supernova explosion has now reached the dense inner ring of material that was ejected from the progenitor star 20,000 years ago and knots of gas are lighting up as they are excited by the impact. Astronomers are on the lookout for a neutron star left behind by SN1987A. So far it has not been found.

13.9.1 Gamma-ray bursts

In 1963, the Nuclear Test Ban Treaty was signed. To monitor compliance with the treaty, four years later the United States launched a series of satellites to detect gamma rays, which are the tell-tale signatures of nuclear explosions. Straight away, these satellites began to detect occasional flashes of gamma rays, or gamma-ray bursts (GRB) as they are known. They were immediately put under investigation.

By 1973 it was clear that the gamma rays originated in space and the research was declassified by the military. GRBs typically last just a few seconds, so uncovering their secrets has proved quite a challenge. Success has depended on the rapid deployment of telescopes to study the afterglow following the detection of a burst of gamma rays by a satellite. This has been possible since the launch in 1996 of the Italian–Dutch satellite BeppoSAX, which was designed to locate GRBs very quickly and precisely, enabling swift follow-up by optical telescopes. The spectral absorption lines in these optical afterglows are highly redshifted and are produced as light from the GRB event passes through a distant intervening galaxy. It is now clear that GRBs are the product of the most violent explosions in the universe. They are extremely rare, but so powerful that we can detect them from the other side of the universe billions of light years away.

GRBs have been divided into two categories. Most of the events last for a few seconds, with a typical duration of 20 seconds, and are known as long GRBs. The other category are the short GRBs that last for less than two seconds. The amount of energy in a GRB is vast. Isotropic emission of so much gamma radiation would require an unfeasible energy source that would vastly exceed even the most luminous supernovae. It is now generally accepted that GRBs are tightly focused beams of radiation. We see a GRB when an intense beam of gamma radiation is fired in our direction from this cataclysmic event. The two categories of GRB are believed to result from two different types of event.

Some of the long GRBs have been identified with examples of the most luminous class of supernovae, known as *hypernovae*, which are at least ten times as bright as normal supernovae. Astrophysicists have concluded that the long GRBs are the result of core collapse of a very massive star to form a black hole. Squeezing an entire star into a black hole is difficult as black holes are tiny by cosmic standards, just a few kilometres across. The collapsing star spins ever faster and while much of the low angular momentum material near the poles readily forms the black hole, the high angular momentum material near the equator creates a rapidly spinning disc. Viscous processes allow the rapid accretion of this disc into the newly formed black hole, but some of the material is ejected at the poles. This is compressed beyond nuclear densities and focused into two jets that shoot outward at almost the speed of light. Electrons within the jets are accelerated and emit synchrotron radiation, generating two intense beams of gamma rays. A civilization on the other side of the universe that happens to be looking down the barrel of this gamma-ray gun sees a brief trace of radiation signalling the death of a mighty star of perhaps $40M_\odot$, and the formation of a black hole.

In some cases the polar gamma ray beams may be unable to punch their way through the collapsing stellar material, so it is likely that not all hypernovae are associated with GRBs. It is also possible that some of the most massive stars collapse to form black holes with a whimper rather than a bang and do not generate a supernova at all. They would simply disappear from view.

Short GRBs have proved even more difficult to study because they are so short-lived and much less powerful. Recently they have been associated with faint visible afterglows that are also billions of light years distant. There is very good evidence that these events are due to the merger of two compact bodies that may be two neutron stars or a neutron star and a black hole.

13.10   The Density–Temperature Diagram

In section 13.4.3 we found that stars obey equation (13.52), which implies that

$$\rho(r) = \frac{1}{(G\mu m_p)^3} \frac{T^3(r)}{M^2}. \tag{13.111}$$

This equation relates the density to the cube of the temperature at each point in the star. Taking logarithms gives $\log \rho = 3 \log T - 2 \log M + \text{const}$, so on a plot of $\log \rho$ against $\log T$ the values appropriate to a star of mass M will fall on a straight line of slope 3, as shown in Figure 13.17. This is the mass track of the star. Equally spaced representative lines are shown for stellar masses of $0.1 M_\odot$, $M_\odot$, $10 M_\odot$ and $100 M_\odot$. (Mass loss during stellar evolution is known to be significant especially for higher-mass stars, but for simplicity the diagram assumes the stellar masses remain constant.)

Much of the content of this chapter may be summarized in this diagram. On the left is the region of white dwarfs where electron degeneracy pressure balances gravitational pressure. Above this region we reach the critical density where the degenerate electrons reach relativistic velocities and the white dwarfs become unstable to collapse into neutron stars. In the region at the bottom right, radiation pressure is greater than thermal pressure within the star. When radiation pressure becomes dominant, a star is unstable. This limits the mass of a star to less than around $120 M_\odot$. The diagonal strip inhabited by the mass tracks of the stars lies between these extremes.

The curved white lines on the diagram show the temperature and density required for the various rounds of nuclear fusion. The leftmost curve represents hydrogen fusion. The low temperature end corresponds to energy production via the proton–proton chain and shows the temperature dependence given in equation (13.59) with exponent $n = 4$. The high temperature end is much steeper corresponding to fusion via the CNO cycle with the temperature dependence given in equation (13.62) with exponent $n = 18$. When a protostar contracts under gravity, the pressure–temperature conditions in its core rise up the appropriate mass track until they reach the hydrogen fusion line and energy generation begins. The core then remains at this point throughout its main sequence lifetime. The density and temperature within the star, as we move outwards from the core through the envelope, follows the mass track back towards the bottom left of the diagram.

When the star depletes the hydrogen fuel in its core, the core contracts and moves further up to the right along the mass track. The trajectory of the track and the fate of the star now depend on its mass. The low-mass tracks, such as for $0.1 M_\odot$, approach the region where

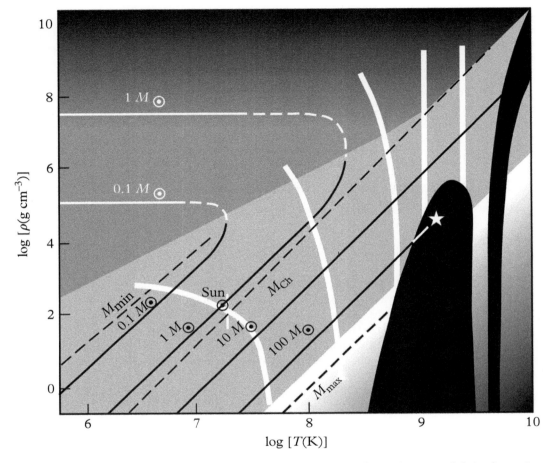

Fig. 13.17 Density–temperature diagram. The dark-grey triangle to the upper left is the region where electron degeneracy pressure balances gravitational pressure. At the top of this region white dwarfs are unstable and collapse to form neutron stars. In the black region at the bottom right, stars are unstable due to radiation pressure. Stable stars exist in the diagonal strip between these extremes.

the equation of state stiffens dramatically and equation (13.52) no longer applies. These tracks curve into the region of white dwarfs, whose stellar cores reach a maximum density supported by electron degeneracy pressure. They then gradually cool down at a constant density and move left horizontally.

As the depleted core of a solar mass star contracts, it follows the $M_{\odot}$ track until it intersects the second curved white line representing the conditions suitable for helium fusion. The core then ceases to contract until the helium fuel is depleted. While helium is burning in the core, conditions in a shell further out in the star may be suitable for hydrogen burning, where the mass track intersects the hydrogen fusion line. When the helium is exhausted, the star again follows the mass track into the white dwarf zone.

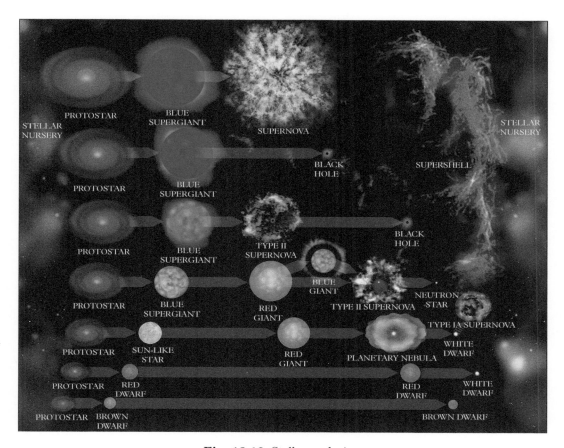

Fig. 13.18 Stellar evolution.

The cores of more massive stars reach the temperatures necessary for nuclear fusion at lower densities. After the depletion of hydrogen and helium in their cores, these stars progress up their mass tracks until new rounds of nuclear fusion are triggered. Such stars may burn nuclear fuel in a number of shells in addition to the fusion in their cores. The cores of these stars eventually reach conditions of instability where they undergo supernova explosions, leaving behind neutron stars or black holes. The fates of stars of varying masses are illustrated in Figure 13.18.

13.11   Further Reading

For an introduction to the theory of stars, see

R.J. Tayler, *The Stars: Their Structure and Evolution (2nd ed.)*, Cambridge: CUP, 1994.

D. Prialnik, *An Introduction to the Theory of Stellar Structure and Evolution (2nd ed.)*, Cambridge: CUP, 2010.

For an exhaustive treatment of nucleosynthesis in stars and supernovae, see

D. Arnett, *Supernovae and Nucleosynthesis: An Investigation of the History of Matter, from the Big Bang to the Present*, Princeton: PUP, 1996.

For a comprehensive survey of exotic stars, including white dwarfs, neutron stars and black holes, see

M. Camenzind, *Compact Objects in Astrophysics: White Dwarfs, Neutron Stars and Black Holes*, Berlin, Heidelberg: Springer, 2007.

For an account of supernovae and Gamma Ray Bursts, see

P. Podsiadlowski, *Supernovae and Gamma-Ray Bursts* in *Planets, Stars and Stellar Systems: Vol. 4, Stellar Structure and Evolution*, eds. T.D. Oswalt and M.A. Barstow, pp 693-733, Dordrecht: Springer, 2013.

14

Cosmology

14.1 Einstein's Universe

Einstein launched the modern era of cosmology in 1917 when he investigated the implications of general relativity for the structure of the universe as a whole. Einstein's starting point was that we do not have a privileged position within the universe and that on the very largest scales the universe is filled with matter of a uniform density. He also assumed that the universe is eternal and unchanging on cosmic timescales. He soon realized that to find a model that fitted this last assumption, he needed to amend the field equation of general relativity by the addition of an extra term. This term has the form $\Lambda g_{\mu\nu}$, where Λ is known as the *cosmological constant* and $g_{\mu\nu}$ is the metric of spacetime. Like the Einstein tensor, $g_{\mu\nu}$ is a covariantly constant symmetric rank 2 tensor. It is the only term that can be added to the field equation (6.47) without ruining its covariance. The revised Einstein equation is

$$G_{\mu\nu} - \Lambda g_{\mu\nu} = 8\pi G T_{\mu\nu}\,. \tag{14.1}$$

In the context of Newtonian gravity, the cosmological term introduces an extra force between any two objects throughout space. This is an additional universal force that is proportional to distance. Depending on the sign of Λ this force may be either attractive like gravity or repulsive. Furthermore, as the force does not decay with the distance squared, like the usual gravitational force, it can affect the structure of the universe on cosmic scales even if it is completely undetectable on much shorter length scales.

Einstein found a solution of his revised equation that describes a static eternal universe. The cosmological term, with Λ positive, provides a repulsive force that balances the tendency for the universe to collapse under the gravitational attraction of its component masses. However, this model is unstable. Any region that is slightly denser than average will undergo gravitational collapse. Any region that is slightly less dense than average will expand without limit. Einstein's static universe is not the universe that is observed. Instead, non-static solutions of the basic equation (14.1) give the best models we have for describing the universe on the largest, cosmic scales.

14.2 The Distance–Redshift Relationship

Just a century ago the Milky Way galaxy was believed to constitute the entire universe and any fuzzy patches in the night sky, such as the Andromeda Nebula, and the Large and Small Magellanic Clouds, were assumed to be gas clouds within our galaxy. This view could only be challenged when it became possible to determine distances to objects more precisely. Now we know that the Andromeda Nebula is a galaxy rather like the Milky Way, at a distance about 25 times the diameter of the Milky Way.

The Physical World. Nicholas Manton and Nicholas Mee, Oxford University Press (2017).
© Nicholas Manton and Nicholas Mee. DOI 10.1093/oso/9780198795933.001.0001

During the 20th century, a huge effort was expended on building the *distance ladder* to the furthest galaxies. The first step of the ladder beyond the solar system is to use parallax. If we can measure the shift in the apparent position of a star due to the changing position of the Earth as it orbits the Sun, then it is a simple exercise in geometry to calculate the distance to the star. If the parallax shift is one second of arc, the distance to the object is called 1 parsec (1 pc), and this corresponds to about 3.26 light years. The nearest stars, such as α Centauri and Sirius are a few parsecs away. Between 1989 and 1993, the distances to 120,000 of the nearest stars were determined by the satellite Hipparcos using this method, as discussed in section 13.2. Then we can move out into the depths of space by using *standard candles*, astronomical objects whose intrinsic luminosity L can be calculated. The intrinsic luminosities of these standard candles may be calibrated using the accurate distance measurements in the Hipparcos catalogue. The distances d to more remote examples are found by measuring their observed luminosity I and then using the formula (13.5),

$$I = \frac{L}{4\pi d^2}.$$

(14.2)

To understand the large-scale structure of the universe it is necessary to measure cosmic distances. These are reckoned in Mpc (megaparsecs). Henrietta Leavitt discovered the most important of the standard candles while working at Harvard College Observatory in 1912. These are very bright stars known as Cepheid variables whose luminosity varies in a regular cycle and for which the period of the cycle is related to the peak intrinsic luminosity of the star. This means that by measuring the length of the cycle it is possible to determine the star's intrinsic brightness and hence determine its distance. The nearest Cepheid variable is the star Polaris, which is about 120 pc away, but Cepheid variables are bright enough to be visible in the nearest galaxies, which means that they can be used to extend the distance ladder to our galactic neighbours. Leavitt discovered the critical period–luminosity relationship by analysing Cepheid variables in the nearby dwarf galaxy known as the Small Magellanic Cloud. Its stars are all situated at essentially the same distance from us, so differences in their observed luminosity reflect true differences in their intrinsic luminosity.

Edwin Hubble used Leavitt's discovery to establish that the universe is much larger than had previously been thought. He showed that our galaxy is just one among many and went on to estimate the distances to numerous relatively nearby galaxies. He also determined the rate at which these galaxies are moving towards or away from us by measuring the Doppler shift of spectral lines in their light. Suppose $\Delta\lambda = \lambda_o - \lambda$ is the shift in wavelength, where λ_o is the observed wavelength of a spectral line in the light reaching Earth from a distant galaxy and λ is the wavelength of the same spectral line produced by excited atoms in a laboratory on Earth. Then the redshift z of the galaxy is defined to be the fractional shift in wavelength $\frac{\Delta\lambda}{\lambda}$, and it is related to the velocity v at which the galaxy is receding from us by

$$z \equiv \frac{\Delta\lambda}{\lambda} = v.$$

(14.3)

Hubble soon concluded that the rate of recession of the galaxies is proportional to their distance,

$$v = H_0 d,$$

(14.4)

where d is the distance to a galaxy and H_0 is the constant of proportionality, now known as the *Hubble constant*. In 1929, Hubble made the momentous announcement that the entire universe is expanding. This observation has formed the bedrock of cosmology ever since.

The modern figure for the Hubble constant is $H_0 = 68$ km s^{-1} Mpc^{-1}, which means that an object at a distance of 1 Mpc $\simeq 3.26$ million light years is receding from us at 68 km s^{-1} due to the expansion of the universe. H_0 is a strange unit, as both 1 Mpc and 1 km are length units. More naturally, H_0 has units of inverse time, and it is approximately the inverse of 14 billion years. Thus Hubble's constant is a measure of the age of the universe, because if the expanding universe is run backwards in time, with H_0 held constant, then all galaxies coalesce about 14 billion years ago.

14.3   Friedmann–Robertson–Walker Cosmology

The standard model of cosmology is known as the *Friedmann–Robertson–Walker* or *FRW cosmology*. It is a highly symmetric solution of the Einstein equation (14.1), named after Alexander Friedmann who first derived it in 1922, and Howard Robertson and Arthur Walker who studied it during the 1930s. Unlike either the Schwarzschild or the Kerr metrics considered in Chapter 6, the FRW cosmology evolves with time. It is built on the assumption that space is perfectly *homogeneous* and *isotropic* on the largest scales.

Homogeneity means that 3-dimensional space is the same everywhere. It follows that cosmic, 4-dimensional spacetime can be neatly sliced into a sequence of 3-spaces parametrized by a time coordinate on which all observers agree. This is known as *cosmic time*. Cosmic time is similar to time in Newtonian physics, with the difference that cosmic spacetime has a Lorentzian metric due to the finite speed of light. Homogeneity implies that at any moment of cosmic time, all spatial points are equivalent geometrically. This is the modern equivalent of the Copernican principle, which says that the Earth is not in a privileged position such as the centre of the universe. Similarly, we now say that the Milky Way (or any other galaxy) does not have a privileged position in the cosmos. (Spacetime is not homogeneous, because the universe evolves with time.)

Isotropy means that from our location (or from any other) space looks the same in all directions. This implies that the universe is not rotating, because an axis of rotation would violate isotropy, so the spacetime metric cannot include any cross-terms between time and space. Also, all the space components of the metric must evolve in the same way.

The critical assumption of homogeneity on which the FRW models are built appears to be justified by observations. We will return to the observational evidence later. The assumption of isotropy is directly verified by astronomers, who observe roughly equal concentrations of galaxies in all directions.

Together, homogeneity and isotropy dramatically constrain the possible geometries of the universe and result in an enormous simplification of the physics. They imply that the spatial Riemann curvature tensor must take the following simple form:

$$^{(3)}R_{abcd} = C(h_{ac}h_{bd} - h_{ad}h_{bc}),\tag{14.5}$$

where h_{ab} is the 3-dimensional metric tensor. This Riemann tensor, derived in Chapter 5, is the most general one for a space of constant curvature. If C is positive then the cosmos is spherical and of finite size, if $C = 0$ the cosmos is flat, and if C is negative it is hyperbolic. If C is negative or zero, the cosmos is infinitely large.

Combining the possible spatial geometries with the cosmic time coordinate produces the following 4-dimensional FRW metrics in polar coordinates,

$$d\tau^2 = dt^2 - a^2(t)\big(d\chi^2 + f^2(\chi)(d\vartheta^2 + \sin^2\vartheta\, d\varphi^2)\big)\,, \tag{14.6}$$

where $f^2(\chi) = \sin^2\chi$ for the spherical spatial geometry, $f^2(\chi) = \chi^2$ for the flat geometry and $f^2(\chi) = \sinh^2\chi$ for the hyperbolic geometry. The only remaining freedom is the scale parameter $a(t)$, which is a function of the cosmic time t. In a spherical universe $a(t)$ is the time-dependent radius, and the spatial curvature is $K(t) = \frac{1}{a^2(t)}$. In the hyperbolic case $K(t) = -\frac{1}{a^2(t)}$. Four-dimensional spacetime is curved even in the spatially flat case. (In most of what follows, for brevity, we will write a for $a(t)$.)

As the universe expands or contracts, each galaxy g retains the same coordinate labels $(\chi_\mathrm{g}, \vartheta_\mathrm{g}, \varphi_\mathrm{g})$ from epoch to epoch. These are known as *comoving coordinates*. The assumption of homogeneity means that the matter within the universe cannot have any localized motion. Put another way, the random relative motion of the galaxies is negligible. We can choose our own location to be $\chi = 0$. Relative to us, χ is a radial coordinate, and ϑ, φ are spherical polar coordinates on the sky. The distance to another galaxy depends on both χ and a.

It is also natural to express the FRW geometry in spatial Cartesian coordinates. All points are equivalent, but in these coordinates we can place ourselves at the origin. The metric is then

$$d\tau^2 = dt^2 - a^2(t)\frac{dx^2 + dy^2 + dz^2}{\left(1 + \frac{k}{4}(x^2 + y^2 + z^2)\right)^2}\,, \tag{14.7}$$

where the curvature characteristic is respectively $k = +1, 0$ or -1 in the cases of spherical, flat or hyperbolic geometry. Here we have used the expression (5.74) for the 3-sphere metric, and its flat and hyperbolic analogues. We will concentrate on the flat geometry with $k = 0$, as this is the simplest case and appears to have most physical significance. The metric tensor $g_{\mu\nu}$ and its inverse $g^{\mu\nu}$ then take the simple forms

$$g_{\mu\nu} = \mathrm{diag}\left(1, -a^2, -a^2, -a^2\right)\,, \quad g^{\mu\nu} = \mathrm{diag}\left(1, -\frac{1}{a^2}, -\frac{1}{a^2}, -\frac{1}{a^2}\right)\,. \tag{14.8}$$

14.3.1   Einstein's equation and the FRW metric

Thus far, we have deduced the possible geometries of the universe solely from our assumptions of homogeneity and isotropy. We need to show these geometries are consistent with general relativity and satisfy Einstein's equation. This is a straightforward and instructive exercise. We assume for the moment that $k = 0$ and also $\Lambda = 0$.

The only non-zero derivatives of the metric (14.8) are $g_{xx,t}$, $g_{yy,t}$ and $g_{zz,t}$, which all equal $-2a\dot{a}$, where the dot indicates a time derivative. The non-zero Christoffel symbols (5.50) therefore all have one time index and two identical space indices.

They are

$$\Gamma^t_{\ yy} = \Gamma^t_{\ zz} = \Gamma^t_{\ xx} \quad = \quad \frac{1}{2}g^{t\sigma}(g_{x\sigma,x} + g_{\sigma x,x} - g_{xx,\sigma}) = \frac{1}{2}g^{tt}(g_{xt,x} + g_{tx,x} - g_{xx,t})$$

$$= \quad -\frac{1}{2}g^{tt}g_{xx,t} = a\dot{a}\,, \tag{14.9}$$

$$\Gamma^y_{\ ty} = \Gamma^y_{\ yt} = \Gamma^z_{\ tz} = \Gamma^z_{\ zt} = \Gamma^x_{\ tx} = \Gamma^x_{\ xt} \quad = \quad \frac{1}{2}g^{x\sigma}(g_{x\sigma,t} + g_{\sigma t,x} - g_{xt,\sigma})$$

$$= \quad \frac{1}{2}g^{xx}g_{xx,t} = \frac{1}{2}\left(-\frac{1}{a^2}\right)(-2a\dot{a})$$

$$= \quad \frac{\dot{a}}{a}\,. \tag{14.10}$$

The Ricci tensor (6.36) is

$$R_{\mu\nu} = \Gamma^\rho_{\ \mu\nu,\rho} - \Gamma^\rho_{\ \rho\nu,\mu} + \Gamma^\alpha_{\ \mu\nu}\Gamma^\rho_{\ \alpha\rho} - \Gamma^\alpha_{\ \rho\nu}\Gamma^\rho_{\ \alpha\mu}\,, \tag{14.11}$$

and its only non-zero components are

$$\begin{aligned}
R_{tt} &= \Gamma^\rho_{\ tt,\rho} - \Gamma^\rho_{\ \rho t,t} + \Gamma^\alpha_{\ tt}\Gamma^\rho_{\ \alpha\rho} - \Gamma^\alpha_{\ \rho t}\Gamma^\rho_{\ \alpha t} \\
&= -\Gamma^\rho_{\ \rho t,t} - \Gamma^\alpha_{\ \rho t}\Gamma^\rho_{\ \alpha t} \\
&= -\Gamma^x_{\ xt,t} - \Gamma^y_{\ yt,t} - \Gamma^z_{\ zt,t} - \Gamma^x_{\ xt}\Gamma^x_{\ xt} - \Gamma^y_{\ yt}\Gamma^y_{\ yt} - \Gamma^z_{\ zt}\Gamma^z_{\ zt} \\
&= -3\Gamma^x_{\ xt,t} - 3\left(\Gamma^x_{\ xt}\right)^2 \\
&= -3\left(\frac{\ddot{a}}{a} - \frac{\dot{a}^2}{a^2}\right) - 3\frac{\dot{a}^2}{a^2} \\
&= -3\frac{\ddot{a}}{a}\,, \tag{14.12}
\end{aligned}$$

and

$$\begin{aligned}
R_{yy} = R_{zz} = R_{xx} &= \Gamma^\rho_{\ xx,\rho} - \Gamma^\rho_{\ \rho x,x} + \Gamma^\alpha_{\ xx}\Gamma^\rho_{\ \alpha\rho} - \Gamma^\alpha_{\ \rho x}\Gamma^\rho_{\ \alpha x} \\
&= \Gamma^t_{\ xx,t} + \Gamma^t_{\ xx}\Gamma^\rho_{\ t\rho} - \Gamma^x_{\ tx}\Gamma^t_{\ xx} - \Gamma^t_{\ xx}\Gamma^x_{\ tx} \\
&= \Gamma^t_{\ xx,t} + 3\Gamma^t_{\ xx}\Gamma^x_{\ tx} - 2\Gamma^x_{\ tx}\Gamma^t_{\ xx} \\
&= \Gamma^t_{\ xx,t} + \Gamma^t_{\ xx}\Gamma^x_{\ tx} = (a\ddot{a} + \dot{a}^2) + a\dot{a}\left(\frac{\dot{a}}{a}\right) \\
&= a\ddot{a} + 2\dot{a}^2\,. \tag{14.13}
\end{aligned}$$

The Ricci scalar is then

$$\begin{aligned}
R = g^{\mu\nu}R_{\mu\nu} &= R_{tt} - \frac{1}{a^2}(R_{xx} + R_{yy} + R_{zz}) \\
&= -3\frac{\ddot{a}}{a} - \frac{3}{a^2}(a\ddot{a} + 2\dot{a}^2) \\
&= -6\left(\frac{\ddot{a}}{a} + \frac{\dot{a}^2}{a^2}\right)\,. \tag{14.14}
\end{aligned}$$

This leads to the following components of the Einstein tensor (6.46):

$$G_{tt} = R_{tt} - \frac{1}{2}Rg_{tt} = -3\frac{\ddot{a}}{a} + 3\left(\frac{\ddot{a}}{a} + \frac{\dot{a}^2}{a^2}\right)$$

$$= 3\frac{\dot{a}^2}{a^2},$$

$$G_{yy} = G_{zz} = G_{xx} = R_{xx} - \frac{1}{2}Rg_{xx} = a\ddot{a} + 2\dot{a}^2 + 3\left(\frac{\ddot{a}}{a} + \frac{\dot{a}^2}{a^2}\right)(-a^2)$$

$$= a\ddot{a} + 2\dot{a}^2 - 3a\ddot{a} - 3\dot{a}^2$$

$$= -2a\ddot{a} - \dot{a}^2. \tag{14.15}$$

We can model the matter content of the universe with the energy–momentum tensor of an ideal fluid (6.28)

$$T_{\mu\nu} = (\rho + P)\,v_\mu v_\nu - Pg_{\mu\nu}, \tag{14.16}$$

where ρ is the energy density and P is the pressure. Homogeneity implies that ρ and P are functions only of time t. The energy density characterizes the matter in gas clouds, stars and galaxies, and also the energy of the radiation in the universe. The matter has no organized motion with respect to the comoving coordinates, and the absence of random motion means that it exerts negligible pressure. The pressure P is mainly due to the radiation. Therefore, in a comoving frame, $v_\mu = (1, 0, 0, 0)$ and the metric is as in equation (14.8), so the energy–momentum tensor has the simple form $T_{\mu\nu} = \mathrm{diag}(\rho, a^2P, a^2P, a^2P)$ and Einstein's equation reduces to

$$\begin{pmatrix} G_{tt} & 0 & 0 & 0 \\ 0 & G_{xx} & 0 & 0 \\ 0 & 0 & G_{yy} & 0 \\ 0 & 0 & 0 & G_{zz} \end{pmatrix} = 8\pi G \begin{pmatrix} \rho & 0 & 0 & 0 \\ 0 & a^2P & 0 & 0 \\ 0 & 0 & a^2P & 0 \\ 0 & 0 & 0 & a^2P \end{pmatrix}. \tag{14.17}$$

There is one equation for G_{tt},

$$3\frac{\dot{a}^2}{a^2} = 8\pi G\rho, \tag{14.18}$$

and another for G_{xx},

$$-2\frac{\ddot{a}}{a} - \frac{\dot{a}^2}{a^2} = 8\pi GP. \tag{14.19}$$

The equations for G_{yy} and G_{zz} are identical to the equation for G_{xx}. Einstein's equation normally relates two symmetric rank 2 tensors, so it consists of ten coupled equations for the ten components of the metric tensor, but the symmetry underlying the FRW geometry reduces Einstein's equation to just two equations that determine how the scale factor a varies with time.

The reward for all this algebra is a very important conclusion. Eliminating $\frac{\dot{a}^2}{a^2}$ between equations (14.18) and (14.19) gives

$$\frac{\ddot{a}}{a} = -\frac{4\pi G}{3}(\rho + 3P). \tag{14.20}$$

On the very reasonable physical assumption that the energy density $\rho > 0$ and the pressure $P \geq 0$, this implies that $\ddot{a} < 0$, which means that the universe cannot be static; it must be dynamic. This is why Einstein introduced the cosmological constant Λ into his field equation. According to Gamow writing in 1960, Einstein described this as the biggest blunder of his life. If Einstein had pursued the consequences of his original field equation with no cosmological constant to its logical conclusion, he could have predicted the expansion (or contraction) of the universe, which would, perhaps, have been the greatest prediction of any scientist of any age.

If the universe is expanding and $\ddot{a} < 0$, then the rate of expansion is decreasing. This is not so surprising as gravity is an attractive force. The expansion is slowing due to the gravitational attraction of the matter and radiation.

14.3.2 The general FRW cosmological solutions

The general FRW cosmology has a non-zero cosmological constant Λ and may be spherical or hyperbolic, so k need not be zero but can be ± 1. Equations (14.18) and (14.19) generalize to

$$3\frac{\dot{a}^2}{a^2} + 3\frac{k}{a^2} = 8\pi G\rho + \Lambda \,, \tag{14.21}$$

$$-2\frac{\ddot{a}}{a} - \frac{\dot{a}^2}{a^2} - \frac{k}{a^2} = 8\pi G P - \Lambda \,. \tag{14.22}$$

The FRW model, which is the basis for all modern studies of cosmology, boils down to these two simple equations. In addition, one needs to specify the type of matter and energy in the cosmos, to determine the relationship between P and ρ.

For most of its history the energy in the universe has mainly been locked up as the rest mass of matter—the matter from which the stars and galaxies are composed. In this matter-dominated limit the universe is a collection of non-interacting, slowly moving particles with $P = 0$ and energy–momentum tensor $T_{\mu\nu} = \mathrm{diag}(\rho, 0, 0, 0)$. However, during its first few hundred thousand years, most of the universe's energy was in the form of radiation or relativistic particles. In Chapter 10, we showed that $\rho = \frac{E}{V} = 3P$ for black body radiation. This is the relationship between pressure and energy density in a radiation-dominated universe.

The simplest solution to these FRW equations is static, with $a = a_0$, a constant. This is the Einstein static universe. If it is matter-dominated, and $P = 0$, then the two equations require that

$$\Lambda = \frac{k}{a_0^2} \quad \text{and} \quad \Lambda = 4\pi G\rho \,. \tag{14.23}$$

As ρ is positive, Λ is positive, so $k = 1$. The Einstein static universe is therefore a finite, spherical universe with cosmological constant $\Lambda_{\mathrm{E}} = 4\pi G\rho$ and radius $a_0 = \Lambda_{\mathrm{E}}^{-\frac{1}{2}}$.

However, a static universe is not what is observed, so we need time-dependent solutions of the FRW equations. Taking the time derivative of the first equation and multiplying by $\frac{a}{\dot{a}}$ we find

$$6\frac{\ddot{a}}{a} - 6\frac{\dot{a}^2}{a^2} - 6\frac{k}{a^2} = 8\pi G\frac{a\dot{\rho}}{\dot{a}} \,, \tag{14.24}$$

whereas adding the first and second equation gives

$$-2\frac{\ddot{a}}{a} + 2\frac{\dot{a}^2}{a^2} + 2\frac{k}{a^2} = 8\pi G(\rho + P) \,. \tag{14.25}$$

The left-hand sides differ only by a factor of -3 so the right-hand sides differ by the same factor. Therefore

$$\frac{a\dot{\rho}}{\dot{a}} = -3(\rho + P)\,, \tag{14.26}$$

which simplifies to

$$a\frac{d\rho}{da} = -3(\rho + P)\,. \tag{14.27}$$

This is called the *continuity equation*, and it is a consequence of the conservation of energy and momentum. It is valid for any values of Λ and k.

We will now assume that the relationship between P and ρ is

$$P = w\rho \tag{14.28}$$

with w a constant. The value $w = 0$ describes non-interacting matter, a *dust* with $P = 0$, and $w = \frac{1}{3}$ describes radiation. The continuity equation becomes

$$a\frac{d\rho}{da} = -3(1 + w)\rho\,, \tag{14.29}$$

which is a (separable) differential equation whose solution is

$$\rho a^{3(1+w)} = c\,, \tag{14.30}$$

with c a constant. For dust the power of a is 3, and for radiation the power is 4. The constancy of ρa^3 for dust is just a statement of conservation of mass. The density decreases as the volume of the universe increases. The constancy of ρa^4 for radiation is also readily understood. Just as with matter, if the volume of the universe increases, the number density of photons decreases in proportion, but there is an additional decrease in the energy density, because the energy ε of a photon of wavelength λ is equal to

$$\varepsilon = \frac{2\pi\hbar}{\lambda}\,. \tag{14.31}$$

As the universe expands, the wavelength of each photon increases with a due to its redshift, as we will see, and therefore its energy decreases. When this is taken into account, the energy density of the radiation is proportional to a^{-4}.

Substituting for ρ in terms of a, the first FRW equation (14.21) reduces to

$$\dot{a} = \left(\frac{8\pi G}{3}ca^{-1-3w} - k + \frac{\Lambda}{3}a^2\right)^{\frac{1}{2}}\,. \tag{14.32}$$

This differential equation can be integrated, numerically if necessary. The second FRW equation (14.22) is automatically satisfied because we have solved the continuity equation (14.29). Possible cosmological solutions of the FRW equations are summarized in Figure 14.1.

There are special cases where the solution has a simple closed form, for example, a dust in a flat cosmos with no cosmological constant. We will derive some of these solutions in section 14.6 on the Big Bang. The FRW equations can also be solved for a mixture of dust and radiation, and for materials with exotic pressure–density relations.

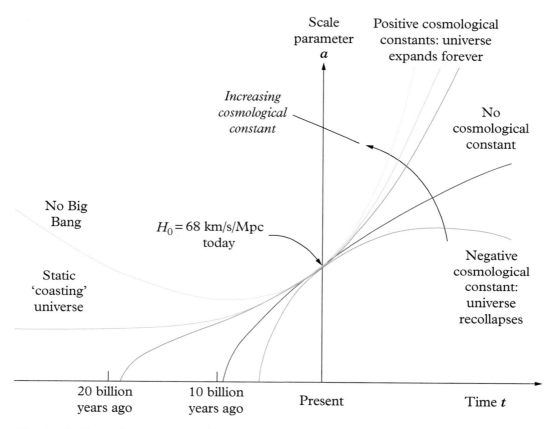

Fig. 14.1 The scale parameter $a(t)$ plotted for flat FRW cosmologies, with $w = 0$ and the cosmological constant taking a range of values. With a negative cosmological constant the universe recollapses. With no cosmological constant the expansion of the universe slows and comes to a halt asymptotically. With a positive cosmological constant the expansion of the universe accelerates. With the Hubble parameter H_0 held fixed at its currently observed value, the age of the universe increases as the cosmological constant increases.

Hubble's law (14.4) is a natural feature of the FRW cosmology, at least as an approximate result. In an FRW universe, galaxies cannot have any peculiar motion, such as a flow towards a neighbouring cluster of galaxies. The velocities of galaxies are simply due to the expansion of the universe. If the distance to a remote galaxy is $d = a\chi$, where a is the scale parameter, then d increases with a, but χ is fixed, so the rate at which the galaxy recedes from us is $v = \dot{a}\chi$. If we define

$$\frac{v}{d} = \frac{\dot{a}}{a} \equiv H_0(t) \,, \tag{14.33}$$

then, by definition, equation (14.4) is satisfied, and we have related Hubble's constant to the rate of change of the scale factor. However, as we have seen, usually $\ddot{a} < 0$, so the Hubble 'constant' $H_0(t)$ is really a time-dependent parameter. It is only for relatively nearby galaxies that it is reasonable to make a linear approximation and take the Hubble parameter to be a constant. The rate at which the expansion of the universe is slowing can be expressed in terms of a dimensionless parameter known as the deceleration parameter that is defined as

$$q_0 = -\frac{\ddot{a}a}{\dot{a}^2} \,. \tag{14.34}$$

14.4 Cosmological Redshift

How does the FRW cosmology tie in with Hubble's observations about the redshift of distant galaxies? Consider first a source of discrete pulses of radiation such as a pulsar, which emits a pulse of radio waves in our direction once every rotation. In its rest frame, the period of the pulses equals the period of the pulsar's rotation. However, if the pulsar is receding from us, the time intervals between our detection of pulses is greater than the rotational period of the pulsar, as each successive pulse travels further than the last. The frequency of the observed pulse train is therefore decreased and its wavelength is increased. In other words, the pulse train is redshifted. Similarly, each successive pulse from a pulsar at a cosmological distance travels further than the last due to the expansion of the universe, so the interval between pulses, as detected here, is greater than the rotational period of the pulsar, i.e. the pulse train undergoes a cosmological redshift due to the expansion of the universe.

In comoving polar coordinates, a pulse of light from a pulsar in a distant galaxy travels to us on a radial null curve,

$$d\tau^2 = 0 = dt^2 - a^2(t)\, d\chi^2 \,, \tag{14.35}$$

starting at some comoving radius χ and arriving at our location $\chi = 0$. Along this curve, $|d\chi| = \frac{dt}{a(t)}$. The pulse is emitted at time t_e and observed at time t_o. The actual distance travelled by each pulse is found by integrating $a(t)\,|d\chi|$ along the null curve, and this clearly increases with time. But in the FRW model the comoving coordinates of each galaxy remain fixed for the entire history of the universe, so each pulse travels the same comoving distance

$$\chi = \int_{t_e}^{t_o} \frac{dt}{a(t)} \,. \tag{14.36}$$

The next pulse emitted a short time δt_e later, and observed δt_o later, travels the same comoving distance, so

$$\int_{t_e+\delta t_e}^{t_o+\delta t_o} \frac{dt}{a(t)} = \chi = \int_{t_e}^{t_o} \frac{dt}{a(t)} \,, \tag{14.37}$$

and the difference between these integrals is zero. These integrals coincide for almost their entire intervals, so the only contribution to their difference is from close to their endpoints. Therefore

$$\int_{t_o}^{t_o+\delta t_o} \frac{dt}{a(t)} - \int_{t_e}^{t_e+\delta t_e} \frac{dt}{a(t)} = 0 \,. \tag{14.38}$$

As the scale parameter $a(t)$ changes negligibly between pulses, these remaining integrals reduce to products of their integrands and integration intervals, so

$$\frac{\delta t_o}{a(t_o)} - \frac{\delta t_e}{a(t_e)} = 0 \,. \tag{14.39}$$

A simple rearrangement then gives

$$\frac{\delta t_e}{\delta t_o} = \frac{a(t_e)}{a(t_o)} = \frac{\omega_o}{\omega_e}, \tag{14.40}$$

where ω_e and ω_o represent the frequency of the emission and observation of a train of pulses emitted at regular intervals of δt_e and observed at intervals of δt_o. Equivalently, if δt_e and δt_o are the time intervals between wave peaks of an emitted and observed light wave, then ω_e and ω_o are the frequencies of the emitted and observed light.

In an expanding universe, this corresponds to a redshift $z = \frac{\Delta\lambda}{\lambda}$, where

$$1 + z = \frac{\lambda_o}{\lambda_e} = \frac{\omega_e}{\omega_o} = \frac{a(t_o)}{a(t_e)}, \tag{14.41}$$

so

$$z = \frac{a(t_o) - a(t_e)}{a(t_e)}. \tag{14.42}$$

Thus, in the FRW model there is a simple relationship between the redshift and the scale factor of the universe $a(t)$. It is often said that the expansion of the universe stretches the wavelength of light as it travels to us from a distant galaxy. In a sense this is true, as the change in wavelength and frequency is a direct consequence of the distortion of space, but it would be wrong to infer that space is physically acting on the light wave in some way. A discrete train of pulses undergoes precisely the same redshift and in this case there is nothing between the pulses for space to act on. In general, it is safest to consider the change in frequency as due to differences in the metric between the point of emission and the point of observation of the light. This reflects the difference in time and space measurements for a comoving emitter and observer at those two points.

14.5   Newtonian Interpretation of the FRW Cosmology

Let us consider an FRW cosmology with $\Lambda = 0$, but with k not fixed. If we multiply equation (14.21) by $\frac{a^2}{6}$ we obtain

$$\frac{1}{2}\dot{a}^2 + \frac{1}{2}k = \frac{4\pi}{3}G\rho a^2. \tag{14.43}$$

This has a simple Newtonian interpretation.

Imagine a comoving ball of uniform density ρ and radius $d = \sigma a$, in a flat Newtonian space. Its mass is $M = \frac{4}{3}\pi\rho\sigma^3 a^3$. As long as the radius is small, the relative velocities are non-relativistic, so the physics is well described in Newtonian terms. Consider the energy of a grain of material within a thin shell surrounding the comoving ball. As the universe is assumed to be isotropic and therefore spherically symmetric, the grain will feel no gravitational attraction due to the material outside the sphere. In accordance with Birkhoff's theorem, discussed in section 6.7, the grain will only be affected gravitationally by the material within the ball.[1] The gravitational energy per unit mass of the grain is therefore

$$-\frac{GM}{d} = -\frac{4\pi}{3}\frac{G\rho\sigma^3 a^3}{\sigma a} = -\frac{4\pi}{3}G\rho\sigma^2 a^2. \tag{14.44}$$

---

[1] This is slightly curious in a homogeneous cosmos, but makes sense for this ball of matter.

As the ball is comoving, the velocity of the grain relative to the centre is $\sigma\dot{a}$ and its kinetic energy per unit mass is $\frac{1}{2}\sigma^2\dot{a}^2$, so the total energy per unit mass is

$$\frac{1}{2}\sigma^2\dot{a}^2 - \frac{4\pi}{3}G\rho\sigma^2a^2 \, . \tag{14.45}$$

If we set this equal to the constant $-\frac{1}{2}k\sigma^2$, we obtain equation (14.43), which shows that this equation simply represents the conservation of energy and $-\frac{1}{2}k\sigma^2$ is the total energy per unit mass of the grain.

If $k < 0$, then the sum of the grain's kinetic energy and gravitational potential energy is positive and the grain will eventually escape to infinity. (But there is nothing special about this grain or the comoving sphere relative to which it has been defined. The sphere's radius falls out of the calculation, so the result applies to all the material in the universe.) $k < 0$ corresponds to a hyperbolic universe, and such a universe will expand forever. If $k > 0$, the sum of the grain's kinetic and potential energy is negative, so the grain is gravitationally bound, along with the rest of the material in the universe. This corresponds to a spherical universe and such a universe will eventually undergo gravitational collapse. Finally, if $k = 0$, the kinetic and potential energies exactly balance and we have the flat cosmology. This is the borderline between eventual re-contraction and eternal expansion. It occurs if

$$\frac{4\pi}{3}G\rho\sigma^2a^2 = \frac{1}{2}\sigma^2\dot{a}^2 \, . \tag{14.46}$$

The critical density is therefore

$$\rho_{\text{crit}} = \frac{3}{8\pi G}\frac{\dot{a}^2}{a^2} = \frac{3H_0^2}{8\pi G} \, , \tag{14.47}$$

depending just on H_0, the current Hubble parameter.

Much effort has gone into measuring ρ to determine the ultimate fate of the universe. ρ is often expressed in terms of the critical density as

$$\Omega = \frac{\rho}{\rho_{\text{crit}}} \, . \tag{14.48}$$

If $\Omega > 1$, then $k > 0$ and the universe is spherical and will eventually contract. If $\Omega = 1$, then $k = 0$ and the universe is flat. If $\Omega < 1$, then $k < 0$ and the universe is hyperbolic and will expand forever. Recent analysis shows that Ω is very close to 1, but only when the cosmological constant is taken into account. This will be discussed in section 14.9.

14.6 The Big Bang

If the universe is expanding now, then the galaxies and all the intergalactic matter must have been much closer together in the past. The energy density of the universe was greater then, and the temperature was higher. It appears that the universe was initially compressed to a point or into a very small region, and that it has been expanding from this point ever since.

The beginning of the universe has been given a catchy name—the Big Bang. It is sometimes represented as an explosion within the universe and this is definitely not correct. It suggests that the universe was a pre-existing container from within which the material forming the stars and galaxies burst forth. This leads to the misconception that the Big Bang happened at a particular place. In fact, if the Big Bang happened anywhere, it happened everywhere at once. The idea is that, in accordance with the FRW model, the universe in its entirety—space, time and matter—began at the Big Bang.

It helps to consider the analogy of a balloon that is being blown up, as depicted in Figure 14.2. The main difference is that the surface of a balloon is 2-dimensional whereas space is 3-dimensional. As the balloon expands, every point on its surface moves away from every other point, and the further apart two points are, the faster they recede from each other, just like the galaxies in the real universe. We can run the expansion backwards until every point on the balloon coalesces into a single point which represents the beginning of the universe. From this perspective we can see that every point of the balloon universe is equally distant from its origin, and the balloon Big Bang happened everywhere simultaneously.

Fig. 14.2 Balloon universe.

14.6.1   The age of the universe

Let us again suppose for the moment that $\Lambda = k = 0$. In a matter-dominated universe, with density ρ and zero pressure, $\rho a^3 = c$. Substituting for ρ in equation (14.18), we find successively

$$\dot{a}^2 \propto a^{-1}, \quad \dot{a} \propto a^{-\frac{1}{2}}, \quad a^{\frac{3}{2}} \propto t, \tag{14.49}$$

so the scale factor varies as $a \propto t^{\frac{2}{3}}$. Here, the origin of time is the time of the Big Bang, when $a = 0$. This solution allows us to estimate the age of the universe. Assuming that the universe has been matter-dominated for almost all its history,

$$H_0(t) = \frac{\dot{a}}{a} = \frac{\frac{2}{3} t^{-\frac{1}{3}}}{t^{\frac{2}{3}}} = \frac{2}{3} t^{-1}. \tag{14.50}$$

Taking the Hubble parameter now to be $H_0 = 68$ km s^{-1} Mpc^{-1}, this gives the age of the universe as $\frac{2}{3} \frac{1}{H_0}$, which is about 10 billion years, as indicated by the curve marked *No cosmological constant* in Figure 14.1.

However, there is now abundant evidence that this figure is too low by quite a significant margin. For instance, the ages of the oldest stars, typically found in globular clusters, are known to be greater than this. The explanation is that the cosmological constant

Λ is positive, and it is this rather than the matter density that provides the dominant contribution to the expansion of the universe in the present epoch. This conclusion has been reached by combining data from gravitational lensing, galactic clustering, the brightness–redshift relation for distant supernovae, and most significantly the anisotropies in the cosmic microwave background. The current best figure for the time since the Big Bang derived from these observations is 13.8 billion years. We will return to this subject in section 14.9.

The cosmological term is independent of the scale factor $a(t)$ of the universe. Travelling back in time we reach an epoch when energy in the form of matter dominated the cosmological term. Much further back in time, in the very early universe, the dominant form of energy was radiation. This is because in a radiation-dominated universe, $\rho = 3P$, and equations (14.28) and (14.30) imply that $\rho \propto \frac{1}{a^4}$, so as the universe shrinks the energy content of the radiation increases faster than the energy content of the matter ($\rho \propto \frac{1}{a^3}$). Substituting $\rho \propto \frac{1}{a^4}$ in equation (14.18), we find that for a radiation-dominated universe

$$\dot{a}^2 \propto a^{-2}, \quad \dot{a} \propto a^{-1}, \quad a^2 \propto t. \tag{14.51}$$

We conclude that in the very early universe the scale factor varied as $a \propto t^{\frac{1}{2}}$, where again $t = 0$ represents the time of the Big Bang. This conclusion is not significantly affected by non-zero Λ or k, because provided there is any positive amount of matter or radiation, the term $8\pi G\rho$ dominates the terms involving Λ and k as the scale parameter a approaches zero.

There is extremely good evidence for the reality of the Big Bang. First, as Hubble discovered, the motion of distant galaxies shows that the universe is expanding, but there is also independent observational evidence. The universe would have been much hotter and denser in the past. For the first couple of minutes it would have been a nuclear furnace in which fusion reactions created deuterium, helium and traces of other very light elements, such as lithium. (These conditions would not have persisted long enough for any heavier nuclei to be synthesized. All the heavier elements were created much later in stars and supernova explosions.) It is possible to measure the amounts of deuterium, helium and other light elements in the universe and the observations closely match the amounts deduced from modelling the Big Bang. The amount of deuterium, in particular, is very sensitive to the conditions in the early universe and enables astrophysicists to determine the energy density in this epoch as well as the proportion of it that was in the form of matter and the proportion of it that was radiation. This has a bearing on estimates of the current density of matter in the universe, a question that we will now consider.

14.7 Dark Matter

There are two ways to determine the amount of material in the universe. One is to measure its gravitational effects, and the other is to measure the amount of light that is being emitted by luminous objects. All objects have a gravitational attraction, but not all objects emit or scatter light, so we would expect the first measure to give an answer that is bigger than the second, and it does. Even so, it is rather surprising that most of the material in the universe does not emit or even scatter light. Astronomers call this invisible stuff *dark matter* simply because we cannot see it.

Several methods have been used to estimate the density of the material that forms the cosmos. For instance, the rate of rotation of a spiral galaxy viewed edge-on can be measured from the Doppler shift of the light from its two edges. The starlight from one edge is redshifted as the stars move away from us, while the starlight from the other edge is blueshifted. From the rate of rotation we can deduce the total mass necessary to gravitationally bind the galaxy together. It is clear from these measurements that the rate of rotation of such galaxies is so high that they would fly apart if they were solely composed of their visible material.

Our galaxy, and many comparable ones, are accompanied by around 100 globular clusters. These are tightly packed swarms of up to a million stars that are gravitationally bound to the galaxy and reside in a spherical halo. Studies of the velocity distribution of globular clusters show that the host galaxy must again contain a large amount of invisible material or the globular clusters would escape. Other studies have analysed the motions of galaxies within galactic clusters and have reached a similar conclusion. A completely different technique is to estimate the mass of a galaxy cluster from its gravitational lensing effect on more distant galaxies, as described in section 6.9. These studies all agree that the universe contains a great deal of dark matter.

This begs a big question: what is it? There have been many suggestions. One class of proposed objects are the MACHOs (Massive Compact Halo Objects). These are massive bodies that are too faint to be seen and include the remnants of burnt-out stars, such as white dwarfs, neutron stars or black holes, and very faint stars, such as brown dwarfs. If the galaxy really is filled with huge quantities of MACHOs then there is an effect that would betray their presence. The chance alignment of a MACHO with a background star would occasionally produce a sharp spike in the star's brightness due to gravitational lensing. Astronomers have undertaken systematic searches for these microlensing events in the halo of our galaxy and have concluded that they are too rare for MACHOs to account for a significant proportion of the dark matter. Another possibility is that dark matter consists of dark gas clouds that have yet to condense into stars. This possibility can also be ruled out. Cosmologists refer to ordinary matter such as this as *baryonic* matter, as it consists mainly of protons and neutrons. The observed abundance of deuterium places strict limits on the density of baryonic matter, as the primordial nucleosynthesis of deuterium is very dependent on the density in the early universe. If dark matter were composed of any form of baryonic matter, then this would ruin the agreement between observation and models of primordial nucleosynthesis.

This leaves particles at the other end of the mass spectrum. Many physicists now believe that dark matter consists of vast quantities of stable particles that only interact very weakly with baryonic matter, which is why they are not found mixed with ordinary matter and why they have not yet been identified in cosmic rays. Rather whimsically, such particles are known as WIMPs (Weakly Interacting Massive Particles). Here, weakly interacting is a general term that does not simply refer to the weak force in the Standard Model. WIMPs do not interact via the electromagnetic or strong forces or they would have been detected long ago; they may interact via the weak force, or via even weaker forces, as yet unknown, or solely through gravity.

Neutrinos were created in vast quantities in the early universe and are still being produced by stars and supernovae. It was once thought that they might be the dark matter. However, we now know that the masses of the neutrinos are too small ($m_\nu < 1$ eV) to account for all of it. Because of their small masses, neutrinos move relativistically and are classed as *hot dark matter*.

In section 14.10 we will look at galaxy formation in the early universe. Computer simulations have shown that complex structures such as galaxies and galaxy clusters can only form in a universe dominated by *cold dark matter* (CDM). This favours WIMPs moving at non-relativistic speeds, and such particles must be much more massive than neutrinos.

In summary, dark matter is believed to consist of vast numbers of an as yet unknown type of stable particle, produced in the very early universe. Identifying such a particle is one of the main targets of the research at the Large Hadron Collider. There are also ongoing attempts to track down these elusive particles in neutrino observatories and cosmic ray detectors. A number of possible relic particles have been suggested, many of which are found in proposed extensions of the Standard Model. One of the leading candidates is the lightest exotic particle predicted by the theory known as *supersymmetry*, which is discussed in section 15.5.2.

14.8   The Cosmic Microwave Background

The most conclusive evidence for the Big Bang was discovered in 1964 by Arno Penzias and Robert Wilson who were building a very sensitive antenna in New Jersey for Bell Labs. Their equipment was plagued with background noise that they initially assumed was due to a fault. Eventually, the explanation was provided by the Princeton astrophysicists Robert Dicke, Jim Peebles and David Wilkinson, who were preparing to search for microwaves from the early universe. Penzias and Wilson had discovered the *cosmic microwave background* (CMB). Its existence had been predicted as early as 1946 by Gamow and his team.

So where do all these microwaves come from? After the era of primordial nucleosynthesis, the expanding universe consisted of a hot plasma of charged particles composed largely of hydrogen and helium nuclei and free electrons. This plasma was awash with photons bouncing around and scattering off the nuclei and electrons. As the radiation was in thermal equilibrium with the matter, it had a black body spectrum (10.108) at a temperature equal to the ambient temperature of the universe.

As the universe expanded the energy density dropped and the temperature fell. After around 380,000 years of expansion the plasma cooled to 3100 K, which was cool enough for hydrogen atoms to form. Cosmologists refer to this as *recombination*, even though it was the epoch when atoms first formed. Prior to this, any electron that combined with a proton to form an atom would quickly have been kicked out again by a passing photon. However, with the expansion of the universe, the redshift in wavelengths meant that most photons now had insufficient energy. Some atoms were still in excited states, but a negligible fraction were ionized. Just as hydrogen gas is transparent, so now the universe was transparent, but it was still bathed in photons with a black body spectrum at 3100 K.

After this, the photons continued to race across the universe for billions of years and the universe continued to expand. These are the CMB photons responsible for the noise that Penzias and Wilson were able to detect. Each photon last interacted with an electron or other charged particle just after the Big Bang, and since then the universe has expanded in size by around 1100 times, so the radiation has been redshifted by a factor of around $z = 1100$, as discussed in section 14.4. What set out a vast distance away in the early universe as visible light is now detectable with a wavelength in the microwave range. The radiation has retained its black body spectrum but now corresponds to a much reduced temperature. The observed CMB has a temperature of just 2.7 K, around 1100th of its original temperature. The present universe contains approximately 4.1×10^8 CMB photons per cubic metre and there are approximately 10^9 CMB photons for every proton. The energy density of the CMB is much greater than the average energy density due to starlight. It is only because the photons of the CMB are in the microwave region of the spectrum that the night sky appears so dark to us.

14.8.1   Precision measurements of the CMB

In 1989, NASA launched the probe COBE (Cosmic Background Explorer) to map the cosmic microwave background over the entire sky. COBE showed that it is distributed uniformly in all directions and has the most perfect black body spectrum ever measured. (The distribution of galaxies is fairly uniform when averaged over very large distances, but it is much more clumpy than the almost perfectly uniform CMB.) A comparison between the calculated spectrum and the observations by COBE is shown in Figure 14.3. The CMB has an almost perfectly constant temperature of 2.726 K right across the entire sky. This is the best evidence we have that in the very early universe space was extremely homogeneous and uniform in temperature, so this fundamental assumption of the FRW model seems to be a very good one.

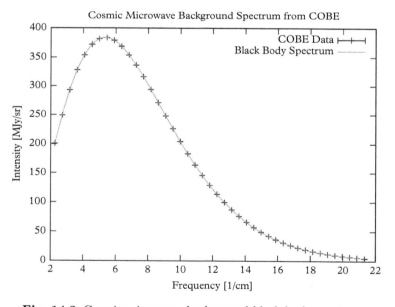

Fig. 14.3 Cosmic microwave background black body spectrum.

The comoving coordinates of the FRW cosmology define the rest frame of the CMB. It is possible to measure the (peculiar) motion of the Earth with respect to this background. Indeed, we now know that the solar system is travelling at a velocity of 370 km s^{-1} in a direction that is almost towards the Virgo cluster of galaxies. The Sun has an orbital velocity of about 250 km s^{-1} around the centre of our galaxy in an almost opposite direction. When these velocities are combined, it gives a velocity for our galaxy relative to the microwave background of 627 ± 22 km s^{-1} in a direction between the Hydra and Centaurus galaxy clusters.

Any slight variation in the temperature of the CMB across the sky contains critical information about the structure of the very early universe. In order to seek these variations, COBE has been followed up by WMAP (Wilkinson Microwave Anisotropy Probe), which was launched in 2001 and has greatly increased the resolution in the measurements. Figure 14.4 shows a map produced by WMAP covering the entire sky and showing regions where the microwave background is very slightly cooler or very slightly warmer than average. Blue corresponds to 0.0002 K cooler and red corresponds to 0.0002 K warmer. These extremely small temperature variations correspond to very slight variations in the density of the universe just 380,000 years after the Big Bang. The denser regions are the seeds that are thought to have eventually grown into clusters of galaxies as the universe evolved. A great deal of information about the structure of the universe has been teased out of these tiny temperature differences or *anisotropies*, as they are called. In 2008, the European Space Agency launched the Planck probe that has studied the microwave background at even higher resolution.

Fig. 14.4 Map of the slight variations in the CMB across the whole sky from the data collected by WMAP, after subtracting the effect of the Earth's peculiar motion through the microwave background.

14.9 The Cosmological Constant

The biggest surprise in cosmology in recent decades has been the discovery that the cosmological constant Λ that Einstein introduced, then rejected, is not zero after all. This was first discovered through distance measurements made using Type Ia supernovae, which are excellent standard candles and are extremely bright so they can be observed at enormous distances, as mentioned in section 13.9. At large redshifts $z \simeq 1$, Type Ia supernovae are fainter than expected. They appear to be further away than they would be in a flat FRW universe with no cosmological term. This implies that the universe was expanding more slowly in the distant past; in other words, the expansion of the universe is accelerating, and therefore our universe has a positive cosmological constant. The implications of a non-zero cosmological constant are illustrated in Figure 14.1.

The interpretation of the cosmological term depends on which side of Einstein's equation (14.1) it is added to. On the left it represents a modification of the Einstein tensor. On the right it can be interpreted as an additional contribution to the energy density of the universe that is somehow built into the fabric of spacetime. In a universe with $\Lambda \neq 0$ and no matter or radiation, Einstein's equation takes the form

$$\begin{pmatrix} G_{tt} & 0 & 0 & 0 \\ 0 & G_{xx} & 0 & 0 \\ 0 & 0 & G_{yy} & 0 \\ 0 & 0 & 0 & G_{zz} \end{pmatrix} = \begin{pmatrix} \Lambda & 0 & 0 & 0 \\ 0 & -a^2\Lambda & 0 & 0 \\ 0 & 0 & -a^2\Lambda & 0 \\ 0 & 0 & 0 & -a^2\Lambda \end{pmatrix}. \tag{14.52}$$

Comparison with equation (14.17), where there is an energy–momentum term, shows that the cosmological term mimics material with energy density $\rho_\Lambda = \frac{1}{8\pi G}\Lambda$ and a negative pressure $P = -\rho$. This negative pressure means that $w = -1$ in equation (14.28).

To draw a parallel with dark matter, the observed cosmological term has been named *dark energy*. Its origin and precise nature remains a mystery. The parallel is not particularly accurate, however. Dark matter is so named because it emits no light, not simply because its composition is unknown. Dark energy is also referred to as vacuum energy, as one possible source is the energy of the vacuum state of quantum fields, as described in Chapter 12.

Analysis of the anisotropies in the cosmic microwave background has enabled cosmologists to make precise measurements of the cosmological parameters for the first time and confirm that Λ is positive. We now know that the age of the universe is 13.798 ± 0.037 billion years. The energy density of the universe has also been accurately determined. It is divided into three components:

$$\Omega_B = \frac{\rho_B(t_0)}{\rho_{\text{crit}}}, \quad \Omega_D = \frac{\rho_D(t_0)}{\rho_{\text{crit}}}, \quad \Omega_\Lambda = \frac{\rho_\Lambda(t_0)}{\rho_{\text{crit}}}, \tag{14.53}$$

corresponding to baryonic matter, dark matter and dark energy respectively. The values of these components at the present cosmic time t_0 are

$$\Omega_B = 0.047, \quad \Omega_D = 0.233, \quad \Omega_\Lambda = 0.72, \tag{14.54}$$

so baryonic matter forms less than 5% of the energy density of the universe, whereas almost a quarter of the energy density is accounted for by dark matter. Most remarkably, over 70% is due to dark energy. There are also contributions from photons and neutrinos, but they are negligible currently, with values $\Omega_\gamma \sim \Omega_\nu \sim 10^{-4}$. This gives the important result that

$$\Omega = \Omega_B + \Omega_D + \Omega_\Lambda = 1, \tag{14.55}$$

which means that the geometry of the universe is flat: $k = 0$.

These parameters are fundamental components of FRW cosmology, and we have reached a remarkable point in the history of cosmology where their values have been accurately determined by observation. The FRW model fits the observational evidence very well, providing a description of the entire cosmos, but, as yet, there is no fundamental theory that explains why the parameters take the observed values or even what dark matter is. The origin of dark energy remains a complete mystery. If cosmology and astrophysics continue to advance at the current rate, these questions may be answered in the coming decades.

14.10 Galaxy Formation

The FRW model of the universe is built on the assumption that the universe is homogeneous on the very largest scales. The CMB radiation offers very good evidence that this assumption is valid, especially in the early universe. Looking out into deep space, however, there is structure on all length scales; we see galaxies, galaxy clusters and galaxy superclusters. We need to explain the origin of these structures and how they evolved from an initial state of almost perfect homogeneity. This is an extremely difficult nonlinear problem that can only be investigated with large-scale numerical simulations.

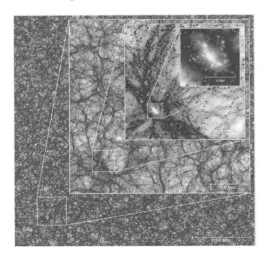 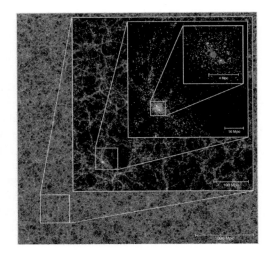

Fig. 14.5 Left: The mass density field in the Millennium-XXL simulation at the current era. Each inset zooms by a factor of 8 from the previous one; the side-length varies from 4.1 Gpc down to 8.1 Mpc. All these images are projections of a thin slice through the simulation of thickness 8 Mpc. (1 Mpc $= 3.26 \times 10^6$ light years.) Right: The predicted galaxy distribution corresponding to the mass density field on the left.

Astrophysicists have designed software that models the evolution of the mass distribution of the universe with the aim of testing detailed cosmological theories. In these models, tiny inhomogeneities in the early universe grow through gravitational clumping, leading to the creation of galaxies, supermassive black holes and quasars and their subsequent interactions and evolution. The origin of the inhomogeneities is not well understood, although one possibility is that they arise from quantum fluctuations of either the matter density or geometry in the early universe. One of these simulations was set up by a collaboration known as the Virgo Consortium; it has run on one of the world's fastest supercomputers and has generated 100 terabytes of data. In 2010, their Millennium-XXL simulation modelled the gravitational interactions of $6720^3 \simeq 3 \times 10^{11}$ massive 'particles' over the course of 13 billion years within an expanding universe. Each particle represents a mass of $7 \times 10^9 M_\odot$. In accordance with the CMB data, the simulation assumes the presence of cold dark matter and dark energy, represented by the cosmological constant Λ. This now standard formulation is known as the ΛCDM model of the universe. Selected results are shown in Figures 14.5 and 14.6. The observed large-scale features of the universe are accounted for very well. It is clear from such simulations that the complex features that we see in the universe can only form in a universe that contains a great deal of cold dark matter. This agrees with the analysis of the CMB and other observations.

The output from the simulations is now being used as a virtual observatory to further refine our understanding of the early universe. Astrophysicists are comparing the data to observations of the real universe, including the most accurate measurements of the CMB to date made by the Planck satellite, as well as galaxy cluster data inferred from gravitational lensing measurements. Further large-scale observational data will soon be available from the Panoramic Survey Telescope and Rapid Response System (PANSTARRS) based in Hawaii. This telescope with its 1.4 gigapixel camera will map large areas of the sky with great sensitivity; every month one-sixth of the sky will be surveyed at five wavelengths.

Fig. 14.6 Expanded view of a galaxy cluster in the Millennium-XXL simulation. (h is defined by the relation: $H_0 = 100h$ km s^{-1} Mpc^{-1}. Its value is believed to lie in the range $0.6 < h < 0.9$.)

14.11   The Inflationary Universe

The evidence for the FRW cosmology all stacks up. However, the universe has a couple of odd features that it does not account for. If the universe is expanding like a balloon, then it should appear curved like the surface of a balloon, but the universe seems to be flat, with $k = 0$. We are familiar with the fact that when we survey our surroundings, the Earth looks flat in our vicinity, even though we know it is spherical. This is because the Earth is large. Similarly, if the universe is spatially flat or very close to flat, then it must be much larger than the region that we can see. Why is this?

When we look out in the night sky, the universe appears the same in every direction. Allowing for our motion relative to the Virgo galaxy cluster, the CMB on one side of the sky looks exactly the same as the CMB on the other. This uniformity indicates that the early universe was very homogeneous and in thermal equilibrium. Although it has taken 13.8 billion years for this radiation to reach us, according to conventional cosmology, there cannot have been sufficient time for opposite regions where the CMB was produced to ever have been in causal contact. So why do they have the same temperature? This is a serious puzzle for cosmologists. To get a handle on the problem, we must investigate the *particle horizon*, also known as the *causal* or *cosmological horizon*. This is the furthest that we can hope to see and is the cosmological equivalent of a black hole's event horizon.

14.11.1 Particle horizons

In a flat Minkowski spacetime it is possible to see to the ends of the universe. There is no horizon. This is because Minkowski spacetime extends an indefinite time into the past, so there is sufficient time for light to reach us from even the most remote regions of space. By contrast, in an expanding universe time only extends a finite period into the past. This being the case, can we still expect to see the whole of the universe?

As time $t = 0$ is approached, the k-dependent term in equation (14.21) becomes negligible and the expansion rate of both spherical and hyperbolic universes is comparable to that of the flat universe. Let us therefore consider the simpler, spatially flat FRW metric (14.7) with $k = 0$. We can transform to a new conformal time coordinate, such that

$$dt' = \frac{dt}{a(t)}, \quad \text{and} \quad t' = \int \frac{dt}{a(t)}. \tag{14.56}$$

With this new time coordinate the metric takes the form

$$d\tau^2 = a^2(t')(dt'^2 - dx^2 - dy^2 - dz^2), \tag{14.57}$$

which is a Minkowski metric multiplied by the time-dependent conformal factor $a^2(t')$. This new version of the metric is *conformally flat* with light signals travelling along the radial lines $r = \pm t' + $ const. Changing coordinates does not alter the causal structure of spacetime. Null geodesics remain null, time-like curves remain time-like and space-like curves remain space-like, but the transformed metric is more convenient for answering questions about the propagation of light signals and causal effects. The state of the universe at our present location has only been causally affected by events inside and on our past lightcone.

The maximum comoving distance that light can travel between an initial time t_i and a later time t is

$$r_{\mathrm{h}}(t) = t' - t_i' = \int_{t_i}^{t} \frac{dt}{a(t)}. \tag{14.58}$$

$r_{\mathrm{h}}(t)$ is called the radius of the particle horizon, and the physical proper distance to the horizon is $d_{\mathrm{h}}(t) = a(t)r_{\mathrm{h}}(t)$. In Minkowski spacetime the time coordinate extends back to minus infinity, so there is always sufficient time for a light signal to pass between any two points, irrespective of the distance between them, and there is no horizon. Using our new coordinates, FRW spacetime has the same causal structure as Minkowski spacetime, so if t' also extends back to minus infinity, then FRW spacetime has no horizon. Now, the original time coordinate t extends back to the origin of the universe at $t_i = 0$. Setting the lower limit of the integral (14.58) to $t_i = 0$, we see that r_{h} is finite and a horizon exists if the integral converges, whereas no horizon exists if the integral diverges. In fact, the integral converges if

$$a(t) \propto t^n \tag{14.59}$$

with $n < 1$. However, we have seen in section 14.6.1 that in a matter-dominated universe $a \propto t^{\frac{2}{3}}$, and in a radiation-dominated universe $a \propto t^{\frac{1}{2}}$. In both cases the integral converges, so there *is* a horizon. There are regions of space too distant for their light, or any particles

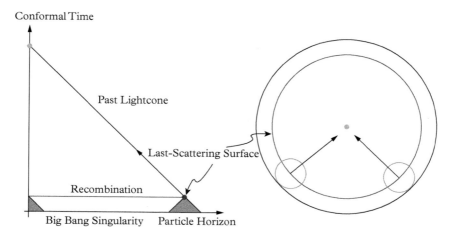

Fig. 14.7 Left: Conformal diagram showing the particle horizons in an FRW universe. The CMB is received from points on the past lightcone where it intersects the last scattering (recombination) surface. Right: Spatial slice of the universe, showing that two causally disconnected regions contribute to the CMB as seen at the centre.

moving more slowly, to reach us since the Big Bang. We must therefore accept that the universe may be much larger than the region we can see, and most of the universe is invisible. In compensation, we will be able to observe an ever larger region of the universe as time proceeds.

Figure 14.7 (left) shows the conformal diagram of the FRW universe. The history of the universe stretches back to the singular point where $a = 0$. We can determine the conformal time at which the singularity occurred in a matter-dominated universe by combining equation (14.56) with the expression for the scale parameter $a \propto t^{\frac{2}{3}}$. This expression implies that $da \propto t^{-\frac{1}{3}} dt \propto a^{-\frac{1}{2}} dt$, so

$$dt' = \frac{dt}{a(t)} \propto a^{-\frac{1}{2}} da. \tag{14.60}$$

Therefore

$$t' \propto a^{\frac{1}{2}} \quad \text{and} \quad a \propto t'^2. \tag{14.61}$$

Similarly, in the radiation-dominated case $a \propto t^{\frac{1}{2}}$, which implies that $da \propto t^{-\frac{1}{2}} dt \propto a^{-1} dt$, so

$$dt' = \frac{dt}{a(t)} \propto da, \tag{14.62}$$

and therefore

$$t' \propto a. \tag{14.63}$$

In both the matter-dominated and radiation-dominated cases, $a = 0$ corresponds to $t' = 0$.

Each point in the diagram in Figure 14.7 (left) is at the apex of its past lightcone which defines its entire causal past. Figure 14.7 (right) shows that when we observe the microwave background, according to the FRW model, we are receiving light from a vast number of regions that could never have had causal contact when the radiation was emitted at the time of last scattering (recombination). Yet, judging by the spectrum of the CMB, this emission took place at the same cosmic time in all regions of the visible universe, so the entire visible universe had precisely the same temperature at this time.

14.11.2 Inflation

The FRW model cannot explain the uniformity of the microwave temperature across the whole sky, as we have just seen. In 1980, Alan Guth proposed a model called the *inflationary universe* that offers a possible solution. Guth postulated that the universe was subject to a brief period of highly accelerated expansion. Prior to this inflationary epoch the entire visible universe existed in a minute volume that is presumed to have reached thermal equilibrium. Regions of the universe then became causally separated by the inflationary expansion, but retained equal temperatures.

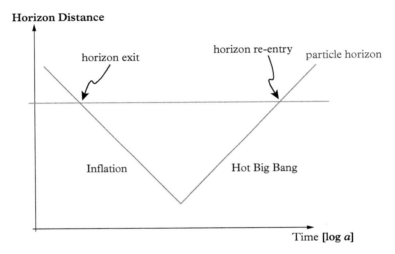

Fig. 14.8 Regions that were in causal contact in the very early universe became causally disconnected during inflation, but may come back into causal contact later. The diagonal lines represent the distance to the cosmological horizon for a representative point during the evolution of the universe. To the left of the figure, during inflation, the horizon distance shrinks. To the right, during the conventional expansion, the horizon distance increases. The horizontal line corresponds to a fixed comoving distance from the representative point. In the very early universe any region at this comoving distance is within the horizon of the point. During inflation, as the horizon shrinks this region passes over the horizon. Then, during the conventional expansion, as the horizon expands the region re-enters the horizon.

This is shown schematically in Figure 14.8. In the earliest moments of the universe, the causal horizon of each point may have encompassed most of the universe. Then during inflation, the causal horizon of each point shrank dramatically. To fit the observational evidence, inflation must have begun around 10^{-36} s after the origin of the universe and lasted for a similar period of time, during which the size of the universe doubled at least 60 times. When inflation turned off, the universe continued to expand, but with the steady expansion of the conventional FRW model. The causal horizon of each point then grew and the portion of the universe open to observation increased. Eventually, regions that lost causal contact in the early universe may have come back into causal contact. By the end of inflation any initial inhomogeneities in the universe would have been inflated out of sight and any spatial curvature would have been stretched until the universe was indistinguishable from a flat universe.

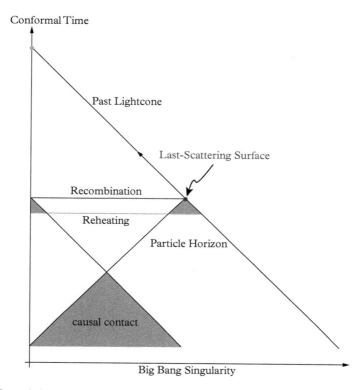

Fig. 14.9 Conformal diagram showing the particle horizons in an inflationary universe (compare Figure 14.7).

What makes Guth's idea physically plausible is that there are solutions of Einstein's equation that have these inflationary properties. What is required is that for a very brief period of time in the early universe the scale factor obeys $\ddot{a} > 0$. From equation (14.20) we see that $\ddot{a} > 0$ requires $\rho + 3P < 0$ and as $\rho > 0$ this produces the condition $P < -\frac{1}{3}\rho$. So a sufficiently negative pressure will produce a universe whose expansion accelerates.

This can be achieved in a universe that contains nothing but a positive cosmological constant Λ. The FRW equations (14.21) and (14.22) in such a universe with $k = 0$ are

$$3\frac{\dot{a}^2}{a^2} = \Lambda, \tag{14.64}$$

$$2\frac{\ddot{a}}{a} + \frac{\dot{a}^2}{a^2} = \Lambda. \tag{14.65}$$

The second of these equations is an automatic consequence of the first, as is easily verified.

The first equation simplifies to

$$\dot{a} = \sqrt{\frac{\Lambda}{3}}\, a\,, \tag{14.66}$$

and has the solution

$$a \propto \exp\left(\sqrt{\frac{\Lambda}{3}}\, t\right)\,. \tag{14.67}$$

This solution is known as *de Sitter space* after Willem de Sitter. As required for inflation, $\ddot{a} = \frac{\Lambda}{3}a > 0$. By combining equation (14.67) with equation (14.56) we see that

$$dt' = \frac{dt}{a(t)} \propto \frac{da}{a^2} \quad \text{so} \quad t' \propto -\frac{1}{a}\,. \tag{14.68}$$

Therefore $a(t') \propto -\frac{1}{t'}$, and the singularity at $a = 0$ corresponds to the conformal time $t' = -\infty$. There is no horizon. The origin of the universe has been pushed back to $-\infty$. As shown in the conformal diagram in Figure 14.9, this means that in contrast with the conventional FRW cosmology, there is ample time in the inflationary cosmology for all regions of the universe to come into thermal equilibrium. It can thereby account for the uniformity of the microwave background.

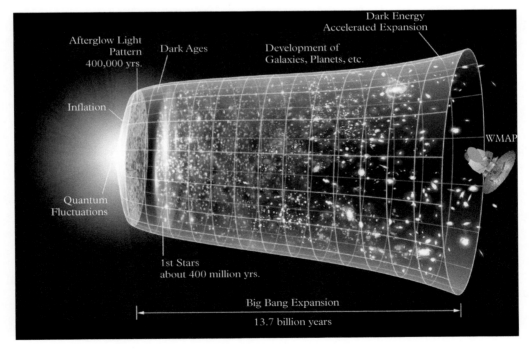

Fig. 14.10 Timeline of the universe.

Inflation offers a mechanism that can, in principle, account for the observed homogeneity of the universe and its apparent flatness. As an added bonus, inflation offers a possible answer to the question of how the small initial inhomogeneities arose that are necessary for galaxy formation. Prior to inflation, there would inevitably have been quantum fluctuations in the energy density of the universe. Any tiny initial regions of greater density would then have inflated to a scale where they might have formed the primordial seeds of inhomogeneity from which galaxies and galaxy clusters would grow. Much effort has been expended devising inflationary models that would result in the creation of a universe that looks like ours. Inflation must be turned on briefly to inflate the universe in its earliest epoch, then the universe undergoes a phase transition and inflation is turned off after the incredibly short time of around 10^{-34} s. The subsequent expansion of the universe is well described by conventional FRW models. Currently this scenario is only possible by postulating the existence of new quantum fields, and ultimately, the appropriate fields should be derived from a unified theory of the forces of nature. There are ongoing efforts to derive suitable quantum fields from string theory.

Our current understanding of the evolution of the universe from the Big Bang to the present epoch is depicted in Figure 14.10. There has been spectacular recent progress in cosmology, nonetheless many puzzles are yet to be solved.

14.12   Further Reading

E. Harrison, *Cosmology: The Science of the Universe (2nd ed.)*, Cambridge: CUP, 2000.

M. Longair, *The Cosmic Century: A History of Astrophysics and Cosmology*, Cambridge: CUP, 2006.

S. Weinberg, *Cosmology*, Oxford: OUP, 2008.

For a review of inflationary cosmology, see

D. Baumann, *TASI Lectures on Inflation*, arXiv: 0907.5424v2 [hep-th], 2012.

15
Frontiers of Physics

Despite the overwhelming successes of modern physics, there are questions that remain to be answered. We will take a look at some of these open questions in this final chapter. They come in several varieties. There are unexplained features of the universe, such as the matter–antimatter asymmetry, the existence of dark matter and the even more mysterious dark energy. There are issues relating to the incompleteness of theory, such as the loose ends of the Standard Model and the need for a quantum theory of gravity that can be integrated into a theory of all the forces. There are also deep philosophical questions to be resolved—what really is a particle, and what is the fundamental nature of space and time? First we will consider the still unresolved issue of how we should interpret quantum mechanics and, indeed, whether quantum mechanics represents the ultimate theory of reality or whether a deeper theory is possible.

15.1 The Interpretation of Quantum Mechanics

As we have seen, quantum mechanics offers an extremely accurate recipe for predicting probabilistic results of experiments. No experiment or observation has ever cast doubt on the fact that quantum mechanics works. However, when we enquire into its metaphysical implications, quantum mechanics suggests that the universe is very different and much stranger than our classical intuitions might lead us to believe.

The standard interpretation of quantum mechanics was thrashed out in 1927 at a meeting of Bohr, Heisenberg and Pauli in Copenhagen. It includes the uncertainty principle, wave–particle duality, the probabilistic interpretation of the wavefunction and the identification of eigenvalues as the measured values of observables. This suite of ideas has become known as the Copenhagen interpretation. The final element of standard quantum mechanics—wavefunction collapse—was added by John von Neumann in his thesis *Mathematical Foundations of Quantum Mechanics* published in 1932 and it is usually considered an integral part of the Copenhagen interpretation. Von Neumann proposed that quantum mechanics consists of two separate processes. First, when not subject to any measurement, a quantum system evolves deterministically according to the time-dependent Schrödinger equation, and during this evolution the state generally consists of a superposition of eigenfunctions of any given observable. Second, when that observable is measured, and immediately afterwards, the wavefunction of the system is the particular eigenfunction corresponding to the measured eigenvalue, and it is the act of making the measurement that projects the system on to this eigenfunction. This projection is known as wavefunction collapse. We have followed the Copenhagen interpretation in all our earlier discussions of quantum mechanics, such as in section 12.9 where we considered neutrino oscillations.

The Physical World. Nicholas Manton and Nicholas Mee, Oxford University Press (2017).
© Nicholas Manton and Nicholas Mee. DOI 10.1093/oso/9780198795933.001.0001

Some of the pioneers of quantum theory, most notably Einstein, Schrödinger and de Broglie, could not accept the Copenhagen interpretation and put considerable effort into finding counterarguments and alternatives. We can illustrate these philosophical differences with a simple example. A radioactive nucleus, such as uranium-238, may exist for billions of years, then suddenly it decays and emits an alpha particle. But the moment of decay cannot be predicted. We can only give a probability that the nucleus will decay within a certain period of time. The same behaviour occurs in fundamental particles such as muons, which decay with a half-life of around one microsecond. In Einstein's view, our inability to determine the moment of decay is simply due to our ignorance of all the relevant variables. He was convinced that behind the quantum description of such processes there must be quasi-classical *hidden variables* that determine exactly when a particle will decay. A uranium nucleus is a complicated object, so we might imagine that its components are shuttling around until they achieve a rather improbable configuration, at which point decay occurs. Muons, however, which are believed to be elementary particles with no substructure, behave in the same way. According to Einstein's viewpoint, even in these cases there must be hidden variables that we are unaware of, determining the moment of decay. If this were true, then it would reduce quantum mechanics to something similar to classical statistical mechanics, a probabilistic account of the 'apparently random' dynamics of a complicated system.

The Copenhagen interpretation of quantum mechanics describes the decay of a particle very differently. According to this view, we must describe a system that includes an unstable particle, such as a muon, by a wavefunction that evolves as a superposition of states. This superposition includes states that describe the undecayed muon and states that include the decay products following the muon decay. The system evolves as a collection of potentialities until a measurement is made, at which point the wavefunction collapses and the system then exists in a state in which the muon has decayed or not, as the case may be. We are only able to predict the probability of the various possible results of the measurement. There is no hidden information.

The Copenhagen interpretation, as championed by Bohr, von Neumann and others, might be consistent with observations, but it raises a number of issues. According to quantum mechanics, muons are identical particles and it is impossible even in principle to distinguish one muon from another muon. It is therefore also impossible to distinguish a muon at this instant from the same muon an instant ago, but suddenly at an unpredictable moment the muon changes and decays. This is rather strange, as it appears to be an effect without a cause, and suggests that we might have to give up determinism. It is also rather odd that the ultimate theory of physics is dependent on a process—wavefunction collapse—that is not described mathematically by the theory. Furthermore, wavefunction collapse is assumed to happen instantaneously, which is contrary to the spirit of relativity at the very least. It also requires some outside intervention—the measurement—and this raises the question of what actually constitutes a measurement.

15.1.1 Schrödinger's cat and Wigner's friend

Einstein was very uncomfortable with the Copenhagen interpretation as he felt that it denies the existence of a commonsense reality. Bohr had argued that we have no direct experience of entities as small as atoms so we should not judge in advance how they might behave. An exchange of letters between Einstein and Schrödinger led to the most famous attempt to demonstrate the absurdity of Bohr's position. The *Schrödinger cat* thought experiment published in 1935 aimed to show that the strange ideas of quantum mechanics cannot be confined to the microworld but must infect the macroworld with results that are counter to our everyday experience.

The set-up of the experiment is as follows. A radioactive atom is connected to an apparatus containing a cat in a closed steel box. The atom has a half-life of one hour and is monitored by a Geiger counter. If the atom undergoes radioactive decay, the Geiger counter triggers a switch and poisonous gas is released that kills the cat. After one hour there is a 50% chance that the atom has decayed and a 50% chance that it has not decayed, but the box containing the cat is closed so it is impossible to know whether or not the atomic decay has occurred. According to the Copenhagen view of quantum mechanics, after one hour the atom must be described by a state that is a superposition of decayed and undecayed, and as no measurement has yet been made, the only way to describe the whole system is to assume that the entire apparatus and the cat are now in a superposition of states. In one of the superposed states the cat is alive, in the other state it is dead, and this superposition of states persists until we perform a measurement by opening the box to take a look inside. At this point the wavefunction collapses to reveal a cat that is either alive or dead.

We might be prepared to accept that an atom exists in a superposition of states, as the microworld of the atom is not directly accessible to us, but we never experience superpositions in everyday life, so can we really accept the possibility that a cat can exist in such a state? We see live cats, and dead cats, but never a superposition of the two. And what if we placed a human in the box instead of a cat?

Eugene Wigner was a proponent of the orthodox view of quantum mechanics and possibly one of its architects as he was a close collaborator of von Neumann at the time when his thesis was written. Wigner devised a thought experiment known as the puzzle of Wigner's friend, which goes as follows: Wigner is busy, so he asks a friend to check the results of an experiment, possibly involving a cat in a box, but more likely the records of LHC collisions.[1] Wigner's friend notes an interesting result, perhaps a rare event, such as the production of a Higgs particle, that was recorded the previous day. She duly reports the result to Wigner. The question is: when did the wavefunction describing the creation of the Higgs particle collapse? When the ATLAS detector at the LHC recorded the event, when Wigner's friend checked the result, or when she reported the news to Wigner?

According to the Copenhagen interpretation, the wavefunction collapsed on that previous day—there was a definite Higgs particle that interacted with the measuring devices and subsequently decayed. In Wigner's friend's mind, however, the wavefunction collapse took place when she looked at the record. Wigner's friend now goes to tell Wigner what she has seen, and it is only then that Wigner's own wavefunction collapses. In other words, up to that moment, Wigner knew there was the possibility that events the previous day might have included a Higgs particle, but it was part of a quantum superposition, in which the amplitude for the Higgs particle is very small. Wigner's wavefunction did not collapse the previous day, and neither did it collapse when his friend examined the records. At that time, all that happened (from Wigner's viewpoint) was that a correlation was set up between the possible Higgs signal and Wigner's friend's mind. Wigner's wavefunction *did* collapse when she told him what happened.

---

[1] Our discussion of the paradox of Wigner's friend is somewhat anachronistic. It was published in 1961, long before the LHC was built.

If we believe in an objective external reality, then only one of these options can be correct, but which one? Wigner took the view that all the components of a detector such as ATLAS are formed of atoms and other particles that obey the rules of quantum mechanics and so must evolve as a superposition of states according to some extremely complicated Schrödinger equation and nothing within the apparatus can cause wavefunction collapse. Wigner knows that when the result of the experiment is reported to him, the wavefunction has collapsed. He could take the view that the collapse occurs at this moment. But this would imply that only he is able to collapse wavefunctions. This would essentially mean that Wigner was the only sentient being and his friend and everyone else were just automatons. This rather catastrophic philosophical position is known as solipsism. Naturally, Wigner rejected this possibility. He decided that it is only reasonable to suppose that his friend is also a sentient being and that she was consciously aware of the results of the experiment before she reported them to Wigner. Wigner's conclusion was that the measurement of an observable is only possible by conscious beings, such as humans, and it is this interaction between consciousness and a quantum system that causes wavefunction collapse.

Does this mean that we, as humans, are the ultimate measuring devices, and only we, and not cats, can collapse wavefunctions? This might appear like an attractive proposal. It means that all those complicated laboratory measuring devices and the physics inside, such as the particle detectors like ATLAS that respond to the outgoing particles and the computers storing long-term records of the collision events, are happily obeying a complicated multi-particle Schrödinger equation, and no wavefunction collapse occurs. The collapse only occurs when *we* look. However, it is this last stage that is now the real puzzle. What is different about humans, placing them outside the rest of physics and obeying different laws?

Wigner's association of consciousness with wavefunction collapse is rather strange. The most basic requirement when 'doing' physics is the belief in an external objective reality. Invoking a role for consciousness in the collapse of the wavefunction threatens to undermine this principle. It suggests the universe is like a virtual reality game whose details are constantly being updated as we explore it further. When we visit a new place or look through a telescope at a distant galaxy, is it really credible that there is a cascade of wavefunction collapses, as nature somehow colludes to construct a consistent universe backwards in time to the Big Bang? Could it really be true that the world existed in an ill-defined superposition of states until the first human became self-aware? Or was Wigner simply conflating two rather different problems that we do not yet understand, the quantum measurement problem and the origin of consciousness? Although consciousness is not at all well understood, it seems reasonable to confer varying degrees of consciousness on all entities from humans to the great apes, dolphins, elephants, cats and so on. Buddhists would have us believe that even trees, amoebae, rocks and fundamental particles share some degree of consciousness. So, which of these entities are able to collapse wavefunctions? Wavefunction collapse must be a discrete process, as the wavefunction either collapses or it doesn't. Can it really be related to a property like consciousness that exists on a continuous spectrum?

A simpler alternative might be a purely mechanical solution to the measurement problem. It is straightforward to imagine that when a wavefunction reaches an appropriate level of complexity, which might be determined by the number of particles or the amount of mass it describes, then it must collapse spontaneously. This would require the inclusion of a new nonlinear term in the Schrödinger equation. Adding a small term that only becomes significant when, say, 10^{10} particles interact, need not undermine the great predictive success of quantum mechanics. One such proposal has been sketched out by Roger Penrose, who has suggested that wavefunction collapse might be triggered when the total mass of entangled particles described by the wavefunction approaches the Planck mass, which is around 10^{-8} kg. This is very large on atomic scales, but very small on human scales.

15.1.2 The many-worlds interpretation

There is another interpretation of quantum mechanics developed by Hugh Everett III in 1957, which attempts to circumvent the problems associated with wavefunction collapse. It is known as the *many-worlds interpretation*. According to this view, wavefunctions do not collapse. Instead, the universe splits whenever particles interact, with or without human intervention. When we make a measurement, the universe splits. We might find ourselves in a universe where the value that we measure is λ_0 and the wavefunction after the measurement is Ψ_0, so if we repeat the measurement, we will again obtain the value λ_0; in another universe the measured value might be λ_1 and the wavefunction after the measurement is Ψ_1, and so on. In this way a single measurement may vastly increase the number of universes, but we avoid unpleasant superpositions. For example, the dead and alive Schrödinger cats exist in different universes. In one universe, the cat has been poisoned; in the other universe, the cat is alive and well.

One problem with the many-worlds interpretation is that it remains unclear what constitutes a measurement. Measurements need not involve macroscopic apparatus, so almost everything that happens is a measurement. If two particles collide, one particle is measuring the position of the other. As a consequence, the universe does not just split on rare occasions; it happens all the time. Such an implausible multiplicity of unobservable universes stretches our credulity. Do we inhabit a single universe? We may think so, but no—we are in many at the same time, and all of them in some sense exist. This produces a dramatic clash with our intuition of the world around us. We experience a single world where, for example, our favoured football team just lost. But according to many-worlds, another 'we' is in a universe where our team won. How do we get there? Why do we only experience a single undivided universe? Many-worlds implies that every interaction since the Big Bang has split the universe into an unclassifiable infinity of universes, as variables such as momentum and scattering angle have a continuum of possible values for quantum systems. The many-worlds interpretation offers no mechanism for universe-splitting and it remains an idea with no quantitative consequences and a large amount of unwanted baggage.

These are rather metaphysical speculations and it is very difficult to come to any firm conclusions about them by pure introspection. Rather remarkably, however, some features of the quantum measurement problem have been tested in the laboratory in recent decades.

15.1.3 The EPR paradox

A thought experiment due to Einstein, Boris Podolsky and Nathan Rosen was published in 1935 and is known as the *EPR paradox*, although the clearest expression of the paradox is due to David Bohm and this is the version that we will consider. Imagine a spin 0 particle X in its rest frame. Particle X decays into two particles A and B that have spin $\frac{1}{2}$ and recede in opposite directions z and $-z$. (These decay products could be an electron and a positron, for instance.) We arrange our apparatus to measure the spin of particle A in a direction x that is perpendicular to z. The measurement determines whether particle A is spin up or spin down in direction x. We know that the total spin of the decay products must be zero, so particle B must have opposite spin to particle A. Measuring the spin state of particle A will therefore simultaneously determine the spin state of particle B.

This may not seem so surprising. We are familiar with situations where we acquire a piece of information about one object and it simultanously reveals a piece of information about a second object. For instance, we might place a black ball and a white ball in a bag and ask someone to retrieve a ball from the bag. If they pull out the white ball, we know that the black ball remains in the bag. We have gained information about the ball in the bag by observing the ball that has been removed. There is no mystery here; we know that the balls have retained their identity throughout this procedure. The ball in the bag has not been affected by our observation.

However, there is a big difference when considering the spin states of quantum particles because, according to the Copenhagen interpretation, these spin states are not defined until we make the measurement. Before the measurement, particles A and B are described as *entangled* and they must be described by a wavefunction ϕ_{AB} that is a superposition of the two possible spin states, such that the total spin is 0. Using the notation in which $\phi_{Ax}^{\uparrow}$ denotes the wavefunction for particle A to be spin up in the x-direction, the spin state of the two particles is described by the wavefunction

$$\phi_{AB} = \frac{1}{\sqrt{2}}(\phi_{Ax}^{\uparrow}\phi_{Bx}^{\downarrow} - \phi_{Ax}^{\downarrow}\phi_{Bx}^{\uparrow}). \tag{15.1}$$

When we make the measurement, the wavefunction collapses and one of the superposed states, either $\phi_{Ax}^{\uparrow}\phi_{Bx}^{\downarrow}$ or $\phi_{Ax}^{\downarrow}\phi_{Bx}^{\uparrow}$, is projected out. This affects the state of both particles. Also, importantly, we could have arranged our apparatus to measure the spin of particle A in the y-direction, or any other direction. This means that the particles are not classical spinning tops whose spin vector was determined at the moment they came into existence following the decay of particle X; the value of their spin is not precisely defined until it is measured. Furthermore, there is no limit to the distance between particles A and B when we perform the measurement. This seems to imply that our interaction with particle A has an instantaneous influence on particle B. If we measure the spin of particle A to be up, then wavefunction collapse implies that immediately afterwards, the spin of particle B is down, but just before that moment the spin of particle B was in a superposition of states. We cannot use this fact to transmit information; nevertheless, it appears to violate the spirit of relativity. Einstein referred to entanglement as *spooky action at a distance*.

John Bell showed that the EPR thought experiment implies that if quantum mechanics is correct then there are correlations between measurements that cannot be explained by any hidden variables theory. The correlations were first tested in a series of experiments by Alain Aspect. These experiments and all the refinements that have followed demonstrate very clearly that the counterintuitive predictions of quantum mechanics are obeyed.

15.1.4 The Aspect experiments

In 1982 a team led by Alain Aspect performed a series of experiments inspired by the EPR paradox to investigate the quantum measurement problem. Their apparatus was constructed in the basement of the Institute for Theoretical and Applied Optics in Paris. These experiments measured the polarization states of entangled photons rather than spin states of particles such as electrons and positrons, but the implications are the same. Calcium atoms have an excited state where the two outer 4s electrons are promoted into a spin 0,

$4p^2$ $^1S_0$ state, and this is the source of the entangled photons used in the experiment. The excited state decays rapidly to the ground state in two steps with the emission of a photon of wavelength 551.3 nm in the green region of the spectrum, followed by the emission of a photon of wavelength 422.7 nm in the blue region of the spectrum. These photons are emitted in opposite directions and, as both the excited state and the ground state have spin 0, the photons have opposite polarizations.

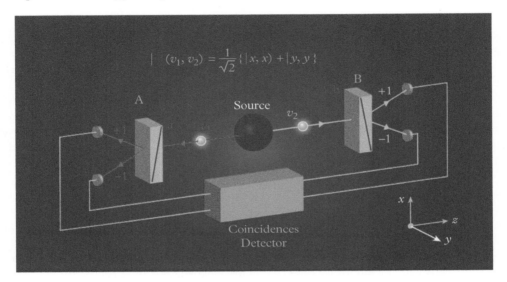

Fig. 15.1 Schematic representation of the apparatus used in the Aspect experiments. (Figure reprinted with permission from *Physics* 8, 123, December 16, 2015. Copyright (2015) the American Physical Society.)

In the Aspect experiments a collimated beam of calcium atoms with a relatively low density of around 3×10^{16} atoms per cubic metre passes into a chamber where the atoms are exposed to two extremely high-power laser beams of wavelengths 406 nm and 581 nm, thus exciting the atoms into the spin 0, $4p^2$ $^1S_0$ state by simultaneous double-photon absorption. (This technique was devised to avoid any confusion between the photons used to excite the calcium atoms and those produced by the decays of the atoms.) The light emitted by the calcium atoms is monitored, as depicted in Figure 15.1. To the left, a filter selects green photons labelled A and to the right, a filter selects blue photons labelled B. These photons then pass into polarization analysers, which are cubes formed from two prisms glued together with a dielectric coating on their common faces. Vertically polarized photons are directly transmitted when they impinge on this face and horizontally polarized photons are reflected at 90°. The photons then pass to photo-multiplier tubes where they are detected. In this way the polarization of any photon that passes through the apparatus may be determined.

Each polarization analyser is mounted on a platform allowing it to be rotated around its optical axis so that the relative orientation of the two analysers can be altered. The half-life for the emission of the second photon from the excited calcium atoms is around 5 ns, so the electronics are set to look for coincidences in the arrival of photons A and B within 20 ns, which is sufficient to collect the true coincidences but short enough to make the probability of overlap between distinct pairs of photons very low. The measured coincidence rates vary in the range of 0–40 per second depending on the angle between the analysers.

We can obtain some insight into what happens within the Aspect apparatus by considering a thin plastic polarizer, usually known as a polaroid. A polaroid sheet contains needle-like crystals, all aligned in the same direction. When randomly polarized light impinges on a polaroid, it splits into the two orthogonal polarization states, parallel and perpendicular to the crystals. These are referred to as the vertical $(+)$ and horizontal $(-)$ polarizations with respect to the polaroid. Only vertically polarized light is transmitted and its intensity is $\frac{1}{2}I_0$, where I_0 is the original intensity. A second polaroid positioned in the path of the transmitted light re-projects the light into vertical and horizontal components. If the angle between the polarization directions of the two polaroids is α, the electric field E of the light transmitted by the first polaroid will project into a vertical component $E\cos\alpha$ that is transmitted by the second polaroid, and a horizontal component $E\sin\alpha$ that is extinguished. The intensity of the light transmitted through both polaroids is $\frac{1}{2}I_0\cos^2\alpha$. This observational result is known as Malus' law. If the two polaroids are perpendicular, then $\alpha = \frac{\pi}{2}$ and no light is transmitted. A remarkable consequence of quantum mechanics is that some light will be transmitted if one more polaroid is placed *between* two perpendicular polaroids. If the angle between the first and second polaroid is α and the angle between the second and third polaroid is $\frac{\pi}{2} - \alpha$, then the intensity of the transmitted light is $\frac{1}{2}I_0\cos^2\alpha\sin^2\alpha = \frac{1}{8}I_0\sin^2 2\alpha$, which has a maximum value $\frac{1}{8}I_0$ when $\alpha = \frac{\pi}{4}$.

Now we can consider what happens to the photons passing through the Aspect apparatus. When an excited calcium atom decays, it emits a green photon followed almost immediately by a blue photon. According to the Copenhagen interpretation of quantum mechanics, these two photons are emitted in a superposition of states with total spin zero. When the green photon interacts with polarization analyser PA1, it is projected into a vertical or horizontal polarization (with probability $\frac{1}{2}$ for each). This wavefunction collapse simultaneously affects the blue photon, which instantaneously projects into a state with the same polarization relative to PA1, but with its phase shifted by π relative to the green photon. The blue photon then meets PA2, which is oriented at an angle α to PA1. This photon now undergoes a second projection into a vertical or horizontal polarization relative to PA2. If the blue photon was vertically polarized with respect to PA1, then the probability of projecting vertically with respect to PA2 is $\cos^2\alpha$ and the probability of projecting horizontally is $\sin^2\alpha$. In total there are four possible outcomes for each pair of photons that are detected: $(++), (--), (+-), (-+)$, where the first sign represents the polarization of the green photon relative to PA1 and the second sign represents the polarization of the blue photon relative to PA2. The probabilities for each outcome are the product of the probabilities for the two projections: $P_{++} = \frac{1}{2}\cos^2\alpha$, $P_{--} = \frac{1}{2}\cos^2\alpha$, $P_{+-} = \frac{1}{2}\sin^2\alpha$, $P_{-+} = \frac{1}{2}\sin^2\alpha$. The results can be summarized with the coincidence function

$$
\begin{aligned}
E(\alpha) &= P_{++} + P_{--} - P_{+-} - P_{-+} \\
&= \frac{1}{2}\cos^2\alpha + \frac{1}{2}\cos^2\alpha - \frac{1}{2}\sin^2\alpha - \frac{1}{2}\sin^2\alpha \\
&= \cos 2\alpha,
\end{aligned}
\tag{15.2}
$$

which varies with the angle α between the two polarization analysers. Aspect's team measured the coincidence function at angles $\alpha = 0, \frac{\pi}{8}, \frac{\pi}{6}, \frac{\pi}{4}, \frac{\pi}{3}, \frac{3\pi}{8}, \frac{\pi}{2}$ and the results fall on the curved line shown in Figure 15.2. There is perfect agreement with the predictions of quantum mechanics.

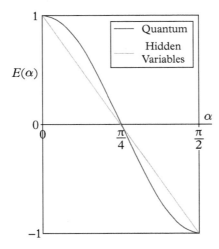

Fig. 15.2 The curved line shows the quantum mechanical prediction for $E(\alpha)$. The straight line shows the prediction of the simple hidden variable model described in the text. The results of the Aspect and other quantum entanglement experiments lie on the curved line and rule out local hidden variables theories.

What would we expect the results of an alternative hidden variables theory to show? In such a theory the polarization of the photons is determined when the calcium atom decays. We will consider a simple theory where this information is stored as a single random variable ϑ that represents the direction in which a photon is polarized. If the green photon is emitted with its polarization in a direction at angle ϑ relative to PA1, then the blue photon is polarized at angle $\vartheta + \pi$ relative to the same direction. We will assume that when a photon described by this theory passes through a polarization analyser it projects vertically or horizontally, depending on which of these directions is closest; that is, it becomes horizontally polarized for $\frac{\pi}{4} < \vartheta < \frac{3\pi}{4}$ and $\frac{5\pi}{4} < \vartheta < \frac{7\pi}{4}$ and vertically polarized for $\frac{3\pi}{4} < \vartheta < \frac{5\pi}{4}$ and $\frac{7\pi}{4} < \vartheta < \frac{\pi}{4}$. Crucially, we don't expect the blue photon to be affected by the projection of the green photon. It is straightforward to compute the results according to this theory. There is a probability of $\frac{1}{2}$ that the green photon will be projected vertically by PA1. If PA2 is set at an angle of α to PA1, then the probability that the blue photon will be projected vertically by PA2 is $\frac{1}{\pi} \times (\pi - 2\alpha)$ and the probability that it will be projected horizontally is $\frac{1}{\pi} \times 2\alpha$. Including all four possible outcomes leads to the following expression for the coincidence function:

$$\begin{aligned}
E(\alpha) &= P_{++} + P_{--} - P_{+-} - P_{-+} \\
&= \frac{1}{2} \times \frac{1}{\pi}\big((\pi - 2\alpha) + (\pi - 2\alpha) - 2\alpha - 2\alpha\big) \\
&= \frac{\pi - 4\alpha}{\pi}.
\end{aligned} \tag{15.3}$$

This function is shown as the straight line in Figure 15.2.

In the Aspect experiments the polarization analysers are positioned 13 metres apart. It would take just over 40 ns to transmit a signal from one analyser to the other at the speed of light. This is twice as long as the maximum period allowed in the detection of pairs of entangled photons, so the experiment rules out the possibility of collusion between the photons due to an unknown subluminal interaction. In one variant of the experiment, known as the delayed choice experiment, two settings are available for the angle of each polarization analyser and the choice is made on which to use while the photons are in mid-flight. This rules out any hidden correlation between the decaying atom and the polarization analysers. In the decades since the original Aspect experiments, further efforts have been made to devise quantum entanglement experiments that close any remaining loopholes that could conceivably allow explanations in terms of local hidden variables. These improvements include ensuring that there is a space-like separation of the detectors, as well as the polarization analysers. Detector efficiency has also been greatly improved to ensure that the experiment does not suffer from unfair sampling and removes the possibility that the system is only able to detect an atypical subset of particles that show quantum correlations. Also, quantum random number generators have been used to determine the orientation for each measurement to ensure that the results cannot be influenced by any memory within the system of previous configurations. In 2015, a team at the University of Delft reported the first loophole-free experiment, measuring the entanglement of two electrons each trapped in a vacancy in a diamond lattice and separated by 1.3 km. It is now generally agreed that the results of these experiments rule out all possible local hidden variables theories.

It seems to be an inescapable fact that quantum mechanics is underpinned by non-local influences between particles and, as demonstrated in the Aspect experiments, these influences are not limited by the speed of light. It is worth noting, however, that no signal can be transmitted in this way. A physicist monitoring PA2 can set the polarizer at any angle and for each blue photon that enters PA2 there is always a probability of $\frac{1}{2}$ that it will be vertically polarized and a probability of $\frac{1}{2}$ that it will be horizontally polarized. This offers no information whatsoever about the orientation of PA1 or the polarization of the green photon. It is only when the results from both PA1 and PA2 are brought together and compared, that the correlation between these statistical results is apparent. So although we might be shocked by the results of these experiments, they offer no evidence for causal superluminal interactions. Furthermore, if the measurements at PA1 and PA2 have a space-like separation, then in some frames the measurement at PA2 will occur before the measurement at PA1. In such frames the results have a rather different interpretation, but this does not produce any inconsistency. Now we interpret the measurement at PA2 as producing the wavefunction collapse and the measurement at PA1 as acting on the resulting eigenstate. The results of the measurements are exactly the same and the same correlations exist between them.

Quantum entanglement is already the basis for a new technology, quantum cryptography, and it holds out the promise of further innovations, such as quantum computing, although considerable technological hurdles remain to be overcome.

At the heart of these quantum mysteries is the notion of wave–particle duality. In some situations the entities that we are interested in, be they electrons, photons or neutrinos, behave like waves, and in other situations they behave like particles. So what are they? Particles are often treated as point-like, but this cannot literally be true, as point-like particles are highly singular. Perhaps a better description of particles would elucidate some of the issues surrounding the interpretation of quantum mechanics.

15.2 The Problem of Point Particles

The ideal classical picture of a particle is that it is a geometrical point, tracing out a worldline in spacetime. This leads to differential equations of motion, including Newton's laws of motion and the geodesic equation of motion for a particle in Einstein's curved spacetime. These equations are well established. The fundamental particles of the Standard Model behave as though they are point-like at the distances that have been explored by the LHC, which is down to around 10^{-18} m. However, a particle has a finite mass and often a non-zero electric charge, which means that if it were genuinely point-like, then it would have infinite mass density, infinite charge density and infinite electrostatic energy. Also, it seems impossible to determine precisely and consistently the classical motion of a point-like particle, taking into account the radiation that it emits. We must therefore consider the point-like nature of particles as an idealization, valid in situations where the distances between particles are much greater than the particle sizes.

The point-like model of particles is retained in standard non-relativistic quantum mechanics. The squared modulus of the wavefunction, $|\psi(\mathbf{x})|^2$, determines the probability of finding the particle at $\mathbf{x}$. $\psi(\mathbf{x})$ is usually smooth and spread out, but the particle itself is not. It is assumed that the particle remains point-like, and that the wavefunction can be arbitrarily narrowly peaked, at least at some initial time, meaning that the position uncertainty is negligible. The wavefunction later, after evolving according to the Schrödinger equation, is much broader. This is because of the uncertainty principle. A precisely localized particle has a great momentum uncertainty, so (heuristically) the particle moves rapidly away from its initial position in all directions. The position probability density therefore spreads rapidly too.

In quantum mechanics, the point-like model must again be regarded as an idealization, which is only approximately true. In quantum field theory, which combines quantum mechanics with relativity, there is a real limitation on how small a particle can be, although it is not very precise. The idea is that if a particle is too highly localized, then it has a significant probability of having a large momentum, and therefore much greater kinetic energy than its rest mass. This energy can reappear as a new particle, or particle–antiparticle pair. With more than one particle around, we no longer know precisely where the original particle is. Suppose the particle is localized within distance L. Then its momentum is at least $\frac{2\pi\hbar}{L}$. The relativistic energy–momentum relation is $E^2 = p^2 + m^2$, and the particle number starts to become uncertain when p is of order m. This occurs if $\frac{2\pi\hbar}{L}$ is of order m, or equivalently, if L is of order $\frac{2\pi\hbar}{m}$. For this reason, $\frac{2\pi\hbar}{m}$ is called the *Compton wavelength* of a particle of mass m, and the particle cannot sensibly be localized within a smaller radius than this. The Compton wavelength of a proton is about 1 fm (10^{-15} m), and for an electron it is 1836 times larger, of order 10^{-12} m. Both of these distances are much smaller than an atomic radius.

These arguments give important limits on how small a particle can be, but they do not give any insight into the intrinsic structure or precise size of the particle. Current understanding is that the electron does not have intrinsic structure at its Compton wavelength. In fact, experiments have not detected any spatial structure in the electron down to about 10^{-18} m. Similarly, individual quarks appear to have no substructure. The proton, on the other hand, is constructed from three quarks. In simple models, the quarks have a spatial wavefunction that gives the proton an intrinsic size almost the same as its Compton wavelength.

Physicists still do not have a convincing picture of the ultimate structure of fundamental particles. Arguments that rely on quantum uncertainty and explanations that treat them as composed of ever smaller subunits are not entirely satisfactory. There is a different approach, which we will consider next, that makes further use of the nonlinearity of field theory. A linear field theory, like Klein–Gordon theory, is fundamentally a theory of waves that have no spatial localization at all. The quantum theory has states obeying $E^2 = p^2 + m^2$, and from this alone one infers that there are particles of mass m. On the other hand, nonlinear field theories often have more localized solutions that are particle-like even before the field is quantized. These solutions are known as *solitons*, and they are not point-like. They offer a radically different model of fundamental particles.

15.2.1 Solitons

The excitations that we have so far considered in field theories are like the waves along a long rope or in an elastic medium. Only when the waves are quantized do we find particle states. Solitons are different. They are particle-like solutions of the original classical field equations. An appropriate analogy for a soliton is a twist in a Möbius strip. A soliton is intrinsically localized, and smooth rather than point-like, with a size that depends on the parameters of the theory. Its classical energy is identified with its rest mass as a particle. When the field theory is quantized, the properties of the soliton are not affected very much.

We shall first consider the *sine–Gordon soliton*. This is a soliton in 1-dimensional space, so it is not a realistic physical particle. It arises, however, in a mathematically elegant field theory that can be analysed, both classically and quantum mechanically, in great detail. The name of this model is a pun on Klein–Gordon. The second soliton we shall discuss is the *Skyrmion* in 3-dimensional space. This is a realistic model for a proton or neutron with some physical interest, although it is not believed to be as fundamental as the QCD model of these particles.

There are many other types of soliton. An example is a solitonic magnetic monopole, with a non-zero magnetic charge. Particles with magnetic charge are not possible in Maxwell theory because they violate the equation $\nabla \cdot \mathbf{B} = 0$, but they are found in some of the more complicated Yang–Mills theories, where Maxwell's equations are combined with equations for other fields. The Standard Model itself does not have any monopoles, which is fortunate as monopoles have never been observed. Solitons also exist in classical wave contexts, including water waves and waves in optical fibres, and in many-body quantum systems. For example, certain magnetic materials admit a 2-dimensional analogue of a Skyrmion. Much of the recent literature on Skyrmions now refers to these objects, but they are not fundamental particles.

The sine–Gordon field theory is a version of the Klein–Gordon theory in one dimension, with a particular interaction term. The theory is Lorentz invariant in 2-dimensional spacetime. It has a Lagrangian for one real scalar field $\phi(x, t)$, from which one derives the sine–Gordon field equation

$$\frac{\partial^2 \phi}{\partial t^2} - \frac{\partial^2 \phi}{\partial x^2} + \sin \phi = 0 \,. \tag{15.4}$$

(We have fixed time and energy units to put this in its simplest form.) The $\sin \phi$ interaction term gives the model its name. If expanded, for small ϕ, the sine–Gordon equation becomes

$$\frac{\partial^2 \phi}{\partial t^2} - \frac{\partial^2 \phi}{\partial x^2} + \phi - \frac{1}{6}\phi^3 + \cdots = 0. \tag{15.5}$$

The ϕ term gives the field a mass, and the ϕ^3 term generates interactions that could be represented by a Feynman diagram vertex, as discussed in Chapter 12.

Recall from our discussion of the Higgs mechanism that the vacuum of a field theory need not be unique. Here, in sine–Gordon theory, a vacuum can be built on any stable, uniform field satisfying the field equation. One vacuum solution is $\phi = 0$, but the solution $\phi = 2\pi N$ for any positive or negative integer N is also a vacuum.

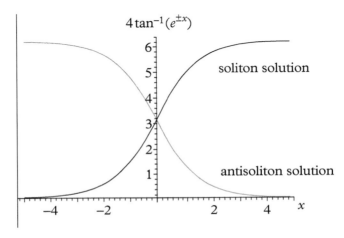

Fig. 15.3 The soliton and antisoliton solutions of the 1-dimensional sine-Gordon theory.

There are plenty of time-dependent, wave-like solutions of the sine–Gordon equation, and remarkably, an infinite number of these can be written in closed form. The solution that interests us, however, is the soliton. This is a localized static solution that approaches different vacua at spatial infinity on the left and on the right, and has finite energy. To find it, we note that if a static field satisfies

$$\frac{d\phi}{dx} = 2\sin\frac{1}{2}\phi, \tag{15.6}$$

then it satisfies the sine–Gordon equation, as

$$\frac{d^2\phi}{dx^2} = \left(\cos\frac{1}{2}\phi\right)\frac{d\phi}{dx} = 2\cos\frac{1}{2}\phi \sin\frac{1}{2}\phi = \sin\phi. \tag{15.7}$$

The first order equation (15.6) is easily solved, and has a solution $\log\tan\frac{1}{4}\phi = x$, or equivalently

$$\phi(x) = 4\tan^{-1}(e^x). \tag{15.8}$$

This soliton solution is shown in Figure 15.3.

The field variable ϕ of the soliton has one unit of winding. As $x \to -\infty$, $e^x \to 0$ and we can choose the value of $\tan^{-1}(e^x)$ so that $\phi \to 0$. As $x \to \infty$, $e^x \to \infty$ and $\tan^{-1}(e^x) \to \frac{1}{2}\pi$, so $\phi \to 2\pi$. The soliton approaches a vacuum at both ends of the spatial line, but has a non-trivial and irremovable topological character due to the increase of ϕ by 2π along the line, which we refer to as unit winding. For this reason, the sine–Gordon soliton is an example of a topological soliton.

Often, the sine–Gordon field ϕ is regarded as an angular variable, in which case the vacua we have discussed are actually indistinguishable. There is just one vacuum after all, because shifting ϕ by 2π has no physical effect. However, the soliton still has unit winding, like a pendulum making one complete circuit, and it cannot be continuously deformed to a constant vacuum solution.

The sine–Gordon soliton clearly has a smooth character and finite width. It has several variants. We can shift the soliton to the left or right (using the arbitrary constant of integration when solving equation (15.6)). We can also boost the soliton to any speed v less than the speed of light. A solution of the full field equation (15.4) is

$$\phi(x,t) = 4\tan^{-1}\left(e^{\gamma(x-vt)}\right), \tag{15.9}$$

where $\gamma = (1-v^2)^{-\frac{1}{2}}$ is the usual relativistic gamma factor. The field energy and momentum of the soliton can also be calculated. The static soliton has energy $E = 8$, and when it is moving it has the relativistic energy $E = 8\gamma$ and momentum $p = 8\gamma v$. The soliton is therefore interpreted as a particle with rest mass 8.

The equation $\frac{d\phi}{dx} = -2\sin\frac{1}{2}\phi$, with reversed sign, also implies that $\frac{d^2\phi}{dx^2} = \sin\phi$, and its solution is an antisoliton with a negative unit of winding but the same energy as the soliton. Both soliton and antisoliton are shown in Figure 15.3.

The sine–Gordon soliton is a new type of particle, but there is also the basic scalar particle of the theory that arises when one quantizes waves close to the vacuum $\phi = 0$. Because the linearized field has mass parameter 1, this particle has mass $\hbar$, a much lighter mass than the soliton mass 8 if $\hbar$ (in our units) is small. Both masses acquire further quantum corrections, but these are small if $\hbar$ is small. The soliton is not just heavier; it has a topological stability because of its winding, and it cannot decay into a collection of the lighter particles.

Much more is known about the sine–Gordon theory and its solitons. One can construct classical solutions with multiple winding, although none of them are static. These solutions correspond to several solitons in interaction. The forces between solitons can be calculated, and the classical and quantized scattering of solitons is also computable. The quantized antisoliton behaves as the antiparticle of the soliton, and there are bound states of the soliton and antisoliton that provide a localized picture of the basic scalar particle of mass $\hbar$ with no winding.

15.2.2 Skyrmions

A more realistic soliton, because it occurs in a theory in 3-dimensional space, is the Skyrmion. The theory was proposed by Tony Skyrme around 1960. It is a development of the Yukawa theory of pions and nucleons. The basic fields are the three scalar pion fields, but there are no explicit Dirac fields for the nucleons. Skyrme's idea was to combine the pion fields in a nonlinear way that allows for a winding number and a topologically stable soliton. Just as the sine–Gordon field is an angular field with values on a circle, so the Skyrme field takes values on a 3-dimensional sphere. This is realized by introducing four fields $\sigma, \pi_1, \pi_2, \pi_3$ and imposing the constraint

$$\sigma^2 + \pi_1^2 + \pi_2^2 + \pi_3^2 = 1. \tag{15.10}$$

Locally, σ can be eliminated, and the physical fields are the pion fields, π_1, π_2, π_3, closely related to the pion particles, π^-, π^0, π^+. The vacuum solution has $\sigma = 1$ everywhere in spacetime, and the pion fields are zero. In addition, there are wave-like solutions close to the vacuum where the pion fields are of small magnitude everywhere, and σ is close to 1. Quantizing these topologically trivial wave fields gives a fairly realistic model of the spin 0 pion particles including their strong interactions. Electromagnetic and weak interaction effects can be added to the theory if desired.

Fig. 15.4 The Skyrmion.

The most interesting classical solution in this theory is the Skyrmion, which is a static soliton. The Skyrmion approaches the vacuum at spatial infinity (in all directions), but in a region surrounding some central point, the field winds entirely around the 3-sphere defined by the constraint equation (15.10). The solution is given pictorially in Figure 15.4. The arrows indicate the values of the pion field (π_1, π_2, π_3), presented as a vector. σ approaches 1 at infinity but has the value -1 at the Skyrmion centre. The unit winding means that, within this theory, the Skyrmion is absolutely stable and cannot decay to pion waves.

Static and dynamic solutions with multiple windings are also possible. The winding number is identified with baryon number, B. This was Skyrme's great idea. Baryon number conservation is a law of nature, but it is not understood in a fundamental way. In Skyrme's model it is a topological conservation law.

Skyrme's theory is not so well understood as the sine–Gordon theory. Approximate quantum results can be calculated for low energy phenomena, and they are very interesting, but ultimately, if we are to model physical reality, it seems that the theory must be replaced by QCD at length scales shorter than those of the Skyrmion. Classically, the $B = 1$ Skyrmion can rotate around any axis through its centre. For topological reasons, the lowest energy quantum states of the rotating Skyrmion have spin $\frac{1}{2}$. The states also have isospin $\frac{1}{2}$ because the orientation of the pion vectors rotates with the Skyrmion. So there are four fundamental quantum states of the Skyrmion with essentially equal energies, each with baryon number $B = 1$. Two states represent a proton, with spin up or down, and two represent a neutron with spin up or down. Higher energy states with spin $\frac{3}{2}$ and isospin $\frac{3}{2}$ represent the Delta resonances Δ^{++}, Δ^{+}, Δ^{0}, Δ^{-}. There are also states of the antiSkyrmion with opposite winding, representing antibaryons.

Remarkably, from one field theory with just three scalar fields we have obtained two types of particle—pions and nucleons—and furthermore, although the nucleons have spin $\frac{1}{2}$, we have not needed a fundamental Dirac equation. This possibility is a fascinating aspect of quantum field theory when the interactions give rise to topological solitons.

As in sine–Gordon theory, one can construct many more solutions in the Skyrme theory than just the static soliton with unit baryon number, but to do this requires numerical assistance. The solutions include bound configurations of Skyrmions with multiple baryon number B that have been used to model nuclei. The quantum states describe nuclei in their ground states and also some excited states. The finite size of the Skyrmion has an important effect here. Usually the protons and neutrons in nuclei are treated as point particles with strong repulsive forces that keep them separated by at least 1 fm. In the Skyrme theory, the repulsive forces are present automatically, but the Skyrmions deform considerably as they approach each other. So Skyrme theory gives a different picture of nuclei than the familiar one of hard, touching yellow and red balls representing protons and neutrons, as in Figure 11.9.

In summary, soliton models of particles are promising alternatives to more orthodox models of particles in quantum field theory. They rely in an essential way on the nonlinear interactions, and to be stable they require some topological structure in the theory and its field equations that one would not easily recognize from Feynman diagrams. The soliton paradigm provides a unified model of the internal structure of particles, as illustrated in Figures 15.3 and 15.4, as well as a description of how the particles interact and scatter. Soliton models do have their limitations. We only have an approximate model of protons and neutrons, using Skyrmions, and there is as yet no successful soliton model of electrons and the other leptons. As far as we know, the leptons are completely structureless.

For now we must set aside our qualms relating to the finite size of particles and return to the best particle physics theory that we have—the Standard Model.

15.3   Critique of the Standard Model

It is quite remarkable that all phenomena relating to electromagnetic, weak and strong interactions, essentially the whole of non-gravitational physics, are encompassed by the Standard Model, a single consistent theory. Furthermore, the theory can be used to make detailed quantitative predictions and every experiment ever performed agrees with these predictions, in some cases to an extraordinary precision. Nevertheless, the Standard Model certainly cannot be the final word in particle physics.

The Standard Model includes a number of free parameters that cannot currently be calculated by theorists and must be measured in the laboratory as inputs to the theory. They are the rest masses of the twelve fundamental fermions, along with the masses of the W, Z and Higgs bosons. There are also the three angles and the phase of the CKM matrix that control the mixing of the quark flavours in weak interactions, and similarly the three angles and phase of the PMNS matrix that control the mixing of neutrino flavours. Finally, there are the coupling strengths of the electromagnetic and strong interactions. (The coupling of the weak interaction is related to that of the electromagnetic interaction by the ratio of the W and Z masses.) These fifteen masses, eight mixing parameters and two coupling strengths give a total of 25 free parameters that are unexplained by the theory.

There is also no explanation of why the Standard Model takes the form that we find. The exchange bosons of the theory reflect the symmetries that underlie the interactions. The symmetry group of the weak force is called SU(2), and the symmetry group of the colour force is SU(3). It is this SU(3) symmetry that determines that there are eight gluons that mediate the force. But why SU(2) and SU(3) and not some other symmetry groups? Similar questions can be asked about the other features of the Standard Model. Why are there three generations of matter particles? Why is it that only left-handed particles interact via the weak force? Why does the Higgs potential take the form that we observe, which is just right to spontaneously break the symmetry of the electroweak force?

How can we build a theory that is even better than the Standard Model that resolves these issues? As yet, no-one knows, but in the next few sections we take a look at some routes towards possible answers.

15.4   Topology and the Standard Model

There are a number of important features of the universe that are a challenge to explain. These include the presence of large quantities of dark matter and the observed matter–antimatter asymmetry. We discussed the evidence for dark matter in Chapter 14. We will now consider the implications of the matter–antimatter asymmetry. We can define baryon number B, such that protons and neutrons have baryon number $B = 1$ and antiprotons and antineutrons have baryon number $B = -1$. Baryon number is conserved in the usual Standard Model interactions, so we might expect that the universe should contain equal amounts of matter and antimatter. However, if there were regions of the universe where antimatter was dominant, then matter–antimatter annihilation would occur at the interface between matter-dominated regions and antimatter-dominated regions with the emission of gamma rays. This would produce a background of high-energy gamma radiation, but no such background is observed. We can be confident therefore that the entire observable universe is dominated by matter.

In 1967, Andrei Sakharov listed three conditions that are necessary if a universe that began in a state with $B = 0$ is to evolve into a matter–antimatter asymmetric state, such as we now see, where the baryon number is $B \gg 0$. First, the universe must be out of thermodynamic equilibrium, otherwise forward and back reactions would occur at equal rates and nothing would change, and B would remain zero. It is not hard to imagine that in its earliest moments, a rapidly expanding universe would be out of thermal equilibrium. Second, there must be interactions that violate baryon number conservation. Third, the symmetries C and CP must be violated, as otherwise, for every process that created an extra baryon, there would be an equivalent process creating an extra antibaryon. We know that in the Standard Model, C as well as P is violated almost maximally in weak interactions. CP is also violated, but this is a much smaller effect that is probably insufficient to explain the observed matter–antimatter asymmetry of the universe. A stronger CP-violating mechanism may be required, involving physics beyond the Standard Model.

Remarkably, the violation of baryon number conservation is also a feature of the Standard Model, although it depends on a topologically non-trivial process that has never yet been seen in accelerator experiments. The Standard Model has some topological structure because of the symmetry groups SU(2) and SU(3) and the way the Higgs mechanism operates. It does not have any stable solitons, but it does have an unstable static solution with topological significance, called a *sphaleron*. The sphaleron is a smooth and localized field configuration sitting at a mountain pass in the energy landscape of field configurations. It is unstable because the energy decreases in one particular direction (and the opposite direction) but increases in all other directions.

Just thinking about the SU(2) gauge and Higgs fields, there is a passage from the vacuum over the sphaleron mountain pass and down the other side to the vacuum again, but this passage significantly changes the state of the fermionic particles. Some fermions are created and some are destroyed. This is because of the interplay of the gauge and Higgs field topology with the Dirac sea of each fermion. The Dirac sea behaves like an 'infinity hotel'. An infinity hotel has rooms numbered $1, 2, 3, \ldots$ up to infinity, and all rooms are occupied. When another guest requests a room, the manager asks all occupants to move to the room next door with the next higher number. They can all do this. This creates a vacancy in room 1 which can be occupied by the new guest.

Similarly, the passage from vacuum to vacuum via the sphaleron creates vacancies in some fermions' Dirac sea, because all energy levels are pushed down. The hole created in the Dirac sea is observed as an antiparticle. Going the other way, the energy levels rise and one fermion which had negative energy ends up in a positive energy state, as illustrated in Figure 15.5. This is observed as a particle. The net effect is that some particles are produced and some antiparticles are produced. They are not of the same type, and in fact three quarks and one lepton can be produced for each generation of Standard Model fermions. This creates, in total, three baryons and three leptons. Baryon number B and lepton number L change by three, but $B - L$ is conserved. Also there is no net change in the electric charge. The rate for the reverse process, creating antibaryons and antileptons, may be slightly different because of CP-violation. So random processes can favour the production of matter over antimatter. These processes do require genuine energy to create the fermions, but it is small compared with the energy temporarily needed to create a sphaleron and go over the mountain pass.

The energy of the sphaleron can be calculated and is found to be about 9 TeV, comparable with a compact collection of about one hundred W, Z and Higgs particles. This energy is just within reach of proton–proton collisions at the LHC, but it is believed that the rate of sphaleron production and decay near this energy is unobservably small, because these collisions do not create large numbers of W, Z and Higgs particles in a coherent way. Recently it has been suggested that a collision involving just two quarks with energy greater than 9 TeV may make sphaleron-mediated processes more likely. Physicists are fairly confident that sphaleron processes were more frequent in the very high temperature conditions of the early universe, and together with some version of CP-violation, may have been partly reponsible for creating the matter–antimatter asymmetry we observe today.

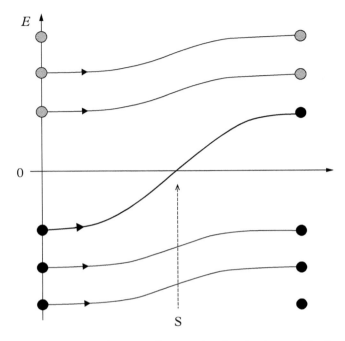

Fig. 15.5 Sphaleron-mediated production of a massive fermionic particle. In the background of the sphaleron S, the fermion has a zero-energy state.

15.5 Beyond the Standard Model

In section 12.9 we discussed neutrino oscillations in which electron, muon and tauon neutrinos are transformed into each other over distances of thousands of kilometres. The neutrino observatory at Gran Sasso in Italy has been investigating the possibility that there might be at least one additional neutrino that mixes with the three established neutrino types. Such a particle is known as a sterile neutrino as it would not interact via the weak force or any other process described by the Standard Model. These short-range neutrino oscillation experiments use an intense electron neutrino source composed of the cerium isotope ^{144}Ce positioned within a few metres of the Borexino liquid scintillator detector, which is surrounded by photomultiplier tubes. Precision timing of the arrival of photons enables the location of neutrino interactions within Borexino to be determined. The experiment is designed to measure any variation with distance from the neutrino source in the number of electron neutrinos detected. The experimental system is too small for significant oscillations into muon or tauon neutrinos to occur, so if there is a shortfall in the number of electron neutrinos detected, then it will signal mixing with a fourth neutrino. The mass of this new neutrino can then be deduced from the wavelength of the oscillations. If a sterile neutrino is discovered in this way, it will be a major breakthrough, as it would represent a sign of physics beyond the Standard Model. It would also be significant, as sterile neutrinos might have been formed in large quantities immediately after the Big Bang and might now form a component of dark matter.

15.5.1 Grand Unified Theories

The Standard Model consists of the combination of the GWS model and QCD, which account for the electroweak and colour forces respectively, as described in Chapter 12. The GWS model combines the electromagnetic and weak force into a single theory that relies on the Higgs mechanism to break some of the gauge symmetry at energies in the region of 100 GeV. Almost as soon as the Standard Model was established the next step in unifying the forces seemed reasonably clear. In 1974 Sheldon Glashow and Howard Georgi proposed that there might be two stages of symmetry breaking. At extremely high energies, the colour force and the electroweak force would be unified and described by a single Yang–Mills theory. Such theories are known as *Grand Unified Theories* or GUTs. At the GUT scale, which is in the region of 10^{15} to 10^{16} GeV, a first round of symmetry breaking produces the distinct colour force and the electroweak force. This is followed at very much lower energies (100 GeV) by the electroweak symmetry breaking of the Standard Model. The simplest such theory and the first to be proposed is described by a symmetry group known as SU(5).

The exchange bosons of the GUT come in a single matrix. In the case of the SU(5) GUT this is a 5 × 5 matrix, with 24 independent components. Twelve of these are the exchange bosons of the Standard Model (eight gluons plus the W^+, W^-, Z and the photon). The other twelve are new, and are usually described as X and Y bosons. The first round of symmetry breaking gives the X and Y bosons a very large mass of the order of the GUT scale, while the other exchange bosons all remain massless (until we reach the electroweak breaking scale). The theory therefore requires a matrix of Higgs bosons with masses in the region of the GUT scale.

Similarly, the matter particles, the leptons and quarks, combine into GUT multiplets and this produces an interesting and testable prediction because it means that the GUT force is able to convert antileptons into quarks and quarks into antileptons, which is impossible in the Standard Model, except possibly through the sphaleron process. Such interactions would be mediated by the X and Y bosons in the manner shown in Figure 15.6. These interactions violate baryon number B, but again $B - L$ is conserved, where L is lepton number.

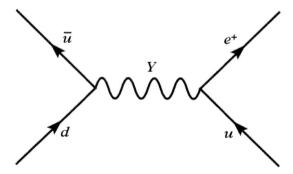

Fig. 15.6 The hypothetical Y boson could mediate processes that would result in proton decay. (The charge on the Y boson is $+\frac{1}{3}$.) The half-life for this process is of order M_Y^4 and is therefore very long.

If this is correct, then the proton *duu* might be unstable to the decay

$$p \quad \rightarrow \quad e^+ \quad + \quad \pi^0. \qquad (15.11)$$

As the X and Y bosons are so massive, such interactions would be incredibly weak, so the half-life of the proton would be very long. For the original SU(5) GUT the proton lifetime is estimated to be in the region of 10^{30} years.

Although the forces and particles of the Standard Model seem to fit together quite naturally in some of the simpler GUTs, these models suffer from serious problems that are both theoretical and experimental. For instance, proton decay has never been observed. Results from the Super-Kamiokande detector give a minimum lifetime for the proton in the region of 10^{34} years, which rules out the simplest GUTs, such as the original SU(5) theory. GUTs also predict the existence of magnetic monopoles. These have never been observed and their existence would create problems for our understanding of cosmology.

The motivation for GUTs is to unify the strong and electroweak forces. If this can be achieved then it would remove one of the undetermined parameters of the Standard Model by establishing a relationship between the strong and electroweak couplings, but this is at the expense of introducing many new particles—exchange bosons, Higgs bosons and usually additional fermions—whose masses are not pinned down by the theory, so overall the number of undetermined parameters is greatly increased.

Even more serious is a theoretical issue related to the idea of having two rounds of symmetry breaking based on the Higgs mechanism. We know that spin 1 particles must be massless if there is a corresponding unbroken gauge symmetry. Spin $\frac{1}{2}$ particles must also be massless if their left-handed and right-handed components carry different charges of an unbroken gauge symmetry. This is known as chiral asymmetry. There is no such principle that guarantees the masslessness of spin 0 particles. We should therefore expect any spin 0 particles to have a mass that is comparable to the natural mass scale of the theory. For GUTs the appropriate scale is in the region of 10^{15} GeV. It is therefore rather mysterious that a GUT could include the Standard Model Higgs boson with a mass of just 125 GeV. This is known as the *hierarchy problem*.

A number of possible solutions to the hierarchy problem have been proposed. One suggestion is that perhaps the Higgs boson is not a fundamental scalar, but a bound state of two new spin $\frac{1}{2}$ particles. Such particles are known as techniquarks and theories of this type are known as *technicolor*. The proposal is that the techniquarks interact via an unknown gauge force that is similar to QCD and that the Higgs boson is a type of technimeson, analogous to a pion. However, the construction of consistent technicolor theories that do not obviously conflict with existing experimental results is a formidable challenge that requires the introduction of many new particles and undetermined parameters.

The most popular solution to the hierarchy problem is to postulate the existence of a symmetry between fermions and bosons. Such a symmetry, which is known as *supersymmetry*, would guarantee the existence of massless spin 0 particles, as long as the theory included massless spin $\frac{1}{2}$ particles. The existence of low mass Higgs bosons would therefore be possible in a world such as ours that contains low mass spin $\frac{1}{2}$ particles. We will take a look at supersymmetry and its implications next.

15.5.2   Supersymmetry

For some years, theorists have been captivated by the possibility of supersymmetric quantum field theories, which possess a symmetry between their fermionic and bosonic fields. These theories include equal numbers of bosonic and fermionic wave modes, which means that the zero-point energies of these fields cancel. We saw in section 12.2.2 that the two components of each mode of a complex scalar field have zero-point energy $\frac{1}{2}\hbar\omega$, and in section 12.2.1 we showed that the Dirac field, with its two spin states, includes negative energy excitations with energy $-\hbar\omega$. For these modes of the Dirac field the zero-point energy is $-\frac{1}{2}\hbar\omega$. So the sum of the zero-point energies of a complex scalar field and a Dirac field is zero. This is an attractive feature of such theories and suggests that supersymmetry might offer a good starting point for quantum field theories, especially if our ultimate aim is to incorporate gravity into such a theory.

In supersymmetric theories, fermions and bosons come in pairs. For instance, the electron (which is a fermion) must have a partner that is a boson, and the photon (which is a boson) must have a partner that is a fermion. The known fundamental particles cannot be paired up in this way, so if the universe really is supersymmetric there must be a lot of new particles awaiting discovery. If supersymmetry were a perfect unbroken symmetry, these pairs of particles would have the same mass, in which case the partner particles would have been discovered long ago. So, if supersymmetry plays any role in particle physics then it must be spontaneously broken, like the electroweak force. The simplest supersymmetric extension of the Standard Model is known as the minimal supersymmetric Standard Model or MSSM.

The hypothetical new particles predicted by supersymmetry have already been named by theorists. The partner of the electron is known as the selectron. In general, the name of the bosonic partner of a fermion is produced by adding the prefix s for supersymmetry to the name of the fermion, so that the partner of a neutrino is known as a sneutrino and the partners of the quarks are known as squarks. The supersymmetric partner of the photon is a fermion. Theorists call it the photino. In general, the supersymmetric partners of bosons all end in the suffix ino, which gives us Winos, Zinos, gluinos and Higgsinos.

The LHC is searching for signs of these new particles. If supersymmetry is a true symmetry of the universe, then there should be a wonderful harvest of new particles to be reaped, and their discovery could answer one of the biggest mysteries of cosmology. Almost all fundamental particles are unstable, decaying rapidly into other lighter particles. Matter is formed from the few types of particle that are stable. A very important consequence of supersymmetry is that the lightest supersymmetry partner particle would be completely stable. This is expected to be a spin $\frac{1}{2}$ particle with zero charge known as a *neutralino*. (It is the lightest of four particles formed of combinations of the photino, Zino and two neutral Higgsinos.) This particle would have been produced in great profusion in the earliest moments of the universe and as it is stable it will be as abundant now as it always was. This makes the neutralino another candidate for explaining dark matter.

None of these attempts to push physics beyond the Standard Model are particularly compelling. If we measure the elegance of a theory by the number of undetermined parameters it includes, then each of these theories—GUTs, technicolor and supersymmetry—requires at least 100 additional parameters beyond those of the Standard Model. Even more serious is the lack of physical support and even downright incompatibility with the observational evidence. For forty years, the Standard Model has been king, winning success after success. What comes next is something of a mystery.

15.6 String Theory

There is an alternative to the bottom-up approach to fundamental physics that we have been considering. Rather than starting with the Standard Model and building a more sophisticated theory around it, we could look for a top-down route towards the Standard

Model. The ultimate goal of such investigations is to discover a unique, consistent theory of the universe and to show that it contains the Standard Model as its low energy limit. String theory is the first serious candidate for this ultimate theory. The aim of string theory is to encapsulate all the forces and particles from which the universe is constructed. The new feature of the theory is that instead of treating particles as zero-dimensional, point-like entities, the fundamental objects are 1-dimensional *strings*. This apparently simple idea has very deep consequences with a wealth of interesting repercussions and has led theorists in many remarkable directions in recent decades. Rather than postulating the existence of multitudes of different particles, there is just one fundamental object—the string—which can vibrate in many ways and each vibration mode represents a different type of particle. For instance, one mode might be an electron, another mode might be a quark and a third mode might be a photon. String theory has turned out to be far richer and more surprising than anyone could have imagined, and the full meaning of the theory is still far from understood. Experimental support is lacking, so the jury is still out regarding whether it has any connection to the real world. It remains a purely speculative search for the ultimate theory of nature. We can only give a flavour of this incredibly rich theory in this very brief survey. So what is string theory all about and why is it so important?

String theory offers the possibility of a consistent quantum theory of gravity for the first time. General relativity is built on fundamentally different principles from quantum mechanics. Nevertheless, it is generally agreed that when considered as a quantum field theory, gravity is mediated by a massless spin 2 particle known as the *graviton*. Gravitons have spin 2 because they are the quantum excitations from which gravitational waves are formed, and gravitational waves have two distinct quadrupole polarizations each of which is symmetrical under a rotation by 180° around the direction of the wavevector $\mathbf{k}$, as can be seen from Figures 6.13 and 6.14. Physicists struggled for many years to devise a consistent quantum theory of gravity without success. The reason this is such a difficult problem seems to lie with the behaviour of such theories on very short length scales. In a quantum theory of gravity we can expect the geometry of spacetime to undergo violent, virtual fluctuations at extremely short distances. This makes the theory very difficult to define. We can estimate the scale at which quantum effects become important in gravity by combining the fundamental constants $\hbar, G$ and c to obtain a fundamental quantity with the dimensions of length known as the *Planck length*, $l_{\mathrm{P}} = \sqrt{\frac{\hbar G}{c^3}} \simeq 1.6 \times 10^{-35}$ m.

Strings may be open or closed. An open string is a curve with two ends, a closed string is a loop. The fundamental mode of vibration of a closed string looks identical to a massless spin 2 particle. This is the graviton. Its presence implies that string theory includes a theory of gravity. If strings are to explain gravity then it is reasonable to expect their size to be in the region of the Planck length. The fundamental mode of vibration of the string corresponds to the lowest mass particles and the string harmonics correspond to more massive particles; the higher the harmonic the greater the mass of the particle. For Planck length strings, these harmonics give rise to particles with masses that are multiples of the *Planck mass*, $m_{\mathrm{P}} = \sqrt{\frac{\hbar c}{G}} \simeq 2.2 \times 10^{-8}$ kg. This is equal to 10^{19} GeV, which is not much beyond the GUT scale but is still around 10^{15} times the energy released by the proton collisions at the LHC. The string excitations form a tower of states that come in multiples of the Planck mass. We would not expect strings to be found in these excited states except in the most extreme of circumstances, such as the centre of a black hole or in the immediate aftermath of the Big

Bang, but they are essential for the consistency of the theory. If string theory is correct, then the whole of our everyday physics is due to the lowest modes of vibration of the strings, which correspond to massless particles.

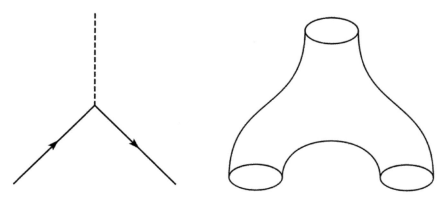

Fig. 15.7 Left: Feynman diagram vertex in particle physics. Right: String theory vertex.

The potential problems that arise in quantum gravity at very short distances seem to be alleviated in string theory. As a string moves through spacetime, it sweeps out a 2-dimensional surface known as the string worldsheet. Interactions between strings can be represented by Feynman diagrams by analogy with point particles. For instance, an interaction in which two closed strings combine to form a single closed string is shown in Figure 15.7. This is sometimes known as a trouser diagram. The interactions in particle physics occur at points, related to the vertices where particle lines meet in Feynman diagrams. This produces a singular result in quantum field theory on very short length scales. By contrast, the string theory Feynman diagrams are smooth everywhere, which makes the theory much better behaved. This is only relevant, of course, if it is possible to define a quantum theory of strings, which is not a foregone conclusion.

Complications arise when one attempts to construct a quantum theory of strings. Although it might be possible to write down a classical theory that includes a particular symmetry, sometimes the symmetry is lost in the quantum version of the theory. When this happens, it is referred to as an *anomaly*. If the lost symmetry is a gauge symmetry or Lorentz invariance, which is vital for the theory's consistency, then we must conclude that a consistent quantum version of the theory does not exist. This is the case for almost all string theories that one might consider. (The symmetry that is lost due to the anomaly is Lorentz invariance.) Only in spacetimes with a critical number of dimensions is a consistent anomaly-free quantum theory of strings possible. In the simplest theory known as the bosonic string the critical number of spacetime dimensions is 26, but this theory is purely bosonic, and does not include any spin $\frac{1}{2}$ particles, so it is not considered as a candidate for a realistic theory of physics. In the supersymmetric string theory, which includes bosons and fermions, the critical spacetime dimension is ten. The *superstring*, as it is called, has all the necessary ingredients to describe the particles and forces of the real world, but the requirement of ten spacetime dimensions means that if it is to explain genuine physics in four spacetime dimensions, then there must be six additional spatial dimensions of which we are not normally aware. Einstein showed us that gravity determines the shape of spacetime. String theorists believe that in a similar way, the string field moulds the background spacetime that the strings move through. This determines the shape of the hidden dimensions and these extra dimensions control the form taken by the non-gravitational forces. Potentially, this offers a complete geometrization of fundamental physics.

String theory is not simply a theory of gravity. The quantized lowest mode of vibration of an open string is a massless spin 1 particle that we would recognize as a photon. If colour charges are attached to the ends of the string, then the theory becomes even more interesting. Now, instead of a single zero-energy mode representing the photon, the theory includes a number of massless spin 1 particles corresponding to the different charge combinations of the string. These are particles that we would recognize as gluons. In the critical dimension, string theory automatically incorporates the symmetries of a Yang–Mills force like QCD. More accurately, it includes a supersymmetric version of a Grand Unified Theory.[2]

By comparison with the Planck energy m_P, the whole of known physics is considered as low energy physics. If strings are to describe established physics, then it must be in terms of the zero-energy states of the string. The full complement of zero-energy states of the superstring, interpreted in ordinary 4-dimensional spacetime, correspond to massless particles with spin 0, $\frac{1}{2}$, 1, $\frac{3}{2}$ and 2. The unique, massless spin 2 particle is the graviton, which gives us gravity. The massless spin $\frac{3}{2}$ particles, of which there can be up to eight types, are known as gravitinos. One gravitino is the supersymmetry partner of the graviton and its presence implies that the low energy limit of the superstring includes the supersymmetric version of quantum gravity known as *supergravity*. The massless spin 1 particles are the exchange bosons of a GUT. The superstring also includes massless spin $\frac{1}{2}$ particles, which are necessary to describe the fermions of the GUT, and hence of the Standard Model, and massless spin 0 particles, which are necessary if the theory is to include Higgs bosons. So string theory includes all the ingredients that are required to describe low energy physics. What about the extra dimensions demanded by the theory?

15.6.1 Compactification

String theorists assume that the six extra spatial dimensions are *compactified*, which means they are rolled up much smaller than the three macroscopic spatial dimensions that we are familiar with. They are so minuscule that we are completely unaware of their spatial extension. It is usually assumed that they are comparable in size to the Planck length, which is far too small for their structure to be probed in the laboratory. However, the precise way in which these dimensions are rolled up is critical for low energy physics. Strings can wrap themselves around the hidden dimensions, which affects the energy of the strings and the symmetry between different string states. This means that the hidden dimensions should determine the spectrum of particles observed in particle accelerators and the nature of the non-gravitational forces. The challenge for string theorists is to show that string compactification leads to a particle spectrum that matches observations.

The starting point for most compactification schemes is to assume the existence of supersymmetry in the low energy physics. If supersymmetry survives, then the six extra dimensions must form a hypersurface or manifold of a type known as a *Calabi–Yau manifold*. Theorists have explored the possible Calabi–Yau geometries in the hope of finding a unique

---

[2] Even in the critical dimension of the superstring there are just two possible anomaly-free GUTs known as the SO(32) and $E_8 \times E_8$ theories.

solution that resembles established physics. Compactification schemes have been discovered that come close to producing the minimal supersymmetric Standard Model, with the correct symmetry group and the correct number of generations of fermions, but unfortunately, these solutions are far from unique. The 6-dimensional Calabi–Yau manifolds have not yet been classified, so no-one knows how many there are. There are certainly thousands, and possibly an infinite number. String theorists are unable to solve the string equations to determine the correct vacuum state, so they have no idea how nature might choose between the myriad possibilities. What makes one manifold more suitable for the creation of a universe than any of the other options? No-one knows. There is a whole landscape of possibilities. In fact, the question of explaining how nature chooses the appropriate vacuum is known as the *landscape problem*. It has cast doubt on whether theorists will ever be able to match the predictions of string theory to physics that can be tested in the laboratory.

There are some generic predictions of string theory that might be confirmed by the LHC. Supersymmetry could exist without string theory, but its discovery at the LHC would encourage string theorists to believe that they are right to seek even greater symmetries in the structure of the laws of nature, and as supersymmetry fits so naturally within string theory, any sign of supersymmetry would be seen as a vindication of their efforts. Most compactification schemes lead to extra forces in addition to the electroweak and strong forces. Physicists at the LHC are on the lookout for new bosonic particles that would mediate these forces. Again, it is possible that forces are at play in particle physics beyond those described by the Standard Model without the existence of string theory. Extra spatial dimensions may also exist without string theory. In the absence of a unique compelling theory, experimental results are needed to guide future theoretical work.

15.7 Further Reading

Fundamental issues in the interpretation of quantum mechanics are considered in

J. Baggott, *Beyond Measure: Modern Physics, Philosophy and the Meaning of Quantum Mechanics*, Oxford: OUP, 2004.

J.S. Bell, *Speakable and Unspeakable in Quantum Mechanics (2nd ed.)*, Cambridge: CUP, 2004.

For the mathematical theory and physical applications of solitons, including Skyrmions, see

N. Manton and P. Sutcliffe, *Topological Solitons*, Cambridge: CUP, 2004.

T. Dauxois and M. Peyrard, *Physics of Solitons*, Cambridge: CUP, 2006.

For a gentle introduction to string theory, see

B. Zwiebach, *A First Course in String Theory*, Cambridge: CUP, 2004.

Picture Credits

Fig. 11.4 – Wikimedia—Benjamin Bannier, Woods–Saxon potential for A=50, in units of fm, relative to V_0, $a = 0.5$fm

Fig. 11.6 – Wikimedia—Numerical data from Katharina Lodders (2003), Solar System Abundances and Condensation Temperatures of the Elements, The Astrophysical Journal 591: 1220–1247. Composed by Orionus

Fig. 11.19 – ITER Organisation

Fig. 12.1 – Wikimedia—MissMJ, PBS NOVA, Fermilab, Office of Science, United States Department of Energy, Particle Data Group

Fig. 12.3 – L. Alvarez-Gaume et al., Introductory lectures on quantum field theory, arXiv:hep-th/0510040 CERN-PH-TH-2005-147, CERN-PH-TH-2009-257

Fig. 12.4 – Wikimedia—Carl D. Anderson (1933), The Positive Electron, Physical Review 43 (6): 491–494. DOI:10.1103/PhysRev.43.491

Fig. 12.12 – F. Halzen and A.D. Martin, Quarks and Leptons: An Introductory Course in Modern Particle Physics, Wiley 1984

Fig. 12.16 – Left: Nicholas Mee, Right: CERN

Fig. 12.17 – Robert G. Arns, Detecting the Neutrino, Phys. Perspect. 3 (2001) 314–334, Courtesy of the Regents of the University of California, operators of Los Alamos National Laboratory

Fig. 12.26 – from Phys. Rept. 427 (2006) 257–454. Precision Electroweak Measurements on the Z Resonance, The ALEPH Collaboration, the DELPHI Collaboration, the L3 Collaboration, the OPAL Collaboration, the SLD Collaboration, the LEP Electroweak Working Group, the SLD electroweak, heavy flavour groups, arXiv:hep-ex/0509008

Fig. 13.2 – Wikimedia—Michael Perryman, Hipparcos satellite: path of star on the sky

Fig. 13.3 – CSIRO

Fig. 13.7 – Wikimedia—Borb, Diagram of the CNO Cycle

Fig. 13.8 – Necklace and Cat's Eye Nebula: NASA, ESA, HEIC, and The Hubble Heritage Team (STScI/AURA). Ring Nebula and IC418: NASA and The Hubble Heritage Team (STScI/AURA). Hour Glass Nebula: NASA, R.Sahai, J.Trauger (JPL), and The WFPC2 Science Team. Eskimo Nebula: NASA, A.Fruchter and the ERO Team (STScI)

Fig. 13.14 – I.R. Seitenzahl, F.X. Timmes and G. Magkotsios, The light curve of SN 1987A revisited: constraining production masses of radioactive nuclides, arXiv:1408.5986 [astro-ph.SR]

Fig. 13.15 – Left: Wikimedia—ESA/Hubble & NASA, Right: ESO/L. Calada, An artist's impression of the rings around SN 1987A

Fig. 13.16 – Supernova 1987A 1994-2003 Hubble Space Telescope – WFPC2 – ACS, NASA and R.Kirshner (Harvard Center for Astrophysics) STScI-PRC04-09b

Fig. 13.17 – Dina Prialnik

Fig. 13.18 – NASA/C210CXC/M.Weiss

Figs. 14.5, 14.6 – The Millennium-XXL Project, Simulating the Galaxy Population in Dark Energy Universes

Figs. 14.7, 14.8, 14.9 – Daniel Baumann, Inflation, arXiv:0907.5424 [hep-th] TASI-2009

Fig. 14.10 – NASA/WMAP Science Team

Fig. 15.1 – Figure reprinted with permission from *Physics* 8, 123, December 16, 2015. Copyright (2015) the American Physical Society

Index

Page numbers indicating figures and tables are given in italics; a page number followed by n indicates a footnote; a page number indicating a definition is given in bold.

Biographies

Nicholas Manton is Professor of Mathematical Physics in the Department of Applied Mathematics and Theoretical Physics at the University of Cambridge. His research is mainly in particle and nuclear physics, and especially the theory and application of localized field structures known as solitons. He has led recent developments of the theory of nuclei based on solitons called Skyrmions. He is also the co-discoverer of the electroweak sphaleron, a twisted topological structure in the Higgs field that may have helped generate the matter-antimatter asymmetry of the universe. He has published more than 100 papers, and the monograph *Topological Solitons* with Paul Sutcliffe. He teaches courses for both maths and science students in Cambridge and has supervised more than 25 PhD students. He was awarded the Whitehead prize of the London Mathematical Society and is a Fellow of the Royal Society.

Nicholas Mee studied theoretical physics and mathematics at the University of Cambridge. He achieved a top distinction in Part III of the Mathematical Tripos of the University of Cambridge and gained his PhD there in theoretical particle physics, with the thesis *Supersymmetric Quantum Mechanics and Geometry*. He is Director of software company Virtual Image and the author of over 50 multimedia titles including *The Code Book on CD-ROM* with Simon Singh and *Connections in Space* with John Barrow, Martin Kemp and Richard Bright. He has played key roles in numerous science and art projects including the *Symbolic Sculpture* project with John Robinson, the European SCIENAR project, and the 2012 *Henry Moore and Stringed Surfaces* exhibition at the Royal Society. He is author of the award-winning popular science books *Higgs Force*: *Cosmic Symmetry Shattered* and *Gravity: Cracking the Cosmic Code*.